Student's Study Guide / Solutions Manual

Volume 1: Chapters 1–20

Sears & Zemansky's

UNIVERSITY PHYSICS

15TH EDITION

WAYNE ANDERSON
A. LEWIS FORD

With contributions by Joshua Ridley

330 Hudson Street, NY NY 10013

Director Physical Science Portfolio: Jeanne Zalesky
Physics and Astronomy Portfolio Analyst: Ian Desrosiers
Senior Content Producer: Martha Steele
Managing Producer: Kristen Flathman
Courseware Editorial Assistant: Sabrina Marshall
Director, Mastering Physics Content Development:
Amir Said

Associate Producer, Science/ECS: Kristen Sanchez
Senior Content Developer (Physics/Astronomy):
David Hoogewerff
Rich Media Content Producer: Keri Rand
Full-Service Vendor: SPi Global
Compositor: SPi Global
Manufacturing Buyer: Stacey Weinberger

This work is solely for the use of instructors and administrators for the purpose of teaching courses and assessing student learning. Unauthorized dissemination, publication or sale of the work, in whole or in part (including posting on the internet) will destroy the integrity of the work and is strictly prohibited.

Cover Photo Credit: John Woodworth/Alamy Stock Photo

Copyright ©2020, 2016, 2012, 2008 Pearson Education, Inc. All Rights Reserved. Printed in the United States of America. This publication is protected by copyright, and permission should be obtained from the publisher prior to any prohibited reproduction, storage in a retrieval system, or transmission in any form or by any means, electronic, mechanical, photocopying, recording, or otherwise. For information regarding permissions, request forms and the appropriate contacts within the Pearson Education Global Rights & Permissions department, please visit www.pearsoned.com/permissions/.

PEARSON, ALWAYS LEARNING and Mastering™ Physics are exclusive trademarks in the U.S. and/or other countries owned by Pearson Education, Inc. or its affiliates.

Unless otherwise indicated herein, any third-party trademarks that may appear in this work are the property of their respective owners and any references to third-party trademarks, logos or other trade dress are for demonstrative or descriptive purposes only. Such references are not intended to imply any sponsorship, endorsement, authorization, or promotion of Pearson's products by the owners of such marks, or any relationship between the owner and Pearson Education, Inc. or its affiliates, authors, licensees or distributors.

Printed in the United States of America.
1 2019

ISBN 10:   0-135-21695-8
ISBN 13: 978-0-135-21695-8

Contents

	Preface	viii
Part I	**Mechanics**	
	Chapter 1 Units, Physical Quantities, and Vectors	
	Study Guide	SG 1-1
	Solutions Manual	SM 1-1
	Chapter 2 Motion Along a Straight Line	
	Study Guide	SG 2-1
	Solutions Manual	SM 2-1
	Chapter 3 Motion in Two or Three Dimensions	
	Study Guide	SG 3-1
	Solutions Manual	SM 3-1
	Chapter 4 Newton's Laws of Motion	
	Study Guide	SG 4-1
	Solutions Manual	SM 4-1
	Chapter 5 Applying Newton's Laws	
	Study Guide	SG 5-1
	Solutions Manual	SM 5-1
	Chapter 6 Work and Kinetic Energy	
	Study Guide	SG 6-1
	Solutions Manual	SM 6-1
	Chapter 7 Potential Energy and Energy Conservation	
	Study Guide	SG 7-1
	Solutions Manual	SM 7-1

Chapter 8	Momentum, Impulse, and Collisions	
	Study Guide	SG 8-1
	Solutions Manual	SM 8-1
Chapter 9	Rotation of Rigid Bodies	
	Study Guide	SG 9-1
	Solutions Manual	SM 9-1
Chapter 10	Dynamics of Rotational Motion	
	Study Guide	SG 10-1
	Solutions Manual	SM 10-1
Chapter 11	Equilibrium and Elasticity	
	Study Guide	SG 11-1
	Solutions Manual	SM 11-1
Chapter 12	Fluid Mechanics	
	Study Guide	SG 12-1
	Solutions Manual	SM 12-1
Chapter 13	Gravitation	
	Study Guide	SG 13-1
	Solutions Manual	SM 13-1
Chapter 14	Periodic Motion	
	Study Guide	SG 14-1
	Solutions Manual	SM 14-1

Part II **Waves/Acoustics**

Chapter 15	Mechanical Waves	
	Study Guide	SG.15-1
	Solutions Manual	SM 15-1
Chapter 16	Sound and Hearing	
	Study Guide	SG 16-1
	Solutions Manual	SM 16-1

Part III **Thermodynamics**

Chapter 17	Temperature and Heat	
	Study Guide	SG 17-1
	Solutions Manual	SM 17-1

© Copyright 2020 Pearson Education, Inc. All rights reserved. This material is protected under all copyright laws as they currently exist. No portion of this material may be reproduced, in any form or by any means, without permission in writing from the publisher.

Chapter 18	Thermal Properties of Matter	
	Study Guide	SG 18-1
	Solutions Manual	SM 18-1
Chapter 19	The First Law of Thermodynamics	
	Study Guide	SG 19-1
	Solutions Manual	SM 19-1
Chapter 20	The Second Law of Thermodynamics	
	Study Guide	SG 20-1
	Solutions Manual	SM 20-1

Part IV   Electromagnetism

Chapter 21	Electric Charge and Electric Field	
	Study Guide	SG 21-1
	Solutions Manual	SM 21-1
Chapter 22	Gauss's Law	
	Study Guide	SG 22-1
	Solutions Manual	SM 22-1
Chapter 23	Electric Potential	
	Study Guide	SG 23-1
	Solutions Manual	SM 23-1
Chapter 24	Capacitance and Dielectrics	
	Study Guide	SG 24-1
	Solutions Manual	SM 24-1
Chapter 25	Current, Resistance, and Electromotive Force	
	Study Guide	SG 25-1
	Solutions Manual	SM 25-1
Chapter 26	Direct-Current Circuits	
	Study Guide	SG 26-1
	Solutions Manual	SM 26-1

	Chapter 27	Magnetic Field and Magnetic Forces	
		Study Guide	SG 27-1
		Solutions Manual	SM 27-1
	Chapter 28	Sources of Magnetic Field	
		Study Guide	SG 28-1
		Solutions Manual	SM 28-1
	Chapter 29	Electromagnetic Induction	
		Study Guide	SG 29-1
		Solutions Manual	SM 29-1
	Chapter 30	Inductance	
		Study Guide	SG 30-1
		Solutions Manual	SM 30-1
	Chapter 31	Alternating Current	
		Study Guide	SG 31-1
		Solutions Manual	SM 31-1
	Chapter 32	Electromagnetic Waves	
		Study Guide	SG 32-1
		Solutions Manual	SM 32-1
Part V	**Optics**		
	Chapter 33	The Nature and Propagation of Light	
		Study Guide	SG 33-1
		Solutions Manual	SM 33-1
	Chapter 34	Geometric Optics	
		Study Guide	SG 34-1
		Solutions Manual	SM 34-1
	Chapter 35	Interference	
		Study Guide	SG 35-1
		Solutions Manual	SM 35-1
	Chapter 36	Diffraction	
		Study Guide	SG 36-1
		Solutions Manual	SM 36-1

Part VI   **Modern Physics**

    Chapter 37   Relativity

        Study Guide .... SG 37-1

        Solutions Manual .... SM 37-1

    Chapter 38   Photons: Light Waves Behaving as Particles

        Study Guide .... SG 38-1

        Solutions Manual .... SM 38-1

    Chapter 39   Particles Behaving as Waves

        Study Guide .... SG 39-1

        Solutions Manual .... SM 39-1

    Chapter 40   Quantum Mechanics I: Wave Functions

        Study Guide .... SG 40-1

        Solutions Manual .... SM 40-1

    Chapter 41   Quantum Mechanics II: Atomic Structure

        Study Guide .... SG 41-1

        Solutions Manual .... SM 41-1

    Chapter 42   Molecules and Condensed Matter

        Study Guide .... SG 42-1

        Solutions Manual .... SM 42-1

    Chapter 43   Nuclear Physics

        Study Guide .... SG 43-1

        Solutions Manual .... SM 43-1

    Chapter 44   Particle Physics and Cosmology

        Study Guide .... SG 44-1

        Solutions Manual .... SM 44-1

PREFACE

This student resource to accompany Volume 1 (Chapters 1-20) of the fifteenth edition of *University Physics* by Roger Freedman and Hugh Young has been updated to combine two separate resources into one text: the *Student's Study Guide and Solutions Manual*.

The *Study Guide* summarizes the essential information in each chapter and provides additional problems for the student to solve, reinforcing the text's emphasis on problem-solving strategies and student misconceptions. It now also includes the new Key Concept Statements frorn each Worked Example in the text and tags these concepts to the Conceptual Questions and Problems presented in the *Student's Study Guide*. Finally, it includes the Key Example Variation Problems from the text, new to the Fifteenth Edition.

The *Solutions Manual* contains detailed solutions for all of the new Key Example Variation Problems and approximately two-thirds of the odd-numbered Exercises and Problems in Chapters 1-20. The Exercise and Problems included were not selected at random but rather were carefully chosen to include at least one representative example of each problem type. The solutions are intended to be used as models to following in working physics problems, and are worked out in the manner and style in which you should carry out your own problem solutions. The remaining Exercise and Problems, for which solutions are not given here, constitute an ample set of problems for you to tackle on your own.

The *Student's Study Guide and Solutions Manual* Volume 2 (Chapters 21–37) and Volume 3 (Chapters 37–44) are also available from your college bookstore.

Wayne Anderson
Lewis Ford
Sacramento, CA

STUDY GUIDE FOR UNITS, PHYSICAL QUANTITIES, AND VECTORS

Summary

Physics is the study of natural phenomena. In physics, we build theories based on observations of nature, and those theories evolve into physical laws. We often seek simplicity: Models are simplified versions of physical phenomena that allow us to gain insight into a physical process. This chapter covers foundational material that we'll use throughout our study. We begin with measurements that include units, conversions, precision, significant figures, estimates, orders of magnitude, and scientific notation. We'll also examine physical quantities. Scalar quantities, such as temperature, are described by a single number. Vector quantities, such as velocity, require both a magnitude and a direction for a complete description. We'll delve deeper into vectors, as they are used throughout physics. We'll also include a summary of critical mathematics skills you'll apply throughout your physics career. Techniques developed in this chapter will be used throughout our investigation of physics.

Objectives

After studying this chapter, you'll understand
- The process of experimentation and its relation to theory and laws.
- The SI units for length, mass, and time and common metric prefixes.
- How to express results in proper units and how to convert between different sets of units.
- Measurement uncertainties and how significant figures express precision.
- The use and meaning of scalar and vector quantities.
- Various ways to represent vectors, including graphical, component, and unit vector representations.
- How to add and subtract vectors, both graphically and componentwise.

- How to multiply vectors (the dot product and the cross product).
- The six most critical mathematics techniques you'll encounter in physics.

Concepts and Equations

Term	Description
Physical Law	A physical law is a well-established description of a physical phenomenon.
Model	A model is a simplified version of a physical system that focuses on its most important features.
Système International (SI)	The Système International (SI) is the system of units based on metric measures. It established refined definitions of units, including definitions of the second, meter, and kilogram.
Significant Figures	The accuracy of a measurement is indicated by the number of significant figures, or the number of meaningful digits, in a value. In multiplying or dividing, the number of significant figures in the result is no greater than in the factor with the fewest significant figures. In adding or subtracting, the result can have no more decimal places than the term with the fewest decimal places.
Scalar Quantity	A scalar quantity is expressed by a single number. Examples include temperature, mass, length, and time.
Vector Quantity	A vector quantity is expressed by both a magnitude and a direction and is often shown as an arrow in sketches. Vectors are frequently represented as single letters with arrows above them or in boldface type. Common examples include velocity, displacement, and force.
Component of a Vector	The vector $\vec{A}$ lying in the xy- plane has components A_x parallel to the x-axis and A_y parallel to the y-axis; A_x and A_y are the x- and y-component vectors of $\vec{A}$. Vector $\vec{A}$ can be described by unit vectors—vectors that have unity magnitude and that align along a particular axis. The unit vectors $\hat{i}$, $\hat{j}$, and $\hat{k}$ respectively align along the x-, y-, and z-axes of the rectangular coordinate system. Here, $$\vec{A} = A_x\hat{i} + A_y\hat{j}.$$
Magnitude of a Vector	The magnitude of a vector is the length of the vector. Magnitude is a scalar quantity that is always positive. It has several representations, including $$\text{Magnitude of } \vec{A} = A = |\vec{A}|.$$ The magnitude can be found from the component vectors as $$A = \sqrt{A_x^2 + A_y^2}.$$
Vector Addition and Subtraction	Two vectors, $\vec{A}$ and $\vec{B}$, are added graphically by placing the tail of $\vec{A}$ at the tip of $\vec{B}$:

	$\vec{A} + \vec{B}$ (triangle addition diagram)
	Vector $\vec{B}$ is subtracted from vector $\vec{A}$ by reversing the direction of $\vec{B}$ and then adding it to $\vec{A}$:
	$\vec{A} - \vec{B} = \vec{A} + (-\vec{B})$ (diagrams)
	Vector addition can also be done with component vectors. For components A_x and A_y of the vector $\vec{A}$ and components B_x and B_y of the vector $\vec{B}$, the components R_x and R_y of the resultant vector $\vec{R}$ are given by $$R_x = A_x + B_x \text{ and } R_y = A_y + B_y.$$
Scalar Product	The scalar, or dot, product $C = \vec{A} \cdot \vec{B}$ of two vectors $\vec{A}$ and $\vec{B}$ is a scalar quantity. It can be expressed in terms of the magnitudes of $\vec{A}$ and $\vec{B}$ and the angle ϕ between the two vectors—that is, $$\vec{A} \cdot \vec{B} = AB\cos\phi = |\vec{A}||\vec{B}|\cos\phi,$$ or in terms of the components of $\vec{A}$ and $\vec{B}$—that is, $$\vec{A} \cdot \vec{B} = A_x B_x + A_y B_y + A_z B_z.$$ The scalar product of two perpendicular vectors is zero.
Vector Product	The vector, or cross, product $\vec{C} = \vec{A} \times \vec{B}$ of two vectors $\vec{A}$ and $\vec{B}$ is a vector quantity. The magnitude of $\vec{A} \times \vec{B}$ depends on the magnitudes of $\vec{A}$ and $\vec{B}$ and the angle ϕ between the two vectors. The direction of $\vec{A} \times \vec{B}$ is perpendicular to the plane in which vectors $\vec{A}$ and $\vec{B}$ lie and is given by the right-hand rule. The magnitude of $\vec{C} = \vec{A} \times \vec{B}$ is $$C = AB\sin\phi$$ and the components are $$C_x = A_y B_z - A_z B_y$$ $$C_y = A_z B_x - A_x B_z$$ $$C_z = A_x B_y - A_y B_x.$$ $\vec{A} \times \vec{B}$ is perpendicular to the plane of $\vec{A}$ and $\vec{B}$. (Magnitude of $\vec{A} \times \vec{B}$) $= AB\sin\phi$ The vector product of two parallel or antiparallel vectors is zero.

Key Concept 1: To convert units, multiply by an appropriate unit multiplier.

Key Concept 2: If the units of a quantity are a product of simpler units, such as $m^3 = m \times m \times m$, use a product of unit multipliers to convert these units.

Key Concept 3: When you are multiplying (or dividing) quantities, the result can have no more significant figures than the quantity with the fewest significant figures.

Key Concept 4: To decide whether the numerical value of a quantity is reasonable, assess the quantity in terms of other quantities that you can estimate, even if only roughly.

Key Concept 5: In every problem involving vector addition, draw the two vectors being added as well as the vector sum. The head-to-tail arrangement shown in Figs. 1.11a and 1.11b is easiest. This will help you to visualize the vectors and understand the direction of the vector sum. Drawing the vectors is equally important for problems involving vector subtraction (see Fig. 1.14).

Key Concept 6: When you are finding the components of a vector, always use a diagram of the vector and the coordinate axes to guide your calculations.

Key Concept 7: When you are adding vectors, the x-component of the vector sum is equal to the sum of the x-components of the vectors being added, and likewise for the y-component. Always use a diagram to help determine the direction of the vector sum.

Key Concept 8: By using unit vectors, you can write a single equation for vector addition that incorporates the x-, y-, and z-components.

Key Concept 9: The scalar product $\vec{A} \cdot \vec{B}$ is a scalar (a number) that equals the sum of the products of the x-components, y-components, and z-components of $\vec{A}$ and $\vec{B}$.

Key Concept 10: You can find the angle φ between two vectors $\vec{A}$ and $\vec{B}$ whose components are known by first finding their scalar product, then using the equation $\vec{A} \cdot \vec{B} = AB \cos\varphi$.

Key Concept 11: The vector product $\vec{A} \times \vec{B}$ of two vectors is a third vector that is perpendicular to both $\vec{A}$ and $\vec{B}$. You can find the vector product either from the magnitudes of the two vectors, the angle between them, and the right-hand rule, or from the components of the two vectors.

Mathematics Review: Top Six Math Skills You Will Need in Introductory Physics

Mathematics is the main language of physics. You'll rely on mathematics throughout your study of physics and therefore must become comfortable with mathematical techniques. Here, we present the six most important mathematical techniques you'll use throughout your physics career. We strongly encourage you to review these materials thoroughly. When your knowledge of mathematics becomes second nature, your understanding of physics will blossom.

Math 1: Trigonometry

We'll use trigonometry throughout physics; problems involving objects tossed into the air, ramps, velocities, and forces all require trigonometry. The basic trigonometric functions relate the lengths of the sides of a right triangle to the inside angle. We define $\sin\theta$, $\cos\theta$, and $\tan\theta$ for the right triangle shown:

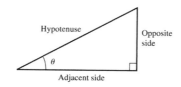

From Chapter 1 of Student's Study Guide to accompany *University Physics with Modern Physics, Volume 1*, Fifteenth Edition. Hugh D. Young and Roger A. Freedman. Copyright © 2020 by Pearson Education, Inc. All rights reserved.

$$\sin\theta = \frac{\text{opposite side}}{\text{hypotenuse}}, \cos\theta = \frac{\text{adjacent side}}{\text{hypotenuse}}, \tan\theta = \frac{\text{opposite side}}{\text{adjacent side}}.$$

Often, the triangle is formed in an *xy*-coordinate system, as is seen in Figure 1. Note that the two inside angles complement each other (add to 90°), so two sets of relations can be used:

$$\cos\theta = \frac{x}{r}, \sin\theta = \frac{y}{r}, \text{ and } \tan\theta = \frac{y}{x}$$

$$\cos\phi = \frac{y}{r}, \sin\phi = \frac{x}{r}, \text{ and } \tan\theta = \frac{x}{y}.$$

Figure 1 *xy*-coordinate system.

CAUTION **Watch sines and cosines!** One common mistake is to automatically associate the *x*-component with cosine and the *y*-component with sine. As you can see from Figure 1, this association does not always hold. One of the most common mistakes encountered in physics is confusing components of sines and cosines. By checking components every time, you avoid this mistake.

We'll also need to manipulate trigonometric relations, so we'll use common trigonometric identities, including the following:

$$\tan\theta = \frac{\sin\theta}{\cos\theta}$$
$$\sin^2\theta + \cos^2\theta = 1$$
$$\sin 2a = 2\sin a \cos a$$
$$\cos 2a = \cos^2 a - \sin^2 a$$

Math 2: Derivatives

Physics often investigates changes and rates of changes of various quantities. Derivatives provide the instantaneous rate of change of a quantity. Derivatives are therefore the natural choice for finding rates of change in physics. They're especially useful when we have functional definitions of quantities. Speed is the rate of change of position. If we're given speed in terms of a position function, we can easily find the speed by taking the derivative of position. The most common derivatives you'll encounter in physics are given in Table 1.

TABLE 1: Common derivatives.

$\frac{d}{dx}x^n = nx^{n-1}$
$\frac{d}{dx}\sin ax = a\cos ax$

$\dfrac{d}{dx}\cos ax = -a\sin ax$
$\dfrac{d}{dx}e^{ax} = ae^{ax}$
$\dfrac{d}{dx}\ln ax = \dfrac{1}{x}$

Math 3: Integrals

In physics, we also need to sum various quantities—quantities that are often given in terms of functions. Integrals sum functions and therefore are used to sum physical quantities. For example, one may find the total mass of an object by integrating its density function. The most common integrals you'll encounter in physics are given in Table 2. In addition, you may want to review integration by parts and trigonometric substitution for integrals in order to solve the more complicated ones.

TABLE 2: Common integrals.

$\int x^n\, dx = \dfrac{x^{n+1}}{n+1} \quad (n \neq -1)$
$\int \dfrac{dx}{x} = \ln x$
$\int \sin ax\, dx = -\dfrac{1}{a}\cos ax$
$\int \cos ax\, dx = \dfrac{1}{a}\sin ax$
$\int e^{ax}\, dx = \dfrac{1}{a}e^{ax}$
$\int \dfrac{dx}{\sqrt{a^2 - x^2}} = \arcsin \dfrac{x}{a}$
$\int \dfrac{dx}{\sqrt{x^2 + a^2}} = \ln\left(x + \sqrt{x^2 + a^2}\right)$
$\int \dfrac{dx}{x^2 + a^2} = \dfrac{1}{a}\arctan \dfrac{x}{a}$
$\int \dfrac{dx}{(x^2 + a^2)^{3/2}} = \dfrac{1}{a^2}\dfrac{x}{\sqrt{x^2 + a^2}}$
$\int \dfrac{x\, dx}{(x^2 + a^2)^{3/2}} = -\dfrac{1}{\sqrt{x^2 + a^2}}$

Math 4: Graphs

Graphs are common in many fields, including physics, in which data or information is plotted as a function of time. However, many students do not gain a full appreciation of graphs, and instructors often take graph interpretation for granted. In physics, graphs provide added insight into complex phenomena. We'll review important features of graphs to help your physics interpretations, as well as to help your interpretation of graphs wherever you encounter them.

Let's begin by examining the graph in Figure 2. Here, position is plotted as a function of time for three different objects. As time increases, the positions of all three objects increase, so the object is moving away from the origin. The slope gives the rate of change of the position—the speed—of the object:

$$\text{slope} = \text{rate of change of position} = \text{speed} = \frac{\Delta(\text{position})}{\Delta(\text{time})}.$$

All three lines are straight lines, indicating that the speed is constant for each object. Both objects A and B start at the same initial position. Object A's line has the greatest slope, so object A moves the fastest or has the greatest speed. Object C starts away from objects A and B and moves away at a slower rate than the other two. Where the lines intersect, objects A and C are at the same position at the same time. After the intersection, object A moves away from object C and so passes object C.

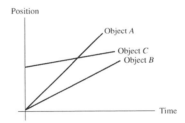

Figure 2 Position-versus-time graph.

Moving on to a more interesting case, we see that the slope in Figure 3 is not constant; calculus will be necessary to interpret this graph. The rate of change of the position varies, so we'll need to consider the instantaneous slope, or the derivative of the position:

$$\text{instantaneous slope} = \lim_{\Delta t \to 0} \frac{\Delta(\text{position})}{\Delta(\text{time})} = \frac{dx}{dt}.$$

Graphically, the instantaneous slope is the tangent to the line in the position-versus-time graph.

If we look at the figure, we see that the object moves away from the initial position, remains at a constant position for a period of time, and then moves toward the initial position. How it moves is found by looking at the slope. The slope of the line increases between point A and point B, indicating that the object begins by moving slowly and then speeding up as it moves away. After point B, the slope decreases until point C, where the slope becomes zero. Thus, the object begins slowing down at point B and stops at point C. At point D, the slope decreases and then becomes constant, indicating that the object moves back toward its initial position. The slope past point D is negative, indicating that the object moves in the direction opposite that of its initial movement. Points E and F show the instantaneous slope at two points on the curve. The slope at point E is greater than the slope at point F, indicating that the object is moving faster at point E.

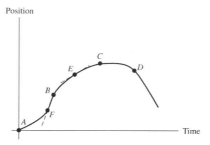

Figure 3 Position-versus-time graph.

The speed of another object is plotted in Figure 4. From this graph, we can use calculus to determine the distance the object travels in a given time interval. The distance traveled by an object between times t_a and t_b is the area under the curve, or the shaded area shown in the figure. The area under the curve is the speed multiplied by the time, which is the distance traveled. To find the area, we will need to sum up, or integrate, the speed over time:

$$\text{distance} = \int_{t_a}^{t_b} (speed)\, dt.$$

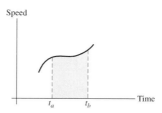

Figure 4 Speed-versus-time graph.

We have now seen somewhat how calculus and certain graphs are intertwined. Derivatives and integrals, respectively, give us the slope and the areas under a curve. We'll see that switching back and forth between graphs and the mathematics of calculus will help develop our physics intuition and skills. When you're confused about integrals, consider using a graph to clarify your understanding.

Math 5: Solving quadratic equations

We'll encounter quadratic equations throughout our physics investigations. We can either factor the equation to obtain the solutions or use the quadratic equation. It is often easier to apply the quadratic equation and not even attempt to factor, as the quadratic equation leads directly to the solutions. For a quadratic equation of the form $ax^2 + bx + c = 0$ with real numbers a, b, and c, the solutions are given by the quadratic equation:

$$x = \frac{-b \pm \sqrt{b^2 - 4ac}}{2a}.$$

Math 6: Solving simultaneous equations

Our goal in solving problems is to determine unknown quantities in equations. When we have multiple unknowns, we'll need multiple equations to find solutions. We'll need at least as many equations as we have unknowns to solve any problem.

When we are presented with multiple equations, one option is to rewrite one of the equations, solving for one unknown in terms of the other unknown(s), and substitute the result into the other equation(s) to eliminate one variable. For example, if we need to solve for both x and y, given two equations, we first rewrite one equation to solve for x in terms of y. Then we replace the x terms in the second equation with the solution of the first equation, leaving an equation having only y terms.

A second technique comes from linear algebra. We multiply each equation by a factor and then add or subtract the two equations. By choosing the proper factor, we eliminate one variable in the process. For example, with the equations

$$3x + 2y = 3$$
$$4x - 3y = 7$$

we multiply the top equation by 4 and the bottom equation by 3, leaving

$$12x + 8y = 12$$
$$12x - 9y = 21$$

If we now subtract the two equations, the x variable is eliminated. We can also multiply the top equation by 3 and the bottom equation by 2 and add the two equations to eliminate y. The key in this process is to multiply all terms of each equation by the same factor.

Other mathematical topics

Other mathematical relations that we'll encounter as we cover the material of this course include the following:

- Circumference, area, surface area, and volume of spheres and cylinders
- Exponentials, logarithms, and their identities
- The binomial theorem
- Power series expansions of algebraic, trigonometric, and exponential functions
- Multivariable calculus, including derivatives and integrals in two and three dimensions.

It is best to review the preceding topics as you encounter them in the course. You may also want to consult your mathematics textbooks or the Internet for more information.

CAUTION **Don't be afraid to review math!** Knowing math will let you focus on physics and save time when you solve problems.

Conceptual Questions

1: Sketch the situation

A man uses a cable to drag a trunk up the loading ramp of a mover's truck. The ramp has a slope angle of 20.0°, and the cable makes an angle of 30.0° with the ramp. Make a sketch of this situation.

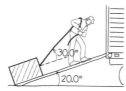

Figure 5 Sketch of trunk being dragged up a loading ramp.

IDENTIFY, SET UP, AND EXECUTE The sketch is shown in Figure 5. The ramp makes an angle of 20° with the ground. The trunk is on the ramp and the cable is attached to the trunk. The cable makes an angle of 30° with respect to the ramp, clearly marked. The mover is shown pulling the trunk up the ramp.

EVALUATE Understanding the physical situation in physics problems is critical for a correct interpretation. You should always draw a diagram (or diagrams) of the physical system you are investigating. Even when a figure is provided, it's often useful to sketch the important aspects. Only after creating a diagram should you proceed to interpret the physics and determine the proper equations to apply.

2: Dimensional analysis practice

Based *only* on consistency of units, which of the following formulas could *not* be correct? In each case, x is distance, v is speed, and t is time.

(a) $t = \sqrt{\dfrac{2x}{9.8 \text{ m/s}^2}}$

(b) $x = vt + (4.9 \text{ m/s}^2)t$

(c) $v = v_0 \sin\theta + \dfrac{(9.8 \text{ m/s}^2)x}{v_0 \cos\theta}$

(d) $x^2 - \dfrac{2v_0^2 \sin\theta}{9.8 \text{ m/s}^2} - \dfrac{2v^2}{9.8 \text{ m/s}^2} = 0$

(e) $v^2 = v_0^2 - 2(9.8 \text{ m/s}^2)\left(v_0 \tan\theta - \dfrac{1}{2}(9.8 \text{ m/s}^2)t^2\right)$

(f) $t = \dfrac{v^2 + (4.9 \text{ m/s}^2)x}{(3.0 \text{ m/s}^2)x}$

IDENTIFY, SET UP, AND EXECUTE For each of the six equations, carefully examine the units of each term in the equation. Equations (*a*), (*c*), and (*d*) are dimensionally correct; however, (*b*), (*e*), and (*f*) are incorrect. The far-right term in equation (*b*) has units of (m/s), while the other two terms have units of (m). In equation (*e*), the left term inside the rightmost set of parentheses (the term $v_0 \tan\theta$) has units of (m/s), which, when combined with the (m/s²) outside of the left parenthesis, would result in units of (m²/s³). The other three terms in the equation have units of (m²/s²). The fraction in equation (*f*) has no units, while the left-hand side has units of (s).

Thus, an error exists in each of the three equations ((b), (e), and (f)), since the units on the two sides of the equation do not agree. The next step would be to recheck our derivation to locate the source of the mistake.

EVALUATE Dimensional analysis is a powerful technique to help keep you from making errors. Catching the three errors in this problem would save time while reducing confusion. Always check your units!

3: Maximum and minimum magnitudes of a vector

Given vector $\vec{A}$ with magnitude 1.3 N and vector $\vec{B}$ with magnitude 3.4 N, what are the minimum and maximum magnitudes of $\vec{A} + \vec{B}$?

IDENTIFY, SET UP, AND EXECUTE The maximum magnitude is achieved when the two vectors are parallel and point in the same direction. The minimum magnitude is achieved when the two vectors are parallel and point in opposite directions. (The vectors are then called antiparallel.)

For parallel vectors, the magnitude is the sum of their magnitudes, 4.7 N in this case. For antiparallel vectors, the magnitude is the difference of their magnitudes, 2.1 N here.

EVALUATE This example helps illustrate the fact that vectors do not add like ordinary scalar numbers. They do not subtract like scalar numbers either. The magnitude of $\vec{A} + \vec{B}$ for any arbitrary alignment of the two vectors must lie between 2.1 N and 4.7 N.

4: Finding parallel and perpendicular vectors

If you are given two vectors, how can you determine whether the vectors are parallel or perpendicular?

IDENTIFY, SET UP, AND EXECUTE The cross product of two parallel vectors is zero. The dot product of two perpendicular vectors is zero. By taking the cross and dot products of the two vectors, you will determine whether they are parallel or perpendicular.

EVALUATE There are circumstances in which you cannot easily identify parallel and perpendicular vectors, such as vectors lying in the *xyz*-plane and that are given in terms of their components. Here, the best way to identify their orientation is to take dot and cross products.

Problems

1: Convert knots to m/s

A yacht is traveling at 18.0 knots. (One knot is 1 nautical mile per hour.) Find the speed of the yacht in m/s.

IDENTIFY AND SET UP We'll use a series of conversion factors to solve this problem. 1 nautical mile = 6080 ft and 1 mi = 5280 ft = 1.609 km. We know that 1 km = 1000 m and that 1 hour = 60 min = 60×(60 s) = 3600 s.

EXECUTE We apply the conversion factors to the speed in knots to solve:

$$18.0 \text{ knots} = \left(\frac{18.0 \text{ nautical miles}}{1 \text{ h}}\right)\left(\frac{6080 \text{ ft}}{1 \text{ nautical mile}}\right)\left(\frac{1.609 \text{ km}}{5280 \text{ ft}}\right)\left(\frac{1000 \text{ m}}{1 \text{ km}}\right)\left(\frac{1 \text{ h}}{3600 \text{ s}}\right) = 9.26 \text{ m/s}.$$

From Chapter 1 of Student's Study Guide to accompany *University Physics with Modern Physics, Volume 1,* Fifteenth Edition. Hugh D. Young and Roger A. Freedman. Copyright © 2020 by Pearson Education, Inc. All rights reserved.

KEY CONCEPT 1 **EVALUATE** Using a combination of several conversion factors, we've found that 18.0 knots is equal to 9.26 m/s. The last four quantities in parentheses are each equal to unity; hence, multiplying 18.0 knots by several factors of unity doesn't change the magnitude of the quantity. Crossing out the units helps prevent mistakes.

Practice Problem: How many fluid pints are in a 2-L bottle of soda? *Answer:* 4.23 pints

Extra Practice: How many gallons is this? *Answer:* 0.528 gallons

2: Finding components of vectors

Find the *x*- and *y*-components of the vector $\vec{A}$ in Figure 6. The magnitude of vector $\vec{A}$ is 26.2 cm.

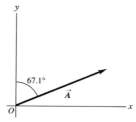

Figure 6 Problem 2.

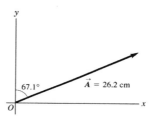

Figure 7 Problem 2 with components.

IDENTIFY AND SET UP We'll find the components of a vector by examining the triangle formed by the vector and the coordinate axes. Figure 7 shows Figure 6 redrawn to include the component vectors.

EXECUTE The *x*-component of $\vec{A}$ is located opposite the 67.1° angle; hence, we'll use the sine function:

$$A_x = A \sin 67.1° = (26.2 \text{ cm}) \sin 67.1° = 24.1 \text{ cm}.$$

The *y*-component of $\vec{A}$ is located adjacent to the 67.1° angle; thus, we'll use the cosine function:

$$A_y = A \cos 67.1° = (26.2 \text{ cm}) \cos 67.1° = 10.2 \text{ cm}.$$

The vector has an *x*-component of 24.1 cm and a *y*-component of 10.2 cm.

KEY CONCEPT 6 **EVALUATE** Finding the components of the vector required applying the sine and cosine functions. Often, but not always, the horizontal components will use cosine and the vertical components will use sine. This example illustrates an exception to that general assertion. It's important to examine a problem carefully in order to identify the proper trigonometric function for each component.

Practice Problem: What is the x-component of a vector that makes a 25° angle with the x-axis and has a magnitude of 18 cm? *Answer:* 16.3 cm

Extra Practice: What if the magnitude is 36 cm? *Answer:* 32.6 cm

3: Vector addition

Find the vector sum $\vec{A} + \vec{B}$ of the two vectors in Figure 8. Express the results in terms of components.

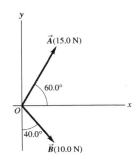

Figure 8 Problem 3.

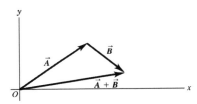

Figure 9 Sketch of Problem 3.

IDENTIFY AND SET UP Figure 9 shows a sketch of the two vectors added together, head to tail. The sketch indicates that we should expect a resultant in the first quadrant, with positive x- and y-components. We'll add the vectors by adding their x- and y-components, using the Cartesian coordinate system provided.

EXECUTE We find the components of the vectors by examining the triangles made by the vectors and their components. For $\vec{A}$,

$$A_x = A\cos 60.0° = (15.0 \text{ N})\cos 60.0° = 7.50 \text{ N},$$
$$A_y = A\sin 60.0° = (15.0 \text{ N})\sin 60.0° = 13.0 \text{ N}.$$

For $\vec{B}$,

$$B_x = B\sin 40.0° = (10.0 \text{ N})\sin 40.0° = 6.43 \text{ N},$$

$$B_y = -B\cos 40.0° = -(10.0\text{ N})\cos 40.0° = -7.66\text{ N}.$$

We can now sum the components:

$$R_x = A_x + B_x = 7.50\text{ N} + 6.43\text{ N} = 13.9\text{ N},$$

$$R_y = A_y + B_y = 13.0\text{ N} - 7.66\text{ N} = 5.34\text{ N}.$$

The resultant vector has an *x*-component of 13.9 N and a *y*-component of 5.34 N.

KEY CONCEPT 5 **EVALUATE** The resultant vector has positive components and resides in the first quadrant, as expected. Note how the components of the two vectors include both sine and cosine terms (i.e., vector *A*'s *x*-component includes the cosine component, and vector *B*'s *x*-component includes the sine component). This results from how the vectors' angles were given: vector *A*'s angle was with respect to the horizontal axis and vector *B*'s angle was with respect to the vertical axis. It's critical not to automatically associate all horizontal components with the cosine and all vertical components with the sine.

Practice Problem: Find the magnitude and direction of the resultant vector. *Answer:* The magnitude is 14.9 N, and its direction is 21.0° above the positive *x*-axis.

Extra Practice: What is the magnitude of the resultant vector if the *y*-component was −5.34 N? *Answer:* 14.9 N

4: Determine displacement on a lake

Marie paddles her canoe around a lake. She first paddles 0.75 km to the east, then paddles 0.50 km 30° north of east, and finally paddles 1.0 km 50° north of west. Find the resulting displacement from her origin.

IDENTIFY Displacement is a vector indicating change in position. The displacement vector points from the starting point of a journey to the endpoint. If we represent each of the three segments of the journey as a vector, the displacement vector is the sum of the three vectors. The goal is to find the sum of the three displacement vectors.

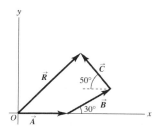

Figure 10 Problem 4.

SET UP Figure 10 shows a sketch of the three displacement segments (labeled $\vec{A}$, $\vec{B}$, and $\vec{C}$) and the resultant displacement vector ($\vec{R}$). We will add the three vectors, using the Cartesian coordinate system in the figure.

EXECUTE We find the components of the vectors by examining the triangles made by the vectors and their components. For $\vec{A}$, there is only a horizontal component:

$$A_x = A = 0.75 \text{ km},$$
$$A_y = 0.$$

For $\vec{B}$,
$$B_x = B\cos 30° = (0.50 \text{ km})\cos 30° = 0.443 \text{ km},$$
$$B_y = B\sin 30° = (0.50 \text{ km})\sin 30° = 0.250 \text{ km}.$$

For $\vec{C}$,
$$C_x = -C\cos 50° = -(1.0 \text{ km})\cos 50° = -0.643 \text{ km},$$
$$C_y = C\sin 50° = (1.0 \text{ km})\sin 50° = 0.766 \text{ km}.$$

The *x*-component is negative, as it points to the west. We can now sum the components:
$$R_x = A_x + B_x + C_x = 0.75 \text{ km} + 0.443 \text{ km} - 0.643 \text{ km} = 0.550 \text{ km},$$
$$R_y = A_y + B_y + C_y = 0 \text{ km} + 0.250 \text{ km} + 0.766 \text{ km} = 1.016 \text{ km}.$$

The resultant displacement vector has an *x*-component of 0.55 km and a *y*-component of 1.02 km. We can express the displacement vector in terms of magnitude and direction. To find the magnitude, we use the Pythagorean theorem:
$$R = \sqrt{R_x^2 + R_y^2} = \sqrt{(0.550 \text{ km})^2 + (1.016 \text{ km})^2} = 1.16 \text{ km}.$$

The inverse tangent gives us the angle:
$$\theta = \tan^{-1}\frac{R_y}{R_x} = \tan^{-1}\frac{1.016 \text{ km}}{0.550 \text{ km}} = 61.6°.$$

The resultant displacement vector has a magnitude of 1.16 km and points 61.6° above the positive *x*-axis.

KEY CONCEPTS 3, 7 **EVALUATE** Marie paddled a total of 2.25 km, only to end up 1.16 km away from her starting point. This shows how the magnitude of a vector sum can be smaller than the sum of the magnitudes of the individual vectors. Note that we carried an extra significant figure through the calculations and rounded off only in the final step.

Practice Problem: John walks a mile north, then a mile west, and then a mile north again. What is the magnitude of his displacement? *Answer:* 2.24 miles

Extra Practice: What if he walked 2 miles for each segment? *Answer:* 4.47 miles

5: Finding the dot product

Find the dot product of vectors $\vec{A}$ and $\vec{B}$ if $\vec{A} = 5.0\hat{i} + 2.3\hat{j} - 6.4\hat{k}$ and $\vec{B} = 12.0\hat{i} - 4.7\hat{j} + 9.3\hat{k}$.

IDENTIFY AND SET UP We're given the *x*-, *y*-, and *z*-components of the two vectors. We'll find the scalar product by using the component form of the dot product relation.

EXECUTE The dot product is the sum of the products of the components of the vectors:
$$\vec{A} \cdot \vec{B} = A_x B_x + A_y B_y + A_z B_z.$$

For our two vectors,

$$\vec{A} \cdot \vec{B} = (5.0)(12.0) + (2.3)(-4.7) + (-6.4)(9.3)$$
$$= 60 - 10.8 - 59.5$$
$$= -10.3.$$

The dot product is -10.3.

KEY CONCEPT 9 **EVALUATE** The result is a negative number, indicating that the projection of one vector onto the other points in the direction opposite that of the other. The vectors are not perpendicular, since the dot product is not zero.

We could've attempted to sketch these vectors, but since they're in three-dimensional xyz space, it's difficult to represent them accurately on a two-dimensional page. Building intuition in two-dimensional space helps us when we work in three-dimensional space.

Practice Problem: What is the dot product of $\vec{A} \cdot (-\vec{B})$? *Answer:* 10.33

Extra Practice: What is the dot product of $(-2\vec{A}) \cdot \vec{B}$? *Answer:* 20.66

6: Finding the cross product

Find the cross product $\vec{A} \times \vec{B}$, given $\vec{A} = 5.0\hat{i} + 2.3\hat{j} - 6.4\hat{k}$ and $\vec{B} = 12.0\hat{i} - 4.7\hat{j} + 9.3\hat{k}$.

IDENTIFY AND SET UP We are given the x-, y-, and z-components of the two vectors. We will find the cross product by using the component form of the cross-product relation.

EXECUTE The three components of the cross product are given by various products of the components of the two vectors:

$$C_x = A_y B_z - A_z B_y$$
$$C_y = A_z B_x - A_x B_z$$
$$C_z = A_x B_y - A_y B_x.$$

For this problem, the components are

$$C_x = (2.3)(9.3) - (-6.4)(-4.7) = (21.4) - (30.1) = -8.7$$
$$C_y = (-6.4)(12.0) - (5.0)(9.3) = (-76.8) - (46.5) = -123.3$$
$$C_z = (5.0)(-4.7) - (2.3)(12.0) = (-23.5) - (27.6) = -51.1.$$

The cross product is $\vec{C} = -8.7\hat{i} - 123.3\hat{j} - 51.1\hat{k}$.

KEY CONCEPT 11 **EVALUATE** The result is a vector, as is expected for the cross product. The vectors are not parallel, since the cross product is not zero.

Practice Problem: Find the magnitude of the resultant vector. *Answer:* The magnitude is 133.8.

Extra Practice: What is the magnitude of $\vec{A} \times \vec{B}$? *Answer:* 1129

7: Review of simultaneous equations

Solve the following expressions for T_A and T_B.

$$27T_A + 13T_B = 0$$
$$32T_A + 52T_B = 22.$$

IDENTIFY AND SET UP Both of the expressions involve two unknowns, so we cannot find a solution by using only one equation. We'll multiply the first equation by 32, multiply the second by 27, and then subtract the second equation from the first.

EXECUTE Multiplying the first expression by 32 and the second expression by 27 gives

$$864T_A + 416T_B = 0$$
$$864T_A + 1404T_B = 594$$

Subtracting the second equation from the first gives

$$864T_A + 416T_B - 864T_A - 1404T_B = 0 - 594,$$
$$988T_B = 594,$$
$$T_B = \frac{594}{988} = 0.601.$$

Substituting the value for T_B back into either expression to find T_A yields

$$32T_A + 52(0.601) = 22,$$
$$T_A = -0.289.$$

The two equations together result in $T_A = -0.29$ and $T_B = 0.60$.

EVALUATE An alternative solution would be to write T_B in terms of T_A, using the first expression, and then substitute for T_B into the second expression. This gives the same result. You may choose the method you prefer and may end up applying both to particular classes of problems.

If we encounter *three* unknowns in an expression, how many equations will we need to solve for each unknown simultaneously? Three equations will be needed to solve for the three unknown quantities.

Practice Problem: Solve the following expressions for x and y: $8x - 2y = -24$ and $-3x + 4y = 22$. Answer: $x = -2, y = 4$

Extra Practice: Solve the following expressions for a and b: $4a - 3b = 13$ and $7a + 8b = -17$. Answer: $a = 1, b = -3$

8: Review of the quadratic formula

The position of a ball tossed in the air depends on the initial speed of the ball and the time elapsed and is given by

$$y = v_0 t - (4.9 \text{ m/s}^2)t^2,$$

where v_0 is the initial speed and t is the time elapsed. For a ball tossed with an initial speed of 30.0 m/s, find the time(s) when the ball is at a height of 12.5 m.

IDENTIFY AND SET UP We recognize that the equation is quadratic, since it has a t^2 term, a t term, and a constant term. If we try to rewrite the equation in terms of t alone, we find that we cannot easily isolate the t term. We'll employ the quadratic formula to solve the problem.

EXECUTE We rewrite the equation, substituting the given values:

$$12.5 \text{ m} = (30.0 \text{ m/s})t - (4.9 \text{ m/s}^2)t^2.$$

The quadratic formula requires that the equation be written as $ax^2 + bx + c = 0$, so we rearrange terms to yield

$$(-4.9 \text{ m/s}^2)t^2 + (30.0 \text{ m/s})t + (-12.5 \text{ m}) = 0.$$

From this rearrangement, we see that $a = -4.9 \text{ m/s}^2$, $b = 30.0 \text{ m/s}$, and $c = -12.5 \text{ m}$. The solutions of the quadratic equation are

$$x = \frac{-b \pm \sqrt{b^2 - 4ac}}{2a}.$$

Substituting our values into the quadratic equation gives

$$t = \frac{-(30.0 \text{ m/s}) \pm \sqrt{(30.0 \text{ m/s})^2 - 4(-4.9 \text{ m/s}^2)(-1.25 \text{ m})}}{2(-4.9 \text{ m/s}^2)}.$$

Multiplying out the terms and canceling the units produces

$$t = \frac{-(30.0 \text{ m/s}) \pm \sqrt{(900.0 - 245.0)(\text{m}^2/\text{s}^2)}}{-9.8 \, (\text{m/s}^2)} = \frac{-30.0 \pm 25.59}{-9.8} \text{ s} = 0.445 \text{ s}, 5.67 \text{ s}.$$

There are two times when the ball is at a height of 12.5 m: 0.445 s and 5.67 s. These are, respectively, when the ball is rising to its maximum height and when it is falling from its maximum height.

EVALUATE You must learn to recognize quadratic equations. Once you identify a quadratic equation, the solution is straightforward (although it requires careful algebra). Quadratic equations result in two solutions, and you must be able to interpret their meanings. In this case, the two solutions corresponded to the upward and downward motion of the ball. You may need only one of the solutions for your situation. If neither solution seems reasonable, then you should check your work.

CAUTION Watch units! It's important to check units every time you write equations. If we found incorrect units when we solved for time, we would've discovered a mistake that would've been quickly corrected.

Practice Problem: Solve for x: $3x^2 + 2x - 1 = 0$. *Answer:* $-1, 1/3$

Extra Practice: Solve for b: $b = \sqrt{b+4}$. *Answer:* $-1.56, 2.56$

Problem Summary

The problems in this chapter represent a foundation that you will use throughout your physics course. Common elements make up good problem-solving techniques, including

- Identifying a procedure to find the solution.
- Making a sketch when no figure is provided.
- Adding appropriate coordinate systems to the sketch.
- Identifying the known and unknown quantities in the problem.
- Finding appropriate equations to solve for the unknown quantities.
- Checking for consistency of units in derived equations.
- Evaluating results to check for inconsistencies.

We'll see how these techniques apply to a wide variety of problems as we progress. Although they may seem cumbersome right now, they'll help you solve the problems you encounter.

Key Example Variation Problems

Solutions to these problems are in Chapter 1 of the Student's Solutions Manual.

Be sure to review EXAMPLE 1.7 (Section 1.8) before attempting these problems.

VP1.7.1 Consider the three vectors $\vec{A}, \vec{B},$ and $\vec{C}$ in Example 1.7. If a fourth vector $\vec{D}$ is added to $\vec{A} + \vec{B} + \vec{C}$, the result is zero: $\vec{A} + \vec{B} + \vec{C} + \vec{D} = 0$. Find the magnitude and direction of $\vec{D}$. State the direction of $\vec{D}$ in terms of an angle measured counterclockwise from the positive *x*-axis, and state in which quadrant this angle lies.

VP1.7.2 Consider the three vectors $\vec{A}, \vec{B},$ and $\vec{C}$ in Example 1.7. Calculate the magnitude and direction of the vector $\vec{S} = \vec{A} - \vec{B} + \vec{C}$. State the direction of $\vec{S}$ in terms of an angle measured counterclockwise from the positive *x*-axis, and state in which quadrant this angle lies. (*Hint:* The components of $-\vec{B}$ are just the negatives of the components of $\vec{B}$.)

VP1.7.3 Consider the three vectors $\vec{A}, \vec{B},$ and $\vec{C}$ in Example 1.7. (a) Find the components of the vector $\vec{T} = \vec{A} - \vec{B} + 2\vec{C}$. (b) Find the magnitude and direction of $\vec{T}$. State the direction of $\vec{T}$ in terms of an angle measured counterclockwise from the positive *x*-axis, and state in which quadrant this angle lies.

VP1.7.4 A hiker undergoes the displacement $\vec{A}$ shown in Example 1.7. The hiker then undergoes a second displacement such that she ends up 38.0 m from her starting point, in a direction from her starting point that is 37.0° west of north. Find the magnitude and direction of this second displacement. State the direction in terms of an angle measured counterclockwise from the positive *x*-axis, and state in which quadrant this angle lies.

Be sure to review EXAMPLES 1.9 and 1.10 (Section 1.10) before attempting these problems.

VP1.10.1 Vector $\vec{A}$ has magnitude 5.00 and is at an angle of 36.9° south of east. Vector $\vec{B}$ has magnitude 6.40 and is at an angle of 20.0° west of north. (a) Choose the positive *x*-direction to the east and the positive *y*-direction to the north. Find the components of $\vec{A}$ and $\vec{B}$. (b) Calculate the scalar product $\vec{A} \cdot \vec{B}$.

VP1.10.2 Vector $\vec{C}$ has magnitude 6.50 and is at an angle of 55.0° measured counterclockwise from the +*x*-axis toward the +*y*-axis. Vector $\vec{D}$ has components $Dx = +4.80$ and $Dy = -8.40$. (a) Calculate the scalar product $\vec{C} \cdot \vec{D}$. (b) Find the angle φ between the vectors $\vec{C}$ and $\vec{D}$.

VP1.10.3 Vector $\vec{A}$ has components $Ax = -5.00, Ay = 3.00,$ and $Az = 0$. Vector $\vec{B}$ has components $Bx = 2.50, By = 4.00,$ and $Bz = -1.50$. Find the angle between the two vectors.

VP1.10.4 If a force $\vec{F}$ acts on an object as that object moves through a displacement $\vec{s}$, the *work* done by that force equals the scalar product of $\vec{F}$ and $\vec{s}$: $W = \vec{F} \cdot \vec{s}$. A certain object moves through displacement $\vec{s} = (4.00 \text{ m})\hat{i} + (5.00 \text{ m})\hat{j}$. As it moves it is acted on by force $\vec{F}$, which has x-component $F_x = -12.0$ N (1 N = 1 newton is the SI unit of force). The work done by this force is $26.0 \text{ N} \cdot \text{m} = 26.0$ J (1 J = 1 joule = 1 newton-meter is the SI unit of work). (a) Find the y-component of $\vec{F}$. (b) Find the angle between $\vec{F}$ and $\vec{s}$.

Student's Solutions Manual for Units, Physical Quantities, and Vectors

VP1.7.1. **IDENTIFY:** We know that the sum of three known vectors and a fourth unknown vector is zero. We want to find the magnitude and direction of the unknown vector.
SET UP: The sum of their x-components and the sum of their y-components must both be zero.
$A_x + B_x + C_x + D_x = 0$
$A_y + B_y + C_y + D_y = 0$
The magnitude of a vector is w $A = \sqrt{A_x^2 + A_y^2}$ and the angle θ it makes with the +x-axis is
$\theta = \arctan \dfrac{A_y}{A_x}$.
EXECUTE: We use the results of Ex. 1.7. See Fig. 1.23 in the textbook.
$A_x = 38.37$ m, $B_x = -46.36$ m, $C_x = 0.00$ m, $A_y = 61.40$ m, $B_y = -33.68$ m, $C_y = -17.80$ m
Adding the x-components gives
38.37 m + (–46.36 m) + 0.00 m + D_x = 0 → D_x = 7.99 m
Adding the y-components gives
61.40 m + (–33.68 m) + (–17.80 m) + D_y = 0 → D_y = –9.92 m
$D = \sqrt{D_x^2 + D_y^2} = \sqrt{(7.99 \text{ m})^2 + (-9.92 \text{ m})^2} = 12.7$ m
$\theta = \arctan \dfrac{D_y}{D_x} = \arctan[(-9.92 \text{ m})/(7.99 \text{ m})] = -51°$
Since $\vec{D}$ has a positive x-component and a negative y-component, it points into the fourth quadrant making an angle of 51° below the +x-axis and an angle of 360° – 51° = 309° counterclockwise with the +x-axis.
EVALUATE: The vector $\vec{D}$ has the same magnitude as the resultant in Ex. 1.7 but points in the opposite direction. This is reasonable because $\vec{D}$ must be opposite to the resultant of the three vectors in Ex. 1.7 to make the resultant of all four vectors equal to zero.

VP1.7.2. **IDENTIFY:** We know three vectors $\vec{A}$, $\vec{B}$, and $\vec{C}$ and we want to find the sum $\vec{S}$ where $\vec{S} = \vec{A} - \vec{B} + \vec{C}$. The components of $-\vec{B}$ are the negatives of the components of $\vec{B}$.
SET UP: The components of $\vec{S}$ are
$S_x = A_x - B_x + C_x$
$S_y = A_y - B_y + C_y$

The magnitude A of a vector $\vec{A}$ is $A = \sqrt{A_x^2 + A_y^2}$ and the angle θ it makes with the $+x$-axis is $\theta = \arctan\dfrac{A_y}{A_x}$.

EXECUTE: Using the components from Ex. 1.7 we have
$S_x = 38.37\text{ m} - (-46.36\text{ m}) + 0.00\text{ m} = 84.73\text{ m}$
$S_y = 61.40\text{ m} - (-33.68\text{ m}) + (-17.80\text{ m}) = 77.28\text{ m}$
$S = \sqrt{S_x^2 + S_y^2} = \sqrt{(84.73\text{ m})^2 + (77.28\text{ m})^2} = 115\text{ m}$
$\theta = \arctan\dfrac{S_y}{S_x} = \arctan[(77.28\text{ m})/(84.73\text{ m})] = 42°$

Since both components of $\vec{S}$ are positive, $\vec{S}$ points into the first quadrant. Therefore it makes an angle of 42° with the $+x$-axis.
EVALUATE:

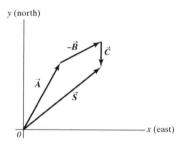

Figure VP1.7.2

The graphical solution shown in Fig. VP1.7.2 shows that our results are reasonable.

VP1.7.3. IDENTIFY: We know three vectors $\vec{A}$, $\vec{B}$, and $\vec{C}$ and we want to find the sum $\vec{T}$ where $\vec{T} = \vec{A} + \vec{B} + 2\vec{C}$.

SET UP: Find the components of vectors $\vec{A}$, $\vec{B}$, and $\vec{C}$ and use them to find the magnitude and direction of $\vec{T}$. The components of $2\vec{C}$ are twice those of $\vec{C}$.

EXECUTE: $S_x = A_x + B_x + 2C_x$ and $S_y = A_y + B_y + 2C_y$

(a) Using the components from Ex. 1.7 gives
$T_x = 38.37\text{ m} + (-46.36\text{ m}) + 2(0.00\text{ m}) = -7.99\text{ m}$
$T_y = 61.40\text{ m} + (-33.68\text{ m}) + 2(-17.80\text{ m}) = -7.88\text{ m}$
(b) $T = \sqrt{T_x^2 + T_y^2} = \sqrt{(-7.99\text{ m})^2 + (-7.88\text{ m})^2} = 11.2\text{ m}$
$\theta = \arctan\dfrac{T_y}{T_x} = \arctan[(-7.88\text{ m})/(-7.99\text{ m})] = 45°$

Both components of $\vec{T}$ are negative, so it points into the third quadrant, making an angle of 45° below the $-x$-axis or $45° + 180° = 225°$ counterclockwise with the $+x$-axis, in the third quadrant.
EVALUATE:

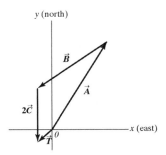

Figure VP1.7.3

The graphical solution shown in Fig. VP1.7.3 shows that this result is reasonable.

VP1.7.4. **IDENTIFY:** The hiker makes two displacements. We know the first one and their resultant, and we want to find the second displacement.

SET UP: Calling $\vec{A}$ the known displacement, $\vec{R}$ the known resultant, and $\vec{D}$ the unknown vector, we know that $\vec{A} + \vec{D} = \vec{R}$. We also know that $R = 38.0$ m and $\vec{R}$ makes an angle $\theta_R = 37.0° + 90° = 127°$ with the $+x$-axis. Fig. VP1.7.4 shows a sketch of these vectors.

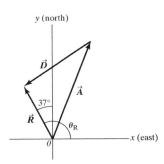

Figure VP1.7.4

EXECUTE: From Ex. 1.7 we have $A_x = 38.37$ m and $A_y = 61.40$ m. The components of $\vec{R}$ are
$R_x = R \cos 127.0° = (38.0 \text{ m}) \cos 127.0° = -22.87$ m
$R_y = R \sin 38.0° = (38.0 \text{ m}) \sin 127.0° = 30.35$ m
$R_x = A_x + D_x$ and $R_y = A_y + D_y$

Using these components, we find the components of $\vec{D}$.
38.37 m + D_x = −22.87 m   → D_x = −22.87 m
61.40 m + D_y = −31.05 m   → D_y = −31.05 m

$D = \sqrt{D_x^2 + D_y^2} = \sqrt{(-61.24 \text{ m})^2 + (-31.05 \text{ m})^2} = 68.7$ m

$\theta = \arctan \dfrac{D_y}{D} = \arctan[(-31.05 \text{ m})/(-61.24 \text{ m})] = 27°$

Both components of $\vec{D}$ are negative, so it points into the third quadrant, making an angle of 27° + 180° = 207° with the $+x$-axis.

EVALUATE: A graphical solution will confirm these results.

VP1.10.1. **IDENTIFY:** We know the magnitude and direction of two vectors. We want to use these to find their components and their scalar product.

SET UP: $A_x = A\cos\theta_A$, $A_y = A\sin\theta_A$, $B_x = B\cos\theta_B$, $B_y = B\sin\theta_B$. We can find the scalar product using the vector components or using their magnitudes and the angle between them. $\vec{A}\cdot\vec{B} = A_x B_x + A_y B_y$ and $\vec{A}\cdot\vec{B} = AB\cos\phi$. Which form you use depends on the information you have.

EXECUTE: (a) $A_x = A\cos\theta_A = (5.00)\cos(360° - 36.9°) = 4.00$
$A_y = A\sin\theta_A = (5.00)\sin(360° - 36.9°) = -3.00$
$B_x = (6.40)\cos(90° + 20.0°) = -2.19$
$B_y = (6.40)\sin(90° + 20.0°) = 6.01$
(b) Using components gives
$\vec{A}\cdot\vec{B} = A_x B_x + A_y B_y = (4.00)(-2.19) + (-3.00)(6.01) = -26.8$

EVALUATE: We check by using $\vec{A}\cdot\vec{B} = AB\cos\phi$.
$\vec{A}\cdot\vec{B} = AB\cos\phi = (5.00)(6.40)\cos(20.0° + 90° + 36.9°) = -26.8$
This agrees with our result in part (b).

VP1.10.2. **IDENTIFY:** We know the magnitude and direction of one vector and the components of another vector. We want to use these to find their scalar product and the angle between them.

SET UP: The scalar product can be expressed as $\vec{A}\cdot\vec{B} = A_x B_x + A_y B_y$ and $\vec{A}\cdot\vec{B} = AB\cos\phi$. Which form you use depends on the information you have.

EXECUTE: (a) $C_x = C\cos\theta_C = (6.50)\cos 55.0° = 3.728$
$C_y = C\sin\theta_C = (6.50)\sin 55.0° = 5.324$
$D_x = 4.80$ and $D_y = -8.40$
Using components gives $\vec{C}\cdot\vec{D} = C_x D_x + C_y D_y = (3.728)(4.80) + (5.324)(-8.40) = -26.8$

(b) $D = \sqrt{D_x^2 + D_y^2} = \sqrt{(4.80)^2 + (-8.40)^2} = 9.675$
$\vec{C}\cdot\vec{D} = CD\cos\phi$, so $\cos\phi = \vec{C}\cdot\vec{D}/CD = (-26.8)/[(6.50)(9.67)] = -0.426$. $\phi = 115°$.

EVALUATE: Find the angle that $\vec{D}$ makes with the $+x$-axis.

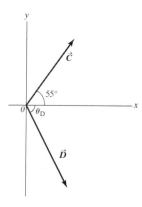

Figure VP1.10.2

$\theta_D = \arctan\dfrac{D_y}{D_x} = \arctan[8.40/(-4.80)] = -60.3°$, which is 60.3° below the $+x$-axis. From Fig. VP1.10.2, we can easily seed that the angle between $\vec{C}$ and $\vec{D}$ is $\phi = 60.3° + 55.0° = 115°$, as we found in (b).

VP1.10.3. IDENTIFY: We know the components of two vectors and want to find the angle between them.
SET UP: The scalar product $\vec{A} \cdot \vec{B} = AB \cos\phi$ involves the angle between two vectors. We can find this product using components from $\vec{A} \cdot \vec{B} = A_x B_x + A_y B_y$. From this result we can find the angle ϕ.
EXECUTE: First find the magnitudes of the two vectors.
$A = \sqrt{A_x^2 + A_y^2 + A_z^2} = \sqrt{(-5.00)^2 + (3.00)^2 + 0^2} = 5.83$
$B = \sqrt{B_x^2 + B_y^2 + B_z^2} = \sqrt{(2.50)^2 + (4.00)^2 + (-1.50)^2} = 4.95$
Now use $\vec{A} \cdot \vec{B} = AB \cos\phi = A_x B_x + A_y B_y$ and solve for ϕ.
$(5.83)(4.95) \cos\phi = (-5.00)(2.50) + (3.00)(4.00) + (0)(-1.50) \rightarrow \phi = 91°$.
EVALUATE: The scalar product is positive, so ϕ must be between 90° and 180°, which agrees with our result.

VP1.10.4. IDENTIFY: We know the scalar product of two vectors. We also know both components of one of them and the x-component of the other one. We want to find the y-component of the other one and the angle between the two vectors. The scalar product involves the angle between two vectors.
SET UP: We use $\vec{A} \cdot \vec{B} = A_x B_x + A_y B_y$ and $\vec{A} \cdot \vec{B} = AB \cos\phi$.
EXECUTE: (a) Use $\vec{F} \cdot \vec{s} = F_x s_x + F_y s_y$ to find F_y.
$26.0 \text{ N} \cdot \text{m} = (-12.0 \text{ N})(4.00 \text{ m}) + F_y(5.00 \text{ m}) \rightarrow F_y = 14.8 \text{ N}$.
(b) Use $\vec{F} \cdot \vec{s} = Fs \cos\phi$ and $A = \sqrt{A_x^2 + A_y^2}$ to find the magnitudes of the two vectors.
$\sqrt{(-12.0 \text{ N})^2 + (14.8 \text{ N})^2} \sqrt{(4.00 \text{ m})^2 + (5.00 \text{ m})^2} \cos\phi = 26.0 \text{ N} \cdot \text{m} \rightarrow \phi = 77.7°$.
EVALUATE: The work is positive, so the angle between $\vec{F}$ and $\vec{s}$ must be between 0° and 90°, which agrees with our result in part (b).

1.3. IDENTIFY: We know the speed of light in m/s. $t = d/v$. Convert 1.00 ft to m and t from s to ns.
SET UP: The speed of light is $v = 3.00 \times 10^8$ m/s. 1 ft = 0.3048 m. 1 s = 10^9 ns.
EXECUTE: $t = \dfrac{0.3048 \text{ m}}{3.00 \times 10^8 \text{ m/s}} = 1.02 \times 10^{-9}$ s = 1.02 ns
EVALUATE: In 1.00 s light travels 3.00×10^8 m = 3.00×10^5 km = 1.86×10^5 mi.

1.7. IDENTIFY: Convert miles/gallon to km/L.
SET UP: 1 mi = 1.609 km. 1 gallon = 3.788 L.
EXECUTE: (a) 55.0 miles/gallon = $(55.0 \text{ miles/gallon})\left(\dfrac{1.609 \text{ km}}{1 \text{ mi}}\right)\left(\dfrac{1 \text{ gallon}}{3.788 \text{ L}}\right) = 23.4$ km/L.
(b) The volume of gas required is $\dfrac{1500 \text{ km}}{23.4 \text{ km/L}} = 64.1$ L. $\dfrac{64.1 \text{ L}}{45 \text{ L/tank}} = 1.4$ tanks.
EVALUATE: 1 mi/gal = 0.425 km/L. A km is very roughly half a mile and there are roughly 4 liters in a gallon, so 1 mi/gal ~ $\frac{2}{4}$ km/L, which is roughly our result.

1.9. IDENTIFY: We know the density and mass; thus we can find the volume using the relation density = mass/volume = m/V. The radius is then found from the volume equation for a sphere and the result for the volume.

SET UP: Density = 19.5 g/cm^3 and $m_{\text{critical}} = 60.0$ kg. For a sphere $V = \frac{4}{3}\pi r^3$.

EXECUTE: $V = m_{\text{critical}}/\text{density} = \left(\dfrac{60.0 \text{ kg}}{19.5 \text{ g/cm}^3}\right)\left(\dfrac{1000 \text{ g}}{1.0 \text{ kg}}\right) = 3080 \text{ cm}^3$.

$r = \sqrt[3]{\dfrac{3V}{4\pi}} = \sqrt[3]{\dfrac{3}{4\pi}(3080 \text{ cm}^3)} = 9.0$ cm.

EVALUATE: The density is very large, so the 130-pound sphere is small in size.

1.15. IDENTIFY: Express 200 kg in pounds. Express each of 200 m, 200 cm, and 200 mm in inches. Express 200 months in years.

SET UP: A mass of 1 kg is equivalent to a weight of about 2.2 lbs. 1 in. = 2.54 cm. 1 y = 12 months.

EXECUTE: **(a)** 200 kg is a weight of 440 lb. This is much larger than the typical weight of a man.

(b) $200 \text{ m} = (2.00 \times 10^4 \text{ cm})\left(\dfrac{1 \text{ in.}}{2.54 \text{ cm}}\right) = 7.9 \times 10^3$ inches. This is much greater than the height of a person.

(c) 200 cm = 2.00 m = 79 inches = 6.6 ft. Some people are this tall, but not an ordinary man.

(d) 200 mm = 0.200 m = 7.9 inches. This is much too short.

(e) $200 \text{ months} = (200 \text{ mon})\left(\dfrac{1 \text{ y}}{12 \text{ mon}}\right) = 17$ y. This is the age of a teenager; a middle-aged man is much older than this.

EVALUATE: None are plausible. When specifying the value of a measured quantity it is essential to give the units in which it is being expressed.

1.17. IDENTIFY: Estimation problem.

SET UP: Estimate that the pile is 18 in. × 18 in. × 5 ft 8 in.. Use the density of gold to calculate the mass of gold in the pile and from this calculate the dollar value.

EXECUTE: The volume of gold in the pile is $V = 18 \text{ in.} \times 18 \text{ in.} \times 68 \text{ in.} = 22{,}000 \text{ in.}^3$. First convert to cm^3:

$V = 22{,}000 \text{ in.}^3 (1000 \text{ cm}^3/61.02 \text{ in.}^3) = 3.6 \times 10^5 \text{ cm}^3$.

The density of gold is 19.3 g/cm^3, so the mass of this volume of gold is

$m = (19.3 \text{ g/cm}^3)(3.6 \times 10^5 \text{ cm}^3) = 6.95 \times 10^6$ g.

The monetary value of one gram is $40, so the gold has a value of

($40/gram)($6.95 \times 10^6$ grams) = 2.8×10^8

or about 300×10^6 (three hundred million dollars).

EVALUATE: This is quite a large pile of gold, so such a large monetary value is reasonable.

1.19. IDENTIFY: Estimate the diameter of a drop and from that calculate the volume of a drop, in m^3. Convert m^3 to L.

SET UP: Estimate the diameter of a drop to be $d = 2$ mm. The volume of a spherical drop is $V = \frac{4}{3}\pi r^3 = \frac{1}{6}\pi d^3$. 10^3 $cm^3 = 1$ L.

EXECUTE: $V = \frac{1}{6}\pi(0.2\text{ cm})^3 = 4 \times 10^{-3}$ cm^3. The number of drops in 1.0 L is
$$\frac{1000 \text{ cm}^3}{4 \times 10^{-3} \text{ cm}^3} = 2 \times 10^5.$$

EVALUATE: Since $V \sim d^3$, if our estimate of the diameter of a drop is off by a factor of 2 then our estimate of the number of drops is off by a factor of 8.

1.23. IDENTIFY: Since she returns to the starting point, the vector sum of the four displacements must be zero.

SET UP: Call the three given displacements $\vec{A}$, $\vec{B}$, and $\vec{C}$, and call the fourth displacement $\vec{D}$. $\vec{A} + \vec{B} + \vec{C} + \vec{D} = 0$.

EXECUTE: The vector addition diagram is sketched in Figure 1.23. Careful measurement gives that $\vec{D}$ is 144 m, 41° south of west.

EVALUATE: $\vec{D}$ is equal in magnitude and opposite in direction to the sum $\vec{A} + \vec{B} + \vec{C}$.

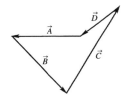

Figure 1.23

1.29. IDENTIFY: If $\vec{C} = \vec{A} + \vec{B}$, then $C_x = A_x + B_x$ and $C_y = A_y + B_y$. Use C_x and C_y to find the magnitude and direction of $\vec{C}$.

SET UP: From Fig. E1.30 in the textbook, $A_x = 0$, $A_y = -8.00$ m and $B_x = +B\sin 30.0° = 7.50$ m, $B_y = +B\cos 30.0° = 13.0$ m.

EXECUTE: (a) $\vec{C} = \vec{A} + \vec{B}$ so $C_x = A_x + B_x = 7.50$ m and $C_y = A_y + B_y = +5.00$ m. $C = 9.01$ m.
$\tan\theta = \frac{C_y}{C_x} = \frac{5.00 \text{ m}}{7.50 \text{ m}}$ and $\theta = 33.7°$.

(b) $\vec{B} + \vec{A} = \vec{A} + \vec{B}$, so $\vec{B} + \vec{A}$ has magnitude 9.01 m and direction specified by 33.7°.

(c) $\vec{D} = \vec{A} - \vec{B}$ so $D_x = A_x - B_x = -7.50$ m and $D_y = A_y - B_y = -21.0$ m. $D = 22.3$ m.
$\tan\phi = \frac{D_y}{D_x} = \frac{-21.0 \text{ m}}{-7.50 \text{ m}}$ and $\phi = 70.3°$. $\vec{D}$ is in the third quadrant and the angle θ counterclockwise from the $+x$-axis is $180° + 70.3° = 250.3°$.

(d) $\vec{B} - \vec{A} = -(\vec{A} - \vec{B})$, so $\vec{B} - \vec{A}$ has magnitude 22.3 m and direction specified by $\theta = 70.3°$.

EVALUATE: These results agree with those calculated from a scale drawing in Problem 1.22.

1.35. IDENTIFY: Find the components of each vector and then use the general equation $\vec{A} = A_x\hat{i} + A_y\hat{j}$ for a vector in terms of its components and unit vectors.

SET UP: $A_x = 0$, $A_y = -8.00$ m. $B_x = 7.50$ m, $B_y = 13.0$ m. $C_x = -10.9$ m, $C_y = -5.07$ m. $D_x = -7.99$ m, $D_y = 6.02$ m.

EXECUTE: $\vec{A} = (-8.00 \text{ m})\hat{j}$; $\vec{B} = (7.50 \text{ m})\hat{i} + (13.0 \text{ m})\hat{j}$; $\vec{C} = (-10.9 \text{ m})\hat{i} + (-5.07 \text{ m})\hat{j}$; $\vec{D} = (-7.99 \text{ m})\hat{i} + (6.02 \text{ m})\hat{j}$.

EVALUATE: All these vectors lie in the xy-plane and have no z-component.

1.39. IDENTIFY: $\vec{A}$ and $\vec{B}$ are given in unit vector form. Find A, B and the vector difference $\vec{A} - \vec{B}$.

SET UP: $\vec{A} = -2.00\hat{i} + 3.00\hat{j} + 4.00\hat{k}$, $\vec{B} = 3.00\hat{i} + 1.00\hat{j} - 3.00\hat{k}$

Use $A = \sqrt{A_x^2 + A_y^2 + A_z^2}$ to find the magnitudes of the vectors.

EXECUTE: (a) $A = \sqrt{A_x^2 + A_y^2 + A_z^2} = \sqrt{(-2.00)^2 + (3.00)^2 + (4.00)^2} = 5.38$

$B = \sqrt{B_x^2 + B_y^2 + B_z^2} = \sqrt{(3.00)^2 + (1.00)^2 + (-3.00)^2} = 4.36$

(b) $\vec{A} - \vec{B} = (-2.00\hat{i} + 3.00\hat{j} + 4.00\hat{k}) - (3.00\hat{i} + 1.00\hat{j} - 3.00\hat{k})$

$\vec{A} - \vec{B} = (-2.00 - 3.00)\hat{i} + (3.00 - 1.00)\hat{j} + (4.00 - (-3.00))\hat{k} = -5.00\hat{i} + 2.00\hat{j} + 7.00\hat{k}$.

(c) Let $\vec{C} = \vec{A} - \vec{B}$, so $C_x = -5.00$, $C_y = +2.00$, $C_z = +7.00$

$C = \sqrt{C_x^2 + C_y^2 + C_z^2} = \sqrt{(-5.00)^2 + (2.00)^2 + (7.00)^2} = 8.83$

$\vec{B} - \vec{A} = -(\vec{A} - \vec{B})$, so $\vec{A} - \vec{B}$ and $\vec{B} - \vec{A}$ have the same magnitude but opposite directions.

EVALUATE: A, B, and C are each larger than any of their components.

1.41. IDENTIFY: $\vec{A} \cdot \vec{B} = AB \cos\phi$

SET UP: For $\vec{A}$ and $\vec{B}$, $\phi = 150.0°$. For $\vec{B}$ and $\vec{C}$, $\phi = 145.0°$. For $\vec{A}$ and $\vec{C}$, $\phi = 65.0°$.

EXECUTE: (a) $\vec{A} \cdot \vec{B} = (8.00 \text{ m})(15.0 \text{ m})\cos 150.0° = -104 \text{ m}^2$

(b) $\vec{B} \cdot \vec{C} = (15.0 \text{ m})(12.0 \text{ m})\cos 145.0° = -148 \text{ m}^2$

(c) $\vec{A} \cdot \vec{C} = (8.00 \text{ m})(12.0 \text{ m})\cos 65.0° = 40.6 \text{ m}^2$

EVALUATE: When $\phi < 90°$ the scalar product is positive and when $\phi > 90°$ the scalar product is negative.

1.43. IDENTIFY: For all of these pairs of vectors, the angle is found from combining $\vec{A} \cdot \vec{B} = AB \cos\phi$ and $\vec{A} \cdot \vec{B} = A_x B_x + A_y B_y + A_z B_z$, to give the angle ϕ as

$$\phi = \arccos\left(\frac{\vec{A} \cdot \vec{B}}{AB}\right) = \arccos\left(\frac{A_x B_x + A_y B_y}{AB}\right).$$

SET UP: $\vec{A} \cdot \vec{B} = A_x B_x + A_y B_y + A_z B_z$ shows how to obtain the components for a vector written in terms of unit vectors.

EXECUTE: (a) $\vec{A} \cdot \vec{B} = -22$, $A = \sqrt{40}$, $B = \sqrt{13}$, and so $\phi = \arccos\left(\frac{-22}{\sqrt{40}\sqrt{13}}\right) = 165°$.

(b) $\vec{A} \cdot \vec{B} = 60$, $A = \sqrt{34}$, $B = \sqrt{136}$, $\phi = \arccos\left(\dfrac{60}{\sqrt{34}\sqrt{136}}\right) = 28°$.

(c) $\vec{A} \cdot \vec{B} = 0$ and $\phi = 90°$.

EVALUATE: If $\vec{A} \cdot \vec{B} > 0$, $0 \le \phi < 90°$. If $\vec{A} \cdot \vec{B} < 0$, $90° < \phi \le 180°$. If $\vec{A} \cdot \vec{B} = 0$, $\phi = 90°$ and the two vectors are perpendicular.

1.45. IDENTIFY: $\vec{A} \times \vec{D}$ has magnitude $AD \sin \phi$. Its direction is given by the right-hand rule.

SET UP: $\phi = 180° - 53° = 127°$

EXECUTE: **(a)** $|\vec{A} \times \vec{D}| = (8.00 \text{ m})(10.0 \text{ m})\sin 127° = 63.9 \text{ m}^2$. The right-hand rule says $\vec{A} \times \vec{D}$ is in the $-z$-direction (into the page).

(b) $\vec{D} \times \vec{A}$ has the same magnitude as $\vec{A} \times \vec{D}$ and is in the opposite direction.

EVALUATE: The component of $\vec{D}$ perpendicular to $\vec{A}$ is $D_\perp = D\sin 53.0° = 7.99$ m. $|\vec{A} \times \vec{D}| = AD_\perp = 63.9 \text{ m}^2$, which agrees with our previous result.

1.47. IDENTIFY: This problem involves the vector product of two vectors.

SET UP: The magnitude is $|\vec{A} \times \vec{B}| = AB \sin \phi$ and the right-hand rule gives the direction. Since $\vec{A} \times \vec{B}$ is in the $+z$-direction, both $\vec{A}$ and $\vec{B}$ must lie in the xy-plane.

EXECUTE: $\vec{A}$ has no y-component and $\vec{B}$ has no x-component, so they must be perpendicular to each other. Since $\vec{A} \times \vec{B}$ is in the $+z$-direction, the right-hand rule tells us that $\vec{B}$ must point in the $-y$-direction. $|\vec{A} \times \vec{B}| = AB \sin \phi = (8.0 \text{ m})B \sin 90° = 16.0 \text{ m}^2$, so $B = 2.0$ m.

EVALUATE: In unit vector notation, $\vec{B} = -2.0 \text{ m } \hat{j}$.

1.51. IDENTIFY: The density relates mass and volume. Use the given mass and density to find the volume and from this the radius.

SET UP: The earth has mass $m_E = 5.97 \times 10^{24}$ kg and radius $r_E = 6.37 \times 10^6$ m. The volume of a sphere is $V = \frac{4}{3}\pi r^3$. $\rho = 1.76 \text{ g/cm}^3 = 1760 \text{ km/m}^3$.

EXECUTE: **(a)** The planet has mass $m = 5.5 m_E = 3.28 \times 10^{25}$ kg.

$$V = \frac{m}{\rho} = \frac{3.28 \times 10^{25} \text{ kg}}{1760 \text{ kg/m}^3} = 1.86 \times 10^{22} \text{ m}^3.$$

$$r = \left(\frac{3V}{4\pi}\right)^{1/3} = \left(\frac{3[1.86 \times 10^{22} \text{ m}^3]}{4\pi}\right)^{1/3} = 1.64 \times 10^7 \text{ m} = 1.64 \times 10^4 \text{ km}$$

(b) $r = 2.57 r_E$

EVALUATE: Volume V is proportional to mass and radius r is proportional to $V^{1/3}$, so r is proportional to $m^{1/3}$. If the planet and earth had the same density its radius would be $(5.5)^{1/3} r_E = 1.8 r_E$. The radius of the planet is greater than this, so its density must be less than that of the earth.

1.55. **IDENTIFY:** We are dealing with unit vectors, which must have magnitude 1. We will need to use the scalar product and to express vectors using the unit vectors.

SET UP: $A = \sqrt{A_x^2 + A_y^2 + A_z^2}$, $\vec{A} \cdot \vec{B} = A_x B_x + A_y B_y + A_z B_z$. If two vectors are perpendicular, their scalar product is zero.

EXECUTE: **(a)** If we divide a vector by its magnitude, the result will have magnitude 1 but still point in the same direction as the original vector, so it will be a unit vector. First find the magnitude of the given vector. $A = \sqrt{A_x^2 + A_z^2} = \sqrt{(3.0)^2 + (-4.0)^2} = 5.0$. Therefore $\dfrac{3.0\hat{i} - 4.0\hat{k}}{5} = 0.60\hat{i} - 0.80\hat{k}$ is a unit vector that is parallel to $\vec{A}$.

(b) Reversing the direction of the unit vector in (a) will make it antiparallel to $\vec{A}$, so the unit vector is $-0.60\hat{i} + 0.80\hat{k}$.

(c) Call $\vec{B}$ the unknown unit vector. Since it has no y-component, we can express it as $\vec{B} = B_x \hat{i} + B_z \hat{k}$. Since $\vec{A}$ and $\vec{B}$ are perpendicular, $\vec{A} \cdot \vec{B} = 0$, so $A_x B_x + A_y B_y + A_z B_z = 0$. This gives $(3.0)B_x - (4.0)B_z = 0 \rightarrow B_z = 0.75 B_x$. Since $\vec{B}$ is a unit vector, we have $B_x^2 + B_z^2 = B_x^2 + (0.75 B_x)^2 = 1$. Solving gives $B_x = \pm 0.80$. Therefore $B_z = \pm(0.75 B_x) = \pm(0.75)(0.80) = \pm 0.60$. Therefore $\vec{B} = \pm(0.80\hat{i} + 0.60\hat{k})$, so the two unit vectors are $\vec{B}_+ = 0.80\hat{i} + 0.60\hat{k}$ and $\vec{B}_- = -0.80\hat{i} - 0.60\hat{k}$.

EVALUATE: $\vec{A} \cdot \vec{B}_+ = (3.0)(0.80) + (-4.0)(6.0) = 0$ and $\vec{A} \cdot \vec{B}_- = (3.0)(-0.80) + (-4.0)(-6.0) = 0$, so the two vectors are perpendicular to $\vec{A}$. Their magnitudes are $\sqrt{(\pm 0.80)^2 + (\pm 0.60)^2} = 1$, so they are unit vectors.

1.57. **IDENTIFY:** Vector addition. Target variable is the fourth displacement.

SET UP: Use a coordinate system where east is in the $+x$-direction and north is in the $+y$-direction.

Let $\vec{A}, \vec{B},$ and $\vec{C}$ be the three displacements that are given and let $\vec{D}$ be the fourth unmeasured displacement. Then the resultant displacement is $\vec{R} = \vec{A} + \vec{B} + \vec{C} + \vec{D}$. And since she ends up back where she started, $\vec{R} = 0$.
$0 = \vec{A} + \vec{B} + \vec{C} + \vec{D}$, so $\vec{D} = -(\vec{A} + \vec{B} + \vec{C})$
$D_x = -(A_x + B_x + C_x)$ and $D_y = -(A_y + B_y + C_y)$

EXECUTE:

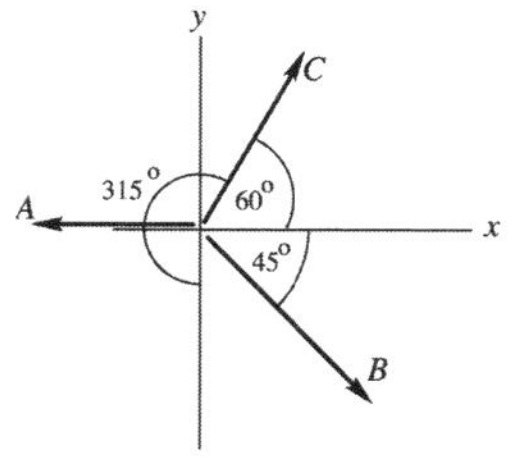

$A_x = -180$ m, $A_y = 0$
$B_x = B\cos 315° = (210\text{ m})\cos 315° = +148.5$ m
$B_y = B\sin 315° = (210\text{ m})\sin 315° = -148.5$ m
$C_x = C\cos 60° = (280\text{ m})\cos 60° = +140$ m
$C_y = C\sin 60° = (280\text{ m})\sin 60° = +242.5$ m

Figure 1.57a

$D_x = -(A_x + B_x + C_x) = -(-180\text{ m} + 148.5\text{ m} + 140\text{ m}) = -108.5$ m
$D_y = -(A_y + B_y + C_y) = -(0 - 148.5\text{ m} + 242.5\text{ m}) = -94.0$ m

Student's Solutions Manual for Units, Physical Quantities, and Vectors

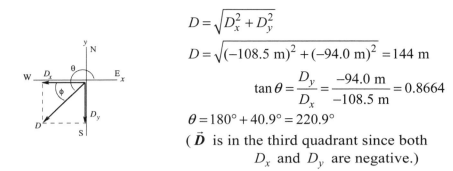

$$D = \sqrt{D_x^2 + D_y^2}$$

$$D = \sqrt{(-108.5 \text{ m})^2 + (-94.0 \text{ m})^2} = 144 \text{ m}$$

$$\tan\theta = \frac{D_y}{D_x} = \frac{-94.0 \text{ m}}{-108.5 \text{ m}} = 0.8664$$

$$\theta = 180° + 40.9° = 220.9°$$

($\vec{D}$ is in the third quadrant since both D_x and D_y are negative.)

Figure 1.57b

The direction of $\vec{D}$ can also be specified in terms of $\phi = \theta - 180° = 40.9°$; $\vec{D}$ is 41° south of west.

EVALUATE: The vector addition diagram, approximately to scale, is

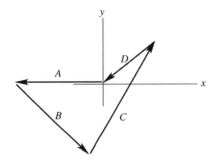

Vector $\vec{D}$ in this diagram agrees qualitatively with our calculation using components.

Figure 1.57c

1.59. **IDENTIFY:** This problem requires vector addition. We can find the components of the given vectors and then use them to find the magnitude and direction of the resultant vector.

SET UP: $A_x = A\cos\theta$, $A_y = A\sin\theta$, $\theta = \arctan\dfrac{A_y}{A_x}$, $A = \sqrt{A_x^2 + A_y^2 + A_z^2}$, $R_x = A_x + B_x$, and $R_y = A_y + B_y$. Sketch the given vectors to help find the components (see Fig. 1.59).

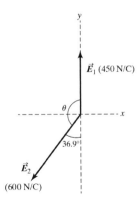

Figure 1.59

EXECUTE: From Figure 1.59 we can see that the components are
$E_{1x} = 0$ and $E_{1y} = 450$ N/C
$E_{2x} = E_2 \cos\theta = (600 \text{ N/C}) \cos 233.1° = -360.25$ N/C

$E_{2y} = E_2 \sin\theta = (600 \text{ N/C}) \sin 233.1° = -479.81$ N/C.
Now find the components of the resultant field:
$E_x = E_{1x} + E_{2x} = 0 + (-360.25 \text{ N/C}) = -360.25$ N/C
$E_y = E_{1y} + E_{2y} = 450 \text{ N/C} + (-479.81 \text{ N/C}) = -29.81$ N/C
Now find the magnitude and direction of $\vec{E}$:
$E = \sqrt{E_x^2 + E_y^2} = \sqrt{(-360.25 \text{ N/C})^2 + (-29.81 \text{ N/C})^2} = 361$ N/C.
$\theta = \arctan\dfrac{A_y}{A_x} = \theta = \arctan\left(\dfrac{-29.81 \text{ N/C}}{-360.25 \text{ N/C}}\right) = 4.73°$. Both components of $\vec{E}$ are negative, so it must point into the third quadrant. Therefore the angle below the $-x$-axis is 4.73°. The angle with the $+x$-axis is $180° + 4.73° = 184.73°$.
EVALUATE: Make a careful graphical sum to check your answer.

1.63. **IDENTIFY:** We have two known vectors and a third unknown vector, and we know the resultant of these three vectors.

SET UP: Use coordinates for which $+x$ is east and $+y$ is north. The vector displacements are: $\vec{A} = 23.0$ km at 34.0° south of east; $\vec{B} = 46.0$ km due north; $\vec{R} = 32.0$ km due west; $\vec{C}$ is unknown.
EXECUTE: $C_x = R_x - A_x - B_x = -32.0 \text{ km} - (23.0 \text{ km})\cos 34.0° - 0 = -51.07$ km;
$C_y = R_y - A_y - B_y = 0 - (-23.0 \text{ km})\sin 34.0° - 46.0 \text{ km} = -33.14$ km;
$C = \sqrt{C_x^2 + C_y^2} = 60.9$ km

Calling θ the angle that $\vec{C}$ makes with the $-x$-axis (the westward direction), we have $\tan\theta = C_y/C_x = \dfrac{33.14}{51.07}$; $\theta = 33.0°$ south of west.
EVALUATE: A graphical vector sum will confirm this result.

1.65. **IDENTIFY:** We want to find the resultant of three known displacement vectors: $\vec{R} = \vec{A} + \vec{B} + \vec{C}$.

SET UP: Let $+x$ be east and $+y$ be north and find the components of the vectors.
EXECUTE: The magnitudes are $A = 20.8$ m, $B = 38.0$ m, $C = 18.0$ m. The components are
$A_x = 0$, $A_y = 28.0$ m, $B_x = 38.0$ m, $B_y = 0$,
$C_x = -(18.0 \text{ m})(\sin 33.0°) = -9.804$ m, $C_y = -(18.0 \text{ m})(\cos 33.0°) = -15.10$ m
$R_x = A_x + B_x + C_x = 0 + 38.0 \text{ m} + (-9.80 \text{ m}) = 28.2$ m
$R_y = A_y + B_y + C_y = 20.8 \text{ m} + 0 + (-15.10 \text{ m}) = 5.70$ m
$R = \sqrt{R_x^2 + R_y^2} = 28.8$ m is the distance you must run. Calling θ_R the angle the resultant makes with the $+x$-axis (the easterly direction), we have
$\tan\theta_R = R_y/R_x = (5.70 \text{ km})/(28.2 \text{ km})$; $\theta_R = 11.4°$ north of east.
EVALUATE: A graphical sketch will confirm this result.

1.67. **IDENTIFY:** We know the resultant of two vectors and one of the vectors, and we want to find the second vector.

SET UP: Let the westerly direction be the $+x$-direction and the northerly direction be the $+y$-direction. We also know that $\vec{R} = \vec{A} + \vec{B}$ where $\vec{R}$ is the vector from you to the truck. Your GPS tells you that you are 122.0 m from the truck in a direction of 58.0° east of south, so a vector from the truck to you is 122.0 m at 58.0° east of south. Therefore the vector from you to the truck is 122.0 m at 58.0° west of north. Thus $\vec{R} = 122.0$ m at 58.0° west of north and $\vec{A}$ is 72.0 m due west. We want to find the magnitude and direction of vector $\vec{B}$.

EXECUTE: $B_x = R_x - A_x = (122.0 \text{ m})(\sin 58.0°) - 72.0 \text{ m} = 31.462 \text{ m}$

$B_y = R_y - A_y = (122.0 \text{ m})(\cos 58.0°) - 0 = 64.450 \text{ m}$; $B = \sqrt{B_x^2 + B_y^2} = 71.9 \text{ m}$.

$\tan \theta_B = B_y / B_x = \dfrac{64.650 \text{ m}}{31.462 \text{ m}} = 2.05486$; $\theta_B = 64.1°$ north of west.

EVALUATE: A graphical sum will show that the results are reasonable.

1.69. IDENTIFY: The sum of the four displacements must be zero. Use components.

SET UP: Call the displacements $\vec{A}$, $\vec{B}$, $\vec{C}$, and $\vec{D}$, where $\vec{D}$ is the final unknown displacement for the return from the treasure to the oak tree. Vectors $\vec{A}$, $\vec{B}$, and $\vec{C}$ are sketched in Figure 1.69a. $\vec{A} + \vec{B} + \vec{C} + \vec{D} = 0$ says $A_x + B_x + C_x + D_x = 0$ and $A_y + B_y + C_y + D_y = 0$. $A = 825$ m, $B = 1250$ m, and $C = 1000$ m. Let $+x$ be eastward and $+y$ be north.

EXECUTE: (a) $A_x + B_x + C_x + D_x = 0$ gives
$D_x = -(A_x + B_x + C_x) = -[0 - (1250 \text{ m})\sin 30.0° + (1000 \text{ m})\cos 32.0°] = -223.0$ m.
$A_y + B_y + C_y + D_y = 0$ gives
$D_y = -(A_y + B_y + C_y) = -[-825 \text{ m} + (1250 \text{ m})\cos 30.0° + (1000 \text{ m})\sin 32.0°] = -787.4$ m. The fourth displacement $\vec{D}$ and its components are sketched in Figure 1.69b.
$D = \sqrt{D_x^2 + D_y^2} = 818.4$ m. $\tan \phi = \dfrac{|D_x|}{|D_y|} = \dfrac{223.0 \text{ m}}{787.4 \text{ m}}$ and $\phi = 15.8°$. You should head 15.8° west of south and must walk 818 m.

(b) The vector diagram is sketched in Figure 1.69c. The final displacement $\vec{D}$ from this diagram agrees with the vector $\vec{D}$ calculated in part (a) using components.

EVALUATE: Note that $\vec{D}$ is the negative of the sum of $\vec{A}$, $\vec{B}$, and $\vec{C}$, as it should be.

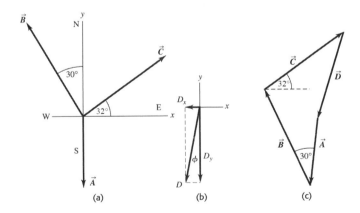

Figure 1.69

1.71. **IDENTIFY:** We are given the resultant of three vectors, two of which we know, and want to find the magnitude and direction of the third vector.

SET UP: Calling $\vec{C}$ the unknown vector and $\vec{A}$, and $\vec{B}$ the known vectors, we have $\vec{A} + \vec{B} + \vec{C} = \vec{R}$. The components are $A_x + B_x + C_x = R_x$ and $A_y + B_y + C_y = R_y$.

EXECUTE: The components of the known vectors are $A_x = 12.0$ m, $A_y = 0$, $B_x = -B\sin 50.0° = -21.45$ m, $B_y = B\cos 50.0° = +18.00$ m, $R_x = 0$, and $R_y = -10.0$ m. Therefore the components of $\vec{C}$ are $C_x = R_x - A_x - B_x = 0 - 12.0$ m $- (-21.45$ m$) = 9.45$ m and $C_y = R_y - A_y - B_y = -10.0$ m $- 0 - 18.0$ m $= -28.0$ m.

Using these components to find the magnitude and direction of $\vec{C}$ gives $C = 29.6$ m and $\tan\theta = \dfrac{9.45}{28.0}$ and $\theta = 18.6°$ east of south.

EVALUATE: A graphical sketch shows that this answer is reasonable.

1.73. **IDENTIFY:** If the vector from your tent to Joe's is $\vec{A}$ and from your tent to Karl's is $\vec{B}$, then the vector from Karl's tent to Joe's tent is $\vec{A} - \vec{B}$.

SET UP: Take your tent's position as the origin. Let $+x$ be east and $+y$ be north.
EXECUTE: The position vector for Joe's tent is
$[(21.0 \text{ m})\cos 23°]\hat{i} - [(21.0 \text{ m})\sin 23°]\hat{j} = (19.33 \text{ m})\hat{i} - (8.205 \text{ m})\hat{j}$.
The position vector for Karl's tent is
$[(32.0 \text{ m})\cos 37°]\hat{i} + [(32.0 \text{ m})\sin 37°]\hat{j} = (25.56 \text{ m})\hat{i} + (19.26 \text{ m})\hat{j}$.
The difference between the two positions is
$(19.33 \text{ m} - 25.56 \text{ m})\hat{i} + (-8.205 \text{ m} - 19.25 \text{ m})\hat{j} = -(6.23 \text{ m})\hat{i} - (27.46 \text{ m})\hat{j}$. The magnitude of this vector is the distance between the two tents: $D = \sqrt{(-6.23 \text{ m})^2 + (-27.46 \text{ m})^2} = 28.2$ m.

EVALUATE: If both tents were due east of yours, the distance between them would be 32.0 m $- 21.0$ m $= 11.0$ m. If Joe's was due north of yours and Karl's was due south of yours, then the distance between them would be 32.0 m $+ 21.0$ m $= 53.0$ m. The actual distance between them lies between these limiting values.

1.77. **IDENTIFY:** We know the scalar product and the magnitude of the vector product of two vectors and want to know the angle between them.

SET UP: The scalar product is $\vec{A} \cdot \vec{B} = AB\cos\theta$ and the vector product is $|\vec{A} \times \vec{B}| = AB\sin\theta$.

EXECUTE: $\vec{A} \cdot \vec{B} = AB\cos\theta = -6.00$ and $|\vec{A} \times \vec{B}| = AB\sin\theta = +9.00$. Taking the ratio gives $\tan\theta = \dfrac{9.00}{-6.00}$, so $\theta = 124°$.

EVALUATE: Since the scalar product is negative, the angle must be between 90° and 180°.

1.79. **IDENTIFY:** This problem involves the vector product and the scalar product of two vectors. It is best to use components.

SET UP: $\vec{A} \cdot \vec{B} = A_x B_x + A_y B_y + A_z B_z$, the components of $\vec{A} \times \vec{B}$ are shown in Eq. (1.25) in the textbook.

EXECUTE: (a) $\vec{A} \cdot \vec{B} = A_x B_x + A_y B_y + A_z B_z = a(0) + (0)(-c) + (-b)(d) = -bd$.

Realizing that $A_y = 0$ and $B_x = 0$, Eq. (1.25) gives the components of $\vec{A} \times \vec{B}$.

$(\vec{A} \times \vec{B})_x = -A_z B_y = -(-b)(-c) = -bc$

$(\vec{A} \times \vec{B})_y = -A_x B_z = -(a)(d) = -ad$

$(\vec{A} \times \vec{B})_z = A_x B_y = (a)(-c) = -ac$

$\vec{A} \times \vec{B} = -bc\,\hat{i} - ad\,\hat{j} - ac\,\hat{k}$.

(b) If $c = 0$, $\vec{A} \cdot \vec{B} = -bd$ and $\vec{A} \times \vec{B} = -ad\,\hat{j}$. The magnitude of $\vec{A} \times \vec{B}$ is ad and its direction is $-\hat{j}$ (which is in the $-y$-direction). Figure 1.79 shows a sketch of $\vec{A}$ and $\vec{B}$ in the xy-plane. In this figure, the $+y$-axis would point into the paper. By the right-hand rule, $\vec{A} \times \vec{B}$ points out of the paper, which is in the $-y$-direction (or the $-\hat{j}$-direction), which agrees with our results.

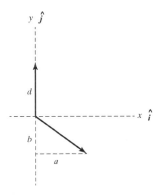

Figure 1.79

From Figure 1.79 we see that the component of $\vec{A}$ that is parallel to $\vec{B}$ is $-b$. So the product of B with the component of $\vec{A}$ that is parallel to $\vec{B}$ is $d(-b) = -bd$, which agrees with our result. From the same figure we see that the component of $\vec{A}$ that is perpendicular to $\vec{B}$ is a. So the product of B and the component of $\vec{A}$ that is perpendicular to $\vec{B}$ is da, which is the magnitude of the vector product we found above.

EVALUATE: The geometric interpretations of $\vec{A} \cdot \vec{B}$ and $\vec{A} \times \vec{B}$ can be reversed in the sense that $\vec{A} \cdot \vec{B}$ equals A times the component of $\vec{B}$ that is parallel to $\vec{A}$ and $|\vec{A} \times \vec{B}|$ equals A times the component of $\vec{B}$ that is perpendicular to $\vec{A}$.

1.83. **IDENTIFY:** We know the scalar product of two vectors, both their directions, and the magnitude of one of them, and we want to find the magnitude of the other vector.

SET UP: $\vec{A} \cdot \vec{B} = AB\cos\phi$. Since we know the direction of each vector, we can find the angle between them.

EXECUTE: The angle between the vectors is $\theta = 79.0°$. Since $\vec{A} \cdot \vec{B} = AB\cos\phi$, we have

$B = \dfrac{\vec{A} \cdot \vec{B}}{A\cos\phi} = \dfrac{48.0 \text{ m}^2}{(9.00 \text{ m})\cos 79.0°} = 28.0 \text{ m}.$

EVALUATE: Vector $\vec{B}$ has the same units as vector $\vec{A}$.

1.85. **IDENTIFY and SET UP:** The target variables are the components of $\vec{C}$. We are given $\vec{A}$ and $\vec{B}$. We also know $\vec{A}\cdot\vec{C}$ and $\vec{B}\cdot\vec{C}$, and this gives us two equations in the two unknowns C_x and C_y.

EXECUTE: $\vec{A}$ and $\vec{C}$ are perpendicular, so $\vec{A}\cdot\vec{C}=0$. $A_x C_x + A_y C_y = 0$, which gives $5.0 C_x - 6.5 C_y = 0$.

$\vec{B}\cdot\vec{C} = 15.0$, so $3.5 C_x - 7.0 C_y = 15.0$

We have two equations in two unknowns C_x and C_y. Solving gives $C_x = -8.0$ and $C_y = -6.1$.

EVALUATE: We can check that our result does give us a vector $\vec{C}$ that satisfies the two equations $\vec{A}\cdot\vec{C} = 0$ and $\vec{B}\cdot\vec{C} = 15.0$.

1.89. **IDENTIFY:** Use the x- and y-coordinates for each object to find the vector from one object to the other; the distance between two objects is the magnitude of this vector. Use the scalar product to find the angle between two vectors.

SET UP: If object A has coordinates (x_A, y_A) and object B has coordinates (x_B, y_B), the vector $\vec{r}_{AB}$ from A to B has x-component $x_B - x_A$ and y-component $y_B - y_A$.

EXECUTE: **(a)** The diagram is sketched in Figure 1.89.

(b) (i) In AU, $\sqrt{(0.3182)^2 + (0.9329)^2} = 0.9857$.

(ii) In AU, $\sqrt{(1.3087)^2 + (-0.4423)^2 + (-0.0414)^2} = 1.3820$.

(iii) In AU, $\sqrt{(0.3182-1.3087)^2 + (0.9329-(-0.4423))^2 + (0.0414)^2} = 1.695$.

(c) The angle between the directions from the earth to the Sun and to Mars is obtained from the dot product. Combining Eqs. (1.16) and (1.19),

$$\phi = \arccos\left(\frac{(-0.3182)(1.3087-0.3182)+(-0.9329)(-0.4423-0.9329)+(0)}{(0.9857)(1.695)}\right) = 54.6°.$$

(d) Mars could not have been visible at midnight, because the Sun–Mars angle is less than 90°.

EVALUATE: Our calculations correctly give that Mars is farther from the Sun than the earth is. Note that on this date Mars was farther from the earth than it is from the Sun.

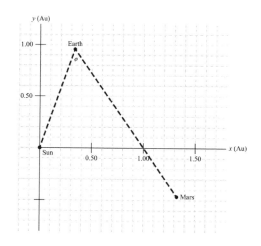

Figure 1.89

1.91. **IDENTIFY:** Draw the vector addition diagram for the position vectors.

SET UP: Use coordinates in which the Sun to Merak line lies along the x-axis. Let $\vec{A}$ be the position vector of Alkaid relative to the Sun, $\vec{M}$ is the position vector of Merak relative to the Sun, and $\vec{R}$ is the position vector for Alkaid relative to Merak. $A = 138$ ly and $M = 77$ ly.

EXECUTE: The relative positions are shown in Figure 1.91. $\vec{M} + \vec{R} = \vec{A}$. $A_x = M_x + R_x$ so $R_x = A_x - M_x = (138 \text{ ly})\cos 25.6° - 77 \text{ ly} = 47.5 \text{ ly}$. $R_y = A_y - M_y = (138 \text{ ly})\sin 25.6° - 0 = 59.6 \text{ ly}$. $R = 76.2$ ly is the distance between Alkaid and Merak.

(b) The angle is angle ϕ in Figure 1.91. $\cos\theta = \dfrac{R_x}{R} = \dfrac{47.5 \text{ ly}}{76.2 \text{ ly}}$ and $\theta = 51.4°$. Then $\phi = 180° - \theta = 129°$.

EVALUATE: The concepts of vector addition and components make these calculations very simple.

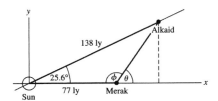

Figure 1.91

Study Guide for Motion along a Straight Line

Summary

We'll introduce *kinematics*, the study of an object's motion, or change of position with time, in this chapter. Motion includes *displacement*, the change in position of an object; *velocity*, the rate of change of position with respect to time; and *acceleration*, the rate of change of velocity with respect to time. We introduce average velocity and average acceleration as changes over a time interval and instantaneous velocity and instantaneous acceleration as changes over an infinitely short time interval. We'll learn relationships between displacement, velocity, and acceleration and see how they are modified for freely falling objects. In this chapter, we'll restrict ourselves to motion along a straight line, or one-dimensional motion and this is our first step into understanding mechanics.

Objectives

After studying this chapter, you'll understand
- The definitions of kinematic variables for position, velocity, and acceleration.
- How to calculate and interpret average and instantaneous velocities.
- How to calculate and interpret average and instantaneous accelerations.
- How to apply the equations of motion for constant acceleration.
- How to apply equations of constant acceleration to freely falling objects.
- How to analyze motion when acceleration is not constant.

Concepts and Equations

Term	Description
Average Velocity	A particle's average x-velocity v_{av-x} over a time interval Δt is its displacement Δx divided by the time interval Δt: $$v_{av-x} = \frac{x_2 - x_1}{t_2 - t_1} = \frac{\Delta x}{\Delta t}.$$ The SI unit of velocity is meters per second (m/s).
Instantaneous Velocity	A particle's instantaneous velocity is the limit of the average velocity as Δt goes to zero, or the derivative of position with respect to time. The x-component is defined as $$v_x = \lim_{t \to \infty} \frac{\Delta x}{\Delta t} = \frac{dx}{dt}.$$ The term *velocity* refers to the instantaneous velocity.
Average Acceleration	The average x-acceleration of a particle over a time interval Δt is the change in the x-component of velocity, $\Delta v_x = v_{2x} - v_{1x}$, divided by the time interval Δt: $$a_{av-x} = \frac{v_{2x} - v_{1x}}{t_2 - t_1} = \frac{\Delta v_x}{\Delta t}.$$ The SI unit of acceleration is meters per second per second (m/s²).
Instantaneous Acceleration	A particle's instantaneous acceleration is the limit of the average acceleration as Δt goes to zero, or the derivative of the velocity with respect to time. The x-component is defined as $$a_x = \lim_{t \to \infty} \frac{\Delta v_x}{\Delta t} = \frac{dv_x}{dt}.$$ The term *acceleration* refers to the instantaneous acceleration.
Motion with Constant Acceleration	When the x-acceleration is constant, position, x-velocity, acceleration, and time are related by $$x = x_0 + v_{0x} t + \tfrac{1}{2} a_x t^2$$ $$v_x = v_{0x} + a_x t$$ $$v_x^2 = v_{0x}^2 + 2a_x (x - x_0)$$ $$x - x_0 = \left(\frac{v_{0x} + v_x}{2} \right) t.$$
Freely Falling Object	A freely falling object is an object that moves under the influence of the gravity. The acceleration due to gravity is denoted by g, is directed downward, and has a value of 9.8 m/s² near the surface of the earth.
Motion with Varying Acceleration	When the acceleration is not constant, we can find the position and velocity as a function of time by integrating the acceleration function: $$x = x_0 + \int_0^t v_x dt$$ $$v_x = v_{0x} + \int_0^t a_x dt.$$

Key Concept 1: To calculate the average velocity of an object in straight-line motion, first find its displacement (final coordinate minus initial coordinate) during a time interval. Then divide by the time interval. To calculate the object's instantaneous velocity (its average velocity over an infinitesimally short time interval), take the derivative of its position with respect to time.

Key Concept 2: To calculate the average acceleration of an object in straight-line motion, first find the change in its velocity (final velocity minus initial velocity) during a time interval. Then divide by the time interval.

Key Concept 3: To calculate an object's instantaneous acceleration (its average acceleration over an infinitesimally short time interval), take the derivative of its velocity with respect to time.

Key Concept 4: By using one or more of the four equations in Table 2.5, you can solve any problem involving straight-line motion with constant acceleration.

Key Concept 5: In straight-line motion, one object meets or passes another at the time when the two objects have the same coordinate x (and so their x-t graphs cross). The objects can have different velocities at that time.

Key Concept 6: By using one or more of the four equations in Table 2.5 with x replaced by y, the positive y-direction chosen to be upward, and acceleration $a_y = -g$, you can solve any free-fall problem.

Key Concept 7: If a freely falling object passes a given point at two different times, once moving upward and once moving downward, its speed will be the same at both times.

Key Concept 8: When the acceleration of an object is constant, its position as a function of time is given by a quadratic equation. Inspect the solutions to this equation to determine what they tell you about the problem you're solving.

Key Concept 9: If the acceleration in straight-line motion is not constant but is a known function of time, you can find the velocity and position as functions of time by integration.

Conceptual Questions

1: Velocity and acceleration at the top of a ball's path

A ball is tossed vertically upward. (a) Describe the velocity and acceleration of the ball just before it reaches the top of its flight. (b) Describe the velocity and acceleration of the ball at the instant it reaches the top of its flight. (c) Describe the velocity and acceleration of the ball just after it reaches the top of its flight.

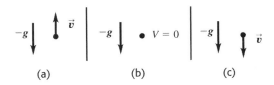

Figure 1 Question 1.

IDENTIFY, SET UP, AND EXECUTE Figure 1 shows the three time frames we'll examine. During its flight, the ball undergoes acceleration due to gravity. The initial velocity is directed upward, slowing to zero at the top of the flight. Then the velocity increases downward.

PART (A) The velocity is directed upward and is very small just before the top of the flight. The acceleration due to gravity is directed downward.

PART (B) The velocity is zero at the top of the flight. The acceleration due to gravity remains constant and is directed downward. The acceleration has caused the velocity to decrease from its small positive value in part (a) to zero.

PART (C) The velocity is directed downward and is very small just after the top of the flight. The acceleration due to gravity remains constant and directed downward. The acceleration has caused the velocity to increase downward from zero in part (b).

EVALUATE The acceleration due to gravity causes a change in the velocity of the ball during its flight. The ball starts with an upward velocity, which slows, drops to zero, and then increases downward. The acceleration due to gravity is constant throughout the motion. The velocity is zero for an instant at the top, changing from slightly upward to slightly downward around this instant.

CAUTION Gravity doesn't turn off! A common misconception is that there's no acceleration due to gravity at the top of an object's trajectory, but how would gravity know to turn off at that instant?

2: Comparing two cyclists

The position-versus-time graph depicting the paths of two cyclists is shown in Figure 2. (a) Do the cyclists start from the same position? (b) Are there any times that they have the same velocity? (c) What is happening at the intersection of lines A and B?

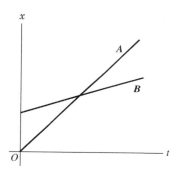

Figure 2 Question 2.

IDENTIFY, SET UP, AND EXECUTE PART (A) We find the starting location by examining the position when time is zero (i.e., by looking at the x-intercept). At $t = 0$, the two cyclists are at different locations.

PART (B) The velocity is found by examining the slope of the position-versus-time graph. The slopes of the two lines are different; hence, the cyclists never have the same velocity.

PART (C) At the intersection of lines A and B, both cyclists are at the same position at the same time. At this point, cyclist A is passing cyclist B, since cyclist A started closer to the origin and has a higher velocity.

KEY CONCEPT 5 **EVALUATE** These three questions show only a small part of what can be learned from graphs, which offer a parallel representation of physical phenomena. The interpretation of graphs is an important tool in physics and, indeed, science in general.

3: Interpreting a position-versus-time graph

Figure 3 shows a position-versus-time graph of the motion of a car. Describe the velocity and acceleration during segments *OA*, *AB*, and *BC*.

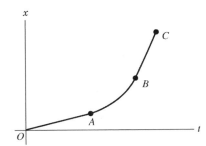

Figure 3 Question 3.

IDENTIFY, SET UP, AND EXECUTE Velocity is the change in position with respect to time and acceleration is the change in velocity with respect to time. We can describe the velocity by examining the slope of the position-versus-time graph, and we can describe the acceleration by noting how the velocity changes.

In segment *OA*, the slope is positive and constant, indicating that the velocity is positive and constant. With constant velocity, there's no acceleration.

In segment *AB*, the slope is increasing smoothly, indicating that the velocity is increasing. There must be acceleration in order for the velocity to increase.

In segment *BC*, the slope is again constant, indicating that the velocity is constant. This velocity is greater in magnitude than the velocity in segment *OA*, since the slope is larger. With constant velocity, there's no acceleration.

EVALUATE This question illustrates how we can describe the velocity and acceleration from the position-versus-time graph.

4: A falling ball

A ball falls from the top of a building. If the ball takes time t_A to fall halfway from the top of the building to the ground, is the time it takes to fall the remaining distance to the ground equal to, greater than, or smaller than t_A?

IDENTIFY, SET UP, AND EXECUTE We can break the problem up into two segments: the first half and the second half. In the first segment, the falling ball starts with an initial velocity of zero. In the second segment, the ball has acquired velocity, so it has an initial velocity. The time to complete the second segment must be shorter than t_A.

EVALUATE If you watch a ball fall, you should be able to see that it spends more time in the first half of the motion than in the second half. We can also look at the equation for the position of a falling body:

$$y = y_0 + v_{0y} t + \tfrac{1}{2} a_y t^2.$$

For the first half of the motion, the velocity term is zero; for the second half, it's not zero. Given equal time and equal acceleration, a segment with an initial velocity will cover a larger distance, or cover the same distance in a shorter time.

Problems

1: Throwing a ball upward

Robert throws a ball vertically upward from the edge of a 150-m-tall building. The ball falls to the ground 9.5 s after leaving Robert's hand. Assume that the ball leaves Robert's hand when it is 2.0 m above the roof of the building. Find the initial velocity of the ball and the time the ball reaches its maximum height.

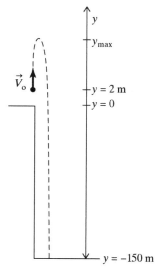

Figure 4 Problem 1.

IDENTIFY The ball undergoes constant acceleration due to gravity, so we'll use the constant-acceleration kinematics equation to solve the problem.

SET UP Figure 4 shows a sketch of the problem. Once thrown, the ball has an initial upward velocity and will undergo downward gravitational acceleration.

We ignore effects due to the air. A vertical coordinate system is shown in the diagram, with the origin located at the edge of the building and positive values directed upward.

EXECUTE We first determine the initial velocity of the ball. We know the initial and final positions, times, and accelerations of the ball; therefore, we use the equation for position as a function of time:

$$y = y_0 + v_{0y} t + \tfrac{1}{2} a_y t^2.$$

The initial position of the ball (y_0) is $+2.0$ m, the final position (y) is -150 m (the ground is below the edge of the building), the acceleration is $-g$, and the time is 9.5 s. Solving for the initial velocity v_{0y} gives

$$v_{0y} = \frac{y - y_0 - \tfrac{1}{2} a_y t^2}{t}.$$

Substituting the given values yields

$$v_{0y} = \frac{(-150 \text{ m}) - (2.0 \text{ m}) - \frac{1}{2}(-9.8 \text{ m/s}^2)(6.5 \text{ s})^2}{(6.5 \text{ s})} = 8.5 \text{ m/s}.$$

The initial velocity of the ball is 8.5 m/s. The value is positive, indicating that the initial velocity is directed upward. To find the time taken to reach the maximum height, we know that the velocity at that height is momentarily zero, so we can use the equation for velocity as a function of time:

$$v_y = v_{0y} + a_y t.$$

We now solve for the time t when the velocity v_y is zero

$$t = \frac{v_y - v_{0y}}{a_y} = \frac{(0 - 8.5 \text{ m/s})}{(-9.8 \text{ m/s}^2)} = 0.87 \text{ s}.$$

The ball reaches its maximum height 0.87 s after leaving Robert's hand.

KEY CONCEPT 6 **EVALUATE** This is a straightforward application of constant-acceleration kinematics. We identified the known and unknown quantities and substituted into appropriate equations to find the unknown quantities.

Practice Problem: Find the maximum height y_{max} of the ball. *Answer:* $y_{max} = 5.7$ m above the top of the building.

Extra Practice: How much time does it take to return from maximum height and pass Robert's hand? *Answer:* 0.87 s

2: Dropping a stone from a moving helicopter

A helicopter is ascending at a constant rate of 18 m/s. A stone falls from the helicopter 12 s after it leaves the ground. How long does it take for the stone to reach the ground?

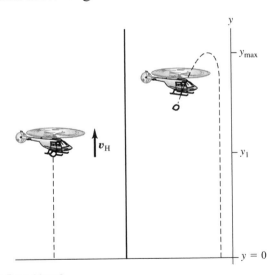

Figure 5 Problem 2.

IDENTIFY There are two segments of the stone's motion: (1) moving upward with the helicopter at constant velocity and (2) free fall after the stone breaks free of the helicopter. To solve the problem, we'll apply the constant-acceleration kinematics equations to the two segments, using the final quantities from the first segment as the initial quantities in the second.

SET UP The two segments of the stone's motion are sketched in Figure 5. As it moves upward with the helicopter, the stone has a constant velocity v_{y0}. When it breaks loose and begins to fall freely, the stone has an initial velocity that is the same as the helicopter's and undergoes acceleration due to gravity. We need to know the position, velocity, and time at the end of the first segment to solve for the second segment. The velocity and time are given in the statement of the problem.

We ignore effects due to the air. A vertical coordinate system is shown in the diagram, with the origin located on the ground and positive values directed upward.

EXECUTE The position is found from the equation for position as a function of time with zero acceleration:

$$y = y_0 + v_{0y} t.$$

The initial position is zero (the helicopter starts at the ground) and the helicopter is ascending, so v_{y0} is +18 m/s and the time is 12 s. Substituting yields

$$y = y_0 + v_{0y} t = 0 + (18 \text{ m/s})(12 \text{ s}) = 216 \text{ m}.$$

For the second segment, the initial position is 216 m, the initial velocity is +18 m/s, the final position is zero, and the acceleration is $-g$. The equation for position as a function of time with constant acceleration can be used to find the time:

$$y = y_0 + v_{0y} t + \tfrac{1}{2} a_y t^2.$$

Substituting values gives

$$0 = (216 \text{ m}) + (18 \text{ m/s}) t + \tfrac{1}{2}(-9.8 \text{ m/s}^2) t^2.$$

We cannot solve this equation directly for t, so we resort to the quadratic equation. In this case, $a = -4.9 \text{ m/s}^2$, $b = 18 \text{ m/s}$, and $c = 216 \text{ m}$. The result is given by

$$t = \frac{-b \pm \sqrt{b^2 - 4ac}}{2a}.$$

Substituting and solving yields

$$t = \frac{-(18 \text{ m/s}) \pm \sqrt{(18 \text{ m/s})^2 - 4(-4.9 \text{ m/s}^2)(216 \text{ m})}}{2(-4.9 \text{ m/s}^2)} = -5.1 \text{ s}, \; +8.7 \text{ s}.$$

The positive solution, 8.7 s, corresponds to the time the stone hits the ground. The stone hits the ground 8.7 s after falling from the helicopter, or 20.7 s after the helicopter originally left the ground.

KEY CONCEPT 8 **EVALUATE** We applied the equations for motion with constant acceleration to each of the two segments in this problem, using the results from the first part as input into the second part. The negative solution of the quadratic equation corresponds to the time the stone would have left the ground, assuming that it was thrown from the ground. Because this aspect of the motion doesn't apply to our problem, we ignore that solution.

Practice Problem: How fast is the stone moving when it reaches the ground? *Answer:* 67.5 m/s

Extra Practice: What is the maximum height of the stone? *Answer:* 233 m

3: Avoiding a ticket

A speed trap is made by placing two pressure-sensitive tracks across a highway, 150 m apart. Suppose you're driving and you notice the speed trap and begin slowing down the instant you cross the first track. If you're moving at a rate of 42 m/s and the speed limit is 35 m/s, what must your minimum acceleration be in order for you to have an average speed within the speed limit by the time your car crosses the second track?

IDENTIFY For the average speed over the interval to be under the speed limit, the final speed at the second track must be less than the speed limit. We'll use the kinematic equations to find the acceleration necessary to avoid a speeding ticket.

SET UP A sketch of the problem is shown in Figure 6. We determine the final speed by writing the average speed in terms of the initial and final speeds, setting the average speed to the speed limit, and solving for the final speed. Once we determine the final speed, we find the acceleration from the kinematics equations.

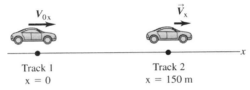

Figure 6 Problem 3.

EXECUTE The average speed (for constant acceleration) is

$$v_{av,x} = \frac{v_{0x} + v_x}{2}.$$

Substituting and solving for the final speed gives

$$v_x = 2 v_{av,x} - v_{0x} = 2(35 \text{ m/s}) - (42 \text{ m/s}) = 28 \text{ m/s}.$$

The final speed must be 28 m/s in order for the average speed to be 35 m/s. We use the equation for velocity as a function of position with constant acceleration, or

$$v_x^2 = v_{0x}^2 + 2 a_x (x - x_0).$$

In our coordinate system, the difference between the final and initial positions is 150 m. Substituting and solving for the acceleration gives

$$a_x = \frac{v_x^2 - v_{0x}^2}{2(x - x_0)} = \frac{(28 \text{ m/s})^2 - (42 \text{ m/s})^2}{2(150 \text{ m})} = -3.3 \text{ m/s}^2.$$

You'll need to accelerate at a rate of -3.3 m/s^2 to avoid a ticket, with the minus indicating that you'll need to slow down.

KEY CONCEPT 4 **EVALUATE** The challenge in this problem was to recognize that we needed a final velocity that would result in the correct average velocity. A common mistake is to take the desired *average* velocity as the *final* velocity. Understanding the difference can help you avoid errors (and a ticket).

Practice Problem: How much time does it take to slow down? *Answer:* 4.24 s

Extra Practice: How much time would it take to travel the 150 m if you did not slow down?
Answer: 3.57 s

4: Graphical solution to an accelerating car

A car undergoing constant acceleration moves 250 m in 8.5 s. If the speed at the end of the 250 m segment is 33 m/s, what was the car's speed at the beginning of the segment?

IDENTIFY We can approach this problem in two ways. First, we can use the kinematics equations to solve the problem, but doing so will require several equations. Second, we can use the velocity-versus-time plot and solve the problem graphically. We'll choose the graphical method in this case.

SET UP A sketch of the problem is shown in Figure 7. We realize that there is no single kinematics equation that ties these quantities together, so we construct the velocity-versus-time graph. The graph must start with initial velocity v_0 and result in final velocity v_1. The slope of the line between the two velocities must be constant, since the car exhibits constant acceleration. The time interval between the two velocities must be the given 8.5 s, so we construct the graph shown in Figure 8.

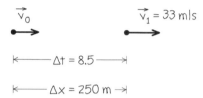

Figure 7 Problem 4 sketch.

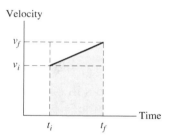

Figure 8 Problem 4 velocity-vs.-time graph.

Examining the graph, we realize that the area under the velocity line is the distance the car travels. We can therefore find the initial velocity by calculating the area under the curve.

EXECUTE The area under the curve is the sum of the area of the rectangle and the area of the triangle shown in the figure. We find these areas by multiplying the time by the velocities:

$$\text{Area} = v_0 t + \tfrac{1}{2}(v_1 - v_0)t.$$

The area under the curve is just the distance the car travels, 250 m. We rewrite the equation to solve for the initial velocity v_0:

$$v_0 = \frac{2\,\text{Area} - v_1 t}{t} = \frac{2((250 \text{ m}) - (33 \text{ m/s})(8.5 \text{ s}))}{(8.5 \text{ s})} = 25.8 \text{ m/s}.$$

The initial velocity is 25.8 m/s.

EVALUATE The example illustrates how graphical analysis can lead to a straightforward solution. The key was to realize that the area under the curve is the distance the car traveled, or its displacement. If we were to solve the problem with kinematics equations, we would've had two unknowns (initial velocity and acceleration), requiring us to utilize two equations in the solution.

Practice Problem: Use kinematic equations to find the car's acceleration. *Answer:* 0.844 m/s²

Extra Practice: Use kinematic equations to solve for the initial velocity. *Answer:* 25.8 m/s

5: Two objects falling from a building

A ball is dropped from the top of a tall building. One second later, another ball is thrown from the top of the building with a velocity of 30 m/s directed vertically downward. Will the balls ever meet? If so, when and where?

IDENTIFY Both balls undergo acceleration due to gravity after being dropped or thrown, so we will apply constant-acceleration kinematics. We'll use separate sets of kinematic equations for the two balls; ball 1 will be the dropped ball and ball 2 will be the thrown ball.

SET UP A sketch of the problem is shown in Figure 9. We are interested in where the balls meet, so we'll use the position equations, setting their positions equal to each other to find out whether they meet at any point in time. The coordinate system is shown in the sketch, with the origin at the top of the building and positive values directed downward.

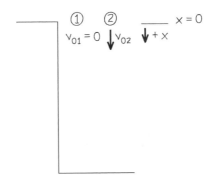

Figure 9 Problem 5.

EXECUTE The equation for the position as a function of time for ball 1, the dropped ball, is
$$x_1 = x_{01} + v_{01}t + \tfrac{1}{2}a_x t^2 = \tfrac{1}{2}gt^2.$$
The equation for the position as a function of time for ball 2, the thrown ball, is
$$x_2 = x_{02} + v_{02}t' + \tfrac{1}{2}a_x t'^2 = v_{02}(t-1\text{ s}) + \tfrac{1}{2}a_x(t-1\text{ s})^2,$$
where we have included the initial velocity v_{02} and the replaced the time t' with $(t-1\text{ s})$. For the two balls to meet, the two positions must be the same. We set the two equations equal to each other and solve for the time they meet:
$$\tfrac{1}{2}a_x t^2 = v_{02}(t-1\text{ s}) + \tfrac{1}{2}a_x(t-1\text{ s})^2$$
$$\tfrac{1}{2}gt^2 = v_{02}t - v_{02}1\text{ s} + \tfrac{1}{2}gt^2 + \tfrac{1}{2}g(-2t\text{ s}) + \tfrac{1}{2}g(1\text{ s})^2.$$

The t^2 term cancels, leaving

$$0 = v_{02} t - v_{02} \, 1 \text{ s} + \tfrac{1}{2} g (-2t \text{ s}) + \tfrac{1}{2} g (1 \text{ s})^2.$$

Solving for t gives

$$t = \frac{\tfrac{1}{2} g (\text{s}) - v_{02}}{g(\text{s}) - v_{02}} = \frac{\tfrac{1}{2}(9.8 \text{ m/s}^2)(\text{s}) - (30 \text{ m/s})}{(9.8 \text{ m/s}^2)(\text{s}) - (30 \text{ m/s})} = 1.24 \text{ s}.$$

The balls meet 1.24 s after the first ball is dropped. The position of the balls at this time is

$$x_1 = \tfrac{1}{2} g t^2 = \tfrac{1}{2}(9.8 \text{ m/s}^2)(1.24 \text{ s})^2 = 7.57 \text{ m}.$$

The balls meet 7.57 m below the top of the building.

EVALUATE We check our results and see that the balls meet 1.24 s after the first ball is dropped, or 0.24 s after the second ball is thrown. Since the balls meet after the second ball is thrown, we conclude these times represent a reasonable result. They meet 7.57 m below the top of the building. (A positive value indicates that they meet below the top of the building.) We also see that the building must be at least 7.57 m tall.

Practice Problem: What is the speed of the dropped ball when they meet? *Answer:* 12.2 m/s

Extra Practice: What is the speed of the thrown ball when they meet? *Answer:* 32.4 m/s

6: Nonconstant acceleration

A particle has an initial velocity of 12.0 m/s and starts at $x = 14.2$ m. It moves along the x-axis and possesses an acceleration given by

$$a_x = bt^2,$$

where b is a constant equal to 3.5 m/s^4. Find the particle's velocity and position as a function of time.

IDENTIFY AND SET UP The acceleration varies with time, so we cannot use the constant-acceleration kinematic equations. Instead we integrate the acceleration to find the velocity as a function of time and we integrate the velocity to find the position as a function of time. Both the initial position and initial velocity are zero.

EXECUTE We begin by integrating the acceleration to find the velocity:

$$v_x = v_{0x} + \int_0^t a_x \, dt.$$

For our problem, v_{0x} is 12.0 m/s and a_x is given. Substituting and solving, we obtain

$$v_x = (12.0 \text{ m/s}) + \int_0^t bt^2 \, dt.$$

$$= (12.0 \text{ m/s}) + \tfrac{1}{3} b t^3 \big|_0^t$$

$$= (12.0 \text{ m/s}) + \tfrac{1}{3} b t^3.$$

To find the position as a function of time, we integrate the velocity:

$$x = x_0 + \int_0^t v_x \, dt.$$

Substituting $x_0 = 14.2$ m as given and the value of v_x that we found and solving yields

Study Guide for Motion along a Straight Line

$$x = (14.2 \text{ m}) + \int_0^t \left[(12.0 \text{ m/s}) + \tfrac{1}{3}bt^3\right]dt$$
$$= (14.2 \text{ m}) + \left[(12.0 \text{ m/s})t + \tfrac{1}{12}bt^4\right]\Big|_0^t$$
$$= (14.2 \text{ m}) + (12.0 \text{ m/s})t + \tfrac{1}{12}bt^4.$$

We have found both the position and velocity as a function of time.

KEY CONCEPT 9   **EVALUATE:** We can check our integration by taking derivatives of our results. When we do, we find the original acceleration.

Practice Problem: What is the particle's position at t = 3 s? *Answer:* 73.8 m

Extra Practice: What is the particle's velocity at t = 3 s? *Answer:* 43.5 m/s

Try It Yourself!

Learning physics requires that you practice problems without having solutions next to the problem. To help you prepare for homework problems and exams, we have included sample problems with checkpoints to help you through them. We encourage you to try these problems on your own and refer to the checkpoints only when you get stuck. So go ahead and *Try It Yourself!*

1: Police chase

A speeder traveling at a constant speed of 100 km/hr passes a waiting police car that immediately starts from rest and accelerates at a constant 2.5 m/s². (a) How long will it take for the police car to catch the speeder? (b) How fast will the police car be traveling when it catches the speeder? (c) How far will the police car have traveled when it catches the speeder?

Solution Checkpoints

IDENTIFY AND SET UP Constant-acceleration kinematics are appropriate in this problem. Two separate sets of kinematics equations should be used to represent the police car and the speeder. We set appropriate values equal to each other to solve the problem.

EXECUTE The positions of the police car and speeder must be the same when the police car catches the speeder:

$$v_{0,\text{ speeder}}\, t = \tfrac{1}{2}at^2.$$

This leads to the conclusion that the speeder is caught 22.2 s after passing the police car. Kinematics then indicates that the police car had a velocity of 200 km/hr and a position of 616 m.

KEY CONCEPT 5   **EVALUATE** Would you expect the police car to accelerate at a constant rate or the speeder not to slow down? How would these changes affect the result? $t = 0$ s is also a solution of the equation. To what event does $t = 0$ s correspond?

Practice Problem: How long would it take if the car were traveling at 200 km/hr? *Answer:* 44.4 s

Extra Practice: How fast (in MPH) would the police car need to travel in this case? *Answer:* 250 MPH

2: Don't hit the truck

A car traveling 100 km/hr is 200 m away from a truck traveling 50 km/hr in the same direction. What minimum acceleration must the car have in order to avoid hitting the truck? Assume constant acceleration during braking.

Solution Checkpoints

IDENTIFY AND SET UP Constant-acceleration kinematics are valid in this problem. Two sets of kinematics equations should be used to represent the car and truck separately. Choose an appropriate coordinate system.

EXECUTE To avoid the collision with the minimum acceleration, the car and truck will meet at the same point at the same time and with the same velocity. Setting the car's and truck's position equations equal to each other gives

$$v_{0\,car}\, t + \tfrac{1}{2} a t^2 = x_{0\,truck} + v_{0\,truck}\, t.$$

This equation has two unknowns, so you'll need to set the velocities equal to each other and solve by using both relations. You should find that the magnitude of the acceleration is 0.48 m/s^2.

KEY CONCEPT 4   **EVALUATE** What is the sign of the acceleration you found? Is it what you would expect for a car slowing down?

Practice Problem: How much time passes until they meet? *Answer:* 28.8 s

Extra Practice: How much distance is traveled by the car when they meet?
Answer: 600 m

3: Throwing a ball upward

A ball is thrown vertically upward from a 125-m-high building with an initial velocity of 45 m/s. (a) What is the ball's maximum height? (b) What is its velocity as it passes the top of the building on its way down? (c) How long does it take the ball to reach the ground?

Solution Checkpoints

IDENTIFY AND SET UP Constant-acceleration kinematics are valid in this problem. You'll need equations for position as a function of time and velocity as a function of time. Choose an appropriate coordinate system.

EXECUTE At the maximum height, the velocity is zero. Solving will give a height of 103 m above the building.
(b) On the way down, the velocity of the ball when it passes the top of the building will have the same magnitude as the initial velocity, but will be opposite in direction.
(c) The equation for position as a function of time can be used to find the time taken for the ball to hit the ground. The position equation will lead to a quadratic equation and a result of 11.4 s.

KEY CONCEPT 7   **EVALUATE** The quadratic equation of part (c) had two roots. Why did you omit one root? What is the physical interpretation of the omitted root?

Practice Problem: How fast is the ball moving when it reaches the ground? *Answer: 66.9 m/s*

Extra Practice: How long would it take to reach the ground if you threw it down instead of up? *Answer: 2.23 s*

Key Example Variation Problems

Solutions to these problems are in Chapter 2 of the Student's Solutions Manual.

Be sure to review EXAMPLE 2.5 (Section 2.4) before attempting these problems.

VP2.5.1 A sports car starts from rest at an intersection and accelerates toward the east on a straight road at 8.0 m/s^2. Just as the sports car starts to move, a bus traveling east at a constant 18 m/s on the same straight road passes the sports car. When the sports car catches up with and passes the bus, (a) how much time has elapsed and (b) how far has the sports car traveled?

VP2.5.2 A car is traveling on a straight road at a constant 30.0 m/s, which is faster than the speed limit. Just as the car passes a police motorcycle that is stopped at the side of the road, the motorcycle accelerates forward in pursuit. The motorcycle passes the car 12.5 s after starting from rest. (a) What is the acceleration of the motorcycle (assumed to be constant)? (b) How far does the motorcycle travel before it passes the car?

VP2.5.3 A police car is traveling north on a straight road at a constant 20.0 m/s. An SUV traveling north at 30.0 m/s passes the police car. The driver of the SUV suspects he may be exceeding the speed limit, so just as he passes the police car he lets the SUV slow down at a constant 18.0 m/s^2. (a) How much time elapses from when the SUV passes the police car to when the police car passes the SUV? (b) What distance does the SUV travel during this time? (c) What is the speed of the SUV when the police car passes it?

VP2.5.4 At $t = 0$ a truck starts from rest at $x = 0$ and speeds up in the positive x-direction on a straight road with acceleration a_T. At the same time, $t = 0$, a car is at $x = 0$ and traveling in the positive x-direction with speed v_C. The car has a constant negative x-acceleration: $a_{\text{car-}x} = -a_C$,

where a_C is a positive quantity. (a) At what time does the truck pass the car? (b) At what x-coordinate does the truck pass the car?

Be sure to review EXAMPLE 2.7 (Section 2.5) before attempting these problems.

VP2.7.1 You throw a ball straight up from the edge of a cliff. It leaves your hand moving at 12.0 m/s. Air resistance can be neglected. Take the positive y-direction to be upward, and choose $y = 0$ to be the point where the ball leaves your hand. Find the ball's position and velocity (a) 0.300 s after it leaves your hand and (b) 2.60 s after it leaves your hand. At each time state whether the ball is above or below your hand and whether it is moving upward or downward.

VP2.7.2 You throw a stone straight down from the top of a tall tower. It leaves your hand moving at 8.00 m/s. Air resistance can be neglected. Take the positive y-direction to be upward, and choose $y = 0$ to be the point where the stone leaves your hand.

(a) Find the stone's position and velocity 1.50 s after it leaves your hand. (b) Find the stone's velocity when it is 8.00 m below your hand.

VP2.7.3 You throw a football straight up. Air resistance can be neglected. (a) When the football is 4.00 m above where it left your hand, it is moving upward at 0.500 m/s. What was the speed of the football when it left your hand? (b) How much time elapses from when the football leaves your hand until it is 4.00 m above your hand?

VP2.7.4 You throw a tennis ball straight up. Air resistance can be neglected. (a) The maximum height above your hand that the ball reaches is H. At what speed does the ball leave your hand? (b) What is the speed of the ball when it is a height $H/2$ above your hand? Express your answer as a fraction of the speed at which it left your hand. (c) At what height above your hand is the speed of the ball half as great as when it left your hand? Express your answer in terms of H.

Be sure to review EXAMPLE 2.8 (Section 2.5) before attempting these problems.

VP2.8.1 You throw a rock straight up from the edge of a cliff. It leaves your hand at time $t = 0$ moving at 12.0 m/s. Air resistance can be neglected. (a) Find both times at which the rock is 4.00 m above where it left your hand. (b) Find the time when the rock is 4.00 m below where it left your hand.

VP2.8.2 You throw a basketball straight down from the roof of a building. The basketball leaves your hand at time $t = 0$ moving at 9.00 m/s. Air resistance can be neglected. Find the time when the ball is 5.00 m below where it left your hand.

VP2.8.3 You throw an apple straight up. The apple leaves your hand at time $t = 0$ moving at 5.50 m/s. Air resistance can be neglected. (a) How many times (two, one, or none) does the apple pass through a point 1.30 m above your hand? If the apple does pass through this point, at what times t does it do so, and is the apple moving upward or downward at each of these times? (b) How many times (two, one, or none) does the apple pass through a point 1.80 m above your hand? If the apple does pass through this point, at what times t does it do so, and is the apple moving upward or downward at each of these times?

VP2.8.4 You throw an orange straight up. The orange leaves your hand at time $t = 0$ moving at speed v_0. Air resistance can be neglected. (a) At what time(s) is the orange at a height $v_0^2/2g$ above the point where it left your hand? At these time(s) is the orange moving upward, downward, or neither? (b) At what time(s) is the orange at a height $3v_0^2/8g$ above the point where it left your hand? At these time(s) is the orange moving upward, downward, or neither?

STUDENT'S SOLUTIONS MANUAL FOR MOTION ALONG A STRAIGHT LINE

VP2.5.1. **IDENTIFY:** The bus and the car leave the same point at the same time. The bus has constant velocity, but the car starts from rest with constant acceleration. So the constant-acceleration formulas apply. We want to know how long it takes for the car to catch up to the bus and how far they both travel during that time.
SET UP: When they meet, x is the same for both of them and they have traveled for the same time. The formulas $x = x_0 + v_{0x}t + \frac{1}{2}a_x t^2$ and $v_x = v_{0x} + a_x t$ both apply.
EXECUTE: (a) When the car and bus meet, they have traveled the same distance in the same time. We apply the formula $x = x_0 + v_{0x}t + \frac{1}{2}a_x t^2$ to each of them, with the origin at their starting point, which makes $x_0 = 0$ for both of them. The bus has no acceleration and the car has no initial velocity. The equation reduces to $\frac{1}{2}a_{car}t^2 = v_{bus}t \rightarrow t = 2v_{bus}/a_{car}$.
$t = 2(18 \text{ m/s})/(8.0 \text{ m/s}^2) = 4.5 \text{ s}$.
(b) The bus has zero acceleration, so $v_x = v_{0x} + a_x t$ reduces to $x_{bus} = v_{bus}t$
$x_{bus} = (18 \text{ m/s})(4.5 \text{ s}) = 81 \text{ m}$.
EVALUATE: To check, use the car's motion to find the distance.
$x_{car} = \frac{1}{2}a_{car}t^2 = \frac{1}{2}(8.0 \text{ m/s}^2)(4.5 \text{ s})^2 = 81 \text{ m}$, which agrees with our result in part (b).

VP2.5.2. **IDENTIFY:** This is very similar to VP2.5.1 and VP2.5.2. The motorcycle and the SUV leave the same point at the same time. The motorcycle has a constant velocity, but the SUV has an initial velocity and a constant acceleration. So the constant-acceleration formulas apply.
SET UP: When they meet, x is the same for both of them and they have traveled for the same time. The formulas $x = x_0 + v_{0x}t + \frac{1}{2}a_x t^2$ and $v_x = v_{0x} + a_x t$ both apply.
EXECUTE: (a) When the SUV and motorcycle meet, they have traveled the same distance in the same time. We apply the formula $x = x_0 + v_{0x}t + \frac{1}{2}a_x t^2$ to each of them, with the origin at their starting point, which makes $x_0 = 0$ for both of them. The motorcycle has no acceleration and the SUV has an initial velocity and an acceleration. The acceleration is *opposite* to the velocity of the SUV. If we take the x-axis to be in the direction of motion, a_{SUV} is negative. The equation reduces to

$v_{\text{motorcycle}}t = v_{\text{SUV}}t + \frac{1}{2}a_{\text{SUV}}t^2$. We want the time. Putting in the numbers gives

$(20.0 \text{ m/s})t = (30.0 \text{ m/s})t + \frac{1}{2}(-1.80 \text{ m/s}^2)t^2$

$t = 0$ s and $t = 11.1$ s. The $t = 0$ s solution is when they both of them leave the same point, and the $t = 11.1$ s is the time when the motorcycle passes the SUV.
(b) Both have traveled the same distance when they meet. For the motorcycle this gives
$x_{\text{motorcycle}} = v_{\text{motorcycle}}t = (20.0 \text{ m/s})(11.1 \text{ s}) = 222$ m.
(c) The equation $v_x = v_{0x} + a_xt$ gives the speed of the SUV when they meet in 11.1 s.
$v_x = 30.0$ m/s $+ (-1.80 \text{ m/s}^2)(11.1 \text{ s}) = 10.0$ s.
EVALUATE: Use $x = v_{\text{av}}t$ to find the distance the SUV has traveled in 11.1 s. For constant acceleration, the average velocity is $v_{\text{av}} = (v_1 + v_2)/2$, which gives us
$x = [(30.0 \text{ m/s} + 10.0 \text{ m/s})/2](11.1 \text{ s}) = 222$ m, which agrees with our previous result.

VP2.5.4. **IDENTIFY:** The truck and car have constant (but different) accelerations and the car has an initial velocity but the truck starts from rest. They leave from the same place at the same time and the truck eventually passes the car. The constant-acceleration equation $x = x_0 + v_{0x}t + \frac{1}{2}a_xt^2$ applies.

SET UP: (a) The truck and car have traveled the same distance in the same time when the truck reaches the car to pass. They start at the same place so x_0 is the same for both and the truck has no initial velocity.

EXECUTE: Use the equation $x = x_0 + v_{0x}t + \frac{1}{2}a_xt^2$ for each of them, which gives

$\frac{1}{2}a_Tt^2 = v_Ct - \frac{1}{2}a_Ct^2 \rightarrow t = \frac{2v_C}{a_T + a_C}$.

(b) Looking at the truck gives $x_T = \frac{1}{2}a_Tt^2 = \frac{1}{2}a_T\left(\frac{2v_C}{a_T + a_C}\right)^2 = \frac{2a_Tv_C^2}{(a_T + a_C)^2}$.

EVALUATE: We can calculate the distance the car travels using $x_C = v_Ct - \frac{1}{2}a_Ct^2$ and the value of t we found in part (a). Doing this and simplifying the result gives the same answer as in part (b).

VP2.7.1. **IDENTIFY:** The ball is in free fall so its acceleration is g downward and the constant-acceleration equations apply.

SET UP: Calling the y-axis vertical, the formulas $y = y_0 + v_{0y}t + \frac{1}{2}a_yt^2$ and $v_y = v_{0y} + a_yt$ apply to the motion of the ball. We know that $a_y = 9.80 \text{ m/s}^2$ downward and $v_{0y} = 12.0$ m/s upward.

EXECUTE: (a) At time $t = 0.300$ s, the vertical coordinate of the ball is given by
$y = y_0 + v_{0y}t + \frac{1}{2}a_yt^2$, where $y_0 = 0$ at the location of the hand.

$y = 0 + (12.0 \text{ m/s})(0.300 \text{ s}) + \frac{1}{2}(-9.80 \text{ m/s}^2)(0.300 \text{ s})^2 = 3.16$ m. Since y is positive, the ball is above the hand. The vertical velocity is given by $v_y = v_{0y} + a_yt$.
$v_y = 12.0$ m/s $+ (-9.80 \text{ m/s}^2)(0.300 \text{ s}) = 9.06$ m/s. Since v_y is positive, the ball is moving upward.
(b) At $t = 2.60$ s, $y = (12.0 \text{ m/s})(2.60 \text{ s}) + \frac{1}{2}(-9.80 \text{ m/s}^2)(2.60 \text{ s})^2 = -1.92$ m. Since y is negative, the ball is now below the hand. The ball must be moving downward since it is now below the hand.

EVALUATE: Check with v_y: $v_y = v_{0y} + a_y t = 12.0$ m/s $+ (-9.80$ m/s$^2)(2.60$ s$) = -13.5$ m/s. Since v_y is negative, the ball is moving downward, as we saw above.

VP2.7.2. **IDENTIFY:** The stone is in free fall so its acceleration is g downward and the constant-acceleration equations apply.

SET UP: Calling the y-axis vertical, the formulas $y = y_0 + v_{0y}t + \frac{1}{2}a_y t^2$, $v_y = v_{0y} + a_y t$, and $v_y^2 = v_{0y}^2 + 2a_y(y - y_0)$ apply to the motion of the stone. We know that $a_y = 9.80$ m/s^2 downward, $v_{0y} = 8.00$ m/s downward, and $y_0 = 0$.

EXECUTE: (a) The equation $y = y_0 + v_{0y}t + \frac{1}{2}a_y t^2$ gives

$y = 0 + (-8.00$ m/s$)(1.50$ s$) + \frac{1}{2}(-9.80$ m/s$^2)(1.50$ s$)^2 = -23.0$ m. The minus sign means that the stone is below your hand. The velocity of the stone is given by
$v_y = v_{0y} + a_y t = -8.00$ m/s $+ (-9.80$ m/s$^2)(1.50$ s$) = -22.7$ m/s. The minus sign tells us it is moving downward.

(b) We know the stone's position and acceleration and want its velocity. The equation $v_y^2 = v_{0y}^2 + 2a_y(y - y_0)$ gives

$v_y^2 = (-8.00$ m/s$)^2 + 2(-9.80$ m/s$^2)(-8.00$ m$)$

$v_y = \pm 14.9$ m/s. The stone must be moving downward, so $v_y = -14.9$ m/s.

EVALUATE: When the stone returned to the level of your hand, its speed was the same as its initial speed of 8.00 m/s. But as the stone has continued to accelerate downward since then, its speed must be greater than its initial speed, which is what we found.

VP2.7.3. **IDENTIFY:** The football is in free fall so its acceleration is g downward and the constant-acceleration equations apply.

SET UP: Calling the y-axis vertical, the formulas $y = y_0 + v_{0y}t + \frac{1}{2}a_y t^2$, $v_y = v_{0y} + a_y t$, and $v_y^2 = v_{0y}^2 + 2a_y(y - y_0)$ apply. We know that $a_y = 9.80$ m/s^2 downward, $v_y = 0.500$ m/s upward when $y = 4.00$ m, and $y_0 = 0$.

EXECUTE: (a) We know the speed, acceleration, and position of the ball and want its initial speed, so we use the equation $v_y^2 = v_{0y}^2 + 2a_y(y - y_0)$ to find its initial speed v_{0y}.

$(0.500$ m/s$)^2 = v_{0y}^2 + 2(-9.80$ m/s$^2)(4.00$ m$) \rightarrow v_{0y} = 8.87$ m/s.

(b) Use the result from (a) in the equation $v_y = v_{0y} + a_y t$ to find the time.

0.500 m/s $= 8.87$ m/s $+ (-9.80$ m/s$^2)t \rightarrow t = 0.854$ s.

EVALUATE: Calculate y using the time from (b) and compare it with the given value of 4.00 m.

$y = y_0 + v_{0y}t + \frac{1}{2}a_y t^2 = 0 + (8.87$ m/s$)(0.854$ s$) + \frac{1}{2}(-9.80$ m/s$^2)(0.854$ s$)^2 = 4.00$ m, which agrees with the given value.

VP2.7.4. **IDENTIFY:** The tennis ball is in free fall so its acceleration is g downward and the constant-acceleration equations apply.

SET UP: When the ball is at its highest point, its vertical velocity is zero. The equation $v_y^2 = v_{0y}^2 + 2a_y(y - y_0)$ applies.

EXECUTE: (a) At the highest point, $v_y = 0$. Use $v_y^2 = v_{0y}^2 + 2a_y(y - y_0)$ to find v_{0y}.

$0 = v_{0y}^2 + 2(-g)(H) \rightarrow v_{0y} = \sqrt{2gH}$

(b) We now know H, v_{0y} and want v_y. The same equation gives
$v_y^2 = v_{0y}^2 + 2(-g)(H/2) = 2gH - 2gH/2 = 2gH/2$

$v_y = \sqrt{\dfrac{2gH}{2}} = \dfrac{\sqrt{2gH}}{\sqrt{2}} = \dfrac{v_0}{\sqrt{2}}$.

(c) We want $y - y_0$, we know v_0, a, and v. The same equation gives
$v_y^2 = v_{0y}^2 + 2a_y(y - y_0)$

$\left(\dfrac{v_0}{2}\right)^2 = v_0^2 + 2(-g)(y - y_0)$

$y - y_0 = \dfrac{3v_0^2}{8g} = \dfrac{3(2gH)}{8g} = \dfrac{3H}{4}$.

EVALUATE: When the ball is half way to the top, $v \neq v_0/2$ because the motion equations involve the *squares* of quantities such as v^2 and t^2.

VP2.8.1. **IDENTIFY:** The rock is in free fall so its acceleration is g downward and the constant-acceleration equations apply.

SET UP: The equation $y = y_0 + v_{0y}t + \dfrac{1}{2}a_y t^2$ applies.

EXECUTE: (a) With the origin at the hand and the y-axis positive upward, $y = 4.00$ m and $y_0 = 0$. We want the time at which this occurs. The equation $y = y_0 + v_{0y}t + \dfrac{1}{2}a_y t^2$ gives

4.00 m $= (12.0$ m/s$)t + \dfrac{1}{2}(-9.80$ m/s$^2)t^2$. Solving this quadratic equation for t gives two answers:
$t = 0.398$ s and $t = 2.05$ s.

(b) Use the same procedure as in (a) except that $y = -4.00$ m. This gives
-4.00 m $= (12.0$ m/s$)t + \dfrac{1}{2}(-9.80$ m/s$^2)t^2$. This quadratic equation has two solutions, $t = 2.75$ s and $t = -0.297$ s. The negative answer is not physical, so $t = 2.75$ s.

EVALUATE: The ball is at 4.00 m above your hand twice, when it is going up and when it is going down, so we get two answers. It is at 4.00 m below the hand only once, when it is going down, so we have just one answer.

VP2.8.2. **IDENTIFY:** The ball is in free fall so its acceleration is g downward and the constant-acceleration equations apply.

SET UP: Calling the y-axis vertical with the origin at the hand, the formulas $y = y_0 + v_{0y}t + \dfrac{1}{2}a_y t^2$, $v_y = v_{0y} + a_y t$, and $v_y^2 = v_{0y}^2 + 2a_y(y - y_0)$ apply. We know that $a_y = 9.80$ m/s^2 downward and v_y is initially 9.00 m/s downward. Since all the quantities are downward, it is convenient to call the $+y$-axis downward.

EXECUTE: First find v_y when $y = 5.00$ m. Then use this result to find the time for the ball to reach this height.

$v_y^2 = v_{0y}^2 + 2a_y(y - y_0) = (9.00 \text{ m/s})^2 + 2(9.80 \text{ m/s}^2)(5.00 \text{ m})$

$v_y = 13.38$ m/s

Now use $v_y = v_{0y} + a_y t$ to find the time t.

$13.38 \text{ m/s} = 9.00 \text{ m/s} + (9.80 \text{ m/s}^2)t \rightarrow t = 0.447$ s.

EVALUATE: Check using the equation $y = y_0 + v_{0y}t + \frac{1}{2}a_y t^2$ when $t = 0.447$ s.

$y = (9.00 \text{ m/s})(0.447 \text{ s}) + \frac{1}{2}(9.80 \text{ m/s}^2)(0.447 \text{ s})^2 = 5.00$ m, which agrees with the given value.

VP2.8.3. IDENTIFY: The apple is in free fall so its acceleration is g downward and the constant-acceleration equations apply.

SET UP: Calling the y-axis vertically upward with the origin at the hand, the formulas $y = y_0 + v_{0y}t + \frac{1}{2}a_y t^2$, $v_y = v_{0y} + a_y t$, and $v_y^2 = v_{0y}^2 + 2a_y(y - y_0)$ apply. We know that $a_y = 9.80$ m/s^2 downward and v_y is initially 5.50 m/s upward.

EXECUTE: (a) Using $y = y_0 + v_{0y}t + \frac{1}{2}a_y t^2$ gives

$1.30 \text{ m} = (5.50 \text{ m/s})t + \frac{1}{2}(-9.80 \text{ m/s}^2)(0.447 \text{ s})^2$

Solving using the quadratic formula gives $t = 0.338$ s and $t = 0.784$ s. The apple passes through this point twice: going up at 0.338 s and going down at 0784 s.

(b) Use the same approach as in (a).

$1.80 \text{ m} = (5.50 \text{ m/s})t + \frac{1}{2}(-9.80 \text{ m/s}^2)t^2$

This equation has no real solutions, so the apple *never* reaches a height of 1.80 m.

EVALUATE: The highest point the apple reaches is when $v_y = 0$. Use $v_y^2 = v_{0y}^2 + 2a_y(y - y_0)$ to find the maximum height.

$0 = (5.50 \text{ m/s})^2 + 2(-9.80 \text{ m/s}^2)(y - y_0)$

$y - y_0 = 1.54$ m, which is *less than* 1.80 m. This is why we had no solutions to the quadratic equation in part (b).

VP2.8.4. IDENTIFY: The orange is in free fall so its acceleration is g downward and the constant-acceleration equations apply.

SET UP: Calling the y-axis vertically upward with the origin at the hand, the formula $y = y_0 + v_{0y}t + \frac{1}{2}a_y t^2$ applies. We know that $a_y = g$ downward and v_y is initially v_0 upward.

EXECUTE: We want the time when $y = \frac{v_0^2}{2g}$, so we use $y = y_0 + v_{0y}t + \frac{1}{2}a_y t^2$.

$\frac{v_0^2}{2g} = v_0 t - \frac{1}{2}gt^2$. Solving this quadratic equation gives $t = v_0/g$. There is only one solution, so the orange reaches this height only once.

(b) Use the same approach as in part (a). $y = y_0 + v_{0y}t + \frac{1}{2}a_y t^2$ gives

$\dfrac{3v_0^2}{8g} = v_0 t - \dfrac{1}{2}gt^2$. The quadratic formula gives two solutions: $t = v_0/2g$ and $t = 3v_0/2g$.

The orange is going up at the smaller solution and going down at the larger solution.

EVALUATE: We got only one solution for part (a) because $y = \dfrac{v_0^2}{2g}$ is the highest point the orange reaches, and that occurs only once because the orange stops there. In (b) the height $\dfrac{3v_0^2}{8g}$ is less than the maximum height, so the orange reaches this height twice, once going up and once going down.

2.3. IDENTIFY: Target variable is the time Δt it takes to make the trip in heavy traffic. Use Eq. (2.2) that relates the average velocity to the displacement and average time.

SET UP: $v_{av-x} = \dfrac{\Delta x}{\Delta t}$ so $\Delta x = v_{av-x}\Delta t$ and $\Delta t = \dfrac{\Delta x}{v_{av-x}}$.

EXECUTE: Use the information given for normal driving conditions to calculate the distance between the two cities, where the time is 1 h and 50 min, which is 110 min:

$\Delta x = v_{av-x}\Delta t = (105 \text{ km/h})(1 \text{ h}/60 \text{ min})(110 \text{ min}) = 192.5 \text{ km}$.

Now use v_{av-x} for heavy traffic to calculate Δt; Δx is the same as before:

$\Delta t = \dfrac{\Delta x}{v_{av-x}} = \dfrac{192.5 \text{ km}}{70 \text{ km/h}} = 2.75 \text{ h} = 2 \text{ h and } 45 \text{ min}$.

The additional time is (2 h and 45 min) – (1 h and 50 min) = (1 h and 105 min) – (1 h and 50 min) = 55 min.

EVALUATE: At the normal speed of 105 km/s the trip takes 110 min, but at the reduced speed of 70 km/h it takes 165 min. So decreasing your average speed by about 30% adds 55 min to the time, which is 50% of 110 min. Thus a 30% reduction in speed leads to a 50% increase in travel time. This result (perhaps surprising) occurs because the time interval is inversely proportional to the average speed, not directly proportional to it.

2.5. IDENTIFY: Given two displacements, we want the average velocity and the average speed.

SET UP: The average velocity is $v_{av-x} = \dfrac{\Delta x}{\Delta t}$ and the average speed is just the total distance walked divided by the total time to walk this distance.

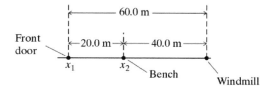

Figure 2.5

EXECUTE: **(a)** Let $+x$ be eastward with the origin at the front door. The trip begins at the front door and ends at the bench as shown in Figure 2.5. Therefore $x_1 = 0.00$ m and $x_2 = 20.0$ m. $\Delta x = x_2 - x_1 = 20.0 \text{ m} - 0.00 \text{ m} = 20.0 \text{ m}$. The total time is $\Delta t = 28.0 \text{ s} + 36.0 \text{ s} = 64.0 \text{ s}$. So

$v_{av-x} = \dfrac{\Delta x}{\Delta t} = \dfrac{20.0 \text{ m}}{64.0 \text{ s}} = 0.313$ m/s.

(b) Average speed $= \dfrac{60.0 \text{ m} + 40.0 \text{ m}}{64.0 \text{ s}} = 1.56 \text{ m/s}$.

EVALUATE: The average speed is much greater than the average velocity because the total distance walked is much greater than the magnitude of the displacement vector.

2.7. (a) IDENTIFY: Calculate the average velocity using $v_{av-x} = \dfrac{\Delta x}{\Delta t}$.

SET UP: $v_{av-x} = \dfrac{\Delta x}{\Delta t}$ so use $x(t)$ to find the displacement Δx for this time interval.

EXECUTE: $t = 0$: $x = 0$

$t = 10.0$ s: $x = (2.40 \text{ m/s}^2)(10.0 \text{ s})^2 - (0.120 \text{ m/s}^3)(10.0 \text{ s})^3 = 240 \text{ m} - 120 \text{ m} = 120 \text{ m}$.

Then $v_{av-x} = \dfrac{\Delta x}{\Delta t} = \dfrac{120 \text{ m}}{10.0 \text{ s}} = 12.0 \text{ m/s}$.

(b) IDENTIFY: Use $v_x = \dfrac{dx}{dt}$ to calculate $v_x(t)$ and evaluate this expression at each specified t.

SET UP: $v_x = \dfrac{dx}{dt} = 2bt - 3ct^2$.

EXECUTE: (i) $t = 0$: $v_x = 0$

(ii) $t = 5.0$ s: $v_x = 2(2.40 \text{ m/s}^2)(5.0 \text{ s}) - 3(0.120 \text{ m/s}^3)(5.0 \text{ s})^2 = 24.0 \text{ m/s} - 9.0 \text{ m/s} = 15.0 \text{ m/s}$.

(iii) $t = 10.0$ s: $v_x = 2(2.40 \text{ m/s}^2)(10.0 \text{ s}) - 3(0.120 \text{ m/s}^3)(10.0 \text{ s})^2 = 48.0 \text{ m/s} - 36.0 \text{ m/s} = 12.0 \text{ m/s}$.

(c) IDENTIFY: Find the value of t when $v_x(t)$ from part **(b)** is zero.

SET UP: $v_x = 2bt - 3ct^2$

$v_x = 0$ at $t = 0$.

$v_x = 0$ next when $2bt - 3ct^2 = 0$

EXECUTE: $2b = 3ct$ so $t = \dfrac{2b}{3c} = \dfrac{2(2.40 \text{ m/s}^2)}{3(0.120 \text{ m/s}^3)} = 13.3$ s

EVALUATE: $v_x(t)$ for this motion says the car starts from rest, speeds up, and then slows down again.

2.9. IDENTIFY: The average velocity is given by $v_{av-x} = \dfrac{\Delta x}{\Delta t}$. We can find the displacement Δt for each constant velocity time interval. The average speed is the distance traveled divided by the time.

SET UP: For $t = 0$ to $t = 2.0$ s, $v_x = 2.0$ m/s. For $t = 2.0$ s to $t = 3.0$ s, $v_x = 3.0$ m/s. In part (b), $v_x = -3.0$ m/s for $t = 2.0$ s to $t = 3.0$ s. When the velocity is constant, $\Delta x = v_x \Delta t$.

EXECUTE: (a) For $t = 0$ to $t = 2.0$ s, $\Delta x = (2.0 \text{ m/s})(2.0 \text{ s}) = 4.0$ m. For $t = 2.0$ s to $t = 3.0$ s, $\Delta x = (3.0 \text{ m/s})(1.0 \text{ s}) = 3.0$ m. For the first 3.0 s, $\Delta x = 4.0 \text{ m} + 3.0 \text{ m} = 7.0$ m. The distance traveled is also 7.0 m. The average velocity is $v_{av-x} = \dfrac{\Delta x}{\Delta t} = \dfrac{7.0 \text{ m}}{3.0 \text{ s}} = 2.33$ m/s. The average speed is also 2.33 m/s.

(b) For $t = 2.0$ s to 3.0 s, $\Delta x = (-3.0 \text{ m/s})(1.0 \text{ s}) = -3.0$ m. For the first 3.0 s, $\Delta x = 4.0 \text{ m} + (-3.0 \text{ m}) = +1.0$ m. The ball travels 4.0 m in the $+x$-direction and then 3.0 m in the $-x$-

direction, so the distance traveled is still 7.0 m. $v_{av-x} = \dfrac{\Delta x}{\Delta t} = \dfrac{1.0 \text{ m}}{3.0 \text{ s}} = 0.33$ m/s. The average speed is $\dfrac{7.00 \text{ m}}{3.00 \text{ s}} = 2.33$ m/s.

EVALUATE: When the motion is always in the same direction, the displacement and the distance traveled are equal and the average velocity has the same magnitude as the average speed. When the motion changes direction during the time interval, those quantities are different.

2.13. IDENTIFY and SET UP: Use $v_x = \dfrac{dx}{dt}$ and $a_x = \dfrac{dv_x}{dt}$ to calculate $v_x(t)$ and $a_x(t)$.

EXECUTE: $v_x = \dfrac{dx}{dt} = 2.00$ cm/s $- (0.125$ cm/s$^2)t$

$a_x = \dfrac{dv_x}{dt} = -0.125$ cm/s^2

(a) At $t = 0$, $x = 50.0$ cm, $v_x = 2.00$ cm/s, $a_x = -0.125$ cm/s^2.
(b) Set $v_x = 0$ and solve for t: $t = 16.0$ s.
(c) Set $x = 50.0$ cm and solve for t. This gives $t = 0$ and $t = 32.0$ s. The turtle returns to the starting point after 32.0 s.
(d) The turtle is 10.0 cm from starting point when $x = 60.0$ cm or $x = 40.0$ cm.
Set $x = 60.0$ cm and solve for t: $t = 6.20$ s and $t = 25.8$ s.
At $t = 6.20$ s, $v_x = +1.23$ cm/s.
At $t = 25.8$ s, $v_x = -1.23$ cm/s.
Set $x = 40.0$ cm and solve for t: $t = 36.4$ s (other root to the quadratic equation is negative and hence nonphysical).
At $t = 36.4$ s, $v_x = -2.55$ cm/s.
(e) The graphs are sketched in Figure 2.13.

Figure 2.13

EVALUATE: The acceleration is constant and negative. v_x is linear in time. It is initially positive, decreases to zero, and then becomes negative with increasing magnitude. The turtle initially moves farther away from the origin but then stops and moves in the $-x$-direction.

2.15. IDENTIFY: The average acceleration is $a_{av-x} = \dfrac{\Delta v_x}{\Delta t}$. Use $v_x(t)$ to find v_x at each t. The instantaneous acceleration is $a_x = \dfrac{dv_x}{dt}$.

SET UP: $v_x(0) = 3.00$ m/s and $v_x(5.00$ s$) = 5.50$ m/s.

EXECUTE: (a) $a_{av-x} = \dfrac{\Delta v_x}{\Delta t} = \dfrac{5.50 \text{ m/s} - 3.00 \text{ m/s}}{5.00 \text{ s}} = 0.500$ m/s^2

(b) $a_x = \dfrac{dv_x}{dt} = (0.100$ m/s$^3)(2t) = (0.200$ m/s$^3)t$. At $t = 0$, $a_x = 0$. At $t = 5.00$ s, $a_x = 1.00$ m/s^2.

(c) Graphs of $v_x(t)$ and $a_x(t)$ are given in Figure 2.15.

EVALUATE: $a_x(t)$ is the slope of $v_x(t)$ and increases as t increases. The average acceleration for $t = 0$ to $t = 5.00$ s equals the instantaneous acceleration at the midpoint of the time interval, $t = 2.50$ s, since $a_x(t)$ is a linear function of t.

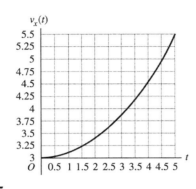
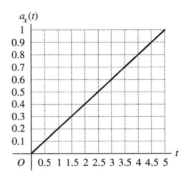

Figure 2.15

2.17. **IDENTIFY:** $v_x(t) = \dfrac{dx}{dt}$ and $a_x(t) = \dfrac{dv_x}{dt}$

SET UP: $\dfrac{d}{dt}(t^n) = nt^{n-1}$ for $n \geq 1$.

EXECUTE: **(a)** $v_x(t) = (9.60 \text{ m/s}^2)t - (0.600 \text{ m/s}^6)t^5$ and $a_x(t) = 9.60 \text{ m/s}^2 - (3.00 \text{ m/s}^6)t^4$. Setting $v_x = 0$ gives $t = 0$ and $t = 2.00$ s. At $t = 0$, $x = 2.17$ m and $a_x = 9.60 \text{ m/s}^2$. At $t = 2.00$ s, $x = 15.0$ m and $a_x = -38.4 \text{ m/s}^2$.

(b) The graphs are given in Figure 2.17.

EVALUATE: For the entire time interval from $t = 0$ to $t = 2.00$ s, the velocity v_x is positive and x increases. While a_x is also positive the speed increases and while a_x is negative the speed decreases.

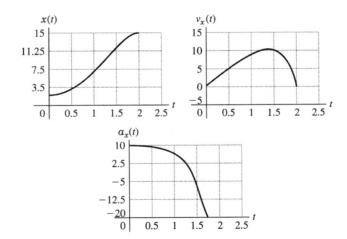

Figure 2.17

2.23. IDENTIFY: Assume that the acceleration is constant and apply the constant-acceleration kinematic equations. Set $|a_x|$ equal to its maximum allowed value.

SET UP: Let $+x$ be the direction of the initial velocity of the car. $a_x = -250$ m/s^2. 105 km/h = 29.17 m/s.

EXECUTE: $v_{0x} = 29.17$ m/s. $v_x = 0$. $v_x^2 = v_{0x}^2 + 2a_x(x - x_0)$ gives

$$x - x_0 = \frac{v_x^2 - v_{0x}^2}{2a_x} = \frac{0 - (29.17 \text{ m/s})^2}{2(-250 \text{ m/s}^2)} = 1.70 \text{ m}.$$

EVALUATE: The car frame stops over a shorter distance and has a larger magnitude of acceleration. Part of your 1.70 m stopping distance is the stopping distance of the car and part is how far you move relative to the car while stopping.

2.29. (a) IDENTIFY and SET UP: The acceleration a_x at time t is the slope of the tangent to the v_x versus t curve at time t.

EXECUTE: At $t = 3$ s, the v_x versus t curve is a horizontal straight line, with zero slope. Thus $a_x = 0$.

At $t = 7$ s, the v_x versus t curve is a straight-line segment with slope $\dfrac{45 \text{ m/s} - 20 \text{ m/s}}{9 \text{ s} - 5 \text{ s}} = 6.3$ m/s^2.

Thus $a_x = 6.3$ m/s^2.

At $t = 11$ s the curve is again a straight-line segment, now with slope $\dfrac{-0 - 45 \text{ m/s}}{13 \text{ s} - 9 \text{ s}} = -11.2$ m/s^2.

Thus $a_x = -11.2$ m/s^2.

EVALUATE: $a_x = 0$ when v_x is constant, $a_x > 0$ when v_x is positive and the speed is increasing, and $a_x < 0$ when v_x is positive and the speed is decreasing.

(b) IDENTIFY: Calculate the displacement during the specified time interval.

SET UP: We can use the constant-acceleration equations only for time intervals during which the acceleration is constant. If necessary, break the motion up into constant-acceleration segments and apply the constant acceleration equations for each segment. For the time interval $t = 0$ to $t = 5$ s the acceleration is constant and equal to zero. For the time interval $t = 5$ s to $t = 9$ s the acceleration is constant and equal to 6.25 m/s^2. For the interval $t = 9$ s to $t = 13$ s the acceleration is constant and equal to -11.2 m/s^2.

EXECUTE: During the first 5 seconds the acceleration is constant, so the constant-acceleration kinematic formulas can be used.

$v_{0x} = 20$ m/s $a_x = 0$ $t = 5$ s $x - x_0 = ?$

$x - x_0 = v_{0x}t$ ($a_x = 0$ so no $\tfrac{1}{2}a_x t^2$ term)

$x - x_0 = (20 \text{ m/s})(5 \text{ s}) = 100$ m; this is the distance the officer travels in the first 5 seconds.

During the interval $t = 5$ s to 9 s the acceleration is again constant. The constant-acceleration formulas can be applied to this 4-second interval. It is convenient to restart our clock so the interval starts at time $t = 0$ and ends at time $t = 4$ s. (Note that the acceleration is *not* constant over the entire $t = 0$ to $t = 9$ s interval.)

$v_{0x} = 20$ m/s $a_x = 6.25$ m/s^2 $t = 4$ s $x_0 = 100$ m $x - x_0 = ?$

$x - x_0 = v_{0x}t + \tfrac{1}{2}a_x t^2$

$x - x_0 = (20 \text{ m/s})(4 \text{ s}) + \tfrac{1}{2}(6.25 \text{ m/s}^2)(4 \text{ s})^2 = 80$ m $+ 50$ m $= 130$ m.

Thus $x - x_0 + 130 \text{ m} = 100 \text{ m} + 130 \text{ m} = 230 \text{ m}$.

At $t = 9$ s the officer is at $x = 230$ m, so she has traveled 230 m in the first 9 seconds. During the interval $t = 9$ s to $t = 13$ s the acceleration is again constant. The constant-acceleration formulas can be applied for this 4-second interval but *not* for the whole $t = 0$ to $t = 13$ s interval. To use the equations restart our clock so this interval begins at time $t = 0$ and ends at time $t = 4$ s.

$v_{0x} = 45$ m/s (at the start of this time interval)

$a_x = -11.2 \text{ m/s}^2 \quad t = 4 \text{ s} \quad x_0 = 230 \text{ m} \quad x - x_0 = ?$

$x - x_0 = v_{0x}t + \frac{1}{2}a_x t^2$

$x - x_0 = (45 \text{ m/s})(4 \text{ s}) + \frac{1}{2}(-11.2 \text{ m/s}^2)(4 \text{ s})^2 = 180 \text{ m} - 89.6 \text{ m} = 90.4 \text{ m}$.

Thus $x = x_0 + 90.4 \text{ m} = 230 \text{ m} + 90.4 \text{ m} = 320 \text{ m}$.

At $t = 13$ s the officer is at $x = 320$ m, so she has traveled 320 m in the first 13 seconds.

EVALUATE: The velocity v_x is always positive so the displacement is always positive and displacement and distance traveled are the same. The average velocity for time interval Δt is $v_{av-x} = \Delta x / \Delta t$. For $t = 0$ to 5 s, $v_{av-x} = 20$ m/s. For $t = 0$ to 9 s, $v_{av-x} = 26$ m/s. For $t = 0$ to 13 s, $v_{av-x} = 25$ m/s. These results are consistent with the figure in the textbook.

2.35. **IDENTIFY:** A ball on Mars that is hit directly upward returns to the same level in 8.5 s with a constant downward acceleration of $0.379g$. How high did it go and how fast was it initially traveling upward?

SET UP: Take $+y$ upward. $v_y = 0$ at the maximum height. $a_y = -0.379g = -3.71 \text{ m/s}^2$. The constant-acceleration formulas $v_y = v_{0y} + a_y t$ and $y = y_0 + v_{0y}t + \frac{1}{2}a_y t^2$ both apply.

EXECUTE: Consider the motion from the maximum height back to the initial level. For this motion $v_{0y} = 0$ and $t = 4.25$ s. $y = y_0 + v_{0y}t + \frac{1}{2}a_y t^2 = \frac{1}{2}(-3.71 \text{ m/s}^2)(4.25 \text{ s})^2 = -33.5 \text{ m}$. The ball went 33.5 m above its original position.

(b) Consider the motion from just after it was hit to the maximum height. For this motion $v_y = 0$ and $t = 4.25$ s. $v_y = v_{0y} + a_y t$ gives $v_{0y} = -a_y t = -(-3.71 \text{ m/s}^2)(4.25 \text{ s}) = 15.8 \text{ m/s}$.

(c) The graphs are sketched in Figure 2.35.

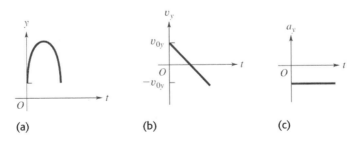

Figure 2.35

EVALUATE: The answers can be checked several ways. For example, $v_y = 0$, $v_{0y} = 15.8$ m/s, and $a_y = -3.71 \text{ m/s}^2$ in $v_y^2 = v_{0y}^2 + 2a_y(y - y_0)$ gives

$y - y_0 = \dfrac{v_y^2 - v_{0y}^2}{2a_y} = \dfrac{0 - (15.8 \text{ m/s})^2}{2(-3.71 \text{ m/s}^2)} = 33.6$ m, which agrees with the height calculated in (a).

2.37. **IDENTIFY:** A rock is thrown upward, so its acceleration is downward and uniform. Therefore the constant-acceleration equations apply. We want to know the rock's velocity at times 1.0 s and 3.0 s after it is thrown.

SET UP: The formula $v_y = v_{0y} + a_y t$ applies. Call the $+y$-axis upward, with the origin at the point where the rock leaves the hand; this makes $y_0 = 0$, $v_{0y} = 24.0$ m/s, and $a_y = -9.80$ m/s^2.

EXECUTE: **(a)** At 1.0 s: $v_y = v_{0y} + a_y t = 24.0$ m/s $+ (-9.80$ m/s$^2)(1.0$ s$) = +14.2$ m/s. The acceleration is downward. The velocity is upward but the speed is decreasing because the acceleration is downward.

(b) At 3.0 s: $v_y = v_{0y} + a_y t = 24.0$ m/s $+ (-9.80$ m/s$^2)(3.0$ s$) = -5.40$ m/s. The acceleration is downward. The velocity is downward and the speed is increasing because the acceleration is also downward. The rock has passed its highest point and is now coming down.

EVALUATE: If only gravity acts on an object, its acceleration is always downward with a magnitude of 9.80 m/s^2.

2.41. **IDENTIFY:** When the only force is gravity the acceleration is 9.80 m/s^2, downward. There are two intervals of constant-acceleration and the constant-acceleration equations apply during each of these intervals.

SET UP: Let $+y$ be upward. Let $y = 0$ at the launch pad. The final velocity for the first phase of the motion is the initial velocity for the free-fall phase.

EXECUTE: **(a)** Find the velocity when the engines cut off.

$y - y_0 = 525$ m, $a_y = 2.25$ m/s^2, $v_{0y} = 0$. $v_y^2 = v_{0y}^2 + 2a_y(y - y_0)$ gives

$v_y = \sqrt{2(2.25 \text{ m/s}^2)(525 \text{ m})} = 48.6$ m/s.

Now consider the motion from engine cut off to maximum height:
$y_0 = 525$ m, $v_{0y} = +48.6$ m/s, $v_y = 0$ (at the maximum height),

$a_y = -9.80$ m/s^2. $v_y^2 = v_{0y}^2 + 2a_y(y - y_0)$ gives $y - y_0 = \dfrac{v_y^2 - v_{0y}^2}{2a_y} = \dfrac{0 - (48.6 \text{ m/s})^2}{2(-9.80 \text{ m/s}^2)} = 121$ m and

$y = 121$ m $+ 525$ m $= 646$ m.

(b) Consider the motion from engine failure until just before the rocket strikes the ground:

$y - y_0 = -525$ m, $a_y = -9.80$ m/s^2, $v_{0y} = +48.6$ m/s. $v_y^2 = v_{0y}^2 + 2a_y(y - y_0)$ gives

$v_y = -\sqrt{(48.6 \text{ m/s})^2 + 2(-9.80 \text{ m/s}^2)(-525 \text{ m})} = -112$ m/s. Then $v_y = v_{0y} + a_y t$ gives

$t = \dfrac{v_y - v_{0y}}{a_y} = \dfrac{-112 \text{ m/s} - 48.6 \text{ m/s}}{-9.80 \text{ m/s}^2} = 16.4$ s.

(c) Find the time from blast off until engine failure: $y - y_0 = 525$ m, $v_{0y} = 0$, $a_y = +2.25$ m/s^2.

$y - y_0 = v_{0y} t + \tfrac{1}{2} a_y t^2$ gives $t = \sqrt{\dfrac{2(y - y_0)}{a_y}} = \sqrt{\dfrac{2(525 \text{ m})}{2.25 \text{ m/s}^2}} = 21.6$ s. The rocket strikes the launch

pad 21.6 s $+ 16.4$ s $= 38.0$ s after blast off. The acceleration a_y is $+2.25$ m/s^2 from $t = 0$ to $t = 21.6$ s. It is -9.80 m/s^2 from $t = 21.6$ s to 38.0 s. $v_y = v_{0y} + a_y t$ applies during each constant-acceleration segment, so the graph of v_y versus t is a straight line with positive slope of 2.25 m/s^2 during the blast off phase and with negative slope of -9.80 m/s^2 after engine failure.

During each phase $y - y_0 = v_{0y}t + \frac{1}{2}a_y t^2$. The sign of a_y determines the curvature of $y(t)$. At $t = 38.0$ s the rocket has returned to $y = 0$. The graphs are sketched in Figure 2.41.

EVALUATE: In part (b) we could have found the time from $y - y_0 = v_{0y}t + \frac{1}{2}a_y t^2$, finding v_y first allows us to avoid solving for t from a quadratic equation.

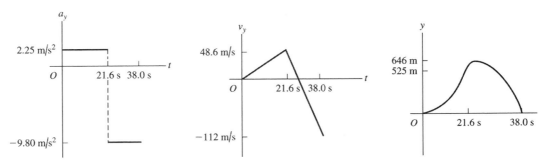

Figure 2.41

2.47. IDENTIFY: The rock has a constant downward acceleration of 9.80 m/s². The constant-acceleration kinematics formulas apply.

SET UP: The formulas $y = y_0 + v_{0y}t + \frac{1}{2}a_y t^2$ and $v_y^2 = v_{0y}^2 + 2a_y(y - y_0)$ both apply. Call +y upward. First find the initial velocity and then the final speed.

EXECUTE: (a) 6.00 s after it is thrown, the rock is back at its original height, so $y = y_0$ at that instant. Using $a_y = -9.80$ m/s² and $t = 6.00$ s, the equation $y = y_0 + v_{0y}t + \frac{1}{2}a_y t^2$ gives $v_{0y} = 29.4$ m/s.

When the rock reaches the water, $y - y_0 = -28.0$ m. The equation $v_y^2 = v_{0y}^2 + 2a_y(y - y_0)$ gives $v_y = -37.6$ m/s, so its speed is 37.6 m/s.

EVALUATE: The final speed is greater than the initial speed because the rock accelerated on its way down below the bridge.

2.49. IDENTIFY: The acceleration is not constant, but we know how it varies with time. We can use the definitions of instantaneous velocity and position to find the rocket's position and speed.

SET UP: The basic definitions of velocity and position are $v_y(t) = v_{0y} + \int_0^t a_y dt$ and $y - y_0 = \int_0^t v_y dt$.

EXECUTE: (a) $v_y(t) = \int_0^t a_y dt = \int_0^t (2.80 \text{ m/s}^3) t\, dt = (1.40 \text{ m/s}^3) t^2$

$y - y_0 = \int_0^t v_y dt = \int_0^t (1.40 \text{ m/s}^3) t^2 dt = (0.4667 \text{ m/s}^3) t^3$. For $t = 10.0$ s, $y - y_0 = 467$ m.

(b) $y - y_0 = 325$ m so $(0.4667 \text{ m/s}^3) t^3 = 325$ m and $t = 8.864$ s. At this time $v_y = (1.40 \text{ m/s}^3)(8.864 \text{ s})^2 = 110$ m/s.

EVALUATE: The time in part (b) is less than 10.0 s, so the given formulas are valid.

2.51. (a) IDENTIFY: Integrate $a_x(t)$ to find $v_x(t)$ and then integrate $v_x(t)$ to find $x(t)$.

SET UP: $v_x = v_{0x} + \int_0^t a_x\, dt$, $a_x = At - Bt^2$ with $A = 1.50$ m/s³ and $B = 0.120$ m/s⁴.

EXECUTE: $v_x = v_{0x} + \int_0^t (At - Bt^2)\, dt = v_{0x} + \frac{1}{2}At^2 - \frac{1}{3}Bt^3$

At rest at $t = 0$ says that $v_{0x} = 0$, so

$v_x = \frac{1}{2}At^2 - \frac{1}{3}Bt^3 = \frac{1}{2}(1.50 \text{ m/s}^3)t^2 - \frac{1}{3}(0.120 \text{ m/s}^4)t^3$

$v_x = (0.75 \text{ m/s}^3)t^2 - (0.040 \text{ m/s}^4)t^3$

SET UP: $x - x_0 + \int_0^t v_x\, dt$

EXECUTE: $x = x_0 + \int_0^t (\frac{1}{2}At^2 - \frac{1}{3}Bt^3)\, dt = x_0 + \frac{1}{6}At^3 - \frac{1}{12}Bt^4$

At the origin at $t = 0$ says that $x_0 = 0$, so

$x = \frac{1}{6}At^3 - \frac{1}{12}Bt^4 = \frac{1}{6}(1.50 \text{ m/s}^3)t^3 - \frac{1}{12}(0.120 \text{ m/s}^4)t^4$

$x = (0.25 \text{ m/s}^3)t^3 - (0.010 \text{ m/s}^4)t^4$

EVALUATE: We can check our results by using them to verify that $v_x(t) = \frac{dx}{dt}$ and $a_x(t) = \frac{dv_x}{dt}$.

(b) IDENTIFY and SET UP: At time t, when v_x is a maximum, $\frac{dv_x}{dt} = 0$. (Since $a_x = \frac{dv_x}{dt}$, the maximum velocity is when $a_x = 0$. For earlier times a_x is positive so v_x is still increasing. For later times a_x is negative and v_x is decreasing.)

EXECUTE: $a_x = \frac{dv_x}{dt} = 0$ so $At - Bt^2 = 0$

One root is $t = 0$, but at this time $v_x = 0$ and not a maximum.

The other root is $t = \frac{A}{B} = \frac{1.50 \text{ m/s}^3}{0.120 \text{ m/s}^4} = 12.5$ s

At this time $v_x = (0.75 \text{ m/s}^3)t^2 - (0.040 \text{ m/s}^4)t^3$ gives

$v_x = (0.75 \text{ m/s}^3)(12.5 \text{ s})^2 - (0.040 \text{ m/s}^4)(12.5 \text{ s})^3 = 117.2 \text{ m/s} - 78.1 \text{ m/s} = 39.1 \text{ m/s}$.

EVALUATE: For $t < 12.5$ s, $a_x > 0$ and v_x is increasing. For $t > 12.5$ s, $a_x < 0$ and v_x is decreasing.

2.53. IDENTIFY: The sprinter's acceleration is constant for the first 2.0 s but zero after that, so it is not constant over the entire race. We need to break up the race into segments.

SET UP: When the acceleration is constant, the formula $x - x_0 = \left(\frac{v_{0x} + v_x}{2}\right)t$ applies. The average velocity is $v_{av-x} = \frac{\Delta x}{\Delta t}$.

EXECUTE: (a) $x - x_0 = \left(\frac{v_{0x} + v_x}{2}\right)t = \left(\frac{0 + 10.0 \text{ m/s}}{2}\right)(2.0 \text{ s}) = 10.0$ m.

(b) (i) 40.0 m at 10.0 m/s so time at constant speed is 4.0 s. The total time is 6.0 s, so

$v_{av-x} = \frac{\Delta x}{\Delta t} = \frac{50.0 \text{ m}}{6.0 \text{ s}} = 8.33$ m/s.

(ii) He runs 90.0 m at 10.0 m/s so the time at constant speed is 9.0 s. The total time is 11.0 s, so

$v_{av-x} = \frac{100 \text{ m}}{11.0 \text{ s}} = 9.09$ m/s.

(iii) He runs 190 m at 10.0 m/s so time at constant speed is 19.0 s. His total time is 21.0 s, so
$$v_{av-x} = \frac{200 \text{ m}}{21.0 \text{ s}} = 9.52 \text{ m/s}.$$
EVALUATE: His average velocity keeps increasing because he is running more and more of the race at his top speed.

2.55. IDENTIFY: In time t_S the S-waves travel a distance $d = v_S t_S$ and in time t_P the P-waves travel a distance $d = v_P t_P$.

SET UP: $t_S = t_P + 33 \text{ s}$

EXECUTE: $\frac{d}{v_S} = \frac{d}{v_P} + 33 \text{ s}.$ $d\left(\frac{1}{3.5 \text{ km/s}} - \frac{1}{6.5 \text{ km/s}}\right) = 33 \text{ s}$ and $d = 250 \text{ km}$.

EVALUATE: The times of travel for each wave are $t_S = 71 \text{ s}$ and $t_P = 38 \text{ s}$.

2.57. IDENTIFY: The average velocity is $v_{av-x} = \frac{\Delta x}{\Delta t}$.

SET UP: Let $+x$ be upward.

EXECUTE: (a) $v_{av-x} = \frac{1000 \text{ m} - 63 \text{ m}}{4.75 \text{ s}} = 197 \text{ m/s}$

(b) $v_{av-x} = \frac{1000 \text{ m} - 0}{5.90 \text{ s}} = 169 \text{ m/s}$

EVALUATE: For the first 1.15 s of the flight, $v_{av-x} = \frac{63 \text{ m} - 0}{1.15 \text{ s}} = 54.8 \text{ m/s}$. When the velocity isn't constant the average velocity depends on the time interval chosen. In this motion the velocity is increasing.

2.61. IDENTIFY: When the graph of v_x versus t is a straight line the acceleration is constant, so this motion consists of two constant-acceleration segments and the constant-acceleration equations can be used for each segment. Since v_x is always positive the motion is always in the $+x$ -direction and the total distance moved equals the magnitude of the displacement. The acceleration a_x is the slope of the v_x versus t graph.

SET UP: For the $t = 0$ to $t = 10.0$ s segment, $v_{0x} = 4.00$ m/s and $v_x = 12.0$ m/s. For the $t = 10.0$ s to 12.0 s segment, $v_{0x} = 12.0$ m/s and $v_x = 0$.

EXECUTE: (a) For $t = 0$ to $t = 10.0$ s, $x - x_0 = \left(\frac{v_{0x} + v_x}{2}\right)t = \left(\frac{4.00 \text{ m/s} + 12.0 \text{ m/s}}{2}\right)(10.0 \text{ s}) = 80.0 \text{ m}.$

For $t = 10.0$ s to $t = 12.0$ s, $x - x_0 = \left(\frac{12.0 \text{ m/s} + 0}{2}\right)(2.00 \text{ s}) = 12.0 \text{ m}.$ The total distance traveled is 92.0 m.

(b) $x - x_0 = 80.0 \text{ m} + 12.0 \text{ m} = 92.0 \text{ m}$

(c) For $t = 0$ to 10.0 s, $a_x = \frac{12.0 \text{ m/s} - 4.00 \text{ m/s}}{10.0 \text{ s}} = 0.800 \text{ m/s}^2$. For $t = 10.0$ s to 12.0 s,

$a_x = \frac{0 - 12.0 \text{ m/s}}{2.00 \text{ s}} = -6.00 \text{ m/s}^2$. The graph of a_x versus t is given in Figure 2.61.

EVALUATE: When v_x and a_x are both positive, the speed increases. When v_x is positive and a_x is negative, the speed decreases.

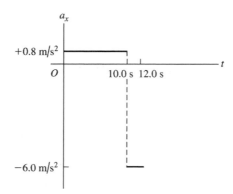

Figure 2.61

2.63. IDENTIFY and SET UP: Apply constant-acceleration kinematics equations.

Find the velocity at the start of the second 5.0 s; this is the velocity at the end of the first 5.0 s. Then find $x - x_0$ for the first 5.0 s.

EXECUTE: For the first 5.0 s of the motion, $v_{0x} = 0$, $t = 5.0$ s.

$v_x = v_{0x} + a_x t$ gives $v_x = a_x (5.0 \text{ s})$.

This is the initial speed for the second 5.0 s of the motion. For the second 5.0 s:
$v_{0x} = a_x (5.0 \text{ s})$, $t = 5.0$ s, $x - x_0 = 200$ m.

$x - x_0 = v_{0x} t + \frac{1}{2} a_x t^2$ gives $200 \text{ m} = (25 \text{ s}^2) a_x + (12.5 \text{ s}^2) a_x$ so $a_x = 5.333 \text{ m/s}^2$.

Use this a_x and consider the first 5.0 s of the motion:

$x - x_0 = v_{0x} t + \frac{1}{2} a_x t^2 = 0 + \frac{1}{2} (5.333 \text{ m/s}^2)(5.0 \text{ s})^2 = 67$ m.

EVALUATE: The ball is speeding up so it travels farther in the second 5.0 s interval than in the first.

2.65. IDENTIFY: Apply constant-acceleration equations to each object.

Take the origin of coordinates to be at the initial position of the truck, as shown in Figure 2.65a. Let d be the distance that the car initially is behind the truck, so $x_0(\text{car}) = -d$ and $x_0(\text{truck}) = 0$. Let T be the time it takes the car to catch the truck. Thus at time T the truck has undergone a displacement $x - x_0 = 60.0$ m, so is at $x = x_0 + 60.0 \text{ m} = 60.0$ m. The car has caught the truck so at time T is also at $x = 60.0$ m.

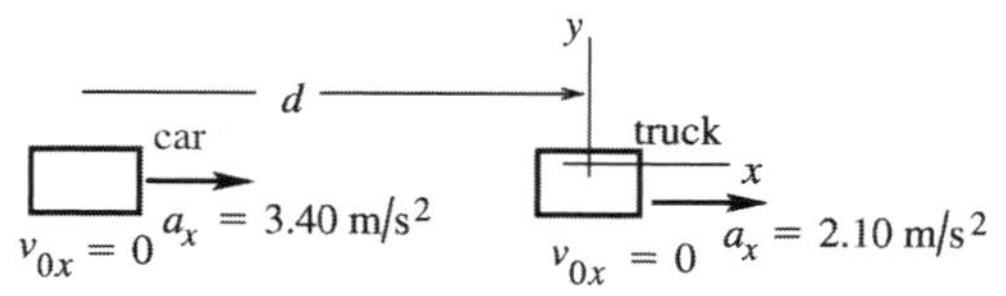

Figure 2.65a

(a) SET UP: Use the motion of the truck to calculate T:

$x - x_0 = 60.0$ m, $v_{0x} = 0$ (starts from rest), $a_x = 2.10 \text{ m/s}^2$, $t = T$

$x - x_0 = v_{0x} t + \frac{1}{2} a_x t^2$

Since $v_{0x} = 0$, this gives $t = \sqrt{\dfrac{2(x-x_0)}{a_x}}$

EXECUTE: $T = \sqrt{\dfrac{2(60.0 \text{ m})}{2.10 \text{ m/s}^2}} = 7.56$ s

(b) SET UP: Use the motion of the car to calculate d:

$x - x_0 = 60.0 \text{ m} + d$, $v_{0x} = 0$, $a_x = 3.40 \text{ m/s}^2$, $t = 7.56$ s

$x - x_0 = v_{0x}t + \dfrac{1}{2}a_x t^2$

EXECUTE: $d + 60.0 \text{ m} = \dfrac{1}{2}(3.40 \text{ m/s}^2)(7.56 \text{ s})^2$

$d = 97.16 \text{ m} - 60.0 \text{ m} = 37.2 \text{ m}$.

(c) car: $v_x = v_{0x} + a_x t = 0 + (3.40 \text{ m/s}^2)(7.56 \text{ s}) = 25.7$ m/s

truck: $v_x = v_{0x} + a_x t = 0 + (2.10 \text{ m/s}^2)(7.56 \text{ s}) = 15.9$ m/s

(d) The graph is sketched in Figure 2.65b.

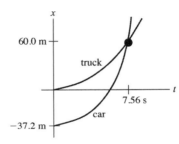

Figure 2.65b

EVALUATE: In part (c) we found that the auto was traveling faster than the truck when they came abreast. The graph in part (d) agrees with this: at the intersection of the two curves the slope of the x-t curve for the auto is greater than that of the truck. The auto must have an average velocity greater than that of the truck since it must travel farther in the same time interval.

2.67. IDENTIFY: The runner has constant acceleration in each segment of the dash, but it is not the same acceleration in the two segments. Therefore we must solve this problem in two segments.

SET UP: In the first segment, her acceleration is a constant a_x for 3.0 s, and in the next 9.0 s her speed remains constant. The total distance she runs is 100 m. A sketch as in Figure 2.67 helps to organize the information. The constant-acceleration equations apply to each segment. We want a_x.

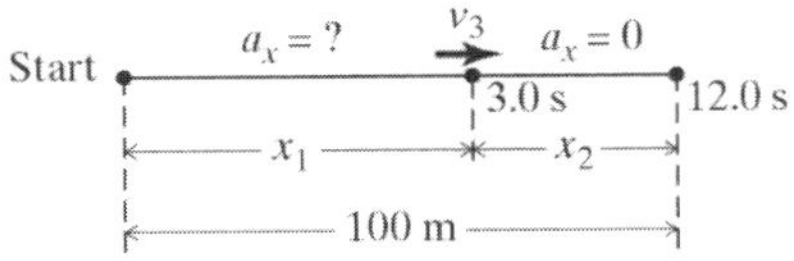

Figure 2.67

EXECUTE: Calling v_3 the velocity at the end of the first 3.0 s and using the notation shown in the figure, we look at one segment at a time. We realize that $v_0 = 0$ and v_3 is the constant speed during the second segment because she is not accelerating during that segment.

First segment: $v_x = v_{0x} + a_x t$ gives $v_3 = (3.0 \text{ s})a_x$

$x = x_0 + v_{0x}t + \dfrac{1}{2}a_x t^2$ gives $x_1 = \dfrac{1}{2}a_x(3.0 \text{ s})^2 = (4.5 \text{ s}^2)a_x$.

Second segment: She has no acceleration, so $x_2 = v_3 t_2 = [(3.0 \text{ s})a_x](9.0 \text{ s}) = (27 \text{ s}^2)a_x$.
The dash is 100 m long, so $x_1 + x_2 = 100$ m, so $(4.5 \text{ s}^2)a_x + (27 \text{ s}^2)a_x = 100$ m $\rightarrow$ $a_x = 3.2 \text{ m/s}^2$.
EVALUATE: We cannot solve this problem in a single step by averaging the accelerations because the accelerations (a_x during the first 3.0 s and zero during the last 9.0 s) do not last for the same time.

2.69. IDENTIFY: In this problem the acceleration is not constant, so the constant-acceleration equations do not apply. We need to go to the basic definitions of velocity and acceleration and use calculus.

SET UP: We know the object starts at $x = 0$ and its velocity is given by $v_x = \alpha t - \beta t^3$. We want to find out when it returns to the origin and what are its velocity and acceleration at that instant. We need to use the definitions $v_x = dx/dt$ and $a_x = dv_x/dt$.

EXECUTE: (a) To find when the object returns to the origin we need to find $x(t)$ and use it to find t when $x = 0$. Using $v_x = dx/dt$ gives $dx = v_x dt$. Using $v_x = \alpha t - \beta t^3$, we integrate to find $x(t)$.

$x - x_0 = \int v_x dt = \int (\alpha t - \beta t^3) dt = \frac{\alpha t^2}{2} - \frac{\beta t^4}{4}$. The object starts from the origin, so $x_0 = 0$ and when $t = 0$, $x = 0$. This gives $\frac{\alpha t^2}{2} - \frac{\beta t^4}{4} = 0$. Solving for t gives $t = \sqrt{\frac{2\alpha}{\beta}} = \sqrt{\frac{2(8.0 \text{ m/s}^2)}{4.0 \text{ m/s}^4}} = 2.0$ s.

(b) When $t = 2.0$ s, $v_x = \alpha t - \beta t^3 = (8.0 \text{ m/s}^2)(2.0 \text{ s}) - (4.0 \text{ m/s}^4)(2.0 \text{ s})^3 = -16$ m/s. The minus sign tells us the object is moving in the $-x$-direction. To find the acceleration, we $a_x = dv_x/dt$.

$a_x = \frac{d(\alpha t - \beta t^3)}{dt} = \alpha - 3\beta t^2 = 8.0 \text{ m/s}^2 - 3(4.0 \text{ m/s}^4)(2.0 \text{ s})^2 = -40 \text{ m/s}^2$, in the $-x$-direction.

EVALUATE: The standard constant-acceleration formulas do not apply when the acceleration is a function of time.

2.71. (a) IDENTIFY and SET UP: Integrate $a_x(t)$ to find $v_x(t)$ and then integrate $v_x(t)$ to find $x(t)$. We know $a_x(t) = \alpha + \beta t$, with $\alpha = -2.00 \text{ m/s}^2$ and $\beta = 3.00 \text{ m/s}^3$.

EXECUTE: $v_x = v_{0x} + \int_0^t a_x\, dt = v_{0x} + \int_0^t (\alpha + \beta t)\, dt = v_{0x} + \alpha t + \tfrac{1}{2}\beta t^2$

$x = x_0 + \int_0^t v_x\, dt = x_0 + \int_0^t (v_{0x} + \alpha t + \tfrac{1}{2}\beta t^2)\, dt = x_0 + v_{0x}t + \tfrac{1}{2}\alpha t^2 + \tfrac{1}{6}\beta t^3$

At $t = 0$, $x = x_0$.

To have $x = x_0$ at $t_1 = 4.00$ s requires that $v_{0x} t_1 + \tfrac{1}{2}\alpha t_1^2 + \tfrac{1}{6}\beta t_1^3 = 0$.

Thus $v_{0x} = -\tfrac{1}{6}\beta t_1^2 - \tfrac{1}{2}\alpha t_1 = -\tfrac{1}{6}(3.00 \text{ m/s}^3)(4.00 \text{ s})^2 - \tfrac{1}{2}(-2.00 \text{ m/s}^2)(4.00 \text{ s}) = -4.00$ m/s.

(b) With v_{0x} as calculated in part (a) and $t = 4.00$ s,

$v_x = v_{0x} + \alpha t + \tfrac{1}{2}\beta t^2 = -4.00 \text{ m/s} + (-2.00 \text{ m/s}^2)(4.00 \text{ s}) + \tfrac{1}{2}(3.00 \text{ m/s}^3)(4.00 \text{ s})^2 = +12.0$ m/s.

EVALUATE: $a_x = 0$ at $t = 0.67$ s. For $t > 0.67$ s, $a_x > 0$. At $t = 0$, the particle is moving in the $-x$-direction and is speeding up. After $t = 0.67$ s, when the acceleration is positive, the object slows down and then starts to move in the $+x$-direction with increasing speed.

2.73. IDENTIFY: The watermelon is in free fall so it has a constant downward acceleration of g. The constant-acceleration equations apply to its motion.

SET UP: The melon starts from rest at the top of a building. You observe that 1.50 s after it is 30.0 m above the ground, it hits the ground, and you want to find the height of the building. It is very helpful to make a sketch to organize the information, as in Figure 2.73. In the figure, we see that the height h of the building is $h = y_1 + 30.0$ m, and it takes 1.50 s to travel those last 30.0 m of the fall. We know that $v_{av-y} = \dfrac{\Delta y}{\Delta t}$, and for constant acceleration it is also true that $v_{av-y} = \dfrac{v_1 + v_2}{2}$. All quantities are downward, so choose the $+y$-axis downward.

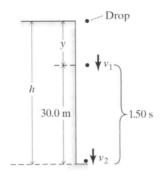

Figure 2.73

EXECUTE: We need to find y_1, but first we need to get v_1. Using $v_{av-y} = \dfrac{\Delta y}{\Delta t}$ during the last 1.50 s of the fall, we have $v_{av-y} = \dfrac{30.0 \text{ m}}{1.50 \text{ s}} = 20.0$ m/s. It is also true that $v_{av-y} = \dfrac{v_1 + v_2}{2}$, so $\dfrac{v_1 + v_2}{2} = 20.0$ m/s, which gives $v_2 = 40.0$ m/s $- v_1$. We can also use $v_y = v_{0y} + a_y t$ to get v_2 in terms of v_1.
$v_2 = v_1 + gt = v_1 + (9.80 \text{ m/s}^2)(1.50 \text{ s}) = v_1 + 14.7$ m/s. Equating our two expressions for v_2 gives $v_1 + 14.7$ m/s $= 40.0$ m/s $- v_1$. Solving for v_1 gives $v_1 = 12.65$ m/s. Now use $v_y^2 = v_{0y}^2 + 2a_y(y - y_0)$ to find y_1. $(12.65 \text{ m/s})^2 = 0 + 2(9.80 \text{ m/s}^2)y_1$ → $y_1 = 8.16$ m. The total height h of the building is $h = y_1 + 30.0$ m $= 8.16$ m $+ 30.0$ m $= 38.2$ m.

EVALUATE: Use our results to find the time to fall the last 30.0 m; it should be 1.50 s if our answer is correct.

Time to fall 8.16 m from rest: $y = y_0 + v_{0y}t + \dfrac{1}{2}a_y t^2$: 8.16 m $= 0 + \dfrac{1}{2}(9.80 \text{ m/s}^2)t^2$ → $t = 1.29$ s.

Time to fall 38.2 m from rest: 38.2 m $= \dfrac{1}{2}(9.80 \text{ m/s}^2)t^2$ → $t = 2.79$ s.

Time to fall the last 30.0 m: 2.79 s $-$ 1.29 s $=$ 1.50 s, which agrees with the observed time.

2.75. (a) IDENTIFY: Consider the motion from when he applies the acceleration to when the shot leaves his hand.

SET UP: Take positive y to be upward. $v_{0y} = 0$, $v_y = ?$, $a_y = 35.0 \text{ m/s}^2$, $y - y_0 = 0.640$ m, $v_y^2 = v_{0y}^2 + 2a_y(y - y_0)$.

EXECUTE: $v_y = \sqrt{2a_y(y - y_0)} = \sqrt{2(35.0 \text{ m/s}^2)(0.640 \text{ m})} = 6.69$ m/s

(b) IDENTIFY: Consider the motion of the shot from the point where he releases it to its maximum height, where $v = 0$. Take $y = 0$ at the ground.

SET UP: $y_0 = 2.20$ m, $y = ?$, $a_y = -9.80$ m/s^2 (free fall), $v_{0y} = 6.69$ m/s (from part (a), $v_y = 0$ at maximum height), $v_y^2 = v_{0y}^2 + 2a_y(y - y_0)$

EXECUTE: $y - y_0 = \dfrac{v_y^2 - v_{0y}^2}{2a_y} = \dfrac{0 - (6.69 \text{ m/s})^2}{2(-9.80 \text{ m/s}^2)} = 2.29$ m, $y = 2.20$ m $+ 2.29$ m $= 4.49$ m.

(c) IDENTIFY: Consider the motion of the shot from the point where he releases it to when it returns to the height of his head. Take $y = 0$ at the ground.

SET UP: $y_0 = 2.20$ m, $y = 1.83$ m, $a_y = -9.80$ m/s^2, $v_{0y} = +6.69$ m/s, $t = ?$ $y - y_0 = v_{0y}t + \tfrac{1}{2}a_yt^2$

EXECUTE: 1.83 m $- 2.20$ m $= (6.69 \text{ m/s})t + \tfrac{1}{2}(-9.80 \text{ m/s}^2)t^2 = (6.69 \text{ m/s})t - (4.90 \text{ m/s}^2)t^2$,

$4.90t^2 - 6.69t - 0.37 = 0$, with t in seconds. Use the quadratic formula to solve for t:

$t = \dfrac{1}{9.80}\left(6.69 \pm \sqrt{(6.69)^2 - 4(4.90)(-0.37)}\right) = 0.6830 \pm 0.7362$. Since t must be positive,

$t = 0.6830$ s $+ 0.7362$ s $= 1.42$ s.

EVALUATE: Calculate the time to the maximum height: $v_y = v_{0y} + a_y t$, so $t = (v_y - v_{0y})/a_y = -(6.69 \text{ m/s})/(-9.80 \text{ m/s}^2) = 0.68$ s. It also takes 0.68 s to return to 2.2 m above the ground, for a total time of 1.36 s. His head is a little lower than 2.20 m, so it is reasonable for the shot to reach the level of his head a little later than 1.36 s after being thrown; the answer of 1.42 s in part (c) makes sense.

2.77. IDENTIFY: Two stones are thrown up with different speeds. (a) Knowing how soon the faster one returns to the ground, how long it will take the slow one to return? (b) Knowing how high the slower stone went, how high did the faster stone go?

SET UP: Use subscripts f and s to refer to the faster and slower stones, respectively. Take $+y$ to be upward and $y_0 = 0$ for both stones. $v_{0f} = 3v_{0s}$. When a stone reaches the ground, $y = 0$. The constant-acceleration formulas $y = y_0 + v_{0y}t + \tfrac{1}{2}a_yt^2$ and $v_y^2 = v_{0y}^2 + 2a_y(y - y_0)$ both apply.

EXECUTE: **(a)** $y = y_0 + v_{0y}t + \tfrac{1}{2}a_yt^2$ gives $a_y = -\dfrac{2v_{0y}}{t}$. Since both stones have the same

a_y, $\dfrac{v_{0f}}{t_f} = \dfrac{v_{0s}}{t_s}$ and $t_s = t_f\left(\dfrac{v_{0s}}{v_{0f}}\right) = \left(\tfrac{1}{3}\right)(10 \text{ s}) = 3.3$ s.

(b) Since $v_y = 0$ at the maximum height, then $v_y^2 = v_{0y}^2 + 2a_y(y - y_0)$ gives $a_y = -\dfrac{v_{0y}^2}{2y}$. Since

both have the same a_y, $\dfrac{v_{0f}^2}{y_f} = \dfrac{v_{0s}^2}{y_s}$ and $y_f = y_s\left(\dfrac{v_{0f}}{v_{0s}}\right)^2 = 9H$.

EVALUATE: The faster stone reaches a greater height so it travels a greater distance than the slower stone and takes more time to return to the ground.

2.83. IDENTIFY and SET UP: Use $v_x = dx/dt$ and $a_x = dv_x/dt$ to calculate $v_x(t)$ and $a_x(t)$ for each car. Use these equations to answer the questions about the motion.

EXECUTE: $x_A = \alpha t + \beta t^2$, $v_{Ax} = \dfrac{dx_A}{dt} = \alpha + 2\beta t$, $a_{Ax} = \dfrac{dv_{Ax}}{dt} = 2\beta$

$x_B = \gamma t^2 - \delta t^3$, $v_{Bx} = \dfrac{dx_B}{dt} = 2\gamma t - 3\delta t^2$, $a_{Bx} = \dfrac{dv_{Bx}}{dt} = 2\gamma - 6\delta t$

(a) IDENTIFY and **SET UP:** The car that initially moves ahead is the one that has the larger v_{0x}.
EXECUTE: At $t = 0$, $v_{Ax} = \alpha$ and $v_{Bx} = 0$. So initially car A moves ahead.
(b) IDENTIFY and **SET UP:** Cars at the same point implies $x_A = x_B$.
$\alpha t + \beta t^2 = \gamma t^2 - \delta t^3$
EXECUTE: One solution is $t = 0$, which says that they start from the same point. To find the other solutions, divide by t: $\alpha + \beta t = \gamma t - \delta t^2$
$\delta t^2 + (\beta - \gamma)t + \alpha = 0$
$t = \dfrac{1}{2\delta}\left(-(\beta-\gamma) \pm \sqrt{(\beta-\gamma)^2 - 4\delta\alpha}\right) = \dfrac{1}{0.40}\left(+1.60 \pm \sqrt{(1.60)^2 - 4(0.20)(2.60)}\right) = 4.00 \text{ s} \pm 1.73 \text{ s}$
So $x_A = x_B$ for $t = 0$, $t = 2.27$ s and $t = 5.73$ s.
EVALUATE: Car A has constant, positive a_x. Its v_x is positive and increasing. Car B has $v_{0x} = 0$ and a_x that is initially positive but then becomes negative. Car B initially moves in the $+x$-direction but then slows down and finally reverses direction. At $t = 2.27$ s car B has overtaken car A and then passes it. At $t = 5.73$ s, car B is moving in the $-x$-direction as it passes car A again.
(c) IDENTIFY: The distance from A to B is $x_B - x_A$. The rate of change of this distance is $\dfrac{d(x_B - x_A)}{dt}$. If this distance is not changing, $\dfrac{d(x_B - x_A)}{dt} = 0$. But this says $v_{Bx} - v_{Ax} = 0$. (The distance between A and B is neither decreasing nor increasing at the instant when they have the same velocity.)
SET UP: $v_{Ax} = v_{Bx}$ requires $\alpha + 2\beta t = 2\gamma t - 3\delta t^2$
EXECUTE: $3\delta t^2 + 2(\beta - \gamma)t + \alpha = 0$
$t = \dfrac{1}{6\delta}\left(-2(\beta-\gamma) \pm \sqrt{4(\beta-\gamma)^2 - 12\delta\alpha}\right) = \dfrac{1}{1.20}\left(3.20 \pm \sqrt{4(-1.60)^2 - 12(0.20)(2.60)}\right)$
$t = 2.667$ s ± 1.667 s, so $v_{Ax} = v_{Bx}$ for $t = 1.00$ s and $t = 4.33$ s.
EVALUATE: At $t = 1.00$ s, $v_{Ax} = v_{Bx} = 5.00$ m/s. At $t = 4.33$ s, $v_{Ax} = v_{Bx} = 13.0$ m/s. Now car B is slowing down while A continues to speed up, so their velocities aren't ever equal again.
(d) IDENTIFY and **SET UP:** $a_{Ax} = a_{Bx}$ requires $2\beta = 2\gamma - 6\delta t$
EXECUTE: $t = \dfrac{\gamma - \beta}{3\delta} = \dfrac{2.80 \text{ m/s}^2 - 1.20 \text{ m/s}^2}{3(0.20 \text{ m/s}^3)} = 2.67$ s.
EVALUATE: At $t = 0$, $a_{Bx} > a_{Ax}$, but a_{Bx} is decreasing while a_{Ax} is constant. They are equal at $t = 2.67$ s but for all times after that $a_{Bx} < a_{Ax}$.

2.85. **IDENTIFY:** A ball is dropped from rest and falls from various heights with constant acceleration. Interpret a graph of the square of its velocity just as it reaches the floor as a function of its release height.

SET UP: Let $+y$ be downward since all motion is downward. The constant-acceleration kinematics formulas apply for the ball.

EXECUTE: **(A)** The equation $v_y^2 = v_{0y}^2 + 2a_y(y - y_0)$ applies to the falling ball. Solving for $y - y_0$ and using $v_{0y} = 0$ and $a_y = g$, we get $y - y_0 = \dfrac{v_y^2}{2g}$. A graph of $y - y_0$ versus v_y^2 will be a straight line with slope $1/2g = 1/(19.6 \text{ m/s}^2) = 0.0510 \text{ s}^2/\text{m}$.

(b) With air resistance the acceleration is less than 9.80 m/s², so the final speed will be smaller.
(c) The graph will not be a straight line because the acceleration will vary with the speed of the ball. For a given release height, v_y with air resistance is less than without it. Alternatively, with air resistance the ball will have to fall a greater distance to achieve a given velocity than without air resistance. The graph is sketched in Figure 2.85.

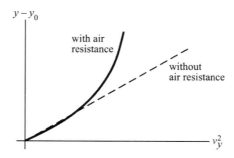

Figure 2.85

EVALUATE: Graphing $y - y_0$ versus v_y^2 for a set of data will tell us if the acceleration is constant. If the graph is a straight line, the acceleration is constant; if not, the acceleration is not constant.

2.89. IDENTIFY: Apply constant acceleration equations to both objects.

SET UP: Let $+y$ be upward, so each ball has $a_y = -g$. For the purpose of doing all four parts with the least repetition of algebra, quantities will be denoted symbolically. That is, let
$y_1 = h + v_0 t - \frac{1}{2} g t^2$, $y_2 = h - \frac{1}{2} g (t - t_0)^2$. In this case, $t_0 = 1.00$ s.

EXECUTE: **(a)** Setting $y_1 = y_2 = 0$, expanding the binomial $(t - t_0)^2$ and eliminating the common term $\frac{1}{2} g t^2$ yields $v_0 t = g t_0 t - \frac{1}{2} g t_0^2$. Solving for t: $t = \dfrac{\frac{1}{2} g t_0^2}{g t_0 - v_0} = \dfrac{t_0}{2} \left(\dfrac{1}{1 - v_0/(g t_0)} \right)$.

Substitution of this into the expression for y_1 and setting $y_1 = 0$ and solving for h as a function of v_0 yields, after some algebra, $h = \frac{1}{2} g t_0^2 \dfrac{\left(\frac{1}{2} g t_0 - v_0 \right)^2}{(g t_0 - v_0)^2}$. Using the given value

$t_0 = 1.00$ s and $g = 9.80$ m/s², $h = 20.0$ m $= (4.9$ m$) \left(\dfrac{4.9 \text{ m/s} - v_0}{9.8 \text{ m/s} - v_0} \right)^2$.

This has two solutions, one of which is unphysical (the first ball is still going up when the second is released; see part (c)). The physical solution involves taking the negative square root before solving for v_0, and yields 8.2 m/s. The graph of y versus t for each ball is given in Figure 2.89.
(b) The above expression gives for (i) 0.411 m and for (ii) 1.15 km.
(c) As v_0 approaches 9.8 m/s, the height h becomes infinite, corresponding to a relative velocity at the time the second ball is thrown that approaches zero. If $v_0 > 9.8$ m/s, the first ball can never catch the second ball.
(d) As v_0 approaches 4.9 m/s, the height approaches zero. This corresponds to the first ball being closer and closer (on its way down) to the top of the roof when the second ball is released. If $v_0 < 4.9$ m/s, the first ball will already have passed the roof on the way down before the second ball is released, and the second ball can never catch up.

EVALUATE: Note that the values of v_0 in parts (a) and (b) are all greater than v_{min} and less than v_{max}.

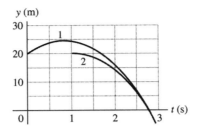

Figure 2.89

STUDY GUIDE FOR MOTION IN TWO OR THREE DIMENSIONS

Summary

In this chapter we expand our kinematics to motion of objects in two or three dimensions. In doing so, we'll find that we can simultaneously apply our one-dimensional kinematics equations to multiple axes independently. Displacement, velocity, and acceleration take on their vector qualities as we expand to more dimensions, requiring us to work with components of each quantity. Our new skills will allow us to investigate projectile motion and the interesting case of uniform circular motion. We'll also learn to analyze motion viewed from different moving reference frames. By the end of this chapter, we'll have laid a strong foundation in kinematics and will be ready to investigate the causes of motion.

Objectives

After studying this chapter, you will understand

- How to describe an object's position, velocity, and acceleration in terms of vector quantities.
- How to apply equations of motion to objects moving in a plane.
- How to describe and analyze the motion of projectiles.
- How to analyze an object in uniform circular motion.
- How to combine components of acceleration that are parallel and perpendicular to an object's path.
- How to relate the velocities of objects to different reference frames.

Concepts and Equations

Term	Description
Position Vector	The position vector $\vec{r}$ of a point P in space is the displacement vector from the origin to P. It has components x, y, and z.
Average Velocity	The average velocity $\vec{v}_{av}$ during a time interval Δt is the displacement $\Delta \vec{r}$ divided by Δt: $$\vec{v}_{av} = \frac{\vec{r}_2 - \vec{r}_1}{t_2 - t_1} = \frac{\Delta \vec{r}}{\Delta t}.$$
Instantaneous Velocity	An object's instantaneous velocity is the derivative of $\vec{r}$ with respect to time: $$\vec{v} = \lim_{t \to \infty} \frac{\Delta \vec{r}}{\Delta t} = \frac{d\vec{r}}{dt}.$$ The instantaneous velocity has components $$v_x = \frac{dx}{dt}, \quad v_y = \frac{dy}{dt}, \quad v_z = \frac{dz}{dt}.$$
Average Acceleration	The average acceleration $\vec{a}_{av}$ during a time interval Δt is the change in velocity $\Delta \vec{v}$ divided by Δt: $$\vec{a}_{av} = \frac{\vec{v}_2 - \vec{v}_1}{t_2 - t_1} = \frac{\Delta \vec{v}}{\Delta t}$$
Instantaneous Acceleration	An object's instantaneous acceleration is the derivative of $\vec{v}$ with respect to time: $$\vec{a} = \lim_{t \to \infty} \frac{\Delta \vec{v}}{\Delta t} = \frac{d\vec{v}}{dt}.$$ The instantaneous velocity has components $$a_x = \frac{dv_x}{dt}, \quad a_y = \frac{dv_y}{dt}, \quad a_z = \frac{dv_z}{dt}.$$ The component of acceleration parallel to the velocity affects the speed of the object. The component of acceleration perpendicular to the velocity affects the object's direction of motion. An object has acceleration if either its speed or direction changes.
Projectile Motion	An object undergoes projectile motion when it is given an initial velocity and then follows a path determined entirely by the effect of a constant gravitational force. The path, or trajectory, is a parabola. The projectile's vertical motion is independent of its horizontal motion. The horizontal acceleration is zero and the vertical acceleration is $-g$. The coordinates and velocities of a projectile with an initial velocity of magnitude v_0 and direction α_0 (measured with respect to the ground) are given as a function of time by $$x = (v_0 \cos \alpha_0) t,$$ $$y = (v_0 \sin \alpha_0) t - \tfrac{1}{2} g t^2,$$ $$v_x = v_0 \cos \alpha_0,$$ $$v_y = v_0 \sin \alpha_0 - gt.$$

Uniform Circular Motion	A particle moving in a circular path of radius R and constant speed v is said to move in uniform circular motion. The particle possesses an acceleration of magnitude $$a_{rad} = \frac{v^2}{R}$$ directed toward the center of the circle. If the particle's speed is not constant in circular motion, then the radial component of acceleration remains as just given and there is also a component parallel to the path of the particle.
Relative Velocity	When an object P moves relative to a reference frame B, and B moves relative to a second reference frame A, the velocity of P relative to B is denoted by $\vec{v}_{P/B}$, the velocity of P relative to A is denoted by $\vec{v}_{P/A}$, and the velocity of B relative to A is denoted by $\vec{v}_{B/A}$. These velocities are related by $$\vec{v}_{P/A} = \vec{v}_{P/B} + \vec{v}_{B/A}.$$

Key Concept 1: To calculate the average velocity vector of an object, first find its displacement vector during a time interval. Then divide by the time interval. To calculate the object's instantaneous velocity vector (its average velocity vector over an infinitesimally short time interval), take the derivative of its position vector with respect to time.

Key Concept 2: To calculate the average acceleration vector of an object, first find the change in its velocity vector (final velocity minus initial velocity) during a time interval. Then divide by the time interval. To calculate the object's instantaneous acceleration vector (its average velocity vector over an infinitesimally short time interval), take the derivative of its velocity vector with respect to time.

Key Concept 3: If an object's *speed* is changing, there is a component of its acceleration vector *parallel* to its velocity vector. If an object's *direction of motion* is changing—that is, it is turning—there is a component of its acceleration vector *perpendicular* to its velocity vector and toward the inside of the turn.

Key Concept 4: If a moving object is turning (changing direction), its acceleration vector points ahead of the normal to its path if it is speeding up, behind the normal if it is slowing down, and along the normal if its speed is instantaneously not changing.

Key Concept 5: No matter how complicated the acceleration of a particle before it becomes a projectile, its acceleration as a projectile is given by $a_x = 0$, $a_y = -g$.

Key Concept 6: The motion of a projectile is a combination of motion with constant velocity in the horizontal x-direction and motion with constant downward acceleration in the vertical y-direction.

Key Concept 7: You can solve most projectile problems by using the equations for x, y, v_x, and v_y as functions of time. The highest point of a projectile's motion occurs at the time its vertical component of velocity is zero.

Key Concept 8: When you solve physics problems in general, and projectile problems in particular, it's best to use symbols rather than numbers as far into the solution as possible. This allows you to better explore and understand your result.

Key Concept 9: A projectile's vertical coordinate y as a function of time is given by a quadratic equation, which in general has more than one solution. Take care to select the solution that's appropriate for the problem you're solving.

Key Concept 10: It can be useful to think of a projectile as following a straight-line path that's pulled downward by gravity a distance $1/2\, gt^2$ in a time t.

Key Concept 11: For uniform circular motion at a given speed, decreasing the radius increases the centripetal acceleration.

Key Concept 12: For uniform circular motion with a given radius, decreasing the period increases the speed and the centripetal acceleration.

Key Concept 13: To solve problems involving relative velocity along a line, use Eq. (3.32) and pay careful attention to the subscripts for the frames of reference in the problem.

Key Concept 14: To solve problems involving relative velocity in a plane or in space, use Eq. (3.35). Pay careful attention to the subscripts for the frames of reference in the problem.

Key Concept 15: The vector equation for relative velocity in a plane, Eq. (3.35), allows you to solve for two unknowns, such as an unknown vector magnitude and an unknown direction.

Conceptual Questions

1: Velocity and acceleration at the top of a projectile's path

A projectile is launched with initial nonzero x and y velocities. (a) Describe the velocity and acceleration just before the projectile reaches the top of its trajectory. (b) Describe the velocity and acceleration at the instant the projectile reaches the top of its trajectory. (c) Describe the velocity and acceleration just after the projectile reaches the top of its trajectory.

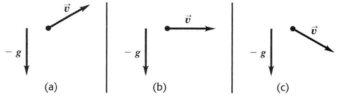

Figure 1 Question 1.

IDENTIFY AND SET UP Figure 1 shows the three time frames we'll examine. During the flight, the projectile undergoes acceleration due to gravity. The projectile's initial velocity has both x- and y-components; the x-component remains constant while the y-component is accelerated by gravity. We have to consider each component of velocity separately.

EXECUTE PART (A) The x-component of velocity is constant and directed to the right, and the y-component of velocity is directed upward and is very small just before the top of the flight. The acceleration due to gravity is directed downward.

PART (B) The x-component of velocity is constant and directed to the right, and the y-component of velocity is zero at the top of the flight. The acceleration due to gravity remains constant and directed downward.

PART (C) The x-component of velocity is constant and directed to the right, and the y-component of velocity is directed downward and is very small just after the top of the flight. The acceleration due to gravity remains constant and directed downward.

EVALUATE The acceleration due to gravity causes a change in velocity during the flight, but affects only the vertical component of velocity. The ball starts with a nonzero velocity, which decreases, drops to a minimum value at the top, and then increases downward. The acceleration due to gravity is constant throughout the motion. The velocity is changing throughout the motion.

2: Launching a marble off the edge of a table

A marble is launched off the edge of a horizontal table and lands on the floor. Draw the trajectory of the ball from the table to the floor. Draw a second line showing the trajectory of the marble if it were given a smaller initial velocity. Draw a third line showing the trajectory if the marble were given a larger initial velocity than the original initial velocity.

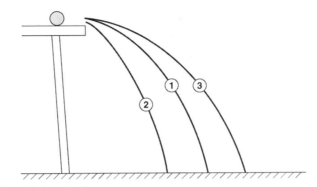

Figure 2 Question 2.

IDENTIFY, SET UP, AND EXECUTE The table and marble are sketched in Figure 2. The initial trajectory is shown and labeled "1." The marble follows a parabolic path, starting with an initial nonzero horizontal velocity. For the smaller initial velocity, the marble also follows a parabolic path, but with a termination point closer to the edge of the table. This path is shown in the figure and is labeled "2." For the larger initial velocity, the marble again follows a parabolic path, but with a termination point farther from the edge of the table. This path is shown in the figure and is labeled "3."

KEY CONCEPT 6 **EVALUATE** The paths are similar; their differences owe to the different initial velocities. How does the time the marble spends in the air compare for the three paths? All three take the same amount of time to reach the ground, as they all start with zero initial vertical velocity and fall the same distance. Since they spend the same time in the air, those with larger initial velocities reach greater horizontal distances.

3: Comparing projectiles

Figure 3 graphs the paths of two projectiles in the xy-plane. If we ignore air resistance, how do the initial velocities compare (magnitude and direction)?

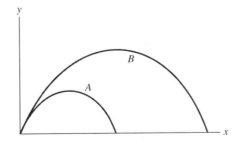

Figure 3 Question 3.

IDENTIFY, SET UP, AND EXECUTE At the origin, we see that the two paths begin identically. This indicates that the initial directions of the velocities of both projectiles are the same.

The graph does not include a time axis, so we need to look to other clues to compare the magnitudes of the initial velocities. Trajectory B reaches a greater height, indicating that its initial vertical component of velocity was larger than trajectory A's initial vertical component. Therefore, trajectory B has a greater magnitude of initial velocity.

EVALUATE Without a time axis, examine the x motion and we cannot assume that trajectory B has a greater magnitude of initial velocity.

How do the horizontal components of the initial velocities compare? Both projectiles have the same initial launch angle; therefore, the ratio of their velocity components must be the same. If the vertical component of B's velocity is larger, so must B's horizontal velocity component be larger.

4: Comparing projectiles again

Figure 4 shows the graph of the paths of two projectiles in the xy-plane. If we ignore air resistance, how do the initial velocities compare (in magnitude and direction)? Which projectile lands first?

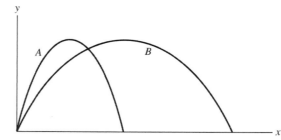

Figure 4 Question 4.

IDENTIFY, SET UP, AND EXECUTE We see that the paths do not coincide at the origin: Projectile A has a larger launch angle.

Again, the graph does not include a time axis, so we need to look to other clues to compare the magnitudes of the initial velocities. Both trajectories reach the same maximum height, indicating that both have the same vertical velocity components. However, their initial directions were different, requiring their initial horizontal components to be different. Trajectory B reaches a greater horizontal distance, so must have a larger initial horizontal velocity. Therefore, trajectory B has a greater magnitude of initial velocity.

Since both trajectories reach the same maximum height and have the same initial vertical velocity, they must end at the same time.

EVALUATE As an alternative analysis, we could have considered the time first and the velocity second. In that case, it might have been easier to see that the initial horizontal velocity of projectile B was larger because it covered more distance in the same time.

5: Falling luggage

A piece of luggage falls out of the cargo door of a airplane flying horizontally at a constant speed. In what direction should the pilot look to follow the luggage to the ground so that it can be recovered?

IDENTIFY, SET UP, AND EXECUTE When the piece of luggage falls from the airplane, its initial velocity is the same as the plane's velocity. As it falls, the luggage accelerates in the vertical direction and its horizontal velocity remains constant (assuming no air resistance). Since the plane and the piece of luggage are moving at the same horizontal velocity, the luggage falls directly below the plane. The pilot should look straight down to see where the luggage will land.

KEY CONCEPT 5 **EVALUATE** You might expect that the piece of luggage would fall behind the airplane. Now, what would cause it to fall behind the plane? For it to fall behind the plane, the luggage would have to slow down, or accelerate in a direction opposite that of its horizontal motion. Air resistance could slow down the bag, because air resistance opposes the motion of a body.

Problems

1: Water Balloon Launch

Your physics professor is walking past the physics building at a constant 3.5 m/s. You're on the third-floor balcony (25 m above the ground) of the building with your new water balloon launcher. The launcher allows you to adjust the speed of the water balloon, but you can launch the balloon only horizontally. What launch speed should be set for the balloon to land on your professor if you launch it just as she passes below? What will be your professor's horizontal distance from the building when the balloon hits her, as measured from a point on the ground directly below you?

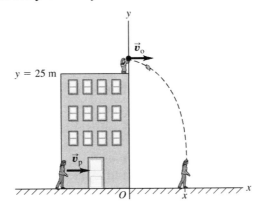

Figure 5 Problem 1.

IDENTIFY Once launched, the water balloon will undergo gravitational acceleration in the vertical direction and continue with constant velocity in the horizontal direction. We'll apply the constant-acceleration kinematics equations separately to the horizontal and vertical components to solve the problem.

SET UP Figure 5 shows a sketch of the situation. We ignore effects due to the air. Your professor is roughly 1.7 m tall, but we'll ignore her height and determine the position where the balloon hits the ground. An xy-coordinate system is shown in the diagram.

We first determine the launch speed. Since there is no acceleration in the horizontal direction, the water balloon must be launched at the same speed as your professor is walking, 3.5 m/s.

EXECUTE To find where the balloon hits her, we find the time from the start of the vertical motion and use that to find the horizontal distance the water balloon travels as it falls. The vertical position for constant acceleration is given by

$$y = y_0 + v_{0y}t + \tfrac{1}{2}at^2.$$

We've set the origin at the ground; therefore, the initial position becomes 25 m and the final position becomes 0. The launcher imparts only a horizontal velocity, so the initial vertical velocity is zero. The acceleration is $-g$, since the positive vertical axis is directed upward. Combining these parameters gives

$$0 = 25 \text{ m} - \tfrac{1}{2}gt^2.$$

Solving for t produces

$$t = \sqrt{\frac{2(25 \text{ m})}{(9.8 \text{ m/s}^2)}} = 2.26 \text{ s}.$$

It takes 2.26 s for the balloon to fall to the ground. During this time, it is traveling with constant horizontal velocity. We find the horizontal distance it travels from the formula

$$x = x_0 + v_{0x}t.$$

Your origin is directly below your position on the balcony ($x_0 = 0$). Substituting the horizontal velocity and time we calculated, we find the horizontal distance:

$$x = v_{0x}t = (3.5 \text{ m/s})(2.26 \text{ s}) = 7.9 \text{ m}.$$

The water balloon will hit your professor a horizontal distance 7.9 m away from your location.

KEY CONCEPT 9 **EVALUATE** This is a straightforward application of two-dimensional kinematics. We solved for one component of the motion and substituted the result into the equation for the other component to arrive at the solution. Note that we solved for the vertical motion and substituted the result into the equation for the horizontal motion, the opposite order of the previous problem. Practicing solving a variety of problems will build proficiency in solving problems involving motion in a plane.

Practice Problem: How much time would it take if the building were 30 m tall? *Answer:* 2.47 s

Extra Practice: Where would the professor get hit? *Answer:* 8.66 m

2: Hitting a Baseball in Fenway Park

You win a chance to try hitting a baseball over the "Green Monster" in Fenway Park. The Green Monster is a 37.2-ft (11.3-m)-high wall in left field of the ballpark. The left end is closest to home plate, 310 ft (94.5m) away. If you give the ball an initial speed of 33 m/s at an initial angle of 47°, by how much does the baseball clear (or miss) the top of the wall?

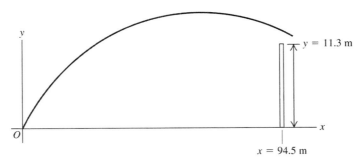

Figure 6 Problem 2 sketch.

IDENTIFY The baseball has a nonzero initial velocity, undergoes acceleration due to gravity in the vertical direction, and has no acceleration in the horizontal direction. Constant-acceleration kinematics equations will be applied separately to the horizontal and vertical components to find the solution.

SET UP We sketch the problem in Figure 6. We ignore effects due to the air. The ball is hit roughly 1 m or so above the ground, but we'll neglect this small distance and set the origin at ground level. An xy-coordinate system that coincides with this choice is shown in the diagram.

We'll solve for the time the baseball arrives at the wall by using the horizontal-position equation. Then we'll substitute into the vertical-position equation to find the vertical position of the ball at the wall.

EXECUTE The horizontal position is given by

$$x = x_0 + v_{0x}t.$$

In this case, we start at the origin $(x_0 = 0)$ and the x-component of velocity includes a cosine term:

$$x = v_{0x}t = v_0 \cos\theta\, t.$$

We wish to find the time t when the baseball is located at the wall $(x = 94.5\text{ m})$:

$$t = \frac{x}{v_0 \cos\theta} = \frac{(94.5\text{ m})}{(33\text{ m/s})\cos 47°} = 4.20\text{ s}.$$

After 4.20 s, the baseball's horizontal position is 94.5 m. We now find the vertical position at that time. The vertical position is given by

$$y = y_0 + v_{0y}t + \tfrac{1}{2}at^2$$

Again, we start at the origin $(y_0 = 0)$, the acceleration is directed downward (negative) and is of magnitude g, and the y-component of velocity includes a sine term:

$$y = v_0 \sin\theta\, t + \tfrac{1}{2}(-g)t^2.$$

We can now substitute our values into the equation to find the height of the ball:

$$y = (33\text{ m/s})(\sin 47°)(4.20\text{ s}) + \tfrac{1}{2}(-9.8\text{ m/s}^2)(4.20\text{ s})^2 = 14.9\text{ m}.$$

At the wall, the ball's height is 14.9 m, or 3.6 m above the 11.3-m high wall. The ball clears the Green Monster by 3.6 m!

KEY CONCEPT 8 **EVALUATE** This is another straightforward application of two-dimensional kinematics. We solved for one component of the motion and substituted the result into the equation for the other component to arrive at the solution. We'll follow this procedure often to solve problems involving motion in a plane.

Practice Problem: Find the *x*- and *y*-components of the baseball's velocity at the wall. *Answer:* $v_x = 22.5$ m/s, $v_y = -17.0$ m/s

Extra Practice: How far would the ball travel if it landed on the ground? *Answer:* 364 ft

CAUTION **Don't Mix *x* and *y*!** It is easy to mix up *x*- and *y*-components for position, velocity, and acceleration. You must label each of these carefully to ensure that you don't make mistakes. Only time is common to both the *x*- and *y*-components.

3: Acceleration of a Propeller Tip

The Wright Brothers' plane had a 2.4-m-long propeller that operated at a constant 350 rpm. Find the acceleration of a particle at the tip of the propeller.

IDENTIFY This is a uniform circular motion problem; the acceleration is determined by the centripetal-acceleration formula.

SET UP We'll need to find the velocity and radius from the information provided. A diagram of the problem is shown in Figure 7.

To find the centripetal acceleration, we need the radius and speed of a particle on the tip of the propeller. We're given the diameter of the propeller, and dividing that in half gives the radius.

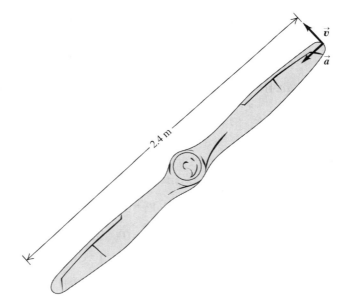

Figure 7 Problem 3.

EXECUTE The speed of a particle at the end of the propeller is found by dividing the circumference at the tip of the propeller $(2\pi r)$ by the time it takes the propeller to make 1 revolution: (T)

Study Guide for Motion in Two or Three Dimensions

$$v = \frac{2\pi r}{T}.$$

The propeller makes 350 revolutions per minute, so we find the time it takes to make 1 revolution by dividing 1 minute by 350 revolutions:

$$T = \frac{1\,\text{min}}{350\,\text{rev}} = \frac{60\,\text{s}}{350\,\text{rev}} = 0.171\,\text{s/rev}.$$

The propeller takes 0.171 s to make 1 revolution. We can now find the velocity:

$$v = \frac{2\pi r}{T} = \frac{2\pi(1.2\,\text{m})}{0.171\,\text{s}} = 44.1\,\text{m/s}.$$

The centripetal acceleration is then

$$a_{rad} = \frac{v^2}{r} = \frac{(44.1\,\text{m/s})^2}{(1.2\,\text{m})} = 1620\,\text{m/s}^2.$$

The centripetal acceleration of a particle on the tip of the propeller is 1620 m/s². This is equivalent to 165 times the acceleration due to gravity!

KEY CONCEPT 12 EVALUATE We've found the magnitude of the acceleration in this problem. Acceleration is a vector, so where does it point? The acceleration is directed toward the center of the propeller, perpendicular to the velocity.

We didn't include gravity in this problem. The particle is affected by attraction toward the ground throughout its motion; however, the resulting acceleration is very small compared with the centripetal acceleration.

Practice Problem: What is the centripetal acceleration if the time of each revolution was 0.15 s? *Answer:* 2106 m/s²

Extra Practice: How many rpm is this? *Answer:* 400 rpm

4: Paddling across a River

You wish to paddle north across a 350-m-wide river. The river has a 1.2 m/s current from east to west, and you can paddle at a steady 1.5 m/s pace. In what direction should you paddle, and how long will it take you to cross the river?

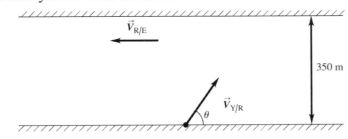

Figure 8 Problem 4.

IDENTIFY You'll need to paddle into the river current to compensate for the river's moving your boat downstream as you cross.

SET UP Figure 8 shows a sketch of the situation. We use relative velocities to solve the problem. The direction in which you must paddle is determined by setting north to be the direction of your resulting relative velocity with respect to the earth.

Figure 9 Problem 4.

EXECUTE Figure 9 combines your velocity with respect to the river ($v_{Y/R}$) with the river current's velocity with respect to the earth ($v_{R/E}$) to form your relative velocity with respect to the earth ($v_{Y/E}$):

$$\vec{v}_{Y/E} = \vec{v}_{Y/R} + \vec{v}_{R/E}.$$

For you to land directly across from your starting point, the direction of $v_{Y/E}$ must be northward. Therefore, the x-component of $v_{Y/R}$ must be equal and opposite to $v_{R/E}$. We find the direction in which you should paddle by equating those two magnitudes:

$$(\vec{v}_{Y/R})_x = v_{R/E},$$
$$(\vec{v}_{Y/R})_x = v_{Y/R} \sin\theta = v_{R/E}.$$

$$\theta = \sin^{-1}\left(\frac{v_{R/E}}{v_{Y/R}}\right) = \sin^{-1}\left(\frac{1.2\,\text{m/s}}{1.5\,\text{m/s}}\right) = 53°.$$

You'll need to paddle 53° east of north to follow a northward path. The time it will take is found from the y-component of the displacement. You're traveling at constant velocity, so the vertical component of the displacement is

$$y - y_0 = (v_{Y/E})_y t.$$

$(v_{Y/E})_y$ is the magnitude of $v_{Y/E}$, because $v_{Y/E}$ has only a y-component. $(v_{Y/E})_y$ must also be equal to the y-component of $v_{Y/R}$. Solving for time gives

$$t = \frac{y - y_0}{(v_{Y/E})_y} = \frac{y - y_0}{v_{Y/E}} = \frac{y - y_0}{(v_{Y/R})_y} = \frac{y - y_0}{v_{Y/E} \cos\theta} = \frac{350\,\text{m}}{(1.5\,\text{m/s})\cos(53°)} = 390\,\text{s}.$$

It'll take you 390 s to paddle across the river.

KEY CONCEPTS 14, 15

EVALUATE When you paddle across a river perpendicular to its flow, your relative velocity with respect to the earth is always less than your velocity with respect to the river. It also takes longer to cross a river that has a current compared with a calm river when your path is perpendicular to the river. The next practice problem lets you compare the time required to cross a calm river with the time you just found in dealing with a river that has a current.

Practice Problem: How long would it take to paddle across the same river with no current? *Answer:* 230 s

Extra Practice: What if the current was 1.4 m/s? *Answer:* 650 s

Try It Yourself!

1: Ball thrown from a cliff

A boy throws a ball from a cliff at an angle of 30.0° above the horizontal with an initial velocity of 10.0 m/s. The ball lands 100.0 m from the base of the cliff. (a) How high is the cliff? (b) What is the time of flight of the ball? (c) What is the velocity of the ball just before impact?

Solution Checkpoints

IDENTIFY AND SET UP Constant-acceleration kinematics equations should be applied separately to the horizontal and vertical components of the ball's motion. The initial velocity should be broken down into x- and y-components. There is acceleration only in the vertical direction.

EXECUTE Equations for the x- and y-components of position and velocity can be found. The x position equation can be solved for time and substituted into the y position equation to find the height of the cliff. Doing this leads to

$$y = x \tan\theta - \tfrac{1}{2} g \left[\frac{x}{v_0 \cos\theta} \right]^2,$$

from which you will find that the cliff is 595 m high. You can then substitute that number into the x position equation to find that the time is 11.5 s.

To find the velocity just before impact, you can find the velocity components from the kinematic equations. Knowing the components, you can find the magnitude and direction of the velocity. The velocity is 108 m/s, directed 85° below the x-axis.

KEY CONCEPT 6 **EVALUATE** We see that the velocity just before impact is almost straight down. Can it ever be exactly straight down?

Practice Problem: What is the height if the initial angle was 0°? *Answer:* 490 m

Extra Practice: What would the speed be just before impact? *Answer:* 98.5 m/s

2: Kicking a soccer ball

A soccer ball is kicked 25 m in the horizontal direction. What is its initial velocity if it reaches a maximum height of 6.0 m?

Solution Checkpoints

IDENTIFY AND SET UP Constant-acceleration kinematics equations should be applied separately to the horizontal and vertical components of the ball's motion. Draw the ball's path and choose an appropriate coordinate system.

EXECUTE The initial y-component of velocity can be found from the formula

$$v_y^2 = v_{0y}^2 + 2a_y \Delta y.$$

This gives an initial vertical velocity of 10.8 m/s. To find the initial horizontal component of velocity, we determine the flight time from the vertical motion and combine that with the horizontal distance traveled. The result is an initial velocity of 15.6 m/s at an angle of 44° above the horizontal.

KEY CONCEPT 7 **EVALUATE** By this point, you have seen many kinematics problems. Can you summarize your problem-solving techniques?

Practice Problem: How long is the ball in the air if the maximum height was 3 m? *Answer:* 1.56s

Extra Practice: What is the initial speed if the maximum height was 3 m? *Answer:* 17.7 m/s

3: Archer and arrow

An archer shoots an arrow into the air at an angle of 30° above the horizontal. It lands on a building 100.0 m away at a height of 20.0 m. What was the initial speed of the arrow?

Solution Checkpoints

IDENTIFY AND SET UP Can constant-acceleration kinematics be used in this case? What assumptions do you need to make?

EXECUTE The x and y position equations can be rearranged to yield

$$y = x \tan\theta - \tfrac{1}{2}g \left[\frac{x}{v_0 \cos\theta} \right]^2.$$

Rewriting the equation and the x and y positions at the building gives $v_0 = 41$ m/s.

KEY CONCEPT 6 **EVALUATE** How can you check your result? How do you know that it is reasonable?

Practice Problem: How long does it take the arrow to land on the building? *Answer:* 2.78 s

Extra Practice: How fast is the arrow moving right before it lands? *Answer:* 36.6 m/s

Key Example Variation Problems

Solutions to these problems are in Chapter 3 of the Student's Solutions Manual.

Be sure to review EXAMPLES 3.6, 3.7, 3.8, and 3.9 (Section 3.3) before attempting these problems. In all problems, ignore air resistance.

VP3.9.1 You launch a projectile from level ground at a speed of 25.0 m/s and an angle of 36.9° above the horizontal. (a) How long after it is launched does the projectile reach its maximum height above the ground, and what is that maximum height? (b) How long after the projectile is launched does it return to ground level, and how far from its launch point does it land?

VP3.9.2 You throw a baseball at an angle of 30.0° above the horizontal. It reaches the highest point of its trajectory 1.05 s later. (a) At what speed does the baseball leave your hand? (b) What is the maximum height above the launch point that the baseball reaches?

VP3.9.3 You toss a walnut at a speed of 15.0 m/s at an angle of 50.0° above the horizontal. The launch point is on the roof of a building that is 20.0 m above the ground. (a) How long after it is launched does the walnut reach the ground? (b) How far does the walnut travel horizontally from launch point to landing point? (c) What are the horizontal and vertical components of the walnut's velocity just before it reaches the ground?

VP3.9.4 You use a slingshot to launch a potato horizontally from the edge of a cliff with speed v_0. The acceleration due to gravity is g. Take the origin at the launch point. (a) How long after you launch the potato has it moved as far horizontally from the launch point as it has moved vertically? What are the coordinates of the potato at this time? (b) How long after you launch the potato is it moving in a direction exactly 45° below the horizontal? What are the coordinates of the potato at this time?

Be sure to review EXAMPLES 3.11 and 3.12 (Section 3.4) before attempting these problems.

VP3.12.1 A cyclist going around a circular track at 10.0 m/s has a centripetal acceleration of 5.00 m/s². What is the radius of the curve?

VP3.12.2 A race car is moving at 40.0 m/s around a circular racetrack of radius 265 m. Calculate (a) the period of the motion and (b) the car's centripetal acceleration.

VP3.12.3 The wheel of a stationary exercise bicycle at your gym makes one rotation in 0.670 s. Consider two points on this wheel: Point P is 10.0 cm from the rotation axis, and point Q is 20.0 cm from the rotation axis. Find (a) the speed of each point on the spinning wheel and (b) the centripetal acceleration of each point.
(c) For points on this spinning wheel, as the distance from the axis increases, does the speed increase or decrease? Does the centripetal acceleration increase or decrease?

VP3.12.4 The planets Venus, Earth, and Mars all move in approximately circular orbits around the sun. Use the data in the table to find (a) the speed of each planet in its orbit and (b) the centripetal acceleration of each planet. (c) As the size of a planet's orbit increases, does the speed increase or decrease? Does the centripetal acceleration increase or decrease?

Planet	Orbital radius (m)	Orbital period (days)
Venus	1.08×10^{11}	225
Earth	1.50×10^{11}	365
Mars	2.28×10^{11}	687

VP3.12.5 Object A is moving at speed v in a circle of radius R. Object B is moving at speed $2v$ in a circle of radius $R/2$. (a) What is the ratio of the period of object A to the period of object B? (b) What is the ratio of the centripetal acceleration of object A to the centripetal acceleration of object B?

Be sure to review EXAMPLES 3.13, 3.14, and 3.15 (Section 3.5) before attempting these problems.

VP3.15.1 A police car in a high-speed chase is traveling north on a two-lane highway at 35.0 m/s. In the southbound lane of the same highway, an SUV is moving at 18.0 m/s. Take the positive x-direction to be toward the north. Find the x-velocity of (a) the police car relative to the SUV and (b) the SUV relative to the police car.

VP3.15.2 Race cars A and B are driving on the same circular racetrack at the same speed of 45.0 m/s. At a given instant car A is on the north side of the track moving eastward and car B is on the south side of the track moving westward. Find the velocity vector (magnitude and direction) of (a) car A relative to car B and (b) car B relative to car A. (c) Does the relative velocity have a component along the line connecting the two cars? Are the two cars approaching each other, moving away from each other, or neither?

VP3.15.3 Two vehicles approach an intersection: a truck moving eastbound at 16.0 m/s and an SUV moving southbound at 20.0 m/s. Find the velocity vector (magnitude and direction) of (a) the truck relative to the SUV and (b) the SUV relative to the truck.

VP3.15.4 A jet is flying due north relative to the ground. The speed of the jet relative to the ground is 155 m/s. The wind at the jet's altitude is 40.0 m/s toward the northeast (45.0° north of east). Find the speed of the jet relative to the air (its airspeed) and the direction in which the pilot of the jet must point the plane so that it travels due north relative to the ground.

STUDENT'S SOLUTIONS MANUAL FOR MOTION IN TWO OR THREE DIMENSIONS

VP3.9.1. **IDENTIFY:** This is a projectile problem with only vertical acceleration.

SET UP: The formulas $y = (v_0 \sin \alpha_0)t - \frac{1}{2}gt^2$, $v_x = v_0 \cos \alpha_0$, $v_y = v_0 \sin \alpha_0 - gt$, and $v_y^2 = v_{0y}^2 + 2a_y(y - y_0)$ apply.

EXECUTE: We know the launch speed and launch angle of the projectile, and want to know its maximum height, maximum range, and the time it is in the air.

(a) At the maximum height, $v_y = 0$. We use $v_y = v_0 \sin \alpha_0 - gt$ to find the time.

$0 = (25.0 \text{ m/s})(\sin 36.9°) - (9.80 \text{ m/s}^2)t \rightarrow t = 1.53$ s. Now use $v_y^2 = v_{0y}^2 + 2a_y(y - y_0)$.

$0 = [(25.0 \text{ m/s})(\sin 36.9°)]^2 - 2(9.80 \text{ m/s}^2)(y - y_0) \rightarrow y - y_0 = 11.5$ m.

(b) When the projectile returns to ground level, $v_y = -v_0 \sin \alpha_0$. So

$v_y = v_0 \sin \alpha_0 - gt = -v_0 \sin \alpha_0$

$2v_0 \sin \alpha_0 = gt$

$2(25.0 \text{ m/s})(\sin 36.9°) = (9.80 \text{ m/s}^2)t \rightarrow t = 3.06$ s.

For the horizontal motion, we have

$x = (v_0 \cos \alpha_0)t = (25.0 \text{ m/s})(\cos 36.9°)(3.06 \text{ s}) = 61.2$ m.

EVALUATE: (a) Calculate y using $y = (v_0 \sin \alpha_0)t - \frac{1}{2}gt^2$.

$y = (25.0 \text{ m/s})(\sin 36.9°)(1.53 \text{ s}) - \frac{1}{2}(9.80 \text{ m/s}^2)(1.53 \text{ s})^2 = 11.5$ m, which agrees with our result. (b) The acceleration is constant, so the time for the upward motion is equal to the time for the downward motion. Thus $t_{tot} = 1.53$ s + 1.53 s = 3.06 s, which agrees with our result.

VP3.9.2. **IDENTIFY:** This is a projectile problem with only vertical acceleration.

SET UP: The formulas $v_x = v_0 \cos \alpha_0$, $v_y = v_0 \sin \alpha_0 - gt$, $y = (v_0 \sin \alpha_0)t - \frac{1}{2}gt^2$, and $v_y^2 = v_{0y}^2 + 2a_y(y - y_0)$ apply. At the highest point in the baseball's trajectory, its vertical velocity is zero, but its horizontal velocity is the same as when it left the ground.

EXECUTE: (a) At the highest point, $v_y = 0$, so $v_y = v_0 \sin \alpha_0 - gt$ gives

$0 = v_0 \sin(30.0°) - (9.80 \text{ m/s}^2)(1.05 \text{ s})^2 \rightarrow v_0 = 20.6$ m/s.

(b) Using $v_y^2 = v_{0y}^2 + 2a_y(y - y_0)$ gives

$0 = [(20.6 \text{ m/s})(\sin 30.0°)]^2 - 2(9.80 \text{ m/s}^2)(y - y_0) \rightarrow y - y_0 = 5.40$ m

Chapter 3 of Student's Solutions Manual to accompany *University Physics with Modern Physics, Volume 1*, Fifteenth Edition. Hugh D. Young and Roger A. Freedman. Copyright © 2020 by Pearson Education, Inc. All rights reserved.

EVALUATE: We can check by finding y when $t = 1.05$ s, using $y = (v_0 \sin\alpha_0)t - \frac{1}{2}gt^2$.

$y = (20.6 \text{ m/s})(\sin 30.0°)(1.05 \text{ s}) - \frac{1}{2}(9.80 \text{ m/s}^2)(1.05 \text{ s})^2 = 5.40$, which agrees with our result.

VP3.9.3. **IDENTIFY:** This is a projectile problem with only vertical acceleration.
SET UP: The formulas $v_x = v_0 \cos\alpha_0$, $v_y = v_0 \sin\alpha_0 - gt$, $x = (v_0 \cos\alpha_0)t$, and $y = (v_0 \sin\alpha_0)t - \frac{1}{2}gt^2$ apply. Call the origin the launch point with the +y-axis vertically upward.

EXECUTE: (a) When the walnut reaches the ground, $y = -20.0$ m. Use $y = (v_0 \sin\alpha_0)t - \frac{1}{2}gt^2$.

$-20.0 \text{ m} = (15.0 \text{ m/s})(\sin 50.0°)t - \frac{1}{2}(9.80 \text{ m/s}^2)t^2$. Using the quadratic formula gives two roots, only one of which is positive: $t = 3.51$ s.
(b) There is no horizontal acceleration, so we use $x = (v_0 \cos\alpha_0)t$ to find the distance x.
$x = (v_0 \cos\alpha_0)t = (15.0 \text{ m/s})(\cos 50.0°)(3.51 \text{ s}) = 33.8$ m.
(c) There is no horizontal acceleration, so $v_x = v_0 \cos\alpha_0 = (15.0 \text{ m/s})(\cos 50.0°) = 9.64$ m/s. The vertical velocity is
$v_y = v_0 \sin\alpha_0 - gt = (15.0 \text{ m/s}) \sin 50.0° - (9.80 \text{ m/s}^2)(3.51 \text{ s}) = -22.9$ m/s. The minus sign tells us the walnut is moving downward.
EVALUATE: We cannot use the projectile range formula because the landing point is not on the same level as the launch point.

VP3.9.4. **IDENTIFY:** This is a projectile problem with only vertical acceleration.
SET UP: The formulas $v_x = v_0 \cos\alpha_0$, $v_y = v_0 \sin\alpha_0 - gt$, $y = (v_0 \sin\alpha_0)t - \frac{1}{2}gt^2$, and $x = (v_0 \cos\alpha_0)t$ apply. Call the origin the launch point with the +y-axis vertically upward.
EXECUTE: (a) The horizontal and vertical distances are equal, so $y = -x$, so
$-(v_0 \cos\alpha_0)t = (v_0 \sin\alpha_0)t - \frac{1}{2}gt^2$. Since $\alpha_0 = 0$, we have $v_0 t = \frac{1}{2}gt^2$, which gives $t = 2v_0/g$. At this time $x = v_0 t = v_0(2v_0/g) = 2v_0^2/g$. Since $-x = y$, $y = -2v_0^2/g$.
(b) When the potato is moving at 45° below the horizontal, $v_y = -v_x$. $v_x = v_0$ and $v_y = -v_0 = -gt$.
Therefore $t = v_0/g$. At this time, $x = v_0 t = v_0(v_0/g) = v_0^2/g$, and $y = -\frac{1}{2}gt^2 = -\frac{1}{2}g(v_0/g)^2 = -v_0^2/2g$.
EVALUATE: As a check, solve for y using the time found in part (a).
$y = -\frac{1}{2}gt^2 = -\frac{1}{2}g(2v_0/g)^2 = -2v_0^2/g$, just as we found. Notice in part (b) that $x \neq y$ when the potato is traveling at 45° with the horizontal, but the magnitudes of the velocity are equal. We cannot use the projectile range formula because the landing point is not on the same level as the launch point.

VP3.12.1. **IDENTIFY:** This problem involves circular motion at constant speed. The acceleration of the cyclist is toward the center of the circle.

SET UP: The radial acceleration of the cyclist is $a_{rad} = \frac{v^2}{R}$ toward the center of the circular track. We know the speed and acceleration and want the radius of the circle.

EXECUTE: Solve $a_{rad} = \frac{v^2}{R}$ for R, giving $R = v^2/a_{rad} = (10.0 \text{ m/s})^2/(5.00 \text{ m/s}^2) = 20.0$ m.

EVALUATE: The speed of the cyclist is constant, but not the velocity since it is tangent to the circular path and is always changing direction.

VP3.12.2. **IDENTIFY:** This problem involves circular motion at constant speed. The acceleration of the car is toward the center of the circle.

SET UP: The radial acceleration of the car is $a_{rad} = \dfrac{v^2}{R}$ toward the center of the circular track. In terms of the period of motion, the acceleration is $a_{rad} = \dfrac{4\pi^2 R}{T^2}$. We know the speed of the car and the radius of the circle and want to find the period of the motion and centripetal acceleration of the car.

EXECUTE: (a) Equate the two expressions for a_{rad}, giving $\dfrac{4\pi^2 R}{T^2} = \dfrac{v^2}{R}$. Now solve for T.

$T = \sqrt{\dfrac{4\pi^2 R^2}{v^2}} = 2\pi R/v = 2\pi(265 \text{ m})/(40.0 \text{ m/s}) = 41.6 \text{ s}$.

(b) $a_{rad} = \dfrac{v^2}{R} = (40.0 \text{ m/s})^2/(265 \text{ m}) = 6.04 \text{ m/s}^2$.

EVALUATE: We can find the period by realizing that vT is the circumference of the track, which is $2\pi R$. Therefore $vT = 2\pi R$, so $T = 2\pi R/v = 2\pi(265 \text{ m})/(40.0 \text{ m/s}) = 41.6 \text{ s}$, which is agrees with our answer in (a). The acceleration of the car (and the driver inside) is quite large, over 60% of g.

VP3.12.3. **IDENTIFY:** This problem involves circular motion at constant speed. The radial acceleration is toward the center of the circle.

SET UP: The equation $a_{rad} = \dfrac{4\pi^2 R}{T^2}$ applies and the speed is $v = 2\pi R/T$. The period is the same for all points on the wheel, but the speed (and hence acceleration) is not.

EXECUTE: (a) $v_{10} = 2\pi R/T = 2\pi(10.0 \text{ cm})/(0.670 \text{ s}) = 93.8 \text{ cm/s} = 0.938 \text{ m/s}$. Since R is twice as great at 20.0 cm, $v_{20} = 2v_{10} = 2(0.938 \text{ m/s}) = 1.88 \text{ m/s}$.
(b) $a_{10} = 4\pi^2 R/T^2 = 4\pi^2(0.100 \text{ m})/(0.670 \text{ s})^2 = 8.79 \text{ m/s}^2$. As in (a), we see that $a_{20} = 2a_{10} = 2(8.79 \text{ m/s}^2) = 17.6 \text{ m/s}^2$.
(c) Both the speed and radial acceleration increase as R increases.

EVALUATE: As R increases, the points farther from the center must travel a greater distance than points closer to the center. So the speed and acceleration are greater for those distant points.

VP3.12.4. **IDENTIFY:** This problem involves circular motion at constant speed. The radial acceleration of a planet is toward the center of the circle, which is essentially the sun.

SET UP: The equation $a_{rad} = \dfrac{v^2}{R}$ applies and the speed of a planet is $v = 2\pi R/T$.

EXECUTE: (a) Apply $v = 2\pi R/T$ to each planet using the data given.
$v_V = 2\pi(1.08 \times 10^{11} \text{ m})/[(225)(86,500)\text{s}] = 3.49 \times 10^4 \text{ m/s}$. Similar calculations for the earth and Mars gives $v_E = 2.99. \times .10^4 \text{ m/s}$ and $v_M = 2.41 \times 10^4 \text{ m/s}$.

(b) Use $a_{rad} = \dfrac{v^2}{R}$ with the speeds found in part (a).

$a_V = (3.49 \times 10^4 \text{ m/s})^2/(1.08 \times 10^{11} \text{ m}) = 1.13 \times 10^{-2} \text{ m/s}^2$. Likewise we get $a_E = 5.95 \times 10^{-3} \text{ m/s}^2$ and $a_M = 2.55 \times 10^{-3} \text{ m/s}^2$.
(c) As the size of the orbit increases, both the orbital speed and the radial acceleration decrease.

EVALUATE: Since R increases and v decreases with distance from the sun, it must follow that the radial acceleration also decreases with distance. This is reasonable because the gravitational pull of the sun (to be studied in a later chapter) is weaker for distant planets than for closer ones.

VP3.12.5. **IDENTIFY:** This problem involves circular motion at constant speed.

SET UP: The equation $a_{rad} = \dfrac{v^2}{R}$ applies and the speed of an object moving in a circle is $v = 2\pi R/T$.

EXECUTE: (a) $v = 2\pi R/T$, so $T = 2\pi R/v$. Take the ratio of the two periods, giving $\dfrac{T_A}{T_B} = \dfrac{2\pi R_A / v_A}{2\pi R_B / v_B} = \dfrac{R_A}{R_B}\dfrac{v_B}{v_A}$. Putting in the given values for the speeds and radii gives $\dfrac{T_A}{T_B} = \dfrac{R(2v)}{(R/2)v} = 4.$

(b) Use $a_{rad} = \dfrac{v^2}{R}$, take the ratio of the accelerations, and use the given speeds and radii.

$\dfrac{a_A}{a_B} = \dfrac{v_A^2 / R_A}{v_B^2 / R_B} = \left(\dfrac{v_A}{v_B}\right)^2 \dfrac{R_B}{R_A} = \left(\dfrac{v}{2v}\right)^2 \dfrac{R/2}{R} = \dfrac{1}{8}.$

EVALUATE: Since $a_{rad} = \dfrac{v^2}{R}$, doubling the speed will increase a_{rad} by a factor of 4. Then halving the radius will increase a_{rad} by another factor of 2, giving a total factor of 8. This tells us that the inner object will have an acceleration 8 times that of the outer object, which is what we found in part (b).

VP3.15.1. **IDENTIFY:** This problem is about relative velocities.

SET UP: If object P is moving relative to object B and B is moving relative to A, then the velocity of P relative to A is given by $\vec{v}_{P/A} = \vec{v}_{P/B} + \vec{v}_{B/A}$. Let subscript P denote the police car, S the SUV, and E the earth.

EXECUTE: (a) We have one-dimensional motion, so the relative velocities in the x-direction are given by $v_{P/S} = v_{P/E} + v_{E/S}$. Using the given values gives
$v_{P/S} = 35.0$ m/s $+ 18.0$ m/s $= 53.0$ m/s.
(b) In this case, we want $v_{S/P}$, so
$v_{S/P} = v_{S/E} + v_{E/P} = -18.0$ m/s $+ (-35.0$ m/s$) = -53.0$ m/s.

EVALUATE: A rider in the SUV sees the police car going north at 53.0 m/s, but a rider in the police car sees the SUV going south at 53.0 m/s. Since they are going in opposite directions, their speeds relative to the earth add.

VP3.15.2. **IDENTIFY:** This problem is about relative velocities.

SET UP: If object P is moving relative to object B and B is moving relative to A, then the velocity of P relative to A is given by $\vec{v}_{P/A} = \vec{v}_{P/B} + \vec{v}_{B/A}$. Let subscript A denote the car A, B car B, and E the earth. The cars have only east-west velocities, so we look at velocity components along those directions.

EXECUTE: (a) We want the velocity of car A relative to car B.
$v_{A/B} = v_{A/E} + v_{E/B} = 45.0$ m/s $+ 45.0$ m/sk $= 90.0$ m/s, eastward.
(b) $v_{B/A} = v_{B/E} + v_{E/A} = -45.0$ m/s $+ (-45.0$ m/s$) = -90.0$ m/s, so the magnitude is 90.0 m/s and the direction is westward.
(c) At the points in this problem, the cars have only east-west velocities, so they have no relative velocity component along the line connecting them. Since both cars are traveling at the same speed in the same clockwise sense in a circle, they always remain the same distance apart. So they are neither approaching nor moving away from each other.

EVALUATE: In both cases the cars have east-west velocities in opposite directions, so their velocities relative to the earth add. When A sees B moving eastward at 90.0 m/s, B sees A moving westward at 90.0 m/s.

VP3.15.3. **IDENTIFY:** This problem is about relative velocities.
SET UP: If object P is moving relative to object B and B is moving relative to A, then the velocity of P relative to A is given by $\vec{v}_{P/A} = \vec{v}_{P/B} + \vec{v}_{B/A}$. Let subscript T denote the truck, S the SUV, and E the earth.
EXECUTE: For this case, we have $\vec{v}_{T/S} = \vec{v}_{T/E} + \vec{v}_{E/S}$. We want $\vec{v}_{T/S}$. We know that $\vec{v}_{T/E} = 16.0$ m/s eastbound and $\vec{v}_{S/E} = 20.0$ m/s southbound. Therefore $\vec{v}_{E/S} = 20.0$ m/s northbound. Figure VP3.15.3 illustrates the velocity vectors.

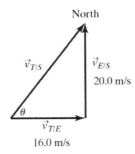

Figure VP3.15.3

Applying $A = \sqrt{A_x^2 + A_y^2}$ for the magnitude of a vector, we have
$v_{T/S} = \sqrt{(16.0 \text{ m/s})^2 + (20.0 \text{ m/s})^2} = 25.6$ m/s. From the figure, we see that
$\theta = \arctan[(20.0 \text{ m/s})/(16.0 \text{ m/s})] = 51.3°$ north of east.
(b) Using $\vec{v}_{S/T} = -\vec{v}_{T/S}$, the speed is 25.6 m/s, and the direction is 51.3° south of west.
EVALUATE: Check (b) by applying the relative velocity formula.
$\vec{v}_{S/T} = \vec{v}_{S/E} + \vec{v}_{E/T} = -20.0 \text{ m/s } \hat{i} + (-16.0 \text{ m/s}) \hat{j} = -(16.0 \text{ m/s } \hat{i} + 20.0 \text{ m/s } \hat{j}) = -\vec{v}_{T/S}$.

VP3.15.4. **IDENTIFY:** This problem is about relative velocities.
SET UP: If object P is moving relative to object B and B is moving relative to A, then the velocity of P relative to A is given by $\vec{v}_{P/A} = \vec{v}_{P/B} + \vec{v}_{B/A}$. Let subscript J denote the jet, A the air, and E the earth.
EXECUTE: The jet's velocity relative to the earth is $\vec{v}_{J/E} = \vec{v}_{J/A} + \vec{v}_{A/E}$. Figure VP3.15.4 illustrates these vectors. We want to find the magnitude of $\vec{v}_{J/A}$ (the airspeed) and its direction (θ in the figure). Therefore we first find the components of $\vec{v}_{J/A}$.

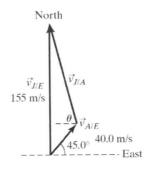

Figure VP3.15.4

Since $\vec{v}_{J/E}$ is due north, it has no east-west component. From the figure, we can therefore see that the east-west components of $\vec{v}_{A/E}$ and $\vec{v}_{J/A}$ must have opposite sign and equal magnitudes. $\vec{v}_{J/A}$ (east component) $= -\vec{v}_{A/E}$ (west component) $= -(40.0 \text{ m/s}) \cos 45.0° = -28.28$ m/s. From the figure, we also see that

$\vec{v}_{J/A}$ (north component) $+ \vec{v}_{A/E}$ (north component) $= \vec{v}_{J/E}$ (north component)

$\vec{v}_{J/A}$ (north component) $+ (40.0 \text{ m/s}) \cos 45.0° = 155$ m/s

$\vec{v}_{J/A}$ (north component) $= 126.7$ m/s. Now find the magnitude of $\vec{v}_{J/A}$ using its components.

$v_{J/A} = \sqrt{(-28.28 \text{ m/s})^2 + (126.7 \text{ m/s})^2} = 130$ m/s. From the figure we see that

$\theta = \arctan[(126.7 \text{ m/s})/(28.28 \text{ m/s})] = 77.4°$. Therefore the airspeed of the jet is 130 m/s and the pilot must point it at 77.4° north of west.

EVALUATE: The pilot points the jet in a direction to offset the wind so the plane is flying directly north at 155 m/s as observed by someone standing on the ground.

3.1. **IDENTIFY and SET UP:** Use $\vec{v}_{av} = \dfrac{\vec{r}_2 - \vec{r}_1}{t_2 - t_1}$ in component form.

EXECUTE: (a) $v_{av-x} = \dfrac{\Delta x}{\Delta t} = \dfrac{x_2 - x_1}{t_2 - t_1} = \dfrac{5.3 \text{ m} - 1.1 \text{ m}}{3.0 \text{ s} - 0} = 1.4$ m/s

$v_{av-y} = \dfrac{\Delta y}{\Delta t} = \dfrac{y_2 - y_1}{t_2 - t_1} = \dfrac{-0.5 \text{ m} - 3.4 \text{ m}}{3.0 \text{ s} - 0} = -1.3$ m/s

(b)

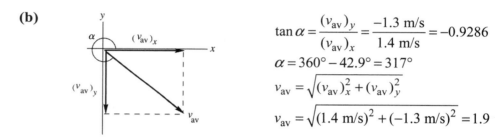

$\tan \alpha = \dfrac{(v_{av})_y}{(v_{av})_x} = \dfrac{-1.3 \text{ m/s}}{1.4 \text{ m/s}} = -0.9286$

$\alpha = 360° - 42.9° = 317°$

$v_{av} = \sqrt{(v_{av})_x^2 + (v_{av})_y^2}$

$v_{av} = \sqrt{(1.4 \text{ m/s})^2 + (-1.3 \text{ m/s})^2} = 1.9$

Figure 3.1

EVALUATE: Our calculation gives that $\vec{v}_{av}$ is in the fourth quadrant. This corresponds to increasing x and decreasing y.

3.3. **(a) IDENTIFY** and **SET UP:** From $\vec{r}$ we can calculate x and y for any t.

Then use $\vec{v}_{av} = \dfrac{\vec{r}_2 - \vec{r}_1}{t_2 - t_1}$ in component form.

EXECUTE: $\vec{r} = [4.0 \text{ cm} + (2.5 \text{ cm/s}^2)t^2]\hat{i} + (5.0 \text{ cm/s})t\hat{j}$

At $t = 0$, $\vec{r} = (4.0 \text{ cm})\hat{i}$.

At $t = 2.0$ s, $\vec{r} = (14.0 \text{ cm})\hat{i} + (10.0 \text{ cm})\hat{j}$.

$v_{av-x} = \dfrac{\Delta x}{\Delta t} = \dfrac{10.0 \text{ cm}}{2.0 \text{ s}} = 5.0$ cm/s.

$v_{av-y} = \dfrac{\Delta y}{\Delta t} = \dfrac{10.0 \text{ cm}}{2.0 \text{ s}} = 5.0$ cm/s.

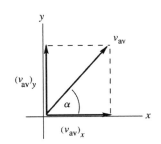

$v_{av} = \sqrt{(v_{av})_x^2 + (v_{av})_y^2} = 7.1$ cm/s

$\tan \alpha = \dfrac{(v_{av})_y}{(v_{av})_x} = 1.00$

$\theta = 45°$.

Figure 3.3a

EVALUATE: Both x and y increase, so $\vec{v}_{av}$ is in the first quadrant.

(b) IDENTIFY and **SET UP:** Calculate $\vec{v}$ by taking the time derivative of $\vec{r}(t)$.

EXECUTE: $\vec{v} = \dfrac{d\vec{r}}{dt} = ([5.0 \text{ cm/s}^2]t)\hat{i} + (5.0 \text{ cm/s})\hat{j}$

$\underline{t = 0}$: $v_x = 0$, $v_y = 5.0$ cm/s; $v = 5.0$ cm/s and $\theta = 90°$

$\underline{t = 1.0 \text{ s}}$: $v_x = 5.0$ cm/s, $v_y = 5.0$ cm/s; $v = 7.1$ cm/s and $\theta = 45°$

$\underline{t = 2.0 \text{ s}}$: $v_x = 10.0$ cm/s, $v_y = 5.0$ cm/s; $v = 11$ cm/s and $\theta = 27°$

(c) The trajectory is a graph of y versus x.

$x = 4.0 \text{ cm} + (2.5 \text{ cm/s}^2)t^2$, $y = (5.0 \text{ cm/s})t$

For values of t between 0 and 2.0 s, calculate x and y and plot y versus x.

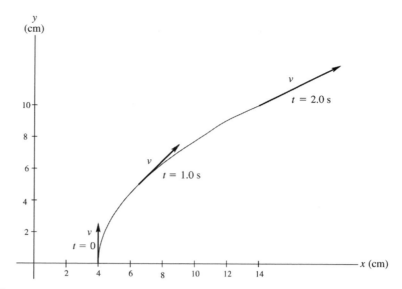

Figure 3.3b

EVALUATE: The sketch shows that the instantaneous velocity at any t is tangent to the trajectory.

3.5. **IDENTIFY and SET UP:** Use Eq. $\vec{a}_{av} = \dfrac{\vec{v}_2 - \vec{v}_1}{t_2 - t_1}$ in component form to calculate a_{av-x} and a_{av-y}.

EXECUTE: **(a)** The velocity vectors at $t_1 = 0$ and $t_2 = 30.0$ s are shown in Figure 3.5a.

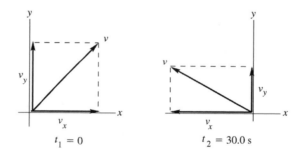

Figure 3.5a

(b) $a_{av-x} = \dfrac{\Delta v_x}{\Delta t} = \dfrac{v_{2x} - v_{1x}}{t_2 - t_1} = \dfrac{-170 \text{ m/s} - 90 \text{ m/s}}{30.0 \text{ s}} = -8.67 \text{ m/s}^2$

$a_{av-y} = \dfrac{\Delta v_y}{\Delta t} = \dfrac{v_{2y} - v_{1y}}{t_2 - t_1} = \dfrac{40 \text{ m/s} - 110 \text{ m/s}}{30.0 \text{ s}} = -2.33 \text{ m/s}^2$

(c)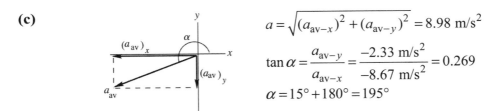

$a = \sqrt{(a_{av-x})^2 + (a_{av-y})^2} = 8.98 \text{ m/s}^2$

$\tan \alpha = \dfrac{a_{av-y}}{a_{av-x}} = \dfrac{-2.33 \text{ m/s}^2}{-8.67 \text{ m/s}^2} = 0.269$

$\alpha = 15° + 180° = 195°$

Figure 3.5b

EVALUATE: The changes in v_x and v_y are both in the negative x- or y-direction, so both components of $\vec{a}_{av}$ are in the third quadrant.

3.7. **IDENTIFY and SET UP:** Use $\vec{v} = \dfrac{d\vec{r}}{dt}$ and $\vec{a} = \dfrac{d\vec{v}}{dt}$ to find v_x, v_y, a_x, and a_y as functions of time. The magnitude and direction of $\vec{r}$ and $\vec{a}$ can be found once we know their components.

EXECUTE: (a) Calculate x and y for t values in the range 0 to 2.0 s and plot y versus x. The results are given in Figure 3.7a.

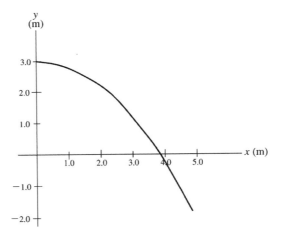

Figure 3.7a

(b) $v_x = \dfrac{dx}{dt} = \alpha \quad v_y = \dfrac{dy}{dt} = -2\beta t$

$a_x = \dfrac{dv_x}{dt} = 0 \quad a_y = \dfrac{dv_y}{dt} = -2\beta$

Thus $\vec{v} = \alpha \hat{i} - 2\beta t \hat{j}$, $\vec{a} = -2\beta \hat{j}$

(c) <u>Velocity:</u> At $t = 2.0$ s, $v_x = 2.4$ m/s, $v_y = -2(1.2 \text{ m/s}^2)(2.0 \text{ s}) = -4.8$ m/s

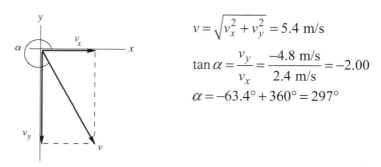

Figure 3.7b

Acceleration: At $t = 2.0$ s, $a_x = 0$, $a_y = -2(1.2 \text{ m/s}^2) = -2.4 \text{ m/s}^2$

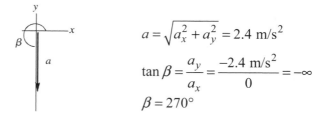

Figure 3.7c

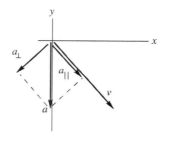

EVALUATE: (d) $\vec{a}$ has a component $a_\parallel$ in the same direction as $\vec{v}$, so we know that v is increasing (the bird is speeding up). $\vec{a}$ also has a component $a_\perp$ perpendicular to $\vec{v}$, so that the direction of $\vec{v}$ is changing; the bird is turning toward the $-y$-direction (toward the right)

Figure 3.7d

$\vec{v}$ is always tangent to the path; $\vec{v}$ at $t = 2.0$ s shown in part (c) is tangent to the path at this t, conforming to this general rule. $\vec{a}$ is constant and in the $-y$-direction; the direction of $\vec{v}$ is turning toward the $-y$-direction.

3.11. **IDENTIFY:** Each object moves in projectile motion.

SET UP: Take $+y$ to be downward. For each cricket, $a_x = 0$ and $a_y = +9.80 \text{ m/s}^2$. For Chirpy, $v_{0x} = v_{0y} = 0$. For Milada, $v_{0x} = 0.950$ m/s, $v_{0y} = 0$.

EXECUTE: Milada's horizontal component of velocity has no effect on her vertical motion. She also reaches the ground in 2.70 s. $x - x_0 = v_{0x}t + \frac{1}{2}a_x t^2 = (0.950 \text{ m/s})(2.70 \text{ s}) = 2.57$ m.

EVALUATE: The x- and y-components of motion are totally separate and are connected only by the fact that the time is the same for both.

3.13. **IDENTIFY:** The car moves in projectile motion. The car travels $21.3 \text{ m} - 1.80 \text{ m} = 19.5 \text{ m}$ downward during the time it travels 48.0 m horizontally.

SET UP: Take $+y$ to be downward. $a_x = 0$, $a_y = +9.80 \text{ m/s}^2$. $v_{0x} = v_0$, $v_{0y} = 0$.

EXECUTE: **(a)** Use the vertical motion to find the time in the air:

$y - y_0 = v_{0y}t + \frac{1}{2}a_y t^2$ gives $t = \sqrt{\frac{2(y-y_0)}{a_y}} = \sqrt{\frac{2(19.5 \text{ m})}{9.80 \text{ m/s}^2}} = 1.995 \text{ s}$

Then $x - x_0 = v_{0x}t + \frac{1}{2}a_x t^2$ gives $v_0 = v_{0x} = \frac{x-x_0}{t} = \frac{48.0 \text{ m}}{1.995 \text{ s}} = 24.1 \text{ m/s}$.

(b) $v_x = 24.06 \text{ m/s}$ since $a_x = 0$. $v_y = v_{0y} + a_y t = -19.55 \text{ m/s}$. $v = \sqrt{v_x^2 + v_y^2} = 31.0 \text{ m/s}$.

EVALUATE: Note that the speed is considerably less than the algebraic sum of the x- and y-components of the velocity.

3.15. **IDENTIFY:** The ball moves with projectile motion with an initial velocity that is horizontal and has magnitude v_0. The height h of the table and v_0 are the same; the acceleration due to gravity changes from $g_E = 9.80 \text{ m/s}^2$ on earth to g_X on planet X.

SET UP: Let $+x$ be horizontal and in the direction of the initial velocity of the marble and let $+y$ be upward. $v_{0x} = v_0$, $v_{0y} = 0$, $a_x = 0$, $a_y = -g$, where g is either g_E or g_X.

EXECUTE: Use the vertical motion to find the time in the air: $y - y_0 = -h$. $y - y_0 = v_{0y}t + \frac{1}{2}a_y t^2$ gives $t = \sqrt{\frac{2h}{g}}$. Then $x - x_0 = v_{0x}t + \frac{1}{2}a_x t^2$ gives $x - x_0 = v_{0x}t = v_0\sqrt{\frac{2h}{g}}$. $x - x_0 = D$ on earth and $2.76D$ on Planet X. $(x - x_0)\sqrt{g} = v_0\sqrt{2h}$, which is constant, so $D\sqrt{g_E} = 2.76D\sqrt{g_X}$.

$g_X = \frac{g_E}{(2.76)^2} = 0.131 g_E = 1.28 \text{ m/s}^2$.

EVALUATE: On Planet X the acceleration due to gravity is less, it takes the ball longer to reach the floor and it travels farther horizontally.

3.19. **IDENTIFY:** Take the origin of coordinates at the point where the quarter leaves your hand and take positive y to be upward. The quarter moves in projectile motion, with $a_x = 0$, and $a_y = -g$. It travels vertically for the time it takes it to travel horizontally 2.1 m.

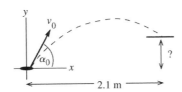

$v_{0x} = v_0 \cos \alpha_0 = (6.4 \text{ m/s}) \cos 60°$
$v_{0x} = 3.20 \text{ m/s}$
$v_{0y} = v_0 \sin \alpha_0 = (6.4 \text{ m/s}) \sin 60°$
$v_{0y} = 5.54 \text{ m/s}$

Figure 3.19

(a) SET UP: Use the horizontal (x-component) of motion to solve for t, the time the quarter travels through the air:

$t = ?$, $x - x_0 = 2.1 \text{ m}$, $v_{0x} = 3.2 \text{ m/s}$, $a_x = 0$

$x - x_0 = v_{0x}t + \frac{1}{2}a_x t^2 = v_{0x}t$, since $a_x = 0$

EXECUTE: $t = \frac{x - x_0}{v_{0x}} = \frac{2.1 \text{ m}}{3.2 \text{ m/s}} = 0.656 \text{ s}$

SET UP: Now find the vertical displacement of the quarter after this time:
$y - y_0 = ?$, $a_y = -9.80 \text{ m/s}^2$, $v_{0y} = +5.54 \text{ m/s}$, $t = 0.656 \text{ s}$
$y - y_0 + v_{0y}t + \tfrac{1}{2} a_y t^2$

EXECUTE: $y - y_0 = (5.54 \text{ m/s})(0.656 \text{ s}) + \tfrac{1}{2}(-9.80 \text{ m/s}^2)(0.656 \text{ s})^2 = 3.63 \text{ m} - 2.11 \text{ m} = 1.5 \text{ m}$.

(b) SET UP: $v_y = ?$, $t = 0.656 \text{ s}$, $a_y = -9.80 \text{ m/s}^2$, $v_{0y} = +5.54 \text{ m/s}$ $v_y = v_{0y} + a_y t$

EXECUTE: $v_y = 5.54 \text{ m/s} + (-9.80 \text{ m/s}^2)(0.656 \text{ s}) = -0.89 \text{ m/s}$.

EVALUATE: The minus sign for v_y indicates that the y-component of $\vec{v}$ is downward. At this point the quarter has passed through the highest point in its path and is on its way down. The horizontal range if it returned to its original height (it doesn't!) would be 3.6 m. It reaches its maximum height after traveling horizontally 1.8 m, so at $x - x_0 = 2.1 \text{ m}$ it is on its way down.

3.21. IDENTIFY: Take the origin of coordinates at the roof and let the +y-direction be upward. The rock moves in projectile motion, with $a_x = 0$ and $a_y = -g$. Apply constant acceleration equations for the x- and y-components of the motion.

SET UP:

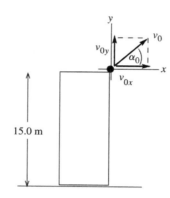

$v_{0x} = v_0 \cos \alpha_0 = 25.2 \text{ m/s}$
$v_{0y} = v_0 \sin \alpha_0 = 16.3 \text{ m/s}$

Figure 3.21a

(a) At the maximum height $v_y = 0$.
$a_y = -9.80 \text{ m/s}^2$, $v_y = 0$, $v_{0y} = +16.3 \text{ m/s}$, $y - y_0 = ?$
$v_y^2 = v_{0y}^2 + 2a_y (y - y_0)$

EXECUTE: $y - y_0 = \dfrac{v_y^2 - v_{0y}^2}{2a_y} = \dfrac{0 - (16.3 \text{ m/s})^2}{2(-9.80 \text{ m/s}^2)} = +13.6 \text{ m}$

(b) SET UP: Find the velocity by solving for its x- and y-components.
$v_x = v_{0x} = 25.2 \text{ m/s}$ (since $a_x = 0$)
$v_y = ?$, $a_y = -9.80 \text{ m/s}^2$, $y - y_0 = -15.0 \text{ m}$ (negative because at the ground the rock is below its initial position), $v_{0y} = 16.3 \text{ m/s}$
$v_y^2 = v_{0y}^2 + 2a_y(y - y_0)$
$v_y = -\sqrt{v_{0y}^2 + 2a_y(y - y_0)}$ (v_y is negative because at the ground the rock is traveling downward.)

EXECUTE: $v_y = -\sqrt{(16.3 \text{ m/s})^2 + 2(-9.80 \text{ m/s}^2)(-15.0 \text{ m})} = -23.7$ m/s

Then $v = \sqrt{v_x^2 + v_y^2} = \sqrt{(25.2 \text{ m/s})^2 + (-23.7 \text{ m/s})^2} = 34.6$ m/s.

(c) SET UP: Use the vertical motion (y-component) to find the time the rock is in the air:
$t = ?$, $v_y = -23.7$ m/s (from part (b)), $a_y = -9.80$ m/s^2, $v_{0y} = +16.3$ m/s

EXECUTE: $t = \dfrac{v_y - v_{0y}}{a_y} = \dfrac{-23.7 \text{ m/s} - 16.3 \text{ m/s}}{-9.80 \text{ m/s}^2} = +4.08$ s

SET UP: Can use this t to calculate the horizontal range:
$t = 4.08$ s, $v_{0x} = 25.2$ m/s, $a_x = 0$, $x - x_0 = ?$

EXECUTE: $x - x_0 = v_{0x}t + \frac{1}{2}a_x t^2 = (25.2 \text{ m/s})(4.08 \text{ s}) + 0 = 103$ m

(d) Graphs of x versus t, y versus t, v_x versus t and v_y versus t:

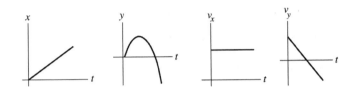

Figure 3.21b

EVALUATE: The time it takes the rock to travel vertically to the ground is the time it has to travel horizontally. With $v_{0y} = +16.3$ m/s the time it takes the rock to return to the level of the roof ($y = 0$) is $t = 2v_{0y}/g = 3.33$ s. The time in the air is greater than this because the rock travels an additional 15.0 m to the ground.

3.27. IDENTIFY: The ball has no vertical acceleration, but it has a horizontal acceleration toward the center of the circle. This acceleration is the radial acceleration.

SET UP: $a_{rad} = \dfrac{4\pi^2 R}{T^2}$, $R = L \sin \theta$.

EXECUTE: Use $a_{rad} = \dfrac{4\pi^2 R}{T^2}$ with $R = L \sin \theta$. $a_{rad} = \dfrac{4\pi^2 L \sin \theta}{T^2}$. Using $L = 0.800$ m, $\theta = 37.0°$, and $T = 0.600$ s, this gives $a_{rad} = 52.8$ m/s^2.

EVALUATE: Compare this acceleration to g: $a_{rad}/g = 52.8/9.80 = 5.39$, so $a_{rad} = 5.39g$, which is fairly large.

3.33. IDENTIFY: Each part of his body moves in uniform circular motion, with $a_{rad} = \dfrac{v^2}{R}$. The speed in rev/s is $1/T$, where T is the period in seconds (time for 1 revolution). The speed v increases with R along the length of his body but all of him rotates with the same period T.

SET UP: For his head $R = 8.84$ m and for his feet $R = 6.84$ m.

EXECUTE: (a) $v = \sqrt{R a_{rad}} = \sqrt{(8.84 \text{ m})(12.5)(9.80 \text{ m/s}^2)} = 32.9$ m/s

(b) Use $a_{rad} = \dfrac{4\pi^2 R}{T^2}$. Since his head has $a_{rad} = 12.5g$ and $R = 8.84$ m,

$T = 2\pi \sqrt{\dfrac{R}{a_{rad}}} = 2\pi \sqrt{\dfrac{8.84\,m}{12.5(9.80\,m/s^2)}} = 1.688\,s.$ Then his feet have $a_{rad} = \dfrac{R}{T^2} = \dfrac{4\pi^2(6.84\,m)}{(1.688\,s)^2} = 94.8\,m/s^2$
$= 9.67\,g$. The difference between the acceleration of his head and his feet is
$12.5g - 9.67g = 2.83g = 27.7\,m/s^2.$

(c) $\dfrac{1}{T} = \dfrac{1}{1.69\,s} = 0.592$ rev/s $= 35.5$ rpm

EVALUATE: His feet have speed $v = \sqrt{Ra_{rad}} = \sqrt{(6.84\,m)(94.8\,m/s^2)} = 25.5$ m/s.

3.35. IDENTIFY: Relative velocity problem. The time to walk the length of the moving sidewalk is the length divided by the velocity of the woman relative to the ground.

SET UP: Let W stand for the woman, G for the ground, and S for the sidewalk. Take the positive direction to be the direction in which the sidewalk is moving.
The velocities are $v_{W/G}$ (woman relative to the ground), $v_{W/S}$ (woman relative to the sidewalk), and $v_{S/G}$ (sidewalk relative to the ground).
The equation for relative velocity becomes $v_{W/G} = v_{W/S} + v_{S/G}$.

The time to reach the other end is given by $t = \dfrac{\text{distance traveled relative to ground}}{v_{W/G}}$

EXECUTE: (a) $v_{S/G} = 1.0$ m/s
$v_{W/S} = +1.5$ m/s
$v_{W/G} = v_{W/S} + v_{S/G} = 1.5$ m/s $+ 1.0$ m/s $= 2.5$ m/s.
$t = \dfrac{35.0\,m}{v_{W/G}} = \dfrac{35.0\,m}{2.5\,m/s} = 14$ s.

(b) $v_{S/G} = 1.0$ m/s
$v_{W/S} = -1.5$ m/s
$v_{W/G} = v_{W/S} + v_{S/G} = -1.5$ m/s $+ 1.0$ m/s $= -0.5$ m/s. (Since $v_{W/G}$ now is negative, she must get on the moving sidewalk at the opposite end from in part (a).)
$t = \dfrac{-35.0\,m}{v_{W/G}} = \dfrac{-35.0\,m}{-0.5\,m/s} = 70$ s.

EVALUATE: Her speed relative to the ground is much greater in part (a) when she walks with the motion of the sidewalk.

3.39. IDENTIFY: The resultant velocity, relative to the ground, is directly southward. This velocity is the sum of the velocity of the bird relative to the air and the velocity of the air relative to the ground.

SET UP: $v_{B/A} = 100$ km/h. $\vec{v}_{A/G} = 40$ km/h, east. $\vec{v}_{B/G} = \vec{v}_{B/A} + \vec{v}_{A/G}$.

EXECUTE: We want $\vec{v}_{B/G}$ to be due south. The relative velocity addition diagram is shown in Figure 3.39.

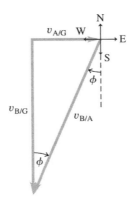

Figure 3.39

(a) $\sin\phi = \dfrac{v_{A/G}}{v_{B/A}} = \dfrac{40 \text{ km/h}}{100 \text{ km/h}}$, $\phi = 24°$, west of south.

(b) $v_{B/G} = \sqrt{v_{B/A}^2 - v_{A/G}^2} = 91.7$ km/h. $t = \dfrac{d}{v_{B/G}} = \dfrac{500 \text{ km}}{91.7 \text{ km/h}} = 5.5$ h.

EVALUATE: The speed of the bird relative to the ground is less than its speed relative to the air. Part of its velocity relative to the air is directed to oppose the effect of the wind.

3.43. IDENTIFY: Use the relation that relates the relative velocities.

SET UP: The relative velocities are the velocity of the plane relative to the ground, $\vec{v}_{P/G}$, the velocity of the plane relative to the air, $\vec{v}_{P/A}$, and the velocity of the air relative to the ground, $\vec{v}_{A/G}$. $\vec{v}_{P/G}$ must be due west and $\vec{v}_{A/G}$ must be south. $v_{A/G} = 80$ km/h and $v_{P/A} = 320$ km/h. $\vec{v}_{P/G} = \vec{v}_{P/A} + \vec{v}_{A/G}$. The relative velocity addition diagram is given in Figure 3.43.

EXECUTE: (a) $\sin\theta = \dfrac{v_{A/G}}{v_{P/A}} = \dfrac{80 \text{ km/h}}{320 \text{ km/h}}$ and $\theta = 14°$, north of west.

(b) $v_{P/G} = \sqrt{v_{P/A}^2 - v_{A/G}^2} = \sqrt{(320 \text{ km/h})^2 - (80.0 \text{ km/h})^2} = 310$ km/h.

EVALUATE: To travel due west the velocity of the plane relative to the air must have a westward component and also a component that is northward, opposite to the wind direction.

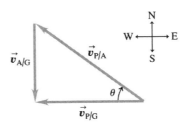

Figure 3.43

3.45. IDENTIFY: $\vec{v} = d\vec{r}/dt$. This vector will make a $45°$ angle with both axes when its x- and y-components are equal.

SET UP: $\dfrac{d(t^n)}{dt} = nt^{n-1}$.

EXECUTE: $\vec{v} = 2bt\hat{i} + 3ct^2\hat{j}$. $v_x = v_y$ gives $t = 2b/3c$.

EVALUATE: Both components of $\vec{v}$ change with t.

3.47. IDENTIFY: Once the rocket leaves the incline it moves in projectile motion. The acceleration along the incline determines the initial velocity and initial position for the projectile motion.

SET UP: For motion along the incline let $+x$ be directed up the incline. $v_x^2 = v_{0x}^2 + 2a_x(x - x_0)$ gives $v_x = \sqrt{2(1.90 \text{ m/s}^2)(200 \text{ m})} = 27.57$ m/s. When the projectile motion begins the rocket has $v_0 = 27.57$ m/s at $35.0°$ above the horizontal and is at a vertical height of $(200.0 \text{ m})\sin 35.0° = 114.7$ m. For the projectile motion let $+x$ be horizontal to the right and let $+y$ be upward. Let $y = 0$ at the ground. Then $y_0 = 114.7$ m, $v_{0x} = v_0 \cos 35.0° = 22.57$ m/s, $v_{0y} = v_0 \sin 35.0° = 15.81$ m/s, $a_x = 0$, $a_y = -9.80$ m/s^2. Let $x = 0$ at point A, so $x_0 = (200.0 \text{ m})\cos 35.0° = 163.8$ m.

EXECUTE: (a) At the maximum height $v_y = 0$. $v_y^2 = v_{0y}^2 + 2a_y(y - y_0)$ gives

$$y - y_0 = \frac{v_y^2 - v_{0y}^2}{2a_y} = \frac{0 - (15.81 \text{ m/s})^2}{2(-9.80 \text{ m/s}^2)} = 12.77 \text{ m}$$ and $y = 114.7$ m $+ 12.77$ m $= 128$ m. The maximum height above ground is 128 m.

(b) The time in the air can be calculated from the vertical component of the projectile motion: $y - y_0 = -114.7$ m, $v_{0y} = 15.81$ m/s, $a_y = -9.80$ m/s^2. $y - y_0 = v_{0y}t + \frac{1}{2}a_y t^2$ gives $(4.90 \text{ m/s}^2)t^2 - (15.81 \text{ m/s})t - 114.7$ m. The quadratic formula gives $t = 6.713$ s for the positive root. Then $x - x_0 = v_{0x}t + \frac{1}{2}a_x t^2 = (22.57 \text{ m/s})(6.713 \text{ s}) = 151.6$ m and $x = 163.8$ m $+ 151.6$ m $= 315$ m. The horizontal range of the rocket is 315 m.

EVALUATE: The expressions for h and R derived in the range formula do not apply here. They are only for a projectile fired on level ground.

3.49. IDENTIFY: The canister moves in projectile motion. Its initial velocity is horizontal. Apply constant acceleration equations for the x- and y-components of motion.

SET UP:

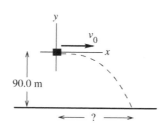

Take the origin of coordinates at the point where the canister is released. Take $+y$ to be upward. The initial velocity of the canister is the velocity of the plane, 64.0 m/s in the $+x$-direction.

Figure 3.49

Use the vertical motion to find the time of fall: $y - y_0 = v_{0y}t + \frac{1}{2}a_y t^2$ where $t = ?$, $v_{0y} = 0$, $a_y = -9.80$ m/s^2, $y - y_0 = -90.0$ m. (When the canister reaches the ground it is 90.0 m *below* the origin.)

EXECUTE: Since $v_{0y} = 0$, $t = \sqrt{\frac{2(y - y_0)}{a_y}} = \sqrt{\frac{2(-90.0 \text{ m})}{-9.80 \text{ m/s}^2}} = 4.286$ s.

SET UP: Then use the horizontal component of the motion to calculate how far the canister falls in this time: $x - x_0 = ?$, $a_x = 0$, $v_{0x} = 64.0$ m/s.

EXECUTE: $x - x_0 = v_0 t + \frac{1}{2} a t^2 = (64.0 \text{ m/s})(4.286 \text{ s}) + 0 = 274$ m.

EVALUATE: The time it takes the canister to fall 90.0 m, starting from rest, is the time it travels horizontally at constant speed.

3.51. IDENTIFY: The person moves in projectile motion. Her vertical motion determines her time in the air.

SET UP: Take $+y$ upward. $v_{0x} = 15.0$ m/s, $v_{0y} = +10.0$ m/s, $a_x = 0$, $a_y = -9.80$ m/s^2.

EXECUTE: **(a)** Use the vertical motion to find the time in the air: $y - y_0 = v_{0y} t + \frac{1}{2} a_y t^2$ with $y - y_0 = -30.0$ m gives $-30.0 \text{ m} = (10.0 \text{ m/s})t - (4.90 \text{ m/s}^2) t^2$. The quadratic formula gives

$t = \frac{1}{2(4.9)}\left(+10.0 \pm \sqrt{(-10.0)^2 - 4(4.9)(-30)}\right)$ s. The positive solution is $t = 3.70$ s. During this time she travels a horizontal distance $x - x_0 = v_{0x} t + \frac{1}{2} a_x t^2 = (15.0 \text{ m/s})(3.70 \text{ s}) = 55.5$ m. She will land 55.5 m south of the point where she drops from the helicopter and this is where the mats should have been placed.

(b) The x-t, y-t, v_x-t and v_y-t graphs are sketched in Figure 3.51.

EVALUATE: If she had dropped from rest at a height of 30.0 m it would have taken her $t = \sqrt{\frac{2(30.0 \text{ m})}{9.80 \text{ m/s}^2}} = 2.47$ s. She is in the air longer than this because she has an initial vertical component of velocity that is upward.

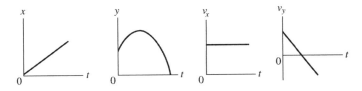

Figure 3.51

3.53. IDENTIFY: Find the horizontal distance a rocket moves if it has a non constant horizontal acceleration but a constant vertical acceleration of g downward.

SET UP: The vertical motion is g downward, so we can use the constant acceleration formulas for that component of the motion. We must use integration for the horizontal motion because the acceleration is not constant. Solving for t in the kinematics formula for y gives $t = \sqrt{\frac{2(y - y_0)}{a_y}}$. In the horizontal direction we must use $v_x(t) = v_{0x} + \int_0^t a_x(t') dt'$ and $x - x_0 = \int_0^t v_x(t') dt'$.

EXECUTE: Use vertical motion to find t. $t = \sqrt{\frac{2(y - y_0)}{a_y}} = \sqrt{\frac{2(30.0 \text{ m})}{9.80 \text{ m/s}^2}} = 2.474$ s.

In the horizontal direction we have

$v_x(t) = v_{0x} + \int_0^t a_x(t') dt' = v_{0x} + (0.800 \text{ m/s}^3) t^2 = 12.0 \text{ m/s} + (0.800 \text{ m/s}^3) t^2$. Integrating $v_x(t)$ gives

$x - x_0 = (12.0 \text{ m/s}) t + (0.2667 \text{ m/s}^3) t^3$. At $t = 2.474$ s, $x - x_0 = 29.69 \text{ m} + 4.04 \text{ m} = 33.7$ m.

EVALUATE: The vertical part of the motion is familiar projectile motion, but the horizontal part is not.

3.55. IDENTIFY: Two-dimensional projectile motion.

SET UP: Let $+y$ be upward. $a_x = 0$, $a_y = -9.80$ m/s^2. With $x_0 = y_0 = 0$, algebraic manipulation of the equations for the horizontal and vertical motion shows that x and y are related by

$$y = (\tan\theta_0)x - \frac{g}{2v_0^2 \cos^2\theta_0}x^2.$$

$\theta_0 = 60.0°$. $y = 8.00$ m when $x = 18.0$ m.

EXECUTE: (a) Solving for v_0 gives $v_0 = \sqrt{\dfrac{gx^2}{2(\cos^2\theta_0)(x\tan\theta_0 - y)}} = 16.6$ m/s.

(b) We find the horizontal and vertical velocity components:

$v_x = v_{0x} = v_0\cos\theta_0 = 8.3$ m/s.

$v_y^2 = v_{0y}^2 + 2a_y(y - y_0)$ gives

$v_y = -\sqrt{(v_0\sin\theta_0)^2 + 2a_y(y - y_0)} = -\sqrt{(14.4 \text{ m/s})^2 + 2(-9.80 \text{ m/s}^2)(8.00 \text{ m})} = -7.1$ m/s

$v = \sqrt{v_x^2 + v_y^2} = 10.9$ m/s. $\tan\theta = \dfrac{|v_y|}{|v_x|} = \dfrac{7.1}{8.3}$ and $\theta = 40.5°$, below the horizontal.

EVALUATE: We can check our calculated v_0.

$t = \dfrac{x - x_0}{v_{0x}} = \dfrac{18.0 \text{ m}}{8.3 \text{ m/s}} = 2.17$ s.

Then $y - y_0 = v_{0y}t + \frac{1}{2}a_yt^2 = (14.4 \text{ m/s})(2.17 \text{ s}) - (4.9 \text{ m/s}^2)(2.17 \text{ s})^2 = 8$ m, which checks.

3.57. IDENTIFY: From the figure in the text, we can read off the maximum height and maximum horizontal distance reached by the grasshopper. Knowing its acceleration is g downward, we can find its initial speed and the height of the cliff (the target variables).

SET UP: Use coordinates with the origin at the ground and $+y$ upward. $a_x = 0$, $a_y = -9.80$ m/s^2. The constant-acceleration kinematics formulas $v_y^2 = v_{0y}^2 + 2a_y(y - y_0)$ and $x - x_0 = v_{0x}t + \frac{1}{2}a_xt^2$ apply.

EXECUTE: (a) $v_y = 0$ when $y - y_0 = 0.0674$ m. $v_y^2 = v_{0y}^2 + 2a_y(y - y_0)$ gives

$v_{0y} = \sqrt{-2a_y(y - y_0)} = \sqrt{-2(-9.80 \text{ m/s}^2)(0.0674 \text{ m})} = 1.15$ m/s. $v_{0y} = v_0\sin\alpha_0$ so

$v_0 = \dfrac{v_{0y}}{\sin\alpha_0} = \dfrac{1.15 \text{ m/s}}{\sin 50.0°} = 1.50$ m/s.

(b) Use the horizontal motion to find the time in the air. The grasshopper travels horizontally $x - x_0 = 1.06$ m. $x - x_0 = v_{0x}t + \frac{1}{2}a_xt^2$ gives $t = \dfrac{x - x_0}{v_{0x}} = \dfrac{x - x_0}{v_0\cos 50.0°} = 1.10$ s. Find the vertical displacement of the grasshopper at $t = 1.10$ s:

$y - y_0 = v_{0y}t + \frac{1}{2}a_yt^2 = (1.15 \text{ m/s})(1.10 \text{ s}) + \frac{1}{2}(-9.80 \text{ m/s}^2)(1.10 \text{ s})^2 = -4.66$ m. The height of the cliff is 4.66 m.

EVALUATE: The grasshopper's maximum height (6.74 cm) is physically reasonable, so its takeoff speed of 1.50 m/s must also be reasonable. Note that the equation $R = \dfrac{v_0^2 \sin 2\alpha_0}{g}$ does *not* apply here since the launch point is not at the same level as the landing point.

3.61. **IDENTIFY:** The snowball moves in projectile motion. In part (a) the vertical motion determines the time in the air. In part (c), find the height of the snowball above the ground after it has traveled horizontally 4.0 m.

SET UP: Let $+y$ be downward. $a_x = 0$, $a_y = +9.80 \text{ m/s}^2$. $v_{0x} = v_0 \cos\theta_0 = 5.36$ m/s, $v_{0y} = v_0 \sin\theta_0 = 4.50$ m/s.

EXECUTE: (a) Use the vertical motion to find the time in the air: $y - y_0 = v_{0y}t + \frac{1}{2}a_y t^2$ with $y - y_0 = 14.0$ m gives $14.0 \text{ m} = (4.50 \text{ m/s})t + (4.9 \text{ m/s}^2)t^2$. The quadratic formula gives

$t = \dfrac{1}{2(4.9)}\left(-4.50 \pm \sqrt{(4.50)^2 - 4(4.9)(-14.0)}\right)$ s. The positive root is $t = 1.29$ s. Then

$x - x_0 = v_{0x}t + \frac{1}{2}a_x t^2 = (5.36 \text{ m/s})(1.29 \text{ s}) = 6.91$ m.

(b) The x-t, y-t, v_x-t and v_y-t graphs are sketched in Figure 3.61.

(c) $x - x_0 = v_{0x}t + \frac{1}{2}a_x t^2$ gives $t = \dfrac{x - x_0}{v_{0x}} = \dfrac{4.0 \text{ m}}{5.36 \text{ m/s}} = 0.746$ s. In this time the snowball travels downward a distance $y - y_0 = v_{0y}t + \frac{1}{2}a_y t^2 = 6.08$ m and is therefore $14.0 \text{ m} - 6.08 \text{ m} = 7.9$ m above the ground. The snowball passes well above the man and doesn't hit him.

EVALUATE: If the snowball had been released from rest at a height of 14.0 m it would have reached the ground in $t = \sqrt{\dfrac{2(14.0 \text{ m})}{9.80 \text{ m/s}^2}} = 1.69$ s. The snowball reaches the ground in a shorter time than this because of its initial downward component of velocity.

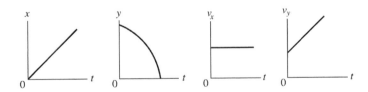

Figure 3.61

3.63. **(a) IDENTIFY:** Projectile motion.

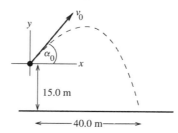

Take the origin of coordinates at the top of the ramp and take $+y$ to be upward.

The problem specifies that the object is displaced 40.0 m to the right when it is 15.0 m below the origin.

Figure 3.63

We don't know t, the time in the air, and we don't know v_0. Write down the equations for the horizontal and vertical displacements. Combine these two equations to eliminate one unknown.

SET UP: *y-component*:

$y - y_0 = -15.0$ m, $a_y = -9.80$ m/s², $v_{0y} = v_0 \sin 53.0°$

$y - y_0 = v_{0y}t + \frac{1}{2}a_y t^2$

EXECUTE: -15.0 m $= (v_0 \sin 53.0°)t - (4.90 \text{ m/s}^2)t^2$

SET UP: *x-component*:

$x - x_0 = 40.0$ m, $a_x = 0$, $v_{0x} = v_0 \cos 53.0°$

$x - x_0 = v_{0x}t + \frac{1}{2}a_x t^2$

EXECUTE: 40.0 m $= (v_0 t)\cos 53.0°$

The second equation says $v_0 t = \dfrac{40.0 \text{ m}}{\cos 53.0°} = 66.47$ m.

Use this to replace $v_0 t$ in the first equation:

-15.0 m $= (66.47 \text{ m})\sin 53° - (4.90 \text{ m/s}^2)t^2$

$t = \sqrt{\dfrac{(66.47 \text{ m})\sin 53° + 15.0 \text{ m}}{4.90 \text{ m/s}^2}} = \sqrt{\dfrac{68.08 \text{ m}}{4.90 \text{ m/s}^2}} = 3.727$ s.

Now that we have t we can use the *x*-component equation to solve for v_0:

$v_0 = \dfrac{40.0 \text{ m}}{t \cos 53.0°} = \dfrac{40.0 \text{ m}}{(3.727 \text{ s})\cos 53.0°} = 17.8$ m/s.

EVALUATE: Using these values of v_0 and t in the $y = y_0 = v_{0y} + \frac{1}{2}a_y t^2$ equation verifies that $y - y_0 = -15.0$ m.

(b) IDENTIFY: $v_0 = (17.8 \text{ m/s})/2 = 8.9$ m/s

This is less than the speed required to make it to the other side, so he lands in the river.
Use the vertical motion to find the time it takes him to reach the water:

SET UP: $y - y_0 = -100$ m; $v_{0y} = +v_0 \sin 53.0° = 7.11$ m/s; $a_y = -9.80$ m/s²

$y - y_0 = v_{0y}t + \frac{1}{2}a_y t^2$ gives $-100 = 7.11t - 4.90t^2$

EXECUTE: $4.90t^2 - 7.11t - 100 = 0$ and $t = \frac{1}{9.80}\left(7.11 \pm \sqrt{(7.11)^2 - 4(4.90)(-100)}\right)$

$t = 0.726$ s ± 4.57 s so $t = 5.30$ s.

The horizontal distance he travels in this time is

$x - x_0 = v_{0x}t = (v_0 \cos 53.0°)t = (5.36 \text{ m/s})(5.30 \text{ s}) = 28.4$ m.

He lands in the river a horizontal distance of 28.4 m from his launch point.
EVALUATE: He has half the minimum speed and makes it only about halfway across.

3.67. IDENTIFY: This is a projectile motion problem. The vertical acceleration is g downward and the horizontal acceleration is zero. The constant-acceleration equations apply.

SET UP: Apply the constant-acceleration formulas. We know that the ball travels 50.0 m horizontally and has a speed of 8.0 m/s at its maximum height. We want to know how long the ball is in the air.

EXECUTE: The horizontal velocity is constant, so $v_x = 8.0$ m/s. The ball moves 50.0 m at this velocity, so $x = v_x t$ gives 50.0 m = (8.0 m/s)t → $t = 6.3$ s.

EVALUATE: The ball's vertical velocity keeps changing, but its horizontal velocity remains constant. At its highest point, the ball does *not stop*. Only its vertical velocity becomes zero there.

3.71. IDENTIFY: Relative velocity problem. The plane's motion relative to the earth is determined by its velocity relative to the earth.

SET UP: Select a coordinate system where $+y$ is north and $+x$ is east.
The velocity vectors in the problem are:
$\vec{v}_{P/E}$, the velocity of the plane relative to the earth.
$\vec{v}_{P/A}$, the velocity of the plane relative to the air (the magnitude $v_{P/A}$ is the airspeed of the plane and the direction of $\vec{v}_{P/A}$ is the compass course set by the pilot).
$\vec{v}_{A/E}$, the velocity of the air relative to the earth (the wind velocity).
The rule for combining relative velocities gives $\vec{v}_{P/E} = \vec{v}_{P/A} + \vec{v}_{A/E}$.

(a) We are given the following information about the relative velocities:
$\vec{v}_{P/A}$ has magnitude 220 km/h and its direction is west. In our coordinates it has components $(v_{P/A})_x = -220$ km/h and $(v_{P/A})_y = 0$.

From the displacement of the plane relative to the earth after 0.500 h, we find that $\vec{v}_{P/E}$ has components in our coordinate system of

$$(v_{P/E})_x = -\frac{120 \text{ km}}{0.500 \text{ h}} = -240 \text{ km/h (west)}$$

$$(v_{P/E})_y = -\frac{20 \text{ km}}{0.500 \text{ h}} = -40 \text{ km/h (south)}$$

With this information the diagram corresponding to the velocity addition equation is shown in Figure 3.71a.

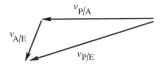

Figure 3.71a

We are asked to find $\vec{v}_{A/E}$, so solve for this vector:
$\vec{v}_{P/E} = \vec{v}_{P/A} + \vec{v}_{A/E}$ gives $\vec{v}_{A/E} = \vec{v}_{P/E} - \vec{v}_{P/A}$.

EXECUTE: The x-component of this equation gives
$(v_{A/E})_x = (v_{P/E})_x - (v_{P/A})_x = -240$ km/h $- (-220$ km/h$) = -20$ km/h.

The y-component of this equation gives
$(v_{A/E})_y = (v_{P/E})_y - (v_{P/A})_y = -40$ km/h.

Now that we have the components of $\vec{v}_{A/E}$ we can find its magnitude and direction.

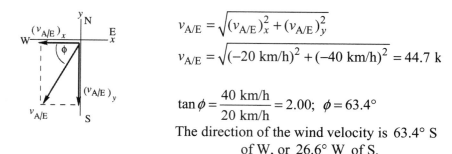

$$v_{A/E} = \sqrt{(v_{A/E})_x^2 + (v_{A/E})_y^2}$$

$$v_{A/E} = \sqrt{(-20 \text{ km/h})^2 + (-40 \text{ km/h})^2} = 44.7 \text{ k}$$

$$\tan\phi = \frac{40 \text{ km/h}}{20 \text{ km/h}} = 2.00; \quad \phi = 63.4°$$

The direction of the wind velocity is 63.4° S of W, or 26.6° W of S.

Figure 3.71b

EVALUATE: The plane heads west. It goes farther west than it would without wind and also travels south, so the wind velocity has components west and south.

(b) SET UP: The rule for combining the relative velocities is still $\vec{v}_{P/E} = \vec{v}_{P/A} + \vec{v}_{A/E}$, but some of these velocities have different values than in part **(a)**.

$\vec{v}_{P/A}$ has magnitude 220 km/h but its direction is to be found.

$\vec{v}_{A/E}$ has magnitude 40 km/h and its direction is due south.

The direction of $\vec{v}_{P/E}$ is west; its magnitude is not given.

The vector diagram for $\vec{v}_{P/E} = \vec{v}_{P/A} + \vec{v}_{A/E}$ and the specified directions for the vectors is shown in Figure 3.71c.

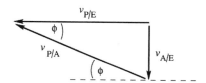

Figure 3.71c

The vector addition diagram forms a right triangle.

EXECUTE: $\sin\phi = \dfrac{v_{A/E}}{v_{P/A}} = \dfrac{40 \text{ km/h}}{220 \text{ km/h}} = 0.1818; \quad \phi = 10.5°.$

The pilot should set her course 10.5° north of west.

EVALUATE: The velocity of the plane relative to the air must have a northward component to counteract the wind and a westward component in order to travel west.

3.75. **IDENTIFY:** We need to use relative velocities.

SET UP: If B is moving relative to M and M is moving relative to E, the velocity of B relative to E is $\vec{v}_{B/E} = \vec{v}_{B/M} + \vec{v}_{M/E}$.

EXECUTE: Let $+x$ be east and $+y$ be north. We have $v_{B/M,x} = 2.50$ m/s, $v_{B/M,y} = -4.33$ m/s, $v_{M/E,x} = 0$, and $v_{M/E,y} = 6.00$ m/s. Therefore $v_{B/E,x} = v_{B/M,x} + v_{M/E,x} = 2.50$ m/s and $v_{B/E,y} = v_{B/M,y} + v_{M/E,y} = -4.33$ m/s $+ 6.00$ m/s $= +1.67$ m/s. The magnitude is $v_{B/E} = \sqrt{(2.50 \text{ m/s})^2 + (1.67 \text{ m/s})^2} = 3.01$ m/s, and the direction is $\tan\theta = \dfrac{1.67}{2.50}$, which gives $\theta = 33.7°$ north of east.

EVALUATE: Since Mia is moving, the velocity of the ball relative to her is different from its velocity relative to the ground or relative to Alice.

3.77. IDENTIFY: The table gives data showing the horizontal range of the potato for various launch heights. You want to use this information to determine the launch speed of the potato, assuming negligible air resistance.

SET UP: The potatoes are launched horizontally, so $v_{0y} = 0$, and they are in free fall, so $a_y = 9.80$ m/s^2 downward and $a_x = 0$. The time a potato is in the air is just the time it takes for it to fall vertically from the launch point to the ground, a distance h.

EXECUTE: **(a)** For the vertical motion of a potato, we have $h = \frac{1}{2} gt^2$, so $t = \sqrt{2h/g}$. The horizontal range R is given by $R = v_0 t = v_0 \sqrt{2h/g}$. Squaring gives $R^2 = \left(\dfrac{2v_0^2}{g}\right) h$. Graphing R^2 versus h will give a straight line with slope $2v_0^2/g$. We can graph the data from the table in the text by hand, or we could use graphing software. The result is shown in Figure 3.77.

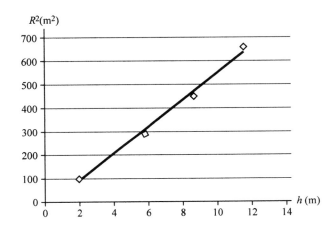

Figure 3.77

(b) The slope of the graph is 55.2 m, so $v_0 = \sqrt{\dfrac{(9.80 \text{ m/s}^2)(55.2 \text{ m})}{2}} = 16.4$ m/s.

(c) In this case, the potatoes are launched and land at ground level, so we can use the range formula with $\theta = 30.0°$ and $v_0 = 16.4$ m/s. The result is $R = \dfrac{v_0^2 \sin(2\theta)}{g} = 23.8$ m.

EVALUATE: This approach to finding the launch speed v_0 requires only simple measurements: the range and the launch height. It would be difficult and would require special equipment to measure v_0 directly.

Study Guide for Newton's Laws of Motion

Summary

We'll define *dynamics*—the study of the relationship of motion to forces—in this chapter. Newton's laws of motion will lay the foundation for our studies and link forces to acceleration. We'll define *force, mass,* and *weight* and apply them to problems. We'll use our knowledge of vectors to better understand forces and construct free-body diagrams. By the end of this chapter, we'll have built a problem-solving framework.

Objectives

After studying this chapter, you'll understand

- The concept of force and why it is a vector quantity.
- How to identify forces acting on a body.
- How to find the resultant force acting on an object by summing multiple forces.
- Newton's three laws of motion.
- The relation between net external force, mass, and acceleration.
- How to recognize an inertial frame of reference, in which Newton's laws are valid.
- How to use a free-body diagram to represent forces acting on an object.
- How to use the free-body diagram as a guide in writing force equations for Newton's laws.

Concepts and Equations

Term	Description
Force	A force is a quantitative measure of the interaction between two objects, represented by a vector. The SI unit of force is the newton (N). One newton equals 1 kilogram-meter per second squared.
Combining Forces	The vector sum of forces acting on a body is the resultant, denoted $\vec{R}$: $$\vec{R} = \vec{F}_1 + \vec{F}_2 + \vec{F}_3 + \ldots = \Sigma \vec{F}.$$ The effect of many forces acting on a body can be captured by the resultant force. This principle is called **superposition of forces**.
Contact Force	A contact force is a force between two objects touching at a surface. A contact force has two components: a component perpendicular to the surface (the normal force) and a component parallel to the surface (the frictional force).
Normal Force	The normal force is the component of a contact force between two objects that is perpendicular to their common surface. The normal force is denoted by $\vec{n}$.
Friction Force	The friction force is the component of a contact force between two objects that is parallel to their common surface. The friction force is denoted by $\vec{f}$. Friction forces often act to resist the sliding of an object.
Tension Force	A tension force is conveyed by the pull of a rope or cord and is denoted by $\vec{T}$.
Newton's First Law	Newton's first law states that when the vector sum of forces acting on an object is zero, the object is in equilibrium and has zero acceleration. The object will remain at rest or move with constant velocity when no net external force acts upon it. The law is valid only in inertial reference frames.
Inertial Reference Frame	An inertial reference frame is a reference frame in which Newton's laws are valid. A common example of a *non*inertial reference frame is that of an accelerating airplane.
Newton's Second Law	Newton's second law of motion states that an object which is not in equilibrium is acted upon by a net external force and accelerates. The acceleration is given in vector form by $$\Sigma \vec{F} = m\vec{a},$$ where m is the object's mass, which characterizes the inertia of the object. Newton's second law can also be written in component form as $$\Sigma F_x = ma_x \quad \Sigma F_y = ma_y \quad \Sigma F_z = ma_z.$$
Weight	An object's weight is the gravitational force exerted on the object by the earth or another astronomical object and is denoted by $\vec{w}$. The magnitude of an object's weight is equal to the product of its mass and the magnitude of the acceleration due to gravity: $$w = mg.$$
Newton's Third Law	Newton's third law states that, for two interacting objects A and B, each exerts a force on the other of equal magnitude and opposite in direction, or $$\vec{F}_{A \text{ on } B} = -\vec{F}_{B \text{ on } A}.$$

From Chapter 4 of Student's Study Guide to accompany *University Physics with Modern Physics, Volume 1*, Fifteenth Edition. Hugh D. Young and Roger A. Freedman. Copyright © 2020 by Pearson Education, Inc. All rights reserved.

Free-Body Diagram	A diagram showing all forces acting *on* an object. The object is represented by a point; the forces are indicated by vectors. A free-body diagram is useful in solving problems involving forces.

Key Concept 1: The net force is the vector sum of all of the individual forces that act on the object. It can be specified by its components or by its magnitude and direction.

Key Concept 2: If the net external force on an object is zero, the object either remains at rest or keeps moving at a constant velocity.

Key Concept 3: If an object either remains at rest or keeps moving at a constant velocity, the net external force on the object is zero.

Key Concept 4: In problems involving forces and acceleration, first identify all of the external forces acting on an object, then choose a coordinate system. Find the vector sum of the external forces, and then set it equal to the mass of the object times the acceleration.

Key Concept 5: In problems involving forces in which you're given velocity, time, and/or displacement data, you'll need to use the equations for motion with constant acceleration as well as Newton's second law.

Key Concept 6: The gravitational force on an object (its weight) does not depend on how the object is moving.

Key Concept 7: In problems involving Newton's second law, make sure that for m you use the mass of the object, *not* its weight.

Key Concept 8: No matter how two interacting objects are moving, the forces that they exert on each other always have the same magnitude and point in opposite directions.

Key Concept 9: The two forces in an action–reaction pair always act on two different objects.

Key Concept 10: In problems that involve more than one object, use Newton's third law to relate the forces that the objects exert on each other.

Key Concept 11: The motion of an object depends on the forces that are exerted on it, not the forces that it exerts on other objects.

Conceptual Questions

1: Winning a Tug-of-War

In a tug-of-war shown in Figure 1, how does the force applied to the rope by the losing team compare with the force applied to the rope by the winning team? Is the magnitude of the force applied by the losing team less than, greater than, or equal to the magnitude of the force applied by the winning team? How does the winning team win?

Figure 1 Question 1.

IDENTIFY, SET UP, AND EXECUTE The free-body diagram in Figure 2 will guide our analysis. Each team experiences four forces: the tension due to the tug-of-war rope (T), a frictional force with the ground (f), the weight of the team (w), and the normal force exerted upward from the ground (*N*). The subscripts indicate the winning (*w*) and losing (*l*) teams. Examining the diagrams, we see that the tension forces must be an action–reaction pair—hence the explicit notation. Therefore, by Newton's third law, the tensions must be equal. The force applied by the losing team on the rope must be of the same magnitude as, but opposite in direction to, the force applied by the winning team on the rope.

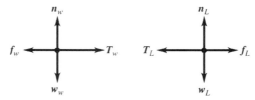

Figure 2 Question 1 free-body diagrams.

The vertical forces will not influence the horizontal interaction, so we look at the remaining force to determine how the winning team wins. The frictional forces must not be equal: The winning team exerts a larger frictional force on the ground than the losing team does in order to accelerate the losing team across the centerline.

KEY CONCEPT 8 **EVALUATE** We see that free-body diagrams and Newton's third law were crucial in our solution. The free-body diagram helped reduce the complexity of the problem and helped show that the frictional force was responsible for the win. After establishing that the tensions were an action–reaction pair, we saw from Newton's third law that the tensions were equal and opposite.

Note that we assumed that the frictional force was between the team and the ground and that each team was able to grip the rope without sliding. There is also a frictional force between the teams' hands and the rope. Differences between the hand and rope frictional forces could also have led to the win.

2: Flying groceries

What force causes a bag of groceries to fly forward when you come to an abrupt stop in a car?

IDENTIFY, SET UP, AND EXECUTE Suppose that, before you come to an abrupt stop, you're moving at a constant velocity. Then no net external force must act on you, the car, or the bag of groceries, according to Newton's first law. As you cause an abrupt stop by hitting the brakes, you increase the frictional force between the car and the road, creating a net external force on the car. When you brake, the force on the bag of groceries doesn't change, so the bag of groceries continues at its initial velocity. (We're assuming that the frictional force between the bag of groceries and the car seat is small.) Therefore, *no force* causes the bag of groceries to fly forward when you come to an abrupt stop in a car.

KEY CONCEPT 2 **EVALUATE** The solution may seem a bit illogical, for consider how the situation would appear to someone outside of the car: The bag of groceries continues moving at a constant velocity after the brakes are applied. This scenario should be more plausible and is a clearer way to imagine the situation.

This is one example of a noninertial frame of reference. The slowing car has negative acceleration and hence is an accelerated frame of reference. Newton's laws don't apply to noninertial frames of reference, so we cannot apply our new force techniques to this problem.

From inside the car, you may try to explain the situation by invoking a "force of inertia." This would be a fictitious force, however, and must be avoided. All of the forces we've encountered (and all of those we will later encounter) arise from known interactions.

3: Does an Apple Accelerate When Placed on a Table?

An apple is placed on a table. Can we describe the apple as having an acceleration of 9.8 m/s^2 toward earth and a second acceleration of 9.8 m/s^2 upward due to the table, thus resulting in a net acceleration of zero?

IDENTIFY, SET UP, AND EXECUTE We have seen how forces can cause accelerations, have heard $F = ma$ often, and know that an object's weight is mg, so it may appear logical to replace forces with mass times acceleration in equations. However, Newton's laws apply to combining forces, not accelerations. Newton's second law states that a net external force on an object will lead to an acceleration equal to the net external force divided by the object's mass.

KEY CONCEPT 3   **EVALUATE** This question points up a common misconception about accelerations and forces. At times, replacing forces with mass times acceleration may lead to the same results as following the correct procedures, but doing so often leads to confusion. An object that is stationary is not accelerating, because there is no *net* external force acting on the object.

4: Forces and Moving Objects

Does a force cause an object to move? Does a moving object "have" a force?

IDENTIFY, SET UP, AND EXECUTE A force does not necessarily cause an object to move. A textbook is acted upon by gravity when placed on a desk, but it does not move. A *net* external force can cause an object to acquire velocity through acceleration.

An object moving at a constant velocity has no net external force acting on it; therefore, the fact that an object is in motion does not indicate that a force is acting upon it. The fact that an object is *accelerating,* however, would certainly indicate that at least one force is acting upon it.

KEY CONCEPT 11   **EVALUATE** Acceleration and motion are *not* equivalent. Acceleration is motion during which the velocity changes over time. An object can also have a *constant* velocity, which is motion without acceleration. You must carefully distinguish between motion and acceleration in order to grasp physics.

5: Definition of equilibrium

Can a moving object be in equilibrium?

IDENTIFY, SET UP, AND EXECUTE Equilibrium occurs when the net external force acting on an object is zero. Newton's first law states that objects with no net external force acting on them remain at rest or continue with constant velocity. An object moving at constant velocity is in equilibrium.

KEY CONCEPT 3   **EVALUATE** *Equilibrium* has a precise definition in physics, even though the word may have connotations of a stationary object. Physics relies upon precise definitions to build representations of physical processes. You must apply physics definitions carefully to build your physics understanding.

Problems

1: Combining several forces to find the resultant

A mover uses a cable to drag a crate across the floor as shown in Figure 3. The mover provides a 300 N force and pulls the cable at an angle of 30.0°. The crate weighs 500 N, and the floor provides a 350 N normal force on the crate and opposes his pull with a 150 N frictional force. Find the resulting force acting on the crate. Will the crate accelerate?

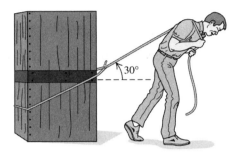

Figure 3 Problem 1.

IDENTIFY We'll combine the forces acting on the crate to find the net external force. If the net external force is not zero, then there'll be acceleration.

SET UP We find the resultant force by adding the forces acting on the crate. Four forces act on the crate: the tension force due to the mover's pull (T), the crate's weight (w), and the normal force (n) and friction force (f) due to the floor. We represent the four forces as vectors in the free-body diagram of the crate in Figure 4.

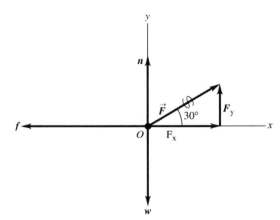

Figure 4 Problem 1 free-body diagram.

We've added an xy-coordinate system to the free-body diagram as the forces act in two dimensions. We've also resolved the tension force into its x- and y-components.

EXECUTE We add the four forces together by adding their components, writing separate equations for the x- and y-components. There are two x-components, due to the horizontal components of the tension force and the friction force:

$$\Sigma F_x = T\cos 30° + (-f).$$

The x-component of the tension force is to the right and is thus assigned a positive value, while the friction force is to the left and is assigned a negative value. We proceed to the y-components. There are three y-components, one due to the normal force, a second due to the weight of the crate, and, finally, the vertical component of the tension force:

$$\Sigma F_y = n + (-w) + T\sin 30°.$$

The y-component of the tension force and the normal force are directed upward and are assigned positive values, while the weight of the crate is directed downward and is assigned a negative value. We now substitute the values for the variables to find the net external force along both axes:

$$\Sigma F_x = T\cos 30° + (-f) = (300\text{ N})\cos 30° + (-150\text{ N}) = +110\text{ N},$$
$$\Sigma F_y = n + (-w) + T\sin 30° = (350\text{ N}) + (-500\text{ N}) + (300\text{ N})\sin 30° = 0\text{ N}.$$

The resultant force on the crate has an x-component of $+110$ N and no y-component, (i.e., the resulting force is horizontal and points to the right). There's also a resulting acceleration of the crate to the right, as there is a net external force.

KEY CONCEPT 1 **EVALUATE:** This is a typical force problem in which we have used our vector addition skills to find the resultant force. We see that there is no net external force in the vertical direction; therefore, the crate remains on the bed of the truck.

Practice Problem: At what rate does the crate accelerate? *Answer*: 2.2 m/s^2

Extra Practice: What tension force is needed to have an acceleration equal to g? *Answer:* 750 N

2: Using Newton's second law to find the mass of a cruise ship

A tugboat pulls a cruise ship out of port. (See Figure 5.) You estimate the acceleration by noting that the tugboat takes 60 s to move the cruise ship 100 m, starting from rest. If the tugboat exerts 3×10^6 N of thrust, what is the mass of the cruise ship? Ignore drag due to the water, and assume that the tugboat accelerates uniformly.

Figure 5 Problem 2.

IDENTIFY We'll use Newton's second law to find the mass of the cruise ship, given the net force and acceleration acting on the ship.

SET UP The problem tells us the net external force provided by the tugboat, and the acceleration can be determined from the kinematics information. We ignore drag or friction with the water, so the only horizontal force acting on the cruise ship is due to the tugboat.

EXECUTE Newton's second law relates the net external force to the mass and resulting acceleration:

$$\sum F_x = ma_x.$$

The net external force acting on the cruise ship is 3×10^6 N. The acceleration is found from the equation for position as a function of time with constant acceleration:

$$x = x_0 + v_0 t + \tfrac{1}{2} a_x t^2.$$

Here, the initial velocity is zero and we take the initial position to be zero. Substituting these values into the equation gives

$$x = \tfrac{1}{2} a_x t^2.$$

Solving for the acceleration yields

$$a_x = \frac{2x}{t^2}.$$

Replacing the distance and time with the given values produces

$$a_x = \frac{2x}{t^2} = \frac{2(100 \text{ m})}{(60 \text{ s})^2} = 0.056 \text{ m/s}^2.$$

We now use Newton's second law to find the mass. Solving for the mass gives

$$m = \frac{F_x}{a_x} = \frac{(3\times10^6 \text{ N})}{(0.056 \text{ m/s}^2)} = 54{,}000{,}000 \text{ kg} = 54 \text{ kilotonnes}.$$

Our estimate shows that the cruise ship has a mass of 54 kilotonnes (1 kilotonne $=10^6$ kg). More correctly, the cruise ship has a mass of 50 kilotonnes, as the values stated in the problem have only one significant figure.

KEY CONCEPT 5 **EVALUATE** This problem shows how we can combine Newton's law with observations to make interesting conclusions about the mass of an object.

Practice Problem: How fast is the ship moving at the end of the 60 s? *Answer:* 3.36 m/s

Extra Practice: How much acceleration would the ship have if it only had a mass of 27 kilotonnes? *Answer:* 0.112 m/s^2

3: Drawing free-body diagrams

Draw a free-body diagram for each of the following situations:

(a) A box slides down a smooth ramp. (See Figure 6.)

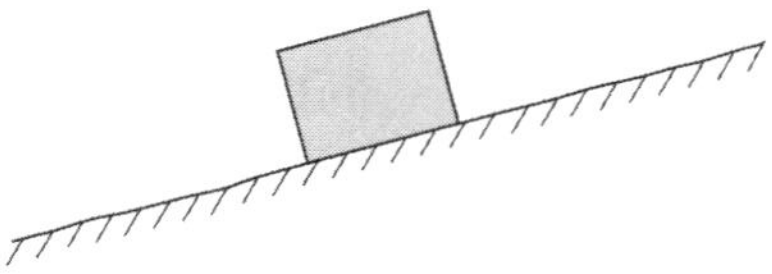

Figure 6 Problem 3a.

(b) A box slides down a rough ramp. (See Figure 7.)

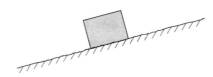

Figure 7 Problem 3b.

(c) A block is placed on top of a crate, and the crate is placed on a horizontal surface. (See Figure 8.) Draw a free-body diagram of the crate.

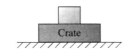

Figure 8 Problem 3c.

(d) A block is placed on top of a crate, and the crate is pulled horizontally across a rough surface. (See Figure 9.) The surface between the crate and the block is rough, and the block is held at rest by a string. Draw a free-body diagram of the crate.

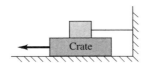

Figure 9 Problem 3d.

IDENTIFY We'll draw free-body diagrams that show all of the forces acting on the object, representing the forces as vectors.

SET UP The first step is to identify the object and then find the forces acting on it. We'll look at the contact tension, normal and frictional forces, and the noncontact gravitational force.

EXECUTE In part (a), there's no friction, since the ramp is smooth. The only contact force acting on the box is the normal force due to the ramp. Gravity also acts on the box. The free-body diagram includes two forces acting on the box: the normal (n) force, directed perpendicular to the ramp; and the weight (w) of the box, directed downward. The free-body diagram of the box is shown in Figure 10.

Figure 10 Problem 3a free-body diagram.

In part (b), there is friction, since the ramp is rough. The contact forces acting on the box are the normal and frictional forces due to the ramp. Gravity also acts on the box. The free-body diagram includes three forces acting on the box: the normal force (*n*), directed perpendicular to the ramp; the frictional force (*f*), directed upward along the ramp (opposing the motion of the box); and the weight (*w*) of the box, directed downward. The free-body diagram of the box is shown in Figure 11.

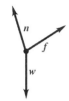

Figure 11 Problem 3b free-body diagram.

In part (c), two contact forces act on the crate: the normal force due to the surface and the normal force due to the block. There are no frictional forces, as neither the crate nor the block is moving. Gravity acts on the box. The free-body diagram includes three forces acting on the crate: the normal force due to the surface (n_{surface}), directed upward; the normal force due to the block (n_{block}), directed downward; and the weight (*w*), of the crate, directed downward. The free-body diagram of the crate is shown in Figure 12.

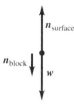

Figure 12 Problem 3c free-body diagram.

In part (d), five contact forces act on the crate: the normal forces due to the surface and the block, the frictional forces due to the surface and the block, and the tension force provided by the pull. The sixth force acting on the crate is gravity. The free-body diagram includes six forces acting on the crate: the normal force due to the surface (n_{surface}), directed upward; the normal force due to the block (n_{block}), directed downward; the frictional forces due to the surface (f_{surface}) and the block (f_{block}), both directed to the right; the tension force (*T*), directed to the left; and the weight (*w*), of the crate, directed downward. The free-body diagram for the crate is shown in Figure 13.

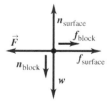

Figure 13 Problem 3d free-body diagram.

KEY CONCEPT 9 **EVALUATE** Drawing free-body diagrams should become second nature to you. Free-body diagrams help catch mistakes by identifying all the forces acting on an object, as well as help identify action–reaction pairs.

4: Tension in a string connecting two blocks

Two blocks are connected by a massless string, as shown in Figure 14. A cable is attached to the upper block and is pulled upward with a 250 N force. Find the tension in the string connecting the two blocks. The upper block has a mass of 7.5 kg and the lower block has a mass of 12 kg.

Figure 14 Problem 4.

IDENTIFY We'll use Newton's laws to solve this problem.

SET UP We cannot determine whether the system is in equilibrium or accelerating from the statement of the problem; therefore, we do not know whether to apply Newton's first law for a system in equilibrium or Newton's second law for an accelerating system. Our first step, therefore, will be to determine whether the system is in equilibrium or accelerating. Then we'll apply the appropriate one of Newton's laws to find the tension in the string.

We'll use free-body diagrams to solve the problem. We can determine whether the blocks are accelerating by considering the two blocks as one system. The left panel of Figure 15 shows a free-body diagram of the system with the two blocks combined. We can find the tension in the string by considering the two blocks separately. The middle and right panels of Figure 15 show the free-body diagrams of the two blocks separately. The top block is designated A, the bottom block B, to reduce confusion.

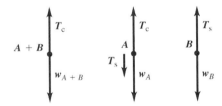

Figure 15 Problem 4 free-body diagram.

The forces in the free-body diagrams are identified by their magnitudes. The combined diagram includes the tension of the cable (T_{cable}) and the weight of the two blocks (w_{A+B}). The other diagrams also include the tension of the string (T_{string}), and the weights of the blocks (w_A and w_B). The upward-pointing vectors will be taken to be positive, the downward-pointing vectors negative.

EXECUTE To determine whether the blocks are accelerating, we examine the net external force acting on them. From the left-hand free-body diagram, we see that there are two forces, the tension of the cable and the weight of the blocks, acting on the combination of blocks:

$$\Sigma F_y = T_{cable} + (-w_{A+B}).$$

The weight is the combined mass of the blocks times the gravitational constant. The net external force is found by replacing the weight and tension by the given values:

$$\Sigma F_y = T_{cable} + (-g(m_A + m_B)) = 250 \text{ N} + (-(9.8 \text{ m/s}^2)(7.5 \text{ kg} + 12 \text{ kg})) = 58.9 \text{ N}.$$

The net external force is not zero; therefore, the blocks are accelerating. We find the acceleration from Newton's second law applied to the combined blocks:

$$\Sigma F_y = (m_A + m_B)a_x.$$

Solving for the acceleration yields

$$a_x = \frac{\Sigma F_y}{(m_A + m_B)} = \frac{58.9 \text{ N}}{(7.5 \text{ kg} + 12 \text{ kg})} = 3.02 \text{ m/s}^2.$$

We now apply Newton's second law to the lower block to find the tension in the string. Two forces are acting on the lower block: the tension due to the string (upward) and gravity (downward). Hence,

$$\Sigma F_y = T_{string} + (-m_B g) = m_B a_y.$$

Solving for the tension in the string and substituting the value for acceleration gives

$$T_{string} = m_B g + m_B a_y = m_B(g + a_y) = (12 \text{ kg})(9.8 \text{ m/s}^2 + 3.02 \text{ m/s}^2) = 150 \text{ N}.$$

The tension in the string is 150 N.

KEY CONCEPT 10 **EVALUATE** We see that the tension force due to the string is less than the tension force due to the cable. This is expected, as the string provides force to accelerate the lower block, whereas the cable provides force to accelerate both blocks. It is important not to assume that tensions are equal in problems; you must consider each tension independently.

Practice Problem: What is the acceleration of the blocks if the second block has a mass of 7.5 kg? *Answer:* 6.87 m/s^2

Extra Practice: What is the tension in the string if the second block has a mass of 7.5 kg? *Answer:* 125 N

Try It Yourself!

1: Rock suspended by wire

A rock of mass 4.0 kg is suspended by a wire. When a horizontal force of 29.4 N is applied to the rock, it moves to the side such that the wire makes an angle θ with the vertical. Find the angle θ and the tension in the wire.

Solution Checkpoints

IDENTIFY AND SET UP The net external force on the rock is zero. Three forces act on the rock. By drawing a free-body diagram, we can see how to set the horizontal and vertical components of force to zero to solve the component force equations.

EXECUTE The net external horizontal and vertical forces acting on the rock are

$$\Sigma F_x = F_H - T \sin\theta = 0$$
$$\Sigma F_y = T \cos\theta - mg = 0.$$

Dividing one equation by the other leads to an angle of 37°. Substituting the angle into either equation leads to a tension of 49 N.

KEY CONCEPT 7 **EVALUATE** We'll often break the net external force into horizontal and vertical components and solve each separately, much as we did in our two-dimensional kinematics problems.

Practice Problem: What horizontal force is required to make a 45° angle? *Answer:* 39.2 N

Extra Practice: What tension results from a 45° angle? *Answer:* 55.4 N

2: Tension in an elevator cable

A 1000.0 kg elevator rises with an upward acceleration equal to g. What is the tension in the supporting cable?

Solution Checkpoints

IDENTIFY AND SET UP There's a net external force on the elevator, so Newton's second law will be used to find the tension.

EXECUTE The net external vertical force acting on the elevator is

$$\Sigma F_y = T - mg = ma_y = mg.$$

Solving for the tension gives 19,600 N.

KEY CONCEPT 4 **EVALUATE** We see that the tension in the cable is larger than the force of gravity on the elevator. Does this make physical sense?

Practice Problem: What is the tension if the acceleration is negative g? *Answer:* 0 N

Extra Practice: What is the tension if the elevator moves with constant velocity? *Answer:* 9,800 N

3: Acceleration in an elevator

A 100.0 kg man stands on a bathroom scale in an elevator. What is the acceleration of the elevator when the scale reading is (a) 1470 N, (b) 980 N, and (c) 490 N?

Solution Checkpoints

IDENTIFY AND SET UP Two forces act on the man: the force of the scale and the force of gravity. Draw a free-body diagram to guide you.

EXECUTE The net external vertical force acting on the man is

$$\Sigma F_y = F_{scale} - mg = ma_y.$$

Solving for the acceleration gives (a) $a = 4.9$ m/s^2, (b) $a = 0$, and (c) $a = -4.9$ m/s^2.

KEY CONCEPT 6 **EVALUATE** In which case(s) does the man feel heavier than normal? In which case(s) does he feel lighter than normal? In which case(s) does he feel as if he has normal weight? What is the significance of the signs in answers (a) and (c)?

Practice Problem: What is the acceleration when the scale reads 0 N? *Answer:* –9.8 m/s^2

Extra Practice: What does the scale read when the acceleration is g? *Answer:* 1,960 N

Key Example Variation Problems

Solutions to these problems are in Chapter 4 of the Student's Solutions Manual.

Be sure to review EXAMPLE 4.1 (Section 4.1) before attempting these problems.

VP4.1.1 Three professional wrestlers are fighting over a champion's belt, and each exerts a force on the belt. Wrestler 1 exerts a force $F_1 = 40.0$ N in the +x-direction, wrestler 2 exerts a force $F_2 = 80.0$ N in the –y-direction, and wrestler 3 exerts a force $F_3 = 60.0$ N at an angle of 36.9° counterclockwise from the +x-direction. Find the x- and y-components of the net external force on the belt, and find the force's magnitude and direction.

VP4.1.2 Three forces act on a statue. Force $\vec{F}_1$ (magnitude 45.0 N) points in the +x-direction, force $\vec{F}_2$ (magnitude 105 N) points in the +y-direction, and force $\vec{F}_3$ (magnitude 235 N) is at an angle of 36.9° from the -x-direction and 53.1° from the +y-direction. Find the x- and y-components of the net external force on the statue, and find the force's magnitude and direction.

VP4.1.3 An eagle descends steeply onto its prey. Its weight (the gravitational force on the eagle), of magnitude 60.0 N, points downward in the –y-direction. The lift force exerted on the eagle's wings by the air, also of magnitude 60.0 N, is at an angle of 20.0° from the vertical (the +y-direction) and 70.0° from the +x-direction. The drag force (air resistance) exerted on the eagle by the air has magnitude 15.0 N and is at an angle of 20.0° from the –x-direction and 70.0° from the +y-direction. Find the x- and y-components of the net external force on the eagle, and find the force's magnitude and direction.

VP4.1.4 A box containing pizza sits on a table. Ernesto, who sits due east of the pizza box, pulls the box toward him with a force of 35.0 N. Kamala, who sits due north of the pizza box, pulls the box toward her with a 50.0 N force. Tsuroku also sits at the table and pulls the box toward her so that the net external force on the box is 24.0 N in a direction 30.0° south of west. Take the +x-direction to be due east and the +y-direction to be due north. Find the x- and y-components of the force that Tsuroku exerts, and find the force's magnitude and direction.

Be sure to review EXAMPLE 4.4 (Section 4.3) before attempting these problems.

VP4.4.1 A box of books with mass 55 kg rests on the level floor of the campus bookstore. The floor is freshly waxed and has negligible friction. A bookstore worker applies a constant horizontal force with magnitude 25 N to the box. What is the magnitude of the acceleration of the box?

VP4.4.2 A block of cheese of mass 2.0 kg sits on a freshly waxed, essentially frictionless table. You apply a constant horizontal force of 0.50 N to the cheese. (a) Name the three external forces that act on the cheese and what exerts each force. (b) What is the magnitude of the acceleration of the cheese?

VP4.4.3 In a game of ice hockey, you use a hockey stick to hit a puck of mass 0.16 kg that slides on essentially frictionless ice. During the hit you exert a constant horizontal force on the puck that gives it an acceleration of 75 m/s^2 for a fraction of a second.

(a) During the hit, what is the magnitude of the horizontal force that you exert on the puck? (b) How does the magnitude of the normal force due to the ice compare to the weight of the puck?

VP4.4.4 A plate of cafeteria food is on a horizontal table. You push it away from you with a constant horizontal force of 14.0 N. The plate has a mass of 0.800 kg, and during the push it has an acceleration of 12.0 m/s² in the direction you are pushing it. (a) What is the magnitude of the net external force on the plate during the push? (b) What are the magnitude and direction of the friction force that the table exerts on the plate during the push?

Be sure to review EXAMPLE 4.5 (Section 4.3) before attempting these problems.

VP4.5.1 On a winter day a child of mass 20.0 kg slides on a horizontal sidewalk covered in ice. Initially she is moving at 3.00 m/s, but due to friction she comes to a halt in 2.25 m. What is the magnitude of the constant friction force that acts on her as she slides?

VP4.5.2 An airliner of mass 1.70×10^5 kg lands at a speed of 75.0 m/s. As it travels along the runway, the combined effects of air resistance, friction from the tires, and reverse thrust from the engines produce a constant force of 2.90×10^5 N opposite to the airliner's motion. What distance along the runway does the airliner travel before coming to a halt?

VP4.5.3 A truck of mass 2.40×10^3 kg is moving at 25.0 m/s. When the driver applies the brakes, the truck comes a stop after traveling 48.0 m. (a) How much time is required for the truck to stop? (b) What is the magnitude of the truck's constant acceleration as it slows down? (c) What is the magnitude of the constant braking force that acts on the truck as it slows down?

VP4.5.4 A car of mass 1.15×10^3 kg is stalled on a horizontal road. You and your friends give the car a constant, forward, horizontal push. There is friction between the car and the road. (a) Name the four external forces that act on the car as you and your friends push it and what exerts each force. (You can regard the combined push from you and your friends as a single force.) (b) The combined force that you and your friends exert has magnitude 8.00×10^2 N, and starting from rest the car reaches a speed of 1.40 m/s after you have pushed it 5.00 m. Find the magnitude of the constant friction force that acts on the car.

STUDENT'S SOLUTIONS MANUAL FOR NEWTON'S LAWS OF MOTION

VP4.1.1. IDENTIFY: This is a problem about vector addition. We know the magnitude and direction of three forces and want to find the magnitude and direction of their resultant force.
SET UP: The components of a vector of magnitude A that make an angle θ with the $+x$-axis are $A_x = A\cos\theta$ and $A_y = A\sin\theta$. The magnitude and direction are $A = \sqrt{A_x^2 + A_y^2}$ and $\theta = \arctan\dfrac{A_y}{A_x}$.
The components of the resultant are $R_x = A_x + B_x + C_x$ and $R_y = A_y + B_y + C_y$.
EXECUTE: For the three given vectors, the components of the resultant are
$R_x = 40.0$ N $+ 0$ N $+ (60.0$ N$)\cos 36.9° = 88.0$ N
$R_y = 0$ N $+ (-80.0$ N$) + (60.0$ N$)\sin 36.9° = -44.0$ N
$R = \sqrt{(88.0 \text{ N})^2 + (-44.0 \text{ N})^2} = 98.4$ N.
$\theta = \arctan[(-44.0 \text{ N})/(88.0 \text{ N})] = -26.6°$. The minus sign tells us that θ is clockwise from the $+x$-axis.
EVALUATE: Since R_x is positive and R_y is negative, the resultant should point into the fourth quadrant, which agrees with our result.

VP4.1.2. IDENTIFY: This is a problem about vector addition. We know the magnitude and direction of three forces and want to find the magnitude and direction of their resultant force.
SET UP: The components of a vector of magnitude A that make an angle θ with the $+x$-axis are $A_x = A\cos\theta$ and $A_y = A\sin\theta$. The magnitude and direction are $A = \sqrt{A_x^2 + A_y^2}$ and $\theta = \arctan\dfrac{A_y}{A_x}$.
The components of the resultant are $R_x = A_x + B_x + C_x$ and $R_y = A_y + B_y + C_y$.
EXECUTE: The components of the resultant are
$R_x = 45.0$ N $+ 0$ N $+ (235$ N$)\cos 143.1° = -143$ N
$R_y = 0$ N $+ 105$ N $+ 235$ N $\sin 143.1° = 246$ N
$R = \sqrt{(-143 \text{ N})^2 + (246 \text{ N})^2} = 285$ N.
$\theta = \arctan[(246 \text{ N})/(-143 \text{ N})] = -60.0°$, so $\theta = 120°$ counterclockwise from the $+x$-axis.
EVALUATE: The resultant has a negative x-component and a positive y-component, so it should point into the second quadrant, which is what our result shows.

VP4.1.3. **IDENTIFY:** This is a problem about vector addition. We know the magnitude and direction of three forces and want to find the magnitude and direction of their resultant force.
SET UP: The components of a vector of magnitude A that make an angle θ with the $+x$-axis are $A_x = A\cos\theta$ and $A_y = A\sin\theta$. The magnitude and direction are $A = \sqrt{A_x^2 + A_y^2}$ and $\theta = \arctan\dfrac{A_y}{A_x}$.
The components of the resultant are $R_x = A_x + B_x + C_x$ and $R_y = A_y + B_y + C_y$.
EXECUTE: The components of the resultant are
$R_x = 0\text{ N} + (60.0\text{ N})\cos 70.0° + (15.0\text{ N})\cos 160.0° = 6.4\text{ N}$
$R_y = -60.0\text{ N} + (60.0\text{ N})\sin 70.0° + (15.0\text{ N})\sin 160.0° = 1.5\text{ N}$
$R = \sqrt{(6.4\text{ N})^2 + (1.5\text{ N})^2} = 6.6\text{ N}$
$\theta = \arctan[(1.5\text{ N})/(64\text{ N})] = 13°$ counterclockwise from the $+x$-axis.
EVALUATE: Since all the components of the resultant are positive, it should point into the first quadrant, which is what we found.

VP4.1.4. **IDENTIFY:** We know the resultant of three vectors, and we know two of them. We want to find the magnitude and direction of the unknown third vector.
SET UP: The components of a vector of magnitude A that make an angle θ with the $+x$-axis are $A_x = A\cos\theta$ and $A_y = A\sin\theta$. The magnitude and direction are $A = \sqrt{A_x^2 + A_y^2}$ and $\theta = \arctan\dfrac{A_y}{A_x}$.
The components of the resultant are $R_x = A_x + B_x + C_x$ and $R_y = A_y + B_y + C_y$.
EXECUTE: Let T refer to Ernesto's force, K for Kamala's force, and T for Tsuroku's unknown force. The components of the resultant force are
$R_x = 35.0\text{ N} + 0\text{ N} + T_x = (24.0\text{ N})\cos 210° \rightarrow T_x = -55.8\text{ N}$
$R_y = 0\text{ N} + 50.0\text{ N} + T_y = (24.0\text{ N})\sin 210° \rightarrow T_y = -62.0\text{ N}$
$T = \sqrt{(-55.8\text{ N})^2 + (-62.0\text{ N})^2} = 83.4\text{ N}$
$\theta = \arctan[(-62.0\text{ N})/(-55.8\text{ N})] = 48.0°$
Since both components of Tsuroku's force are negative, its direction is 48.0° below the $-x$-axis, which is 48.0° south of west.
EVALUATE: A graphical sum will confirm this result.

VP4.4.1. **IDENTIFY:** Apply Newton's second law to the box.
SET UP: Take the x-axis along the floor. Use $\sum F_x = ma_x$ to find a_x. The only horizontal force acting on the box is the force due to the worker.
EXECUTE: $\sum F_x = ma_x$ gives $25\text{ N} = (55\text{ kg})a_x \rightarrow a_x = 0.45\text{ m/s}^2$. This acceleration is in the direction of the worker's force.
EVALUATE: Other forces act on the box, such as gravity downward and the upward push of the floor. But these do not affect the horizontal acceleration since they have no horizontal components.

VP4.4.2. **IDENTIFY:** Apply Newton's second law to the cheese.
SET UP: Take the x-axis along the surface of the table. Use $\sum F_x = ma_x$ to find a_x. The only horizontal force acting on the box is the 0.50-N force.
EXECUTE: **(a)** The forces are gravity acting vertically downward, the normal force due to the tabletop acting vertically upward, and the 0.50-N force due to your hand acting horizontally.
(b) $\sum F_x = ma_x$ gives $0.50\text{ N} = (2.0\text{ kg})a_x \rightarrow a_x = 0.25\text{ m/s}^2$.
EVALUATE: The vertical forces do not affect the horizontal motion, so only the 0.50-N force causes the acceleration.

VP4.4.3. **IDENTIFY:** Apply Newton's second law to the puck.
SET UP: Take the x-axis along the direction of the horizontal hit on the puck. We know the acceleration of the puck, so use $\sum F_x = ma_x$ to find the force of the hit. The only horizontal force acting on the puck is the hit. For the vertical forces, use $\sum F_y = ma_y$.
EXECUTE: (a) With the x-axis horizontal, $\sum F_x = ma_x$ gives
$F_x = (0.16 \text{ kg})(75 \text{ m/s}^2) = 12$ N.
(b) The vertical forces are gravity (the weight w) and the normal force n due to the ice. Using $\sum F_y = ma_y$, we have $n - w = 0$ since $a_y = 0$. So the normal force must be equal to the weight of the puck.
EVALUATE: The vertical forces do not affect the horizontal motion.

VP4.4.4. **IDENTIFY:** We apply Newton's second law to the plate. We know its horizontal acceleration and mass and one of the horizontal forces acting on it, so we can find the friction force, which is horizontal.
SET UP: Apply $\sum F_x = ma_x$ with the x-axis horizontal. The two horizontal forces are friction f and the push P.
EXECUTE: (a) $\sum F_x = ma_x = (0.800 \text{ kg})(12.0 \text{ m/s}^2) = 9.60$ N. This is the net force.
(b) $F_{net} = P - f$, so 9.60 N $= 14.0$ N $- f$, so $f = 4.4$ N. The direction is opposite to the push.
EVALUATE: Friction is less than the push, which it should be since the plate accelerates in the direction of the push.

VP4.5.1. **IDENTIFY:** The forces on the child are constant, so her acceleration is constant. Thus we can use the constant-acceleration motion equations to find her acceleration. Then apply Newton's second law to find the friction force.
SET UP: The formulas $v_x^2 = v_{0x}^2 + 2a_x(x - x_0)$ and $\sum F_x = ma_x$ both apply.
EXECUTE: First find the girl's acceleration using $v_x^2 = v_{0x}^2 + 2a_x(x - x_0)$. Putting in the known numbers gives $0^2 = (3.00 \text{ m/s})^2 + 2a_x(2.25 \text{ m})$, giving $a_x = -2.00 \text{ m/s}^2$. The minus sign means that a_x is opposite to v_x. Now use $\sum F_x = ma_x$ to find the friction force. Since friction is the only horizontal force acting on her, it must be in the same direction as her acceleration. This gives $f = ma_x = (20.0 \text{ kg})(-2.00 \text{ m/s}^2) = -40.0$ N. The magnitude is 40.0 N and the direction is the same as the acceleration, which is opposite to the velocity.
EVALUATE: The other forces (gravity and the normal force due to the ice) are vertical, so they do not affect the horizontal motion.

VP4.5.2. **IDENTIFY:** This problem involves Newton's second law and motion with uniform acceleration. Thus we can use the constant-acceleration motion equations.
SET UP: First use $\sum F_x = ma_x$ to find the plane's acceleration. Then use $v_x^2 = v_{0x}^2 + 2a_x(x - x_0)$ to find how far it travels while stopping.
EXECUTE: Using $\sum F_x = ma_x$ gives 2.90 N $= (1.70 \times 10^5 \text{ kg}) a_x \rightarrow a_x = 1.706 \text{ m/s}^2$. Now use $v_x^2 = v_{0x}^2 + 2a_x(x - x_0)$ to find $x - x_0$. Call the +x-direction to be that of the velocity, so a_x will be negative. Thus
$0^2 = (75.0 \text{ m/s})^2 + 2(-1.706 \text{ m/s}^2)(x - x_0) \rightarrow x - x_0 = 1650$ m $= 1.65$ km.
EVALUATE: This distance is about a mile, which is not so unreasonable for stopping a large plane landing at a fairly high speed.

VP4.5.3. IDENTIFY: This problem involves Newton's second law and motion with uniform acceleration. Thus we can use the constant-acceleration motion equations. We know the truck's mass, its initial speed, and the distance it travels while stopping. We want to find how long it takes to stop, its acceleration while stopping, and the braking force while stopping.

SET UP: The braking force is opposite to the truck's velocity but is in the same direction as the truck's acceleration. The equations $v_x = v_{0x} + a_x t$ and $\sum F_x = ma_x$ apply. In addition, the average velocity is $v_{av-x} = \dfrac{\Delta x}{\Delta t}$, and for uniform acceleration, it is also true that $v_{av} = \dfrac{v_0 + v}{2}$.

EXECUTE: (a) Combining the two equations for v_{av-x} gives

$$\Delta t = \dfrac{\Delta x}{v_{av-x}} = \dfrac{\Delta x}{\left(\dfrac{v_0 + v}{2}\right)} = \dfrac{48.0 \text{ m}}{\dfrac{25.0 \text{ m/s} + 0}{2}} = 3.84 \text{ s}.$$

(b) Using $v_x = v_{0x} + a_x t$ gives $0 = 25.0 \text{ m/s} + a_x (3.84 \text{ s})$
$a_x = -6.51 \text{ m/s}^2$. The minus sign means that a_x is opposite to v_x. The magnitude is 6.51 m/s².
(c) $\sum F_x = ma_x = (2400 \text{ kg})(6.51 \text{ m/s}^2) = 1.56 \times 10^4$ N.
EVALUATE: This may seem like a large force, but it is the only force stopping a massive object with a large acceleration.

VP4.5.4. IDENTIFY: This problem involves Newton's second law and motion with uniform acceleration. Thus we can use the constant-acceleration motion equations.
SET UP: The equations $\sum F_x = ma_x$ and $v_x^2 = v_{0x}^2 + 2a_x(x - x_0)$ apply.
EXECUTE: (a) Gravity acts downward and the normal force due to the road acts upward. The horizontal forces are the push and the friction force from the road. The friction force is opposite to the push.
(b) Use $v_x^2 = v_{0x}^2 + 2a_x(x - x_0)$ to find the acceleration of the car, giving
$(1.40 \text{ m/s})^2 = 0 + 2a_x(5.00 \text{ m}) \rightarrow a_x = 0.1960 \text{ m/s}^2$. Now apply $\sum F_x = ma_x$.
$P - f = ma_x \rightarrow 8.00 \times 10^2 \text{ N} - f = (1.15 \times 10^3 \text{ kg})(0.1960 \text{ m/s}^2) \rightarrow f = 575$ N.
EVALUATE: The push is 800 N and the opposing friction force is 575 N, so the car accelerates in the direction of the push.

4.5. IDENTIFY: Add the two forces using components.
SET UP: $F_x = F\cos\theta$, $F_y = F\sin\theta$, where θ is the angle $\vec{F}$ makes with the $+x$-axis.
EXECUTE: (a) $F_{1x} + F_{2x} = (9.00 \text{ N})\cos 120° + (6.00 \text{ N})\cos(233.1°) = -8.10$ N
$F_{1y} + F_{2y} = (9.00 \text{ N})\sin 120° + (6.00 \text{ N})\sin(233.1°) = +3.00$ N.
(b) $R = \sqrt{R_x^2 + R_y^2} = \sqrt{(8.10 \text{ N})^2 + (3.00 \text{ N})^2} = 8.64$ N.
EVALUATE: Since $F_x < 0$ and $F_y > 0$, $\vec{F}$ is in the second quadrant.

4.7. IDENTIFY: Friction is the only horizontal force acting on the skater, so it must be the one causing the acceleration. Newton's second law applies.
SET UP: Take $+x$ to be the direction in which the skater is moving initially. The final velocity is $v_x = 0$, since the skater comes to rest. First use the kinematics formula $v_x = v_{0x} + a_x t$ to find the acceleration, then apply $\sum \vec{F} = m\vec{a}$ to the skater.

EXECUTE: $v_x = v_{0x} + a_x t$ so $a_x = \dfrac{v_x - v_{0x}}{t} = \dfrac{0 - 2.40 \text{ m/s}}{3.52 \text{ s}} = -0.682 \text{ m/s}^2$. The only horizontal force on the skater is the friction force, so $f_x = ma_x = (68.5 \text{ kg})(-0.682 \text{ m/s}^2) = -46.7 \text{ N}$. The force is 46.7 N, directed opposite to the motion of the skater.

EVALUATE: Although other forces are acting on the skater (gravity and the upward force of the ice), they are vertical and therefore do not affect the horizontal motion.

4.11. IDENTIFY and SET UP: Use Newton's second law in component form to calculate the acceleration produced by the force. Use constant acceleration equations to calculate the effect of the acceleration on the motion.

EXECUTE: (a) During this time interval the acceleration is constant and equal to
$$a_x = \dfrac{F_x}{m} = \dfrac{0.250 \text{ N}}{0.160 \text{ kg}} = 1.562 \text{ m/s}^2$$
We can use the constant acceleration kinematic equations from Chapter 2.
$x - x_0 = v_{0x}t + \tfrac{1}{2}a_x t^2 = 0 + \tfrac{1}{2}(1.562 \text{ m/s}^2)(2.00 \text{ s})^2 = 3.12 \text{ m}$, so the puck is at $x = 3.12$ m.
$v_x = v_{0x} + a_x t = 0 + (1.562 \text{ m/s}^2)(2.00 \text{ s}) = 3.12 \text{ m/s}$.

(b) In the time interval from $t = 2.00$ s to 5.00 s the force has been removed so the acceleration is zero. The speed stays constant at $v_x = 3.12$ m/s. The distance the puck travels is $x - x_0 = v_{0x}t = (3.12 \text{ m/s})(5.00 \text{ s} - 2.00 \text{ s}) = 9.36$ m. At the end of the interval it is at $x = x_0 + 9.36 \text{ m} = 12.5$ m.

In the time interval from $t = 5.00$ s to 7.00 s the acceleration is again $a_x = 1.562 \text{ m/s}^2$. At the start of this interval $v_{0x} = 3.12$ m/s and $x_0 = 12.5$ m.
$x - x_0 = v_{0x}t + \tfrac{1}{2}a_x t^2 = (3.12 \text{ m/s})(2.00 \text{ s}) + \tfrac{1}{2}(1.562 \text{ m/s}^2)(2.00 \text{ s})^2$.
$x - x_0 = 6.24 \text{ m} + 3.12 \text{ m} = 9.36 \text{ m}$.
Therefore, at $t = 7.00$ s the puck is at $x = x_0 + 9.36 \text{ m} = 12.5 \text{ m} + 9.36 \text{ m} = 21.9$ m.
$v_x = v_{0x} + a_x t = 3.12 \text{ m/s} + (1.562 \text{ m/s}^2)(2.00 \text{ s}) = 6.24$ m/s.

EVALUATE: The acceleration says the puck gains 1.56 m/s of velocity for every second the force acts. The force acts a total of 4.00 s so the final velocity is $(1.56 \text{ m/s})(4.0 \text{ s}) = 6.24$ m/s.

4.13. IDENTIFY: The force and acceleration are related by Newton's second law.

SET UP: $\Sigma F_x = ma_x$, where ΣF_x is the net force. $m = 4.50$ kg.

EXECUTE: (a) The maximum net force occurs when the acceleration has its maximum value. $\Sigma F_x = ma_x = (4.50 \text{ kg})(10.0 \text{ m/s}^2) = 45.0$ N. This maximum force occurs between 2.0 s and 4.0 s.

(b) The net force is constant when the acceleration is constant. This is between 2.0 s and 4.0 s.

(c) The net force is zero when the acceleration is zero. This is the case at $t = 0$ and $t = 6.0$ s.

EVALUATE: A graph of ΣF_x versus t would have the same shape as the graph of a_x versus t.

4.15. IDENTIFY: The net force and the acceleration are related by Newton's second law. When the rocket is near the surface of the earth the forces on it are the upward force $\vec{F}$ exerted on it because of the burning fuel and the downward force $\vec{F}_{grav}$ of gravity. $F_{grav} = mg$.

SET UP: Let $+y$ be upward. The weight of the rocket is $F_{grav} = (8.00 \text{ kg})(9.80 \text{ m/s}^2) = 78.4 \text{ N}$.

EXECUTE: (a) At $t = 0$, $F = A = 100.0 \text{ N}$. At $t = 2.00 \text{ s}$, $F = A + (4.00 \text{ s}^2)B = 150.0 \text{ N}$ and $B = \dfrac{150.0 \text{ N} - 100.0 \text{ N}}{4.00 \text{ s}^2} = 12.5 \text{ N/s}^2$.

(b) (i) At $t = 0$, $F = A = 100.0 \text{ N}$. The net force is $\Sigma F_y = F - F_{grav} = 100.0 \text{ N} - 78.4 \text{ N} = 21.6 \text{ N}$. $a_y = \dfrac{\Sigma F_y}{m} = \dfrac{21.6 \text{ N}}{8.00 \text{ kg}} = 2.70 \text{ m/s}^2$. (ii) At $t = 3.00 \text{ s}$, $F = A + B(3.00 \text{ s})^2 = 212.5 \text{ N}$. $\Sigma F_y = 212.5 \text{ N} - 78.4 \text{ N} = 134.1 \text{ N}$. $a_y = \dfrac{\Sigma F_y}{m} = \dfrac{134.1 \text{ N}}{8.00 \text{ kg}} = 16.8 \text{ m/s}^2$.

(c) Now $F_{grav} = 0$ and $\Sigma F_y = F = 212.5 \text{ N}$. $a_y = \dfrac{212.5 \text{ N}}{8.00 \text{ kg}} = 26.6 \text{ m/s}^2$.

EVALUATE: The acceleration increases as F increases.

4.19. IDENTIFY and SET UP: $w = mg$. The mass of the watermelon is constant, independent of its location. Its weight differs on earth and Jupiter's moon. Use the information about the watermelon's weight on earth to calculate its mass:

EXECUTE: (a) $w = mg$ gives that $m = \dfrac{w}{g} = \dfrac{44.0 \text{ N}}{9.80 \text{ m/s}^2} = 4.49 \text{ kg}$.

(b) On Jupiter's moon, $m = 4.49 \text{ kg}$, the same as on earth. Thus the weight on Jupiter's moon is $w = mg = (4.49 \text{ kg})(1.81 \text{ m/s}^2) = 8.13 \text{ N}$.

EVALUATE: The weight of the watermelon is less on Io, since g is smaller there.

4.23. IDENTIFY: The system is accelerating so we use Newton's second law.

SET UP: The acceleration of the entire system is due to the 250-N force, but the acceleration of box B is due to the force that box A exerts on it. $\Sigma F = ma$ applies to the two-box system and to each box individually.

EXECUTE: For the two-box system: $a_x = \dfrac{250 \text{ N}}{25.0 \text{ kg}} = 10.0 \text{ m/s}^2$. Then for box B, where F_A is the force exerted on B by A, $F_A = m_B a = (5.0 \text{ kg})(10.0 \text{ m/s}^2) = 50 \text{ N}$.

EVALUATE: The force on B is less than the force on A.

4.27. IDENTIFY: Identify the forces on each object.

SET UP: In each case the forces are the noncontact force of gravity (the weight) and the forces applied by objects that are in contact with each crate. Each crate touches the floor and the other crate, and some object applies $\vec{F}$ to crate A.

EXECUTE: (a) The free-body diagrams for each crate are given in Figure 4.27. F_{AB} (the force on m_A due to m_B) and F_{BA} (the force on m_B due to m_A) form an action-reaction pair.

(b) Since there is no horizontal force opposing F, any value of F, no matter how small, will cause the crates to accelerate to the right. The weight of the two crates acts at a right angle to the horizontal, and is in any case balanced by the upward force of the surface on them.
EVALUATE: Crate B is accelerated by F_{BA} and crate A is accelerated by the net force $F - F_{AB}$. The greater the total weight of the two crates, the greater their total mass and the smaller will be their acceleration.

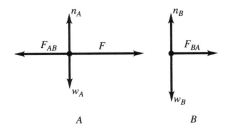

Figure 4.27

4.33. **IDENTIFY:** Apply Newton's second law to the bucket and constant-acceleration kinematics.

SET UP: The minimum time to raise the bucket will be when the tension in the cord is a maximum since this will produce the greatest acceleration of the bucket.
EXECUTE: Apply Newton's second law to the bucket: $T - mg = ma$. For the maximum acceleration, the tension is greatest, so $a = \dfrac{T - mg}{m} = \dfrac{75.0 \text{ N} - (5.60 \text{ kg})(9.8 \text{ m/s}^2)}{5.60 \text{ kg}} = 3.593 \text{ m/s}^2$.

The kinematics equation for $y(t)$ gives $t = \sqrt{\dfrac{2(y - y_0)}{a_y}} = \sqrt{\dfrac{2(12.0 \text{ m})}{3.593 \text{ m/s}^2}} = 2.58 \text{ s}$.

EVALUATE: A shorter time would require a greater acceleration and hence a stronger pull, which would break the cord.

4.35. **IDENTIFY:** If the box moves in the $+x$-direction it must have $a_y = 0$, so $\Sigma F_y = 0$.

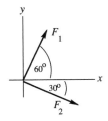

The smallest force the child can exert and still produce such motion is a force that makes the y-components of all three forces sum to zero, but that doesn't have any x-component.

Figure 4.35

SET UP: $\vec{F}_1$ and $\vec{F}_2$ are sketched in Figure 4.35. Let $\vec{F}_3$ be the force exerted by the child.
$\Sigma F_y = ma_y$ implies $F_{1y} + F_{2y} + F_{3y} = 0$, so $F_{3y} = -(F_{1y} + F_{2y})$.
EXECUTE: $F_{1y} = +F_1 \sin 60° = (100 \text{ N}) \sin 60° = 86.6 \text{ N}$
$F_{2y} = +F_2 \sin(-30°) = -F_2 \sin 30° = -(140 \text{ N}) \sin 30° = -70.0 \text{ N}$
Then $F_{3y} = -(F_{1y} + F_{2y}) = -(86.6 \text{ N} - 70.0 \text{ N}) = -16.6 \text{ N}$; $F_{3x} = 0$
The smallest force the child can exert has magnitude 17 N and is directed at 90° clockwise from the $+x$-axis shown in the figure.

(b) IDENTIFY and SET UP: Apply $\sum F_x = ma_x$. We know the forces and a_x so can solve for m. The force exerted by the child is in the $-y$-direction and has no x-component.

EXECUTE: $F_{1x} = F_1 \cos 60° = 50$ N

$F_{2x} = F_2 \cos 30° = 121.2$ N

$\sum F_x = F_{1x} + F_{2x} = 50$ N $+ 121.2$ N $= 171.2$ N

$m = \dfrac{\sum F_x}{a_x} = \dfrac{171.2 \text{ N}}{2.00 \text{ m/s}^2} = 85.6$ kg

Then $w = mg = 840$ N.

EVALUATE: In part (b) we don't need to consider the y-component of Newton's second law. $a_y = 0$ so the mass doesn't appear in the $\sum F_y = ma_y$ equation.

4.37. IDENTIFY: Use Newton's second law to relate the acceleration and forces for each crate.

(a) SET UP: Since the crates are connected by a rope, they both have the same acceleration, 2.50 m/s².

(b) The forces on the 4.00 kg crate are shown in Figure 4.37a.

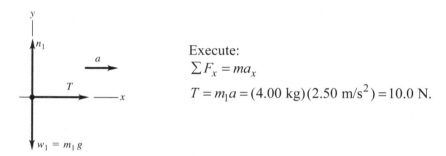

Execute:
$\sum F_x = ma_x$
$T = m_1 a = (4.00 \text{ kg})(2.50 \text{ m/s}^2) = 10.0$ N.

Figure 4.37a

(c) SET UP: Forces on the 6.00 kg crate are shown in Figure 4.37b.

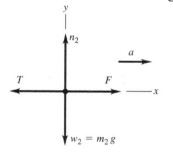

The crate The crate accelerates to the right,
so the net force is to the right.
F must be larger than T.

Figure 4.37b

(d) EXECUTE: $\sum F_x = ma_x$ gives $F - T = m_2 a$

$F = T + m_2 a = 10.0$ N $+ (6.00 \text{ kg})(2.50 \text{ m/s}^2) = 10.0$ N $+ 15.0$ N $= 25.0$ N

EVALUATE: We can also consider the two crates and the rope connecting them as a single object of mass $m = m_1 + m_2 = 10.0$ kg. The free-body diagram is sketched in Figure 4.37c.

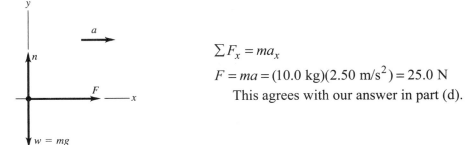

$\sum F_x = ma_x$

$F = ma = (10.0 \text{ kg})(2.50 \text{ m/s}^2) = 25.0 \text{ N}$

This agrees with our answer in part (d).

Figure 4.37c

4.39. **IDENTIFY and SET UP:** Take derivatives of $x(t)$ to find v_x and a_x. Use Newton's second law to relate the acceleration to the net force on the object.

EXECUTE:

(a) $x = (9.0 \times 10^3 \text{ m/s}^2)t^2 - (8.0 \times 10^4 \text{ m/s}^3)t^3$

$x = 0$ at $t = 0$

When $t = 0.025$ s, $x = (9.0 \times 10^3 \text{ m/s}^2)(0.025 \text{ s})^2 - (8.0 \times 10^4 \text{ m/s}^3)(0.025 \text{ s})^3 = 4.4$ m.

The length of the barrel must be 4.4 m.

(b) $v_x = \dfrac{dx}{dt} = (18.0 \times 10^3 \text{ m/s}^2)t - (24.0 \times 10^4 \text{ m/s}^3)t^2$

At $t = 0$, $v_x = 0$ (object starts from rest).

At $t = 0.025$ s, when the object reaches the end of the barrel,

$v_x = (18.0 \times 10^3 \text{ m/s}^2)(0.025 \text{ s}) - (24.0 \times 10^4 \text{ m/s}^3)(0.025 \text{ s})^2 = 300$ m/s

(c) $\sum F_x = ma_x$, so must find a_x.

$a_x = \dfrac{dv_x}{dt} = 18.0 \times 10^3 \text{ m/s}^2 - (48.0 \times 10^4 \text{ m/s}^3)t$

(i) At $t = 0$, $a_x = 18.0 \times 10^3 \text{ m/s}^2$ and $\sum F_x = (1.50 \text{ kg})(18.0 \times 10^3 \text{ m/s}^2) = 2.7 \times 10^4$ N.

(ii) At $t = 0.025$ s, $a_x = 18 \times 10^3 \text{ m/s}^2 - (48.0 \times 10^4 \text{ m/s}^3)(0.025 \text{ s}) = 6.0 \times 10^3 \text{ m/s}^2$ and

$\sum F_x = (1.50 \text{ kg})(6.0 \times 10^3 \text{ m/s}^2) = 9.0 \times 10^3$ N.

EVALUATE: The acceleration and net force decrease as the object moves along the barrel.

4.41. **IDENTIFY:** You observe that your weight is different from your normal weight in an elevator, so you must have acceleration. Apply $\sum \vec{F} = m\vec{a}$ to your body inside the elevator.

SET UP: The quantity $w = 683$ N is the force of gravity exerted on you, independent of your motion. Your mass is $m = w/g = 69.7$ kg. Use coordinates with $+y$ upward. Your free-body diagram is shown in Figure 4.41, where n is the scale reading, which is the force the scale exerts on you. You and the elevator have the same acceleration.

Figure 4.41

EXECUTE: $\sum F_y = ma_y$ gives $n - w = ma_y$ so $a_y = \dfrac{n-w}{m}$.

(a) $n = 725$ N, so $a_y = \dfrac{725 \text{ N} - 683 \text{ N}}{69.7 \text{ kg}} = 0.603$ m/s^2. a_y is positive so the acceleration is upward.

(b) $n = 595$ N, so $a_y = \dfrac{595 \text{ N} - 683 \text{ N}}{69.7 \text{ kg}} = -1.26$ m/s^2. a_y is negative so the acceleration is downward.

EVALUATE: If you appear to weigh less than your normal weight, you must be accelerating downward, but not necessarily *moving* downward. Likewise if you appear to weigh more than your normal weight, you must be acceleration upward, but you could be *moving* downward.

4.43. IDENTIFY: The ball changes velocity, so it has acceleration. Therefore Newton's second law applies to it.

SET UP: Apply $\sum F_x = ma_x$ to the ball. Assume that the acceleration is constant, so we can use the constant-acceleration equation $v_x = v_{0x} + a_x t$. Call the x-axis horizontal with +x in the direction of the ball's original velocity.

EXECUTE: First use $v_x = v_{0x} + a_x t$ to find the acceleration.
-50.0 m/s $= 40.0$ m/s $+ a_x(8.00 \times 10^{-3}$ s$) \rightarrow a_x = -1.125 \times 10^4$ m/s^2.
Now apply $\sum F_x = ma_x$ to the ball. Only the bat exerts a horizontal force on the ball.
$F_{\text{bat}} = ma_x = (0.145 \text{ kg})(-1.125 \times 10^4 \text{ m/s}^2) = -1630$ N. The minus sign tells us that the force and the acceleration are directed opposite to the original velocity of the ball, which is away from the batter.

EVALUATE: Compare the acceleration to g: $a/g = (1.125 \times 10^4$ m/s$^2)/(9.80$ m/s$^2) = 1150$, so the acceleration is $1150g$—a huge acceleration but for only a very brief time.

4.45. IDENTIFY: The system is accelerating, so we apply Newton's second law to each box and can use the constant acceleration kinematics for formulas to find the acceleration.

SET UP: First use the constant acceleration kinematics for formulas to find the acceleration of the system. Then apply $\sum F = ma$ to each box.

EXECUTE: **(a)** The kinematics formula $y - y_0 = v_{0y}t + \tfrac{1}{2}a_y t^2$ gives
$a_y = \dfrac{2(y-y_0)}{t^2} = \dfrac{2(12.0 \text{ m})}{(4.0 \text{ s})^2} = 1.5$ m/s^2. For box B, $mg - T = ma$ and
$m = \dfrac{T}{g-a} = \dfrac{36.0 \text{ N}}{9.8 \text{ m/s}^2 - 1.5 \text{ m/s}^2} = 4.34$ kg.

(b) For box A, $T + mg - F = ma$ and $m = \dfrac{F-T}{g-a} = \dfrac{80.0 \text{ N} - 36.0 \text{ N}}{9.8 \text{ m/s}^2 - 1.5 \text{ m/s}^2} = 5.30$ kg.

EVALUATE: The boxes have the same acceleration but experience different forces because they have different masses.

4.47. **IDENTIFY:** The rocket engines reduce the speed of the rocket, so Newton's second law applies to the rocket. Two vertical forces act on the rocket: gravity and the force F of the engines. The constant-acceleration equations apply because both F and gravity are constant, which makes the acceleration constant.

SET UP: The rocket needs to reduce its speed from 30.0 m/s to zero while traveling 80.0 m. We can find the acceleration from this information using $v_y^2 = v_{0y}^2 + 2a_y(y-y_0)$. Then we can apply $\Sigma F_y = ma_y$ to find the force F. Choose the $+y$-axis upward with the origin at the ground.

EXECUTE: We know that $v_y = 0$ as the rocket reaches the ground. Using $v_y^2 = v_{0y}^2 + 2a_y(y-y_0)$ we get $0 = (-30.0 \text{ m/s})^2 + 2a_y(0 - 80.0 \text{ m}) \rightarrow a_y = 5.625 \text{ m/s}^2$. Now apply $\Sigma F_y = ma_y$.
$F - mg = ma_y \rightarrow F - (20.0 \text{ kg})(9.80 \text{ m/s}^2) = (20.0 \text{ kg})(5.625 \text{ m/s}^2) \rightarrow F = 309$ N.

EVALUATE: The weight of this rocket is $w = mg = (20.0 \text{ kg})(9.80 \text{ m/s}^2) = 196$ N, so we see that $F > w$. This is reasonable because the engine force must oppose gravity to reduce the rocket's speed. Notice that the rocket is moving downward but its acceleration is upward. This is reasonable because the rocket is slowing down, so its acceleration must be opposite to its velocity.

4.49. **IDENTIFY:** The rocket accelerates due to a variable force, so we apply Newton's second law. But the acceleration will not be constant because the force is not constant.

SET UP: We can use $a_x = F_x/m$ to find the acceleration, but must integrate to find the velocity and then the distance the rocket travels.

EXECUTE: Using $a_x = F_x/m$ gives $a_x(t) = \dfrac{(16.8 \text{ N/s})t}{45.0 \text{ kg}} = (0.3733 \text{ m/s}^3)t$. Now integrate the acceleration to get the velocity, and then integrate the velocity to get the distance moved.
$v_x(t) = v_{0x} + \int_0^t a_x(t')dt' = (0.1867 \text{ m/s}^3)t^2$ and $x - x_0 = \int_0^t v_x(t')dt' = (0.06222 \text{ m/s}^3)t^3$. At $t = 5.00$ s, $x - x_0 = 7.78$ m.

EVALUATE: The distance moved during the next 5.0 s would be considerably greater because the acceleration is increasing with time.

4.51. **IDENTIFY:** Kinematics will give us the average acceleration of each car, and Newton's second law will give us the average force that is accelerating each car.

SET UP: The cars start from rest and all reach a final velocity of 60 mph (26.8 m/s). We first use kinematics to find the average acceleration of each car, and then use Newton's second law to find the average force on each car.

EXECUTE: **(a)** We know the initial and final velocities of each car and the time during which this change in velocity occurs. The definition of average acceleration gives $a_{\text{av}} = \dfrac{\Delta v}{\Delta t}$.

Then $F = ma$ gives the force on each car. For the Alpha Romeo, the calculations are $a_{\text{av}} = (26.8 \text{ m/s})/(4.4 \text{ s}) = 6.09 \text{ m/s}^2$. The force is $F = ma = (895 \text{ kg})(6.09 \text{ m/s}^2) = 5.451 \times 10^3$ N $= 5.451$ kN, which we should round to 5.5 kN for 2 significant figures. Repeating this calculation for the other cars and rounding the force to 2 significant figures gives:
Alpha Romeo: $a = 6.09 \text{ m/s}^2$, $F = 5.5$ kN
Honda Civic: $a = 4.19 \text{ m/s}^2$, $F = 5.5$ kN
Ferrari: $a = 6.88 \text{ m/s}^2$, $F = 9.9$ kN
Ford Focus: $a = 4.97 \text{ m/s}^2$, $F = 7.3$ kN
Volvo: $a = 3.72 \text{ m/s}^2$, $F = 6.1$ kN
The smallest net force is on the Alpha Romeo and Honda Civic, to two-figure accuracy. The largest net force is on the Ferrari.

(b) The largest force would occur for the largest acceleration, which would be in the Ferrari. The smallest force would occur for the smallest acceleration, which would be in the Volvo.

(c) We use the same approach as in part (a), but now the final velocity is 100 mph (44.7 m/s).

$a_{av} = (44.7 \text{ m/s})/(8.6 \text{ s}) = 5.20 \text{ m/s}^2$, and $F = ma = (1435 \text{ kg})(5.20 \text{ m/s}^2) = 7.5$ kN. The average force is considerably smaller in this case. This is because air resistance increases with speed.

(d) As the speed increases, so does the air resistance. Eventually the air resistance will be equal to the force from the roadway, so the new force will be zero and the acceleration will also be zero, so the speed will remain constant.

EVALUATE: The actual forces and accelerations involved with auto dynamics can be quite complicated because the forces (and hence the accelerations) are not constant but depend on the speed of the car.

4.53. **IDENTIFY:** A block is accelerated upward by a force of magnitude F. For various forces, we know the time for the block to move upward a distance of 8.00 m starting from rest. Since the upward force is constant, so is the acceleration. Newton's second law applies to the accelerating block.

SET UP: The acceleration is constant, so $y - y_0 = v_{0y}t + \frac{1}{2}a_y t^2$ applies, and $\sum F_y = ma_y$ also applies to the block.

EXECUTE: **(a)** Using the above formula with $v_{0y} = 0$ and $y - y_0 = 8.00$ m, we get $a_y = (16.0 \text{ m})/t^2$. We use this formula to calculate the acceleration for each value of the force F. For example, when $F = 250$ N, we have $a = (16.0 \text{ m})/(3.3 \text{ s})^2 = 1.47 \text{ m/s}^2$. We make similar calculations for all six values of F and then graph F versus a. We can do this graph by hand or using graphing software. The result is shown in Figure 4.53.

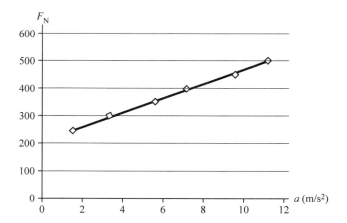

Figure 4.53

(b) Applying Newton's second law to the block gives $F - mg = ma$, so $F = mg + ma$. The equation of our best-fit graph in part (a) is $F = (25.58 \text{ kg})a + 213.0$ N. The slope of the graph is the mass m, so the mass of the block is $m = 26$ kg. The y intercept is mg, so $mg = 213$ N, which gives $g = (213 \text{ N})/(25.58 \text{ kg}) = 8.3 \text{ m/s}^2$ on the distant planet.

EVALUATE: The acceleration due to gravity on this planet is not too different from what it is on earth.

4.55. **IDENTIFY:** The force on the block is a function of time, so the acceleration will also be a function of time. Therefore we cannot use the constant-acceleration formulas, but instead must use the basic velocity and acceleration definitions. We also need to apply Newton's second law.

SET UP: Using $\sum F_x = ma_x$ we can determine the acceleration and use it to find the velocity and position of the block. We use $a_x = dv_x/dt$ and $v_x = dx/dt$. We know that $F(t) = \beta - \alpha t$ and that the object starts from rest.

EXECUTE: **(a)** We want the largest positive value of x. Using $\sum F_x = ma_x$ we can find $a_x(t)$, and from that we can find $v_x(t)$ and then find $x(t)$. First find a_x: $a_x(t) = F_x/m = (\beta - \alpha t)/m$.

Now use $a_x = dv_x/dt$ to find $v_x(t)$. $v_x = \int a_x dt = \int \dfrac{\beta - \alpha t}{m} dt = \dfrac{\beta t - \dfrac{\alpha t^2}{2}}{m}$, where we have used $v_x = 0$ when $t = 0$. Now use $v_x = dx/dt$ to find $x(t)$. $x(t) = \int v_x dt = \int \dfrac{\beta t - \dfrac{\alpha t^2}{2}}{m} dt = \dfrac{\dfrac{\beta t^2}{2} - \dfrac{\alpha t^3}{6}}{m}$, where we have used $x = 0$ when $t = 0$. The largest value of x will occur when $v_x = 0$ because the block will go no further. Equating v_x to zero gives $v_x = \dfrac{\beta t - \dfrac{\alpha t^2}{2}}{m} = 0$. Solving for t and calling it t_{max} gives $t_{max} = 2\beta/\alpha$.

Now use t_{max} in our equation for $x(t)$ to find the largest value of x. This gives

$x(t_{max}) = \dfrac{\dfrac{\beta(2\beta/\alpha)^2}{2} - \dfrac{\alpha(2\beta/\alpha)^3}{6}}{m} = \dfrac{2\beta^3}{3m\alpha^2}$.

The force at t_{max} is $F_x = \beta - \alpha t_{max} = \beta - \alpha\left(\dfrac{2\beta}{\alpha}\right) = -\beta$.

Putting in the values for α and β gives the following results:
$t_{max} = 2\beta/\alpha = 2(4.00\text{ N})/(6.00\text{ N/s}) = 1.33$ s.

$x_{max} = \dfrac{2\beta^3}{3m\alpha^2} = \dfrac{2(4.00\text{ N})^3}{3(2.00\text{ kg})(6.00\text{ N/s})^2} = 0.593$ m.

$F_{max} = -\beta = -4.00$ N, but we only want the magnitude, so $F_{max} = 4.00$ N.

(b) When $x = 0$, the block has returned to where it started. Using our equation for $x(t)$, we get

$x(t) = \dfrac{\dfrac{\beta t^2}{2} - \dfrac{\alpha t^3}{6}}{m} = 0$, which gives $t = 3\beta/\alpha = 3(4.00\text{ N})/(6.00\text{ N/s}) = 2.00$ s.

Using our equation for $v_x(t)$ and evaluating it at $t = 2.00$ s gives

$v_x = \dfrac{\beta t - \dfrac{\alpha t^2}{2}}{m} = \dfrac{(4.00\text{ N})(2.00\text{ s}) - \dfrac{(6.00\text{ N/s})(2.00\text{ s})^2}{2}}{(2.00\text{ kg})} = -2.00$ m/s. Its speed is 2.00 m/s.

EVALUATE: The constant-acceleration equations would be of no help with this type of time-dependent force.

4.57. **IDENTIFY:** Newton's second law applies to the dancer's head.

SET UP: We use $a_{av} = \dfrac{\Delta v}{\Delta t}$ and $\vec{F}_{net} = m\vec{a}$.

EXECUTE: First find the average acceleration: a_{av} = (4.0 m/s)/(0.20 s) = 20 m/s². Now apply Newton's second law to the dancer's head. Two vertical forces act on the head: $F_{neck} - mg = ma$, so $F_{neck} = m(g + a)$, which gives F_{neck} = (0.094)(65 kg)(9.80 m/s² + 20 m/s²) = 180 N, which is choice (d).

EVALUATE: The neck force is not simply ma because the neck must balance her head against gravity, even if the head were not accelerating. That error would lead one to incorrectly select choice (c).

5

STUDY GUIDE FOR APPLYING NEWTON'S LAWS

Summary

In this chapter, we'll apply Newton's laws of motion to objects that are in *equilibrium* (at rest or in uniform motion) and to objects that are *not in equilibrium* (in accelerated motion). We'll develop a consistent problem-solving strategy that utilizes a free-body diagram to identify the relevant forces acting on an object. We'll also expand our catalog of forces by quantifying contact forces and friction forces, as well as examine forces on objects in uniform circular motion. By the end of the chapter, we'll have built a foundation for solving equilibrium and nonequilibrium problems involving any combination of forces.

Objectives

After studying this chapter, you'll understand
- How to efficiently represent forces acting on an object by using a free-body diagram.
- How to use the free-body diagram as a guide in writing force equations for Newton's laws.
- How to use Newton's first law to solve problems involving objects in equilibrium.
- How to use Newton's second law to solve problems involving accelerating objects.
- How to apply contact forces and various friction forces to a variety of situations.
- How to recognize action–reaction pairs and use Newton's third law to quantify their magnitudes.
- How to apply Newton's laws of motion to objects moving in uniform circular motion.
- How to use Newton's laws to solve problems proficiently.

Concepts and Equations

Term	Description
Using Newton's First Law	An object in equilibrium (either at rest or moving with constant velocity) is acted upon by no net external force: The vector sum of the forces acting on the object must be zero according to Newton's first law of motion, $\Sigma \vec{F} = 0$. When solving equilibrium problems, one starts with free-body diagrams, finds the net external forces along two perpendicular components, and then solves by using the equations $$\Sigma F_x = 0 \quad \Sigma F_y = 0.$$
Applying Newton's Second Law	An object that is acted upon by a nonzero net force accelerates. The acceleration is given by Newton's second law of motion, $$\Sigma \vec{F} = m\vec{a}.$$ When solving nonequilibrium problems, one identifies the forces acting on the object with free-body diagrams, finds the net forces along two perpendicular components, and determines the equations of motion given by $$\Sigma F_x = ma_x \quad \Sigma F_y = ma_y$$ In problems with multiple interacting objects, it may be necessary to apply Newton's laws to each object individually and solve the equations simultaneously.
Friction Force	A friction force is that component of the contact force between two objects which is parallel to the surfaces in contact. Friction forces, denoted by $\vec{f}$, are generally proportional to the normal force and include kinetic friction forces (when the objects move relative to each other), static friction forces (when there is no motion between the objects), viscosity and drag forces (for motion involving liquids and gases), and rolling frictional forces (for rolling objects).
Kinetic Friction Force	The kinetic friction force is the friction force between two objects moving relative to each other and is generally proportional to the normal force between the objects. The proportionality constant is the coefficient of kinetic friction (μ_k), which depends on the objects' surface characteristics and has no units. The direction of the kinetic friction force is always opposite the direction of motion. Mathematically, $$f_k = \mu_k n.$$
Static Friction Force	The static friction force is the friction force between two objects that are not moving relative to each another. The maximum static friction is generally proportional to the normal force between the objects, where the proportionality constant is the coefficient of static friction (μ_s). Often, μ_s is greater than μ_k for a given surface. The static friction can vary from zero to the maximum value; its magnitude depends on the component of the applied forces parallel to the surface. The direction of the static frictional force is opposite that of the parallel component of the net applied force. Mathematically, $$f_s \leq \mu_s n.$$
Forces in Circular Motion	For an object in uniform circular motion, the acceleration is directed toward the center of the circle. The motion is determined by Newton's second law, $$\Sigma \vec{F} = m\vec{a}.$$

	The object's acceleration is $$a_{\text{rad}} = \frac{v^2}{R} = \frac{4\pi^2 R}{T^2}.$$

Key Concept 1: The sum of all the external forces on an object in equilibrium is zero. The tension has the same value at either end of a rope or string of negligible mass.

Key Concept 2: If there's more than one object in a problem that involves Newton's laws, the safest approach is to treat each object separately.

Key Concept 3: In two-dimensional problems that involve forces, always write *two* force equations for each object: one for the x-components of the forces and one for the y-components of the forces.

Key Concept 4: In two-dimensional equilibrium problems, choose the coordinate axes so that as many forces as possible lie along either the x-axis or the y-axis.

Key Concept 5: If there's more than one object in a problem involving forces, you are free to choose different x- and y-axes for each object to make it easier to find force components.

Key Concept 6: For problems in which an object is accelerating, it's usually best to choose one positive axis to be in the direction of the acceleration.

Key Concept 7: When friction is present, the friction force on an object is always in the direction that opposes sliding.

Key Concept 8: If an object suspended from a cable (or rope or string) is accelerating vertically, the tension in the cable is *not* equal to the weight of the object.

Key Concept 9: When you are riding in an accelerating vehicle such as an elevator, your apparent weight (the normal force that the vehicle exerts on you) is in general *not* equal to your actual weight.

Key Concept 10: In problems with an object on an incline, it's usually best to take the positive x-direction for that object to be down the incline. The force of gravity will then have both an x-component and a y-component.

Key Concept 11: When two objects are touching each other, the free-body diagram for each object must include the force exerted on it by the other object.

Key Concept 12: If two objects are connected by a string under tension, both objects have the same magnitude of acceleration but may accelerate in different directions. Choose the positive x-direction for each object to be in the direction of its acceleration.

Key Concept 13: For any object, the *maximum* magnitude of the *static* friction force and the magnitude of the *kinetic* friction force are proportional to the magnitude of the normal force on that object.

Key Concept 14: The magnitude f_s of the static friction force on an object at rest does *not* have to equal the maximum magnitude $\mu_s n$. The actual value of f_s depends on the other forces acting on the object; you can find this value using Newton's first law.

Key Concept 15: In problems that involve kinetic friction, you'll always need at least three equations: two from Newton's first or second law in component form, and Eq. (5.3) for kinetic friction, $f_k = \mu_k n$.

Key Concept 16: When kinetic friction of magnitude $f_k = \mu_k n$ is present for an object on an incline, the magnitude n of the normal force on the object is *not* equal to the object's weight.

Key Concept 17: The magnitude $f_k = \mu_k n$ of the kinetic friction force is the same whether or not the object is accelerating.

Key Concept 18: A falling object reaches its terminal speed when the upward force of fluid resistance equals the downward force of gravity. Depending on the object's speed, use either Eq. (5.8) or Eq. (5.12) to find the terminal speed.

Key Concept 19: In problems that involve forces on an object in uniform circular motion, take the positive x-direction to be toward the center of the circle. The net force component in that direction is equal to the object's mass times its radial acceleration.

Key Concept 20: In uniform circular motion, any kind of force (or component of a force) can produce the radial acceleration.

Key Concept 21: For a vehicle following a curved path on a level road, the radial acceleration is produced by the static friction force exerted by the road on the vehicle.

Key Concept 22: The normal force exerted by the road on a vehicle can provide the radial acceleration needed for the vehicle to follow a curved path, provided the road is banked at the correct angle for the vehicle's speed.

Key Concept 23: Even when an object moves with varying speed along a circular path, at any point along the path the net force component toward the center of the circle equals the object's mass times its radial acceleration.

Conceptual Questions

1: Finding errors in a free-body diagram

Two weights are suspended from the ceiling and each other by ropes as shown in Figure 1a. A free-body diagram is shown in Figure 1b for the upper block A. Find the error in the free-body diagram and draw the correct diagram.

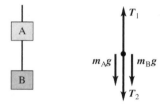

Figure 1 Question 1.

IDENTIFY, SET UP, AND EXECUTE Three forces act on block A: two tension forces due to the ropes and the gravitational force on block A. The gravitational force on block B has been incorrectly included in the diagram. Block B is not in direct contact with block A; only the rope is in contact with block A. The corrected free-body diagram is shown in Figure 2.

Figure 2 Question 1 corrected free-body diagram.

EVALUATE When drawing free-body diagrams, one must include only those forces acting *on* the object. Identifying the forces acting on an object is necessary to apply Newton's laws correctly.

2: Investigation of the normal force

For which of the following figures is the normal force not equal to the object's weight?

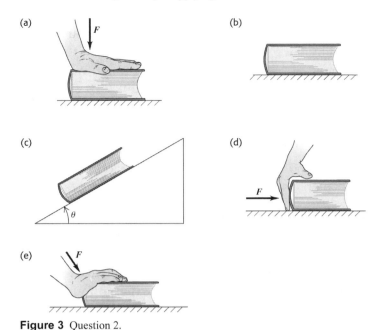

Figure 3 Question 2.

IDENTIFY, SET UP, AND EXECUTE The normal force is not equal to the object's weight in figures (a), (c), and (e). In (a), the normal force is equal to the book's weight plus the force pushing down on the book. In (c), the book's weight is directed downward and the normal force is directed upward and to the left, perpendicular to the ramp's surface. Here, the normal force is equal to the weight multiplied by the cosine of the ramp angle. In (d), the normal force is equal to the book's weight, but the applied force is along the surface; thus, it does not affect the vertical forces. In (e), a component of the applied force is parallel to the normal force, thus increasing the normal force by the amount of that component. The two forces are directed downward.

EVALUATE Often, the normal force is not equal to an object's weight. A common mistake initially encountered in force problems is assuming that the normal force is always equal to some object's weight. You must analyze all problems carefully to determine the proper normal force.

3: Acceleration and tension in blocks connected by a rope

Consider the situation shown in Figure 4. Cart A is placed on a table and is connected to block B by a rope that passes over a frictionless pulley.

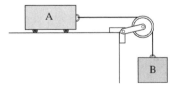

Figure 4 Question 3.

(a) How does the acceleration of cart A compare with that of block B?

IDENTIFY, SET UP, AND EXECUTE Both objects must accelerate at the same rate, since they are connected by the rope (as long as the rope doesn't stretch). To get a better intuitive grasp of this statement, note that cart A will move 10 cm when block B moves 10 cm. If block B moves the 10 cm in 1 second, then cart A moves 10 cm in 1 second; their velocities are the same. If block B's velocity changes by 2 m/s in 1 second, then block A's velocity must change by 2 m/s in 1 second; their accelerations are the same. We say that the rope constrains both objects to accelerate at the same rate.

(b) How does the tension force acting on cart A compare with the weight of block B as the system accelerates?

IDENTIFY, SET UP, AND EXECUTE The tension force in the string is constant along the string, so the tension force is the same on block B as it is on cart A. Therefore, we compare the tension at block B with block B's weight. Newton's second law tells us that the net external force on an object is equal to its mass times its acceleration. Two forces act on block B: B's weight and the tension force. The net external force on block B is its weight minus the tension force. This net external force must be equivalent to the acceleration multiplied by block B's mass; therefore, the tension must be less than the weight. The tension force acting on cart A is less than the weight of block B.

KEY CONCEPT 8 **EVALUATE** Solving this problem gives us two important results that we'll apply repeatedly to later problems: First, objects connected by a rope are constrained to have the same magnitude of acceleration; second, the tension force in a rope connected to an object is not always equal to the object's weight if the object is accelerating.

4: What can a hanging ball indicate

A ball hangs on a string attached to the top of a box, as shown in Figure 5. The box is placed on a horizontal truck bed and the truck moves over a flat roadway. You observe the ball and find that it remains in the position shown in the figure for a long time. By looking only inside the box, what can be determined about the truck's motion?

Figure 5 Question 4.

IDENTIFY, SET UP, AND EXECUTE We see that the ball has swung to the left, so we may suspect that the truck is moving. Let's look at the forces acting on the ball: to investigate the motion. Two forces act on the ball: gravity and the tension force due to the string. Figure 6 shows the free-body diagram.

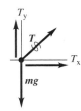

Figure 6 Question 4 free-body diagram.

We see that the tension force has components in both the vertical and horizontal direction. The vertical component of the tension force must be equivalent to the force of gravity, since the box moves horizontally and there is no vertical acceleration. There is only one horizontal force—the horizontal component of the tension force—so there is horizontal acceleration to the right.

We can conclude that the truck is accelerating to the right in the horizontal direction and not at all in the vertical direction. We cannot determine the velocity of the truck; the horizontal velocity components could be zero or nonzero. For example, the truck could be moving to the left with decreasing velocity or it could be accelerating to the right from rest. Both of these motions would result in the ball's being swung to the left.

EVALUATE This problem shows that the absence of a net external force results in zero acceleration of an object and the presence of a net external force results in a nonzero acceleration of an object. We cannot determine the velocity of an object by knowing only its acceleration; we need additional information.

Practice Problem: How would the ball appear if the box moved with constant velocity?
Answer: The ball would hang vertically, since no net external force would act on it.

5: Motion of a box on a rough surface

A constant horizontal force is applied to a box on a rough floor. With a 15 N applied force, the box begins to slide. What is the motion of the box after it begins to slide, assuming that the applied force remains constant?

IDENTIFY, SET UP, AND EXECUTE Before the box slides, there is static friction. Once it begins to slide, the static friction becomes kinetic friction. Kinetic friction is smaller in magnitude than static friction; therefore, the applied force must be larger than the kinetic friction force, and the box accelerates.

EVALUATE This problem helps illustrate the fact that kinetic friction is generally less than static friction. The reason is that the coefficient of kinetic friction is less than the coefficient of static friction.

6: Frictional forces

A box is placed on a rough floor. When you push horizontally against the box with a 20 N force, the box just begins to slide. What is the magnitude of the frictional force when you push against the box with a 10 N force? With a 15 N force?

IDENTIFY, SET UP, AND EXECUTE Since the box just begins to slide when the 20 N force is applied, the maximum static friction force is 20 N. When you push against the box with a 10 N force, you are pushing with less than the maximum static friction force. By Newton's third law, the box must push back with the same force; therefore, the static friction force must have a magnitude of 10 N. For the same reason, when you push with a 15 N force, the static friction force has a magnitude of 15 N.

KEY CONCEPT 14 EVALUATE Static friction varies from zero to its maximum value. The static friction force equals the net external force acting against the friction force. Be careful not to assume that static friction is always at its maximum.

7: A vertical frictional force

A block is placed against the vertical front of an accelerating cart as shown in Figure 7. What condition must hold in order to keep the block from falling?

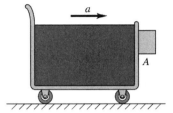

Figure 7 Question 7.

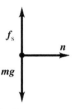

Figure 8 Question 7 free-body diagram.

IDENTIFY, SET UP, AND EXECUTE The free-body diagram shown in Figure 8 indicates that three forces act on the box: gravity (mg, downward), the normal force due to the cart (n, to the right), and the friction force at the block–cart surface (f_s). To keep the block from falling, the friction force must be equal and opposite to the gravitational force. The condition for equilibrium in the vertical direction gives

$$\Sigma F_y = f_s - mg = 0, \qquad f_s = mg.$$

The friction force must be due to static friction, in order to prevent the block from moving. The static friction force is given by

$$f_s \leq \mu_s n.$$

The gravitational force can then be related to the normal force and the coefficient of static friction:

$$mg \leq \mu_s n.$$

The block accelerates to the right with acceleration a_x, so we can use Newton's second law to find an expression for the normal force:

$$\Sigma F_x = n = ma_x.$$

Replacing the normal force with mass and acceleration gives

$$mg \leq \mu_s\, ma_x, \qquad g \leq \mu_s a_x.$$

$$\mu_s \geq \frac{g}{a_x}.$$

The last inequality tells us that the coefficient of static friction must be equal to or greater than the gravitational constant divided by the cart's acceleration.

KEY CONCEPT 3 **EVALUATE** Problems that initially appear complicated often have relatively straightforward solutions. Following a consistent problem-solving procedure helps identify the key points that you'll need to solve the problem.

8: Turning while riding in a car

As you make a right turn in your car, what pushes you against the car door?

IDENTIFY, SET UP, AND EXECUTE As your car turns, your body tends to continue moving in a straight line; therefore, you push up against the car door. So you can say that *no force* actually pushes you against the door and your body tries to maintain a constant velocity due to Newton's first law. Once the car door comes in contact with you, it pushes you in the direction of the turn, accelerating you to the right.

KEY CONCEPT 20 **EVALUATE** From within the car, you may wonder what force pushes you to the side. However, since the car is turning to the right, it is accelerating, and the car is not an inertial reference frame. Therefore, we cannot apply Newton's laws inside the car. If we consider how the situation would appear to someone outside of the car (in an inertial reference frame), we can apply Newton's laws: You would appear to continue moving in a straight line while the car moves to the right.

9: Free-body diagram for a car on a hill

Draw a free-body diagram for a car going over the top of a round hill at a constant speed. Is there a nonzero net external force acting on the car if it is moving at constant speed?

SET UP AND SOLVE Two forces act on the car at the top of the hill: the normal force due to the road and gravity. The normal force is directed upward and gravity is directed downward. Figure 9 shows the free-body diagram.

Figure 9 Question 9 free-body diagram.

The force vectors in the free-body diagram are not drawn to have equal length: The gravitational force is larger than the normal force. This is because there is a nonzero net external force acting on the car: The car's velocity is changing direction, so the car has a centripetal acceleration. The centripetal acceleration is downward, so the net external force must be downward.

There is a net external force acting on the car even though the car is moving at constant speed.

EVALUATE Constant speed does not necessarily imply constant velocity. You must carefully interpret problems involving circular motion and constant speed.

Problems

1: Equilibrium in two dimensions

A 322 kg block hangs from two cables as shown in Figure 10. Find the tension in cables A and B.

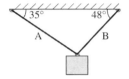

Figure 10 Problem 1.

IDENTIFY The block is in equilibrium, so we can apply Newton's first law to it. The cables have tensions in two dimensions, so we will have to apply the first law to two axes. The target variables are the two tension forces (labeled T_A for cable A and T_B for cable B).

SET UP The free-body diagram of the block is shown in Figure 11. Three forces act on the block: the two tension forces (T_A and T_B) and gravity (mg). The tensions act in two dimensions and gravity acts in the vertical direction.

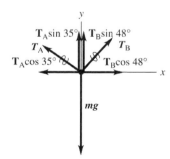

Figure 11 Problem 1 free-body diagram.

We've added an xy-coordinate system to the figure to illustrate in what directions the forces act. We've also resolved the two tension forces into their x- and y-components.

EXECUTE We apply the equilibrium conditions to the block, writing separate equations for the x- and y-components:

$$\Sigma F_x = 0, \quad T_B \cos 48° + (-T_A \cos 35°) = 0,$$
$$\Sigma F_y = 0, \quad T_B \sin 48° + T_A \sin 35° + (-mg) = 0.$$

Note that components directed to the left and downward are negative, consistent with our coordinate system. We can rewrite the first equation as

$$T_B = T_A \frac{\cos 35°}{\cos 48°}.$$

Substituting for T_B in the second equation

$$T_A \frac{\cos 35°}{\cos 48°} \sin 48° + T_A \sin 35° - mg = 0$$

Solving for T_A yields

$$T_A (\cos 35° \tan 48° + \sin 35°) = mg,$$

$$T_A = \frac{mg}{(\cos 35° \tan 48° + \sin 35°)} = \frac{(322 \text{ kg})(9.80 \text{ m/s}^2)}{(\cos 35° \tan 48° + \sin 35°)} = 2130 \text{ N}.$$

Substituting the value for T_A into the first equation gives

$$T_B = T_A \frac{\cos 35°}{\cos 48°} = 2130 \text{ N} \frac{\cos 35°}{\cos 48°} = 2600 \text{ N}.$$

The tension in cable A is 2130 N and the tension in cable B is 2600 N.

KEY CONCEPT 3 **EVALUATE** The sum of the magnitudes of the two tension forces (4730 N) is larger than the weight of the block (3160 N). This is consistent because the tension forces are in two dimensions and their magnitudes are greater than their components. In addition, we check that the two horizontal components of the tension forces are equal in magnitude. Substituting into the terms in the first equation gives a magnitude of 1740 N for each component, each having opposite signs.

Practice Problem: If the block is 1.5 m from the ceiling, then how long are the cables? *Answer:* $A = 2.62$ m, $B = 2.02$ m

Extra Practice: What would the tensions be if the cables had equal length? *Answer:* $T_A = T_B = 2231$ N

2: Accelerated motion of a block on an inclined plane

A block with mass 3.00 kg is placed on a frictionless inclined plane inclined at 35.0° above the horizontal and is connected to a second hanging block with mass 7.50 kg by a cord passing over a small, frictionless pulley (See Figure 12). Find the acceleration (magnitude and direction) of the 3.00 kg block.

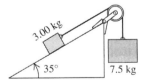

Figure 12 Problem 2.

IDENTIFY The blocks are accelerating, so we apply Newton's second law to both blocks. The target variable is the acceleration of the blocks (a).

SET UP Both blocks accelerate, so we'll apply Newton's second law to each block to find two equations and solve these equations simultaneously to determine the acceleration. The free-body diagrams of both blocks are shown in Figure 13.

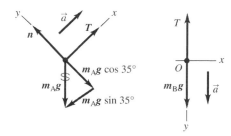

Figure 13 Problem 2 free-body diagrams.

The forces are identified by their magnitudes. Acting on the left-hand block (block *A*) are gravity ($m_A g$), the normal force (*n*), and the tension force (*T*). The right-hand block (block *B*) is acted upon only by the tension force (*T*) and gravity ($m_B g$). The tension forces must be equal in magnitude and the accelerations must be equal in magnitude, since the cord connects the two blocks. (See Conceptual Question 3.) As block *B* accelerates downward, block *A* accelerates up the ramp. We've added an *xy*- coordinate system separately to each free-body diagram, with the positive axes aligned with the direction of acceleration. Using two different coordinate systems is preferred in these situations. The coordinate system for the right-hand block is rotated to coincide with the inclined plane. A rotated axis simplifies the analysis for ramp problems. This rotated axis requires resolving the gravity force into two components, one parallel, and one perpendicular, to the incline.

EXECUTE We apply Newton's second law to each block. Block *A* (with mass m_A) accelerates in the *x*-direction (along the ramp), so

$$\Sigma F_x = T + (-m_A g \sin 35°) = m_A a.$$

Block *B* (with mass m_B) accelerates at the same rate in the *y*-direction; hence,

$$\Sigma F_y = m_B g + (-T) = m_B a.$$

Both equations include the tension force, so we solve for the tension force in the second equation and substitute into the first. Our second equation becomes

$$T = m_B g - m_B a.$$

Replacing the tension force in the first equation yields

$$(m_B g - m_B a) + (-m_A g \sin 35°) = m_A a,$$

Solving for the acceleration gives

$$m_B g + (-m_A g \sin 35°) = (m_A + m_B) a,$$

$$a = \frac{g(m_B - m_A \sin 35°)}{(m_A + m_B)} = \frac{(9.80 \text{ m/s}^2)((7.50 \text{ kg}) - (3.00 \text{ kg}) \sin 35°)}{((7.50 \text{ kg}) + (3.00 \text{ kg}))} = 5.39 \text{ m/s}^2.$$

The 3.00 kg block accelerates up the ramp at 5.39 m/s². The positive value of acceleration confirms the block's acceleration up the ramp.

KEY CONCEPTS **EVALUATE** The value of acceleration is less than *g*, consistent with expectations. If the cord
2, 5, 6 were cut, block *B* would accelerate at *g*. When block *B* is connected to block *A* through the cord, block *B* accelerates with an acceleration less than *g*. We say that block *B* has *additional inertia* when connected to block *A*.

What would have happened if we chose the direction of acceleration incorrectly? We would have found a negative acceleration, indicating that the acceleration was down the incline. In this problem, the forces do not depend on the direction of motion, and a negative acceleration would not indicate an error. It does, however, serve as a checkpoint for our calculation: A negative result with our choice of axes would cause suspicion because the right mass is larger and we expect it to accelerate downward.

Practice Problem: What mass must block A have for the system to remain at rest? Will that mass simply be 7.50 kg? *Answer:* $m_1 = 13.1$ kg, no.

Extra Practice: What is the tension in the cord? *Answer:* 33.1 N

3: Frictional force on an accelerating block

Two blocks are connected to each other by a light cord passing over a small, frictionless pulley as shown in Figure 14. Block A has mass 5.00 kg and block B has mass 4.00 kg. If block B descends at a constant acceleration of 2.00 m/s² when set in motion, find the coefficient of kinetic friction between block A and the table.

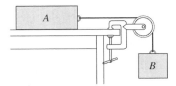

Figure 14 Problem 3.

IDENTIFY The target variable is the coefficient of kinetic friction. The blocks are accelerating, so we apply Newton's second law to both blocks. We'll find the friction force and determine the coefficient from that.

SET UP Both blocks accelerate, so we'll apply Newton's second law to each block to find two equations and solve those equations simultaneously. The free-body diagrams of the two blocks are shown in Figure 15.

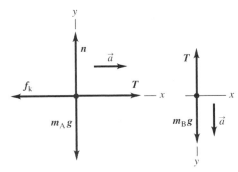

Figure 15 Problem 3 free-body diagram.

The forces are identified by their magnitudes. For block A, there is kinetic friction (f_k), gravity ($m_A g$), the normal force (n), and the tension force (T). For block B there is the tension force (T) and gravity ($m_B g$). The tension forces are equal in magnitude and the magnitudes of the acceleration are equal, as we've seen. As block B accelerates downward, block

A accelerates to the right. To ensure that the acceleration of each block is in the positive direction, we added an *xy*-coordinate system separately to each free-body diagram, with the positive axes aligned with the direction of acceleration. All forces act along the coordinate axes, so we won't need to break the forces into components.

EXECUTE We now apply Newton's second law to each block to find the friction force. Block *A* (with mass m_A) accelerates in the *x*-direction, so

$$\Sigma F_x = T + (-f_k) = m_A a.$$

Block *B* (with mass m_B) accelerates at the same rate in the *y*-direction; thus,

$$\Sigma F_y = m_B g + (-T) = m_B a.$$

Both equations include the tension force, so we solve for the tension force in the second equation and substitute into the first. Our second equation is then

$$T = m_B g - m_B a = m_B (g - a).$$

Replacing the tension force in the first equation gives

$$m_B (g - a) + (-f_k) = m_A a.$$

Solving for the friction force yields

$$f_k = m_B (g - a) - m_1 a = (4.00 \text{ kg})(9.80 \text{ m/s}^2 - 2.00 \text{ m/s}^2) - (5.00 \text{ kg})(2.00 \text{ m/s}^2) = 21.2 \text{ N}.$$

The friction force is related to the coefficient of kinetic friction through the normal force. We find the normal force by examining the vertical components of the forces acting on block *A*. There is no acceleration in the vertical direction for block *A*, so we can apply the equilibrium condition to block *A*:

$$\Sigma F_y = n - m_A g = 0, \quad n = m_A g.$$

Since there are no other vertical forces acting on block *A*, the normal force equals the weight of block *A*. The kinetic frictional force is given by

$$f_k = \mu_k n,$$

which we can solve for μ_k:

$$\mu_k = \frac{f_k}{n} = \frac{f_k}{m_A g} = \frac{(21.2 \text{ N})}{(5.00 \text{ kg})(9.80 \text{ m/s}^2)} = 0.43.$$

We find the coefficient of kinetic friction between the block and the table to be 0.43.

KEY CONCEPTS 7, 8 **EVALUATE** A coefficient of kinetic friction equal to 0.43 compares reasonably well with values we've seen previously for smooth surfaces. Note that the tension is not equal to the weight of block *B*. That it is a common misconception arising from examining the free-body diagram for block *B* without realizing that the block is accelerating. If we look at the rearranged second equation, the relation between the tension and the weight of block *B* becomes clearer:

$$T = m_B (g - a).$$

The tension force is equal to the weight only when block *B*'s acceleration is zero. Calculating the tension for this problem, we obtain a value of 31.2 N, 20% less than the weight of block *B* (39.2 N).

Practice Problem: What is the tension in the cord? *Answer:* 31.2 N

Extra Practice: What does the frictional force need to be to have zero acceleration? *Answer:* 39.2 N

4: Motion of a crate up a *rough* inclined plane at constant velocity

A student pushes a crate up a rough inclined plane as shown in Figure 16. Find the magnitude of the horizontal force the student must apply for the crate to move up the incline at constant velocity. The crate has a mass of 15.0 kg, the incline is sloped at 30.0°, and the coefficient of kinetic friction between the crate and the incline is 0.600.

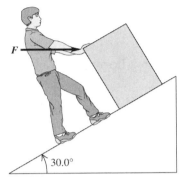

Figure 16 Problem 4.

IDENTIFY There is no acceleration, so we apply the equilibrium condition to the crate to find the applied force.

SET UP The free-body diagram of the crate is shown in Figure 17. The forces are identified by their magnitudes: kinetic frictional force (f_k), gravity (mg), the normal force (n), and the applied force (F). Kinetic friction opposes the motion up the incline and thus is directed down the incline. The rotated xy-coordinate system is indicated in the diagram. This rotated axis requires resolving the gravity and applied forces into components parallel and perpendicular to the incline, as shown in the diagram.

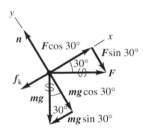

Figure 17 Problem 4 free-body diagram.

EXECUTE We apply the equilibrium condition to the crate. In the x-direction, along the incline, we have

$$\Sigma F_x = F\cos 30° + (-f_k) + (-mg\sin 30°) = 0.$$

We must apply the equilibrium condition in the y-direction to find the normal force in order to quantify the friction force. Thus,

$$\Sigma F_y = n + (-mg\cos 30°) + (-F\sin 30°) = 0$$
$$n = mg\cos 30° + F\sin 30°.$$

The kinetic friction is then

$$f_k = \mu_k n = \mu_k mg\cos 30° + \mu_k F\sin 30°.$$

We now substitute this result into the first equation:
$$F\cos 30° + (-\mu_k mg\cos 30° - \mu_k F\sin 30°) + (-mg\sin 30°) = 0.$$
Solving for the applied force gives
$$F\cos 30° - \mu_k F\sin 30° = \mu_k mg\cos 30° + mg\sin 30°$$
$$F(\cos 30° - \mu_k \sin 30°) = mg(\mu_k \cos 30° + \sin 30°)$$
$$F = \frac{mg(\mu_k \cos 30° + \sin 30°)}{(\cos 30° - \mu_k \sin 30°)} = \frac{(15.0\text{ kg})(9.80\text{ m/s}^2)((0.600)\cos 30° + \sin 30°)}{(\cos 30° - (0.600)\sin 30°)} = 265\text{ N}.$$
The student must push with a horizontal force of 265 N to move the crate up the incline at constant velocity.

KEY CONCEPTS 15, 16, 17 **EVALUATE** A force of 265 N is roughly equivalent to the weight of a 27 kg object. Would it be easier to push the crate up the incline by pushing parallel to the incline? Yes, it would be easier to push along the ramp. In this problem, the component of the applied force directed into the incline ($F\sin 30°$) does nothing to move the crate up the ramp. In fact, this component increases the normal force and therefore the friction force.

Practice Problem: If the student pushed along the incline, the problem would be simplified. What force would be necessary along the incline to maintain the crate at constant velocity? *Answer:* 150 N, 56% of the required horizontal force.

Extra Practice: If the student lets go, what is the acceleration of the crate down the ramp? *Answer:* 0 m/s^2

5: Two blocks suspended by a pulley

Two blocks are connected by a rope that passes over a small, frictionless pulley as shown in Figure 18. Find the tension in the rope and the acceleration of the blocks. Block 1 has a mass of 15.0 kg and block 2 has a mass of 8.0 kg.

Figure 18 Problem 5.

IDENTIFY Our target variables are the tension (T) and acceleration (a). we'll apply Newton's second law to the problem to find the tension and acceleration.

SET UP The two blocks are separate objects, so we draw free-body diagrams of each block, shown in Figure 19. Two forces act on each block: gravity (mg) and the tension force (T). There is no friction in the pulley and the string is considered to be massless, so the tension in the rope is the same throughout.

We assume that the rope doesn't stretch, so the magnitudes of the two accelerations are the same, but the directions are opposite. Included in the diagrams are separate

xy-coordinate axes with the positive y-axis in the direction of the acceleration for both blocks (upward for block 2 and downward for block 1). This choice of axes simplifies our analysis. All of the forces act along the y-axes, so we won't need to break the forces into components.

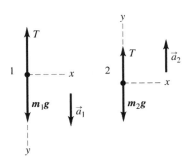

Figure 19 Problem 5 free-body diagrams.

EXECUTE Applying Newton's second law to both blocks gives
$$\Sigma F_y = m_1 g - T = m_1 a \qquad \text{(Block 1)},$$
$$\Sigma F_y = T - m_2 g = m_2 a \qquad \text{(Block 2)}.$$

We add both equations to eliminate T, leaving
$$m_1 g - m_2 g = m_1 a + m_2 a$$
$$a = \frac{(m_1 - m_2)g}{m_1 + m_2} = \frac{((15.0 \text{ kg}) - (8.0 \text{ kg}))(9.8 \text{ m/s}^2)}{(15.0 \text{ kg}) + (8.0 \text{ kg})} = 2.98 \text{ m/s}^2.$$

We find the tension by substituting into the second-law equation for block 2, giving
$$T = m_2(g + a) = (8.0 \text{ kg})(9.8 \text{ m/s}^2 + 2.98 \text{ m/s}^2) = 102.2 \text{ N}.$$

Both blocks accelerate at 2.98 m/s^2, block 1 downward and block 2 accelerates upward. The tension in the rope is 102.2 N.

KEY CONCEPTS **EVALUATE** We check that we get the same value for the tension by using the second-law
8, 12 equation for block 1. We find that the equation gives a tension of 102 N, so the result is the same. We also see that the acceleration is less than 9.8 m/s^2, as is expected, since the net external force on either block is less than its weight.

Practice Problem: How long will it take block 1 to drop 5 m, assuming it starts at rest?
Answer: 1.83 s

Extra Practice: What is the tension if both blocks have a mass of 8 kg? *Answer:* 78.4 N

6: Friction force between two boxes

Two boxes, one on top of the other, are being pulled up a ramp at constant speed by an applied force, as shown in Figure 20. The coefficient of kinetic friction between box B and the ramp is 0.35, and the coefficient of static friction between the two boxes is 0.80. Box A has a mass of 3.00 kg and box B has a mass of 8.00 kg. What is the applied force?

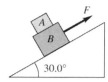

Figure 20 Problem 6.

IDENTIFY Our target variable is the applied force F. We'll use Newton's first law to find the forces acting on the two boxes and then solve for F.

SET UP The two boxes are separate objects, so we draw free-body diagrams of each box, shown in Figure 21. Three forces act on box A: gravity ($m_A g$), the normal force due to box B ($n_{B \text{ on } A}$), and static friction (f_s). Static friction must point up the ramp in order for the net external force on box A to be zero. Six forces act on box B: the applied force (F), gravity ($m_B g$), the normal force due to box A ($n_{A \text{ on } B}$), the normal force due to the ramp (n_{ramp}), kinetic friction (f_k), and static friction (f_s). Static friction and the normal force due to box A are action–reaction pairs; we set their directions opposite those of the forces acting on box A.

Included in the diagrams are a separate xy-coordinate axes with the positive y-axis in the direction of the motion of both boxes (up along the ramp). We'll need to break the forces into components to solve the problem.

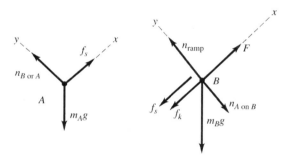

Figure 21 Problem 6 free-body diagrams.

EXECUTE Applying Newton's first law along both axes to both boxes gives

$\Sigma F_x = f_s - m_A g \sin 30.0° = 0$ \qquad (box A)

$\Sigma F_y = n_{B \text{ on } A} - m_A g \cos 30.0° = 0$ \qquad (box A)

$\Sigma F_x = F - f_k - f_s - m_B g \sin 30.0° = 0$ \qquad (box B)

$\Sigma F_y = n_{\text{ramp}} - n_{A \text{ on } B} - m_B g \cos 30.0° = 0$ \qquad (box B)

We have four equations and five unknowns. To solve these equations we need the magnitude of the kinetic friction, which is

$$f_k = \mu_k n_{\text{ramp}}.$$

We begin by solving the first two equations for f_s and $n_{B \text{ on } A}$,

$$f_s = m_A g \sin 30.0° = (3.0 \text{ kg})(9.8 \text{ m/s}^2)\sin 30.0° = 14.7 \text{ N}$$

$$n_{B \text{ on } A} = m_A g \cos 30.0° = (3.0 \text{ kg})(9.8 \text{ m/s}^2)\cos 30.0° = 25.5 \text{ N}.$$

Next, we use the equation for the net external force along the y-axis for box B to solve for the normal force due to the ramp, giving

$$n_{\text{ramp}} = n_{A \text{ on } B} + m_B g \cos 30.0° = (25.5 \text{ N}) + (8.0 \text{ kg})(9.8 \text{ m/s}^2)\cos 30.0° = 93.4 \text{ N}.$$

We can finally find the applied force by using the equation for the net external force along the x-axis for box B, along with the kinetic friction. This gives

$$F = f_k + f_s + m_B g \sin 30.0°$$
$$= \mu_k n_{\text{ramp}} + f_s + m_B g \sin 30.0°$$
$$= (0.35)(93.4 \text{ N}) + (14.7 \text{ N}) + (8.0 \text{ kg})(9.8 \text{ m/s}^2)\sin 30.0° = 86.6 \text{ N}.$$

The applied force is 86.6 N.

KEY CONCEPT 11 EVALUATE Is the coefficient of static friction large enough to keep box A from sliding off box B? We can find out by computing the maximum static friction

$$f_s^{\max} = \mu_s n_{B \text{ on } A} = (0.80)(25.5 \text{ N}) = 20.4 \text{ N}.$$

We see that the applied static friction force (14.7 N) is less than the maximum static friction force (20.4 N), so the box remains in place.

We saw how Newton's third law force pairs can be useful in identifying forces and their directions. In this problem, we used those pairs to find the directions of forces on box B. There are also cases in which the force pairs help us identify missing force pairs. For example, the normal force acting on box B due to box A is often omitted in solutions. If we omitted $n_{A \text{ on } B}$, but had labeled the force pairs carefully, we would've realized that there was a missing normal force.

Practice Problem: What force is needed to accelerate the boxes up the ramp with an acceleration of 0.5 m/s²? *Answer:* 92.1 N

Extra Practice: What acceleration causes box A to slip? *Answer:* 1.9 m/s²

7: Acceleration in a two-pulley system

A mass is attached to a rope that is connected to the ceiling and passes through two light, frictionless pulleys. One pulley is attached to the ceiling and a second mass is attached to the other pulley, as shown in Figure 22. Mass 1 is 20.0 kg and mass 2 is 5.0 kg. Find the acceleration of each mass.

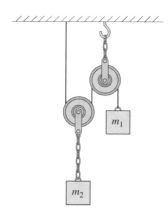

Figure 22 Problem 7.

IDENTIFY Our target variables are the two accelerations, a_1 and a_2, of mass 1 and mass 2, respectively. We'll apply Newton's second law to find these accelerations.

SET UP The free-body diagrams for each mass are shown in Figure 23. Two forces act on mass 1: gravity ($m_1 g$) and the tension in the rope (T). Three forces act on mass 2: gravity ($m_1 g$) and the tension in the rope on both sides of the pulley (two factors of T).

Included in the diagrams are xy-coordinate axes with the positive y-axis in the direction of motion for each mass (upward for mass 2 and downward for mass 1). Also shown are the accelerations a_1 and a_2 of the two masses.

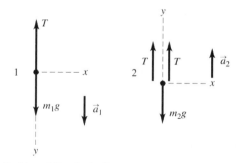

Figure 23 Problem 7 free-body diagrams.

EXECUTE Applying Newton's second law to each mass gives

$$\Sigma F_y = m_1 g - T = m_1 a_1 \quad \text{(mass 1)}$$
$$\Sigma F_y = T + T - m_2 g = m_2 a_2 \quad \text{(mass 2)}$$

We have three unknowns in these two equations; thus, we'll need more information to solve the problem. Let's examine the accelerations. As mass 1 moves down a distance L, mass 2 moves up a distance $L/2$. The change in position of mass 1 is twice the change in position of mass 2, so the velocity and acceleration of mass 1 are twice the velocity and acceleration of mass 2; therefore,

$$a_1 = 2a_2.$$

We now have enough information to solve the system of equations. We begin by replacing a_1 by a_2:

Study Guide for Applying Newton's Laws

$$m_1 g - T = m_1 2a_2$$
$$2T - m_2 g = m_2 a_2.$$

Doubling the first equation and adding the two equations together gives

$$2m_1 g - m_2 g = 4m_1 a_2 + m_2 a_2$$

$$a_2 = \frac{(2m_1 - m_2)g}{4m_1 + m_2} = \frac{(2(20.0 \text{ kg}) - (5.0 \text{ kg}))(9.8 \text{ m/s}^2)}{4(20.0 \text{ kg}) + (5.0 \text{ kg})} = 4.04 \text{ m/s}^2,$$

from which we obtain

$$a_1 = 2a_2 = 2(4.04 \text{ m/s}^2) = 8.08 \text{ m/s}^2.$$

Mass 1 accelerates downward at 8.08 m/s^2 and mass 2 accelerates upward at 4.04 m/s^2.

KEY CONCEPTS **EVALUATE** This problem illustrates the fact that not all accelerations are equal when objects
2, 6, 8 are connected by ropes. We must always evaluate the situation carefully to determine the relation between the accelerations.

Practice Problem: What does mass 2 need to be to make the acceleration zero? *Answer:* 40 kg

Extra Practice: What acceleration do you get if you reverse the masses? *Answer:* -2.45 m/s^2 and -4.9 m/s^2

8: Coefficient of friction in a banked curve

A circular section of road with a radius of 150 m is banked at an angle of 12°. What should be the minimum coefficient of friction between the tires and the road if the roadway is designed for a speed of 25 m/s ?

IDENTIFY Our target variable is the coefficient of static friction, μ_s. We will use Newton's second law to find μ_s by finding the friction force.

SET UP Figure 24 is a free-body diagram of the car tire on the road, showing the three forces acting on the tire: the normal force due to the road (n), static friction with the road (f), and gravity (mg). For the tire not to slip, the vertical forces must be in equilibrium and there must be a net external horizontal force toward the center.

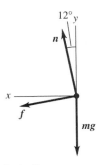

Figure 24 Problem 8 free-body diagram.

We've added an *xy*-coordinate system to the figure, since the forces act in two dimensions. Note that we aligned the axes horizontally and vertically to coincide with the directions of the net external forces.

EXECUTE We apply the force equations to the tire, writing separate equations for the x- and y-components. In the vertical direction, there is no net force:

$$\Sigma F_y = 0, \quad n\cos 12° - f\sin 12° - mg = 0.$$

In the horizontal direction, we use Newton's second law with centripetal acceleration:

$$\Sigma F_x = ma_{\text{rad}}, \quad n\sin 12° + f\cos 12° = m\frac{v^2}{r}.$$

Note that the downward components are negative, consistent with our coordinate system. The static friction force can be replaced with $\mu_s n$ in our two equations:

$$n\cos 12° - \mu_s n\sin 12° - mg = 0,$$

$$n\sin 12° + \mu_s n\cos 12° = m\frac{v^2}{r}.$$

We now rewrite the first equation in terms of n and substitute into the second equation:

$$n = \frac{mg}{\cos 12° - \mu_s \sin 12°},$$

$$\frac{mg}{\cos 12° - \mu_s \sin 12°}\sin 12° + \mu_s \frac{mg}{\cos 12° - \mu_s \sin 12°}\cos 12° = m\frac{v^2}{r}.$$

The mass cancels and we can solve for μ_s:

$$\mu_s = \frac{v^2 \cos 12° - rg\sin 12°}{v^2 \sin 12° + rg\cos 12°} = \frac{(25 \text{ m/s})^2 \cos 12° - (150 \text{ m})(9.8 \text{ m/s}^2)\sin 12°}{(25 \text{ m/s})^2 \sin 12° + (150 \text{ m})(9.8 \text{ m/s}^2)\cos 12°} = 0.20.$$

The minimum coefficient of static friction between the tire and road is 0.20.

KEY CONCEPTS 19, 22 **EVALUATE** The technique for solving this problem is similar to those set forth earlier in this chapter. The differences here were the inclusion of centripetal acceleration and the choice of axes that corresponded to our knowledge of the net external forces.

Practice Problem: What speed would require no frictional force? *Answer:* $v = 18$ m/s.

Extra Practice: If there is no friction, then what banking is required to maintain the speed of 25 m/s? *Answer:* 23°

9: Normal force on a roller coaster

A roller coaster has a vertical loop of radius 45 m. If the roller coaster operates at a constant speed of 35 m/s while in the loop, what normal force does the seat provide for a 75 kg passenger at the top of the loop? The roller coaster is upside down at the top of the loop.

IDENTIFY Our target variable is the normal force n. We'll apply Newton's second law to find the force.

SET UP Figure 25 shows a free-body diagram of the passenger on the roller coaster. The figure shows the two forces acting on the passenger: the normal force due to the seat (n) and gravity (mg). At the top of the loop, the net external force is downward and the person is accelerating toward the center. We've added an xy-coordinate system to the figure, with positive forces directed downward.

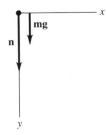

Figure 25 Problem 9 free-body diagram.

EXECUTE The net external force on the passenger is directed downward. Newton's second law with centripetal acceleration gives

$$\Sigma F_y = ma_{rad}, \qquad n + mg = m\frac{v^2}{r}.$$

Solving for the normal force, we obtain

$$n = m\frac{v^2}{r} - mg = (75 \text{ kg})\frac{(35 \text{ m/s})^2}{(45 \text{ m})} - (75 \text{ kg})(9.8 \text{ m/s}^2) = 1300 \text{ N}.$$

The seat exerts a force of 1300 N on the passenger.

KEY CONCEPT 19 EVALUATE We see that the seat provides a force nearly twice the passenger's weight. A seat belt would not be needed to prevent a fall from the roller coaster at the top of the loop. Amusement park rides get their reputation for excitement from their ability to rapidly change the magnitudes and directions of forces applied to passengers. The normal force of the seat on the passenger is even larger at the bottom of the loop.

Practice Problem: What normal force does the seat provide at the bottom of the loop?
Answer: $N = 2800$ N.

Extra Practice: What is the normal force at the bottom of the loop when the speed is doubled?
Answer: 8400 N

10: Investigating a tetherball

A tetherball is attached to a vertical pole with a 2.0 m length of rope, as shown in Figure 26. If the rope makes an angle of 25.0° with the vertical pole, find the time required for one revolution of the tetherball.

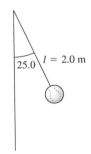

Figure 26 Problem 10.

IDENTIFY Our target variable is the period of 1 revolution, T. The period will be found from the ball's velocity and circumference. The velocity will be found by applying Newton's second law to the ball as it undergoes centripetal acceleration.

SET UP The free-body diagram of the tetherball is shown in Figure 27. Two forces act on the ball: gravity (mg) and the tension in the rope (T). An xy-coordinate axis is included in the diagram.

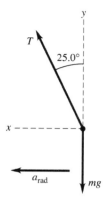

Figure 27 Problem 10 free-body diagram.

EXECUTE The tetherball undergoes centripetal acceleration in the horizontal direction and no net external force in the vertical direction. Applying Newton's first law in the vertical direction gives

$$\Sigma F_y = T\cos 25.0° - mg = 0$$

$$T = \frac{mg}{\cos 25.0°}.$$

Applying Newton's second law in the horizontal direction gives

$$\Sigma F_x = T\sin 25.0° = ma_{\text{rad}} = m\frac{v^2}{R}$$

$$v^2 = T\frac{R}{m}\sin 25.0° = \frac{mg}{\cos 25.0°}\frac{R}{m}\sin 25.0° = Rg\tan 25.0°.$$

We now have the velocity in terms of the radius, angle, and g. To find the radius, we use the length of the rope and the sine:

$$R = l\sin 25.0° = (2.0\text{ m})\sin 25.0° = 0.845\text{ m}.$$

Thus,

$$v = \sqrt{Rg\tan 25.0°} = \sqrt{(0.845\text{ m})(9.8\text{ m/s}^2)\tan 25.0°} = 1.97\text{ m/s}.$$

The period is the time required for 1 revolution. The ball travels the circumference of a circle of radius R in one period. The period is then

$$T = \frac{2\pi R}{v} = \frac{2\pi(0.845\text{ m})}{(1.97\text{ m/s})} = 2.69\text{ s}.$$

The ball completes 1 revolution in 2.69 s.

KEY CONCEPTS 13, 19

EVALUATE This problem illustrates the inclusion of centripetal acceleration into force problems. We see that Newton's laws remain unchanged. We simply set the acceleration equal to the radial acceleration in these problems.

Practice Problem: Does the period increase or decrease with larger angles? Find the period when the rope makes an angle of 50.0° with respect to the vertical. *Answer:* 2.28 s; the period decreased slightly.

Extra Practice: How many rpm is the ball moving when the length of the rope is reduced to 1 m? (Assume the rope makes a 25° angle.) *Answer:* 115 rpm

11: A rotating mass

A mass attached to a vertical post by two strings rotates in a circle of constant velocity v. (See Figure 28.) At high velocities, both strings are taut. Below a critical velocity, the lower string slackens. Find the critical velocity. The mass is 1.0 kg and is 1.5 m from the post.

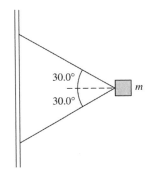

Figure 28 Problem 11.

IDENTIFY Our target variable is the critical velocity, at which the tension in the lower string is zero. We'll find the tension in the lower string and the velocity by applying Newton's second law. Then we'll set the tension in the lower string to zero and solve for the critical velocity.

SET UP The free-body diagram of the mass is shown in Figure 29. Three forces act on the mass: gravity (mg), the tension in the upper rope (T_1), and the tension in the lower rope (T_2). An xy-coordinate axis is included in the diagram.

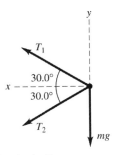

Figure 29 Problem 11 free-body diagram.

EXECUTE The mass undergoes centripetal acceleration in the horizontal direction and no net external force in the vertical direction. Applying Newton's first law in the vertical direction gives

$$\Sigma F_y = T_1 \sin 30° - T_2 \sin 30° - mg = 0.$$

Applying Newton's second law in the horizontal direction gives

$$\Sigma F_x = T_1 \cos 30° + T_2 \cos 30° = ma_{\text{rad}} = m\frac{v^2}{R}.$$

We can solve for T_1 by multiplying the first equation by cos 30°, multiplying the second equation by sin 30°, and adding the two resulting equations. Doing this gives

$$2T_1 \sin 30° \cos 30° = mg \cos 30° + m\frac{v^2}{R} \sin 30°,$$

which reduces to

$$T_1 = \frac{mg}{2\sin 30°} + \frac{mv^2}{2R\cos 30°}.$$

Solving for T_2 produces

$$T_2 = \frac{-mg}{2\sin 30°} + \frac{mv^2}{2R\cos 30°}.$$

At the critical velocity, T_2 is zero. Solving for v, we obtain

$$\frac{mg}{2\sin 30°} = \frac{mv^2}{2R\cos 30°},$$

$$v = \sqrt{\frac{gR}{\tan 30°}} = \sqrt{\frac{(9.8 \text{ m/s}^2)(1.5 \text{ m})}{\tan 30°}} = 5.04 \text{ m/s}.$$

The critical velocity is 5.04 m/s.

KEY CONCEPTS 3, 19 **EVALUATE** For velocities less than the critical velocity, the bottom string is slack. Strings always have positive tensions. When they are slack, they provide no support.

Practice Problem: What is the tension in the upper string at the critical velocity? *Answer:* 20 N.

Extra Practice: What is the length of the ropes? *Answer:* 1.73 m

Try It Yourself!

1: Constant velocity on an incline

Two weights are attached by a light cord that passes over a light, frictionless pulley as shown in Figure 30. The left weight moves up a rough ramp. Find the weight w_2 necessary to keep w_1 (15.0 N) moving up the ramp at a constant rate once it is put in motion. The coefficient of kinetic friction is 0.25 and the ramp is inclined at 30.0°.

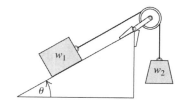

Figure 30 Try it yourself 1.

Solution Checkpoints

IDENTIFY AND SET UP To move at constant velocity, the net external force on each weight must be zero. The rough surface indicates that there is friction between the weight and the ramp. Draw a free-body diagram and apply Newton's first law to solve.

EXECUTE The net external force acting on weight 1 along the ramp is

$$\Sigma F_x = T - f - w_1 \sin\theta = 0.$$

The net external force acting on weight 2 in the vertical direction is

$$\Sigma F_y = T - w_2 = 0.$$

These two equations have three unknowns. You need to find another expression to solve for w_2. The expression will lead to

$$w_2 = \mu_k w_1 \cos\theta + w_1 \sin\theta = 10.7 \text{ N}$$

KEY CONCEPT 5 **EVALUATE** Can you explain why w_2 has less weight than w_1?

Practice Problem: What is the acceleration if $w_2 = w_1$? *Answer:* 1.39 m/s²

Extra Practice: What is w_2 if there is no incline and no acceleration? *Answer:* 3.25 N

2: Box sliding across rough floor

A box is kicked, giving it an initial velocity of 2.0 m/s. It slides across a rough, horizontal floor and comes to rest 1.0 m from its initial position. Find the coefficient of friction.

Solution Checkpoints

IDENTIFY AND SET UP Begin with a sketch and free-body diagram. In the horizontal direction, the only force acting on the box is friction. Use the acceleration of the box and Newton's second law to find the friction.

EXECUTE The net external horizontal force acting on the box is

$$\Sigma F_x = f = ma.$$

To find the acceleration, apply

$$v^2 = v_0^2 + 2a\Delta x,$$

one of the kinematics relations for constant acceleration. You should find that the coefficient of kinetic friction is 0.20.

KEY CONCEPT 15 **EVALUATE** We see how we can combine kinematics with our force problems in this problem. Did we find a reasonable coefficient of kinetic friction?

Practice Problem: How long did it take to stop? *Answer:* 1 s

Extra Practice: What force is required to keep it moving at 2 m/s? *Answer:* 7.8 N

3: Three connected masses

Three blocks are attached to each other with ropes that pass over pulleys as shown in Figure 31. The masses of the ropes and pulleys can be ignored, and there is no friction on the surface over which the blocks slide or in the pulleys. Find the tension in the two ropes and the acceleration of the blocks.

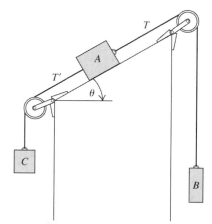

Figure 31 Try it yourself 2.

Solution Checkpoints

IDENTIFY AND SET UP Three objects are in motion, so you must draw three free-body diagrams. How does the acceleration of all three objects compare? You can apply Newton's laws to solve. Assume that the system accelerates clockwise.

EXECUTE The net external forces acting in the direction of the acceleration of the blocks are

$$\Sigma F_A = T - T' - m_A g \sin\theta = m_A a$$
$$\Sigma F_B = m_B g - T = m_B a$$
$$\Sigma F_C = T' - m_C g = m_C a.$$

We can also find the net external force acting on block A normal to the ramp. Using these equations to solve for the tensions gives

$$T = \left(\frac{2m_C m_B + m_B m_A (1+\sin\theta)}{m_A + m_B + m_C} \right) g$$

$$T' = \left(\frac{2m_C m_B + m_C m_A (1-\sin\theta)}{m_A + m_B + m_C} \right) g.$$

Solving for the acceleration yields

$$a = \left(\frac{m_B - m_A \sin\theta - m_C}{m_A + m_B + m_C} \right) g.$$

KEY CONCEPTS **EVALUATE** How do we interpret the results if we find that the acceleration is negative? Do
5, 8, 12 we need to rework the solution? Is the acceleration greater or less than g?

Practice Problem: What is the acceleration when the incline is 30° and block B is twice as massive as both A and C? *Answer:* 1.23 m/s²

Extra Practice: What is the acceleration when the incline is 30° and all blocks have the same mass? *Answer:* −1.63 m/s²

4: Tension along a heavy rope

A heavy rope of mass 10.0 kg and length 5.0 m lies on a frictionless horizontal surface. If a certain horizontal force is applied, the rope accelerates at 1.5 m/s². Find the tension in the rope at any point along its length.

Solution Checkpoints

IDENTIFY AND SET UP To find the tension at any point in the rope, you must break the rope into many pieces and find the tension of any piece. It is easiest to pick a piece of the rope at the end where the force is applied, as shown in Figure 32. By applying Newton's second law to this piece, you can find the tension anywhere in the rope.

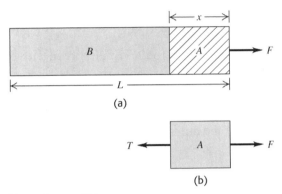

Figure 32 Try it yourself 3.

EXECUTE The net external force required to accelerate the rope is 15.0 N. The mass of the length x of the rope is

$$m_x = \frac{x}{L}m.$$

The net external force on the piece of rope is

$$\Sigma F_x = F - T = m_x a = \frac{x}{L}ma.$$

Solving for T to find the tension as a function of x:

$$T = F\left(1 - \frac{x}{L}\right) = (10.0 \text{ N})\left(1 - \frac{x}{5.0 \text{ m}}\right).$$

KEY CONCEPT 11 EVALUATE This example illustrates how to apply Newton's second law to more complicated problems. Does the tension in the rope increase or decrease as you move away from the end where the force is applied?

Practice Problem: What is the tension at the midpoint of the rope? *Answer:* 7.5 N

Extra Practice: What is the tension at the end of the rope? *Answer:* 0 N

5: Riding a carousel

A 100.0 kg man stands on the outer edge of a carousel of 4.0 m radius. The coefficient of static friction between his shoes and the carousel is 0.30. What is the minimum period of rotation required for the man to remain on the carousel?

Solution Checkpoints

IDENTIFY AND SET UP Three forces act on the man: gravity, the normal force, and friction. The condition required for him to slip is that the centripetal force exceed the static friction. The period is determined from the velocity.

EXECUTE The net external horizontal force acting on the man is

$$\Sigma F_x = f = ma_{\text{rad}} = m\frac{v^2}{R}.$$

This yields the critical velocity when the friction force is equal to the centripetal force. For velocities larger than the critical velocity, the man will slide off. The period is found from the velocity, given that the time required for 1 revolution is the distance traveled divided by the velocity. The minimum period is 7.4 s.

KEY CONCEPT 19 EVALUATE We see that the period and velocity are inversely related: Larger velocities result in shorter periods. How does the friction force vary when the period is greater than 7.4 s?

Practice Problem: What is the minimum period if he doubles his mass? *Answer:* 7.4 s

Extra Practice: What is the minimum period if he stands 2 m closer to the center? *Answer:* 5.2 s

Problem Summary

This chapter has examined a variety of problems with applied forces in diverse applications, but it shares a common problem-solving foundation. For all problems, we

- Identified the general procedure for finding the solution.
- Sketched the situation when no figure was provided.
- Identified the forces acting on the objects of interest.
- Drew free-body diagrams of forces acting *on* the objects.
- Added appropriate coordinate systems to the free-body diagrams.
- Applied the equilibrium condition, Newton's second law, or both to the objects in order to find relations among the forces, masses, and accelerations.
- Solved the equations through algebra, trigonometry, and calculus.
- Reflected on the results, thus checking for inconsistencies.

This problem-solving foundation can be applied to all problems involving forces. Following this procedure enables one to master Newton's laws.

Key Example Variation Problems

Solutions to these problems are in Chapter 5 of the Student's Solutions Manual.

Be sure to review EXAMPLE 5.5 (Section 5.1) before attempting these problems. In all problems, ignore air resistance.

VP5.5.1 In a modified version of the cart and bucket in Fig. 5.5a, the angle of the slope is 36.9° and the bucket weighs 255 N. The cart moves up the incline and the bucket moves downward, both at constant speed. The cable has negligible mass, and there is no friction. (a) What is the weight of the cart? (b) What is the tension in the cable?

VP5.5.2 You increase the angle of the slope in Fig. 5.5a to 25.0° and use a different cart and a different bucket. You observe that the cart and bucket remain at rest when released and that the tension in the cable of negligible mass is 155 N. There is no friction. (a) What is the weight of the cart? (b) What is the *combined* weight of the cart and bucket?

VP5.5.3 You construct a version of the cart and bucket in Fig. 5.5a, but with a slope whose angle can be adjusted. You use a cart of mass 175 kg and a bucket of mass 65.0 kg. The cable has negligible mass, and there is no friction. (a) What must be the angle of the slope so that the cart moves downhill at a constant speed and the bucket moves upward at the same constant speed? (b) With this choice of angle, what will be the tension in the cable?

VP5.5.4 In the situation shown in Fig. 5.5a, let θ be the angle of the slope and suppose there *is* friction between the cart and the track. You find that if the cart and bucket each have the same weight w, they remain at rest when released. In this case, what is the magnitude of the friction force on the cart? Is it less than, greater than, or equal to w?

Be sure to review EXAMPLES 5.13, 5.14, and 5.15 (Section 5.3) before attempting these problems.

VP5.15.1 You pull on a crate using a rope as in Fig. 5.21a, except the rope is at an angle of 20.0° above the horizontal. The weight of the crate is 325 N, and the coefficient of kinetic friction between the crate and the floor is 0.250. (a) What must be the tension in the rope to make the crate move at a constant velocity? (b) What is the normal force that the floor exerts on the crate?

VP5.15.2 You pull on a large box using a rope as in Fig. 5.21a, except the rope is at an angle of 15.0° *below* the horizontal. The weight of the box is 325 N, and the coefficient of kinetic friction between the box and the floor is 0.250. (a) What must be the tension in the rope to make the box move at a constant velocity? (b) What is the normal force that the floor exerts on the box?

VP5.15.3 You are using a lightweight rope to pull a sled along level ground. The sled weighs 475 N, the coefficient of kinetic friction between the sled and the ground is 0.200, the rope is at an angle of 12.0° above the horizontal, and you pull on the rope with a force of 125 N. (a) Find the normal force that the ground exerts on the sled. (b) Find the acceleration of the sled. Is the sled speeding up or slowing down?

VP5.15.4 A large box of mass m sits on a horizontal floor. You attach a lightweight rope to this box, hold the rope at an angle θ above the horizontal, and pull. You find that the minimum tension you can apply to the rope in order to make the box start moving is T_{min}. Find the coefficient of static friction between the floor and the box.

Be sure to review EXAMPLES 5.20, 5.21, and 5.22 (Section 5.4) before attempting these problems.

VP5.22.1 You make a conical pendulum (see Fig. 5.32a) using a string of length 0.800 m and a bob of mass 0.250 kg. When the bob is moving in a circle at a constant speed, the string is at an angle of 20.0° from the vertical. (a) What is the radius of the circle around which the bob moves? (b) How much time does it take the bob to complete one circle? (c) What is the tension in the string?

VP5.22.2 A competition cyclist rides at a constant 12.5 m/s around a curve that is banked at 40.0°. The cyclist and her bicycle have a combined mass of 64.0 kg. (a) What must be the radius of her turn if there is to be no friction force pushing her either up or down the banked curve? (b) What is the magnitude of her acceleration? (c) What is the magnitude of the normal force that the surface of the banked curve exerts on the bicycle?

VP5.22.3 An aerobatic airplane flying at a constant 80.0 m/s makes a horizontal turn of radius 175 m. The pilot has mass 80.0 kg. (a) What is the bank angle of the airplane? (b) What is the pilot's apparent weight during the turn? How many times greater than his actual weight is this?

VP5.22.4 A sports car moves around a banked curve at just the right constant speed v so that no friction is needed to make the turn. During the turn, the driver (mass m) feels as though she weighs x times her actual weight. (a) Find the magnitude of the *net* force on the driver during the turn in terms of m, g, and x. (b) Find the radius of the turn in terms of v, g, and x.

STUDENT'S SOLUTIONS MANUAL FOR APPLYING NEWTON'S LAWS

VP5.5.1. **IDENTIFY:** The cart and bucket move with constant speed, so their acceleration is zero, which means that the forces on each of them must balance.
SET UP: Apply $\Sigma F_x = 0$ and $\Sigma F_y = 0$ to the cart and the bucket. For the cart, take the $+x$-axis parallel to the surface of the incline pointing upward. For the bucket, take the $+y$-axis vertically upward.
EXECUTE: **(a)** Isolate the bucket and apply $\Sigma F_y = 0$. $T - w = 0 \rightarrow T = w = 255$ N.
Now apply $\Sigma F_x = 0$ to the cart. $T - w \sin 36.9° = 0$. 255 N $- w \sin 36.9° = 0$. $w = 425$ N.
(b) From above, $T = 255$ N.
EVALUATE: The bucket can balance a much heavier cart because it only needs to balance the component of the cart's weight that is parallel to the surface of the incline.

VP5.5.2. **IDENTIFY:** The cart and bucket are at rest, which means that the forces on each of them must balance.
SET UP: Apply $\Sigma F_x = 0$ and $\Sigma F_y = 0$ to the cart and the bucket. For the cart, take the $+x$-axis parallel to the surface of the incline pointing upward. For the bucket, take the $+y$-axis vertically upward.
EXECUTE: **(a)** Apply $\Sigma F_x = 0$ to the cart, giving $T - w_C \sin 25.0° = 0$.
155 N $- w_C \sin 25.0°$, so $w_C = 367$ N.
(b) Isolate the bucket and apply $\Sigma F_y = 0$. $T - w_B = 0$, so $w_B = T = 367$ N.
The total weight is 367 N $+ 155$ N $= 522$ N.
EVALUATE: A light bucket can balance a heavy cart because it must balance only the weight component of the cart that is parallel to the surface of the incline.

VP5.5.3. **IDENTIFY:** The cart and bucket move with constant speed, so their acceleration is zero, which means that the forces on each of them must balance.
SET UP: Apply $\Sigma F_x = 0$ and $\Sigma F_y = 0$ to the cart and the bucket. For the cart, take the $+x$-axis parallel to the surface of the incline pointing upward. For the bucket, take the $+y$-axis vertically upward. We want to find the angle θ of the slope. Call T the tension in the cable.

EXECUTE: **(a)** Applying $\Sigma F_y = 0$ to the bucket gives $T = w_B$. Applying $\Sigma F_x = 0$ to the cart gives $T = w_C \sin\theta$. Equating the two expressions for T gives $w_B = w_C \sin\theta$, which tells us $\sin\theta = \dfrac{w_B}{w_C} = \dfrac{m_B g}{m_C g} = \dfrac{65.0 \text{ kg}}{175 \text{ kg}} = 0.371$, so $\theta = 21.8°$.

(b) From part (a), $T = w_B = (65.0 \text{ kg})(9.80 \text{ m/s}^2) = 637 \text{ N}$.

EVALUATE: It must also be true that $T = w_C \sin\theta$. $T = (175 \text{ kg})(9.80 \text{ m/s}^2) \sin 21.8° = 637 \text{ N}$, which agrees with our answer in (b).

VP5.5.4. **IDENTIFY:** The cart and bucket remain at rest, so the forces on each of them must balance.

SET UP: Apply $\Sigma F_x = 0$ and $\Sigma F_y = 0$ to the cart and the bucket. For the cart, take the $+x$-axis parallel to the surface of the incline pointing upward. For the bucket, take the $+y$-axis vertically upward. We want to find the angle θ of the slope. Call T the tension in the cable and f the friction force. If there were no friction, the cart would slide *up* the incline because $w > w\sin\theta$. Since friction opposes motion, it must act *down* the incline.

EXECUTE: Applying $\Sigma F_y = 0$ to the bucket gives $T = w_B = w$. Applying $\Sigma F_x = 0$ to the cart gives $T - w_C \sin\theta - f = 0$, which becomes $T - w_C \sin\theta - f = 0$. Combining the results gives

$f = w(1 - \sin\theta)$. Since $\sin\theta < 1$, we know that $f < w$.

EVALUATE: The weight of the bucket must now balance *two* forces: the weight of the cart acting down the incline and the friction force down the incline.

VP5.15.1. **IDENTIFY:** The crate moves at constant velocity, so the forces on it must balance.
SET UP: Apply $\Sigma F_x = 0$ and $\Sigma F_y = 0$ to the crate. Take the x-axis horizontal and the y-axis vertical. Make a free-body diagram as in Fig. VP5.15.1.

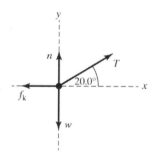

Figure VP5.15.1

EXECUTE: **(a)** Using the notation in Fig. VP5.15.1, $\Sigma F_x = 0$ gives $T \cos 20.0° - f_k = 0$ and $\Sigma F_y = 0$ gives $T \sin 20.0° + n - w = 0$, so $n = w - T \sin 20.0°$. For sliding friction $f_k = \mu_k n$. Combining these results gives $T \cos 20.0° - \mu_k (w - T \sin 20.0°) = 0$. Putting in the numbers:

$T \cos 20.0° - (0.250)(325 \text{ N}) + (0.250) T \sin 20.0° = 0 \rightarrow T = 79.3 \text{ N}$.

(b) $n = w - T \sin 20.0° = 325 \text{ N} - (79.3 \text{ N}) \sin 20.0° = 298 \text{ N}$.

EVALUATE: We find that $n < w$. This is reasonable because the upward component of the tension balances part of the weight of the crate.

VP5.15.2. **IDENTIFY:** The crate moves at constant velocity, so the forces on it must balance.

SET UP: Apply $\Sigma F_x = 0$ and $\Sigma F_y = 0$ to the crate. Take the *x*-axis horizontal and the *y*-axis vertical. Make a free-body diagram as in Fig. 5.15.2.

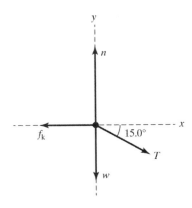

Figure VP5.15.2

EXECUTE: The procedure is exactly the same as for VP5.15.1 *except* that the tension is now directed at 15.0° *below* the horizontal.

(a) $\Sigma F_x = 0$ gives $n - w - T \sin 15.0° = 0$, so $n = w + T \sin 15.0°$.

$\Sigma F_y = 0$ gives $T \cos 15.0° - f_k = T \cos 15.0° - \mu_k n = 0$.

Combining these equations and solving for T gives $T = 90.2$ N.

(b) $n = w + T \sin 15.0° = 325$ N $+ (90.2$ N$) \sin 15.0° = 348$ N.

EVALUATE: The normal force must balance the downward component of the tension in addition to the weight of the crate, so it is greater than the weight.

VP5.15.3. **IDENTIFY:** The sled is accelerated horizontally, so Newton's second law applies to it.

SET UP: $\Sigma F_x = ma_x$ applies to the horizontal motion and $\Sigma F_y = 0$ applies to the vertical motion. Use $\Sigma F_y = 0$ to find the normal force n. The kinetic friction force is $f_k = \mu_k n$. Call P the magnitude of the pull and w the weight of the sled. Take the +*x*-axis to be horizontal in the direction of the horizontal component of the pull.

EXECUTE: (a) $\Sigma F_y = 0$: $T \sin 12.0° + n - w = 0$.

$n = w - T \sin 12.0° = 475$ N $- (125$ N$) \sin 12.0° = 449$ N.

(b) $\Sigma F_x = ma_x$: $P \cos 12.0° - f_k = ma_x = P \cos 12.0° - \mu_k n$

$(125$ N$) \cos 12.0° - (0.200)(449$ N $= [(475$ N$)/(9.80$ m/s$^2)] a_x \to a_x = 0.670$ m/s^2. We have chosen the +*x*-axis in the same direction as the horizontal component of the pull. Since a_x is positive, the acceleration is in the same direction as that pull, so the sled is *speeding up*.

EVALUATE: If the sled were slowing down, that would mean that friction was greater than the horizontal component of the pull. In that case, the sled could never have started moving in the first place, so our answer is reasonable.

VP5.15.4. IDENTIFY: Before it slides, the forces on the box must balance.
SET UP: For the minimum pull, the box is just ready to slide, so static friction is at its maximum, which is $f_s = \mu_s n$. Apply $\Sigma F_x = 0$ and $\Sigma F_y = 0$ to the box just as it is ready to slide.

EXECUTE: $\Sigma F_x = 0$: $T_{min} \cos\theta - f_s = 0 \rightarrow T_{min} \cos\theta - \mu_s n = 0$.

$\Sigma F_y = 0$: $T_{min} \sin\theta + n - mg = 0 \rightarrow n = mg - T_{min} \sin\theta$.

Combine these results and solve for μ_s.

$$T_{min} \cos\theta - \mu_s(mg - T_{min} \sin\theta) = 0 \rightarrow \mu_s = \frac{T_{min} \cos\theta}{mg - T_{min} \sin\theta}.$$

EVALUATE: If the box is *not* just ready to slide, our analysis is not valid. The forces still balance, but $f_s \neq \mu_s n$ in that case.

VP5.22.1. IDENTIFY: The pendulum bob is moving in a horizontal circle at constant speed. Therefore it has horizontal acceleration toward the center of the circle, but it has no vertical acceleration. The vertical forces on it must balance, but we need to use Newton's second law for the horizontal motion.
SET UP: Vertically $\Sigma F_y = 0$ and horizontally $\Sigma F_x = ma_x$, where a_x is the radial acceleration $a_{rad} = v^2/R$. Therefore horizontally we use $\Sigma F = m\frac{v^2}{R}$. The speed is $v = 2\pi R/t$, where t is the time for one cycle (do not use T for the period to avoid confusion with the tension T). Make a free-body diagram like Fig. 5.32b in the text.

EXECUTE: **(a)** $R = L \sin\beta = (0.800 \text{ m}) \sin 20.0° = 0.274$ m.

(b) $\Sigma F = m\frac{v^2}{R}$ gives $T \sin\beta = m\frac{v^2}{R}$. Using $v = 2\pi R/t$, this becomes $T \sin\beta = \frac{m}{R}\left(\frac{2\pi R}{t}\right)^2$.

Solving for t we get $t = \sqrt{\frac{4\pi^2 Rm}{T \sin\beta}}$, so we need to find T.

(c) $\Sigma F_y = 0$ gives $T \cos\beta = W = mg \rightarrow T = (0.250 \text{ kg})(9.80 \text{ m/s}^2)/(\cos 20.0°) = 2.61$ N.

Now return to part (b) to find the time $t = \sqrt{\frac{4\pi^2 Rm}{T \sin\beta}}$. Putting in the numbers gives

Ita $t = 2\pi\sqrt{\frac{(0.274 \text{ m})(0.250 \text{ kg})}{(2.61 \text{ N})(\sin 20.0°)}} = 1.74$ s.

EVALUATE: Even though the bob has constant speed, the horizontal forces do not balance because its *velocity* is changing direction, so it has acceleration.

VP5.22.2. **IDENTIFY:** The cyclist is moving in a horizontal circle at constant speed. Therefore she has horizontal acceleration toward the center of the circle, but no vertical acceleration. The vertical forces on her must balance, but we need to use Newton's second law for her horizontal motion.

SET UP: Vertically $\sum F_y = 0$ and horizontally $\sum F_x = ma_x$, where a_x is the radial acceleration $a_{rad} = v^2/R$. Therefore horizontally we use $\sum F = m\dfrac{v^2}{R}$. Make a free-body diagram as shown in Fig. VP5.22.2.

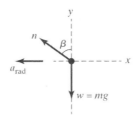

Figure VP5.22.2

EXECUTE: (a) $\sum F_x = ma_x = n \sin \beta = mv^2/R$

$\sum F_y = 0 = n \cos \beta - mg$.

Combining these two equations gives $\tan \beta = v^2/Rg$, which gives

$R = v^2/(g \tan \beta) = (12.5 \text{ m/s})^2/[(9.80 \text{ m/s}^2)(\tan 40.0°)] = 19.0$ m.

(b) $a_{rad} = v^2/R = (12.5 \text{ m/s})^2/(19.0 \text{ m}) = 8.22 \text{ m/s}^2$.

(c) Using $n \cos \beta - mg = 0$, we get $n = mg/\cos \beta = (64.0 \text{ kg})(9.80 \text{ m/s}^2)/(\cos 40.0°) = 819$ N.

EVALUATE: We found that $n > w$. This is reasonable since only the vertical component of n balances her weight, which means that n has to be greater then w.

VP5.22.3. **IDENTIFY:** The plane is moving in a horizontal circle at constant speed. Therefore it has horizontal acceleration toward the center of the circle, but no vertical acceleration. The vertical forces on it must balance, but we need to use Newton's second law for its horizontal motion.

SET UP: Vertically $\sum F_y = 0$ and horizontally $\sum F_x = ma_x$, where a_x is the radial acceleration $a_{rad} = v^2/R$. Therefore horizontally we use $\sum F = m\dfrac{v^2}{R}$. Make a free-body diagram like Fig. 5.35 in the text.

EXECUTE: (a) $\sum F_x = ma_x = n \sin \beta = mv^2/R$

$\sum F_y = 0 = n \cos \beta - mg$.

Combining these two equations gives $\tan \beta = v^2/Rg = (80.0 \text{ m/s})^2/[(175 \text{ m})(9.80 \text{ m/s}^2)]$, which gives $\beta = 75.0°$.

(b) The pilot's apparent weight will be the force n due to the seat. Using $\sum F_y = 0$ gives

$w = n \cos \beta$, so $n = w/\cos \beta = (80.0 \text{ kg})(9.80 \text{ m/s}^2)/(\cos 75.0°) = 3.03 \times 10^3$ N.

$w_{apparent}/w_{actual} = (3.03 \times 10^3 \text{ N})/[(80.0 \text{ kg})(9.80 \text{ m/s}^2)] = 3.86$, which means that is apparent weight is 3.86 times great than his actual weight.

EVALUATE: Notice that $\tan \beta \propto v^2$, so a large speed means a large bank angle. We also saw that $n = w/\cos \beta$, so as v gets larger and larger, β gets closer and closer to 90°, and $\cos \beta$ gets closer and closer to zero. Therefore n gets larger and larger. This can be dangerous for pilots in high speed turns. The effects from such turns can cause a pilot to black out if the speed is great enough.

VP5.22.4. IDENTIFY: The driver is moving in a horizontal circle at constant speed. Therefore she has horizontal acceleration toward the center of the circle, but no vertical acceleration. The vertical forces on her must balance, but we need to use Newton's second law for her horizontal motion.

SET UP: Vertically $\Sigma F_y = 0$ and horizontally $\Sigma F_x = ma_x$, where a_x is the radial acceleration $a_{rad} = v^2/R$. Therefore horizontally we use $\Sigma F = m\dfrac{v^2}{R}$. Make a free-body diagram like Fig. 5.34b in the textbook. Her apparent weight x times her actual weight, which means that the normal force n on her due to the seat is $n = xmg$.

EXECUTE: (a) Using $\Sigma F = m\dfrac{v^2}{R}$, we see that the net force on her is $F_{net} = ma_x = mv^2/R$. From the free-body diagram, we see that $F_{rad} = n \sin \beta$. So $F_{net} = F_{rad} = n \sin \beta = xmg \sin \beta$. Use $\Sigma F_y = 0$ to find β: $n \cos \beta - mg = 0 \to \cos \beta = mg/n = mg/xmg = 1/x$. This means that $\sin \beta = \dfrac{\sqrt{x^2-1}}{x}$. We showed that $F_{net} = xmg \sin \beta$, so $F_{net} = xmg\dfrac{\sqrt{x^2-1}}{x} = mg\sqrt{x^2-1}$.

(b) $F_{rad} = F_{net} = mv^2/R$. Use the result from (a) for F_{net}.

$mg\sqrt{x^2-1} = mv^2/R \to R = \dfrac{v^2}{g\sqrt{x^2-1}}$.

EVALUATE: Our result in (b) says that as x gets larger and larger, R gets smaller and smaller. This is reasonable, since the larger x is, the greater the apparent weight of the driver. So for a given speed, a sharper turn will produce a greater apparent weight than a wide turn.

5.7. IDENTIFY: Apply $\Sigma \vec{F} = m\vec{a}$ to the object and to the knot where the cords are joined.

SET UP: Let $+y$ be upward and $+x$ be to the right.

EXECUTE: (a) $T_C = w$, $T_A \sin 30° + T_B \sin 45° = T_C = w$, and $T_A \cos 30° - T_B \cos 45° = 0$. Since $\sin 45° = \cos 45°$, adding the last two equations gives $T_A(\cos 30° + \sin 30°) = w$, and so $T_A = \dfrac{w}{1.366} = 0.732w$. Then, $T_B = T_A \dfrac{\cos 30°}{\cos 45°} = 0.897w$.

(b) Similar to part (a), $T_C = w$, $-T_A \cos 60° + T_B \sin 45° = w$, and $T_A \sin 60° - T_B \cos 45° = 0$. Adding these two equations, $T_A = \dfrac{w}{(\sin 60° - \cos 60°)} = 2.73w$, and $T_B = T_A \dfrac{\sin 60°}{\cos 45°} = 3.35w$.

EVALUATE: In part (a), $T_A + T_B > w$ since only the vertical components of T_A and T_B hold the object against gravity. In part (b), since T_A has a downward component T_B is greater than w.

5.9. IDENTIFY: Since the velocity is constant, apply Newton's first law to the piano. The push applied by the man must oppose the component of gravity down the incline.

SET UP: The free-body diagrams for the two cases are shown in Figure 5.9. $\vec{F}$ is the force applied by the man. Use the coordinates shown in the figure.

EXECUTE: **(a)** $\Sigma F_x = 0$ gives $F - w\sin 19.0° = 0$ and
$F = (180 \text{ kg})(9.80 \text{ m/s}^2) \sin 19.0° = 574 \text{ N}$.

(b) $\Sigma F_y = 0$ gives $n\cos 19.0° - w = 0$ and $n = \dfrac{w}{\cos 19.0°}$. $\Sigma F_x = 0$ gives $F - n\sin 19.0° = 0$ and $F = \left(\dfrac{w}{\cos 19.0°}\right) \sin 19.0° = w\tan 19.0° = 607 \text{ N}$.

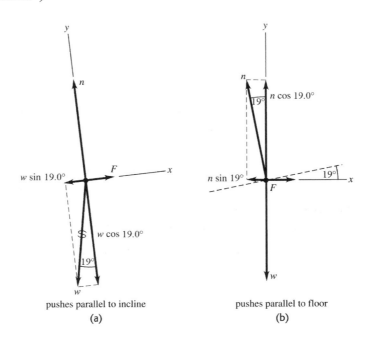

Figure 5.9

EVALUATE: When pushing parallel to the floor only part of the push is up the ramp to balance the weight of the piano, so you need a larger push in this case than if you push parallel to the ramp.

5.11. IDENTIFY: We apply Newton's second law to the rocket and the astronaut in the rocket. A constant force means we have constant acceleration, so we can use the standard kinematics equations.

SET UP: The free-body diagrams for the rocket (weight w_r) and astronaut (weight w) are given in Figure 5.11. F_T is the thrust and n is the normal force the rocket exerts on the astronaut. The speed of sound is 331 m/s. We use $\Sigma F_y = ma_y$ and $v = v_0 + at$.

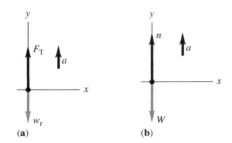

Figure 5.11

EXECUTE: (a) Apply $\Sigma F_y = ma_y$ to the rocket: $F_T - w_r = ma$. $a = 4g$ and $w_r = mg$, so
$F = m(5g) = (2.25 \times 10^6 \text{ kg})(5)(9.80 \text{ m/s}^2) = 1.10 \times 10^8$ N.

(b) Apply $\Sigma F_y = ma_y$ to the astronaut: $n - w = ma$. $a = 4g$ and $m = \dfrac{w}{g}$, so
$n = w + \left(\dfrac{w}{g}\right)(4g) = 5w.$

(c) $v_0 = 0$, $v = 331$ m/s and $a = 4g = 39.2$ m/s^2. $v = v_0 + at$ gives
$t = \dfrac{v - v_0}{a} = \dfrac{331 \text{ m/s}}{39.2 \text{ m/s}^2} = 8.4$ s.

EVALUATE: The 8.4 s is probably an unrealistically short time to reach the speed of sound because you would not want your astronauts at the brink of blackout during a launch.

5.15. IDENTIFY: Apply $\Sigma \vec{F} = m\vec{a}$ to the load of bricks and to the counterweight. The tension is the same at each end of the rope. The rope pulls up with the same force (T) on the bricks and on the counterweight. The counterweight accelerates downward and the bricks accelerate upward; these accelerations have the same magnitude.

(a) SET UP: The free-body diagrams for the bricks and counterweight are given in Figure 5.15.

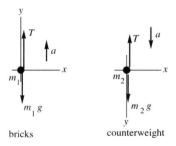

Figure 5.15

(b) EXECUTE: Apply $\Sigma F_y = ma_y$ to each object. The acceleration magnitude is the same for the two objects. For the bricks take $+y$ to be upward since $\vec{a}$ for the bricks is upward. For the counterweight take $+y$ to be downward since $\vec{a}$ is downward.

bricks: $\Sigma F_y = ma_y$
$T - m_1 g = m_1 a$
counterweight: $\Sigma F_y = ma_y$
$m_2 g - T = m_2 a$

Add these two equations to eliminate T:
$$(m_2 - m_1)g = (m_1 + m_2)a$$
$$a = \left(\frac{m_2 - m_1}{m_1 + m_2}\right)g = \left(\frac{28.0 \text{ kg} - 15.0 \text{ kg}}{15.0 \text{ kg} + 28.0 \text{ kg}}\right)(9.80 \text{ m/s}^2) = 2.96 \text{ m/s}^2$$

(c) $T - m_1 g = m_1 a$ gives $T = m_1(a + g) = (15.0 \text{ kg})(2.96 \text{ m/s}^2 + 9.80 \text{ m/s}^2) = 191 \text{ N}$
As a check, calculate T using the other equation.
$m_2 g - T = m_2 a$ gives $T = m_2(g - a) = 28.0 \text{ kg}(9.80 \text{ m/s}^2 - 2.96 \text{ m/s}^2) = 191 \text{ N}$, which checks.

EVALUATE: The tension is 1.30 times the weight of the bricks; this causes the bricks to accelerate upward. The tension is 0.696 times the weight of the counterweight; this causes the counterweight to accelerate downward. If $m_1 = m_2$, $a = 0$ and $T = m_1 g = m_2 g$. In this special case the objects don't move. If $m_1 = 0$, $a = g$ and $T = 0$; in this special case the counterweight is in free fall. Our general result is correct in these two special cases.

5.17. IDENTIFY: Apply $\Sigma \vec{F} = m\vec{a}$ to each block. Each block has the same magnitude of acceleration a.

SET UP: Assume the pulley is to the right of the 4.00 kg block. There is no friction force on the 4.00 kg block; the only force on it is the tension in the rope. The 4.00 kg block therefore accelerates to the right and the suspended block accelerates downward. Let $+x$ be to the right for the 4.00 kg block, so for it $a_x = a$, and let $+y$ be downward for the suspended block, so for it $a_y = a$.

EXECUTE: **(a)** The free-body diagrams for each block are given in Figures 5.17a and b.
(b) $\Sigma F_x = m a_x$ applied to the 4.00 kg block gives $T = (4.00 \text{ kg})a$ and
$$a = \frac{T}{4.00 \text{ kg}} = \frac{15.0 \text{ N}}{4.00 \text{ kg}} = 3.75 \text{ m/s}^2.$$
(c) $\Sigma F_y = m a_y$ applied to the suspended block gives $mg - T = ma$ and
$$m = \frac{T}{g - a} = \frac{15.0 \text{ N}}{9.80 \text{ m/s}^2 - 3.75 \text{ m/s}^2} = 2.48 \text{ kg}.$$
(d) The weight of the hanging block is $mg = (2.48 \text{ kg})(9.80 \text{ m/s}^2) = 24.3 \text{ N}$. This is greater than the tension in the rope; $T = 0.617 mg$.

EVALUATE: Since the hanging block accelerates downward, the net force on this block must be downward and the weight of the hanging block must be greater than the tension in the rope. Note that the blocks accelerate no matter how small m is. It is not necessary to have $m > 4.00$ kg, and in fact in this problem m is less than 4.00 kg.

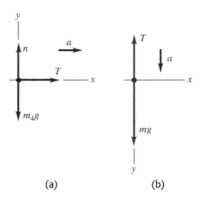

Figure 5.17

5.19. IDENTIFY: While the person is in contact with the ground, he is accelerating upward and experiences two forces: gravity downward and the upward force of the ground. Once he is in the air, only gravity acts on him so he accelerates downward. Newton's second law applies during the jump (and at all other times).

SET UP: Take $+y$ to be upward. After he leaves the ground the person travels upward 60 cm and his acceleration is $g = 9.80 \text{ m/s}^2$, downward. His weight is w so his mass is w/g. $\Sigma F_y = ma_y$ and $v_y^2 = v_{0y}^2 + 2a_y(y - y_0)$ apply to the jumper.

EXECUTE: (a) $v_y = 0$ (at the maximum height), $y - y_0 = 0.60$ m, $a_y = -9.80$ m/s^2. $v_y^2 = v_{0y}^2 + 2a_y(y - y_0)$ gives $v_{0y} = \sqrt{-2a_y(y-y_0)} = \sqrt{-2(-9.80 \text{ m/s}^2)(0.60 \text{ m})} = 3.4$ m/s.

(b) The free-body diagram for the person while he is pushing up against the ground is given in Figure 5.19.

(c) For the jump, $v_{0y} = 0$, $v_y = 3.4$ m/s (from part (a)), and $y - y_0 = 0.50$ m. $v_y^2 = v_{0y}^2 + 2a_y(y-y_0)$ gives $a_y = \dfrac{v_y^2 - v_{0y}^2}{2(y-y_0)} = \dfrac{(3.4 \text{ m/s})^2 - 0}{2(0.50 \text{ m})} = 11.6 \text{ m/s}^2$.

$\Sigma F_y = ma_y$ gives $n - w = ma$. $n = w + ma = w\left(1 + \dfrac{a}{g}\right) = 2.2w$.

Figure 5.19

EVALUATE: To accelerate the person upward during the jump, the upward force from the ground must exceed the downward pull of gravity. The ground pushes up on him because he pushes down on the ground.

5.21. IDENTIFY: We know the external forces on the box and want to find the distance it moves and its speed. The force is not constant, so the acceleration will not be constant, so we cannot use the standard constant-acceleration kinematics formulas. But Newton's second law will apply.

SET UP: First use Newton's second law to find the acceleration as a function of time: $a_x(t) = \frac{F_x}{m}$. Then integrate the acceleration to find the velocity as a function of time, and next integrate the velocity to find the position as a function of time.

EXECUTE: Let +x be to the right. $a_x(t) = \frac{F_x}{m} = \frac{(-6.00 \text{ N/s}^2)t^2}{2.00 \text{ kg}} = -(3.00 \text{ m/s}^4)t^2$. Integrate the acceleration to find the velocity as a function of time: $v_x(t) = -(1.00 \text{ m/s}^4)t^3 + 9.00 \text{ m/s}$. Next integrate the velocity to find the position as a function of time: $x(t) = -(0.250 \text{ m/s}^4)t^4 + (9.00 \text{ m/s})t$. Now use the given values of time.

(a) $v_x = 0$ when $(1.00 \text{ m/s}^4)t^3 = 9.00 \text{ m/s}$. This gives $t = 2.08$ s. At $t = 2.08$ s, $x = (9.00 \text{ m/s})(2.08 \text{ s}) - (0.250 \text{ m/s}^4)(2.08 \text{ s})^4 = 18.72 \text{ m} - 4.68 \text{ m} = 14.0 \text{ m}$.

(b) At $t = 3.00$ s, $v_x(t) = -(1.00 \text{ m/s}^4)(3.00 \text{ s})^3 + 9.00 \text{ m/s} = -18.0 \text{ m/s}$, so the speed is 18.0 m/s.

EVALUATE: The box starts out moving to the right. But because the acceleration is to the left, it reverses direction and v_x is negative in part (b).

5.25. (a) IDENTIFY: Constant speed implies $a = 0$. Apply Newton's first law to the box. The friction force is directed opposite to the motion of the box.

SET UP: Consider the free-body diagram for the box, given in Figure 5.25a. Let $\vec{F}$ be the horizontal force applied by the worker. The friction is kinetic friction since the box is sliding along the surface.

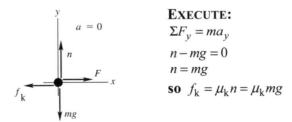

EXECUTE:
$\Sigma F_y = ma_y$
$n - mg = 0$
$n = mg$
so $f_k = \mu_k n = \mu_k mg$

Figure 5.25a

$\Sigma F_x = ma_x$
$F - f_k = 0$
$F = f_k = \mu_k mg = (0.20)(16.8 \text{ kg})(9.80 \text{ m/s}^2) = 33$ N

(b) IDENTIFY: Now the only horizontal force on the box is the kinetic friction force. Apply Newton's second law to the box to calculate its acceleration. Once we have the acceleration, we can find the distance using a constant acceleration equation. The friction force is $f_k = \mu_k mg$, just as in part (a).

SET UP: The free-body diagram is sketched in Figure 5.25b.

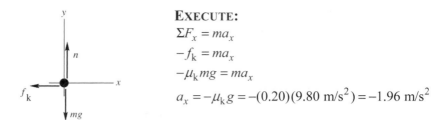

EXECUTE:
$\Sigma F_x = ma_x$
$-f_k = ma_x$
$-\mu_k mg = ma_x$
$a_x = -\mu_k g = -(0.20)(9.80 \text{ m/s}^2) = -1.96 \text{ m/s}^2$

Figure 5.25b

Use the constant acceleration equations to find the distance the box travels:
$v_x = 0$, $v_{0x} = 3.50 \text{ m/s}$, $a_x = -1.96 \text{ m/s}^2$, $x - x_0 = ?$
$v_x^2 = v_{0x}^2 + 2a_x(x - x_0)$
$x - x_0 = \dfrac{v_x^2 - v_{0x}^2}{2a_x} = \dfrac{0 - (3.50 \text{ m/s})^2}{2(-1.96 \text{ m/s}^2)} = 3.1 \text{ m}$

EVALUATE: The normal force is the component of force exerted by a surface perpendicular to the surface. Its magnitude is determined by $\Sigma \vec{F} = m\vec{a}$. In this case n and mg are the only vertical forces and $a_y = 0$, so $n = mg$. Also note that f_k and n are proportional in magnitude but perpendicular in direction.

5.27. IDENTIFY: Apply $\Sigma \vec{F} = m\vec{a}$ to the crate. $f_s \leq \mu_s n$ and $f_k = \mu_k n$.

SET UP: Let $+y$ be upward and let $+x$ be in the direction of the push. Since the floor is horizontal and the push is horizontal, the normal force equals the weight of the crate: $n = mg = 441$ N. The force it takes to start the crate moving equals max f_s and the force required to keep it moving equals f_k.

EXECUTE: **(a)** max $f_s = 313$ N, so $\mu_s = \dfrac{313 \text{ N}}{441 \text{ N}} = 0.710$. $f_k = 208$ N, so
$\mu_k = \dfrac{208 \text{ N}}{441 \text{ N}} = 0.472$.

(b) The friction is kinetic. $\Sigma F_x = ma_x$ gives $F - f_k = ma$ and
$F = f_k + ma = 208 \text{ N} + (45.0 \text{ kg})(1.10 \text{ m/s}^2) = 258 \text{ N}$.

(c) (i) The normal force now is $mg = 72.9$ N. To cause it to move,
$F = \max f_s = \mu_s n = (0.710)(72.9 \text{ N}) = 51.8 \text{ N}$.

(ii) $F = f_k + ma$ and $a = \dfrac{F - f_k}{m} = \dfrac{258 \text{ N} - (0.472)(72.9 \text{ N})}{45.0 \text{ kg}} = 4.97 \text{ m/s}^2$

EVALUATE: The kinetic friction force is independent of the speed of the object. On the moon, the mass of the crate is the same as on earth, but the weight and normal force are less.

Using the notation in Fig. 5.28, $\Sigma F_y = 0$ gives $n - w \cos \beta = 0$, so $n = w \cos \beta$.

$\Sigma F_x = 0$ gives $P - f_k - w \sin \beta = 0$. We also have $f_k = \mu_k n$, so the last equation becomes

$P - \mu_k w \cos \beta - w \sin \beta = 0$. Solve for w: $w = \dfrac{P}{\mu_k \cos \beta + \sin \beta} = \dfrac{75 \text{ lb}}{(0.50)\cos 60° + \sin 60°}$,

so

$w = 67$ lb.

EVALUATE: This weight is less than you can push on a level surface because you need to balance the weight component down the ramp in addition to the friction force.

5.31. IDENTIFY: A 10.0-kg box is pushed on a ramp, causing it to accelerate. Newton's second law applies.

SET UP: Choose the x-axis along the surface of the ramp and the y-axis perpendicular to the surface. The only acceleration of the box is in the x-direction, so $\Sigma F_x = ma_x$ and $\Sigma F_y = 0$. The external forces acting on the box are the push P along the surface of the ramp, friction f_k, gravity mg, and the normal force n. The ramp rises at 55.0° above the horizontal, and $f_k = \mu_k n$. The friction force opposes the sliding, so it is directed up the ramp in part (a) and down the ramp in part (b).

EXECUTE: (a) Applying $\Sigma F_y = 0$ gives $n = mg \cos(55.0°)$, so the force of kinetic friction is $f_k = \mu_k n = (0.300)(10.0 \text{ kg})(9.80 \text{ m/s}^2)(\cos 55.0°) = 16.86$ N. Call the +x-direction down the ramp since that is the direction of the acceleration of the box. Applying $\Sigma F_x = ma_x$ gives $P + mg \sin(55.0°) - f_k = ma$. Putting in the numbers gives (10.0 kg)a = 120 N + (98.0 N)(sin 55.0°) − 16.86 N; $a = 18.3$ m/s².

(b) Now P is up the up the ramp and f_k is down the ramp, but the other force components are unchanged, so $f_k = 16.86$ N as before. We now choose +x to be up the ramp, so $\Sigma F_x = ma_x$ gives

$P - mg \sin(55.0°) - f_k = ma$. Putting in the same numbers as before gives $a = 2.29$ m/s².

EVALUATE: Pushing up the ramp produces a much smaller acceleration than pushing down the ramp because gravity helps the downward push but opposes the upward push.

5.35. IDENTIFY: Use $\Sigma \vec{F} = m\vec{a}$ to find the acceleration that can be given to the car by the kinetic friction force. Then use a constant acceleration equation.

SET UP: Take +x in the direction the car is moving.
EXECUTE: (a) The free-body diagram for the car is shown in Figure 5.35. $\Sigma F_y = ma_y$ gives $n = mg$.

$\Sigma F_x = ma_x$ gives $-\mu_k n = ma_x$. $-\mu_k mg = ma_x$ and $a_x = -\mu_k g$. Then $v_x = 0$ and

$v_x^2 = v_{0x}^2 + 2a_x(x - x_0)$ gives $(x - x_0) = -\dfrac{v_{0x}^2}{2a_x} = +\dfrac{v_{0x}^2}{2\mu_k g} = \dfrac{(28.7 \text{ m/s})^2}{2(0.80)(9.80 \text{ m/s}^2)} = 52.5$ m.

(b) $v_{0x} = \sqrt{2\mu_k g(x - x_0)} = \sqrt{2(0.25)(9.80 \text{ m/s}^2)52.5 \text{ m}} = 16.0$ m/s

EVALUATE: For constant stopping distance $\dfrac{v_{0x}^2}{\mu_k}$ is constant and v_{0x} is proportional to $\sqrt{\mu_k}$. The answer to part (b) can be calculated as $(28.7 \text{ m/s})\sqrt{0.25/0.80} = 16.0$ m/s.

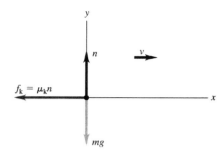

Figure 5.35

5.39. **IDENTIFY:** Apply $\Sigma \vec{F} = m\vec{a}$ to each block. The target variables are the tension T in the cord and the acceleration a of the blocks. Then a can be used in a constant acceleration equation to find the speed of each block. The magnitude of the acceleration is the same for both blocks.

SET UP: The system is sketched in Figure 5.39a.

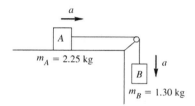

For each block take a positive coordinate direction to be the direction of the block's acceleration.

Figure 5.39a

Block on the table: The free-body is sketched in Figure 5.39b (next page).

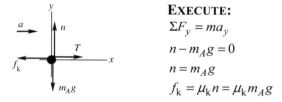

EXECUTE:
$\Sigma F_y = ma_y$
$n - m_A g = 0$
$n = m_A g$
$f_k = \mu_k n = \mu_k m_A g$

Figure 5.39b

$\Sigma F_x = ma_x$
$T - f_k = m_A a$
$T - \mu_k m_A g = m_A a$

SET UP: Hanging block: The free-body is sketched in Figure 5.39c.

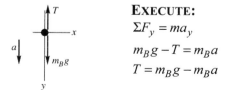

EXECUTE:
$\Sigma F_y = ma_y$
$m_B g - T = m_B a$
$T = m_B g - m_B a$

Figure 5.39c

(a) Use the second equation in the first
$$m_B g - m_B a - \mu_k m_A g = m_A a$$
$$(m_A + m_B)a = (m_B - \mu_k m_A)g$$

$$a = \frac{(m_B - \mu_k m_A)g}{m_A + m_B} = \frac{(1.30 \text{ kg} - (0.45)(2.25 \text{ kg}))(9.80 \text{ m/s}^2)}{2.25 \text{ kg} + 1.30 \text{ kg}} = 0.7937 \text{ m/s}^2$$

SET UP: Now use the constant acceleration equations to find the final speed. Note that the blocks have the same speeds. $x - x_0 = 0.0300$ m, $a_x = 0.7937$ m/s^2, $v_{0x} = 0$, $v_x = ?$

$$v_x^2 = v_{0x}^2 + 2a_x(x - x_0)$$

EXECUTE: $v_x = \sqrt{2a_x(x - x_0)} = \sqrt{2(0.7937 \text{ m/s}^2)(0.0300 \text{ m})} = 0.218$ m/s $= 21.8$ cm/s.

(b) $T = m_B g - m_B a = m_B(g - a) = 1.30 \text{ kg}(9.80 \text{ m/s}^2 - 0.7937 \text{ m/s}^2) = 11.7$ N

Or, to check, $T - \mu_k m_A g = m_A a$.

$T = m_A(a + \mu_k g) = 2.25 \text{ kg}(0.7937 \text{ m/s}^2 + (0.45)(9.80 \text{ m/s}^2)) = 11.7$ N, which checks.

EVALUATE: The force T exerted by the cord has the same value for each block. $T < m_B g$ since the hanging block accelerates downward. Also, $f_k = \mu_k m_A g = 9.92$ N. $T > f_k$ and the block on the table accelerates in the direction of T.

5.41. (a) IDENTIFY: Apply $\Sigma \vec{F} = m\vec{a}$ to the crate. Constant v implies $a = 0$. Crate moving says that the friction is kinetic friction. The target variable is the magnitude of the force applied by the woman.

SET UP: The free-body diagram for the crate is sketched in Figure 5.41.

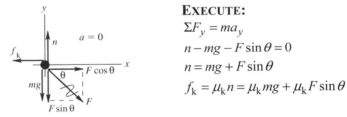

EXECUTE:
$\Sigma F_y = ma_y$
$n - mg - F\sin\theta = 0$
$n = mg + F\sin\theta$
$f_k = \mu_k n = \mu_k mg + \mu_k F \sin\theta$

Figure 5.41

$\Sigma F_x = ma_x$
$F\cos\theta - f_k = 0$
$F\cos\theta - \mu_k mg - \mu_k F\sin\theta = 0$
$F(\cos\theta - \mu_k \sin\theta) = \mu_k mg$
$$F = \frac{\mu_k mg}{\cos\theta - \mu_k \sin\theta}$$

(b) IDENTIFY and SET UP: "Start the crate moving" means the same force diagram as in part (a), except that μ_k is replaced by μ_s. Thus $F = \dfrac{\mu_s mg}{\cos\theta - \mu_s \sin\theta}$.

EXECUTE: $F \to \infty$ if $\cos\theta - \mu_s \sin\theta = 0$. This gives $\mu_s = \dfrac{\cos\theta}{\sin\theta} = \dfrac{1}{\tan\theta}$.

EVALUATE: $\vec{F}$ has a downward component so $n > mg$. If $\theta = 0$ (woman pushes horizontally), $n = mg$ and $F = f_k = \mu_k mg$.

5.45. IDENTIFY: Apply $\Sigma \vec{F} = m\vec{a}$ to the car. It has acceleration $\vec{a}_{rad}$, directed toward the center of the circular path.

SET UP: The analysis is the same as in Example 5.23.

EXECUTE: (a) $F_A = m\left(g + \dfrac{v^2}{R}\right) = (1.60 \text{ kg})\left(9.80 \text{ m/s}^2 + \dfrac{(12.0 \text{ m/s})^2}{5.00 \text{ m}}\right) = 61.8$ N.

(b) $F_B = m\left(g - \dfrac{v^2}{R}\right) = (1.60 \text{ kg})\left(9.80 \text{ m/s}^2 - \dfrac{(12.0 \text{ m/s})^2}{5.00 \text{ m}}\right) = -30.4$ N, where the minus sign indicates that the track pushes down on the car. The magnitude of this force is 30.4 N.

EVALUATE: $|F_A| > |F_B|$. $|F_A| - 2mg = |F_B|$.

5.49. IDENTIFY: Apply Newton's second law to the car in circular motion, assume friction is negligible.

SET UP: The acceleration of the car is $a_{\text{rad}} = v^2/R$. As shown in the text, the banking angle β is given by $\tan \beta = \dfrac{v^2}{gR}$. Also, $n = mg/\cos\beta$. 65.0 mi/h = 29.1 m/s.

EXECUTE: (a) $\tan \beta = \dfrac{(29.1 \text{ m/s})^2}{(9.80 \text{ m/s}^2)(225 \text{ m})}$ and $\beta = 21.0°$. The expression for $\tan \beta$ does not involve the mass of the vehicle, so the truck and car should travel at the same speed.

(b) For the car, $n_{\text{car}} = \dfrac{(1125 \text{ kg})(9.80 \text{ m/s}^2)}{\cos 21.0°} = 1.18 \times 10^4$ N and $n_{\text{truck}} = 2n_{\text{car}} = 2.36 \times 10^4$ N, since $m_{\text{truck}} = 2m_{\text{car}}$.

EVALUATE: The vertical component of the normal force must equal the weight of the vehicle, so the normal force is proportional to m.

5.51. IDENTIFY: Apply $\Sigma \vec{F} = m\vec{a}$ to the composite object of the person plus seat. This object moves in a horizontal circle and has acceleration a_{rad}, directed toward the center of the circle.

SET UP: The free-body diagram for the composite object is given in Figure 5.51. Let $+x$ be to the right, in the direction of $\vec{a}_{\text{rad}}$. Let $+y$ be upward. The radius of the circular path is $R = 7.50$ m. The total mass is $(255 \text{ N} + 825 \text{ N})/(9.80 \text{ m/s}^2) = 110.2$ kg. Since the rotation rate is 28.0 rev/min = 0.4667 rev/s, the period T is $\dfrac{1}{0.4667 \text{ rev/s}} = 2.143$ s.

EXECUTE: $\Sigma F_y = ma_y$ gives $T_A \cos 40.0° - mg = 0$ and

$T_A = \dfrac{mg}{\cos 40.0°} = \dfrac{255 \text{ N} + 825 \text{ N}}{\cos 40.0°} = 1410$ N. $\Sigma F_x = ma_x$ gives $T_A \sin 40.0° + T_B = ma_{\text{rad}}$ and

$T_B = m\dfrac{4\pi^2 R}{T^2} - T_A \sin 40.0° = (110.2 \text{ kg})\dfrac{4\pi^2 (7.50 \text{ m})}{(2.143 \text{ s})^2} - (1410 \text{ N})\sin 40.0° = 6200$ N

The tension in the horizontal cable is 6200 N and the tension in the other cable is 1410 N.

EVALUATE: The weight of the composite object is 1080 N. The tension in cable A is larger than this since its vertical component must equal the weight. The tension in cable B is less than ma_{rad} because part of the required inward force comes from a component of the tension in cable A.

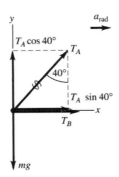

Figure 5.51

5.53. IDENTIFY: The acceleration due to circular motion is $a_{rad} = \dfrac{4\pi^2 R}{T^2}$.

SET UP: $R = 400$ m. $1/T$ is the number of revolutions per second.
EXECUTE: **(a)** Setting $a_{rad} = g$ and solving for the period T gives
$$T = 2\pi\sqrt{\dfrac{R}{g}} = 2\pi\sqrt{\dfrac{400 \text{ m}}{9.80 \text{ m/s}^2}} = 40.1 \text{ s},$$
so the number of revolutions per minute is $(60 \text{ s/min})/(40.1 \text{ s}) = 1.5$ rev/min.
(b) The lower acceleration corresponds to a longer period, and hence a lower rotation rate, by a factor of the square root of the ratio of the accelerations,
$T' = (1.5 \text{ rev/min}) \times \sqrt{3.70/9.8} = 0.92$ rev/min.

EVALUATE: In part (a) the tangential speed of a point at the rim is given by $a_{rad} = \dfrac{v^2}{R}$, so $v = \sqrt{R a_{rad}} = \sqrt{Rg} = 62.6$ m/s; the space station is rotating rapidly.

5.55. IDENTIFY: Newton's second law applies to the rock moving in a vertical circle of radius L. Gravity and the tension in the string are the forces acting on it.

SET UP: $\Sigma F = m\dfrac{v^2}{R}$, where $R = L$ in this case.

EXECUTE: **(a)** Apply $\Sigma F = m\dfrac{v^2}{R}$ at the top of the circle. Both gravity and the tension act downward, so $T + mg = mv^2/L$. The smallest that T can be is zero, in which case $mg = mv^2/L$, so $v = \sqrt{Lg}$.

(b) In this case, $v = 2\sqrt{Lg}$. $\Sigma F = m\dfrac{v^2}{R}$ gives $T - mg = \dfrac{mv^2}{L} = \dfrac{m(2\sqrt{Lg})^2}{L} = 4mg$, so $T = 5mg$.

EVALUATE: Note in (a) that the rock does *not stop* at the top of the circle. If it did, it would just fall down and not complete the rest of the circle.

5.63. IDENTIFY: Newton's second law applies to the block. Gravity, kinetic friction, and the normal force due to the board act upon it.

SET UP: Apply $\Sigma F_x = ma_x$ and $\Sigma F_y = 0$ to the block. Choose the $+x$-axis along the surface of the board pointing downward. At the maximum angle α_0 just before slipping,

$\tan\alpha_0 = \mu_s = 0.600$, so $\alpha_0 = 30.96°$. Fig. 5.63 shows a free-body diagram of the block after it has slipped.

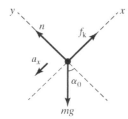

Figure 5.63

EXECUTE: First find the acceleration using Newton's laws. $\Sigma F_y = 0$ gives $n = mg \cos\alpha_0$ $\Sigma F_x = ma_x$ gives

$mg \sin\alpha_0 - f_k = ma_x$

$mg \sin\alpha_0 - \mu_k mg \cos\alpha_0 = ma_x$

$a_x = g(\sin\alpha_0 - \mu_k \cos\alpha_0) = (9.80 \text{ m/s}^2)[\sin 30.96° - (0.400)\cos 30.96°] = 1.681 \text{ m/s}^2$.
Now use $v_x^2 = v_{0x}^2 + 2a_x(x - x_0)$ to find the speed at the bottom of the board.

$v_x^2 = 0 + 2(1.681 \text{ m/s}^2)(3.00 \text{ m}) \rightarrow v_x = 3.18$ m/s.

EVALUATE: We chose the +x-axis to be downward because that is the direction of the acceleration. In most cases, it is easiest to make that choice if the direction of the acceleration is known.

5.65. **IDENTIFY:** Newton's second law applies to the accelerating box. The forces acting on it are the force F, gravity, the normal force due to the surface, and kinetic friction.

SET UP: Apply $\Sigma F_y = 0$ and $\Sigma F_x = ma_x$. Fig. 5.65 shows a free-body diagram of the box. Choose the +x-axis in the direction of the acceleration.

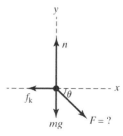

Figure 5.65

EXECUTE: **(a)** We want to find the magnitude of the force F. First use
$v_x^2 = v_{0x}^2 + 2a_x(x - x_0)$ to find a_x. $(6.00 \text{ m/s})^2 = 0 + 2a_x(8.00 \text{ m}) \rightarrow a_x = 2.250 \text{ m/s}^2$.

Now apply $\Sigma F_x = ma_x$: $F\cos\theta - f_k = ma_x$

We also have $f_k = \mu_k n$

Apply $\Sigma F_y = 0$: $n - mg - F\sin\theta = 0$

Combine these three results and solve for F: $F = \dfrac{m(a_x + \mu_k g)}{\cos\theta - \mu_k \sin\theta}$. Using $m = 12.0$ kg, $a_x = 2.250$ m/s^2, $\mu_k = 0.300$, and $\theta = 37.0°$, we get $F = 101$ N.

(b) If $f_k = 0$, we have $F\cos\theta = ma_x \rightarrow F\cos 37.0° = (12.0\text{ kg})(2.250\text{ m/s}^2) \rightarrow F = 33.8$ N.

(c) If F is horizontal, $\theta = 0°$, $n = mg$, and $f_k = \mu_k mg$, so $F = m(a_x + \mu_k g) = 62.3$ N.

EVALUATE: We see that F is least when there is no friction and greatest when it has a downward component. These results are reasonable since a downward component increases the normal force which increases friction. And the fact that it has a downward component means that there is less horizontal component to cause acceleration.

5.67. IDENTIFY: Kinematics will give us the acceleration of the person, and Newton's second law will give us the force (the target variable) that his arms exert on the rest of his body.

SET UP: Let the person's weight be W, so $W = 680$ N. Assume constant acceleration during the speeding up motion and assume that the body moves upward 15 cm in 0.50 s while speeding up. The constant-acceleration kinematics formula $y - y_0 = v_{0y}t + \tfrac{1}{2}a_y t^2$ and $\Sigma F_y = ma_y$ apply. The free-body diagram for the person is given in Figure 5.67. F is the force exerted on him by his arms.

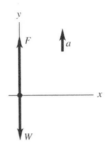

Figure 5.67

EXECUTE: $v_{0y} = 0$, $y - y_0 = 0.15$ m, $t = 0.50$ s. $y - y_0 = v_{0y}t + \tfrac{1}{2}a_y t^2$ gives $a_y = \dfrac{2(y-y_0)}{t^2} = \dfrac{2(0.15\text{ m})}{(0.50\text{ s})^2} = 1.2$ m/s^2. $\Sigma F_y = ma_y$ gives $F - W = ma$. $m = \dfrac{W}{g}$, so $F = W\left(1 + \dfrac{a}{g}\right) = 1.12W = 762$ N.

EVALUATE: The force is greater than his weight, which it must be if he is to accelerate upward.

5.69. IDENTIFY: We know the forces on the box and want to find information about its position and velocity. Newton's second law will give us the box's acceleration.

SET UP: $a_y(t) = \dfrac{\Sigma F_y}{m}$. We can integrate the acceleration to find the velocity and the velocity to find the position. At an altitude of several hundred meters, the acceleration due to gravity is essentially the same as it is at the earth's surface.

EXECUTE: Let $+y$ be upward. Newton's second law gives $T - mg = ma_y$, so $a_y(t) = (12.0 \text{ m/s}^3)t - 9.8 \text{ m/s}^2$. Integrating the acceleration gives $v_y(t) = (6.00 \text{ m/s}^3)t^2 - (9.8 \text{ m/s}^2)t$.

(a) (i) At $t = 1.00$ s, $v_y = -3.80$ m/s. (ii) At $t = 3.00$ s, $v_y = 24.6$ m/s.

(b) Integrating the velocity gives $y - y_0 = (2.00 \text{ m/s}^3)t^3 - (4.9 \text{ m/s}^2)t^2$. $v_y = 0$ at $t = 1.63$ s. At $t = 1.63$ s, $y - y_0 = 8.71 \text{ m} - 13.07 \text{ m} = -4.36$ m.

(c) Setting $y - y_0 = 0$ and solving for t gives $t = 2.45$ s.

EVALUATE: The box accelerates and initially moves downward until the tension exceeds the weight of the box. Once the tension exceeds the weight, the box will begin to accelerate upward and will eventually move upward, as we saw in part (b).

5.73. IDENTIFY: Newton's second law applies to the box.

SET UP: $f_k = \mu_k n$, $\Sigma F_x = ma_x$, and $\Sigma F_y = ma_y$ apply to the box. Take the $+x$-axis down the surface of the ramp and the $+y$-axis perpendicular to the surface upward.

EXECUTE: $\Sigma F_y = ma_y$ gives $n + F\sin(33.0°) - mg\cos(33.0°) = 0$, which gives $n = 51.59$ N. The friction force is $f_k = \mu_k n = (0.300)(51.59 \text{ N}) = 15.48$ N. Parallel to the surface we have $\Sigma F_x = ma_x$ which gives $F\cos(33.0°) + mg\sin(33.0°) - f_k = ma$, which gives $a = 6.129 \text{ m/s}^2$. Finally the velocity formula gives us $v_x = v_{0x} + a_x t = 0 + (6.129 \text{ m/s}^2)(2.00 \text{ s}) = 12.3$ m/s.

EVALUATE: Even though F is horizontal and mg is vertical, it is best to choose the axes as we have done, rather than horizontal-vertical, because the acceleration is then in the x-direction. Taking x and y to be horizontal-vertical would give the acceleration x- and y-components, which would complicate the solution.

5.75. IDENTIFY: Newton's second law applies, as do the constant-acceleration kinematics equations.

SET UP: Call the $+x$-axis horizontal and to the right and the $+y$-axis vertically upward. $\Sigma F_y = ma_y$ and $\Sigma F_x = ma_x$ both apply to the book.

EXECUTE: The book has no horizontal motion, so $\Sigma F_x = ma_x = 0$, which gives us the normal force n:
$n = F\cos(60.0°)$. The kinetic friction force is $f_k = \mu_k n = (0.300)(96.0 \text{ N})(\cos 60.0°) = 14.4$ N. In the vertical direction, we have $\Sigma F_y = ma_y$, which gives $F\sin(60.0°) - mg - f_k = ma$. Solving for a gives us
$a = [(96.0 \text{ N})(\sin 60.0°) - 49.0 \text{ N} - 14.4 \text{ N}]/(5.00 \text{ kg}) = 3.948 \text{ m/s}^2$ upward. Now the velocity formula $v_y^2 = v_{0y}^2 + 2a_y(y - y_0)$ gives $v_y = \sqrt{2(3.948 \text{ m/s}^2)(0.400 \text{ m})} = 1.78$ m/s.

EVALUATE: Only the upward component of the force F makes the book accelerate upward, while the horizontal component of T is the magnitude of the normal force.

5.79. IDENTIFY: Apply $\Sigma \vec{F} = m\vec{a}$ to each block. Use Newton's third law to relate forces on A and on B.

SET UP: Constant speed means $a = 0$.

EXECUTE: **(a)** Treat A and B as a single object of weight $w = w_A + w_B = 1.20 \text{ N} + 3.60 \text{ N} = 4.80$ N.

The free-body diagram for this combined object is given in Figure 5.79a. $\Sigma F_y = ma_y$ gives $n = w = 4.80$ N. $f_k = \mu_k n = (0.300)(4.80$ N$) = 1.44$ N. $\Sigma F_x = ma_x$ gives $F = f_k = 1.44$ N.

(b) The free-body force diagrams for blocks A and B are given in Figure 5.79b. n and f_k are the normal and friction forces applied to block B by the tabletop and are the same as in part (a). f_{kB} is the friction force that A applies to B. It is to the right because the force from A opposes the motion of B. n_B is the downward force that A exerts on B. f_{kA} is the friction force that B applies to A. It is to the left because block B wants A to move with it. n_A is the normal force that block B exerts on A. By Newton's third law, $f_{kB} = f_{kA}$ and these forces are in opposite directions. Also, $n_A = n_B$ and these forces are in opposite directions.

$\Sigma F_y = ma_y$ for block A gives $n_A = w_A = 1.20$ N, so $n_B = 1.20$ N.

$f_{kA} = \mu_k n_A = (0.300)(1.20$ N$) = 0.360$ N, and $f_{kB} = 0.360$ N.

$\Sigma F_x = ma_x$ for block A gives $T = f_{kA} = 0.360$ N.

$\Sigma F_x = ma_x$ for block B gives $F = f_{kB} + f_k = 0.360$ N $+ 1.44$ N $= 1.80$ N.

EVALUATE: In part (a) block A is at rest with respect to B and it has zero acceleration. There is no horizontal force on A besides friction, and the friction force on A is zero. A larger force F is needed in part (b), because of the friction force between the two blocks.

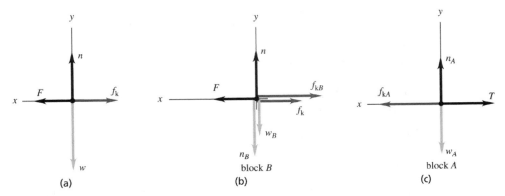

Figure 5.79

5.81. **IDENTIFY:** $a = dv/dt$. Apply $\Sigma \vec{F} = m\vec{a}$ to yourself.

SET UP: The reading of the scale is equal to the normal force the scale applies to you.
EXECUTE: The elevator's acceleration is

$a = \dfrac{dv(t)}{dt} = 3.0$ m/s$^2 + 2(0.20$ m/s$^3)t = 3.0$ m/s$^2 + (0.40$ m/s$^3)t$. At

$t = 4.0$ s, $a = 3.0$ m/s$^2 + (0.40$ m/s$^3)(4.0$ s$) = 4.6$ m/s^2. From Newton's second law, the net force on you is $F_{net} = F_{scale} - w = ma$ and

$F_{scale} = w + ma = (64$ kg$)(9.8$ m/s$^2) + (64$ kg$)(4.6$ m/s$^2) = 920$ N.

EVALUATE: a increases with time, so the scale reading is increasing.

5.83. **IDENTIFY:** The blocks move together with the same acceleration. Newton's second law applies to each of them. The forces acting on each block are gravity downward the upward tension in the rope.

SET UP: $\Sigma F_y = ma_y$ applies to each block. The heavier block accelerates downward while the lighter one accelerates upward, both with acceleration a. Call the +y-axis downward for the heavier block and upward for the lighter block. We want to find the mass of each block.

EXECUTE: Heavier block: First find the acceleration using $y = y_0 + v_{0y}t + \frac{1}{2}a_y t^2$, which gives us $y = \frac{1}{2}at^2$, so $5.00 \text{ m} = \frac{1}{2}a(2.00 \text{ s})^2 \rightarrow a = 2.50 \text{ m/s}^2$. Now apply $\Sigma F_y = ma_y$ to this block, which gives $m_2 g - T = m_2 a$. Now solve for m_2: $m_2 = \frac{T}{g-a} = \frac{16.0 \text{ N}}{9.80 \text{ m/s}^2 - 2.50 \text{ m/s}^2} = 2.19$ kg.

Lighter block: $\Sigma F_y = ma_y$ gives $T - m_1 g = m_1 a$, so $m_1 = \frac{T}{a+g} = 1.30$ kg.

EVALUATE: As a check, consider the two blocks as a singe system The only external force causing the acceleration is $m_2 g - m_1 g$, so $\Sigma F_y = ma_y$ gives $m_2 g - m_1 g = (m_1 + m_2)a$. Solving for a using our results for the two masses gives $a = 2.50$ m/s², which agrees with our result.

5.85. **IDENTIFY:** Apply $\Sigma \vec{F} = m\vec{a}$ to the point where the three wires join and also to one of the balls. By symmetry the tension in each of the 35.0 cm wires is the same.

SET UP: The geometry of the situation is sketched in Figure 5.85a. The angle ϕ that each wire makes with the vertical is given by $\sin\phi = \frac{12.5 \text{ cm}}{47.5 \text{ cm}}$ and $\phi = 15.26°$. Let T_A be the tension in the vertical wire and let T_B be the tension in each of the other two wires. Neglect the weight of the wires. The free-body diagram for the left-hand ball is given in Figure 5.85b and for the point where the wires join in Figure 5.85c. n is the force one ball exerts on the other.

EXECUTE: **(a)** $\Sigma F_y = ma_y$ applied to the ball gives $T_B \cos\phi - mg = 0$.

$T_B = \frac{mg}{\cos\phi} = \frac{(15.0 \text{ kg})(9.80 \text{ m/s}^2)}{\cos 15.26°} = 152$ N. Then $\Sigma F_y = ma_y$ applied in Figure 5.85c gives $T_A - 2T_B \cos\phi = 0$ and $T_A = 2(152 \text{ N})\cos\phi = 249$ N.

(b) $\Sigma F_x = ma_x$ applied to the ball gives $n - T_B \sin\phi = 0$ and $n = (152 \text{ N})\sin 15.26° = 40.0$ N.

EVALUATE: T_A equals the total weight of the two balls.

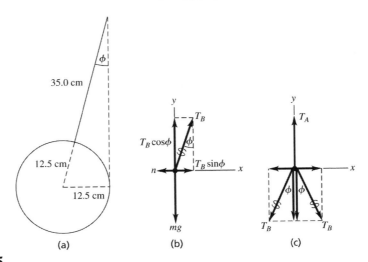

Figure 5.85

5.87. IDENTIFY: Apply $-\left(\dfrac{M+m}{M}\right)\tan\alpha$ to each block. Forces between the blocks are related by Newton's third law. The target variable is the force F. Block B is pulled to the left at constant speed, so block A moves to the right at constant speed and $a=0$ for each block.

SET UP: The free-body diagram for block A is given in Figure 5.87a. n_{BA} is the normal force that B exerts on A. $f_{BA} = \mu_k n_{BA}$ is the kinetic friction force that B exerts on A. Block A moves to the right relative to B, and f_{BA} opposes this motion, so f_{BA} is to the left. Note also that F acts just on B, not on A.

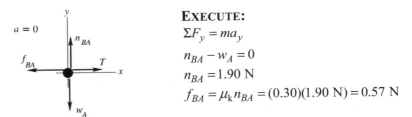

EXECUTE:
$\Sigma F_y = ma_y$
$n_{BA} - w_A = 0$
$n_{BA} = 1.90 \text{ N}$
$f_{BA} = \mu_k n_{BA} = (0.30)(1.90 \text{ N}) = 0.57 \text{ N}$

Figure 5.87a

$\Sigma F_x = ma_x$. $T - f_{BA} = 0$. $T = f_{BA} = 0.57$ N.
SET UP: The free-body diagram for block B is given in Figure 5.87b.

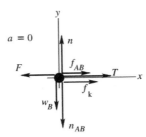

Figure 5.87b

EXECUTE: n_{AB} is the normal force that block A exerts on block B. By Newton's third law n_{AB} and n_{BA} are equal in magnitude and opposite in direction, so $n_{AB} = 1.90$ N. f_{AB} is the kinetic friction force that A exerts on B. Block B moves to the left relative to A and f_{AB} opposes this motion, so f_{AB} is to the right. $f_{AB} = \mu_k n_{AB} = (0.30)(1.90 \text{ N}) = 0.57$ N. n

and f_k are the normal and friction force exerted by the floor on block B; $f_k = \mu_k n$. Note that block B moves to the left relative to the floor and f_k opposes this motion, so f_k is to the right.

$\Sigma F_y = ma_y$: $n - w_B - n_{AB} = 0$. $n = w_B + n_{AB} = 4.20 \text{ N} + 1.90 \text{ N} = 6.10 \text{ N}$. Then $f_k = \mu_k n = (0.30)(6.10 \text{ N}) = 1.83 \text{ N}$. $\Sigma F_x = ma_x$: $f_{AB} + T + f_k - F = 0$.
$F = T + f_{AB} + f_k = 0.57 \text{ N} + 0.57 \text{ N} + 1.83 \text{ N} = 3.0 \text{ N}$.

EVALUATE: Note that f_{AB} and f_{BA} are a third law action-reaction pair, so they must be equal in magnitude and opposite in direction and this is indeed what our calculation gives.

Now use this value of a_x in the equation for block A, giving $P - \mu_s m_A g = m_A \left(\dfrac{P}{m_A + m_B} \right)$.

Solving for P and using $m_A = 2.00$ kg, $m_B = 5.00$ kg, and $\mu_s = 0.400$ gives $P = 11.0$ N.

EVALUATE: If P were to exceed 11.0 N, slipping would occur, the friction force would be kinetic friction, and the blocks would not have the same acceleration.

5.91. **IDENTIFY:** Let the tensions in the ropes be T_1 and T_2.

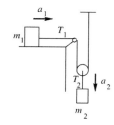

Figure 5.91a

Consider the forces on each block. In each case take a positive coordinate direction in the direction of the acceleration of that block.

SET UP: The free-body diagram for m_1 is given in Figure 5.91b.

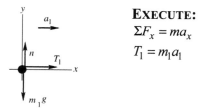

EXECUTE:
$\Sigma F_x = ma_x$
$T_1 = m_1 a_1$

Figure 5.91b

SET UP: The free-body diagram for m_2 is given in Figure 5.91c.

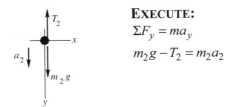

EXECUTE:
$\Sigma F_y = ma_y$
$m_2 g - T_2 = m_2 a_2$

Figure 5.91c

This gives us two equations, but there are four unknowns (T_1, T_2, a_1 and a_2) so two more equations are required.

SET UP: The free-body diagram for the moveable pulley (mass m) is given in Figure 5.91d.

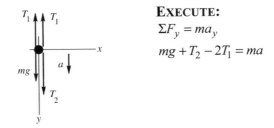

EXECUTE:
$\Sigma F_y = ma_y$
$mg + T_2 - 2T_1 = ma$

Figure 5.91d

But our pulleys have negligible mass, so $mg = ma = 0$ and $T_2 = 2T_1$. Combine these three equations to eliminate T_1 and T_2: $m_2 g - T_2 = m_2 a_2$ gives $m_2 g - 2T_1 = m_2 a_2$. And then with $T_1 = m_1 a_1$ we have $m_2 g - 2m_1 a_1 = m_2 a_2$.

SET UP: There are still two unknowns, a_1 and a_2. But the accelerations a_1 and a_2 are related. In any time interval, if m_1 moves to the right a distance d, then in the same time m_2 moves downward a distance $d/2$. One of the constant acceleration kinematic equations says $x - x_0 = v_{0x} t + \frac{1}{2} a_x t^2$, so if m_2 moves half the distance it must have half the acceleration of m_1: $a_2 = a_1/2$, or $a_1 = 2a_2$.

EXECUTE: This is the additional equation we need. Use it in the previous equation and get
$m_2 g - 2m_1 (2a_2) = m_2 a_2$.
$a_2 (4m_1 + m_2) = m_2 g$
$a_2 = \dfrac{m_2 g}{4m_1 + m_2}$ and $a_1 = 2a_2 = \dfrac{2m_2 g}{4m_1 + m_2}$.

EVALUATE: If $m_2 \to 0$ or $m_1 \to \infty$, $a_1 = a_2 = 0$. If $m_2 \gg m_1$, $a_2 = g$ and $a_1 = 2g$.

5.93. IDENTIFY: Both blocks accelerate together. The force F accelerates the two-block system, and the blocks exert kinetic friction forces on each other. Newton's second law applies to each block as well as the entire system.

SET UP: We apply $\Sigma F_x = ma_x$ and $\Sigma F_y = 0$ to each block. Block A accelerates to the right and B accelerates to the left. Call the direction of acceleration the $+x$-direction in each case, and call each acceleration a. Fig. 5.93 shows free-body diagrams of each block. It is important to realize two things immediately: the normal force that A exerts on B is equal and opposite to the normal force that B exerts on A, and the blocks exert equal but opposite friction forces on each other. Both of these points are due to Newton's third law (action-reaction). We want to find the tension in the cord and the coefficient of kinetic friction between A and B.

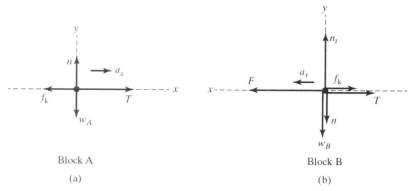

Figure 5.93

EXECUTE: **(a) & (b)** <u>Isolate block A</u>: Call n the normal force that the blocks exert on each other and f_k the kinetic friction force they exert on each other. From Fig. 5.93a, $\Sigma F_y = 0$ gives $n = w_A$ and $\Sigma F_x = ma_x$ gives $T - f_k = m_A a$, which becomes $T - \mu_k w_A = m_A a$. (Eq. 1)

<u>Isolate block B</u>: Call n_t the normal force due to the table. From Fig. 5.93b wee see that $\Sigma F_x = ma_x$ gives $T + f_k - F = m_B a$, which becomes $F - T - \mu_k w_A = m_B a$. (Eq. 2)

Combining Eq. 1 and Eq. 2 and solve for μ_k gives $F - 2\mu_k w_A = (m_A + m_B)a$. Solving for μ_k gives $\mu_k = \dfrac{F - (m_A + m_B)a}{2m_A g}$. Using the given masses, force, and acceleration gives $\mu_k = 0.242$.

Now use Eq. 1 (or Eq. 2) to find T: $T = m_A a + \mu_k m_A g = 7.75$ N.

EVALUATE: The tension is less than the force F, which is reasonable because B could never accelerate if T were greater than F. Also, from Table 5.1 we see that 0.242 is a reasonable coefficient of kinetic friction.

5.95. **IDENTIFY:** Apply $\Sigma \vec{F} = m\vec{a}$ to the block. The cart and the block have the same acceleration. The normal force exerted by the cart on the block is perpendicular to the front of the cart, so is horizontal and to the right. The friction force on the block is directed so as to hold the block up against the downward pull of gravity. We want to calculate the minimum a required, so take static friction to have its maximum value, $f_s = \mu_s n$.

SET UP: The free-body diagram for the block is given in Figure 5.95.

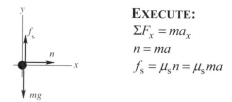

EXECUTE:
$\Sigma F_x = ma_x$
$n = ma$
$f_s = \mu_s n = \mu_s ma$

Figure 5.95

$\Sigma F_y = ma_y$: $f_s - mg = 0$

$\mu_s ma = mg$, so $a = g/\mu_s$.

EVALUATE: An observer on the cart sees the block pinned there, with no reason for a horizontal force on it because the block is at rest relative to the cart. Therefore, such an observer concludes that $n = 0$ and thus $f_s = 0$, and he doesn't understand what holds the block up against the downward force of gravity. The reason for this difficulty is that $\Sigma \vec{F} = m\vec{a}$ does not apply in a coordinate frame attached to the cart. This reference frame is accelerated, and hence not inertial. The smaller μ_s is, the larger a must be to keep the block pinned against the front of the cart.

5.97. IDENTIFY: Apply $\Sigma \vec{F} = m\vec{a}$ to the block and to the plank.

SET UP: Both objects have $a = 0$.

EXECUTE: Let n_B be the normal force between the plank and the block and n_A be the normal force between the block and the incline. Then, $n_B = w\cos\theta$ and $n_A = n_B + 3w\cos\theta = 4w\cos\theta$. The net frictional force on the block is $\mu_k(n_A + n_B) = \mu_k 5w\cos\theta$. To move at constant speed, this must balance the component of the block's weight along the incline, so $3w\sin\theta = \mu_k 5w\cos\theta$, and $\mu_k = \frac{3}{5}\tan\theta = \frac{3}{5}\tan 37° = 0.452$.

EVALUATE: In the absence of the plank the block slides down at constant speed when the slope angle and coefficient of friction are related by $\tan\theta = \mu_k$. For $\theta = 36.9°$, $\mu_k = 0.75$. A smaller μ_k is needed when the plank is present because the plank provides an additional friction force.

5.99. IDENTIFY: Apply $\Sigma \vec{F} = m\vec{a}$ to the person. The person moves in a horizontal circle so his acceleration is $a_{rad} = v^2/R$, directed toward the center of the circle. The target variable is the coefficient of static friction between the person and the surface of the cylinder.

$$v = (0.60 \text{ rev/s})\left(\frac{2\pi R}{1 \text{ rev}}\right) = (0.60 \text{ rev/s})\left(\frac{2\pi(2.5 \text{ m})}{1 \text{ rev}}\right) = 9.425 \text{ m/s}$$

(a) SET UP: The problem situation is sketched in Figure 5.99a.

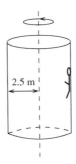

Figure 5.99a

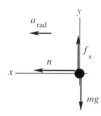

The free-body diagram for the person is sketched in Figure 5.99b. The person is held up against gravity by the static friction force exerted on him by the wall. The acceleration of the person is a_{rad}, directed in toward the axis of rotation.

Figure 5.99b

(b) EXECUTE: To calculate the minimum μ_s required, take f_s to have its maximum value, $f_s = \mu_s n$.

$\Sigma F_y = ma_y$: $f_s - mg = 0$

$\mu_s n = mg$

$\Sigma F_x = ma_x$: $n = mv^2/R$

Combine these two equations to eliminate n: $\mu_s mv^2/R = mg$

$\mu_s = \dfrac{Rg}{v^2} = \dfrac{(2.5 \text{ m})(9.80 \text{ m/s}^2)}{(9.425 \text{ m/s})^2} = 0.28$

(c) EVALUATE: No, the mass of the person divided out of the equation for μ_s. Also, the smaller μ_s is, the larger v must be to keep the person from sliding down. For smaller μ_s the cylinder must rotate faster to make n large enough.

5.101. IDENTIFY: The race car is accelerated toward the center of the circle of the curve. If the car goes at the proper speed for the banking angle, there will be no friction force on it, and will not tend to slide either up or down the road. In this case, it is going faster than the proper speed, so it will tend to slide up the road, so the friction force will be down the road. Newton's second law applies to the car.

SET UP: At the maximum speed so the car will not slide up the road, static friction from the road on the tires is at its maximum value, so $f_s = \mu_s n$ and its direction is down the road. Fig. 5.101 shows a free-body diagram of the car. Call the +x-axis horizontal pointing toward the center of the circular curve, and the y-axis perpendicular to the road surface. This is one of the few times that it is better *not* to take the x-axis parallel to the surface of the incline. The reason is that the acceleration is horizontal, not parallel to the surface. Apply $\Sigma F_x = ma_x$ and $\Sigma F_y = 0$ letting β be the banking angle.

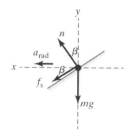

Figure 5.101

EXECUTE: (a) Start with $\Sigma F_x = ma_{\text{rad}}$. From the free-body diagram, we see that

$\Sigma F_x = n \sin \beta + f_s \cos \beta = n \sin \beta + \mu_s n \cos \beta$, so

$ma_{\text{rad}} = n(\sin \beta + \mu_s \cos \beta)$ (Eq. 1)

$\Sigma F_y = 0 = n \cos \beta - f_s \sin \beta - mg = n \cos \beta - \mu_s n \sin \beta - mg = 0$, which gives

$mg = n(\cos \beta - \mu_s \sin \beta)$ (Eq. 2)

Solving for n gives $n = \dfrac{mg}{\cos \beta - \mu_s \sin \beta} = \dfrac{(1200 \text{ kg})(9.80 \text{ m/s}^2)}{\cos 18.0° - (0.400) \sin 18.0°} = 1.42 \times 10^4$ N.

(b) Divide Eq. 1 by Eq. 2, giving $a_{rad} = \left(\dfrac{\sin \beta + \mu_s \cos \beta}{\cos \beta - \mu_s \sin \beta} \right) g$. Using $\beta = 18.0°$ and $\mu_s = 0.400$ gives $a_{rad} = 8.17$ m/s².

(c) $a_{rad} = v^2/R$, so $v = \sqrt{R a_{rad}} = \sqrt{(90.0 \text{ m})(8.17 \text{ m/s}^2)} = 27.1$ m/s.

EVALUATE: Check our result in (b) for a very smooth road in which $\mu_s = 0$. This gives $a_{rad} = g \tan \beta$, so $\dfrac{v^2}{R} = g \tan \beta$, or $\tan \beta = \dfrac{v^2}{Rg}$. This result give the familiar angle for a properly banked road for speed v. If there were no friction, a car would have to go at the proper speed to avoid slipping. So our result in this special case checks out. The proper speed for $\beta = 18.0°$ should be $v = \sqrt{Rg \tan \beta} = \sqrt{(90.0 \text{ m})(9.80 \text{ m/s}^2) \tan 18.0°} = 16.9$ m/s, so the speed we just found is obviously much greater than the proper speed.

5.105. **IDENTIFY:** Apply $\Sigma \vec{F} = m\vec{a}$, with $f = kv$.

SET UP: Follow the analysis that leads to the equation $v_y = v_t [1 - e^{-(k/m)t}]$, except now the initial speed is $v_{0y} = 3mg/k = 3v_t$ rather than zero.

EXECUTE: The separated equation of motion has a lower limit of $3v_t$ instead of zero; specifically,

$$\int_{3v_t}^{v} \dfrac{dv}{v - v_t} = \ln \dfrac{v_t - v}{-2v_t} = \ln \left(\dfrac{v}{2v_t} - \dfrac{1}{2} \right) = -\dfrac{k}{m}t, \text{ or } v = 2v_t \left[\dfrac{1}{2} + e^{-(k/m)t} \right]$$

where $v_t = mg/k$.

EVALUATE: As $t \to \infty$ the speed approaches v_t. The speed is always greater than v_t and this limit is approached from above.

5.107. **IDENTIFY:** Apply $\Sigma \vec{F} = m\vec{a}$ to the circular motion of the bead. Also use $a_{rad} = 4\pi^2 R/T^2$ to relate a_{rad} to the period of rotation T.

SET UP: The bead and hoop are sketched in Figure 5.107a.

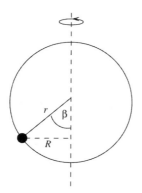

The bead moves in a circle of radius $R = r\sin\beta$. **The normal force exerted on the bead by the hoop is radially inward.**

Figure 5.107a

The free-body diagram for the bead is sketched in Figure 5.107b.

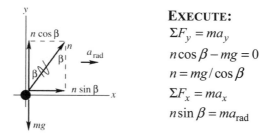

EXECUTE:
$\Sigma F_y = ma_y$
$n\cos\beta - mg = 0$
$n = mg/\cos\beta$
$\Sigma F_x = ma_x$
$n\sin\beta = ma_{\text{rad}}$

Figure 5.107b

Combine these two equations to eliminate n:

$\left(\dfrac{mg}{\cos\beta}\right)\sin\beta = ma_{\text{rad}}$

$\dfrac{\sin\beta}{\cos\beta} = \dfrac{a_{\text{rad}}}{g}$

$a_{\text{rad}} = v^2/R$ and $v = 2\pi R/T$, so $a_{\text{rad}} = 4\pi^2 R/T^2$, where T is the time for one revolution. $R = r\sin\beta$, so $a_{\text{rad}} = \dfrac{4\pi^2 r\sin\beta}{T^2}$

Use this in the above equation: $\dfrac{\sin\beta}{\cos\beta} = \dfrac{4\pi^2 r\sin\beta}{T^2 g}$

This equation is satisfied by $\sin\beta = 0$, so $\beta = 0$, or by $\dfrac{1}{\cos\beta} = \dfrac{4\pi^2 r}{T^2 g}$, which gives

$\cos\beta = \dfrac{T^2 g}{4\pi^2 r}$.

(a) 4.00 rev/s implies $T = (1/4.00)$ s $= 0.250$ s

Then $\cos\beta = \dfrac{(0.250\text{ s})^2(9.80\text{ m/s}^2)}{4\pi^2(0.100\text{ m})}$ and $\beta = 81.1°$.

(b) This would mean $\beta = 90°$. But $\cos 90° = 0$, so this requires $T \to 0$. So β approaches $90°$ as the hoop rotates very fast, but $\beta = 90°$ is not possible.

(c) 1.00 rev/s implies $T = 1.00$ s

The $\cos\beta = \dfrac{T^2 g}{4\pi^2 r}$ equation then says $\cos\beta = \dfrac{(1.00\text{ s})^2(9.80\text{ m/s}^2)}{4\pi^2(0.100\text{ m})} = 2.48$, which is not possible. The only way to have the $\Sigma\vec{F} = m\vec{a}$ equations satisfied is for $\sin\beta = 0$. This means $\beta = 0$; the bead sits at the bottom of the hoop.

EVALUATE: $\beta \to 90°$ as $T \to 0$ (hoop moves faster). The largest value T can have is given by $T^2 g/(4\pi^2 r) = 1$ so $T = 2\pi\sqrt{r/g} = 0.635$ s. This corresponds to a rotation rate of $(1/0.635)$ rev/s $= 1.58$ rev/s. For a rotation rate less than 1.58 rev/s, $\beta = 0$ is the only solution and the bead sits at the bottom of the hoop. Part (c) is an example of this.

5.109. IDENTIFY: The block begins to move when static friction has reached its maximum value. After that, kinetic friction acts and the block accelerates, obeying Newton's second law.

SET UP: $\Sigma F_x = ma_x$ and $f_{s,max} = \mu_s n$, where n is the normal force (the weight of the block in this case).

EXECUTE: (a) & (b) $\Sigma F_x = ma_x$ gives $T - \mu_k mg = ma$. The graph with the problem shows the acceleration a of the block versus the tension T in the cord. So we solve the equation from Newton's second law for a versus T, giving $a = (1/m)T - \mu_k g$. Therefore the slope of the graph will be $1/m$ and the intercept with the vertical axis will be $-\mu_k g$. Using the information given in the problem for the best-fit equation, we have $1/m = 0.182$ kg^{-1}, so $m = 5.4945$ kg and $-\mu_k g = -2.842$ m/s^2, so $\mu_k = 0.290$.

When the block is just ready to slip, we have $f_{s,max} = \mu_s n$, which gives $\mu_s = (20.0\text{ N})/[(5.4945\text{ kg})(9.80\text{ m/s}^2)] = 0.371$.

(c) On the Moon, g is less than on earth, but the mass m of the block would be the same as would μ_k. Therefore the slope ($1/m$) would be the same, but the intercept ($-\mu_k g$) would be less negative.

EVALUATE: Both coefficients of friction are reasonable for ordinary materials, so our results are believable.

5.111. IDENTIFY: A cable pulling parallel to the surface of a ramp accelerates 2170-kg metal blocks up a ramp that rises at 40.0° above the horizontal. Newton's second law applies to the blocks, and the constant-acceleration kinematics formulas can be used.

SET UP: Call the $+x$-axis parallel to the ramp surface pointing upward because that is the direction of the acceleration of the blocks, and let the y-axis be perpendicular to the surface. There is no acceleration in the y-direction. $\Sigma F_x = ma_x$, $f_k = \mu_k n$, and
$x - x_0 = v_{0x}t + \dfrac{1}{2}a_x t^2$.

EXECUTE: (a) First use $x - x_0 = v_{0x}t + \dfrac{1}{2}a_x t^2$ to find the acceleration of a block. Since $v_{0x} = 0$, we have $a_x = 2(x - x_0)/t^2 = 2(8.00\text{ m})/(4.20\text{ s})^2 = 0.9070$ m/s^2. The forces in the y-direction balance, so
$n = mg\cos(40.0°)$, so $f_k = (0.350)(2170\text{ kg})(9.80\text{ m/s}^2)\cos(40.0°) = 5207$ N. Using $\Sigma F_x = ma_x$, we have $T - mg\sin(40.0°) - f_k = ma$. Solving for T gives $T = (2170\text{ kg})(9.80\text{ m/s}^2)\sin(40.0°) + 5207\text{ N} + (2170\text{ kg})(0.9070\text{ m/s}^2) = 2.13 \times 10^4$ N $= 21.3$ kN.
From the table shown with the problem, this tension is greater than the safe load of a ½ inch diameter cable (which is 19.0 kN), so we need to use a 5/8-inch cable.

(b) We assume that the safe load (SL) is proportional to the cross-sectional area of the cable, which means that SL $\propto \pi(D/2)^2 \propto (\pi/4)D^2$, where D is the diameter of the cable. Therefore a graph of SL versus D^2 should give a straight line. We could use the data given in the table with the problem to make the graph by hand, or we could use graphing software. The resulting graph is shown in Figure 5.111 (next page). The best-fit line has a slope of 74.09 kN/in.^{2} and a y-intercept of 0.499 kN. For a cable of diameter $D = 9/16$ in., this equation gives SL = (74.09 kN/in.^{2})(9/16 in.)^2 + 0.499 kN = 23.9 kN.

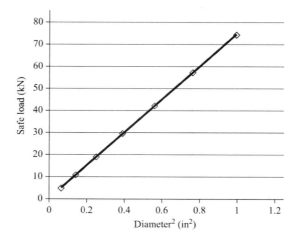

Figure 5.111

(c) The acceleration is now zero, so the forces along the surface balance, giving $T + f_s = mg \sin(40.0°)$. Using the numbers we get $T = 3.57$ kN.
(d) The tension at the top of the cable must accelerate the block and the cable below it, so the tension at the top would be larger. For a 5/8-inch cable, the mass per meter is 0.98 kg/m, so the 9.00-m long cable would have a mass of (0.98 kg/m)(9.00 m) = 8.8 kg. This is only 0.4% of the mass of the block, so neglecting the cable weight has little effect on accuracy.
EVALUATE: It is reasonable that the safe load of a cable is proportional to its cross-sectional area. If we think of the cable as consisting of many tiny strings each pulling, doubling the area would double the number of strings.

Study Guide for Work and Kinetic Energy

Summary

We introduce two new concepts in this chapter: *work* and *energy*. Our investigation begins by learning how work can be used to solve problems with variable forces. We'll see how work is a form of energy transfer, leading us to learn about energy, one of the most important concepts in physics. We'll learn how work can be used to change an object's kinetic energy (the energy of motion), how to determine the work expended in many situations, and how power is the rate of change of work with respect to time.

Objectives

After studying this chapter, you'll understand

- The definition of *work* and how to calculate the work done by a force on an object.
- The definition and interpretation of *kinetic energy*.
- How to apply the work–energy theorem to problems.
- How to use kinetic energy and work in problems involving varying forces applied along curved paths.
- How to analyze springs and the elastic force.
- The definition of *power* and how to calculate power for objects performing work or on which work is performed.

Concepts and Equations

Term	Description
Work Done by a Force	A constant force acting on and displacing an object does work. For a constant force $\vec{F}$ acting on a particle causing a straight-line displacement $\vec{s}$ at an angle ϕ with respect to the force, the work done by the force on the object is $$W = \vec{F} \cdot \vec{s} = Fs\cos\phi.$$ The SI unit of work is 1 joule = 1 newton-meter (1 J = 1 N·m).
Kinetic Energy	Kinetic energy K is the energy of motion of a particle with mass. Kinetic energy is equal to the amount of work required to accelerate a particle from rest to a speed v. A particle of mass m and velocity v has kinetic energy $$K = \tfrac{1}{2}mv^2.$$
Work–Energy Theorem	The work–energy theorem states that the total work done by a net external force on a particle as it undergoes a displacement is equal to the change in kinetic energy of the particle: $$W_{\text{tot}} = K_2 - K_1 = \Delta K.$$
Work Done by a Varying Force along a Curved Path	The work done by a varying force on a particle as it follows a curved path is determined by $$W = \int_{P_1}^{P_2} \vec{F} \cdot d\vec{l} = \int_{P_1}^{P_2} F\cos\phi\, dl = \int_{P_1}^{P_2} F_s\, dl.$$
Elastic Force	An elastic force is a force that restores an object to its original equilibrium position after deformation. For a spring, the deformation is approximately proportional to the applied force, as given by Hooke's law, $$F_{\text{spr}} = kx,$$ where k is the force constant and x is the displacement of the spring from its equilibrium position.
Power	Power is the rate of change of work with respect to time. *Average power* is defined as $$P_{\text{av}} = \frac{\Delta W}{\Delta t},$$ where ΔW is the quantity of work performed during the time interval Δt. Instantaneous power is defined as $$P = \lim_{\Delta t \to 0} \frac{\Delta W}{\Delta t} = \frac{dW}{dt}.$$ For a force acting on a moving particle, the instantaneous power is $$P = \vec{F} \cdot \vec{v}.$$ The SI unit of power is 1 watt = 1 joule/second (1 W = 1 J/s).

Key Concept 1: To find the work W done by a constant force $\vec{F}$ acting on an object that undergoes a straight-line displacement $\vec{s}$, calculate the scalar product of these two vectors: $W = \vec{F} \cdot \vec{s}$.

Key Concept 2: To find the total work done on a moving object, calculate the sum of the amounts of work done by each force that acts on the object. The total work also equals the work done by the *net* force on the object.

Key Concept 3: You can use the work–energy theorem to easily relate the initial and final speeds of an object that moves while being acted on by constant forces.

Key Concept 4: If you know the initial and final speeds of an object that moves over a given straight-line distance, the work–energy theorem lets you calculate the net force that causes the change in speed.

Key Concept 5: The kinetic energy of an object with speed v equals the amount of work you must do to accelerate it from rest to speed v.

Key Concept 6: You must use Eq. (6.10) to calculate the work done by the nonconstant force that a spring exerts.

Key Concept 7: The work–energy theorem also allows you to solve problems with *varying* forces, such as the force exerted by a spring.

Key Concept 8: The work–energy theorem can help you solve problems in which an object follows a curved path. Take care in calculating the work done on such a path.

Key Concept 9: To find the power of a force acting on a moving object, multiply the component of force in the direction of motion by the object's speed.

Key Concept 10: To calculate average power (the average rate of doing work), divide the work done by the time required to do that work.

Conceptual Questions

1: Work done by the normal force

How much work does the normal force do on a box sliding across the floor?

IDENTIFY, SET UP, AND EXECUTE Work is produced when a force acts on an object in the direction the object is displaced. As a box slides across the floor, the normal force is perpendicular to its motion. Therefore, the normal force does no work on the sliding box.

KEY CONCEPT 1 **EVALUATE** Work has a strict definition in physics. The normal force prevents the box from falling into the floor, but it does no work thereby. You also do no work as you carry your backpack, even though your arm tires.

2: Which force does the most work?

Rank the four situations shown in Figure 1 from most to least work done by the force. The displacement is the same in each case.

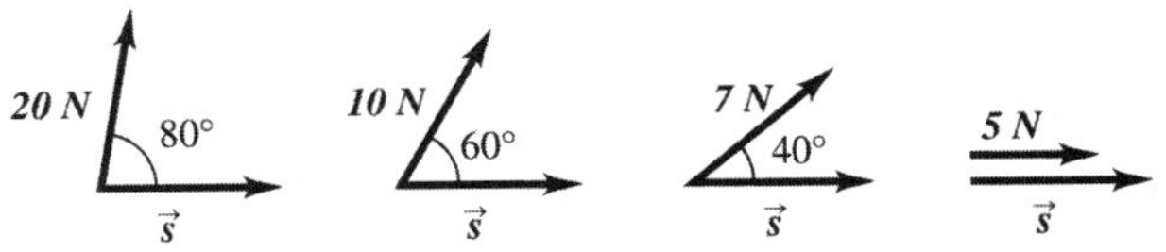

Figure 1 Conceptual Question 2.

IDENTIFY, SET UP, AND EXECUTE Work is the dot product of force and displacement, or displacement times the component of force parallel to the displacement. All four situations have the same displacement, so we rank the components of the force parallel to the displacement. The parallel components of the forces are the magnitudes of the forces times the cosine of the angle between the force and the displacement vectors.

We find that the parallel components for the four situations are, in order, 3.47 N, 4.00 N, 5.36 N, and 5.00 N. Therefore, the ranking from greatest to least amount of work is (c), (d), (b), and (a).

KEY CONCEPT 1 **EVALUATE** We see that larger forces do not necessarily produce more work. Work depends on *both* the magnitude of the force *and* the direction of the force with respect to the displacement. The 20 N force produces much less work than the 5 N force.

3: Ranking stopping distance

Five identical shipping crates slide down a ramp onto a rough horizontal floor. Each crate carries a different mass and has a different velocity at the bottom of the ramp, given in Table 1. Rank the distances required, from least to greatest, for each crate to stop.

TABLE 1: Conceptual Problem 3.

	Mass (kg)	Velocity at bottom of ramp (m/s)
Crate 1	20.0	10.0
Crate 2	30.0	20.0
Crate 3	20.0	15.0
Crate 4	100.0	10.0
Crate 5	200.0	5.0

IDENTIFY, SET UP, AND EXECUTE At the bottom of the ramp, each crate has kinetic energy. As they slow, the crates lose kinetic energy because friction does work on them. The change in kinetic energy is equal to the work done by friction, which in turn is equal to the friction force (μmg) times the displacement (x). Since the crates are identical, their coefficient of friction is the same. Algebraically, the work done is

$$W = \Delta K = K_2 - K_1,$$
$$\mu mgx = 0 - \tfrac{1}{2}mv^2,$$
$$\mu gx = -\tfrac{1}{2}v^2.$$

We see that the displacement is proportional to the velocity squared. To rank the stopping distances, we compare the velocities at the bottom of the ramp.

Crate 5 has the smallest velocity, crates 1 and 4 have the next-largest velocity, crate 3 has the next-largest velocity, and crate 2 has the largest velocity. The ranking of stopping distance for the crates is (5), (1) = (4), (3), and (2).

KEY CONCEPT 3 **EVALUATE** We see that in this case the results do not depend on the mass of the object. Crate 4 has much more mass than crate 1, so crate 4 has more initial kinetic energy. However, the effect of friction on crate 4 is greater than the effect of friction on crate 1. These two effects cancel, resulting in the same stopping distance.

We also see that the work done by friction is negative, because the change in kinetic energy is negative.

4: Work in carnival ride

A swing ride at a carnival consists of chairs attached by a cable to a vertical pole. The vertical pole rotates, causing the chairs to swing in a circle as shown in Figure 2. How much work is done by the tension in the cable as one chair (with a mass of 70 kg) makes one complete revolution? The length of the rope and angle are shown in the figure.

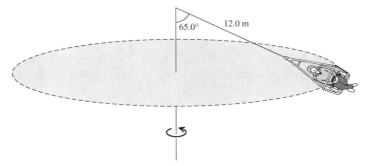

Figure 2 Conceptual Question 4.

IDENTIFY, SET UP, AND EXECUTE As the chair swings, it is displaced along the circle. The tension in the cable is directed perpendicular to the chair's displacement. Since work is the dot product of the force and displacement, the work done by the tension is zero.

EVALUATE Does the kinetic energy change? The chair moves at a constant rate, so the kinetic energy remains constant. No work is needed to keep the chair moving at the same speed.

5: Work with equal forces

A force is applied to several boxes. Rank the five situations shown in Figure 3 from least to most work done by the force. The same magnitude of force is applied to all boxes, and the mass of each box is given.

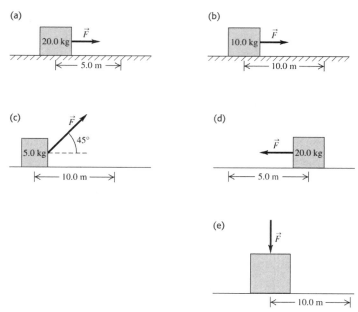

Figure 3 Conceptual Question 5.

IDENTIFY, SET UP, AND EXECUTE Work is the dot product of force and displacement. All five situations have the same magnitude of applied force, so we rank them on the basis of the displacement and the cosine of the angle between the force and the resulting displacement.

In cases (a) and (b), the displacement is in the direction of the force. The box in (b) is displaced further, so more work is done by the force. In case (c), the force is applied at an angle of 45° with respect to the displacement and the box is displaced 10.0 m, resulting in less work than in (b), but more than in (a). The work in case (d) is negative, since the force is opposite the displacement. In case (e), no work is done, since the force is perpendicular to the displacement.

The amount of work done by the force, from least to most, is (d), (e), (a), (c), and (b).

KEY CONCEPT 1 EVALUATE We see that the mass of the box does not influence the results: The work done by the force depends only on the force, the displacement, and the angle between them.

How much work is done on the box in (e)? The box could be moving at constant speed on a frictionless surface, in which case no work is done. Or another force may be acting on the box, creating work. From the information we have, we cannot determine which is correct.

We also see that we can have negative work in problems. Negative work indicates that energy is being removed from the system, perhaps as the box slows to a stop.

Problems

1: Work done in pushing a crate up an inclined plane.

A student pushes a crate 3.50 m up a rough inclined plane with a constant horizontal force of 225 N, starting from rest as shown in Figure 4. Find the work done by the student, the work done by friction, the work done by gravity, and the change in the crate's kinetic energy. How does the work done by the student, friction, and gravity compare with the change in kinetic energy? The crate has a mass of 15.0 kg, the incline is sloped at 30.0°, and the coefficient of kinetic friction between the crate and the incline is 0.400.

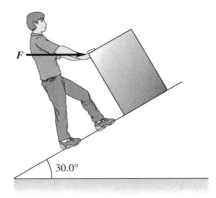

Figure 4 Problem 1.

IDENTIFY Each force is constant and the displacement is along a straight line, so we can find the work and kinetic energy from their definitions.

SET UP We'll begin with a free-body diagram to find the work done by the three forces. Figure 5 shows the free-body diagram with a rotated coordinate system that coincides with the incline to simplify the analysis. The forces are identified by their magnitudes: kinetic friction (f_k), gravity (mg), the normal force (n), and the force applied by the student (F). Kinetic friction opposes the motion up the incline and thus is directed down the incline. Gravity and the applied force are resolved into components parallel and perpendicular to the incline.

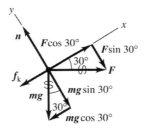

Figure 5 Problem 1 free-body diagram.

EXECUTE The work done by the student pushing the crate is

$$W_{student} = Fs\cos\phi = (225 \text{ N})(3.50 \text{ m})\cos 30° = 682 \text{ J},$$

where the angle between the force and the displacement is 30°. To find the work done by friction, we need to know the friction force. We apply the equilibrium condition in the y-direction to find the normal force in order to quantify the friction force:

$$\Sigma F_y = n + (-mg\cos 30°) + (2\,F\sin 30°) = 0,$$

$$n = mg\cos 30° + F\sin 30° = (15.0 \text{ kg})(9.8 \text{ m/s}^2)(\cos 30°) + (225 \text{ N})(\sin 30°) = 240 \text{ N}.$$

The kinetic friction force is then

$$f_k = \mu_k n = (0.400)(240 \text{ N}) = 95.9 \text{ N}.$$

Friction is directed opposite to the displacement, so the angle between them is 180°. The work done by friction is

$$W_{f_k} = f_k s \cos\phi = (95.9 \text{ N})(3.50 \text{ m})\cos 180° = -336 \text{ J}.$$

The component of gravity along the displacement is also opposite to the displacement. The work done by gravity is

$$W_{\text{grav}} = F_g s \cos\phi = mgs \cos 120° = (15.0 \text{ kg})(9.8 \text{ m/s}^2)(3.50 \text{ m})\cos 120° = -257 \text{ J}.$$

To find the change in kinetic energy, we need the initial and final velocities. The initial velocity is zero. The final velocity is found by applying Newton's second law and kinematics. In the x-direction, along the incline, Newton's second law gives

$$\Sigma F_x = F\cos 30° + (-f_k) + (-mg\sin 30°) = ma_x.$$

The acceleration of the box is then

$$a_x = \frac{F\cos 30° + (-f_k) + (-mg\sin 30°)}{m}$$

$$= \frac{(225 \text{ N})\cos 30° - (95.9 \text{ N}) - (15.0 \text{ kg})(9.8 \text{ m/s}^2)\sin 30°}{(15.0 \text{ kg})} = 1.70 \text{ m/s}^2.$$

Constant-acceleration kinematics gives the final velocity:

$$v^2 = v_{x0}^2 + 2a_x(x - x_0),$$

$$v = \sqrt{0 + 2(1.70 \text{ m/s}^2)(3.50 \text{ m})} = 3.45 \text{ m/s}.$$

The change in kinetic energy is then

$$\Delta K = K_2 - K_1 = \tfrac{1}{2}mv^2 - 0 = \tfrac{1}{2}(15.0 \text{ kg})(3.45 \text{ m/s})^2 = 89.3 \text{ J}.$$

In sum, we found that the student did 682 J of work on the crate, friction did −336 J of work on the crate, gravity did −257 J of work, and the kinetic energy increased by 89 J. When we add the work due to the three forces together, the total work is 89 J. The total work is equal to the change in kinetic energy.

KEY CONCEPT 2 **EVALUATE** This problem affords a thorough investigation of work and kinetic energy. It illustrates how to combine the work due to several forces into the total work and how the total work on a system is used to increase the kinetic energy of the system.

CAUTION **Watch Signs!** You must evaluate the signs carefully when determining work and energy. Negative work indicates that the force is directed opposite to the displacement and often slows the object, as it does in this problem.

Practice Problem: What is the net work done if the student pushes directly up the ramp? *Answer:* 195 J

Extra Practice: What is the net work done if the angle is 0°? *Answer:* 452 J

2: Spring force between two blocks

Two blocks are placed on a horizontal, frictionless surface and attached to each other by a spring with force constant 4500 N/m. If the right-hand block is pulled with a force of 150.0 N, find the displacement of the spring as the blocks accelerate. The left-hand block has a mass of 5.00 kg, and the right-hand block has a mass of 3.00 kg.

IDENTIFY Our target variable is the displacement of the spring. We can solve this problem with Newton's second law.

SET UP The displacement of the spring is proportional to the spring force. We find the spring force by applying Newton's second law to the blocks and then use Hooke's law to find the displacement. The first task is to sketch the situation, as shown in Figure 6.

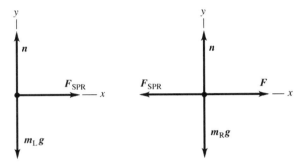

Figure 6 Problem 2.

Examining the sketch, we see the two blocks interact through the spring. We draw free-body diagrams for the two blocks, shown in Figure 7.

Figure 7 Problem 2 free-body diagram.

The forces are identified by their magnitudes: force due to the spring (F_{spr}), gravity (mg), the normal force (n), and the applied force (F). Knowledge of the vertical forces will not be necessary to solve this problem. The blocks are connected to each other; therefore, both accelerate at the same rate and are acted upon by the same magnitude of spring force. The diagrams include a common xy-coordinate axis, with the positive x-axis in the direction of the acceleration. All of the forces act along the coordinate axes, so we won't need to break the forces into components.

EXECUTE We apply Newton's second law to the horizontal forces acting on each block to determine the force due to the spring. For the right-hand block (with mass m_R),

$$\sum F_x = F + (-F_{spr}) = m_R a_x.$$

For the left-hand block (with mass m_L),

$$\sum F_x = F_{spr} = m_L a_x.$$

Examining these two equations, we find two unknowns: F_{spr} and a_x. We wish to find the spring force, so we rewrite the second equation in terms of acceleration:

$$a_x = \frac{F_{spr}}{m_L}.$$

Substituting for the acceleration in the first equation yields

$$F + (-F_{spr}) = m_R \frac{F_{spr}}{m_L},$$

$$F = F_{spr} + F_{spr}\frac{m_R}{m_L} = F_{spr}\left(1 + \frac{m_R}{m_L}\right),$$

$$F_{spr} = \frac{F}{\left(1 + \frac{m_R}{m_L}\right)} = F\frac{m_L}{m_R + m_L} = (150)\frac{(5.00 \text{ kg})}{(3.00 \text{ kg}) + (5.00 \text{ kg})} = 93.8 \text{ N}.$$

This gives us the magnitude of the force due to the spring. The direction is opposite to the displacement. We can now use Hooke's law,

$$F_{spr} = kx,$$

to solve for the displacement:

$$x = \frac{F_{spr}}{k} = \frac{(93.8 \text{ N})}{4500 \text{ N/m}} = 0.0208 \text{ m} = 2.08 \text{ cm}.$$

The spring is displaced 2.08 cm when the blocks are pulled.

EVALUATE We see that the force due to the spring is less than the applied force in this problem. That is reasonable, as the spring must provide force to accelerate only the right-hand mass, while the applied force must accelerate both masses.

Practice Problem: What is the acceleration of the blocks? *Answer:* 18.8 m/s^2

Extra Practice: How much energy is stored up in the spring? *Answer:* 0.973 J

3: Block stopped by spring and friction

A 5.00 kg block is moving along a rough horizontal surface toward a spring with force constant 500 N/m. The velocity of the block just before it contacts the spring is 12.0 m/s. If the coefficient of kinetic friction between the block and the surface is 0.400, what is the maximum compression of the spring?

IDENTIFY We can apply the work–energy theorem to the problem. Two forces (friction and the spring force) do work to slow the block, taking away the kinetic energy from the block. The target variable is the spring's maximum compression.

SET UP Figure 8 shows a sketch of the situation. Just before contacting the spring, the block has kinetic energy and the spring is uncompressed. At maximum compression, the spring has been compressed a distance X, the block is not moving ($K = 0$), and friction has done work on the block as it moves the same distance X. The work–energy theorem tells us that the change in kinetic energy is equal to the total work done.

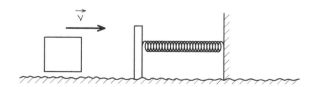

Figure 8 Problem 3 sketch.

Four forces act on the block as it slows: the force due to the spring (F_{spr}), gravity (mg), the normal force (n), and the force of kinetic friction (f_k). These forces are shown in the free-body diagram in Figure 9. The normal force and gravity do no work on the block as it slows.

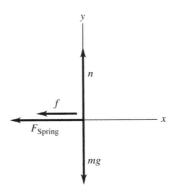

Figure 9 Problem 3 free-body diagram.

EXECUTE The change in kinetic energy of the block is

$$\Delta K = K_2 - K_1 = 0 - \tfrac{1}{2}mv^2 = -\tfrac{1}{2}mv^2.$$

The work done on the spring by the block is

$$W_{\text{Block on Spring}} = \tfrac{1}{2}kx_2^2 - \tfrac{1}{2}kx_1^2 = \tfrac{1}{2}kX^2.$$

The work done by the spring on the block is the negative of this value:

$$W_{\text{Spring}} = -\tfrac{1}{2}kX^2.$$

The work done by friction on the block is

$$W_f = Fs\cos\phi = (\mu mg)s\cos 180° = -\mu mgX.$$

The total work is the work due to the spring and the work due to friction. Setting the sum of these two quantities equal to the change in kinetic energy gives

$$-\tfrac{1}{2}kX^2 - \mu mgX = -\tfrac{1}{2}mv^2.$$

This is a quadratic equation. Solving for X yields

$$X = \frac{-(\mu mg) \pm \sqrt{(\mu mg)^2 - 4\left(\tfrac{1}{2}k\right)\left(-\tfrac{1}{2}mv^2\right)}}{2\left(\tfrac{1}{2}k\right)}$$

$$= \frac{-(19.6) \pm \sqrt{(19.6)^2 + 4(15.0)(22.5)}}{2(15.0)} \text{ m} = 0.735, -2.04 \text{ m}.$$

In this case, we require the positive root, 0.735 m. The spring compresses 0.735 m when the block comes to a momentary stop.

KEY CONCEPT 6 **EVALUATE** This problem illustrates how to use energy to work with varying forces. The analysis would have been more difficult had we used Newton's second law, as we would then have had to integrate the force equation.

CAUTION **Watch signs in work done by a spring!** As you pull on a spring, you do positive work on it and the spring does negative work on you. Understanding the signs for work done on or by a spring will help you understand the proper signs for work and energy.

Practice Problem: How far would the block slide if there were no spring? *Answer:* 18.4 m

Extra Practice: How long would it take to stop the block if there were no spring? *Answer:* 3.06 s

4: A novel spring

Suppose you have invented a novel spring that will slow a 2.5 kg toy car. The spring exerts a force that depends on position such that $F_x = [10.0 \text{ N} + (16.0 \text{ N/m}^2)x^2]$. What maximum speed of the toy car will the spring stop with a compression of 0.50 m?

IDENTIFY The maximum work due to the spring occurs when the displacement is directed along the force, so we'll take that direction as the direction of displacement. Our target variable is the speed of the toy car, and we'll find it by using the work–energy theorem.

SET UP Only the force due to the novel spring acts on the toy car. Initially, there is just the kinetic energy of the car. At the maximum compression, there is no kinetic energy.

EXECUTE The force of the spring is in the direction of the motion. The work that the spring does on the toy car as the spring is compressed a distance X is

$$W = \int_0^X F_x \, dx.$$

Substituting for the force and finding the work when $X = 0.50$ m yields

$$W = \int_0^X \left[10.0 \text{ N} + \left(16.0 \frac{\text{N}}{\text{m}^2}\right) x^2 \right] dx$$

$$= \left[(10.0 \text{ N})x + \left(16.0 \frac{\text{N}}{\text{m}^2}\right) \frac{x^3}{3} \right]_0^X$$

$$= \left[(10.0 \text{ N})X + \left(16.0 \frac{\text{N}}{\text{m}^2}\right) \frac{X^3}{3} \right]$$

$$= \left[(10.0 \text{ N})(0.50 \text{ m}) + \left(16.0 \frac{\text{N}}{\text{m}^2}\right) \frac{(0.50 \text{ m})^3}{3} \right]$$

$$= 6.33 \text{ J}.$$

The spring can stop a toy car with up to 6.33 J of energy. The velocity of that car is determined from the kinetic energy:

$$K = \tfrac{1}{2}mv^2,$$
$$v = \sqrt{\frac{2K}{m}} = \sqrt{\frac{2(6.33 \text{ J})}{(2.5 \text{ kg})}} = 2.25 \text{ m/s}.$$

The maximum velocity that the spring can stop is 2.25 m/s.

KEY CONCEPT 7 **EVALUATE** In this problem, the work is negative because the force due to the spring acts opposite to the direction of the displacement in order to reduce the kinetic energy. The change in kinetic energy is also negative, so the work–energy theorem is satisfied.

Practice Problem: How much force does the spring provide at maximum compression? *Answer:* 14 N

Extra Practice: What spring constant is needed for a normal spring to stop the car in 0.5 m? *Answer:* 51 N/m

5: Average power to run an escalator

What average power does an escalator require to lift twenty 100.0 kg people 3.0 m high in 1 minute?

IDENTIFY The target variable is the power, or the amount of work done per unit time.

SET UP The escalator provides work equivalent to the amount of work due to gravity as the people are lifted. We'll find the work done by gravity and divide by the time.

EXECUTE The work done by gravity is the weight of the people, multiplied by their change in height:

$$W = mgh = (20 \times 100 \text{ kg})(9.8 \text{ m/s}^2)(3.0 \text{ m}) = 58{,}800 \text{ J}.$$

The power is the work of gravity divided by time:

$$P = \frac{\Delta W}{\Delta t} = \frac{(58{,}800 \text{ J})}{(60 \text{ s})} = 980 \text{ W}.$$

The power needed to run the escalator is 980 W.

KEY CONCEPT 10 **EVALUATE** We see that the power is independent of the angle of the escalator. This is due to the fact that the gravitational work depends only on the change in vertical elevation.

Practice Problem: How fast are the people moving if the angle of the escalator is 45°? *Answer:* 0.707 m/s

Extra Practice: What is the average force provided if this were an elevator? *Answer:* 19,600 N

Try It Yourself!

1: Box on a smooth incline

A 10.0 kg box is pushed 2.0 m up a smooth inclined plane of angle 30.0° by a 100 N horizontal force, starting from rest. Find the work done on the box and the change in kinetic energy.

Study Guide for Work and Kinetic Energy

Solution Checkpoints

IDENTIFY AND SET UP Start with a free-body diagram to identify the forces acting on the box. The work due to each force is equal to the dot product of the force and the displacement. The change in kinetic energy is equal to the total work done.

EXECUTE Three forces act on the box. The work done by each is

$$W_{applied} = Fs\cos\phi = 173 \text{ J},$$
$$W_{normal} = 0,$$
$$W_g = -98 \text{ J}.$$

The change in kinetic energy is 75 J.

KEY CONCEPT 5 **EVALUATE** You can check your results by using Newton's second law to find the acceleration, which leads to the final velocity of the box and final kinetic energy of the box. Do they agree?

Practice Problem: How fast is the box moving at the top? *Answer:* 3.87 m/s

Extra Practice: How fast is the box moving at the top if the applied force is up the ramp? *Answer:* 4.52 m/s

2: Object sliding on a rough surface

An object slides on a rough surface. If the object is given an initial velocity of 3.0 m/s, it comes to a stop in 1.0 m. Find the coefficient of kinetic friction between the object and the surface.

Solution Checkpoints

IDENTIFY AND SET UP Use the work–energy theorem to relate the change in kinetic energy to the work done by friction. Only one force acts in the direction of motion.

EXECUTE Set the change in kinetic energy equal to the work done by friction. This results in a coefficient of kinetic friction equal to 0.46.

KEY CONCEPT 4 **EVALUATE** Is the change in kinetic energy and in the work negative or positive? Why?

Practice Problem: How long does it take to stop the box? *Answer:* 0.67 s

Extra Practice: How fast is it moving after traveling 0.5 m? *Answer:* 2.12 m/s

3: Drag on an automobile

An automobile has a 150 hp engine and a top speed of 100 mph. If you assume that half of the power of the engine is delivered to the tires on the road, find the net drag (air resistance and other dissipative forces) on the automobile.

Solution Checkpoints

IDENTIFY AND SET UP At constant velocity, the net external force on the car must be zero and the force acting on the tires must be equal to the drag forces. Force is related to power.

EXECUTE The power is equivalent to the force multiplied by the velocity. The units need to be converted to solve the problem:

$$100 \text{ mph} = 147 \text{ ft/s}.$$

The force is

$$F = \frac{\frac{1}{2}P}{v} = 281 \text{ lb} = 1250 \text{ N}.$$

KEY CONCEPT 9 **EVALUATE** Why was the power divided by 2 to get the force?

Practice Problem: Assume you lift a 5 kg box 1.5 m vertically in 0.5 seconds. What was your average power output? *Answer:* 147 W

Extra Practice: If you exert that power by using a 70 N force, then how fast can the object move? *Answer:* 2.1 m/s

Key Example Variation Problems

Solutions to these problems are in Chapter 6 of the Student's Solutions Manual.

Be sure to review EXAMPLES 6.1 and 6.2 (Section 6.1) before attempting these problems.

VP6.2.1 As a football player moves in a straight line [displacement $(3.00 \text{ m})\hat{i} - (6.50 \text{ m})\hat{j}$], an opponent exerts a constant force $(126 \text{ N})\hat{i} - (168 \text{ N})\hat{j}$ on him. (a) How much work does the opponent do on the football player? (b) How much work does the football player do on the opponent?

VP6.2.2 You push a stalled car with a constant force of 215 N as it moves a distance 8.40 m in a straight line. The amount of work that you do in this process is 1.47×10^3 J. What is the angle between the direction of your push and the direction of the car's motion?

VP6.2.3 A block of mass 15.0 kg slides down a ramp inclined at 28.0° above the horizontal. As it slides, a kinetic friction force of 30.0 N parallel to the ramp acts on it. If the block slides for 3.00 m along the ramp, find (a) the work done on the block by friction, (b) the work done on the block by the force of gravity, (c) the work done on the block by the normal force, and (d) the total work done on the block.

VP6.2.4 Three students are fighting over a T-shirt. Student 1 exerts a constant force $\vec{F}_1 = F_0 \hat{i}$ on the shirt, student 2 exerts a constant force $\vec{F}_2 = -3F_0 \hat{j}$ and student 3 exerts a constant force $\vec{F}_3 = -4F_0 \hat{i} + G\hat{j}$. (In these expressions F_0 and G are positive constants with units of force.) As the three students exert these forces, the T-shirt undergoes a straight-line displacement $2d\hat{i} + d\hat{j}$ where d is a positive constant with units of distance. (a) Find the work done on the T-shirt by each student. (b) What must be the value of G in order for the total work to be equal to zero?

Be sure to review EXAMPLES 6.3 and 6.4 (Section 6.2) before attempting these problems.

VP6.4.1   A nail is partially inserted into a block of wood, with a length of 0.0300 m protruding above the top of the block. To hammer the nail in the rest of the way, you drop a 20.0 kg metal cylinder onto it. The cylinder rides on vertical tracks that exert an upward friction force of 16.0 N on the cylinder as it falls. You release the cylinder from rest at a height of 1.50 m above the top of the nail. The cylinder comes to rest on top of the block of wood, with the nail fully inside the block. Use the work–energy theorem to find (a) the speed of the cylinder just as it hits the nail and (b) the average force the cylinder exerts on the nail while pushing it into the block. Ignore the effects of the air.

VP6.4.2   You are using a rope to lift a 14.5 kg crate of fruit. Initially you are lifting the crate at 0.500 m/s. You then increase the tension in the rope to 175 N and lift the crate an additional 1.25 m. During this 1.25 m motion, how much work is done on the crate by (a) the tension force, (b) the force of gravity, and (c) the net force? (d) What are the kinetic energy and speed of the crate after being lifted the additional 1.25 m?

VP6.4.3   A helicopter of mass 1.40×10^3 kg is descending vertically at 3.00 m/s. The pilot increases the upward thrust provided by the main rotor so that the vertical speed decreases to 0.450 m/s as the helicopter descends 2.00 m. (a) What are the initial and final kinetic energies of the helicopter? (b) As the helicopter descends this 2.00 m distance, how much total work is done on it? How much work is done on it by gravity? How much work is done on it by the upward thrust force? (c) What is the magnitude of the upward thrust force (assumed constant)?

VP6.4.4   A block of mass m is released from rest on a ramp that is inclined at an angle θ from the horizontal. The coefficient of kinetic friction between the block and the ramp is μ_k. The block slides a distance d along the ramp until it reaches the bottom. (a) How much work is done on the block by the force of gravity? (b) How much work is done on the block by the friction force? (c) Use the work–energy theorem to find the speed of the block just as it reaches the bottom of the ramp.

Be sure to review EXAMPLES 6.7 and 6.8 (Section 6.3) before attempting these problems.

VP6.8.1   An air-track glider of mass 0.150 kg is attached to the end of a horizontal air track by a spring with force constant 30.0 N/m (see Fig. 6.22a). Initially the spring is unstretched and the glider is moving at 1.25 m/s to the right. Find the maximum distance d that the glider moves to the right (a) if the air track is turned on, so that there is no friction, and (b) if the air is turned off, so that there is kinetic friction with coefficient $\mu_k = 0.320$.

VP6.8.2   An air-track glider of mass m is attached to the end of a horizontal air track by a spring with force constant k as in Fig. 6.22a. The air track is turned off, so there is friction between the glider and the track. Initially the spring is unstretched, but unlike the situation in Fig. 6.22a, the glider is initially moving to the *left* at speed v_1.

The glider moves a distance *d* to the left before coming momentarily to rest. Use the work–energy theorem to find the coefficient of kinetic friction between the glider and the track.

VP6.8.3 A pendulum is made up of a small sphere of mass 0.500 kg attached to a string of length 0.750 m. The sphere is swinging back and forth between point *A*, where the string is at the maximum angle of 35.0° to the left of vertical, and point *C*, where the string is at the maximum angle of 35.0° to the right of vertical. The string is vertical when the sphere is at point *B*. Calculate how much work the force of gravity does on the sphere (a) from *A* to *B*, (b) from *B* to *C*, and (c) from *A* to *C*.

VP6.8.4 A spider of mass *m* is swinging back and forth at the end of a strand of silk of length *L*. During the spider's swing the strand makes a maximum angle of θ with the vertical. What is the speed of the spider at the low point of its motion, when the strand of silk is vertical?

STUDENT'S SOLUTIONS MANUAL FOR WORK AND KINETIC ENERGY

VP6.2.1. **IDENTIFY:** This problem requires the calculation of work knowing the components of the force and displacement vectors.
SET UP: $W = \vec{F} \cdot \vec{S}$, which in terms of components is $W = F_x s_x + F_y s_y$.
EXECUTE: (a) $W = F_x s_x + F_y s_y = (126 \text{ N})(3.00 \text{ m}) + (168 \text{ N})(-6.50 \text{ m}) = -714 \text{ J}$
(b) The football player's force is equal and opposite to that of the opponent, but his displacement is the same, so the work he does is the negative of the work done by his opponent. $W = -(-714 \text{ J}) = +714 \text{ J}$.
EVALUATE: As we see in this problem, negative work definitely has physical meaning.

VP6.2.2. **IDENTIFY:** This problem requires a calculation using work. We know the work, force, and displacement and want to find the angle between the force and displacement vectors.
SET UP: To find the angle, we write work in the form $W = \vec{F} \cdot \vec{S} = Fs \cos\phi$.
EXECUTE: $W = Fs \cos\phi \rightarrow 1.47 \times 10^3 \text{ J} = (215 \text{ N})(8.40 \text{ m}) \cos\phi \rightarrow \phi = 35.5°$.
EVALUATE: The angle ϕ is less then 90°, so the work should be positive, which it is.

VP6.2.3. **IDENTIFY:** This problem requires the calculation of work for three forces knowing the magnitudes of the forces and displacements and the angles between them. Work is a scalar quantity, so the total work is the algebraic sum of the individual works.
SET UP: Use $W = Fs \cos\phi$ in each case. Then find the sum of the works.
EXECUTE: (a) The friction force is along the surface of the ramp but in the opposite direction from the displacement, so $\phi = 180°$. Thus
$W_f = f_k s \cos\phi° = (30.0 \text{ N})(3.00 \text{ m}) \cos 180° = -90.0 \text{ J}$.
(b) Gravity acts downward and the displacement is along the ramp surface, so $\phi = 62.0°$. Thus $W_g = mgs \cos 62.0° = (15.0 \text{ kg})(9.80 \text{ m/s}^2)(3.00 \text{ m}) \cos 62.0° = 207 \text{ J}$.
(c) The normal force is perpendicular to the displacment, so $\phi = 90°$ and $\cos\phi = 0$, so the normal force does no work.
(d) $W_{tot} = W_f + W_g + W_n = -90.0 \text{ J} + 207 \text{ J} + 0 = 117 \text{ J}$.
EVALUATE: The work is positive because the gravitational force along the ramp is greater than the friction, so the net force does positive work and the displacement is down the ramp.

VP6.2.4. **IDENTIFY:** We want to calculate work using the components of the force and displacement.
SET UP: Since we know the components, we use $W = F_x s_x + F_y s_y$.
EXECUTE: (a) $W_1 = F_0(2d) = 2F_0 d$, $W_2 = (-3F_0)d = -3F_0 d$, $W_3 = (-4F_0)(2d) + Gd = -8F_0 d + Gd$.
(b) $W_{tot} = 2F_0 d + (-3F_0 d) + (-8F_0 d + Gd) = 0 \rightarrow G = 9F_0$.
EVALUATE: The equation $W = Fs \cos\phi$ is still correct, but it would not allow us to easily calculate the work for these forces.

VP6.4.1. **IDENTIFY:** This problem requires the work-energy theorem. As the cylinder falls, work is done on it by gravity and friction. This work changes its kinetic energy.
SET UP: $W_{tot} = \Delta K = K_2 - K_1$, where $K = \frac{1}{2}mv^2$. The work is $W = Fs \cos\phi$. Figure VP6.4.1 illustrates the arrangement.

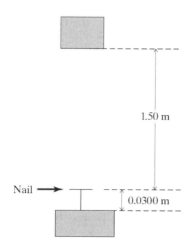

Figure VP6.4.1

EXECUTE: (a) Call point 1 where the cylinder is first released, and point 2 where it is just about to hit the nail. $W_g = mgs$ and $W_f = -fs$, so $W_{tot} = mgs - fs$. $K_1 = 0$ and $K_2 = \frac{1}{2}mv^2$. The work-energy theorem gives $mgs - fs = \frac{1}{2}mv^2$. Using $s = 1.50$ m, $f = 16.0$ N, $m = 20.0$ kg gives $v = 5.20$ m/s.

(b) Now we want to know about the force the cylinder exerts on the nail while hammering it in. Therefore we choose point 1 to be the instant just as the cylinder strikes the nail (that was point 2 in part (a)) and point 2 to be the instant when the nail has first stopped. As the cylinder pushes in the nail, it exerts a force F_N on the nail while pushing 0.0300 m into the block of wood. Likewise the other forces act for the same distance.
$W_{tot} = W_g + W_f + W_N = mgs - fs - F_N s$, $K_1 = \frac{1}{2}mv^2$ (using $v = 5.20$ m/s), $K_2 = 0$. The work-energy theorem gives $mgs - fs - f_N s = 0 - \frac{1}{2}mv^2$. Putting in the numbers, with $s = 0.0300$ m, gives $F_N = 9180$ N.
EVALUATE: The force in (a) is around a thousand pounds. From experience, we know that it takes a large force to hammer in a nail.

VP6.4.2. **IDENTIFY:** This problem requires the work-energy theorem. Gravity and the tension in the rope do work on the crate.

SET UP: $W_{tot} = \Delta K = K_2 - K_1$, where $K = \frac{1}{2}mv^2$. The work is $W = Fs\cos\phi$. Call point 1 when you just increase the tension to 175 N and point 2 just after you have lifted it the additional 1.25 m.

EXECUTE: (a) $W_T = Ts = (175 \text{ N})(1.25 \text{ m}) = 219$ J.
(b) $W_g = mgs\cos\phi = (14.5 \text{ kg})(9.80 \text{ m/s}^2)(1.25 \text{ m})\cos 180° = -178$ J
(c) $F_{net} = T - mg$, so $W_{tot} = F_{net}s = (T - mg)s = W_T + W_g = 219$ J $- 178$ J $= 41$ J.
(d) $W_{tot} = K_2 - K_1 = K_2 - \frac{1}{2}mv_1^2$. Solving for K_2 gives $K_2 = W_{tot} + \frac{1}{2}mv_1^2$, so we get

$K_2 = 41$ J $+ \frac{1}{2}(14.5 \text{ kg})(0.500 \text{ m/s})^2 = 43$ J $= \frac{1}{2}mv_2^2 = \frac{1}{2}(14.5 \text{ kg})v_2^2 \rightarrow v_2 = 2.4$ m/s.

EVALUATE: The tension does positive work and therefore increases the crate's kinetic energy.

VP6.4.3. **IDENTIFY:** This problem requires the work-energy theorem. Gravity and the thrust of the engine do work on the helicopter.

SET UP: $W_{tot} = \Delta K = K_2 - K_1$, where $K = \frac{1}{2}mv^2$. The work is $W = Fs\cos\phi$. Call point 1 when the pilot just increases the thrust and point 2 just as the speed has decreased to 0.450 m/s.

EXECUTE: (a) $K_1 = \frac{1}{2}mv_1^2 = \frac{1}{2}(1400 \text{ kg})(3.00 \text{ m/s})^2 = 6300$ J.

$K_2 = \frac{1}{2}mv_2^2 = \frac{1}{2}(1400 \text{ kg})(0.450 \text{ m/s})^2 = 142$ J.

(b) $W_{tot} = K_2 - K_1 = 142$ J $- 6300$ J $= -6160$ J
$W_g = mg\Delta y = (1400 \text{ kg})(9.80 \text{ m/s}^2)(2.00 \text{ m}) = 27{,}400$ J
$W_{tot} = W_{thrust} + W_g \rightarrow -6160$ J $= W_{thrust} + 27{,}400$ J $\rightarrow W_{thrust} = -3.36\times 10^4$ J.
(c) $W_{thrust} = F_{thrust}s\cos\phi \rightarrow -3.36\times 10^4$ J $= F_{thrust}(2.00 \text{ m})\cos 180° \rightarrow F_{thrust} = 1.68\times 10^4$ N.

EVALUATE: The magnitude of the thrust force is around ten times as great as the force of gravity, so the net work is negative in order to slow down the helicopter.

VP6.4.4. **IDENTIFY:** This problem requires the work-energy theorem.

SET UP: $W_{tot} = \Delta K = K_2 - K_1$, where $K = \frac{1}{2}mv^2$. The work is $W = Fs\cos\phi$. Call point 1 the point where the block is released and point 2 the bottom of the ramp. Let d be the distance it moves along the surface of the ramp. We also use $\Sigma F_y = 0$.

EXECUTE: (a) See Fig. VP6.4.4. $W_g = mgd\cos\phi = mgd\sin\theta$.

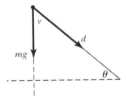

Figure VP6.4.4

(b) $W_f = f_k d\cos 180° = -f_k d = -\mu_k n$. Using $\Sigma F_y = 0$, where the y-axis is perpendicular to the face of the ramp, gives $n = mg\cos\theta$, so $W_f = -\mu_k mgd\cos\theta$.
(c) $W_{tot} = K_2 - K_1 = W_g + W_f + W_n$. $W_n = 0$ and $K_1 = 0$, so we get

$$mgd \sin\theta - \mu_k mgd \cos\theta = \frac{1}{2}mv_2^2 \rightarrow v_2 = \sqrt{2gd(\sin\theta - \mu_k \cos\theta)}.$$

EVALUATE: To check our result, let $\mu_k = 0$ for a frictionless surface. In that case, our result reduces to $v = \sqrt{2gd \sin\theta} = \sqrt{2gh}$. From our study of free-fall, we know that an object falling a distance h from rest has speed $v = \sqrt{2gh}$, which agrees with our result in this special case.

VP6.8.1. IDENTIFY: This problem requires the work-energy theorem and involves the work done by a spring.

SET UP: $W_{tot} = K_2 - K_1$, where $K = \frac{1}{2}mv^2$. The work is $W = Fs \cos\phi$, and the work done to compress a spring is $W = \frac{1}{2}kx^2$. Call point 1 the point just before the glider hits the spring and point 2 to be at maximum compression of the spring.

EXECUTE: **(a)** $W_{tot} = K_2 - K_1$. The final kinetic energy is zero and the work done by the spring on the glider is $-\frac{1}{2}kd^2$, so we get $-\frac{1}{2}kd^2 = -\frac{1}{2}mv_1^2$, which gives
$(30.0 \text{ N/m})d^2 = (0.150 \text{ kg})(1.25 \text{ m/s})^2 \rightarrow d = 0.0884 \text{ m} = 8.84 \text{ cm}.$
(b) Both friction and the spring oppose the glider's motion and therefore do negative work on the glider. The work done by the spring is still $-\frac{1}{2}kd^2$, but d is now different. The work done by kinetic friction is $W_f = -f_k d = -\mu_k mgd$. So the work-energy theorem gives
$-\frac{1}{2}kd^2 - \mu_k mgd = 0 - \frac{1}{2}mv_1^2$. Putting in the numbers and simplifying gives
$(30.0 \text{ N/m})d^2 + 2(0.320)(0.150 \text{ kg})(9.80 \text{ m/s}^2)d - (0.150 \text{ kg})(1.25 \text{ m/s})^2 = 0$. The quadratic formula gives two solutions (one positive and one negative), but we need the positive root, which is $d = 0.0741 \text{ m} = 7.41 \text{ cm}$.

EVALUATE: With no friction, the glider traveled 8.84 cm, but with friction it traveled only 7.41 cm. This is reasonable since both friction and the spring slowed down the glider in the case with friction.

VP6.8.2. IDENTIFY: This problem requires the work-energy theorem and involves the work done by a spring.

SET UP: $W_{tot} = K_2 - K_1$, where $K = \frac{1}{2}mv^2$. The work is $W = Fs \cos\phi$, and the work done to compress a spring is $W = \frac{1}{2}kx^2$. Call point 1 the point just before the glider hits the spring and point 2 to be at maximum compression of the spring.

EXECUTE: $W_{tot} = K_2 - K_1$. The final kinetic energy is zero and the work done by the spring on the glider is $-\frac{1}{2}kd^2$. The work done by kinetic friction is $W_f = -f_k d = -\mu_k mgd$. So the work-energy theorem gives so the work-energy theorem gives $-\frac{1}{2}kd^2 - \mu_k mgd = 0 - \frac{1}{2}mv_1^2$. Solving for μ_k gives $\mu_k = \frac{mv_1^2 - kd^2}{2mgd}$.

EVALUATE: Checking units shows that $\frac{mv_1^2 - kd^2}{2mgd}$ is dimensionless, which is correct for a coefficient of friction.

VP6.8.3. IDENTIFY: This problem involves the work done on the bob of a swinging pendulum.

SET UP: $W = Fs \cos\phi$. Let L be the length of the string and θ the angle it makes with the vertical.

EXECUTE: (a) $W_g = mg\,\Delta y = mgL(1 - \cos\theta)$
$W_g = (0.500 \text{ kg})(9.80 \text{ m/s}^2)(0.750 \text{ m})(1 - \cos 35.0°) = 0.665$ J.
(b) $W_g = mg\,\Delta y$ where Δy is negative, so $W_g = -0.665$ J.
(c) $W_{tot} = W_{AB} + W_{BC} = 0.665$ J $- 0.665$ J $= 0$.

EVALUATE: The total work is zero because gravity does positive work as the bob is moving downward and negative work as it is moving upward.

VP6.8.4. IDENTIFY: This problem requires application of the work-energy theorem to a pendulum.

SET UP: $W_{tot} = K_2 - K_1$, where $K = \frac{1}{2}mv^2$, and work is $W = Fs\cos\phi$. Call point 1 the highest point in its swing point 2 the lowest point.

EXECUTE: The tension in the silk strand does zero work because the tension is perpendicular to the path of the spider, so only gravity does work. $K_1 = 0$ and $K_2 = \frac{1}{2}mv^2$. Therefore

$mgL(1-\cos\theta) = \frac{1}{2}mv^2 \rightarrow v = \sqrt{2gL(1-\cos\theta)}$.

EVALUATE: Using our result, we see that as θ approaches 90°, $\cos\theta$ approaches zero, so v gets larger and larger. This is reasonable because the spider starts swinging from a higher and higher elevation.

6.5. IDENTIFY: The gravity force is constant and the displacement is along a straight line, so $W = Fs\cos\phi$.

SET UP: The displacement is upward along the ladder and the gravity force is downward, so $\phi = 180.0° - 30.0° = 150.0°$. $w = mg = 735$ N.

EXECUTE: (a) $W = (735 \text{ N})(2.75 \text{ m})\cos 150.0° = -1750$ J.
(b) No, the gravity force is independent of the motion of the painter.

EVALUATE: Gravity is downward and the vertical component of the displacement is upward, so the gravity force does negative work.

6.7. IDENTIFY: All forces are constant and each block moves in a straight line, so $W = Fs\cos\phi$. The only direction the system can move at constant speed is for the 12.0 N block to descend and the 20.0 N block to move to the right.

SET UP: Since the 12.0 N block moves at constant speed, $a = 0$ for it and the tension T in the string is $T = 12.0$ N. Since the 20.0 N block moves to the right at constant speed, the friction force f_k on it is to the left and $f_k = T = 12.0$ N.

EXECUTE: (a) (i) $\phi = 0°$ and $W = (12.0 \text{ N})(0.750 \text{ m})\cos 0° = 9.00$ J. (ii) $\phi = 180°$ and $W = (12.0 \text{ N})(0.750 \text{ m})\cos 180° = -9.00$ J.
(b) (i) $\phi = 90°$ and $W = 0$. (ii) $\phi = 0°$ and $W = (12.0 \text{ N})(0.750 \text{ m})\cos 0° = 9.00$ J. (iii) $\phi = 180°$ and $W = (12.0 \text{ N})(0.750 \text{ m})\cos 180° = -9.00$ J. (iv) $\phi = 90°$ and $W = 0$.
(c) $W_{tot} = 0$ for each block.

EVALUATE: For each block there are two forces that do work, and for each block the two forces do work of equal magnitude and opposite sign. When the force and displacement are in opposite directions, the work done is negative.

6.9. IDENTIFY: We want to compare the work done by friction over two different paths.

SET UP: Work is $W = Fs \cos\phi$. Call A the starting point and C the ending point for the first trip, and call B the end of the first segment of the second trip. On the first trip, $W_{AC} = -4.8$ J. We want to find the work done on the second trip, which is $W_{ABC} = W_{AB} + W_{BC}$. Call f_k the friction force, which is the same on all parts of the trips.

EXECUTE: $W_{AC} = -f_k s_{AC} = -f_k(0.500 \text{ m})$. Likewise $W_{ABC} = W_{AB} + W_{BC}$ which gives $W_{ABC} = -f_k(0.300 \text{ m}) - f_k(0.400 \text{ m}) = -f_k(0.700 \text{ m})$. Comparing the work in the two paths gives $\dfrac{W_{ABC}}{W_{AC}} = \dfrac{(-0.700 \text{ m})f_k}{(-0.500 \text{ m})f_k} = 1.40$. So $W_{ABC} = 1.40 W_{AC} = (1.40)(-4.8 \text{ J}) = -6.7$ J.

EVALUATE: We found that friction does more work during the ABC path than during the AC path. This is reasonable because the force is the same but the distance is greater during the ABC path. Notice that the work done by friction between two points depends on the *path* taken between those points.

6.11. IDENTIFY: We need to calculate work and use the work-energy theorem.

SET UP: $W = Fs \cos\phi$, $W_{tot} = K_2 - K_1$, $K = \dfrac{1}{2}mv^2$.

EXECUTE: (a) Using $W = Fs \cos\phi$, we see that to double the work for the same force and displacement, the angle $\cos\phi$ must double. Thus $\cos\phi = 2\cos 60° = 1.00$, so $\phi = 0°$. We should pull horizontally.

(b) The block starts from rest so $K_1 = 0$. Using the work-energy theorem tells us that in this case $W_1 = \dfrac{1}{2}mv^2$. To double v, the work must be $W_2 = \dfrac{1}{2}m(2v)^2 = 4\left(\dfrac{1}{2}mv^2\right) = 4K_1 = 4W_1$. Since only the force F varies, F must increase by a factor of 4 to double the speed.

EVALUATE: To double the speed does not take twice as much work; it takes 4 times as much.

6.13. IDENTIFY: We want the work done by a known force acting through a known displacement.

SET UP: $W = Fs \cos\phi$

EXECUTE: $W = (48.0 \text{ N})(12.0 \text{ m})\cos(173°) = -572$ J.

EVALUATE: The force has a component opposite to the displacement, so it does negative work.

6.15. IDENTIFY: We want the work done by the force, and we know the force and the displacement in terms of their components.

SET UP: We can use either $W = \vec{F} \cdot \vec{s} = F_x s_x + F_y s_y$ or $W = Fs \cos\phi$, depending on what we know.

EXECUTE: (a) We know the magnitudes of the two given vectors and the angle between them, so $W = Fs \cos\phi = (30.0 \text{ N})(5.00 \text{ m})(\cos 37°) = 120$ J.

(b) As in (a), we have $W = Fs \cos\phi = (30.0 \text{ N})(6.00 \text{ m})(\cos 127°) = -108$ J.

(c) We know the components of both vectors, so we use $W = \vec{F} \cdot \vec{s} = F_x s_x + F_y s_y$.

$W = \vec{F} \cdot \vec{s} = F_x s_x + F_y s_y = (30.0 \text{ N})(\cos 37°)(-2.00 \text{ m}) + (30.00 \text{ N})(\sin 37°)(4.00 \text{ m}) = 24.3$ J.

EVALUATE: We could check parts (a) and (b) using the method from part (c).

6.21. IDENTIFY: We need to calculate work.

SET UP: $W = Fs\cos\phi$. Since d and ϕ are the same in both cases, but F is different.

EXECUTE: If the box travels a distance d in time t starting from rest, $x = x_0 + v_{0x}t + \frac{1}{2}a_x t^2$ tells us that $d = \frac{1}{2}a_1 t^2$. If it travels the same distance in half the time, we have

$d = \frac{1}{2}a_2\left(\frac{t}{2}\right)^2 = \frac{1}{2}a_2\frac{t^2}{4} = \frac{1}{2}\left(\frac{a_2}{4}\right)t^2$. Comparing these two equations for d shows that $a_2/4 = a_1$ so $a_2 = 4a_1$. $\Sigma F_x = ma_x$ tells us that the force F is 4 times as great, and so the work $W = Fs\cos\phi$ done by the force is 4 times as great. Therefore the work is $4W_1$.

EVALUATE: We must be careful of using ratios when quantities are squared.

6.23. IDENTIFY and SET UP: Apply the work-energy theorem $W_{\text{tot}} = K_2 - K_1$ to the box. Let point 1 be at the bottom of the incline and let point 2 be at the skier. Work is done by gravity and by friction. Solve for K_1 and from that obtain the required initial speed.

EXECUTE: $W_{\text{tot}} = K_2 - K_1$
$K_1 = \frac{1}{2}mv_0^2$, $K_2 = 0$
Work is done by gravity and friction, so $W_{\text{tot}} = W_{mg} + W_f$.
$W_{mg} = -mg(y_2 - y_1) = -mgh$
$W_f = -fs$. The normal force is $n = mg\cos\alpha$ and $s = h/\sin\alpha$, where s is the distance the box travels along the incline.
$W_f = -(\mu_k mg\cos\alpha)(h/\sin\alpha) = -\mu_k mgh/\tan\alpha$
Substituting these expressions into the work-energy theorem gives $-mgh - \mu_k mgh/\tan\alpha = -\frac{1}{2}mv_0^2$.
Solving for v_0 then gives $v_0 = \sqrt{2gh(1 + \mu_k/\tan\alpha)}$.

EVALUATE: The result is independent of the mass of the box. As $\alpha \to 90°$, $h = s$ and $v_0 = \sqrt{2gh}$, the same as throwing the box straight up into the air. For $\alpha = 90°$ the normal force is zero so there is no friction.

6.25. IDENTIFY: Apply $W = Fs\cos\phi$ and $W_{\text{tot}} = K_2 - K_1$.

SET UP: $\phi = 0°$

EXECUTE: Use $W = Fs\cos\phi$, $W_{\text{tot}} = K_2 - K_1$, and $K = \frac{1}{2}mv^2$ and solve for F, giving

$F = \frac{\Delta K}{s} = \frac{\frac{1}{2}m(v_2^2 - v_1^2)}{s} = \frac{\frac{1}{2}(12.0\text{ kg})\left[(6.00\text{ m/s})^2 - (4.00\text{ m/s})^2\right]}{(2.50\text{ m})} = 48.0\text{ N}$

EVALUATE: The force is in the direction of the displacement, so the force does positive work and the kinetic energy of the object increases.

6.27. IDENTIFY: Apply $W_{\text{tot}} = \Delta K$.

SET UP: $v_1 = 0$, $v_2 = v$. $f_k = \mu_k mg$ and f_k does negative work. The force $F = 36.0$ N is in the direction of the motion and does positive work.

EXECUTE: (a) If there is no work done by friction, the final kinetic energy is the work done by the applied force, and solving for the speed,

$$v = \sqrt{\frac{2W}{m}} = \sqrt{\frac{2Fs}{m}} = \sqrt{\frac{2(36.0\text{ N})(1.20\text{ m})}{(4.30\text{ kg})}} = 4.48\text{ m/s}.$$

(b) The net work is $Fs - f_k s = (F - \mu_k mg)s$, so

$$v = \sqrt{\frac{2(F-\mu_k mg)s}{m}} = \sqrt{\frac{2(36.0\text{ N} - (0.30)(4.30\text{ kg})(9.80\text{ m/s}^2)(1.20\text{ m})}{(4.30\text{ kg})}} = 3.61\text{ m/s}$$

EVALUATE: The total work done is larger in the absence of friction and the final speed is larger in that case.

6.29. IDENTIFY: We use the work-energy theorem.

SET UP: $W_{\text{tot}} = K_2 - K_1$, $K = \frac{1}{2}mv^2$.

EXECUTE: (a) $K_A = \frac{1}{2}m_A v_A^2 = 27$ J and $K_B = \frac{1}{2}\left(\frac{m_A}{4}\right)v_B^2 = 27$ J. Equate both expressions, which gives $\frac{1}{2}m_A v_A^2 = = \frac{1}{2}\left(\frac{m_A}{4}\right)v_B^2$. Solving for v_B we have $v_B = 2v_A$. So B is moving faster and its speed is twice that of A.

(b) Apply the work-energy theorem to A. -18 J $= K_2 - 27$ J, so $K_2 = 9$ J. Take the ratio of its initial and final kinetic energy, giving $\frac{K_2}{K_1} = \frac{\frac{1}{2}mv_2^2}{\frac{1}{2}mv_1^2} = \frac{9\text{ J}}{27\text{ J}} = \frac{1}{3} = \frac{v_2^2}{v_1^2}$. Thus $v_2 = \frac{v_1}{\sqrt{3}}$. We will get the same result for B because both objects have the same initial kinetic energy (27 J) and the same amount of work (–18 J) is done on them.

EVALUATE: The result to part (a) is reasonable since a lighter object must move faster to have the same kinetic energy as a heavier object. Part (b) is reasonable because the kinetic energy decreases since the work done on the objects is negative.

6.31. IDENTIFY: $W_{\text{tot}} = K_2 - K_1$. Only friction does work.

SET UP: $W_{\text{tot}} = W_{f_k} = -\mu_k mgs$. $K_2 = 0$ (car stops). $K_1 = \frac{1}{2}mv_0^2$.

EXECUTE: (a) $W_{\text{tot}} = K_2 - K_1$ gives $-\mu_k mgs = -\frac{1}{2}mv_0^2$. $s = \frac{v_0^2}{2\mu_k g}$.

(b) (i) $\mu_{kb} = 2\mu_{ka}$. $s\mu_k = \frac{v_0^2}{2g} =$ constant so $s_a \mu_{ka} = s_b \mu_{kb}$. $s_b = \left(\frac{\mu_{ka}}{\mu_{kb}}\right)s_a = s_a/2$. The minimum stopping distance would be halved. **(ii)** $v_{0b} = 2v_{0a}$. $\frac{s}{v_0^2} = \frac{1}{2\mu_k g} =$ constant, so $\frac{s_a}{v_{0a}^2} = \frac{s_b}{v_{0b}^2}$.

$s_b = s_a\left(\frac{v_{0b}}{v_{0a}}\right)^2 = 4s_a$. The stopping distance would become 4 times as great. **(iii)** $v_{0b} = 2v_{0a}$,

$\mu_{kb} = 2\mu_{ka}$. $\frac{s\mu_k}{v_0^2} = \frac{1}{2g} =$ constant, so $\frac{s_a\mu_{ka}}{v_{0a}^2} = \frac{s_b\mu_{kb}}{v_{0b}^2}$. $s_b = s_a\left(\frac{\mu_{ka}}{\mu_{kb}}\right)\left(\frac{v_{0b}}{v_{0a}}\right)^2 = s_a\left(\frac{1}{2}\right)(2)^2 = 2s_a$.

The stopping distance would double.

EVALUATE: The stopping distance is directly proportional to the square of the initial speed and indirectly proportional to the coefficient of kinetic friction.

6.35. IDENTIFY: Use the work-energy theorem and the results of Problem 6.36.

SET UP: For $x=0$ to $x=8.0$ m, $W_{tot} = 40$ J. For $x=0$ to $x=12.0$ m, $W_{tot} = 60$ J.

EXECUTE: (a) $v = \sqrt{\dfrac{(2)(40 \text{ J})}{10 \text{ kg}}} = 2.83$ m/s

(b) $v = \sqrt{\dfrac{(2)(60 \text{ J})}{10 \text{ kg}}} = 3.46$ m/s.

EVALUATE: $\vec{F}$ is always in the $+x$-direction. For this motion $\vec{F}$ does positive work and the speed continually increases during the motion.

6.39. IDENTIFY: Apply $\Sigma \vec{F} = m\vec{a}$ to calculate the μ_s required for the static friction force to equal the spring force.

SET UP: (a) The free-body diagram for the glider is given in Figure 6.39.

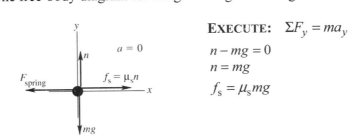

EXECUTE: $\Sigma F_y = ma_y$

$n - mg = 0$

$n = mg$

$f_s = \mu_s mg$

Figure 6.39

$\Sigma F_x = ma_x$

$f_s - F_{spring} = 0$

$\mu_s mg - kd = 0$

$\mu_s = \dfrac{kd}{mg} = \dfrac{(20.0 \text{ N/m})(0.086 \text{ m})}{(0.100 \text{ kg})(9.80 \text{ m/s}^2)} = 1.76$

(b) IDENTIFY and SET UP: Apply $\Sigma \vec{F} = m\vec{a}$ to find the maximum amount the spring can be compressed and still have the spring force balanced by friction. Then use $W_{tot} = K_2 - K_1$ to find the initial speed that results in this compression of the spring when the glider stops.

EXECUTE: $\mu_s mg = kd$

$d = \dfrac{\mu_s mg}{k} = \dfrac{(0.60)(0.100 \text{ kg})(9.80 \text{ m/s}^2)}{20.0 \text{ N/m}} = 0.0294$ m

Now apply the work-energy theorem to the motion of the glider:

$W_{tot} = K_2 - K_1$

$K_1 = \tfrac{1}{2}mv_1^2$, $K_2 = 0$ (instantaneously stops)

$W_{tot} = W_{spring} + W_{fric} = -\tfrac{1}{2}kd^2 - \mu_k mgd$ (as in Example 6.7)

$W_{tot} = -\tfrac{1}{2}(20.0 \text{ N/m})(0.0294 \text{ m})^2 - 0.47(0.100 \text{ kg})(9.80 \text{ m/s}^2)(0.0294 \text{ m}) = -0.02218$ J

Then $W_{tot} = K_2 - K_1$ gives -0.02218 J $= -\tfrac{1}{2}mv_1^2$.

$v_1 = \sqrt{\dfrac{2(0.02218 \text{ J})}{0.100 \text{ kg}}} = 0.67$ m/s.

EVALUATE: In Example 6.7 an initial speed of 1.50 m/s compresses the spring 0.086 m and in part (a) of this problem we found that the glider doesn't stay at rest. In part (b) we found that a smaller displacement of 0.0294 m when the glider stops is required if it is to stay at rest. And we calculate a smaller initial speed (0.67 m/s) to produce this smaller displacement.

6.41. **IDENTIFY and SET UP:** The magnitude of the work done by F_x equals the area under the F_x versus x curve. The work is positive when F_x and the displacement are in the same direction; it is negative when they are in opposite directions.

EXECUTE: (a) F_x is positive and the displacement Δx is positive, so $W > 0$.
$W = \frac{1}{2}(2.0 \text{ N})(2.0 \text{ m}) + (2.0 \text{ N})(1.0 \text{ m}) = +4.0$ J

(b) During this displacement $F_x = 0$, so $W = 0$.

(c) F_x is negative, Δx is positive, so $W < 0$. $W = -\frac{1}{2}(1.0 \text{ N})(2.0 \text{ m}) = -1.0$ J

(d) The work is the sum of the answers to parts (a), (b), and (c), so $W = 4.0 \text{ J} + 0 - 1.0 \text{ J} = +3.0$ J.

(e) The work done for $x = 7.0$ m to $x = 3.0$ m is $+1.0$ J. This work is positive since the displacement and the force are both in the $-x$-direction. The magnitude of the work done for $x = 3.0$ m to $x = 2.0$ m is 2.0 J, the area under F_x versus x. This work is negative since the displacement is in the $-x$-direction and the force is in the $+x$-direction. Thus $W = +1.0 \text{ J} - 2.0 \text{ J} = -1.0$ J.

EVALUATE: The work done when the car moves from $x = 2.0$ m to $x = 0$ is $-\frac{1}{2}(2.0 \text{ N})(2.0 \text{ m}) = -2.0$ J. Adding this to the work for $x = 7.0$ m to $x = 2.0$ m gives a total of $W = -3.0$ J for $x = 7.0$ m to $x = 0$. The work for $x = 7.0$ m to $x = 0$ is the negative of the work for $x = 0$ to $x = 7.0$ m.

6.43. **IDENTIFY and SET UP:** Apply the work-energy theorem. Let point 1 be where the sled is released and point 2 be at $x = 0$ for part (a) and at $x = -0.200$ m for part (b). Use $W = \frac{1}{2}kx^2$ for the work done by the spring and calculate K_2. Then $K_2 = \frac{1}{2}mv_2^2$ gives v_2.

EXECUTE: (a) $W_{tot} = K_2 - K_1$ so $K_2 = K_1 + W_{tot}$
$K_1 = 0$ (released with no initial velocity), $K_2 = \frac{1}{2}mv_2^2$

The only force doing work is the spring force. $W = \frac{1}{2}kx^2$ gives the work done *on* the spring to move its end from x_1 to x_2. The force the spring exerts on an object attached to it is $F = -kx$, so the work the spring does is
$W_{spr} = -\left(\frac{1}{2}kx_2^2 - \frac{1}{2}kx_1^2\right) = \frac{1}{2}kx_1^2 - \frac{1}{2}kx_2^2$. Here $x_1 = -0.375$ m and $x_2 = 0$. Thus
$W_{spr} = \frac{1}{2}(4000 \text{ N/m})(-0.375 \text{ m})^2 - 0 = 281$ J.
$K_2 = K_1 + W_{tot} = 0 + 281 \text{ J} = 281$ J.

Then $K_2 = \frac{1}{2}mv_2^2$ implies $v_2 = \sqrt{\frac{2K_2}{m}} = \sqrt{\frac{2(281 \text{ J})}{70.0 \text{ kg}}} = 2.83$ m/s.

(b) $K_2 = K_1 + W_{tot}$
$K_1 = 0$
$W_{tot} = W_{spr} = \frac{1}{2}kx_1^2 - \frac{1}{2}kx_2^2$. Now $x_2 = -0.200$ m, so
$W_{spr} = \frac{1}{2}(4000 \text{ N/m})(-0.375 \text{ m})^2 - \frac{1}{2}(4000 \text{ N/m})(-0.200 \text{ m})^2 = 281 \text{ J} - 80 \text{ J} = 201$ J

Thus $K_2 = 0 + 201\text{ J} = 201\text{ J}$ and $K_2 = \frac{1}{2}mv_2^2$ gives $v_2 = \sqrt{\dfrac{2K_2}{m}} = \sqrt{\dfrac{2(201\text{ J})}{70.0\text{ kg}}} = 2.40\text{ m/s}$.

EVALUATE: The spring does positive work and the sled gains speed as it returns to $x = 0$. More work is done during the larger displacement in part (a), so the speed there is larger than in part (b).

6.45. IDENTIFY: The force does work on the box, which gives it kinetic energy, so the work-energy theorem applies. The force is variable so we must integrate to calculate the work it does on the box.

SET UP: $W_{\text{tot}} = \Delta K = K_f - K_i = \frac{1}{2}mv_f^2 - \frac{1}{2}mv_i^2$ and $W_{\text{tot}} = \int_{x_1}^{x_2} F(x)\,dx$.

EXECUTE: $W_{\text{tot}} = \int_{x_1}^{x_2} F(x)\,dx = \int_0^{14.0\text{ m}} [18.0\text{ N} - (0.530\text{ N/m})x]\,dx$

$W_{\text{tot}} = (18.0\text{ N})(14.0\text{ m}) - (0.265\text{ N/m})(14.0\text{ m})^2 = 252.0\text{ J} - 51.94\text{ J} = 200.1\text{ J}$. The initial kinetic energy is zero, so $W_{\text{tot}} = \Delta K = K_f - K_i = \frac{1}{2}mv_f^2$. Solving for v_f gives

$v_f = \sqrt{\dfrac{2W_{\text{tot}}}{m}} = \sqrt{\dfrac{2(200.1\text{ J})}{6.00\text{ kg}}} = 8.17\text{ m/s}$.

EVALUATE: We could not readily do this problem by integrating the acceleration over time because we know the force as a function of x, not of t. The work-energy theorem provides a much simpler method.

6.53. IDENTIFY: $P_{\text{av}} = \dfrac{\Delta W}{\Delta t}$. The work you do in lifting mass m a height h is mgh.

SET UP: 1 hp = 746 W

EXECUTE: **(a)** The number per minute would be the average power divided by the work (mgh) required to lift one box, $\dfrac{(0.50\text{ hp})(746\text{ W/hp})}{(30\text{ kg})(9.80\text{ m/s}^2)(0.90\text{ m})} = 1.41/\text{s}$, or 84.6/min.

(b) Similarly, $\dfrac{(100\text{ W})}{(30\text{ kg})(9.80\text{ m/s}^2)(0.90\text{ m})} = 0.378/\text{s}$, or 22.7 / min.

EVALUATE: A 30-kg crate weighs about 66 lbs. It is not possible for a person to perform work at this rate.

6.55. IDENTIFY: To lift the skiers, the rope must do positive work to counteract the negative work developed by the component of the gravitational force acting on the total number of skiers, $F_{\text{rope}} = Nmg\sin\alpha$.

SET UP: $P = F_\parallel v = F_{\text{rope}} v$

EXECUTE: $P_{\text{rope}} = F_{\text{rope}} v = [+Nmg(\cos\phi)]v$.

$P_{\text{rope}} = [(50\text{ riders})(70.0\text{ kg})(9.80\text{ m/s}^2)(\cos 75.0)]\left[(12.0\text{ km/h})\left(\dfrac{1\text{ m/s}}{3.60\text{ km/h}}\right)\right]$.

$P_{\text{rope}} = 2.96 \times 10^4\text{ W} = 29.6\text{ kW}$.

EVALUATE: Some additional power would be needed to give the riders kinetic energy as they are accelerated from rest.

6.57. IDENTIFY: Relate power, work, and time.

SET UP: Work done in each stroke is $W = Fs$ and $P_{\text{av}} = W/t$.

EXECUTE: 100 strokes per second means $P_{av} = 100Fs/t$ with $t = 1.00$ s, $F = 2mg$ and $s = 0.010$ m. $P_{av} = 0.20$ W.

EVALUATE: For a 70-kg person to apply a force of twice his weight through a distance of 0.50 m for 100 times per second, the average power output would be 7.0×10^4 W. This power output is very far beyond the capability of a person.

6.59. IDENTIFY and **SET UP:** Since the forces are constant, $W_F = (F\cos\phi)s$ can be used to calculate the work done by each force. The forces on the suitcase are shown in Figure 6.59a.

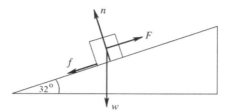

Figure 6.59a

In part (f), the work-energy theorem is used to relate the total work to the initial and final kinetic energy.

EXECUTE: (a) $W_F = (F\cos\phi)s$

Both $\vec{F}$ and $\vec{s}$ are parallel to the incline and in the same direction, so $\phi = 90°$ and $W_F = Fs = (160\text{ N})(3.80\text{ m}) = 608$ J.

(b) The directions of the displacement and of the gravity force are shown in **Figure 6.59b**.

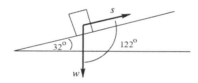

$W_w = (w\cos\phi)s$
$\phi = 122°$, so
$W_w = (196\text{ N})(\cos 122°)(3.80\text{ m})$
$W_w = -395$ J

Figure 6.59b

Alternatively, the component of w parallel to the incline is $w\sin 32°$. This component is down the incline so its angle with $\vec{s}$ is $\phi = 180°$. $W_{w\sin 32°} = (196\text{ N}\sin 32°)(\cos 180°)(3.80\text{ m}) = -395$ J. The other component of w, $w\cos 32°$, is perpendicular to $\vec{s}$ and hence does no work. Thus $W_w = W_{w\sin 25°} = -315$ J, which agrees with the above.

(c) The normal force is perpendicular to the displacement ($\phi = 90°$), so $W_n = 0$.

(d) $n = w\cos 32°$ so $f_k = \mu_k n = \mu_k w\cos 32° = (0.30)(196\text{ N})\cos 32° = 49.87$ N
$W_f = (f_k \cos\phi)x = (49.87\text{ N})(\cos 180°)(3.80\text{ m}) = -189$ J.

(e) $W_{tot} = W_F + W_w + W_n + W_f = +608\text{ J} - 395\text{ J} + 0 - 189\text{ J} = 24$ J.

(f) $W_{tot} = K_2 - K_1$, $K_1 = 0$, so $K_2 = W_{tot}$

$\frac{1}{2}mv_2^2 = W_{tot}$ so $v_2 = \sqrt{\frac{2W_{tot}}{m}} = \sqrt{\frac{2(24\text{ J})}{20.0\text{ kg}}} = 1.5$ m/s.

EVALUATE: The total work done is positive and the kinetic energy of the suitcase increases as it moves up the incline.

6.61. **IDENTIFY:** This problem requires use of the work-energy theorem. Both friction and gravity do work on the block as it slides down the ramp, but the normal force does no work.

SET UP: $W_{tot} = K_2 - K_1$, $W = Fs\cos\phi$, $K = \frac{1}{2}mv^2$, $K_1 = 0$ when the block is released. Fig. 6.61 shows the information in the problem.

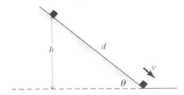

Figure 6.61

EXECUTE: **(a)** We want the work done by friction. $W_{tot} = K_2 - K_1$, where $K_1 = 0$, $K_2 = \frac{1}{2}mv^2$, $W_{tot} = W_f + W_g$. So $W_f = K_2 - W_g = \frac{1}{2}mv^2 - mgh$.

$W_f = \frac{1}{2}(5.00 \text{ kg})(5.00 \text{ m/s})^2 - (5.00 \text{ kg})(9.80 \text{ m/s}^2)(2.00 \text{ m}) = -35.5$ J.

(b) Now the ramp is lowered to 50.0° and the block is released from higher up but still 2.00 m above the bottom. Using $W = Fs\cos\phi$, we have $W_f = -f_k d = -\mu_k nd$. Using $\Sigma F_y = 0$ perpendicular to the surface of the ramp tells us that $n = mg\cos\theta$, and Fig. 6.61 shows that $d = \frac{h}{\sin\theta}$, so

$W_f = -\mu_k mg\cos\theta\left(\frac{h}{\sin\theta}\right) = -\frac{\mu_k mgh}{\tan\theta}$. From this result we see that as θ decreases, $\tan\theta$ decreases, so $1/\tan\theta$ increases. Therefore the magnitude of the work done by friction *increases* as θ decreases providing that h remains the same.

(c) Take the ratio of the works to find $W_{50°}$. $\frac{W_{50°}}{W_{60°}} = \frac{-\frac{\mu_k mgh}{\tan 50.0°}}{-\frac{\mu_k mgh}{\tan 60.0°}} = \frac{\tan 60.0°}{\tan 50.0°} = 1.45$, which gives

$W_{50°} = (1.45)W_{60°} = (1.45)(-35.5 \text{ J}) = -51.6$ J.

EVALUATE: As the slope angle θ decreases, the normal force increases so friction increases, and the distance d that the block slides also increases. So the magnitude of the work $W_f = f_k d$ increases, which agrees with our result.

6.65. **IDENTIFY:** The initial kinetic energy of the head is absorbed by the neck bones during a sudden stop. Newton's second law applies to the passengers as well as to their heads.

SET UP: In part **(a)**, the initial kinetic energy of the head is absorbed by the neck bones, so $\frac{1}{2}mv_{max}^2 = 8.0$ J. For part **(b)**, assume constant acceleration and use $v_f = v_i + at$ with $v_i = 0$, to calculate a; then apply $F_{net} = ma$ to find the net accelerating force.

Solve: (a) $v_{max} = \sqrt{\frac{2(8.0 \text{ J})}{5.0 \text{ kg}}} = 1.8$ m/s = 4.0 mph.

(b) $a = \frac{v_f - v_i}{t} = \frac{1.8 \text{ m/s} - 0}{10.0 \times 10^{-3} \text{ s}} = 180 \text{ m/s}^2 \approx 18g$, and $F_{net} = ma = (5.0 \text{ kg})(180 \text{ m/s}^2) = 900$ N.

EVALUATE: The acceleration is very large, but if it lasts for only 10 ms it does not do much damage.

6.67. IDENTIFY: Calculate the work done by friction and apply $W_{tot} = K_2 - K_1$. Since the friction force is not constant, use $W = \int F_x dx$ to calculate the work.

SET UP: Let x be the distance past P. Since μ_k increases linearly with x, $\mu_k = 0.100 + Ax$. When $x = 12.5$ m, $\mu_k = 0.600$, so $A = 0.500/(12.5 \text{ m}) = 0.0400/\text{m}$.

EXECUTE: (a) $W_{tot} = \Delta K = K_2 - K_1$ gives $-\int \mu_k mg\, dx = 0 - \frac{1}{2}mv_1^2$. Using the above expression for μ_k, $g\int_0^{x_2}(0.100 + Ax)dx = \frac{1}{2}v_1^2$ and $g\left[(0.100)x_2 + A\frac{x_2^2}{2}\right] = \frac{1}{2}v_1^2$.

$(9.80 \text{ m/s}^2)\left[(0.100)x_2 + (0.0400/\text{m})\frac{x_2^2}{2}\right] = \frac{1}{2}(4.50 \text{ m/s})^2$. Solving for x_2 gives $x_2 = 5.11$ m.

(b) $\mu_k = 0.100 + (0.0400/\text{m})(5.11 \text{ m}) = 0.304$

(c) $W_{tot} = K_2 - K_1$ gives $-\mu_k mg x_2 = 0 - \frac{1}{2}mv_1^2$. $x_2 = \frac{v_1^2}{2\mu_k g} = \frac{(4.50 \text{ m/s})^2}{2(0.100)(9.80 \text{ m/s}^2)} = 10.3$ m.

EVALUATE: The box goes farther when the friction coefficient doesn't increase.

6.69. IDENTIFY: Use the work-energy theorem.

SET UP: $W_{tot} = K_2 - K_1$, $K = \frac{1}{2}mv^2$. When a force does work W_D on the box, its speed is V starting from rest, so $W_D = \frac{1}{2}mV^2$.

EXECUTE: (a) We want to find the speed v of the box when half the total work has been done on it. Using $W_{tot} = K_2 - K_1$, we have $\frac{1}{2}W_D = \frac{1}{2}mv^2$. But as we saw above, $W_D = \frac{1}{2}mV^2$. So $\frac{1}{2}\left(\frac{1}{2}mV^2\right) = \frac{1}{2}mv^2$ which gives $v = \frac{V}{\sqrt{2}}$. Since $v \approx 0.707V$, $v > V/2$.

(b) Now we want to find how much work has been done on the box to reach half of its maximum speed, which is $\frac{1}{2}V$. The work-energy theorem gives $W = \frac{1}{2}m\left(\frac{V}{2}\right)^2 = \frac{1}{4}\left(\frac{1}{2}mV^2\right)$. From above, we know that $W_D = \frac{1}{2}mV^2$, so $W = \frac{1}{4}W_D$, which is *less than* $\frac{1}{2}W_D$.

EVALUATE: It takes $\frac{1}{4}$ of the total work to get the box to half its maximum velocity, and $\frac{3}{4}$ of the total work to get the box from $V/2$ to V.

6.71. IDENTIFY and SET UP: Use $\Sigma \vec{F} = m\vec{a}$ to find the tension force T. The block moves in uniform circular motion and $\vec{a} = \vec{a}_{rad}$.

(a) The free-body diagram for the block is given in Figure 6.71.

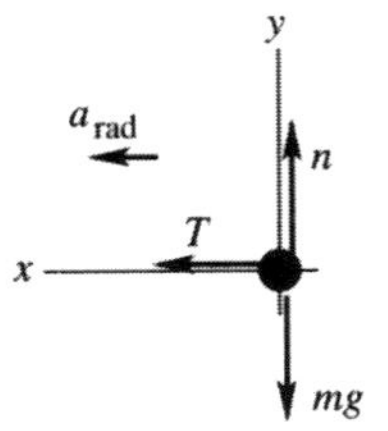

EXECUTE: $\Sigma F_x = ma_x$

$T = m\dfrac{v^2}{R}$

$T = (0.0600 \text{ kg})\dfrac{(0.70 \text{ m/s})^2}{0.40 \text{ m}} = 0.074$ N.

Figure 6.71

(b) $T = m\dfrac{v^2}{R} = (0.0600 \text{ kg})\dfrac{(2.80 \text{ m/s})^2}{0.10 \text{ m}} = 4.7$ N.

(c) SET UP: The tension changes as the distance of the block from the hole changes. We could use $W = \int_{x_1}^{x_2} F_x \, dx$ to calculate the work. But a much simpler approach is to use $W_{tot} = K_2 - K_1$.

EXECUTE: The only force doing work on the block is the tension in the cord, so $W_{tot} = W_T$.

$K_1 = \tfrac{1}{2}mv_1^2 = \tfrac{1}{2}(0.0600 \text{ kg})(0.70 \text{ m/s})^2 = 0.01470$ J,

$K_2 = \tfrac{1}{2}mv_2^2 = \tfrac{1}{2}(0.0600 \text{ kg})(2.80 \text{ m/s})^2 = 0.2352$ J, so

$W_{tot} = K_2 - K_1 = 0.2352 \text{ J} - 0.01470 \text{ J} = 0.22$ J. This is the amount of work done by the person who pulled the cord.

EVALUATE: The block moves inward, in the direction of the tension, so T does positive work and the kinetic energy increases.

6.75. **IDENTIFY and SET UP:** Use the work-energy theorem $W_{tot} = K_2 - K_1$. Work is done by the spring and by gravity. Let point 1 be where the textbook is released and point 2 be where it stops sliding. $x_2 = 0$ since at point 2 the spring is neither stretched nor compressed. The situation is sketched in Figure 6.75.

EXECUTE:

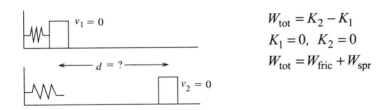

$W_{tot} = K_2 - K_1$
$K_1 = 0, \quad K_2 = 0$
$W_{tot} = W_{fric} + W_{spr}$

Figure 6.75

$W_{spr} = \tfrac{1}{2}kx_1^2$, where $x_1 = 0.250$ m (the spring force is in direction of motion of block so it does positive work).

$W_{fric} = -\mu_k mgd$

Then $W_{tot} = K_2 - K_1$ gives $\tfrac{1}{2}kx_1^2 - \mu_k mgd = 0$

$d = \dfrac{kx_1^2}{2\mu_k mg} = \dfrac{(250 \text{ N/m})(0.250 \text{ m})^2}{2(0.30)(2.50 \text{ kg})(9.80 \text{ m/s}^2)} = 1.1$ m, measured from the point where the block was released.

EVALUATE: The positive work done by the spring equals the magnitude of the negative work done by friction. The total work done during the motion between points 1 and 2 is zero, and the textbook starts and ends with zero kinetic energy.

6.77. **IDENTIFY:** A constant horizontal force pushes a block against a spring on a rough floor. The work-energy theorem and Newton's second law both apply.

SET UP: In part (a), we apply the work-energy theorem $W_{tot} = K_2 - K_1$ to the block. $f_k = \mu_k n$ and $W_{spring} = -\frac{1}{2}kx^2$. In part (b), we apply Newton's second law to the block.

EXECUTE: **(a)** $W_F + W_{spring} + W_f = K_2 - K_1$. $Fx - \frac{1}{2}kx^2 - \mu_k mgx = \frac{1}{2}mv^2 - 0$. Putting in the numbers from the problem gives $(82.0 \text{ N})(0.800 \text{ m}) - (130.0 \text{ N/m})(0.800 \text{ m})^2/2 - (0.400)(4.00 \text{ kg})(9.80 \text{ m/s}^2)(0.800 \text{ m}) = (4.00 \text{ kg})v^2/2$, $v = 2.39$ m/s.

(b) Looking at quantities parallel to the floor, with the positive direction toward the wall, Newton's second law gives $F - f_k - F_{spring} = ma$.
$F - \mu_k mg - kx = ma$: $82.0 \text{ N} - (0.400)(4.00 \text{ kg})(9.80 \text{ m/s}^2) - (130.0 \text{ N/m})(0.800 \text{ m}) = (4.00 \text{ kg})a$
$a = -9.42$ m/s². The minus sign means that the acceleration is away from the wall.

EVALUATE: The force you apply is toward the wall but the block is accelerating away from the wall.

6.81. **IDENTIFY and SET UP:** Apply $W_{tot} = K_2 - K_1$ to the system consisting of both blocks. Since they are connected by the cord, both blocks have the same speed at every point in the motion. Also, when the 6.00-kg block has moved downward 1.50 m, the 8.00-kg block has moved 1.50 m to the right. The target variable, μ_k, will be a factor in the work done by friction. The forces on each block are shown in Figure 6.81.

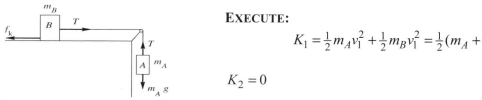

EXECUTE:

$K_1 = \frac{1}{2}m_A v_1^2 + \frac{1}{2}m_B v_1^2 = \frac{1}{2}(m_A +$

$K_2 = 0$

Figure 6.81

The tension T in the rope does positive work on block B and the same magnitude of negative work on block A, so T does no net work on the system. Gravity does work $W_{mg} = m_A g d$ on block A, where $d = 2.00$ m. (Block B moves horizontally, so no work is done on it by gravity.) Friction does work $W_{fric} = -\mu_k m_B g d$ on block B. Thus $W_{tot} = W_{mg} + W_{fric} = m_A g d - \mu_k m_B g d$. Then $W_{tot} = K_2 - K_1$ gives $m_A g d - \mu_k m_B g d = -\frac{1}{2}(m_A + m_B)v_1^2$ and

$$\mu_k = \frac{m_A}{m_B} + \frac{\frac{1}{2}(m_A + m_B)v_1^2}{m_B g d} = \frac{6.00 \text{ kg}}{8.00 \text{ kg}} + \frac{(6.00 \text{ kg} + 8.00 \text{ kg})(0.900 \text{ m/s})^2}{2(8.00 \text{ kg})(9.80 \text{ m/s}^2)(2.00 \text{ m})} = 0.786$$

EVALUATE: The weight of block A does positive work and the friction force on block B does negative work, so the net work is positive and the kinetic energy of the blocks increases as block A descends. Note that K_1 includes the kinetic energy of both blocks. We could have applied the work-energy theorem to block A alone, but then W_{tot} includes the work done on block A by the tension force.

6.83. **IDENTIFY:** Apply the work-energy theorem $W_{tot} = K_2 - K_1$ to the skater.

SET UP: Let point 1 be just before she reaches the rough patch and let point 2 be where she exits from the patch. Work is done by friction. We don't know the skater's mass so can't calculate either friction or the initial kinetic energy. Leave her mass m as a variable and expect that it will divide out of the final equation.

EXECUTE: $f_k = 0.25mg$ so $W_f = W_{tot} = -(0.25mg)s$, where s is the length of the rough patch.
$W_{tot} = K_2 - K_1$

$K_1 = \frac{1}{2}mv_0^2$, $K_2 = \frac{1}{2}mv_2^2 = \frac{1}{2}m(0.55v_0)^2 = 0.3025\left(\frac{1}{2}mv_0^2\right)$

The work-energy relation gives $-(0.25mg)s = (0.3025-1)\frac{1}{2}mv_0^2$.

The mass divides out, and solving gives $s = 1.3$ m.

EVALUATE: Friction does negative work and this reduces her kinetic energy.

6.89. **IDENTIFY** and **SET UP:** For part (a) calculate m from the volume of blood pumped by the heart in one day. For part (b) use W calculated in part (a) in $P_{av} = \dfrac{\Delta W}{\Delta t}$.

EXECUTE: (a) The work to life the blood is $W = mgh$. We need the mass of blood lifted; we are given the volume $V = (7500 \text{ L})\left(\dfrac{1\times 10^{-3} \text{ m}^3}{1 \text{ L}}\right) = 7.50 \text{ m}^3$.

$m = \text{density} \times \text{volume} = (1.05\times 10^3 \text{ kg/m}^3)(7.50 \text{ m}^3) = 7.875\times 10^3$ kg

Then $W = mgh = (7.875\times 10^3 \text{ kg})(9.80 \text{ m/s}^2)(1.63 \text{ m}) = 1.26\times 10^5$ J.

(b) $P_{av} = \dfrac{\Delta W}{\Delta t} = \dfrac{1.26\times 10^5 \text{ J}}{(24 \text{ h})(3600 \text{ s/h})} = 1.46$ W.

EVALUATE: Compared to light bulbs or common electrical devices, the power output of the heart is rather small.

6.91. **IDENTIFY:** We know a spring obeys Hooke's law, and we want to use observations of the motion of a block attached to this spring to determine its force constant and the coefficient of friction between the block and the surface on which it is sliding. The work-energy theorem applies.

SET UP: $W_{tot} = K_2 - K_1$, $W_{spring} = \frac{1}{2}kx^2$.

EXECUTE: (a) The spring force is initially greater than friction, so the block accelerates forward. But eventually the spring force decreases enough so that it is less than the force of friction, and the block then slows down (decelerates).

(b) The spring is initially compressed a distance x_0, and after the block has moved a distance d, the spring is compressed a distance $x = x_0 - d$. Therefore the work done by the spring is $W_{spring} = \dfrac{1}{2}kx_0^2 - \dfrac{1}{2}k(x_0-d)^2$. The work done by friction is $W_f = -\mu_k mgd$.

The work-energy theorem gives $W_{spring} + W_f = K_2 - K_1 = \frac{1}{2}mv^2$. Using our previous results, we get $\dfrac{1}{2}kx_0^2 - \dfrac{1}{2}k(x_0-d)^2 - \mu_k mgd = \dfrac{1}{2}mv^2$. Solving for v^2 gives $v^2 = -\dfrac{k}{m}d^2 + 2d\left(\dfrac{k}{m}x_0 - \mu_k g\right)$, where $x_0 = 0.400$ m.

(c) Figure 6.91 shows the resulting graph of v^2 versus d. Using a graphing program and a quadratic fit gives $v^2 = -39.96d^2 + 16.31d$. The maximum speed occurs when $dv^2/dd = 0$, which gives $(-39.96)(2d) + 16.31 = 0$, so $d = 0.204$ m. For this value of d, we have $v^2 = (-39.96)(0.204 \text{ m})^2 + (16.31)(0.204 \text{ m})$, giving $v = 1.29$ m/s.

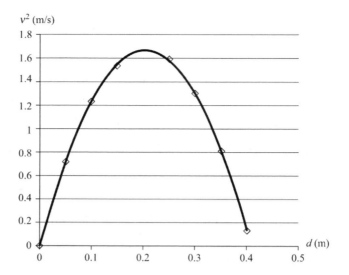

Figure 6.91

(d) From our work in (b) and (c), we know that $-k/m$ is the coefficient of d^2, so $-k/m = -39.96$, which gives $k = (39.96)(0.300 \text{ kg}) = 12.0$ N/m. We also know that $2(kx_0/m - \mu_k g)$ is the coefficient of d. Solving for μ_k and putting in the numbers gives $\mu_k = 0.800$.

EVALUATE: The graphing program makes analysis of complicated behavior relatively easy.

STUDY GUIDE FOR POTENTIAL ENERGY AND ENERGY CONSERVATION

Summary

In this chapter, we'll continue our investigation of energy by defining potential energy and learning about conservation of energy. *Potential energy* is a form of energy storage that applies to gravitational and elastic forces. Conservation of energy is one of the most fundamental concepts in physics, and we'll learn how it can be applied to problems. We'll learn the difference between conservative and nonconservative forces. We'll conclude by learning how to find forces, given a potential-energy function. By the end of the chapter, we'll be able to apply energy concepts to the analysis of problems and be prepared to extend our methods to additional forms of energy.

Objectives

After studying this chapter, you'll understand

- The definition of potential energy.
- How to use gravitational potential energy and elastic potential energy in a variety of problems.
- The definitions of conservative and nonconservative forces.
- How to apply conservation of energy to problems.
- How to find the force, given a potential-energy function.

Concepts and Equations

Term	Description
Gravitational Potential Energy	Gravitational potential energy is the potential energy associated with the position of a particle relative to earth. For a particle of mass m at a vertical distance y above the origin in a uniform gravitational field g, the gravitational potential energy of the system is $$U_{grav} = mgy.$$ Gravitational potential energy does not depend upon the location of the origin; only *differences* in gravitational potential energy are significant.
Elastic Potential Energy	Elastic potential energy is the potential energy associated with an ideal spring. For a spring of force constant k stretched or compressed a distance x from equilibrium, the elastic potential energy is $$U_{el} = \tfrac{1}{2}kx^2.$$
Conservation of Total Mechanical Energy	When only conservative forces act on a particle, the total mechanical energy is constant; that is, $$K_1 + U_1 = K_2 + U_2,$$ where U is the sum of the gravitational and elastic potential energies.
Nonconservation of Total Mechanical Energy	When forces other than gravitation or elastic forces do work on a particle, the work W_{other} done by these other forces equals the change in the total mechanical energy: $$K_1 + U_1 + W_{other} = K_2 + U_2.$$
Conservative Forces and Conservation of Energy	Forces are either conservative or nonconservative. Conservative forces are forces for which the work–kinetic-energy theorem is completely reversible and the work can be represented by potential-energy functions. Work done by nonconservative forces manifests itself as changes in the internal energy of the object. The sum of the kinetic, potential, and internal energy is always conserved: $$\Delta K + \Delta U + \Delta U_{int} = 0.$$
Determining Force from Potential Energy	A conservative force is the negative derivative of its potential-energy function in one, two, or three dimensions: $$F_x(x) = -\frac{dU(x)}{dx},$$ $$F_x = -\frac{\partial U}{\partial x},$$ $$F_y = -\frac{\partial U}{\partial y},$$ $$F_z = -\frac{\partial U}{\partial z},$$ $$\vec{F} = -\left(\frac{\partial U}{\partial x}\hat{i} + \frac{\partial U}{\partial y}\hat{j} + \frac{\partial U}{\partial z}\hat{k}\right).$$

From Chapter 7 of Student's Study Guide to accompany *University Physics with Modern Physics, Volume 1*, Fifteenth Edition. Hugh D. Young and Roger A. Freedman. Copyright © 2020 by Pearson Education, Inc. All rights reserved.

Key Concept 1: Total mechanical energy (the sum of kinetic energy and gravitational potential energy) is conserved when only the force of gravity does work.

Key Concept 2: When a force that cannot be described in terms of potential energy does work W_{other}, the final value of the total mechanical energy equals the initial value of the mechanical energy plus W_{other}.

Key Concept 3: The gravitational potential energy of an object depends on its height, not on the path the object took to reach that height.

Key Concept 4: If one of the forces that acts on a moving object is always perpendicular to the object's path, that force does no work on the object and plays no role in the equation for total mechanical energy.

Key Concept 5: Whether an object's path is straight or curved, the relationship is the same among the initial total mechanical energy, the final total mechanical energy, and the work done by forces other than gravity.

Key Concept 6: For straight-line motion problems in which the forces are constant in each stage of the motion, you can use the total mechanical energy to find the magnitude of an unknown force that does work.

Key Concept 7: You can use elastic potential energy to describe the work done by an ideal spring that obeys Hooke's law.

Key Concept 8: You can solve problems that involve elastic potential energy by using the same steps as for problems that involve gravitational potential energy, even when work is done by other forces.

Key Concept 9: For problems in which you use an energy approach to analyze an object that both changes height and interacts with an ideal spring, you must include both gravitational potential energy and elastic potential energy.

Key Concept 10: The work done by a nonconservative force on an object that moves between two points depends on the path that the object follows. Unlike for conservative forces, you *cannot* express the work done by a nonconservative force in terms of a change in potential energy.

Key Concept 11: The work done on an object that makes a complete trip around a closed path is zero if the force is conservative, but nonzero if the force is nonconservative.

Key Concept 12: In any physical process, energy is never created or destroyed; it is merely converted among the forms of kinetic energy, potential energy, and internal energy.

Key Concept 13: For motion in one dimension, the force associated with a potential-energy function equals the negative derivative of that function with respect to position.

Key Concept 14: For motion in two or three dimensions, the force associated with a potential-energy function equals the negative gradient of that function.

Conceptual Questions

1: Launching a ball

A compressed spring is used to shoot a ball straight up into the air. Compressing the spring a distance of 10 cm results in a maximum height of 3.2 m. How high does the ball go if the spring is compressed 5.0 cm?

IDENTIFY, SET UP, AND EXECUTE The spring stores elastic potential energy that is converted to gravitational potential energy at the top of the ball's flight. Elastic potential energy is proportional to the displacement squared, and gravitational potential energy is proportional to the height. One-half the compression reduces the potential energy of the spring by a factor of four, so the ball reaches one-fourth the height, or 0.8 m.

KEY CONCEPT 9 **EVALUATE** How does the velocity just above the spring compare for the two cases? Just above the spring, the elastic potential energy has been transformed to kinetic energy. Kinetic energy depends on the velocity squared; therefore, the velocity is proportional to the compression. One-half of the compression results in one-half the velocity just above the spring.

2: An accelerating car

A car accelerates from zero to 30 mph in 2.0 s. How long does it take to accelerate from zero to 60 mph?

IDENTIFY, SET UP, AND EXECUTE We assume that the power provided to the wheels is constant and that there is no friction. Kinetic energy is proportional to velocity squared, so a doubling of the final speed requires 4 times the energy. Power is energy per unit time; therefore, the time required to reach the final speed will increase by a factor of four, assuming constant power. The car will take 8.0 s to accelerate from zero to 60 mph.

EVALUATE This problem illustrates how energy principles provide alternative solutions to our earlier kinematics problems. How would you solve the problem by using forces?

3: Multiple routes to bottom of a hill

Figure 1 shows four different routes to the bottom of a hill, all starting from the same initial height. If you and three of your friends slide down the four routes, how do the four speeds at the bottom of the hill compare? Each of the paths is frictionless, and everyone starts from rest.

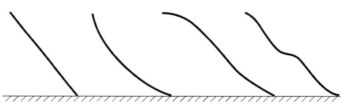

Figure 1 Question 3.

IDENTIFY, SET UP, AND EXECUTE Gravitational potential energy depends only on *changes* in height. Therefore, the change in gravitational potential energy is the same for all four routes, as all four have the same change in height. The kinetic energy at the bottom of the hill will be the same for all of the friends; therefore, the speeds of all four friends will be the same at the bottom.

KEY CONCEPT 3 **EVALUATE** If the four speeds are the same at the bottom, what quantity differs? You can see that the four routes have different lengths and are shaped differently. If you compare path 2 with path 3, you see that path 3 has a steep initial drop-off while path 2 has a shallower initial drop-off. The friend on path 3 will accelerate faster initially than the friend on path 2, will have a larger speed throughout, and will arrive at the bottom first. Therefore, the time to reach the bottom differs for the different paths.

4: Swinging by vines

Tarzan crosses a river gorge by starting from rest and swinging across the gorge on a vine. Can he ever reach a height above his starting point with this method?

IDENTIFY, SET UP, AND EXECUTE From an energy standpoint, Tarzan converts gravitational potential energy into kinetic energy as he swings across the gorge. After he passes the low point of his path, his speed slows as his gravitational potential energy increases. At his starting height, he'll come to a momentary stop after all of his kinetic energy has converted to potential energy. To get to a greater final height, he needs additional energy. He could increase his initial energy by starting with an initial velocity, using a running start, for example.

KEY CONCEPT 12 EVALUATE Without additional energy, Tarzan's final height cannot be greater than his initial height.

Problems

1: Velocity of a mass on a string

A mass m is attached by a string of length l to the ceiling and is released from rest at an angle of 60° from the vertical. Find the velocity as a function of the angle.

IDENTIFY Only gravity and tension act on the mass. Tension does no work, so we can use energy conservation to solve the problem. The target variable is the velocity.

SET UP A sketch of the problem is shown in Figure 2. At the initial angle, the mass has only gravitational potential energy. As the mass falls, the potential energy transforms to kinetic energy. At any point, the sum of the potential and kinetic energy is equal to the initial gravitational potential energy. We'll need to relate the height of the mass to the angle. The origin is placed at the bottom of the mass' path, below the anchor point.

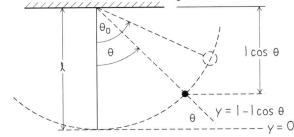

Figure 2 Problem 1.

EXECUTE Energy is conserved, so the initial energy is equal to the energy at any other point:

$$U_1 + K_1 = U_2 + K_2.$$

The initial kinetic energy is zero (the mass starts from rest), and the initial potential energy is

$$U_1 = mgy_1 = mg(l - l\cos\theta_0),$$

where the initial angle $\theta_0 = 60.0°$. At any other point, the kinetic and potential energies are

$$K_2 = \tfrac{1}{2}mv^2,$$
$$U_2 = mgy_2 = mg(l - l\cos\theta).$$

Equating the energies gives

Study Guide for Potential Energy and Energy Conservation

$$mg(l - l\cos\theta_0) = \tfrac{1}{2}mv^2 + mg(l - l\cos\theta).$$

Rearranging to solve for v as a function of the angle gives

$$v = \sqrt{2gl(\cos\theta - \cos\theta_0)}$$
$$= \sqrt{2gl(\cos\theta - \tfrac{1}{2})}.$$

KEY CONCEPT 1 **EVALUATE** What is the maximum velocity? The maximum velocity occurs at the bottom of the swing, where $\theta = 0°$, and is equal to $v = \sqrt{gl}$.

Practice Problem: What is the maximum velocity when the mass is 2 kg and the length is 2.5 m? *Answer:* 4.95 m/s

Extra Practice: What is the potential energy when the angle is 30°? *Answer:* 6.56 J

2: Professor landing on spring platform

Your professor, with a mass of 60.0 kg, falls from a height of 2.50 m onto a platform mounted on a spring. As the springs compresses, she compresses the spring a maximum distance of 0.240 m. What is the force constant of the spring? Assume that the spring and platform have negligible mass.

IDENTIFY Energy is conserved, as the only forces acting on the professor are gravity and the spring force. The target variable is the force constant of the spring.

SET UP Figure 3 shows a sketch of the situation. Initially, the professor has only U_{grav}, since her velocity is zero ($K = 0$) and the spring is uncompressed ($U_{el} = 0$). As she falls to the top of the platform, her

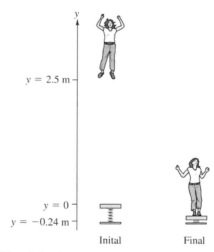

Figure 3 Problem 2 sketch.

kinetic energy increases and gravitational potential energy decreases. As she starts to compress the spring, she slows down as energy is transformed to the spring's elastic potential energy. At maximum compression, she comes to a momentary stop ($K = 0$). At this point, she is below the origin, so she has negative gravitational potential energy. We'll use energy conservation to solve for the spring's force constant.

EXECUTE Energy conservation relates the initial and final energies:

$$K_1 + U_1 = K_2 + U_2.$$

Initially, there is only U_{grav}. At the maximum compression, there are two potential-energy terms: U_{grav} and U_{el}. Also,

$$U_{grav,1} = U_{grav,2} + U_{el,2}.$$

Substituting the expressions for the energies yields

$$mgy_1 = mgy_2 + \tfrac{1}{2}kx^2.$$

The initial height is 2.50 m, and the final height and compression is −0.240 m. Solving for k gives

$$k = \frac{2mg(y_1 - y_2)}{x^2} = \frac{2(60.0 \text{ kg})(9.80 \text{ m/s}^2)(2.50 \text{ m} - (-0.240 \text{ m}))}{(0.240 \text{ m})^2} = 55{,}900 \text{ N/m}.$$

The force constant of the spring is 55,900 N/m.

KEY CONCEPT 9 **EVALUATE** Our choice of origin gave a negative y_2, but only *differences* in gravitational potential energies influence the result. This problem would have been much more challenging to solve with our force techniques, as the force of the spring varies with position.

> **CAUTION** **You set zero for gravitational potential energy!** Only *differences* in gravitational potential energy are useful in energy problems. You may set the zero at any point. It is best to choose one that simplifies the solution.

Practice Problem: How fast was she moving when she hit the spring? *Answer:* 7.0 m/s

Extra Practice: How far would the spring compress if she just stepped on it? *Answer:* 1.05 cm

3: Designing a bungee jump

You're entering the bungee-jumping business and must design the bungee cord. The jump will be from a bridge that is 100.0 m above a river. The design calls for 2.00 seconds of free fall before the cord begins to slow the fall, and the person just touches the water after jumping. Find the force constant and length of the bungee cord for a 100.0 kg person.

IDENTIFY The forces acting on the jumper are gravity and the spring force. There is no mechanical work done on the system, so we'll use energy conservation to solve the problem. Our target variables are the force constant and the length of the bungee cord.

SET UP Figure 4 shows a sketch of the situation with the coordinate origin at the river. On the bridge, the jumper has gravitational potential energy. After he jumps, the energy transforms to kinetic and elastic potential energies. At the river, the jumper momentarily stops and all the energy has transformed into elastic potential energy.

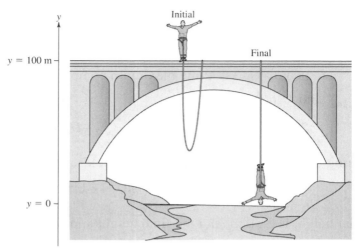

Figure 4 Problem 3.

We'll also need to recall our kinematics for freely falling objects to find the length of the bungee cord. The length of the bungee cord is found by determining its length when it becomes taut. We know that the cord becomes taut after 2.00 s, so we can use free-fall kinematics to solve for the distance the person falls in 2.00 s, which is equal to the length of the bungee cord. We'll ignore air resistance and any friction in the bungee cord.

EXECUTE Energy conservation relates the initial and final energies:

$$K_1 + U_1 = K_2 + U_2.$$

Initially, there is only U_{grav} at the bridge. At the river, there is only U_{el}. Hence,

$$U_{grav} = U_{el}.$$

Replacing the energies gives

$$mgy_1 = \tfrac{1}{2}kx^2,$$

where m is the mass of the jumper, y_1 is height of the bridge, x is the stretch length of the bungee cord, and k is the force constant of the bungee cord. We need the amount of stretch in the bungee cord, so we first use constant acceleration kinematics for freely falling objects to find the position where the bungee cord becomes taut:

$$y_{taut} = y_0 + v_{0y}t + \tfrac{1}{2}a_y t^2.$$

Here, v_{0y} is zero as the person starts from rest, $y_0 = 100$ m, $a_y = -g$, and t is 2.00 s. Solving for the position where the bungee cord becomes taut, we have

$$y_{taut} = 100 \text{ m} + \tfrac{1}{2}(-9.8 \text{ m/s}^2)(2.00 \text{ s})^2 = 80.4 \text{ m}.$$

Thus, $y_{taut} = 80.4$ m is the vertical position above the river where the bungee cord becomes taut. The starting point was at $y_0 = 100$ m, so the length of the bungee cord is the difference between y_0 and y_{taut}: 100 m − 80.4 m = 19.6 m. The stretch length is how much the cord is stretched from its original length, or 80.4 m in this case (i.e., the distance from where the cord becomes taut to the river). Substituting into our energy expression to solve for the force constant yields

$$k = \frac{2mg(y_0)}{x^2} = \frac{2(100.0 \text{ kg})(9.8 \text{ m/s}^2)(100.0 \text{ m})}{(80.4 \text{ m})^2} = 30.3 \text{ N/m}.$$

You'll need a bungee cord that is 19.6 m long with a 30.3 N/m force constant.

KEY CONCEPT 9 **EVALUATE** The spring constant was found to be relatively small, indicating that the person will be slowed gently. What will happen to a person with a mass of less than 100 kg? What about a person with a mass greater than 100 kg? The lighter person has less initial energy and so stops above the river. The heavier person has more initial energy and so stops under the surface of the river. (So the cord should be changed!)

Practice Problem: How much energy is stored in the cord at the river surface? *Answer:* 98,000 J

Extra Practice: How far above the river would an 80 kg person stop? *Answer:* 8.5 m

4: Toy car loop-the-loop

A toy car is released from a spring launcher onto a horizontal track that leads to a vertical loop-the-loop, as shown in Figure 5. What is the minimum compression needed for the launcher so that, when released, the car remains on the track throughout the loop? The mass of the car is 10.0 g, the force constant of the launcher is 20.0 N/m, the loop has a radius of 20.0 cm, and you may assume that the car moves along the track without friction.

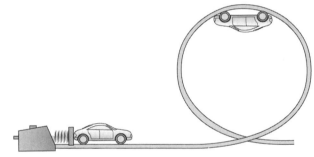

Figure 5 Problem 4.

IDENTIFY We'll use both energy conservation and Newton's second law to solve the problem. The target variable is minimum compression of the spring.

SET UP The forces acting on the car are gravity, the spring force of the launcher while the car is in contact with the launcher, and the normal force of the ground or loop. There is no mechanical work done on the system, so energy is conserved.

For the car to remain in contact with the track at the top of the loop, it must have sufficient velocity to maintain centripetal force. A free-body diagram for the car at the top of the loop is shown in Figure 6.

Figure 6 Problem 4 free-body diagram.

We place the origin at ground level. Our initial point (1) will be when the car is at rest and the launcher is compressed, storing all the energy in the spring. After the car is released, the elastic potential energy is transformed into kinetic energy and then into a combination of kinetic and gravitational potential energy when the car enters the loop. The final point (2) will be at the top of the loop, where there are both kinetic and gravitational potential energies. We'll use energy conservation to solve for the spring's compression.

EXECUTE Energy conservation relates the initial and final energies:
$$K_1 + U_1 = K_2 + U_2.$$

Initially, there is only U_{el} stored in the spring. At the top of the loop, both K and U_{grav} are stored, and
$$U_{el} = U_{grav} + K.$$

Substituting with the expressions for the energies gives
$$\tfrac{1}{2}kx^2 = mgy_2 + \tfrac{1}{2}mv^2,$$

where x is the spring compression, k is the force constant of the spring, m is the mass of the car, y_2 is twice the radius of the loop (the height at the top of the loop), and v is the speed of the car at the top of the loop. We find the velocity at the top of the loop by applying Newton's second law. To find the minimum compression of the spring, we need the minimum velocity at the top of the loop. The minimum velocity corresponds to the minimum force on the car at the top; therefore, the only force acting on the car at the top is gravity, so
$$\Sigma F_y = mg = ma_{rad} = \frac{mv^2}{r}.$$

Solving for v yields the velocity at the top of the loop:
$$v = \sqrt{gr} = \sqrt{(9.80 \text{ m/s}^2)(0.200 \text{ m})} = 1.40 \text{ m/s}.$$

Combining the results and solving for the displacement of the spring gives
$$x = \sqrt{\frac{m(4gr+v^2)}{k}} = \sqrt{\frac{0.0100 \text{ kg}(4(9.80 \text{ m/s}^2)(0.200 \text{ m}) + (1.40 \text{ m/s})^2)}{20.0 \text{ N/m}}} = 0.0700 \text{ m}.$$

The minimum spring compression necessary to keep the car on the track throughout the loop is 7.00 cm.

EVALUATE We found the minimum compression of the spring. Additional compression would have resulted in greater total energy after the car is launched, which would also keep the car on the track.

KEY CONCEPT 4 This problem illustrates how we'll sometimes combine our knowledge of previous materials (e.g., forces) with our current topics. As we progress through the text, add to our knowledge base and not merely exchange one concept for another.

Practice Problem: What is the velocity at the bottom of the loop? *Answer:* 3.13 m/s

Extra Practice: How much energy does the car have the top of the loop? *Answer:* 0.049 J

5: Losing contact with the hill

A frictionless puck slides down a large, round dome of radius 2.0 m. If the puck starts at the top of the dome with a very small initial velocity, how far below the starting point does the puck lose contact with the dome?

IDENTIFY Gravitation and the normal force are the only forces acting on the puck. The normal force does no work, so we'll use energy conservation. At the point where the puck loses contact with the dome, the normal force is zero. The target variable is the height at which the puck loses contact.

SET UP A sketch of the problem is shown in Figure 7. The origin is at the center of the dome and the target variable is the change in height, y. No mechanical work is done on the system, so energy is conserved.

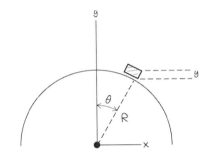

Figure 7 Problem 5.

The forces acting on the puck are gravity and the normal force, and the net external force is a centripetal force directed toward the center of the dome. When the puck loses contact with the dome, the normal force is zero. A free-body diagram of the puck is shown in Figure 8.

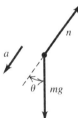

Figure 8 Problem 5 free-body diagram.

Our initial point (1) will be when the puck is at the top of the dome, where there is only gravitational potential energy. (We ignore the small quantity of kinetic energy at the top of the dome, to simplify the solution.) Our second point (2) will be the moment the puck leaves the dome, when there are both kinetic and gravitational potential energies.

EXECUTE We write the change in height in terms of the radius and angle:

$$y = R - R\cos\theta.$$

We need to find the angle at which the puck leaves the dome. We start with the forces acting on the puck. The net external force along the radius at any point is given by Newton's second law:

$$\sum F_{rad} = mg\cos\theta - n = ma_{rad} = m\frac{v^2}{R}.$$

When the puck loses contact, the normal force goes to zero, producing the following relation between the angle and the velocity:

$$v^2 = gR\cos\theta.$$

We now have θ in terms of velocity. The v^2 reminds us of energy, so we write the expression for energy conservation:

$$U_1 + K_1 = U_2 + K_2.$$

The potential energy at the top of the dome is mgR and the kinetic energy is zero. At a later point, the puck has both kinetic energy and potential energy. Substituting in the expressions for the energies gives

$$mgR = mg(R\cos\theta) + \tfrac{1}{2}mv^2.$$

Substituting the expression from the forces results in

$$mgR = mg(R\cos\theta) + \tfrac{1}{2}mgR\cos\theta.$$

Simplifying this equation yields

$$\tfrac{2}{3}R = R\cos\theta.$$

Solving for y, we obtain

$$y = R - R\cos\theta$$
$$= R - \tfrac{2}{3}R$$
$$= \tfrac{1}{3}R = \tfrac{1}{3}(2.0 \text{ m}) = 0.67 \text{ m}.$$

The puck leaves the dome 0.67 m below the top.

KEY CONCEPT 4 **EVALUATE** We see that the result depends on neither the puck's mass nor gravity. If the puck and dome were transported to the moon, the puck would lose contact at the same position as on earth.

Practice Problem: At what angle does the puck leave the dome? *Answer:* 48.2°

Extra Practice: How fast is the puck moving when it leaves the dome? *Answer:* 3.61 m/s

6: Force from potential-energy function

The potential-energy function of a particle is

$$U(x, y) = axy - 3bx^2 + 2cy^2,$$

where a, b, and c are constants. What is the force on the particle?

IDENTIFY AND SET UP Given the potential-energy function, we find the force by taking partial derivatives. The potential-energy function depends on x and y, so we'll find the negative partial derivative of the potential-energy function with respect to x and y.

EXECUTE The x-component of the force is

$$F_x = -\frac{\partial U}{\partial x}.$$

Substituting the expression U and solving gives

$$F_x = -\frac{\partial}{\partial x}(axy - 3bx^2 + 2cy^2) = ay - 6bx.$$

The y-component of the force is

$$F_y = -\frac{\partial U}{\partial y}.$$

Substituting again and solving yields

$$F_y = -\frac{\partial}{\partial y}(axy - 3bx^2 + 2cy^2) = ax + 4cy.$$

The force is

$$\vec{F} = (ay - 6bx)\hat{i} + (ax + 4cy)\hat{j}.$$

KEY CONCEPT 14 **EVALUATE** This problem illustrates how we can find the force, given a potential-energy function. We see that it is easier to find the force from the potential energy than the potential energy from the force.

CAUTION **Use both forces and energy!** This problem shows how you can combine your knowledge of both forces and energy to solve complex problems. Without using the combined technique, the solution would have been more difficult.

Try It Yourself!

1: Mass on a spring

A 0.5 kg box hangs from a spring whose unstretched length is 1.0 m. The box stretches the spring 0.5 m. The box is pulled 0.5 m from its equilibrium position and is released from rest. Find its maximum velocity and height.

Solution Checkpoints

IDENTIFY AND SET UP Start by finding the spring constant by using information from the first part of the problem. What forces act on the box? Can you use energy conservation to find the position and velocity of the box at any point? The initial kinetic energy of the box is zero when it is released. What is the initial elastic potential energy?

EXECUTE Energy conservation gives

$$\tfrac{1}{2}k(1.0 \text{ m})^2 = mgy + \tfrac{1}{2}mv^2 + \tfrac{1}{2}k(1.0 \text{ m} - y)^2$$

when the gravitational potential energy is initially zero. This gives a maximum velocity of 2.2 m/s when $y = 0.5$ m and a maximum height of 1.0 m when $v = 0$.

KEY CONCEPT 9 **EVALUATE** Do you get the same result when you use a different origin?

Practice Problem: What is the total energy of the system? *Answer:* 4.9 J

Extra Practice: What is the maximum velocity if you replace the 0.5 kg box with a 0.25 kg box? *Answer:* 4.43 m/s

2: Two masses connected by a pulley

For the frictionless system shown in Figure 9, find the velocity of the mass m when it hits the floor if it is released from a height L above the floor.

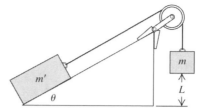

Figure 9 Problem 1.

Solution Checkpoints

IDENTIFY AND SET UP What forces act on the masses? Can you use energy conservation in this case? The initial kinetic energy is zero when mass m is released.

EXECUTE Energy conservation gives
$$mgL = \tfrac{1}{2}(m+m')v_2^2 + m'gL\sin\theta$$
when the origin is at the bottom of the incline. This equation results in a velocity of
$$v_2 = \sqrt{2\frac{(m-m'\sin\theta)}{(m+m')}gL}.$$

KEY CONCEPT 1 **EVALUATE** What is the velocity of m as θ approaches 90° (an Atwood's machine)? What is the velocity as θ approaches 0° (a flat table)?

Practice Problem: If $m' = 2m$, then what is the maximum angle that allows the blocks to move? *Answer:* 30°

Extra Practice: What is the final velocity when the angle is 30°, $m = m'$, and $L = 0.5$ m? *Answer:* 1.57 m/s

3: Force from the potential-energy function

The potential-energy function of a particle is
$$U(x) = ax + \tfrac{1}{2}kx^2.$$

What is the force on the particle? What is the equilibrium position of the particle?

Solution Checkpoints

IDENTIFY AND SET UP How do you find the force, given a potential energy function? The equilibrium position is where the force is zero and is found by setting the force equal to zero and solving for position.

EXECUTE The force on the particle is the negative derivative with respect to position:
$$F_x = -(a + kx).$$
The force is zero when $x = 2\,a/k$.

KEY CONCEPT 14 EVALUATE We see that this force is an elastic force.

Practice Problem: What is the force when $U(y) = 2/3ay^2 - 3/5by^3$? Answer: $F_y = -2ay + by^2$

Extra Practice: Where is the equilibrium position? Answer: $y = 2a/b$

Problem Summary

Energy analysis shares many of the problem-solving principles we have encountered. In these problems, we

- Identified the general procedure to find the solution.
- Sketched the situation when no figure was provided.
- Identified the energies involved and the forces acting in the system.
- Applied energy principles, including conservation of energy (when possible).
- Drew free-body diagrams of the objects when appropriate.
- Applied Newton's laws when appropriate.
- Solved the equations through algebra and substitutions.
- Reflected on the results, checking for inconsistencies.

This problem-solving foundation can be applied to all problems, including those that could be solved by force analysis. Following the procedure set forth here leads to mastery of many physics problems.

Key Example Variation Problems

Solutions to these problems are in Chapter 7 of the Student's Solutions Manual.

Be sure to review EXAMPLES 7.1 and 7.2 (Section 7.1) before attempting these problems.

VP7.2.1 You throw a baseball (mass 0.145 kg) vertically upward. It leaves your hand moving at 12.0 m/s. Air resistance can be neglected. At what height above your hand does the ball have (a) half as much upward velocity, (b) half as much kinetic energy as when it left your hand?

VP7.2.2 You toss a rock of mass m vertically upward. Air resistance can be neglected. The rock reaches a maximum height h above your hand. What is the speed of the rock when it is at height (a) $h/4$ and (b) $3h/4$?

VP7.2.3 You throw a tennis ball (mass 0.0570 kg) vertically upward. It leaves your hand moving at 15.0 m/s. Air resistance cannot be neglected, and the ball reaches a maximum height of 8.00 m. (a) By how much does the total mechanical energy decrease from when the ball leaves your hand to when it reaches its maximum height? (b) What is the magnitude of the average force of air resistance?

VP7.2.4 You catch a volleyball (mass 0.270 kg) that is moving downward at 7.50 m/s. In stopping the ball, your hands and the volleyball descend together a distance of 0.150 m. (a) How much work do your hands do on the volleyball in the process of stopping it? (b) What is the magnitude of the force (assumed constant) that your hands exert on the volleyball?

Be sure to review EXAMPLES 7.4 and 7.5 (Section 7.1) before attempting these problems.

VP7.5.1 A well-greased, essentially frictionless, metal bowl has the shape of a hemisphere of radius 0.150 m. You place a pat of butter of mass 5.00×10^{-3} kg at the rim of the bowl and let it slide to the bottom of the bowl. (a) What is the speed of the pat of butter when it reaches the bottom of the bowl? (b) At the bottom of the bowl, what is the force that the bowl exerts on the pat of butter? How does this compare to the weight of the pat?

VP7.5.2 A snowboarder and her board (combined mass 40.0 kg) are moving at 9.30 m/s at the bottom of a curved ditch. (a) If friction can be ignored, what is the maximum vertical distance that she can travel up the sides of the ditch? Does this answer depend on the shape of the ditch? (b) The snowboarder finds that, due to friction, the maximum vertical distance she can travel up the sides of the ramp is 3.50 m. How much work did the force of friction do on her?

VP7.5.3 A pendulum is made of a small sphere of mass 0.250 kg attached to a lightweight string 1.20 m in length. As the pendulum swings back and forth, the maximum angle that the string makes with the vertical is 34.0°. Friction can be ignored. At the low point of the sphere's trajectory, what are (a) the kinetic energy of the sphere and (b) the tension in the string?

VP7.5.4 You are testing a new roller coaster ride in which a car of mass m moves around a vertical circle of radius R. In one test, the car starts at the bottom of the circle (point A) with initial kinetic energy K_i. When the car reaches the top of the circle (point B), its kinetic energy is $\frac{1}{4}K_i$, and its gravitational potential energy has increased by $\frac{1}{2}K_i$. (a) What was the speed of the car at point A, in terms of g and R? (b) How much work was done on the car by the force of friction as it moved from point A to point B, in terms of m, g, and R? (c) What was the magnitude of the friction force (assumed to be constant throughout the motion), in terms of m and g?

Be sure to review EXAMPLES 7.7, 7.8, and 7.9 (Section 7.2) before attempting these problems.

VP7.9.1 A glider of mass 0.240 kg is on a frictionless, horizontal track, attached to a horizontal spring of force constant 6.00 N/m. Initially the spring (whose other end is fixed) is stretched by 0.100 m and the attached glider is moving at 0.400 m/s in the direction that causes the spring to stretch farther. (a) What is the total mechanical energy (kinetic energy plus elastic potential energy) of the system? (b) When the glider comes momentarily to rest, by what distance is the spring stretched?

VP7.9.2 A glider of mass 0.240 kg is on a horizontal track, attached to a horizontal spring of force constant 6.00 N/m. There is friction between the track and the glider. Initially the spring (whose other end is fixed) is stretched by 0.100 m and the attached glider is moving at 0.400 m/s in the direction that causes the spring to stretch farther. The glider comes momentarily to rest when the spring is stretched by 0.112 m. (a) How much work does the force of friction do on the glider as the stretch of the spring increases from 0.100 m to 0.112 m? (b) What is the coefficient of kinetic friction between the glider and the track?

VP7.9.3 A lightweight vertical spring of force constant k has its lower end mounted on a table. You compress the spring by a distance d, place a block of mass m on top of the compressed spring, and then release the block. The spring launches the block upward, and the block rises to a maximum height some distance above the now-relaxed spring. (a) Find the speed of the block just as it loses contact with the spring. (b) Find the total vertical distance that the block travels from when it is first released to when it reaches its maximum height.

VP7.9.4 A cylinder of mass m is free to slide in a vertical tube. The kinetic friction force between the cylinder and the walls of the tube has magnitude f. You attach the upper end of a lightweight vertical spring of force constant k to the cap at the top of the tube, and attach the lower end of the spring to the top of the cylinder. Initially the cylinder is at rest and the spring is relaxed. You then release the cylinder. What vertical distance will the cylinder descend before it comes momentarily to rest?

STUDENT'S SOLUTIONS MANUAL FOR POTENTIAL ENERGY AND ENERGY CONSERVATION

VP7.2.1. **IDENTIFY:** We use energy conservation. The ball has kinetic energy and gravitational potential energy.
SET UP: $U_1 = m_A g y_{A,1} = (12.0 \text{ kg})(9.80 \text{ m/s}^2)(2.00 \text{ m}) = 235.2$ J. and $U_g = mgy$.
$U_1 + K_1 + W_{\text{other}} = U_2 + K_2$ Call point 1 the place where the ball leaves your hand and point 2 the height where it has the desired speed. In this case, $W_{\text{other}} = 0$.
EXECUTE: **(a)** In this case, $v_2 = v_1/2$ and $U_1 = 0$. $y_1 = (142 \text{ m})\sin 53° = 113$ m. gives $U_2 = 0$.
$\frac{1}{2}mv_2^2 = mgy_2 + \frac{1}{2}m\left(\frac{v_1}{2}\right)^2$. Solve for y_2 gives $y_2 = \frac{3v_1^2}{8g} = \frac{3(12.0 \text{ m/s})^2}{8(9.80 \text{ m/s}^2)} = 5.51$ m.
(b) In this case, $K_2 = K_1/2$ and $U_1 = 0$, so $0 + K_1 + 0 = mgy_2 + K_1/2$. Solving for y_2 gives
$y_2 = \frac{K_1}{2mg} = \frac{\frac{1}{2}mv_1^2}{2mg} = \frac{v_1^2}{4g} = \frac{(12.0 \text{ m/s})^2}{4(9.80 \text{ m/s}^2)} = 3.67$ m.
EVALUATE: Notice that the ball does *not* have half of its initial kinetic energy when it is half-way to the top. Likewise when it has half of its initial kinetic energy, it is *not* half-way to the top.

VP7.2.2. **IDENTIFY:** We use energy conservation. The rock has kinetic energy and gravitational potential energy.
SET UP: $K = \frac{1}{2}mv^2$ and $U_g = mgy$. $U_1 + K_1 + W_{\text{other}} = U_2 + K_2$. Call point 1 the place where the rock leaves your hand and point 2 the height where it has the desired height. In this case, $W_{\text{other}} = 0$ and $U_1 = 0$.
EXECUTE: First find the initial speed in terms of h. $U_1 + K_1 + W_{\text{other}} = U_2 + K_2$ gives
$K_1 = U_2 + K_2 \rightarrow \frac{1}{2}mv_1^2 = mgh \rightarrow v_1^2 = 2gh$.
(a) In this case, $y_2 = h/4$ and we use , so $U_1 + K_1 + W_{\text{other}} = U_2 + K_2$ gives
$\frac{1}{2}mv_1^2 = mg(h/4) + \frac{1}{2}mv_2^2 \rightarrow 2(2gh) - gh = 2v_2^2 \rightarrow v_2 = \sqrt{\frac{3gh}{2}}$.
(b) Follow the same procedure as in (a) *except* that $y_2 = 3h/4$. Energy conservation gives
$\frac{1}{2}mv_1^2 = mg(3h/4) + \frac{1}{2}mv_2^2$. Using $v_1^2 = 2gh$ gives $v_2 = \sqrt{\frac{gh}{2}}$.
EVALUATE: Our result says that the speed when $y = 3h/4$ is less than when $y_2 = h/4$, which is reasonable because the rock is slowing down as it rises.

VP7.2.3. IDENTIFY: We use energy conservation. The ball has kinetic energy and gravitational potential energy.

SET UP: $K = \frac{1}{2}mv^2$ and $U_g = mgy$. $U_1 + K_1 + W_{other} = U_2 + K_2$. Call point 1 the place where the ball leaves your hand and point 2 the height where it reaches its maximum height, so $U_1 = 0$ and $K_2 = 0$. Call E the total mechanical energy of the ball, and use $W = Fs \cos\phi$.

EXECUTE: (a) $E_1 = K_1 = \frac{1}{2}mv_1^2 = \frac{1}{2}(0.0570 \text{ kg})(15.0 \text{ m/s})^2 = 6.4125$ J.

$E_2 = U_2 = mgy_2 = (0.0570 \text{ kg})(9.80 \text{ m/s}^2)(8.00 \text{ m}) = 4.4688$ J.

$\Delta E = E_2 - E_1 = 4.4688 \text{ J} - 6.4125 \text{ J} = -1.94$ J, so the total mechanical energy has *decreased* by 1.94 J.

(b) The loss of mechanical energy is equal to the work done by the friction of air resistance.
$Fs \cos\phi = \Delta E \rightarrow F(8.00 \text{ m}) \cos 180° = -1.94 \text{ J} \rightarrow F = 0.243$ N.

EVALUATE: From our results, we have $E_2/E_1 = (4.4688 \text{ J})/(6.4125 \text{ J}) = 0.700$. We can also calculate this ratio as $E_2/E_1 = \frac{U_2}{U_1} = \frac{mgh}{\frac{1}{2}mv_1^2} = \frac{2gh}{v_1^2} = \frac{2(9.80 \text{ m/s}^2)}{(15.0 \text{ m/s})^2} = 0.700$, so $E_2 = 0.700 E_1$. The mechanical energy has decreased by 30%.

VP7.2.4. IDENTIFY: We use energy conservation. The ball has kinetic energy and gravitational potential energy.

SET UP: $K = \frac{1}{2}mv^2$ and $U_g = mgy$. $U_1 + K_1 + W_{other} = U_2 + K_2$. Call point 1 the place where you catch the ball and point 2 the place where the ball stops, so $U_1 = 0$, $y_2 = -0.150$ m, and $K_2 = 0$. Use $W = Fs \cos\phi$. W_{other} is the work done by your hands in stopping the ball.

EXECUTE: (a) Using $U_1 = 0$, $K_2 = 0$, $U_1 + K_1 + W_{other} = U_2 + K_2$ becomes

$\frac{1}{2}mv_1^2 + W_{hands} = mgy_2$

$W_{hands} = m\left(gy_2 - \frac{v_1^2}{2}\right) = (0.270 \text{ kg})\left[(9.80 \text{ m/s}^2)(-0.150 \text{ m}) - \frac{(7.50 \text{ m/s})^2}{2}\right] = -7.99$ J.

(b) $W = Fs \cos\phi \rightarrow -7.99 \text{ J} = F(0.150 \text{ m}) \cos 180° \rightarrow F = 53.3$ N.

EVALUATE: The work done by gravity is $W_g = mgy = (0.270 \text{ kg})(9.80 \text{ m/s}^2)(0.150 \text{ m}) = 39.7$ J, which is much more than the magnitude of the work your hands do. This is reasonable because your hands also reduce (to zero) the ball's initial kinetic energy.

VP7.5.1. IDENTIFY We use energy conservation. The butter has kinetic energy and gravitational potential energy. Use Newton's second law for circular motion.

SET UP: $K = \frac{1}{2}mv^2$ and $U_g = mgy$. $U_1 + K_1 + W_{other} = U_2 + K_2$. Call point 1 the place at the rim of the bowl and point 2 the bottom of the bowl, so $K_1 = 0$, $y_2 = 0$, and $W_{other} = 0$. Use $\Sigma F_y = ma_y$ at the bottom of the bowl, where $a_y = \frac{v^2}{R}$. At the top of the bowl, $y_1 = R$ (the radius of the bowl).

EXECUTE: (a) Using $K_1 = 0$, $U_2 = 0$, and $W_{other} = 0$, $U_1 + K_1 + W_{other} = U_2 + K_2$ becomes

$mgR = \frac{1}{2}mv_2^2 \rightarrow v_2 = \sqrt{2Rg} = \sqrt{2(0.150 \text{ m})(9.80 \text{ m/s}^2)} = 1.71$ m/s.

(b) At the bottom of the bowl, apply Newton's second law for circular motion.

$$\Sigma F_y = ma_y = \frac{v^2}{R} \rightarrow F_{bowl} - mg = mv^2/R \rightarrow F_{bowl} = m(g + v^2/R)$$

For the numbers here we get
$F_{bowl} = (5.00 \times 10^{-3} \text{ kg})[9.80 \text{ m/s}^2 + (1.71 \text{ m/s})^2/(0.150 \text{ m})] = 0.147$ N.
$w = mg = (5.00 \times 10^{-3} \text{ kg})(9.80 \text{ m/s}^2) = 0.0490$ N.
$F_{bowl}/w = (0.147 \text{ N})/(0.0490 \text{ N}) = 3.00$, so the force due to the bowl is 3 times the weight of the butter.
EVALUATE: The ratio F_{bowl}/w is $m(g + v^2/R)/mg = g + v^2/R$, which is independent of the butter's mass. Therefore *any* object sliding down this bowl under the same conditions would experience a force from the bowl equal to 3 times the weight of the object.

VP7.5.2. **IDENTIFY:** We use energy conservation. The snowboarder has kinetic energy and gravitational potential energy.
SET UP: $K = \frac{1}{2}mv^2$ and $U_g = mgy$. $U_1 + K_1 + W_{other} = U_2 + K_2$. Call point 1 the bottom of the ditch and point 2 the highest point she reaches, so $K_2 = 0$, $U_1 = 0$.
EXECUTE: **(a)** If there is no friction, $W_{other} = 0$, so $U_1 + K_1 + W_{other} = U_2 + K_2$ becomes
$K_1 = U_2 \rightarrow \frac{1}{2}mv^2 = mgh \rightarrow h = v^2/2g = (9.30 \text{ m/s}^2)/[2(9.80 \text{ m/s}^2)] = 4.41$ m. This answer does not depend on the shape of the ditch since there is no friction.
(b) The work done by friction is equal to the loss of mechanical energy, so $W_f = U_2 - K_1$.
$W_f = mgh - \frac{1}{2}mv^2 = m(gh - v^2/2) = (40.0 \text{ kg})[(9.80 \text{ m/s}^2)(3.50 \text{ m}) - (9.30 \text{ m/s})^2/2] = -358$ J.
EVALUATE: The work done by friction is negative because the friction force is opposite to the displacement of the snowboard, which agrees with our result.

VP7.5.3. **IDENTIFY:** We apply energy conservation and Newton's second law to a swinging pendulum. The small sphere (the bob) has kinetic energy and potential energy. Use Newton's second law for circular motion.
SET UP: $K = \frac{1}{2}mv^2$ and $U_g = mgy$. $U_1 + K_1 + W_{other} = U_2 + K_2$. Call point 1 the high point of the swing and point 2 its low point, so $K_1 = 0$, $U_2 = 0$, and $W_{other} = 0$. Use $\Sigma F_y = ma_y$ at the bottom of the swing, where $a_y = \frac{v^2}{R}$. At the top of the swing, $h = L(1 - \cos\theta)$, where L is the length of the string and θ is the largest angle it makes with the vertical.
EXECUTE: **(a)** For conditions here, we find that $U_1 = K_2 = mgh = mgL(1 - \cos\theta)$, so
$K_2 = (0.250 \text{ kg})(9.80 \text{ m/s}^2)(1.20 \text{ m})(1 - \cos 34.0°) = 0.503$ J.
(b) Apply $\Sigma F_y = ma_y = \frac{v^2}{R}$ at the bottom of the swing: $T - mg = \frac{mv^2}{L} = \frac{2}{L}\left(\frac{1}{2}mv^2\right) = \frac{2K}{L}$.
$T = mg + 2K/L = (0.250 \text{ kg})(9.80 \text{ m/s}^2) + 2(0.503 \text{ J})/(1.20 \text{ m}) = 3.29$ m.
EVALUATE: As a check, find v^2 from the known kinetic energy, giving $v^2 = 4.021$ m²/s². Then use $T - mg = mv^2/L$ to find T. The result is the same as we found in (b).

VP7.5.4. **IDENTIFY:** We use energy conservation. The car has kinetic energy and gravitational potential energy.
SET UP: $K = \frac{1}{2}mv^2$ and $U_g = mgy$. $U_1 + K_1 + W_{other} = U_2 + K_2$. Call A point 1 and B point 2, so $U_1 = 0$. In this case, $K_A = K_i$, $K_B = K_i/4$, $U_B = K_i/2$, and $y_B = 2R$.

EXECUTE: (a) $\Delta U_{AB} = mg(2R) = \frac{1}{2}K_i = \frac{1}{2}\left(\frac{1}{2}mv_A^2\right) \rightarrow v_A = \sqrt{8gR}$.

(b) Energy conservation gives $K_A + W_{other} = K_B + (U_B - U_B)$. Since W_{other} is due to friction, this becomes $K_i + W_f = \frac{1}{4}K_i + \frac{1}{2}K_i = \frac{3}{4}K_i$. Solving for W_f gives

$W_f = -\frac{1}{4}K_i = -\frac{1}{4}\left(\frac{1}{2}mv_A^2\right) = -\frac{1}{8}m(8gR) = -mgR.$

(c) Using $W_f = fs \cos\phi$ and the result from (b) gives $-mgR = f(\pi R) \cos 180° = -\pi R f$
$f = mg/\pi$.

EVALUATE: From (c) we see that the heavier the roller coaster, the greater the friction force, which is reasonable since friction depends on the normal force at the surface of contact.

VP7.9.1. IDENTIFY: We use energy conservation. The system has kinetic energy and elastic potential energy in the spring.

SET UP: $K = \frac{1}{2}mv^2$ and $U = \frac{1}{2}kx^2$. The mechanical energy is the kinetic energy plus the potential energy: $E = K + U$.

EXECUTE: (a) $E = K + U = \frac{1}{2}mv_1^2 + \frac{1}{2}kx_1^2$

$E = \frac{1}{2}(0.240 \text{ kg})(0.400 \text{ m/s})^2 + \frac{1}{2}(6.00 \text{ N/m})(0.100 \text{ m})^2 = 0.0492 \text{ J}.$

(b) There is no friction, so E is constant. So when the glider stops, $U = 0.0492$ J. Therefore
$\frac{1}{2}kx_1^2 = 0.0492 \text{ J} \rightarrow \frac{1}{2}(6.00 \text{ N/m}) x^2 = 0.0492 \text{ J} \rightarrow x = 0.128 \text{ m}.$

EVALUATE: During most of the motion, the glider has kinetic energy and potential energy, but the sum of the two is always equal to 0.0492 J.

VP7.9.2. IDENTIFY: We use energy conservation. The system has kinetic energy and elastic potential energy in the spring and there is friction.

SET UP: $K = \frac{1}{2}mv^2$ and $U = \frac{1}{2}kx^2$. $U_1 + K_1 + W_{other} = U_2 + K_2$. $W = Fs \cos\phi$. Call position 1 when the spring is stretched by 0.100 m and position 2 when the glider has instantaneously stopped, so $K_2 = 0$.

EXECUTE: (a) $U_1 + K_1 + W_f = U_2 + K_2$ gives $\frac{1}{2}kx_1^2 + \frac{1}{2}mv_1^2 + W_f = \frac{1}{2}kx_2^2$. Using $k = 6.00$ N/m, $m = 0.240$ kg, $x_1 = 0.100$ m, $x_2 = 0.112$ m, and $v_1 = 0.400$ m/s gives $W_f = -0.0116$ J.

(b) $W_f = fs \cos\phi = \mu_k n(x_2 - x_1) \cos 180° = -\mu_k mg(x_2 - x_1)$. Use the result from (a) for W_f.
$-\mu_k (0.240 \text{ kg})(9.80 \text{ m/s}^2)(0.112 \text{ m} - 0.100 \text{ m}) = -0.0116 \text{ J} \rightarrow \mu_k = 0.410.$

EVALUATE: From Table 5.1 we see that our result is very reasonable for metal-on-metal coefficients of kinetic friction. For example, $\mu_k = 0.47$ for aluminum on steel.

VP7.9.3. IDENTIFY: We use energy conservation. The system has kinetic energy, gravitational potential energy, and elastic potential energy, but there is no friction.

SET UP: $K = \frac{1}{2}mv^2$ and $U = \frac{1}{2}kx^2$. $U_1 + K_1 + W_{other} = U_2 + K_2$, and $W_{other} = 0$.

EXECUTE: (a) Take point 1 to be the the instant the compressed spring is released and point 2 to be just when the block loses contact with the spring, so $K_1 = 0$ and $U_2 = 0$. This gives

$$U_{\text{spring}} + U_g = K_2 \rightarrow \frac{1}{2}kd^2 - mgd = \frac{1}{2}mv_2^2 \rightarrow v_2 = \sqrt{\frac{kd^2}{m} - 2gd}.$$

(b) After the block clears the spring, we reapply $U_1 + K_1 + W_{\text{other}} = U_2 + K_2$. Now call point 1 to be the instant it has left the spring and point 2 to be its maximum height. The v_1 in this part is the v_2 we found in part (a). Let maximum height be h.

$$\frac{1}{2}mv_1^2 = mgh \rightarrow h = \frac{v_1^2}{2g}$$ The *total* vertical distance the block travels is $D = d + h$, so

$D = d + \frac{v_1^2}{2g}$. Using the value of v_2 (which is v_1 for this part) gives

$$D = d + \frac{\frac{kd^2}{m} - 2gd}{2g} = d + \frac{kd^2}{2mg} - d = \frac{kd^2}{2mg}.$$

EVALUATE: Check units to be sure that both answers have the proper dimensions of length.

VP7.9.4. **IDENTIFY:** We use energy conservation. The system has kinetic energy, gravitational potential energy, and elastic potential energy, but there is no friction.

SET UP: $K = \frac{1}{2}mv^2$ and $U = \frac{1}{2}kx^2$. $U_1 + K_1 + W_{\text{other}} = U_2 + K_2$, and $W_{\text{other}} = W_f$. Call point 1 to be at the place where the cylinder is released with the spring relaxed, and point 2 to be at the maximum elongation of the spring, so $K_1 = 0$, $U_1 = 0$, and $K_2 = 0$. Call x the maximum elongation of the spring. The friction force and gravity both act through a distance x, so $W_f = -fx$ and $U_{2\text{-grav}} = -mgx$. This energy is negative because the cylinder is *below* point 1.

EXECUTE: $U_1 + K_1 + W_{\text{other}} = U_2 + K_2$ gives $-fx = \frac{1}{2}kx^2 - mgx$. Solving for x gives

$x = 2(mg - f)/k$.

EVALUATE: Check units to be sure that the answer has the proper dimensions of length.

7.3. **IDENTIFY:** Use the free-body diagram for the bag and Newton's first law to find the force the worker applies. Since the bag starts and ends at rest, $K_2 - K_1 = 0$ and $W_{\text{tot}} = 0$.

SET UP: A sketch showing the initial and final positions of the bag is given in Figure 7.3a.

$\sin\phi = \frac{2.0 \text{ m}}{3.5 \text{ m}}$ and $\phi = 34.85°$. The free-body diagram is given in Figure 7.3b. $\vec{F}$ is the horizontal force applied by the worker. In the calculation of U_{grav} take $+y$ upward and $y = 0$ at the initial position of the bag.

EXECUTE: **(a)** $\Sigma F_y = 0$ gives $T\cos\phi = mg$ and $\Sigma F_x = 0$ gives $F = T\sin\phi$. Combining these equations to eliminate T gives $F = mg\tan\phi = (90.0 \text{ kg})(9.80 \text{ m/s}^2)\tan 34.85° = 610$ N.

(b) (i) The tension in the rope is radial and the displacement is tangential so there is no component of T in the direction of the displacement during the motion and the tension in the rope does no work.
(ii) $W_{\text{tot}} = 0$ so

$W_{\text{worker}} = -W_{\text{grav}} = U_{\text{grav},2} - U_{\text{grav},1} = mg(y_2 - y_1) = (90.0 \text{ kg})(9.80 \text{ m/s}^2)(0.6277 \text{ m}) = 550$ J.

EVALUATE: The force applied by the worker varies during the motion of the bag and it would be difficult to calculate W_{worker} directly.

Figure 7.3

7.5. **IDENTIFY** and **SET UP:** Use $K_1 + U_1 + W_{\text{other}} = K_2 + U_2$. Points 1 and 2 are shown in Figure 7.5.

(a) $K_1 + U_1 + W_{\text{other}} = K_2 + U_2$. Solve for K_2 and then use $K_2 = \frac{1}{2}mv_2^2$ to obtain v_2.

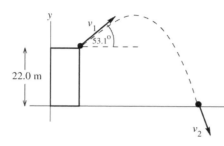

$W_{\text{other}} = 0$ (The only force on the ball while it is in the air is gravity.)
$K_1 = \frac{1}{2}mv_1^2$; $K_2 = \frac{1}{2}mv_2^2$
$U_1 = mgy_1$, $y_1 = 22.0$ m
$U_2 = mgy_2 = 0$, since $y_2 = 0$
for our choice of coordinates.

Figure 7.5

EXECUTE: $\frac{1}{2}mv_1^2 + mgy_1 = \frac{1}{2}mv_2^2$

$v_2 = \sqrt{v_1^2 + 2gy_1} = \sqrt{(12.0 \text{ m/s})^2 + 2(9.80 \text{ m/s}^2)(22.0 \text{ m})} = 24.0$ m/s

EVALUATE: The projection angle of 53.1° doesn't enter into the calculation. The kinetic energy depends only on the magnitude of the velocity; it is independent of the direction of the velocity.
(b) Nothing changes in the calculation. The expression derived in part (a) for v_2 is independent of the angle, so $v_2 = 24.0$ m/s, the same as in part (a).
(c) The ball travels a shorter distance in part (b), so in that case air resistance will have less effect.

7.7. **IDENTIFY:** The take-off kinetic energy of the flea goes into gravitational potential energy.

SET UP: Use $K_1 + U_1 = K_2 + U_2$. Let $y_1 = 0$ and $y_2 = h$ and note that $U_1 = 0$ while $K_2 = 0$ at the maximum height. Consequently, conservation of energy becomes $mgh = \frac{1}{2}mv_1^2$.

EXECUTE: **(a)** $v_1 = \sqrt{2gh} = \sqrt{2(9.80 \text{ m/s}^2)(0.20 \text{ m})} = 2.0$ m/s.

(b) $K_1 = mgh = (0.50 \times 10^{-6} \text{ kg})(9.80 \text{ m/s}^2)(0.20 \text{ m}) = 9.8 \times 10^{-7}$ J. The kinetic energy per kilogram is

$$\frac{K_1}{m} = \frac{9.8 \times 10^{-7} \text{ J}}{0.50 \times 10^{-6} \text{ kg}} = 2.0 \text{ J/kg}.$$

(c) The human can jump to a height of $h_h = h_f \left(\frac{l_h}{l_f}\right) = (0.20 \text{ m})\left(\frac{2.0 \text{ m}}{2.0 \times 10^{-3} \text{ m}}\right) = 200$ m. To attain this height, he would require a takeoff speed of: $v_1 = \sqrt{2gh} = \sqrt{2(9.80 \text{ m/s}^2)(200 \text{ m})} = 63$ m/s.

(d) The human's kinetic energy per kilogram is $\dfrac{K_1}{m} = gh = (9.80 \text{ m/s}^2)(0.60 \text{ m}) = 5.9$ J/kg.

(e) EVALUATE: The flea stores the energy in its tensed legs.

7.9. IDENTIFY: $W_{tot} = K_B - K_A$. The forces on the rock are gravity, the normal force and friction.

SET UP: Let $y = 0$ at point B and let $+y$ be upward. $y_A = R = 0.50$ m. The work done by friction is negative; $W_f = -0.22$ J. $K_A = 0$. The free-body diagram for the rock at point B is given in Figure 7.9. The acceleration of the rock at this point is $a_{rad} = v^2/R$, upward.

EXECUTE: **(a)** (i) The normal force is perpendicular to the displacement and does zero work.

(ii) $W_{grav} = U_{grav,A} - U_{grav,B} = mgy_A = (0.20 \text{ kg})(9.80 \text{ m/s}^2)(0.50 \text{ m}) = 0.98$ J.

(b) $W_{tot} = W_n + W_f + W_{grav} = 0 + (-0.22 \text{ J}) + 0.98 \text{ J} = 0.76$ J. $W_{tot} = K_B - K_A$ gives $\tfrac{1}{2}mv_B^2 = W_{tot}$.

$$v_B = \sqrt{\dfrac{2W_{tot}}{m}} = \sqrt{\dfrac{2(0.76 \text{ J})}{0.20 \text{ kg}}} = 2.8 \text{ m/s}.$$

(c) Gravity is constant and equal to mg. n is not constant; it is zero at A and not zero at B. Therefore, $f_k = \mu_k n$ is also not constant.

(d) $\Sigma F_y = ma_y$ applied to Figure 7.9 gives $n - mg = ma_{rad}$.

$$n = m\left(g + \dfrac{v^2}{R}\right) = (0.20 \text{ kg})\left(9.80 \text{ m/s}^2 + \dfrac{[2.8 \text{ m/s}]^2}{0.50 \text{ m}}\right) = 5.1 \text{ N}.$$

EVALUATE: In the absence of friction, the speed of the rock at point B would be $\sqrt{2gR} = 3.1$ m/s. As the rock slides through point B, the normal force is greater than the weight $mg = 2.0$ N of the rock.

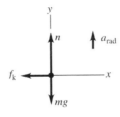

Figure 7.9

7.13. IDENTIFY: This problem involves kinetic energy, gravitational potential energy, and energy conservation.

SET UP: $U_g = mgh$, $K = \tfrac{1}{2}mv^2$, $U_1 + K_1 = U_2 + K_2$.

EXECUTE: **(a)** $\Delta U_6 = m_6 g \Delta y = (6.00 \text{ kg})(9.80 \text{ /s}^2)(0.200 \text{ m}) = 11.8$ J

$\Delta U_8 = m_8 g \Delta y = (8.00 \text{ kg})(9.80 \text{ m/s}^2)(-0.200 \text{ m}) = -15.7$ J.

(b) $W_6 = T\Delta y = (0.200 \text{ m})T$ and $W_8 = T\Delta y = (-0.200 \text{ m})T$.

(c) From part (b), we have $W_T = (0.200 \text{ m})T + (-0.200 \text{ m})T = 0$.

$\Delta U_g = \Delta U_6 + \Delta U_8 = 11.8 \text{ J} - 15.7 \text{ J} = -3.9$ J, and $K_1 = 0$.

$U_1 + K_1 = U_2 + K_2$ gives $K_2 = U_1 - U_2 = -(U_2 - U_1) = -\Delta U = -(-3.9 \text{ J}) = 3.9$ J.

$$K_2 = \tfrac{1}{2}mv_2^2 = -\Delta U \rightarrow v_2 = \sqrt{\dfrac{-2\Delta U}{m}} = \sqrt{\dfrac{-2(-3.9 \text{ J})}{14 \text{ kg}}} = 0.75 \text{ m/s}.$$

EVALUATE: To check, we could find v_2 using Newton's second law. Treating the two masses as a single system gives $m_8 g - m_6 g = (m_8 + m_6)a$. Kinematics gives $a = \dfrac{v_2^2}{2\Delta y}$. Combining these equations and putting in the numbers gives $v_2 = 0.75$ m/s.

7.19. IDENTIFY and **SET UP:** Use energy methods. There are changes in both elastic and gravitational potential energy; elastic; $U = \tfrac{1}{2}kx^2$, gravitational: $kx - mg = ma$,

EXECUTE: (a) $U_{el} = \tfrac{1}{2}kx^2$ so $x = \sqrt{\dfrac{2U_{el}}{k}} = \sqrt{\dfrac{2(1.20 \text{ J})}{800 \text{ N/m}}} = 0.0548$ m $= 5.48$ cm.

(b) The work done by gravity is equal to the gain in elastic potential energy: $W_{grav} = U_{el}$. $mgx = \tfrac{1}{2}kx^2$, so $x = 2mg/k = 2(1.60$ kg$)(9.80$ m/s$^2)/(800$ N/m$) = 0.0392$ m $= 3.92$ cm.

EVALUATE: When the spring is compressed 3.92 cm, it exerts an upward force of 31.4 N on the book, which is greater than the weight of the book (15.6 N). The book will be accelerated upward from this position.

7.21. IDENTIFY: The energy of the book-spring system is conserved. There are changes in both elastic and gravitational potential energy.

SET UP: $U_{el} = \tfrac{1}{2}kx^2$, $U_{grav} = mgy$, $W_{other} = 0$.

EXECUTE: (a) $U = \tfrac{1}{2}kx^2$ so $x = \sqrt{\dfrac{2U}{k}} = \sqrt{\dfrac{2(3.20 \text{ J})}{1600 \text{ N/m}}} = 0.0632$ m $= 6.32$ cm

(b) Points 1 and 2 in the motion are sketched in Figure 7.21. We have $K_1 + U_1 + W_{other} = K_2 + U_2$, where $W_{other} = 0$ (only work is that done by gravity and spring force), $K_1 = 0$, $K_2 = 0$, and $y = 0$ at final position of book. Using $U_1 = mg(h+d)$ and $U_2 = \tfrac{1}{2}kd^2$ we obtain $0 + mg(h+d) + 0 = \tfrac{1}{2}kd^2$. The original gravitational potential energy of the system is converted into potential energy of the compressed spring. Finally, we use the quadratic formula to solve for d: $\tfrac{1}{2}kd^2 - mgd - mgh = 0$, which gives $d = \dfrac{1}{k}\left(mg \pm \sqrt{(mg)^2 + 4\left(\dfrac{1}{2}k\right)(mgh)}\right)$. In our analysis we have assumed that d is positive, so we get

$$d = \dfrac{(1.20 \text{ kg})(9.80 \text{ m/s}^2) + \sqrt{\left[(1.20 \text{ kg})(9.80 \text{ m/s}^2)\right]^2 + 2(1600 \text{ N/m})(1.20 \text{ kg})(9.80 \text{ m/s}^2)(0.80 \text{ m})}}{1600 \text{ N/m}},$$

which gives $d = 0.12$ m $= 12$ cm.

EVALUATE: It was important to recognize that the total displacement was $h + d$; gravity continues to do work as the book moves against the spring. Also note that with the spring compressed 0.12 m it exerts an upward force (192 N) greater than the weight of the book (11.8 N). The book will be accelerated upward from this position.

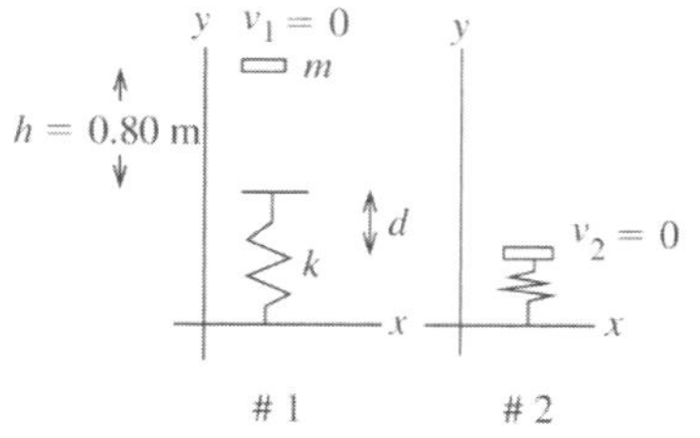

Figure 7.21

7.23. **IDENTIFY:** Only the spring does work and $K_1 + U_1 = K_2 + U_2$ applies. $a = \dfrac{F}{m} = \dfrac{-kx}{m}$, where F is the force the spring exerts on the mass.

SET UP: Let point 1 be the initial position of the mass against the compressed spring, so $K_1 = 0$ and $U_1 = 11.5$ J. Let point 2 be where the mass leaves the spring, so $U_{el,2} = 0$.

EXECUTE: (a) $K_1 + U_{el,1} = K_2 + U_{el,2}$ gives $U_{el,1} = K_2$. $\frac{1}{2}mv_2^2 = U_{el,1}$ and

$$v_2 = \sqrt{\dfrac{2U_{el,1}}{m}} = \sqrt{\dfrac{2(11.5\text{ J})}{2.50\text{ kg}}} = 3.03 \text{ m/s}.$$

K is largest when U_{el} is least and this is when the mass leaves the spring. The mass achieves its maximum speed of 3.03 m/s as it leaves the spring and then slides along the surface with constant speed.

(b) The acceleration is greatest when the force on the mass is the greatest, and this is when the spring has its maximum compression. $U_{el} = \frac{1}{2}kx^2$ so $x = -\sqrt{\dfrac{2U_{el}}{k}} = -\sqrt{\dfrac{2(11.5\text{ J})}{2500\text{ N/m}}} = -0.0959 \text{ m}$.

The minus sign indicates compression. $F = -kx = ma_x$ and

$$a_x = -\dfrac{kx}{m} = -\dfrac{(2500\text{ N/m})(-0.0959\text{ m})}{2.50\text{ kg}} = 95.9 \text{ m/s}^2.$$

EVALUATE: If the end of the spring is displaced to the left when the spring is compressed, then a_x in part (b) is to the right, and vice versa.

7.25. **IDENTIFY:** Apply $K_1 + U_1 + W_{other} = K_2 + U_2$ and $F = ma$.

SET UP: $W_{other} = 0$. There is no change in U_{grav}. $K_1 = 0$, $U_2 = 0$.

EXECUTE: $\frac{1}{2}kx^2 = \frac{1}{2}mv_x^2$. The relations for m, v_x, k and x are $kx^2 = mv_x^2$ and $kx = 5mg$.

Dividing the first equation by the second gives $x = \dfrac{v_x^2}{5g}$, and substituting this into the second gives

$$k = 25\dfrac{mg^2}{v_x^2}.$$

(a) $k = 25\dfrac{(1160\text{ kg})(9.80\text{ m/s}^2)^2}{(2.50\text{ m/s})^2} = 4.46 \times 10^5 \text{ N/m}$

(b) $x = \dfrac{(2.50\text{ m/s})^2}{5(9.80\text{ m/s}^2)} = 0.128 \text{ m}$

EVALUATE: Our results for k and x do give the required values for a_x and v_x:

$$a_x = \dfrac{kx}{m} = \dfrac{(4.46\times 10^5\text{ N/m})(0.128\text{ m})}{1160\text{ kg}} = 49.2 \text{ m/s}^2 = 5.0g \text{ and } v_x = x\sqrt{\dfrac{k}{m}} = 2.5 \text{ m/s}.$$

7.29. **IDENTIFY:** Some of the mechanical energy of the skier is converted to internal energy by the non-conservative force of friction on the rough patch. Use $K_1 + U_1 + W_{other} = K_2 + U_2$.

SET UP: For part (a) use $E_{mech, 2} = E_{mech, 1} - f_k s$ where $f_k = \mu_k mg$. Let $y_2 = 0$ at the bottom of the hill; then $y_1 = 2.50$ m along the rough patch. The energy equation is

$\frac{1}{2}mv_2^2 = \frac{1}{2}mv_1^2 + mgy_1 - \mu_k mgs$. Solving for her final speed gives $v_2 = \sqrt{v_1^2 + 2gy_1 - 2\mu_k gs}$. For part (b), the internal energy is calculated as the negative of the work done by friction: $-W_f = +f_k s = +\mu_k mgs$.

EXECUTE: (a)

$v_2 = \sqrt{(6.50 \text{ m/s})^2 + 2(9.80 \text{ m/s}^2)(2.50 \text{ m}) - 2(0.300)(9.80 \text{ m/s}^2)(4.20 \text{ m})} = 8.16 \text{ m/s}.$

(b) Internal energy $= \mu_k mgs = (0.300)(62.0 \text{ kg})(9.80 \text{ m/s}^2)(4.20 \text{ m}) = 766$ J.

EVALUATE: Without friction the skier would be moving faster at the bottom of the hill than at the top, but in this case she is moving *slower* because friction converted some of her initial kinetic energy into internal energy.

7.31. IDENTIFY: We know the potential energy function and want to find the force causing this energy.

SET UP: $F_x = -\dfrac{dU}{dx}$. The sign of F_x indicates its direction.

EXECUTE: $F_x = -\dfrac{dU}{dx} = -4\alpha x^3 = -4(0.630 \text{ J/m}^4)x^3.$

$F_x(-0.800 \text{ m}) = -4(0.630 \text{ J/m}^4)(-0.80 \text{ m})^3 = 1.29$ N. The force is in the $+x$-direction.

EVALUATE: $F_x > 0$ when $x < 0$ and $F_x < 0$ when $x > 0$, so the force is always directed toward the origin.

7.33. IDENTIFY: From the potential energy function of the block, we can find the force on it, and from the force we can use Newton's second law to find its acceleration.

SET UP: The force components are $F_x = -\dfrac{\partial U}{\partial x}$ and $F_y = -\dfrac{\partial U}{\partial y}$. The acceleration components are $a_x = F_x/m$ and $a_y = F_y/m$. The magnitude of the acceleration is $a = \sqrt{a_x^2 + a_y^2}$ and we can find its angle with the $+x$ axis using $\tan\theta = a_y/a_x$.

EXECUTE: $F_x = -\dfrac{\partial U}{\partial x} = -(11.6 \text{ J/m}^2)x$ and $F_y = -\dfrac{\partial U}{\partial y} = (10.8 \text{ J/m}^3)y^2$. At the point ($x = 0.300$ m, $y = 0.600$ m), $F_x = -(11.6 \text{ J/m}^2)(0.300 \text{ m}) = -3.48$ N and $F_y = (10.8 \text{ J/m}^3)(0.600 \text{ m})^2 = 3.89$ N. Therefore $a_x = \dfrac{F_x}{m} = -87.0 \text{ m/s}^2$ and $a_y = \dfrac{F_y}{m} = 97.2 \text{ m/s}^2$, giving $a = \sqrt{a_x^2 + a_y^2} = 130 \text{ m/s}^2$ and $\tan\theta = \dfrac{97.2}{87.0}$, so $\theta = 48.2°$. The direction is 132° counter-clockwise from the $+x$-axis.

EVALUATE: The force is not constant, so the acceleration will not be the same at other points.

7.35. IDENTIFY and SET UP: Use $F = -dU/dr$ to calculate the force from U. At equilibrium $F = 0$.

(a) EXECUTE: The graphs are sketched in Figure 7.35.

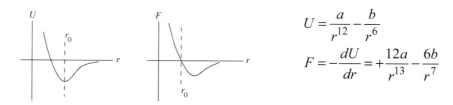

$U = \dfrac{a}{r^{12}} - \dfrac{b}{r^6}$

$F = -\dfrac{dU}{dr} = +\dfrac{12a}{r^{13}} - \dfrac{6b}{r^7}$

Figure 7.35

(b) At equilibrium $F = 0$, so $\dfrac{dU}{dr} = 0$

$F = 0$ implies $\dfrac{+12a}{r^{13}} - \dfrac{6b}{r^7} = 0$

$6br^6 = 12a$; solution is the equilibrium distance $r_0 = (2a/b)^{1/6}$
U is a minimum at this r; the equilibrium is stable.

(c) At $r = (2a/b)^{1/6}$, $U = a/r^{12} - b/r^6 = a(b/2a)^2 - b(b/2a) = -b^2/4a$.
At $r \to \infty$, $U = 0$. The energy that must be added is $-\Delta U = b^2/4a$.

(d) $r_0 = (2a/b)^{1/6} = 1.13 \times 10^{-10}$ m gives that
$2a/b = 2.082 \times 10^{-60}$ m^6 and $b/4a = 2.402 \times 10^{59}$ m^{-6}
$b^2/4a = b(b/4a) = 1.54 \times 10^{-18}$ J
$b(2.402 \times 10^{59}$ m$^{-6}) = 1.54 \times 10^{-18}$ J and $b = 6.41 \times 10^{-78}$ J·m^6.
Then $2a/b = 2.082 \times 10^{-60}$ m^6 gives $a = (b/2)(2.082 \times 10^{-60}$ m$^6) =$
$\tfrac{1}{2}(6.41 \times 10^{-78}$ J·m$^6)(2.082 \times 10^{-60}$ m$^6) = 6.67 \times 10^{-138}$ J·m^{12}

EVALUATE: As the graphs in part (a) show, $F(r)$ is the slope of $U(r)$ at each r. $U(r)$ has a minimum where $F = 0$.

7.37. **IDENTIFY:** Apply $\Sigma \vec{F} = m\vec{a}$ to the bag and to the box. Apply $K_1 + U_1 + W_{\text{other}} = K_2 + U_2$ to the motion of the system of the box and bucket after the bag is removed.

SET UP: Let $y = 0$ at the final height of the bucket, so $y_1 = 2.00$ m and $y_2 = 0$. $K_1 = 0$. The box and the bucket move with the same speed v, so $K_2 = \tfrac{1}{2}(m_{\text{box}} + m_{\text{bucket}})v^2$. $W_{\text{other}} = -f_k d$, with $d = 2.00$ m and $f_k = \mu_k m_{\text{box}} g$. Before the bag is removed, the maximum possible friction force the roof can exert on the box is $(0.700)(80.0 \text{ kg} + 50.0 \text{ kg})(9.80 \text{ m/s}^2) = 892$ N. This is larger than the weight of the bucket (637 N), so before the bag is removed the system is at rest.

EXECUTE: **(a)** The friction force on the bag of gravel is zero, since there is no other horizontal force on the bag for friction to oppose. The static friction force on the box equals the weight of the bucket, 637 N.

(b) Applying $K_1 + U_1 + W_{\text{other}} = K_2 + U_2$ gives $m_{\text{bucket}} g y_1 - f_k d = \tfrac{1}{2} m_{\text{tot}} v^2$, with $m_{\text{tot}} = 145.0$ kg.

$v = \sqrt{\dfrac{2}{m_{\text{tot}}}(m_{\text{bucket}} g y_1 - \mu_k m_{\text{box}} g d)}$.

$v = \sqrt{\dfrac{2}{145.0 \text{ kg}}\left[(65.0 \text{ kg})(9.80 \text{ m/s}^2)(2.00 \text{ m}) - (0.400)(80.0 \text{ kg})(9.80 \text{ m/s}^2)(2.00 \text{ m})\right]} = 2.99$ m/s.

EVALUATE: If we apply $\Sigma \vec{F} = m\vec{a}$ to the box and to the bucket we can calculate their common acceleration a. Then a constant acceleration equation applied to either object gives $v = 2.99$ m/s, in agreement with our result obtained using energy methods.

7.41. **IDENTIFY:** The mechanical energy of the roller coaster is conserved since there is no friction with the track. We must also apply Newton's second law for the circular motion.

SET UP: For part (a), apply conservation of energy to the motion from point A to point B: $U(x)$. with $x = 3.00$ m, Defining $y_B = 0$ and $y_A = 13.0$ m, conservation of energy becomes $\frac{1}{2}mv_B^2 = mgy_A$ or $v_B = \sqrt{2gy_A}$. In part (b), the free-body diagram for the roller coaster car at point B is shown in Figure 7.41. $\Sigma F_y = ma_y$ gives $mg + n = ma_{rad}$, where $K_1 = 0$, Solving for the normal force gives $n = m\left(\dfrac{v^2}{r} - g\right)$.

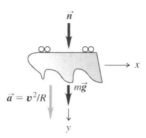

Figure 7.41

EXECUTE: (a) $v_B = \sqrt{2(9.80 \text{ m/s}^2)(13.0 \text{ m})} = 16.0$ m/s.
(b) $x = y$
EVALUATE: The normal force n is the force that the tracks exert on the roller coaster car. The car exerts a force of equal magnitude and opposite direction on the tracks.

7.43. (a) **IDENTIFY:** Use $K_1 + U_1 + W_{other} = K_2 + U_2$ to find the kinetic energy of the wood as it enters the rough bottom.

SET UP: Let point 1 be where the piece of wood is released and point 2 be just before it enters the rough bottom. Let $y_1 = 2.50$ m be at point 2.
EXECUTE: $U_1 = K_2$ gives $K_2 = mgy_1 = 78.4$ J.
IDENTIFY: Now apply $K_1 + U_1 + W_{other} = K_2 + U_2$ to the motion along the rough bottom.
SET UP: Let point 1 be where it enters the rough bottom and point 2 be where it stops.
$K_1 + U_1 + W_{other} = K_2 + U_2$
EXECUTE: $W_{other} = W_f = -\mu_k mgs$, $K_2 = U_1 = U_2 = 0$; $K_1 = 78.4$ J
$78.4 \text{ J} - \mu_k mgs = 0$; solving for s gives θ
The wood stops after traveling 20.0 m along the rough bottom.
(b) Friction does -78.4 J of work.
EVALUATE: The piece of wood stops before it makes one trip across the rough bottom. The final mechanical energy is zero. The negative friction work takes away all the mechanical energy initially in the system.

7.45. **IDENTIFY:** Apply $K_1 + U_1 + W_{other} = K_2 + U_2$ to the motion of the stone.

SET UP: $K_1 + U_1 + W_{other} = K_2 + U_2$. Let point 1 be point A and point 2 be point B. Take $y = 0$ at B.
EXECUTE: $mgy_1 + \frac{1}{2}mv_1^2 = \frac{1}{2}mv_2^2$, with $h = 20.0$ m and $v_1 = 10.0$ m/s,
so $v_2 = \sqrt{v_1^2 + 2gh} = 22.2$ m/s.
EVALUATE: The loss of gravitational potential energy equals the gain of kinetic energy.

(b) IDENTIFY: Apply $K_1 + U_1 + W_{other} = K_2 + U_2$ to the motion of the stone from point B to where it comes to rest against the spring.

SET UP: Use $K_1 + U_1 + W_{other} = K_2 + U_2$, with point 1 at B and point 2 where the spring has its maximum compression x.

EXECUTE: $U_1 = U_2 = K_2 = 0;\ K_1 = \tfrac{1}{2}mv_1^2$ with $v_1 = 22.2$ m/s.

$W_{other} = W_f + W_{el} = -\mu_k mgs - \tfrac{1}{2}kx^2$, with $s = 100$ m $+ x$. The work-energy relation gives

$K_1 + W_{other} = 0.\ \tfrac{1}{2}mv_1^2 - \mu_k mgs - \tfrac{1}{2}kx^2 = 0.$

Putting in the numerical values gives $x^2 + 29.4x - 750 = 0$. The positive root to this equation is $x = 16.4$ m.

EVALUATE: Part of the initial mechanical (kinetic) energy is removed by friction work and the rest goes into the potential energy stored in the spring.

(c) IDENTIFY and SET UP: Consider the forces.

EXECUTE: When the spring is compressed $x = 16.4$ m the force it exerts on the stone is $F_{el} = kx = 32.8$ N. The maximum possible static friction force is

$\max f_s = \mu_s mg = (0.80)(15.0\ \text{kg})(9.80\ \text{m/s}^2) = 118$ N.

EVALUATE: The spring force is less than the maximum possible static friction force so the stone remains at rest.

7.49. IDENTIFY: Use $K_1 + U_1 + W_{other} = K_2 + U_2$. Solve for K_2 and then for v_2.

SET UP: Let point 1 be at his initial position against the compressed spring and let point 2 be at the end of the barrel, as shown in Figure 7.49. Use $F = kx$ to find the amount the spring is initially compressed by the 4400 N force.

$K_1 + U_1 + W_{other} = K_2 + U_2$

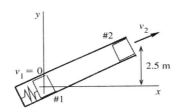

Take $y = 0$ at his initial position.

EXECUTE: $K_1 = 0,\ K_2 = \tfrac{1}{2}mv_2^2$

$W_{other} = W_{fric} = -fs$

$y = y_0 + v_{0y}t + \tfrac{1}{2}a_y t^2$

Figure 7.49

$U_{1,\text{grav}} = 0,\ dK_2/dy = 0.$ where d is the distance the spring is initially compressed.

$U_2 = \tfrac{1}{2}ky^2.$ so $K_1 = 0,$

and $U_g = mgy,$

$x = \dfrac{mg}{k}\left(1 \pm \sqrt{1 + \dfrac{2hk}{mg}}\right).\ U_{2,el} = 0$

Then $K_1 + U_1 + W_{other} = K_2 + U_2$ gives

$8800\ \text{J} - 160\ \text{J} = \tfrac{1}{2}mv_2^2 + 1470\ \text{J}$

$\tfrac{1}{2}mv_2^2 = 7170$ J and $v_2 = \sqrt{\dfrac{2(7170\ \text{J})}{60\ \text{kg}}} = 15.5$ m/s.

EVALUATE: Some of the potential energy stored in the compressed spring is taken away by the work done by friction. The rest goes partly into gravitational potential energy and partly into kinetic energy.

7.51. IDENTIFY: Apply $K_1 + U_1 + W_{other} = K_2 + U_2$ to the system consisting of the two buckets. If we ignore the inertia of the pulley we ignore the kinetic energy it has.

SET UP: $K_1 + U_1 + W_{other} = K_2 + U_2$. Points 1 and 2 in the motion are sketched in Figure 7.51.

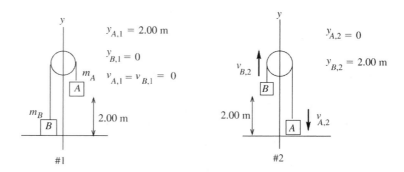

Figure 7.51

The tension force does positive work on the 4.0 kg bucket and an equal amount of negative work on the 12.0 kg bucket, so the net work done by the tension is zero.
Work is done on the system only by gravity, so $W_{other} = 0$ and $K_1 + U_1 + W_{other} = K_2 + U_2$

EXECUTE: $K_1 = 0$, $K_2 = \frac{1}{2}m_A v_{A,2}^2 + \frac{1}{2}m_B v_{B,2}^2$. But since the two buckets are connected by a rope they move together and have the same speed: $v_{A,2} = v_{B,2} = v_2$. Thus

$K_2 = \frac{1}{2}(m_A + m_B)v_2^2 = (8.00 \text{ kg})v_2^2$.

$U_1 = m_A g y_{A,1} = (12.0 \text{ kg})(9.80 \text{ m/s}^2)(2.00 \text{ m}) = 235.2$ J.

$U_2 = m_B g y_{B,2} = (4.0 \text{ kg})(9.80 \text{ m/s}^2)(2.00 \text{ m}) = 78.4$ J.

Putting all this into $K_1 + U_1 + W_{other} = K_2 + U_2$ gives $U_1 = K_2 + U_2$.

$235.2 \text{ J} = (8.00 \text{ kg})v_2^2 + 78.4 \text{ J}$. $v_2 = \sqrt{\dfrac{235.2 \text{ J} - 78.4 \text{ J}}{8.00 \text{ kg}}} = 4.4$ m/s

EVALUATE: The gravitational potential energy decreases and the kinetic energy increases by the same amount. We could apply $K_1 + U_1 + W_{other} = K_2 + U_2$ to one bucket, but then we would have to include in W_{other} the work done on the bucket by the tension T.

7.55. IDENTIFY and SET UP: First apply $\Sigma \vec{F} = m\vec{a}$ to the skier.

Find the angle α where the normal force becomes zero, in terms of the speed v_2 at this point. Then apply the work-energy theorem to the motion of the skier to obtain another equation that relates v_2 and α. Solve these two equations for α.

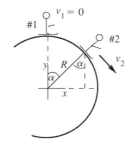

Let point 2 be where the skier loses contact with the snowball, as sketched in Figure 7.55a
Loses contact implies $n \to 0$.
$y_1 = R$, $y_2 = R\cos\alpha$

Figure 7.55a

First, analyze the forces on the skier when she is at point 2. The free-body diagram is given in Figure 7.55b. For this use coordinates that are in the tangential and radial directions. The skier moves in an arc of a circle, so her acceleration is $a_{\text{rad}} = v^2/R$, directed in toward the center of the snowball.

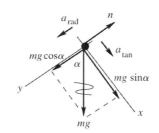

EXECUTE: $\Sigma F_y = ma_y$

$mg\cos\alpha - n = mv_2^2/R$

But $n = 0$ so $mg\cos\alpha = mv_2^2/R$

$v_2^2 = Rg\cos\alpha$

Figure 7.55b

Now use conservation of energy to get another equation relating v_2 to α:
$K_1 + U_1 + W_{\text{other}} = K_2 + U_2$
The only force that does work on the skier is gravity, so $W_{\text{other}} = 0$.
$K_1 = 0$, $K_2 = \tfrac{1}{2}mv_2^2$
$U_1 = mgy_1 = mgR$, $U_2 = mgy_2 = mgR\cos\alpha$
Then $mgR = \tfrac{1}{2}mv_2^2 + mgR\cos\alpha$
$v_2^2 = 2gR(1-\cos\alpha)$
Combine this with the $\Sigma F_y = ma_y$ equation:
$Rg\cos\alpha = 2gR(1-\cos\alpha)$
$\cos\alpha = 2 - 2\cos\alpha$
$3\cos\alpha = 2$ so $\cos\alpha = 2/3$ and $\alpha = 48.2°$

EVALUATE: She speeds up and her a_{rad} increases as she loses gravitational potential energy. She loses contact when she is going so fast that the radially inward component of her weight isn't large enough to keep her in the circular path. Note that α where she loses contact does not depend on her mass or on the radius of the snowball.

7.57. **IDENTIFY and SET UP:**

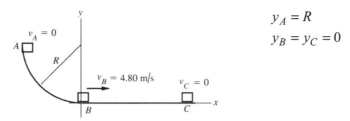

Figure 7.57

$y_A = R$
$y_B = y_C = 0$

(a) Apply conservation of energy to the motion from B to C:
$K_B + U_B + W_{\text{other}} = K_C + U_C$. The motion is described in Figure 7.57.
EXECUTE: The only force that does work on the package during this part of the motion is friction, so
$W_{\text{other}} = W_f = f_k(\cos\phi)s = \mu_k mg(\cos 180°)s = -\mu_k mgs$
$K_B = \frac{1}{2}mv_B^2$, $K_C = 0$
$U_B = 0$, $U_C = 0$
Thus $K_B + W_f = 0$
$\frac{1}{2}mv_B^2 - \mu_k mgs = 0$
$\mu_k = \dfrac{v_B^2}{2gs} = \dfrac{(4.80 \text{ m/s})^2}{2(9.80 \text{ m/s}^2)(3.00 \text{ m})} = 0.392$.

EVALUATE: The negative friction work takes away all the kinetic energy.

(b) IDENTIFY and SET UP: Apply conservation of energy to the motion from A to B:
$$K_A + U_A + W_{\text{other}} = K_B + U_B$$
EXECUTE: Work is done by gravity and by friction, so $W_{\text{other}} = W_f$.
$K_A = 0$, $K_B = \frac{1}{2}mv_B^2 = \frac{1}{2}(0.200 \text{ kg})(4.80 \text{ m/s})^2 = 2.304$ J
$U_A = mgy_A = mgR = (0.200 \text{ kg})(9.80 \text{ m/s}^2)(1.60 \text{ m}) = 3.136$ J, $U_B = 0$
Thus $U_A + W_f = K_B$
$W_f = K_B - U_A = 2.304 \text{ J} - 3.136 \text{ J} = -0.83$ J

EVALUATE: W_f is negative as expected; the friction force does negative work since it is directed opposite to the displacement.

7.59. (a) IDENTIFY: We are given that $F_x = -\alpha x - \beta x^2$, $\alpha = 60.0$ N/m and $\beta = 18.0$ N/m^2. Use $W_{F_x} = \int_{x_1}^{x_2} F_x(x)\,dx$ to calculate W and then use $W = -\Delta U$ to identify the potential energy function $U(x)$.

SET UP: $W_{F_x} = U_1 - U_2 = \int_{x_1}^{x_2} F_x(x)\,dx$

Let $x_1 = 0$ and $U_1 = 0$. Let x_2 be some arbitrary point x, so $U_2 = U(x)$.

EXECUTE: $U(x) = -\int_0^x F_x(x)\,dx = -\int_0^x (-\alpha x - \beta x^2)\,dx = \int_0^x (\alpha x + \beta x^2)\,dx = \frac{1}{2}\alpha x^2 + \frac{1}{3}\beta x^3$.

EVALUATE: If $\beta = 0$, the spring does obey Hooke's law, with $k = \alpha$, and our result reduces to $\frac{1}{2}kx^2$.

(b) IDENTIFY: Apply $K_1 + U_1 + W_{\text{other}} = K_2 + U_2$ to the motion of the object.

SET UP: The system at points 1 and 2 is sketched in Figure 7.59.

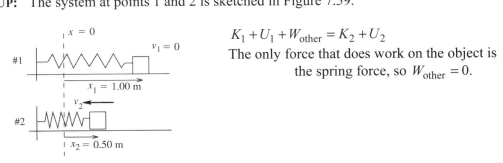

$K_1 + U_1 + W_{\text{other}} = K_2 + U_2$
The only force that does work on the object is the spring force, so $W_{\text{other}} = 0$.

Figure 7.59

EXECUTE: $K_1 = 0$, $K_2 = \frac{1}{2}mv_2^2$

$U_1 = U(x_1) = \frac{1}{2}\alpha x_1^2 + \frac{1}{3}\beta x_1^3 = \frac{1}{2}(60.0 \text{ N/m})(1.00 \text{ m})^2 + \frac{1}{3}(18.0 \text{ N/m}^2)(1.00 \text{ m})^3 = 36.0 \text{ J}$

$U_2 = U(x_2) = \frac{1}{2}\alpha x_2^2 + \frac{1}{3}\beta x_2^3 = \frac{1}{2}(60.0 \text{ N/m})(0.500 \text{ m})^2 + \frac{1}{3}(18.0 \text{ N/m}^2)(0.500 \text{ m})^3 = 8.25 \text{ J}$

Thus $36.0 \text{ J} = \frac{1}{2}mv_2^2 + 8.25 \text{ J}$, which gives $v_2 = \sqrt{\dfrac{2(36.0 \text{ J} - 8.25 \text{ J})}{0.900 \text{ kg}}} = 7.85 \text{ m/s}$.

EVALUATE: The elastic potential energy stored in the spring decreases and the kinetic energy of the object increases.

7.61. IDENTIFY: We have a conservative force, so we can relate the force and the potential energy function. Energy conservation applies.

SET UP: $F_x = -dU/dx$, U goes to 0 as x goes to infinity, and $F(x) = \dfrac{\alpha}{(x+x_0)^2}$.

EXECUTE: (a) Using $dU = -F_x dx$, we get $U_x - U_\infty = -\int_\infty^x \dfrac{\alpha}{(x+x_0)^2}dx = \dfrac{\alpha}{x+x_0}$.

(b) Energy conservation tells us that $U_1 = K_2 + U_2$. Therefore $\dfrac{\alpha}{x_1+x_0} = \dfrac{1}{2}mv_x^2 + \dfrac{\alpha}{x_2+x_0}$. Putting in $m = 0.500$ kg, $\alpha = 0.800$ N·m, $x_0 = 0.200$ m, $x_1 = 0$, and $x_2 = 0.400$ m, solving for v gives $v = 3.27$ m/s.

EVALUATE: The potential energy is not infinite even though the integral in (a) is taken over an infinite distance because the force rapidly gets smaller with increasing distance x.

7.63. IDENTIFY: Apply $K_1 + U_1 + W_{\text{other}} = K_2 + U_2$ to the motion of the block.

SET UP: Let $y = 0$ at the floor. Let point 1 be the initial position of the block against the compressed spring and let point 2 be just before the block strikes the floor.

EXECUTE: With $U_2 = 0$, $K_1 = 0$, $K_2 = U_1$. $\frac{1}{2}mv_2^2 = \frac{1}{2}kx^2 + mgh$. Solving for v_2,

$v_2 = \sqrt{\dfrac{kx^2}{m} + 2gh} = \sqrt{\dfrac{(1900 \text{ N/m})(0.045 \text{ m})^2}{(0.150 \text{ kg})} + 2(9.80 \text{ m/s}^2)(1.20 \text{ m})} = 7.01 \text{ m/s}$.

EVALUATE: The potential energy stored in the spring and the initial gravitational potential energy all go into the final kinetic energy of the block.

7.67. **IDENTIFY:** Only conservative forces (gravity and the spring) act on the fish, so its mechanical energy is conserved.

SET UP: Energy conservation tells us $K_1 + U_1 + W_{other} = K_2 + U_2$, where $W_{other} = 0$. $U_g = mgy$, $K = \frac{1}{2}mv^2$, and $U_{spring} = \frac{1}{2}ky^2$.

EXECUTE: **(a)** $K_1 + U_1 + W_{other} = K_2 + U_2$. Let y be the distance the fish has descended, so $y = 0.0500$ m. $K_1 = 0$, $W_{other} = 0$, $U_1 = mgy$, $K_2 = \frac{1}{2}mv_2^2$, and $U_2 = \frac{1}{2}ky^2$. Solving for K_2 gives

$$K_2 = U_1 - U_2 = mgy - \frac{1}{2}ky^2 = (3.00 \text{ kg})(9.8 \text{ m/s}^2)(0.0500 \text{ m}) - \frac{1}{2}(900 \text{ N/m})(0.0500 \text{ m})^2$$

$K_2 = 1.47$ J $- 1.125$ J $= 0.345$ J. Solving for v_2 gives $v_2 = \sqrt{\frac{2K_2}{m}} = \sqrt{\frac{2(0.345 \text{ J})}{3.00 \text{ kg}}} = 0.480$ m/s.

(b) The maximum speed is when K_2 is maximum, which is when $dK_2/dy = 0$. Using $K_2 = mgy - \frac{1}{2}ky^2$ gives $\frac{dK_2}{dy} = mg - ky = 0$. Solving for y gives

$$y = \frac{mg}{k} = \frac{(3.00 \text{ kg})(9.8 \text{ m/s}^2)}{900 \text{ N/m}} = 0.03267 \text{ m}. \text{ At this } y,$$

$$K_2 = (3.00 \text{ kg})(9.8 \text{ m/s}^2)(0.03267 \text{ m}) - \frac{1}{2}(900 \text{ N/m})(0.03267 \text{ m})^2.$$

$K_2 = 0.9604$ J $- 0.4803$ J $= 0.4801$ J, so $v_2 = \sqrt{\frac{2K_2}{m}} = 0.566$ m/s.

EVALUATE: The speed in part (b) is greater than the speed in part (a), as it should be since it is the maximum speed.

7.71. **IDENTIFY:** We can apply Newton's second law to the block. The only forces acting on the block are gravity downward and the normal force from the track pointing toward the center of the circle. The mechanical energy of the block is conserved since only gravity does work on it. The normal force does no work since it is perpendicular to the displacement of the block. The target variable is the normal force at the top of the track.

SET UP: For circular motion $\Sigma F = m\frac{v^2}{R}$. Energy conservation tells us that $K_A + U_A + W_{other} = K_B + U_B$, where $W_{other} = 0$. $U_g = mgy$ and $K = \frac{1}{2}mv^2$.

EXECUTE: Let point A be at the bottom of the path and point B be at the top of the path. At the bottom of the path, $n_A - mg = m\frac{v^2}{R}$ (from Newton's second law).

$v_A = \sqrt{\frac{R}{m}(n_A - mg)} = \sqrt{\frac{0.800 \text{ m}}{0.0500 \text{ kg}}(3.40 \text{ N} - 0.49 \text{ N})} = 6.82$ m/s. Use energy conservation to find the speed at point B. $K_A + U_A + W_{other} = K_B + U_B$, giving $\frac{1}{2}mv_A^2 = \frac{1}{2}mv_B^2 + mg(2R)$. Solving for v_B gives $v_B = \sqrt{v_A^2 - 4Rg} = \sqrt{(6.82 \text{ m/s})^2 - 4(0.800 \text{ m})(9.8 \text{ m/s}^2)} = 3.89$ m/s. Then at point B,

Newton's second law gives $n_B + mg = m\dfrac{v_B^2}{R}$. Solving for n_B gives $n_B = m\dfrac{v_B^2}{R} - mg =$
$(0.0500 \text{ kg})\left(\dfrac{(3.89 \text{ m/s})^2}{0.800 \text{ m}} - 9.8 \text{ m/s}^2\right) = 0.456 \text{ N}.$

EVALUATE: The normal force at the top is considerably less than it is at the bottom for two reasons: the block is moving slower at the top and the downward force of gravity at the top aids the normal force in keeping the block moving in a circle.

7.73. IDENTIFY: Apply $K_1 + U_1 + W_{other} = K_2 + U_2$ to the motion of the block.

SET UP: The motion from A to B is described in Figure 7.73.

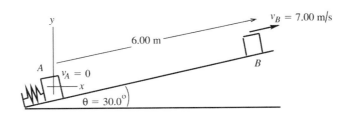

Figure 7.73

The normal force is $n = mg\cos\theta$, so $f_k = \mu_k n = \mu_k mg\cos\theta$. $y_A = 0$;
$y_B = (6.00 \text{ m})\sin 30.0° = 3.00 \text{ m}.$
$K_A + U_A + W_{other} = K_B + U_B$

EXECUTE: Work is done by gravity, by the spring force, and by friction, so $W_{other} = W_f$ and
$U = U_{el} + U_{grav}$
$K_A = 0$, $K_B = \tfrac{1}{2}mv_B^2 = \tfrac{1}{2}(1.50 \text{ kg})(7.00 \text{ m/s})^2 = 36.75 \text{ J}$
$U_A = U_{el,A} + U_{grav,A} = U_{el,A}$, since $U_{grav,A} = 0$
$U_B = U_{el,B} + U_{grav,B} = 0 + mgy_B = (1.50 \text{ kg})(9.80 \text{ m/s}^2)(3.00 \text{ m}) = 44.1 \text{ J}$
$W_{other} = W_f = (f_k \cos\phi)s = \mu_k mg\cos\theta(\cos 180°)s = -\mu_k mg\cos\theta s$
$W_{other} = -(0.50)(1.50 \text{ kg})(9.80 \text{ m/s}^2)(\cos 30.0°)(6.00 \text{ m}) = -38.19 \text{ J}$
Thus $U_{el,A} - 38.19 \text{ J} = 36.75 \text{ J} + 44.10 \text{ J}$, giving $U_{el,A} = 38.19 \text{ J} + 36.75 \text{ J} + 44.10 \text{ J} = 119 \text{ J}.$

EVALUATE: U_{el} must always be positive. Part of the energy initially stored in the spring was taken away by friction work; the rest went partly into kinetic energy and partly into an increase in gravitational potential energy.

7.77. IDENTIFY: The mechanical energy of the system is conserved, and Newton's second law applies. As the pendulum swings, gravitational potential energy gets transformed to kinetic energy.

SET UP: For circular motion, $F = mv^2/r$. $U_{grav} = mgh$.
EXECUTE: (a) Conservation of mechanical energy gives $mgh = \tfrac{1}{2}mv^2 + mgh_0$, where $h_0 = 0.800$ m. Applying Newton's second law at the bottom of the swing gives $T = mv^2/L + mg$. Combining these two equations and solving for T as a function of h gives $T = (2mg/L)h + mg(1 - 2h_0/L)$. In a graph of T versus h, the slope is $2mg/L$. Graphing the data given in the problem, we get the graph shown in Figure 7.77. Using the best-fit equation, we get $T = (9.293 \text{ N/m})h + 257.3 \text{ N}$. Therefore

$2mg/L = 9.293$ N/m. Using $mg = 265$ N and solving for L, we get $L = 2(265\text{ N})/(9.293\text{ N/m}) = 57.0$ m.

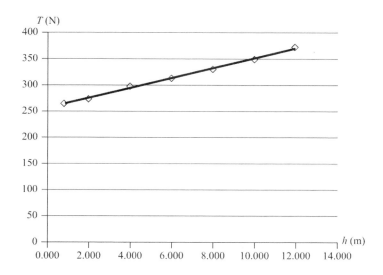

Figure 7.77

(b) $T_{max} = 822$ N, so $T = T_{max}/2 = 411$ N. We use the equation for the graph with $T = 411$ N and solve for h. $411\text{ N} = (9.293\text{ N/m})h + 257.3\text{ N}$, which gives $h = 16.5$ m.
(c) The pendulum is losing energy because negative work is being done on it by friction with the air and at the point of contact where it swings.
EVALUATE: The length of this pendulum may seem extremely large, but it is not unreasonable for a museum exhibit, which can cover a height of several floor levels.

7.79. **IDENTIFY:** For a conservative force, mechanical energy is conserved and we can relate the force to its potential energy function.
SET UP: $F_x = -dU/dx$.
EXECUTE: **(a)** $U + K = E$ = constant. If two points have the same kinetic energy, they must have the same potential energy since the sum of U and K is constant. Since the kinetic energy curve symmetric, the potential energy curve must also be symmetric.
(b) At $x = 0$ we can see from the graph with the problem that $E = K + 0 = 0.14$ J. Since E is constant, if $K = 0$ at $x = -1.5$ m, then U must be equal to 0.14 J at that point.
(c) $U(x) = E - K(x) = 0.14\text{ J} - K(x)$, so the graph of $U(x)$ is like the sketch in Figure 7.79.

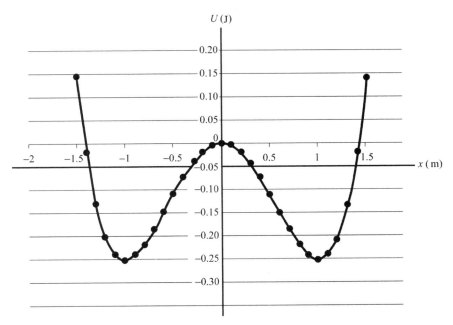

Figure 7.79

(d) Since $F_x = -dU/dx$, $F(x) = 0$ at $x = 0$, $+1.0$ m, and -1.0 m.

(e) $F(x)$ is positive when the slope of the $U(x)$ curve is negative, and $F(x)$ is negative when the slope of the $U(x)$ curve is positive. Therefore $F(x)$ is positive between $x = -1.5$ m and $x = -1.0$ m and between $x = 0$ and $x = 1.0$ m. $F(x)$ is negative between $x = -1.0$ m and 0 and between $x = 1.0$ m and $x = 1.5$ m.

(f) When released from $x = -1.30$ m, the sphere will move to the right until it reaches $x = -0.55$ m, at which point it has 0.12 J of potential energy, the same as at is original point of release.

EVALUATE: Even though we do not have the equation of the kinetic energy function, we can still learn much about the behavior of the system by studying its graph.

Study Guide for Momentum, Impulse, and Collisions

Summary

In this chapter, we'll introduce two new concepts—*momentum* and *impulse*—which together will serve as our third major analysis technique in mechanics. Like energy analysis, momentum and impulse analysis will expand our problem-solving repertoire and allow us to tackle collision problems that would be challenging with Newton's laws. Also, like energy, momentum is a conserved quantity that has important consequences throughout physics. We'll be able to apply momentum and impulse analyses to a wide variety of problems by the end of the chapter, and, when those analyses are combined with force and energy analyses, we'll be able apply a powerful set of tools to investigate many natural phenomena.

Objectives

After studying this chapter, you'll understand

- The definition of *momentum* and the distinction between momentum and velocity.
- The definition of *impulse* and the distinction between impulse and force.
- How to restate Newton's law in terms of momentum and impulse.
- How to use conservation of momentum to solve a variety of collision problems.
- How to identify elastic, inelastic, and totally inelastic collisions and how to apply conservation of energy appropriately to each of these situations.
- The definition of the *center of mass* of a system and how to use the center of mass to solve problems.
- How to apply momentum conservation to rocket propulsion problems in which the rocket's mass changes.

Concepts and Equations

Term	Description
Momentum	The momentum $\vec{p}$ of a particle of mass m moving with velocity $\vec{v}$ is defined as $$\vec{p} = m\vec{v}.$$ Newton's second law states that the net external force on a particle is equal to the rate of change of momentum of the particle: $$\sum \vec{F} = \frac{d\vec{p}}{dt}.$$ The total momentum $\vec{P}$ of a system of particles is the vector sum of the individual momenta: $$\vec{P} = \vec{p}_A + \vec{p}_B + \cdots = m_A \vec{v}_A + m_B \vec{v}_B + \cdots.$$
Impulse	The impulse $\vec{J}$ of the net external force is the product of force and the time interval over which the force acts: $$\vec{J} = \vec{F}(t_2 - t_1) = \vec{F}\Delta t.$$ For net external forces that vary with time, the impulse is the integral of the net external force over the time interval: $$\vec{J} = \int_{t_1}^{t_2} \sum \vec{F} \, dt.$$ The change in a particle's momentum during a certain time interval equals the impulse of the net external force acting on the particle during that interval: $$\vec{J} = \Delta \vec{p} = \vec{p}_2 - \vec{p}_1.$$
Conservation of Momentum	The total momentum of a system is constant when the net external force on the system is zero: $$\vec{P} = \text{constant if } \sum \vec{F} = 0.$$ Each component of momentum is separately conserved.
Elastic Collision	In an elastic collision between two objects, the initial and final kinetic energies are equal and the initial and final relative velocities have equal magnitudes.
Inelastic Collision	In an inelastic collision between two objects, the final kinetic energy is less than the initial kinetic energy. If the two objects have the same final velocity, the collision is completely inelastic.
Center of Mass	The position vector $\vec{r}_{cm}$ for the center of mass of a system of particles is a weighted average of the positions $\vec{r}_1, \vec{r}_2, \ldots$, of the individual particles: $$\vec{r}_{cm} = \frac{m_1 \vec{r}_1 + m_2 \vec{r}_2 + \cdots}{m_1 + m_2 + \cdots} = \frac{\sum_i m_i \vec{r}_i}{\sum_i m_i}$$
Motion of Center of Mass	The total momentum of a system equals its total mass multiplied by the velocity of the center of mass:

	$$\vec{P} = M\vec{v}_{cm} = m_1\vec{v}_1 + m_2\vec{v}_2 + \cdots.$$ The center of mass of a system moves as if all the mass were located at the center of mass: $$\sum \vec{F}_{ext} = M\vec{a}_{cm}.$$ If the net external force on a system is zero, the velocity $\vec{v}_{cm}$ of the center of mass of the system is constant.
Rocket Motion	Any analysis of rocket motion must include the momentum carried away by the spent fuel and the momentum of the rocket itself.

Key Concept 1: The momentum of an object equals the product of its mass and its velocity $\vec{v}$, and also equals the impulse (net external force multiplied by time) needed to accelerate it from rest to $\vec{v}$.

Key Concept 2: The impulse-momentum theorem states that when a net external force acts on an object, the object's momentum changes by an amount equal to the impulse of the net external force. Both momentum and impulse are vector quantities.

Key Concept 3: In two-dimensional problems involving impulse and momentum, you must apply the impulse–momentum theorem separately to the x-components and the y-components.

Key Concept 4: The momentum of a system is conserved if no net external force acts on the system. If the momentum of one part of the system changes by $\Delta\vec{p}$, the momentum of the other parts of the system changes by $-\Delta\vec{p}$ so that the total momentum remains the same.

Key Concept 5: In any collision, momentum is conserved: The total momentum of the colliding objects has the same value just after the collision as just before the collision.

Key Concept 6: In problems that involve a two-dimensional collision, write separate conservation equations for the x-component and y-component of total momentum.

Key Concept 7: In a completely inelastic collision, the colliding objects come together and stick. Momentum is conserved, but kinetic energy is not.

Key Concept 8: Conservation of momentum holds true only when the net external force is zero. In some situations momentum is conserved during part of the motion (such as during a collision) but not during other parts.

Key Concept 9: You can use conservation of momentum in collision problems even though external forces act on the system. That's because the external forces are typically small compared to the internal forces that the colliding objects exert on each other.

Key Concept 10: In an elastic collision both total momentum and total kinetic energy are conserved. The relative velocity of the two colliding objects has the same magnitude after the collision as before, but in the opposite direction.

Key Concept 11: When a particle undergoes a head-on elastic collision with a more massive, initially stationary target, the particle bounces back in the opposite direction. The recoiling target carries away some of the particle's initial kinetic energy.

Key Concept 12: There are three conservation equations for a problem that involves a two-dimensional elastic collision: one for kinetic energy, one for the x-component of momentum, and one for the y-component of momentum.

Key Concept 13: The x-coordinate of the center of mass of a collection of particles is a weighted sum of the x-coordinates of the individual particles, and similarly for the y-coordinate.

Key Concept 14: If there is no net external force on a system of particles, the center of mass of the system maintains the same velocity. As a special case, if the center of mass is at rest, it remains at rest.

Key Concept 15: The thrust provided by a rocket equals the exhaust speed multiplied by the mass of fuel ejected per unit time. The resulting acceleration equals the thrust divided by the rocket's mass.

Key Concept 16: Because the mass of a rocket decreases as it ejects fuel, its acceleration is not constant even if the thrust is constant.

Conceptual Questions

1: Jumping off a wall

If you fall off a 2-m-high wall, would you prefer to land on concrete or grass? Why?

Your speed at the ground will be the same in both cases. The change in your momentum, or impulse, as you come to rest will also be the same. Given the same impulse, the average force will be less if the time interval is longer. Since you will be in contact with the grass longer as you land, grass is the preferred landing material.

KEY CONCEPT 1 This conceptual question illustrates why cushioning is used to reduce the average force exerted in a collision by increasing the duration of the collision. Consider this principle when you buy your next pair of running shoes or feel your car's padded dashboard.

2: Beanbag versus tennis ball

You wish to close your bedroom door from across the room. You can toss either a beanbag or a tennis ball at the door. (Both have the same mass.) Which should you choose?

Consider how much momentum each object can impart to the door. Both the beanbag and the tennis ball have the same mass, and you give each the same velocity, so their initial momenta are the same. After colliding with the door, the beanbag falls to the floor while the tennis ball bounces back toward you. The beanbag ends with zero momentum, whereas the tennis ball has momentum in the direction opposite that of its original momentum. The change in momentum of the tennis ball is larger than the change in momentum for the beanbag, so you should use the tennis ball to close the door.

KEY CONCEPT 4 Even though both objects have the same initial momentum, we see that the *change* in momentum determines the best choice.

3: Getting off the ice

You're standing on a frictionless ice rink. If you toss your physics book vertically upward, will you move?

You will not move, since the book carries away no component of momentum parallel to the ice rink. You should toss your physics book horizontally to move along the ice.

KEY CONCEPT 4 Is momentum conserved in this case? Initially, there is no momentum. When you toss the book, you and the earth must move in the opposite direction. Since the mass of the earth is extremely large, the velocity imparted to the earth is tiny.

4: Car–train collision

If a train engine and a compact car collide, which exhibits the greatest impact force? Which exhibits the greatest change in momentum? Which exhibits the greatest acceleration?

Both the train engine and the compact car exhibit the same impact force, due to Newton's third law. Both also exhibit the same change in momentum, since the momentum is conserved. (The friction force with the ground is very small compared with the impact force and may be ignored.) However, the car exhibits the greatest acceleration, since it has the least mass.

KEY CONCEPT 5 The preferred vehicle to be in during a crash is a larger vehicle, because you'll experience less acceleration (and less force will be acting on *you*). Of course, it is far better to avoid the crash altogether.

5: Collisions on a pool table

A billiard ball can stop when it collides head-on with another ball of the same mass that is at rest. Can the ball stop if the collision is at a slight angle?

KEY CONCEPT 6 If the balls collide head-on, then all of the momentum is in one direction (the direction of the initial velocity) and momentum is conserved if the second ball moves away with the first ball's initial velocity. If the balls collide at an angle, the impact gives the second ball a component of momentum perpendicular to the first ball's initial direction of motion. For the component of momentum perpendicular to the initial velocity to be conserved, the first ball must also have a perpendicular component of momentum and cannot stop.

6: Walking in a canoe

You are standing in a canoe. If you walk to the other end of the canoe, what happens to the canoe? You may neglect the resistance of the water.

No external forces act on you or the canoe (ignoring the resistance of the water). The center of mass of you and the canoe remains at rest, and its location is constant. When you walk to the other end of the canoe, the canoe must move in the opposite direction to preserve the location of the center of mass.

KEY CONCEPT 14 This conceptual question also illustrates how the frictional force is necessary for walking. As you walk, the frictional force between your shoes and the canoe pushes you in one direction while pushing the canoe in the opposite direction.

Problems

1: Tossing bubble gum

You throw your bubble gum at a stationary puck on a frictionless air-hockey table. The gum sticks to the puck, and both move away with a velocity of 1.2 m/s. If the puck has a mass of 0.30 kg and the bubble gum has a mass of 0.020 kg, find the initial speed of the bubble gum.

IDENTIFY There are no horizontal external forces, so the *x*-component of total momentum is the same before and after the collision. The target variable is the bubble gum's initial speed.

SET UP Figure 1 shows the before and after sketches of the situation, including axes. Our system consists of the bubble gum and puck. Before the collision, the gum has a velocity v_{1x} and the puck is stationary. After the collision, the gum and puck move away at velocity v_{2x}. All of the velocities and momenta have only x-components.

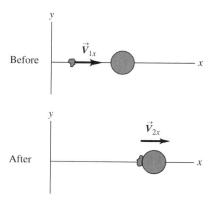

Figure 1 Problem 1.

EXECUTE We solve the problem by using conservation of momentum. The initial momentum is that of the gum:

$$p_{1x} = m_g v_{1x}.$$

The final momentum is that of the gum and puck together:

$$p_{2x} = (m_g + m_p) v_{2x}.$$

Momentum is conserved, so we set the momenta equal to each other:

$$m_g v_{1x} = (m_g + m_p) v_{2x}.$$

Solving for v_{1x} yields

$$v_{1x} = \frac{(m_g + m_p) v_{2x}}{m_g} = \frac{((0.020 \text{ kg}) + (0.30 \text{ kg}))(1.2 \text{ m/s})}{(0.020 \text{ kg})} = 19 \text{ m/s}.$$

The initial speed of the bubble gum is 19 m/s.

EVALUATE This problem illustrates how we can determine a projectile's velocity by examining its collision with a larger object and applying conservation of momentum.

KEY CONCEPT 7 Was the collision elastic? No, the collision was not elastic: The initial kinetic energy was 3.6 J, and the final kinetic energy was 0.23 J. This is an example of a totally inelastic collision, since the masses stuck together after the collision.

Practice Problem: What would the final velocity be if the gum had an initial velocity of 30 m/s? *Answer:* 1.88 m/s

Extra Practice: What would the final velocity be if each object had a mass of 0.25 kg and the gum had an initial velocity of 30 m/s? *Answer:* 15 m/s

2: Ball hits the floor

A golf ball of mass 0.045 kg bounces off a tile floor. The velocity of the ball just before it hits the floor is 6.2 m/s. If the ball is in contact with the floor for 0.012 s and the floor exerts an average force of 40.0 N, find the maximum height of the ball after its impact with the floor.

IDENTIFY We can find the maximum height h by using conservation of energy, but we need to know the velocity of the ball just after impact with the floor. We're given the initial velocity and can determine the impulse imparted by the floor. Since the impulse is the change in momentum, we can find the momentum (and velocity) after the impact.

SET UP The motion is purely vertical, so we'll use a single vertical axis with the positive direction taken to be upward as shown in Figure 2. We'll first use the impulse–momentum theorem to find the velocity v_{2y} just after the ball's impact with the floor. We'll then use energy conservation to find our target variable, the height h the ball reaches. The origin is located at the ground, so the golf ball only has kinetic energy immediately after the bounce. At the maximum height, the ball has pure gravitational potential energy.

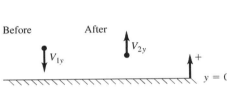

Figure 2 Problem 2.

EXECUTE Impulse is change in momentum, which is equal to the average force times the contact interval:

$$J_y = p_{2y} - p_{1y} = (F_{av})_y \Delta t.$$

We solve this equation for the final momentum:

$$\begin{aligned} p_{2y} &= (F_{av})_y \Delta t + p_{1y} \\ &= (F_{av})_y \Delta t + mv_{1y} \\ &= (40.0\,\text{N})(0.012\,\text{s}) + (0.045\,\text{kg})(-6.2\,\text{m/s}) \\ &= +0.201\,\text{kg} \cdot \text{m/s}. \end{aligned}$$

Note that the initial velocity is downward, or negative, with our choice of axis. The velocity just after impact is

$$v_{2y} = \frac{p_{2y}}{m} = \frac{(1\,0.201\,\text{kg} \cdot \text{m/s})}{(0.045\,\text{kg})} = 4.47\,\text{m/s}.$$

We can now apply conservation of energy to find the maximum height of the ball. We equate the kinetic energy of the ball just after impact with the gravitational potential energy gained by the ball when it reaches maximum height:

$$\tfrac{1}{2} m v_{2y}^2 = mgh,$$

$$h = \frac{\tfrac{1}{2} m v_{2y}^2}{mg} = \frac{v_{2y}^2}{2g} = \frac{(4.47 \text{ m/s})^2}{2(9.8 \text{ m/s}^2)} = 1.02 \text{ m}.$$

The maximum height reached by the golf ball after bouncing off the tile floor is 1.02 m.

KEY CONCEPT 2 **EVALUATE** This problem illustrates the definitions of *momentum* and *impulse* in a relatively straightforward way.

We also see how we must evaluate signs carefully, since momentum is a vector. The initial momentum is directed downward, which is negative, in our example. Had we omitted the negative sign, we would have calculated a final velocity of 16.9 m/s. Clearly, this velocity is nonsensical, since it is greater than the velocity just before the bounce!

Practice Problem: What was the initial height of the ball? *Answer:* 1.96 m

Extra Practice: How long does it take to reach the maximum height? *Answer:* 0.46 s

3: Stepping off a sled

A 60.0 kg sled is traveling across an ice rink at 4.5 m/s. Irene, riding on the sled, jumps off and lands on the ice with a velocity of 1.5 m/s in the opposite direction. What is the velocity of the sled after Irene jumps? Irene has a mass of 45.0 kg.

IDENTIFY There are no external horizontal forces acting on the sled-plus-Irene system, so momentum is conserved.

SET UP We take the x-axis to lie along the direction of motion of the sled, with the positive direction taken to be in the initial direction of the sled as shown in Figure 3. We are given the masses of the sled and Irene, the initial velocity of the sled (with Irene on it), and Irene's final velocity. Our target variable is v_{S2x}, the final velocity of the sled.

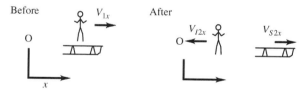

Figure 3 Problem 3.

EXECUTE The x-component of total momentum before Irene jumps off is

$$p_{1x} = m_S v_{S1x} + m_I v_{I1x}$$
$$= (60.0 \text{ kg})(4.5 \text{ m/s}) + (45.0 \text{ kg})(4.5 \text{ m/s})$$
$$= 472.5 \text{ kg} \cdot \text{m/s}.$$

(Note again that both the sled and Irene are moving with the same initial velocity.) After Irene jumps off, the x-component of total momentum has the same value, so

$$p_{2x} = m_S v_{S2x} + m_I v_{I2x}.$$

Solving for v_{S2x}, gives

$$v_{S2x} = \frac{p_{1x} - m_1 v_{12x}}{m_S}$$

$$= \frac{(472.5 \text{ kg} \cdot \text{m}/\text{s}) - (45.0 \text{ kg})(21.5 \text{ m/s})}{(60.0 \text{ kg})}$$

$$= 9.0 \text{ m/s},$$

where Irene's final velocity is negative, since she ends up moving away from the sled. The final velocity of the sled is 9.0 m/s in the positive direction.

KEY CONCEPT 5 **EVALUATE** We check the result by noting that the final velocity of the sled is greater than the initial velocity. For Irene to move away from the sled, she had to give the cart an impulse that resulted in a larger final velocity.

This problem illustrates how we can apply conservation of momentum to problems involving no net external force. Here, there is no explicit collision, but simply someone jumping off a sled.

CAUTION Watch out for signs! In the last two problems, we paid careful attention to the direction, and therefore the sign, of the momentum. Note how an incorrect sign could have led to nonsensical results. Momentum is a vector, and you must always check and recheck the direction to ensure accurate results.

Practice Problem: What impulse was given to the sled? *Answer:* +270 kg

Extra Practice: What force did Irene exert if her jump took 0.03 seconds? *Answer:* 9000 N

4: Colliding pucks in two dimensions

Two pucks collide on a frictionless air-hockey table. Initially, puck A is traveling at 3.50 m/s and puck B is at rest. After the collision, puck A moves away at a speed of 2.50 m/s and an angle of 30.0° from the initial direction. Find the final velocity of puck B. Puck A has a mass of 3.0 kg and puck B has a mass of 5.0 kg.

IDENTIFY There are no horizontal external forces, so both the x-component and the y-component of the total momentum are conserved in the collision.

SET UP Figure 4 shows the before and after sketches of the situation, including axes. Our system consists of the two pucks. Before the collision, puck A moves with velocity v_{A1} to the right. After the collision, puck A moves away at velocity v_{A2} 30.0° above the x-axis and puck B moves away at velocity v_{B2} at an angle θ below the x-axis (to conserve momentum). The velocities are not along a single axis, so we'll have to solve for both the x- and y-components of momentum. Our target variable is the final velocity v_{B2} of puck B.

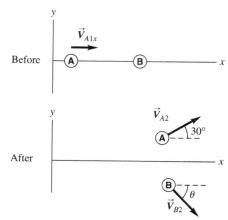

Figure 4 Problem 4.

EXECUTE Starting with the x-components of momentum before and after the collision, we have

$$p_{1x} = m_A v_{A1x},$$

$$p_{2x} = m_A v_{A2x} + m_B v_{B2x} = m_A v_{A2} \cos 30.0° + m_B v_{B2x}.$$

The x-component of momentum is conserved, so we set the expressions for p_{1x} and p_{2x} equal to each other:

$$m_A v_{A1x} = m_A v_{A2} \cos 30.0° + m_B v_{B2x}.$$

Solving for v_{B2x} yields

$$v_{B2x} = \frac{m_A v_{A1x} - m_A v_{A2} \cos 30.0°}{m_B}$$

$$= \frac{(3.0 \text{ kg})(3.50 \text{ m/s}) - (3.0 \text{ kg})(2.50 \text{ m/s})\cos 30.0°}{(5.0 \text{ kg})}$$

$$= 0.801 \text{ m/s}.$$

We follow the same procedure for the y-components of momentum. The initial y-component of momentum is zero. The y-components of the final momentum must sum to zero; that is,

$$0 = p_{2y} = m_A v_{A2y} + m_B v_{B2y} = m_A v_{A2} \sin 30.0° + m_B v_{B2y}.$$

Solving for v_{B2y} gives

$$v_{B2y} = \frac{-m_A v_{A2} \sin 30.0°}{m_B}$$

$$= \frac{-(2.50 \text{ m/s})(3.0 \text{ kg}) \sin 30.0°}{(5.0 \text{ kg})}$$

$$= -0.75 \text{ m/s}.$$

The x- and y-components of puck B's velocity are 0.80 m/s and −0.75 m/s, respectively. Thus, puck B must travel into the fourth quadrant, as expected. We find the magnitude and direction of puck B's velocity from

$$v_{B2} = \sqrt{v_{B2x}^2 + v_{B2y}^2} = \sqrt{(0.80 \text{ m/s})^2 + (-0.75 \text{ m/s})^2} = 1.10 \text{ m/s},$$

$$\phi = \tan^{-1}\frac{v_{B2y}}{v_{B2x}} = \tan^{-1}\frac{(-0.75 \text{ m/s})^2}{(0.80 \text{ m/s})} = -43.2°.$$

Puck B's final velocity is 1.10 m/s, directed at an angle of 43.2° below the positive x-axis.

KEY CONCEPT 6 **EVALUATE** We can check our answer by examining the momentum components before and after the collision. Initially, puck A has an x-component of momentum equal to 10.5 kg m/s. After the collision, the x-component of momentum of puck A is 6.5 kg m/s and that of puck B is 4.0 kg m/s, or a total of 10.5 kg m/s, as expected. There is no initial momentum along the y-axis. After the collision, the y-component of momentum of puck A is 3.75 kg m/s and that of puck B is −3.75 kg m/s, also as expected.

Solving momentum problems in two dimensions follows from the one-dimensional cases. You just need to remember that momentum is a vector that can have multiple components, each of which can be conserved individually.

Practice Problem: What are the x- and y-impulses given to puck A? *Answer:* −4.00 kg m/s and +3.75 kg m/s

Extra Practice: What is the velocity of puck B if puck A moves at a 0° angle? *Answer:* 0.6 m/s

5: Elastic collision in one dimension

Two gliders collide elastically on a frictionless, linear air track. Glider A has a mass of 0.60 kg and initially moves to the right at 3.0 m/s. Glider B has mass of 0.40 kg and initially moves to the left at 4.0 m/s. What are the final velocities of the two gliders after the collision?

IDENTIFY There are no net external forces acting on the system, so the momentum of the system is conserved. The collision is elastic; therefore, energy is conserved as well.

SET UP Figure 5 shows the before and after sketches of the situation including axes. Our system consists of the two gliders. Before the collision, glider A moves with velocity $v_{A1x} = 3.0$ m/s to the right and glider B moves with velocity $v_{B1x} = -4.0$ m/s to the left. After the collision, glider A moves away with velocity v_{A2x} and glider B moves away with velocity v_{B2x}. Our target variables are the final velocities of the two gliders. We'll need to use the relative velocity relation for elastic collisions in our solution.

Before

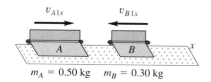

$m_A = 0.50$ kg $m_B = 0.30$ kg

After

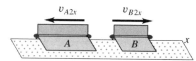

Figure 5 Problem 5.

EXECUTE From conservation of momentum,

$$m_A v_{A1x} + m_B v_{B1x} = m_A v_{A2x} + m_B v_{B2x},$$
$$(0.60 \text{ kg})(3.0 \text{ m/s}) + (0.40 \text{ kg})(-4.0 \text{ m/s}) = (0.60 \text{ kg}) v_{A2x} + (0.40 \text{ kg}) v_{B2x},$$
$$0.20 \text{ m/s} = 0.60 v_{A2x} + 0.40 v_{B2x}.$$

The last equation has two unknowns, and we need more information to solve for the velocities. Since this is an elastic collision, we can apply the relative velocity relation to solve for the velocities:

$$v_{B2x} - v_{A2x} = -(v_{B1x} - v_{A1x})$$
$$v_{B2x} - v_{A2x} = -((-4.0 \text{ m/s}) - (3.0 \text{ m/s})) = 7.0 \text{ m/s}$$
$$v_{B2x} = v_{A2x} + 7.0 \text{ m/s}.$$

Substituting the last expression into the earlier momentum conservation relation gives

$$0.60 v_{A2x} + 0.40 (v_{A2x} + 7.0 \text{ m/s}) = 0.20 \text{ m/s}$$
$$v_{A2x} = (0.20 \text{ m/s}) - (0.40(7.0 \text{ m/s})) = -2.6 \text{ m/s}$$
$$v_{B2x} = v_{A2x} + 7.0 \text{ m/s} = (-2.6 \text{ m/s}) + 7.0 \text{ m/s} = 4.4 \text{ m/s}.$$

After the collision, puck A moves to the left at 2.6 m/s (v_{A2x} is negative) and puck B moves to the right at 4.4 m/s.

KEY CONCEPT 10 EVALUATE Both gliders reversed their directions in this elastic collision. Are the kinetic energies equivalent before and after the collision? The initial kinetic energy is

$$K_i = \tfrac{1}{2} m_A v_{A1x}^2 + \tfrac{1}{2} m_B v_{B1x}^2 = \tfrac{1}{2}(0.60 \text{ kg})(3.0 \text{ m/s})^2 + \tfrac{1}{2}(0.40 \text{ kg})(-4.0 \text{ m/s})^2 = 5.9 \text{ J}.$$

The final kinetic energy is

$$K_f = \tfrac{1}{2} m_A v_{A2x}^2 + \tfrac{1}{2} m_B v_{B2x}^2 = \tfrac{1}{2}(0.60 \text{ kg})(-2.6 \text{ m/s})^2 + \tfrac{1}{2}(0.40 \text{ kg})(4.4 \text{ m/s})^2 = 5.9 \text{ J}.$$

The initial and final kinetic energies are equivalent, as expected.

Practice Problem: What impulse is given to car A? *Answer:* −3.36 kg m/s

Extra Practice: What impulse is given to car B? *Answer:* +3.36 kg m/s

6: Ballistic pendulum

A 6.00 g bullet is shot through a 2.00 kg block suspended on a 1.00-m-long string. If the initial speed of the bullet is 650 m/s and it emerges with a velocity of 175 m/s, find the maximum angle through which the block swings after it is hit.

IDENTIFY We'll analyze this problem in two stages. First, we'll look at the interaction of the bullet with the block. Second, we'll examine the swinging of the block on the string after the bullet passes through the block.

In the first stage, the bullet passes through the block, giving the block an impulse. The bullet passes through the block very quickly, so the block has no time to swing any significant distance from its initial position. The only force acting on the bullet-and-block system is between the bullet and block; there are no appreciable external forces acting on the system. We conclude that, during the first stage, the horizontal component of momentum is conserved.

In the second stage, the block moves away from its initial position. The only forces acting on the block are gravity (a conserved force) and tensions due to the strings (which do no work as the block swings). Total mechanical energy is conserved as the block swings.

SET UP Figure 6 shows sketches of the two stages of the problem. We take the positive x-axis to be to the right and the positive y-axis upward. Our target variable for the first stage is the velocity V_{2x} of the block after the bullet emerges from the block. We'll use momentum conservation to find V_{2x}. Our target variable for the second stage is θ, the angle the string makes with the vertical when the block stops momentarily after swinging to the right. We'll use energy conservation to find the height h the block rises to after the collision and relate h to θ.

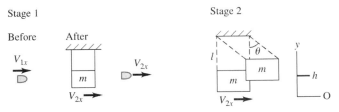

Figure 6 Problem 6.

EXECUTE In the first stage, conservation of momentum gives
$$m v_{1x} = m v_{2x} + M V_{2x}.$$

The velocity of the block is
$$V_{2x} = \frac{m(v_{1x} - v_{2x})}{M}$$
$$= \frac{(0.006 \text{ kg})(650 \text{ m/s} - 175 \text{ m/s})}{(2.0 \text{ kg})}$$
$$= 1.43 \text{ m/s}.$$

At the beginning of the second stage, the block has kinetic energy. Afterward, the block swings up and comes to rest momentarily. At this point, the block's kinetic energy is zero and its gravitational potential energy has increased by *mgh*. Energy conservation gives

$$\tfrac{1}{2}MV_{2x}^2 = Mgh.$$

The block reaches a height

$$h = \frac{V_{2x}^2}{2g} = 0.104 \text{ m}.$$

The angle between the vertical and the string when the block stops momentarily is

$$\theta = \cos^{-1}\frac{l-h}{l} = 26.4°.$$

The maximum angle the block swings to after the collision is 26.4°.

KEY CONCEPT 8 **EVALUATE** Where did most of the energy go? A quick check of the kinetic energies gives an initial energy of the bullet (before striking the block) of 1270 J, a final energy of the bullet of 92 J, and a final energy of the block of 2.0 J. Most of the initial energy was dissipated in the deformation and heating of the bullet and block.

Practice Problem: What force did the bullet apply to the block if it took 0.004 seconds to travel through it? *Answer:* 715 N

Extra Practice: How far did the block travel horizontally? *Answer:* 0.44 m

7: Boat on a lake

Luka is standing in a boat on a calm lake. His position is 5.00 m from the end of the pier. As he walks toward the pier, the boat moves away from it. When Luka stops walking, he finds that the boat has moved 2.00 m away from the pier. If you ignore the boat's resistance to motion in the water, how far from the end of the pier is Luka when he stops walking? The mass of the boat is 80.0 kg, and Luka's mass is 60.0 kg.

IDENTIFY Since we can ignore friction between the boat and the water, the net external force on the boat-and-Luka system is zero. Momentum is therefore conserved. There is no initial motion, so the total momentum is zero and the velocity of the center of mass of the system is zero. The center of mass will remain at rest as Luka walks in the boat.

SET UP We take the origin to be the end of the pier, since the measurements are given with respect to that point. Figure 7 shows a sketch of the situation. Our target variable is the final position of Luka, x_{L2}.

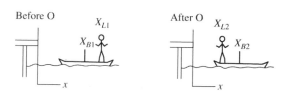

Figure 7 Problem 7.

EXECUTE The initial coordinate of the center of mass of the boat-and-Luka system is given by the formula

$$x_{cm} = \frac{x_{L1}m_L + x_{B1}m_B}{m_L + m_B}.$$

The final coordinate of the center of mass of the system is given by the equation

$$x_{cm} = \frac{x_{L2}m_L + x_{B2}m_B}{m_L + m_B}.$$

Since the center of mass doesn't move, we set the two right-hand expressions equal to each other:

$$\frac{x_{L1}m_L + x_{B1}m_B}{m_L + m_B} = \frac{x_{L2}m_L + x_{B2}m_B}{m_L + m_B}.$$

The boat moves 2.0 m away from the end of the pier, so $x_{B2} = x_{B1} + 2.0$ m. Substituting and solving gives

$$x_{L1}m_L + x_{B1}m_B = x_{L2}m_L + (x_{B1} + 2.0 \text{ m})m_B,$$
$$(5.0 \text{ m})(60.0 \text{ kg}) + x_{B1}(80.0 \text{ kg}) = x_{L2}(60.0 \text{ kg}) + (x_{B1} + 2.0 \text{ m})(80.0 \text{ kg}),$$
$$x_{L2} = 2.3 \text{ m}.$$

KEY CONCEPT 14 **EVALUATE** Since the boat and Luka's mass are similar, Luka moves a distance similar to the distance the boat moves. Luka's mass is less, so he moves a little farther than the boat moves.

Practice Problem: What is the center of mass if the boat's initial position was 1.5 m from the end of the pier? *Answer:* 3.0 m

Extra Practice: What is Luka's final position if he has the same mass as the boat? *Answer:* 3.0 m

8: Falling sand

Sand is dropped from a height of 1.0 m onto a kitchen scale at the uniform rate of 100.0 g/s. Find the force on the scale and its reading if the scale is calibrated in kg.

IDENTIFY We need to find the force on the scale due to the falling sand. We can find the change in momentum of the falling sand and relate it to the force through impulse. The target variable is the force on the scale.

SET UP We'll first find the velocity of the sand as it hits the scale. We'll then use that velocity to find the change in momentum as the sand strikes the scale. We'll incorporate the change in momentum into the impulse formula and solve for the force on the scale. We'll complete the problem by converting the force into a calibrated scale reading. We'll take our vertical axis to be positive downward.

EXECUTE As the sand falls 1.0 m, it acquires a velocity

$$v_{1y} = \sqrt{2gh},$$

according to energy conservation. The sand stops after striking the scale. The change in momentum as the sand strikes the scale is

$$\Delta p = p_{2y} - p_{1y} = -\Delta m v = -\Delta m \sqrt{2gh}.$$

The change in momentum is negative (directed upward) in our coordinate system. The force on the sand due to the scale is directed upward to stop the sand. The impulse imparted to the sand by the scale is

$$J = -F\Delta t = \Delta p = -\Delta m \sqrt{2gh}.$$

Rearranging terms and dividing both sides by Δt gives

$$F = \frac{\Delta m}{\Delta t}\sqrt{2gh}$$
$$= (0.100 \text{ kg/s})\sqrt{2(9.8 \text{ m/s}^2)(1.0 \text{ m})}$$
$$= 0.44 \text{ N}.$$

Here, the rate of change of mass per unit time is the given rate at which the sand falls. The scale is calibrated in terms of mass, so we divide the force by g to get the reading on the scale:

$$m = \frac{F}{g} = \frac{0.44 \text{ N}}{9.8 \text{ m/s}^2} = 0.045 \text{ kg}.$$

The scale reads 0.045 kg as the sand falls.

KEY CONCEPT 2 **EVALUATE** This problem illustrates how we can use force, momentum, and impulse together to solve problems.

Practice Problem: What is the reading on the scale after 5 seconds of falling sand?
Answer: 0.545 kg

Extra Practice: How fast does a piece of sand move right before it hits the scale? *Answer:* 4.43 m/s

Try It Yourself!

1: Rocket motion

A two-stage rocket traveling at 350 m/s through space separates, with one stage having twice the mass of the other. If the final velocity of the larger stage is 120 m/s in a direction opposite its initial direction, find the final velocity of the smaller stage.

Solution Checkpoints

IDENTIFY AND SET UP Confirm that there are no net external forces acting on the rocket. Apply conservation of momentum along one axis. The target variable is the velocity of the smaller rocket stage.

EXECUTE Momentum conservation gives

$$m_A v_{1x} + m_B v_{1x} = m_A v_{A2x} + m_B v_{B2x}.$$

Solving for v_{A2x} yields

$$v_{A2x} = \frac{(m_A + m_B)v_{1x} - m_B v_{B2x}}{m_A} = 1300 \text{ m/s}.$$

KEY CONCEPT 5 **EVALUATE** Do you expect the smaller stage to have a final velocity that is larger or smaller than the other velocities in the problem? Do you expect energy to be conserved in this case?

Practice Problem: How much more energy do the stages have after the separation? *Answer:* 4.5 times more

Extra Practice: If the larger stage had its final velocity in the same direction as its initial direction, then what is the final velocity of the smaller stage? *Answer:* 810 m/s

2: Car collision

A 2000 kg car moving at 30 km/hr collides with a stopped 1000 kg car, and the two lock bumpers. What is their common velocity after the collision? What fraction of the initial energy is dissipated in the collision?

Solution Checkpoints

IDENTIFY AND SET UP Confirm that there are no net external forces acting on the cars just before and just after the collision. Apply conservation of momentum along one axis. The target variables are the velocity of the combined cars and the fraction of energy lost in the collision.

EXECUTE Momentum conservation gives

$$m_A v_{A1x} = (m_A + m_B) v_{2x}.$$

Solving for the final velocity gives 20 km/hr. The initial and final energies are

$$E_1 = \tfrac{1}{2} m_A v_{A1x}^2$$
$$E_2 = \tfrac{1}{2} (m_A + m_B) v_{2x}^2.$$

Dividing the two demonstrates that two-thirds of the initial energy remains kinetic after the collision.

KEY CONCEPT 7 **EVALUATE** We'll contrast this problem with the next one, in which the collision is elastic.

Practice Problem: What is the final velocity if both cars have the same mass? *Answer:* 15 km/hr

Extra Practice: What is the final velocity if the stopped car is twice as massive as the moving car? *Answer:* 10 km/hr

3: Car collision revisited

A 2000 kg car moving at 30 km/hr collides with a stopped 1000 kg car, and the two have spring bumpers so that the collision is perfectly elastic. What is the velocity of each car after the collision?

Solution Checkpoints

IDENTIFY AND SET UP Confirm that there are no net external forces acting on the cars just before and just after the collision. Apply conservation of momentum along one axis. Apply conservation of energy, since the collision is elastic. The target variables are the velocities of the two cars.

EXECUTE Momentum conservation gives

$$m_A v_{A1x} = m_A v_{A2x} + m_B v_{B2x}.$$

Energy conservation gives

$$\tfrac{1}{2} m_A v_{A1x}^2 = \tfrac{1}{2} m_A v_{A2x}^2 + \tfrac{1}{2} m_B v_{B2x}^2.$$

These two equations can be solved to find $v_{A2x} = 10$ km/hr and $v_{2Bx} = 40$ km/hr. You may wish to consult the text for a derivation of velocities in elastic collisions.

KEY CONCEPT 10 **EVALUATE** In this elastic collision, we see that the lighter car ends up with a velocity greater than the heavier car after the collision. You can also check that the initial and final momenta and energies are equal.

Practice Problem: What are their final velocities if their masses are switched? *Answer:* −10 km/hr and 20 km/hr

Extra Practice: What are their final velocities if both cars have the same mass? *Answer:* 0 km/hr and 30 km/hr

4: Putty hitting mass on spring

A 0.75 kg mass of putty is hurled with a velocity of 2.0 m/s against a 0.50 kg mass attached to a spring, as shown in Figure 8. The mass attached to the spring slides across the frictionless horizontal surface, depressing the spring a maximum distance of 12.0 cm. Find the spring constant if the putty sticks to the mass on the spring.

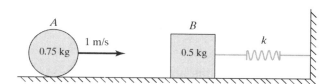

Figure 8 Try It Yourself! 4.

Solution Checkpoints

IDENTIFY AND SET UP Is the collision elastic or inelastic if the putty sticks to the mass? You can apply conservation of momentum to the initial collision (before the spring begins to compress). Then apply conservation of energy to the interval after the spring begins to compress.

EXECUTE Momentum conservation gives

$$m_A v_{A1x} = (m_A + m_B) v_{2x}.$$

The velocity of the putty plus mass is 1.2 m/s just after the collision. Energy conservation applied to the interval after the collision gives

$$\tfrac{1}{2}(m_A + m_B) v_{2x}^2 = \tfrac{1}{2} k x^2.$$

Substituting yields a spring constant of 125 N/m.

KEY CONCEPT 8 **EVALUATE** Energy and momentum conservation do not apply to the whole process; rather, the process must be broken into intervals during which energy and momentum can be applied.

Practice Problem: What would the velocities be after the collision if the putty were replaced by a rubber ball that collides elastically with the mass on the spring? *Answer:* 0.40 m/s and 2.4 m/s

Extra Practice: How far would the spring compress in this case? *Answer:* 0.15 m

5: Collision on a football field

A 135 kg football player traveling at 5.0 m/s collides with an 85 kg player at rest. The two slide 2.0 m on wet grass. Their collision is completely inelastic. What is the coefficient of friction between the grass and the players? How much energy is dissipated in the collision?

Solution Checkpoints

IDENTIFY AND SET UP Confirm that there are minimal external forces acting on the players just before and just after the collision. Apply conservation of momentum to the two players as they collide. Then use the work–energy theorem to find the friction force. The energy dissipated is found by comparing the energy just before and just after the collision.

EXECUTE Momentum conservation applied to the collision gives

$$m_A v_{A1x} = (m_A + m_B) v_{2x}.$$

The work–energy theorem applied to the two players as they slide on the grass after they collide results in

$$-\mu(m_A + m_B)gx = 0 - \tfrac{1}{2}(m_A + m_B)v_{2x}^2.$$

Solving these equations, we obtain $\mu = 0.24$. The change in energy during the collision is

$$\Delta E = \tfrac{1}{2}(m_A + m_B)v_{2x}^2 - \tfrac{1}{2}m_A v_{A1x}^2,$$

or 650 J dissipated during the collision.

KEY CONCEPT 7 **EVALUATE** In this collision, we had to split the process up into two intervals, to which we separately applied conservation of momentum and the work–energy theorem. We also found that almost one-third of the energy is dissipated in the totally inelastic collision of the players.

Practice Problem: How far would they slide if they both had the same mass? *Answer:* 1.33 m

Extra Practice: How fast would the first player need to be moving to slide 2.0 m after the collision in this case? *Answer:* 6.13 m/s

Problem Summary

This chapter has augmented our knowledge of energy, forces, and kinematics with our newly formed knowledge of momentum analysis. Momentum analysis shares many of the problem-solving principles we have encountered. In the problems presented, we

- Identified the general procedure to find the solution.
- Sketched the situation, including before and after views.
- Identified the momenta, energies, and forces in the system.
- Applied conservation of momentum to the system.
- Applied energy principles, including conservation of energy (when possible).
- Drew free-body diagrams of the objects when appropriate.
- Applied Newton's laws when appropriate.
- Solved for the target variable(s) algebraically from the equations we derived.
- Reflected on the results, checking for inconsistencies.

Key Example Variation Problems

Solutions to these problems are in Chapter 8 of the Student's Solutions Manual

Be sure to review EXAMPLES 8.4, 8.5, and 8.6 (Section 8.2) before attempting these problems.

VP8.6.1 You hold glider A of mass 0.125 kg and glider B of mass 0.375 kg at rest on an air track with a compressed spring of negligible mass between them. When you release the gliders, the spring pushes them apart. (a) Once the gliders are no longer in contact with the spring, glider A is moving to the right at 0.600 m/s. What is the velocity (magnitude and direction) of glider B at this time? (b) Glider A moving to the right at 0.600 m/s then collides head-on with a third glider C of mass 0.750 kg that is moving to the left at 0.400 m/s. After this collision, glider C is moving to the left at 0.150 m/s. What is the velocity (magnitude and direction) of glider A after this collision?

VP8.6.2 Two hockey players skating on essentially frictionless ice collide head-on. Madeleine, of mass 65.0 kg, is moving at 6.00 m/s to the east just before the collision and at 3.00 m/s to the west just after the collision. Buffy, of mass 55.0 kg, is moving at 3.50 m/s to the east just after the collision. (a) Find Buffy's velocity (magnitude and direction) just before the collision. (b) What are the changes in the velocities of the two hockey players during the collision? Take east to be the positive direction. Who has the greater magnitude of velocity change: more massive Madeleine or less massive Buffy?

VP8.6.3 A 2.40 kg stone is sliding in the $+x$-direction on a horizontal, frictionless surface. It collides with a 4.00 kg stone at rest. After the collision the 2.40 kg stone is moving at 3.60 m/s at an angle of 30.0° measured from the $+x$-direction toward the $+y$-direction, and the 4.00 kg stone is moving at an angle of 45.0° measured from the $+x$-direction toward the $-y$-direction. (a) What is the y-component of momentum of the 2.40 kg stone after the collision? What must be the y-component of momentum of the 4.00 kg stone after the collision? (b) What is the speed of the 4.00 kg stone after the collision? (c) What is the x-component of the total momentum of the two stones after the collision? (d) What is the speed of the 2.40 kg stone before the collision?

VP8.6.4 A hockey puck of mass m is moving in the $+x$-direction at speed v_{P1} on a frictionless, horizontal surface. It collides with a stone of mass $2m$ that is initially at rest. After the collision the hockey puck is moving at an angle θ measured from the $+x$-

direction toward the +y-direction, and the stone is moving at the same angle θ but measured from the +x-direction toward the −y-direction. (a) In order for the y-component of momentum to be conserved, what must be the ratio of the final speed v_{S2} of the stone to the final speed v_{P2} of the hockey puck? (b) Use conservation of the x-component of momentum to find v_{S2} and v_{P2} in terms of v_{P1} and θ.

Be sure to review EXAMPLES 8.7, 8.8, and 8.9 (Section 8.3) before attempting these problems.

VP8.9.1 Two blocks of clay, one of mass 1.00 kg and one of mass 4.00 kg, undergo a completely inelastic collision. Before the collision one of the blocks is at rest and the other block is moving with kinetic energy 32.0 J. (a) If the 4.00 kg block is initially at rest and the 1.00 kg block is moving, what is the initial speed of the 1.00 kg block? What is the common final speed of the two blocks? How much kinetic energy is lost in the collision? (b) If the 1.00 kg block is initially at rest and the 4.00 kg block is moving, what is the initial speed of the 4.00 kg block? What is the common final speed of the two blocks? How much kinetic energy is lost in the collision? (c) In which case is more of the initial kinetic energy lost in a completely inelastic collision: a moving object collides (i) with a heavier object at rest or (ii) with a lighter object at rest?

VP8.9.2 A 0.500 kg block of cheese sliding on a frictionless tabletop collides with and sticks to a 0.200 kg apple. Before the collision the cheese was moving at 1.40 m/s and the apple was at rest. The cheese and apple then slide together off the edge of the table and fall to the floor 0.600 m below. (a) Find the speed of the cheese and apple just after the collision. In this collision, what is conserved: momentum, total mechanical energy, both, or neither? (b) What is the speed of the cheese and apple just before they hit the floor? During the fall from the tabletop to the floor, what is conserved: momentum, total mechanical energy, both, or neither?

VP8.9.3 A 2.40 kg can of coffee moving at 1.50 m/s in the +x-direction on a kitchen counter collides head-on with a 1.20 kg box of macaroni that is initially at rest. After the collision the can of coffee is moving at 0.825 m/s in the +x-direction.
(a) What is the velocity (magnitude and direction) of the box of macaroni after the collision? (b) What are the kinetic energies of the can before and after the collision, and of the box after the collision? (c) Is this collision elastic, inelastic, or completely inelastic? How can you tell?

VP8.9.4 A block of mass m moving due east at speed v collides with and sticks to a block of mass $2m$ that is moving at the same speed v but in a direction 45.0° north of east. Find the direction in which the two blocks move after the collision.

Be sure to review EXAMPLES 8.13 and 8.14 (Section 8.5) before attempting these problems.

VP8.14.1 Find the x- and y-coordinates of the center of mass of a system composed of the following particles: a 0.500 kg particle at the origin; a 1.25 kg particle at $x = 0.150$ m, $y = 0.200$ m; and a 0.750 kg particle at $x = 0.200$ m, $y = -0.800$ m.

VP8.14.2 Three objects lie along the x-axis. A 3.00 kg object is at the origin, a 2.00 kg object is at $x = 1.50$ m, and a 1.20 kg object is at an unknown position. The center of mass of the system of three objects is at $x = -0.200$ m. What is the position of the 1.20 kg object?

VP8.14.3 You hold a 0.125 kg glider A and a 0.500 kg glider B at rest on an air track with a compressed spring of negligible mass between them. When you release the gliders, the spring pushes them apart so that they move in opposite directions. When glider A has moved 0.960 m to the left from its starting position, how far to the right from its starting position has glider B moved?

VP8.14.4 Three objects each have mass m. Each object feels a force from the other two, but not from any other object. Initially the first object is at $x = -L$, $y = 0$; the second object is at $x = +L$, $y = 0$; and the third object is at $x = 0$, $y = L$. At a later time the first object is at $x = -L/3$, $y = +L/4$; and the second object is at $x = +L/2$, $y = -L$. At this later time, where is the third object?

STUDENT'S SOLUTIONS MANUAL FOR MOMENTUM, IMPULSE, AND COLLISIONS

VP8.6.1. **IDENTIFY:** This problem involves a one-dimensional collision, so we use momentum conservation. The total momentum before the collision must equal the total momentum after the collision.
SET UP: $p_x = mv_x$ and $P_x = p_{1x} + p_{2x} + \ldots$. Call the x-axis positive horizontally to the right.
EXECUTE: **(a)** After the gliders move free of the spring, their total momentum is conserved. Before the gliders were released, they were at rest so their total momentum was zero. This means that their total momentum will be zero later also. So $P_{1x} = P_{2x}$.
$m_A v_{Ax} + m_B v_{Bx} = 0$
$(0.125 \text{ kg})(0.600 \text{ m/s}) + (0.375 \text{ kg}) v_{Bx} \rightarrow v_{Bx} = -0.200 \text{ m/s}$, to the left.
(b) Figure VP8.6.1 shows before and after sketches of the collision. Use momentum conservation.

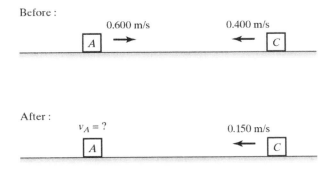

Figure VP8.6.1

$m_A v_{A1x} + m_C v_{C1x} = m_A v_{A2x} + m_C v_{C2x}$
$(0.125 \text{ kg})(0.600 \text{ m/s}) + (0.750 \text{ kg})(-0.400 \text{ m/s}) = (0.125 \text{ kg}) v_{A2x} + (0.750 \text{ kg})(-0.150 \text{ m/s})$
$v_{A2x} = -0.900 \text{ m/s}$, to the left.
EVALUATE: Since momentum is a vector, we need to pay close attention to the *signs* of its components.

VP8.6.2. **IDENTIFY:** This problem involves a one-dimensional collision, so we use momentum conservation. The total momentum before the collision must equal the total momentum after the collision.
SET UP: $p_x = mv_x$ and $P_x = p_{1x} + p_{2x} + ...$. Call the x-axis positive horizontally to the right. Let B stand for Buffy and M for Madeleine. Fig. VP8.6.2 shows before and after sketches.

Figure VP8.6.2

EXECUTE: (a) $P_{1x} = P_{2x}$ → $m_M v_{M1x} + m_B v_{B1x} = m_M v_{M2x} + m_B v_{B2x}$
(65.0 kg)(6.00 m/s) + (55.0 kg)v_{B1x} = (65.0 kg)(–3.00 m/s) + (55.0 kg)(3.50 m/s)
v_{B1x} = –7.14 m/s. The minus sign means it is to the west.
(b) $\Delta v_{Mx} = v_{M2x} - v_{M1x}$ = –3.00 m/s – 6.00 m/s = –9.00 m/s, to the west.
$\Delta v_{Bx} = v_{B2x} - v_{B1x}$ = 3.50 m/s – (–7.14 m/s) = 10.6 m/s, to the east. Buffy has the greater magnitude velocity change.
EVALUATE: It is reasonable that the magnitude of light-weight Buffy's velocity change is greater than that of the heavier Madeleine. Both experience the same magnitude change in *momentum*, but the smaller-mass Buffy needs a larger velocity change than Madeleine so their momentum changes can be equal in magnitude.

VP8.6.3. **IDENTIFY:** This problem involves a two-dimensional collision, so we use momentum conservation. The total momentum before the collision must equal the total momentum after the collision.
SET UP: $p_x = mv_x$ and $P_x = p_{1x} + p_{2x} + ...$ and likewise for y-axis. Call A the 2.40-kg stone and B the 4.00-kg stone. Fig. VP8.6.3 shows before and after sketches.

Figure VP8.6.3

EXECUTE: (a) $m_A v_{A1y}$ = (2.40 kg)(3.60 m/s) sin 30.0° = 4.32 kg·m/s . The initial y-component of the momentum is zero, so the final y-component must be zero. So p_{B2y} = –4.32 kg·m/s .
(b) $p_{B2y} = m_B v_{B2y}$ sin 45.0° → –4.32 kg·m/s = (4.00 kg)v_{B2y} sin 45.0°
v_{B2y} = –1.53 m/s, so its speed is 1.53 m/s.
(c) $P_x = p_{Ax} + p_{Bx}$
P_x = (2.40 kg)(3.60 m/s) cos 30.0° + (4.00 kg)(1.53 m/s) cos 45.0° = 11.8 kg·m/s .

(d) Initially A has all the x-momentum. Since the x-component of the momentum is conserved $(2.40 \text{ kg})v_{Ax} = P_x = 11.8 \text{ kg} \cdot \text{m/s} \rightarrow v_{Ax} = 4.92 \text{ m/s}$.

EVALUATE: In a two-dimensional collision, the x-components of the momentum must *always* be treated separately from the y-components.

VP8.6.4. **IDENTIFY:** This problem involves a two-dimensional collision, so we use momentum conservation. The total momentum before the collision must equal the total momentum after the collision.

SET UP: $p_x = mv_x$ and $P_x = p_{1x} + p_{2x} + \ldots$ and likewise for y-axis. Call P the hockey puck and S the stone. Figure VP8.6.4 shows before and after sketches.

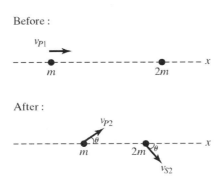

Figure VP8.6.4

EXECUTE: **(a)** The puck and stone must have equal-magnitude y-components of their momentum. $mv_{P2} \sin\theta = (2m)v_{S2} \sin\theta \rightarrow v_{S2}/v_{P2} = \frac{1}{2}$.

(b) The x-components of the momentum give $mv_{P1} = mv_{P2} \cos\theta + (2m)v_{S2} \cos\theta$. From part (a) we have $v_{P2} = 2v_{S2}$. Combining these two equations gives $v_{S2} = \dfrac{v_{P1}}{4\cos\theta}$ and $v_{P2} = \dfrac{v_{P1}}{2\cos\theta}$.

EVALUATE: In a two-dimensional collision, the x-components of the momentum must *always* be treated separately from the y-components.

VP8.9.1. **IDENTIFY:** This problem involves a one-dimensional collision in which the colliding objects stick together. This makes it a completely *inelastic* collision. Momentum is conserved.

SET UP: $p_x = mv_x$ and $P_x = p_{1x} + p_{2x} + \ldots$. $K = \dfrac{1}{2}mv^2$.

EXECUTE: **(a)** $K_1 = \dfrac{1}{2}mv_1^2 = \dfrac{1}{2}(1.00 \text{ kg})v_1^2 = 32.0 \text{ J} \rightarrow v_1 = 8.00 \text{ m/s}$.

Momentum conservation gives $m_1v_1 = (m_1 + m_2)v_2 \rightarrow (1.00 \text{ kg})(8.00 \text{ m/s}) = (5.00 \text{ kg})v_2$, so $v_2 = 1.60 \text{ m/s}$.

The lost kinetic energy is

$K_1 - K_2 = K_1 - \dfrac{1}{2}(m_1 + m_2)v_2^2 = 32.0 \text{ J} - \dfrac{1}{2}(5.00 \text{ kg})(1.60 \text{ m/s})^2 = 25.6 \text{ J}$.

(b) $K = \dfrac{1}{2}mv^2 = \dfrac{1}{2}(4.00 \text{ kg})v^2 = 32 \text{ J} \rightarrow v = 4.00 \text{ m/s}$.

Momentum conservation gives $m_1v_1 = (m_1 + m_2)v_2 \rightarrow (4.00 \text{ kg})(4.00 \text{ m/s}) = (5.00 \text{ kg})v_2$

$v_2 = 3.20 \text{ m/s}$. The loss of kinetic energy is $K_1 - K_2 = K_1 - \dfrac{1}{2}(m_1 + m_2)v^2$

$K_1 - K_2 = 32.0 \text{ J} - \dfrac{1}{2}(5.00 \text{ kg})(3.20 \text{ m/s})^2 = 6.4 \text{ J}$.

(c) More kinetic energy is lost if a light object collides with a stationary heavy object, which is case (i).

EVALUATE: Careful! Even though momentum is always conserved during a collision, kinetic energy may or may not be conserved. The amount of energy lost depends on the nature of the collision and the relative masses of the colliding objects.

VP8.9.2. IDENTIFY: During the collision, momentum is conserved. After the collision, energy is conserved. We must break this problem up into two parts.

SET UP: During the collision, we use $p_x = mv_x$ and $P_x = p_{1x} + p_{2x} + \ldots$ and momentum conservation. After the collision we use $U_1 + K_1 + W_{\text{other}} = U_2 + K_2$, where $W_{\text{tot}} = 0$. Take the x-axis to be horizontal in the direction in which the cheese is originally moving. Figure VP8.9.2 shows the given information.

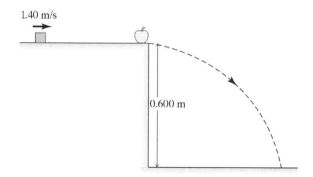

Figure VP8.9.2

EXECUTE: (a) During the collision: $p_{\text{cheese}} = p_{\text{apple + cheese}}$, so $m_c v_c = (m_c + m_a)v$
$(0.500 \text{ kg})(1.40 \text{ m/s}) = (0.700 \text{ kg})v \rightarrow v = 1.00$ m/s. Only momentum is conserved because no net external forces to the apple-cheese system due to the collision. The kinetic energy is *not* conserved because the objects stick together.

(b) Energy conservation after the collision gives $K_1 + U_1 = K_2 + U_2$. Call $y = 0$ at the floor level. $\frac{1}{2}mv_1^2 + mgh = \frac{1}{2}mv_2^2$, which gives $v_2 = \sqrt{v_1^2 + 2gh} = \sqrt{(1.00 \text{ m/s})^2 + 2(9.80 \text{ m/s}^2)(0.600 \text{ m})} =$ 3.57 m/s. Only the total mechanical energy is conserved during the fall. Gravity is an external force acting on the falling object, so its momentum is *not* conserved.

EVALUATE: We cannot do this type of problem in a single step. Some of the initial mechanical energy of the cheese is lost during the collision, and we have no way of knowing how much without examining the collision.

VP8.9.3. IDENTIFY: This is a collision, but we do not know if it's elastic or inelastic from the given information. Momentum is conserved during the collision regardless of whether it is elastic or inelastic.

SET UP: During the collision, we use $p_x = mv_x$ and $P_x = p_{1x} + p_{2x} + \ldots$ and momentum conservation. Let C refer to the coffee can and M refer to the macaroni.

EXECUTE: (a) The momentum before the collision is equal to the momentum after the collision. Using $m_C v_{C1} = m_C v_{C2} + m_M v_{M2}$ gives $(2.40 \text{ kg})(1.50 \text{ m/s}) = (2.40 \text{ kg})(0.825 \text{ m/s}) + (1.20 \text{ kg})v_{M2}$, so $v_{M2} = 1.35$ m/s in the +x direction.

(b) $K_{C1} = \frac{1}{2}m_C v_{C1}^2 = \frac{1}{2}(2.40 \text{ kg})(1.50 \text{ m/s})^2 = 2.70$ J.

$K_{C2} = \frac{1}{2}m_C v_{C2}^2 = \frac{1}{2}(2.40 \text{ kg})(0.825 \text{ m/s})^2 = 0.817$ J.

$K_{M2} = \frac{1}{2}m_M v_{M2}^2 = \frac{1}{2}(1.20 \text{ kg})(1.35 \text{ m/s})^2 = 1.09$ J.

(c) $K_1 = K_{C1} = 2.70$ J
$K_2 = K_{C2} + K_{M2} = 0.817$ J + 1.09 J = 1.91 J

$K_2 < K_1$, so the collision is *inelastic*.
EVALUATE: This collision is inelastic, but it is not *perfectly* elastic because the objects do not stick together.

VP8.9.4. **IDENTIFY:** This is a two-dimensional collision in which the objects stick together, so it is perfectly inelastic. Momentum is conserved during the collision regardless of whether it is elastic or inelastic. We treat the *x*-components of the momentum separately from the *y*-components.
SET UP: During the collision, we use $p_x = mv_x$ and $P_x = p_{1x} + p_{2x} + ...$ and likewise for P_y. Call the *x*-axis positive toward the east and *y*-axis positive toward the north. Let θ be the angle with the +*x*-axis at which the blocks move together after the collision, and let *V* be their common speed. Figure VP8.9.4 shows before and after sketches.

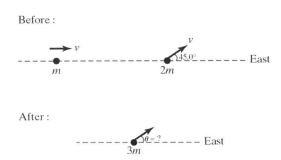

Figure VP8.9.4

EXECUTE: Before the collision the *x*-components of the momentum are mv and $(2m)v \cos 45.0°$ and the *y*-component is $(2m)v \sin 45.0°$. After the collision the *x*-component is $(3m)V \cos \theta$ and the *y*-component is $(3m)V \sin \theta$. These must also be the components after the collision because momentum is conserved.
Easterly components: $mv + 2mv \cos 45.0° = 3mV \cos \theta \rightarrow v(1 + 2 \cos 45.0°) = 3V \cos \theta$
Northerly components: $2mv \sin 45.0° = 3mV \sin \theta \rightarrow 2v \sin 45.0° = 3V \sin \theta$
Dividing these two equations gives $\dfrac{3V \sin \theta}{3V \cos \theta} = \dfrac{2v \sin 45.0°}{v(1 + 2 \cos 45.0°)}$ which simplifies to
$\tan \theta = \dfrac{2 \sin 45.0°}{1 + 2 \cos 45.0°} \rightarrow \theta = 30.4°$ north of east.
EVALUATE: The momentum is conserved because the collision generates no external forces on the system of colliding objects. Kinetic energy is not conserved since the objects stick together.

VP8.14.1. **IDENTIFY:** We want to find the location of the center of mass of a system of particles.
SET UP: The *x*-coordinate of the center of mass is $x_{cm} = \dfrac{m_1 x_1 + m_2 x_2 + m_3 x_3 + \cdots}{m_1 + m_2 + m_3 + \cdots}$ and likewise for the *y*-coordinate.
EXECUTE: **(a)** $x_{cm} = \dfrac{(0.500 \text{ kg})(0) + (1.25 \text{ kg})(0.150 \text{ m}) + (0.750 \text{ kg})(0.200 \text{ m})}{2.50 \text{ kg}} = 0.135$ m.
$y_{cm} = \dfrac{(0.500 \text{ kg})(0) + (1.25 \text{ kg})(0.200 \text{ m}) + (0.750 \text{ kg})(-0.800 \text{ m})}{2.50 \text{ kg}} = -0.140$ m.
(b) Calling *d* the distance, we can calculate the distance between the center of mass and a given particle. For the 0.500-kg particle, we have
$d_{0.500} = \sqrt{(x_{cm} - x_{0.5})^2 + (y_{cm} - y_{0.5})^2} = \sqrt{(0.135 \text{ m} - 0)^2 + (-0.140 \text{ m} - 0)^2} = 0.1945$ m. Likewise for the 1.25-kg particle we find $d_{1.25} = 0.3403$ m. The result is that the center of mass is closest to the 0.500-kg particle.

EVALUATE: A plot on graph paper indicates that the center of mass is closest to the 0.500-kg particle, as we just calculated.

VP8.14.2. IDENTIFY: We know the location of the center of mass of a system of particles and the location of two of them. We want to find the location of the third particle. The particles all lie on the *x*-axis.

SET UP: The *x*-coordinate of the center of mass is $x_{cm} = \dfrac{m_1 x_1 + m_2 x_2 + m_3 x_3}{m_1 + m_2 + m_3}$.

EXECUTE: Putting the known *x*-coordinates into the center of mass formula gives

$-0.200 \text{ m} = \dfrac{(3.00 \text{ kg})(0) + (2.00 \text{ kg})(1.50 \text{ m}) + (1.20 \text{ kg})x}{6.20 \text{ kg}} \rightarrow x = -3.53 \text{ m}.$

EVALUATE: Since x_{cm} is negative, the 1.20-kg object must lie on the –*x* side of the origin and farther from the origin than the center of mass, which it does. So our result is reasonable.

VP8.14.3. IDENTIFY: The force of the spring is internal to the two-glider system, so the center of mass of that system does not move.

SET UP: Use $x_{cm} = \dfrac{m_1 x_1 + m_2 x_2}{m_1 + m_2}$. Take the origin as the original position of the center of mass of the gliders, which makes $x_{cm} = 0$.

EXECUTE: $0 = \dfrac{(0.125 \text{ kg})(-0.960 \text{ m}) + (0.500 \text{ kg})x}{0.625 \text{ kg}} \rightarrow x = 0.240 \text{ m}.$

EVALUATE: Since *B* is more massive than *A*, it should have moved a shorter distance since both of them felt the same force from the spring. This agrees with our result.

VP8.14.4. IDENTIFY: The objects exert forces on each other, but no external forces act on the system of three objects. Therefore the center of mass of this system does not move.

SET UP: The *x*-coordinate of the center of mass is $x_{cm} = \dfrac{m_1 x_1 + m_2 x_2 + m_3 x_3}{m_1 + m_2 + m_3}$ and likewise for the *y*-coordinate. Using the initial conditions, first find the location of the center of mass of the system. Then find the center of mass (which is still the same) after the objects have moved and use it to find the position *x* of the third object.

EXECUTE: Initially: $x_{cm} = \dfrac{m(-L) + m(0) + m(L)}{3m} = 0$ and $y_{cm} = \dfrac{m(0) + m(L) + m(0)}{3m} = \dfrac{L}{3}$. So the center of mass is at $(0, L/3)$ and does not change location.

For the new arrangement: $x_{cm} = \dfrac{m(-L/3) + m(L/2) + mx}{3m} = 0 \rightarrow x = -\dfrac{L}{6}$.

$y_{cm} = \dfrac{m(L/4) + m(-L) + my}{3m} = \dfrac{L}{3} \rightarrow y = \dfrac{7L}{4}$.

EVALUATE: As a check, use the coordinates of the third mass (–*L*/6, 7*L*/4) to calculate the location of the center of mass of the new arrangement. The result should come out (0, *L*/3).

8.3. IDENTIFY and SET UP: We use $p = mv$ and add the respective components.

EXECUTE: (a) $P_x = p_{Ax} + p_{Cx} = 0 + (10.0 \text{ kg})(-3.0 \text{ m/s}) = -30 \text{ kg} \cdot \text{m/s}$
$P_y = p_{Ay} + p_{Cy} = (5.0 \text{ kg})(-11.0 \text{ m/s}) + 0 = -55 \text{ kg} \cdot \text{m/s}$

(b) $P_x = p_{Bx} + p_{Cx} = (6.0 \text{ kg})(10.0 \text{ m/s} \cos 60°) + (10.0 \text{ kg})(-3.0 \text{ m/s}) = 0$
$P_y = p_{By} + p_{Cy} = (6.0 \text{ kg})(10.0 \text{ m/s} \sin 60°) + 0 = 52 \text{ kg} \cdot \text{m/s}$

(c) $P_x = p_{Ax} + p_{Bx} + p_{Cx} = 0 + (6.0 \text{ kg})(10.0 \text{ m/s} \cos 60°) + (10.0 \text{ kg})(-3.0 \text{ m/s}) = 0$
$P_y = p_{Ay} + p_{By} + p_{Cy} = (5.0 \text{ kg})(-11.0 \text{ m/s}) + (6.0 \text{ kg})(10.0 \text{ m/s} \sin 60°) + 0 = -3.0 \text{ kg} \cdot \text{m/s}$

EVALUATE: A has no x-component of momentum so P_x is the same in (b) and (c). C has no y-component of momentum so P_y in (c) is the sum of P_y in (a) and (b).

8.9. IDENTIFY: Use $J_x = p_{2x} - p_{1x}$. We know the initial momentum and the impulse so can solve for the final momentum and then the final velocity.

SET UP: Take the x-axis to be toward the right, so $v_{1x} = +3.00$ m/s. Use $J_x = F_x \Delta t$ to calculate the impulse, since the force is constant.

EXECUTE: (a) $J_x = p_{2x} - p_{1x}$

$$J_x = F_x(t_2 - t_1) = (+25.0 \text{ N})(0.050 \text{ s}) = +1.25 \text{ kg} \cdot \text{m/s}$$

Thus $p_{2x} = J_x + p_{1x} = +1.25 \text{ kg} \cdot \text{m/s} + (0.160 \text{ kg})(+3.00 \text{ m/s}) = +1.73 \text{ kg} \cdot \text{m/s}$

$$v_{2x} = \frac{p_{2x}}{m} = \frac{1.73 \text{ kg} \cdot \text{m/s}}{0.160 \text{ kg}} = +10.8 \text{ m/s (to the right)}$$

(b) $J_x = F_x(t_2 - t_1) = (-12.0 \text{ N})(0.050 \text{ s}) = -0.600 \text{ kg} \cdot \text{m/s}$ (negative since force is to left)

$p_{2x} = J_x + p_{1x} = -0.600 \text{ kg} \cdot \text{m/s} + (0.160 \text{ kg})(+3.00 \text{ m/s}) = -0.120 \text{ kg} \cdot \text{m/s}$

$$v_{2x} = \frac{p_{2x}}{m} = \frac{-0.120 \text{ kg} \cdot \text{m/s}}{0.160 \text{ kg}} = -0.75 \text{ m/s (to the left)}$$

EVALUATE: In part (a) the impulse and initial momentum are in the same direction and v_x increases. In part (b) the impulse and initial momentum are in opposite directions and the velocity decreases.

8.11. IDENTIFY: The force is not constant so $\vec{J} = \int_{t_1}^{t_2} \vec{F} dt$. The impulse is related to the change in velocity by $J_x = m(v_{2x} - v_{1x})$.

SET UP: Only the x-component of the force is nonzero, so $J_x = \int_{t_1}^{t_2} F_x dt$ is the only nonzero component of $\vec{J}$. $J_x = m(v_{2x} - v_{1x})$. $t_1 = 2.00$ s, $t_2 = 3.50$ s.

EXECUTE: (a) $A = \dfrac{F_x}{t^2} = \dfrac{781.25 \text{ N}}{(1.25 \text{ s})^2} = 500 \text{ N/s}^2$.

(b) $J_x = \int_{t_1}^{t_2} At^2 dt = \frac{1}{3} A(t_2^3 - t_1^3) = \frac{1}{3}(500 \text{ N/s}^2)([3.50 \text{ s}]^3 - [2.00 \text{ s}]^3) = 5.81 \times 10^3$ N·s.

(c) $\Delta v_x = v_{2x} - v_{1x} = \dfrac{J_x}{m} = \dfrac{5.81 \times 10^3 \text{ N} \cdot \text{s}}{2150 \text{ kg}} = 2.70$ m/s. The x-component of the velocity of the rocket increases by 2.70 m/s.

EVALUATE: The change in velocity is in the same direction as the impulse, which in turn is in the direction of the net force. In this problem the net force equals the force applied by the engine, since that is the only force on the rocket.

8.13. IDENTIFY: The force is constant during the 1.0 ms interval that it acts, so $\vec{J} = \vec{F} \Delta t$.
$\vec{J} = \vec{p}_2 - \vec{p}_1 = m(\vec{v}_2 - \vec{v}_1)$.

SET UP: Let $+x$ be to the right, so $v_{1x} = +5.00$ m/s. Only the x-component of $\vec{J}$ is nonzero, and $J_x = m(v_{2x} - v_{1x})$.

EXECUTE: **(a)** The magnitude of the impulse is
$J = F\Delta t = (2.50 \times 10^3 \text{ N})(1.00 \times 10^{-3} \text{ s}) = 2.50 \text{ N} \cdot \text{s}$. The direction of the impulse is the direction of the force.

(b) (i) $v_{2x} = \dfrac{J_x}{m} + v_{1x}$. $J_x = +2.50 \text{ N} \cdot \text{s}$. $v_{2x} = \dfrac{+2.50 \text{ N} \cdot \text{s}}{2.00 \text{ kg}} + 5.00 \text{ m/s} = 6.25 \text{ m/s}$. The stone's velocity has magnitude 6.25 m/s and is directed to the right. (ii) Now $J_x = -2.50 \text{ N} \cdot \text{s}$ and
$v_{2x} = \dfrac{-2.50 \text{ N} \cdot \text{s}}{2.00 \text{ kg}} + 5.00 \text{ m/s} = 3.75 \text{ m/s}$. The stone's velocity has magnitude 3.75 m/s and is directed to the right.

EVALUATE: When the force and initial velocity are in the same direction the speed increases, and when they are in opposite directions the speed decreases.

8.17. **IDENTIFY:** Since the rifle is loosely held there is no net external force on the system consisting of the rifle, bullet, and propellant gases and the momentum of this system is conserved. Before the rifle is fired everything in the system is at rest and the initial momentum of the system is zero.

SET UP: Let $+x$ be in the direction of the bullet's motion. The bullet has speed $601 \text{ m/s} - 1.85 \text{ m/s} = 599 \text{ m/s}$ relative to the earth. $P_{2x} = p_{rx} + p_{bx} + p_{gx}$, the momenta of the rifle, bullet, and gases. $v_{rx} = -1.85 \text{ m/s}$ and $v_{bx} = +599 \text{ m/s}$.

EXECUTE: $P_{2x} = P_{1x} = 0$. $p_{rx} + p_{bx} + p_{gx} = 0$.
$p_{gx} = -p_{rx} - p_{bx} = -(2.80 \text{ kg})(-1.85 \text{ m/s}) - (0.00720 \text{ kg})(599 \text{ m/s})$ and
$p_{gx} = +5.18 \text{ kg} \cdot \text{m/s} - 4.31 \text{ kg} \cdot \text{m/s} = 0.87 \text{ kg} \cdot \text{m/s}$. The propellant gases have momentum $0.87 \text{ kg} \cdot \text{m/s}$, in the same direction as the bullet is traveling.

EVALUATE: The magnitude of the momentum of the recoiling rifle equals the magnitude of the momentum of the bullet plus that of the gases as both exit the muzzle.

8.19. **IDENTIFY:** Since drag effects are neglected, there is no net external force on the system of squid plus expelled water, and the total momentum of the system is conserved. Since the squid is initially at rest, with the water in its cavity, the initial momentum of the system is zero. For each object, $K = \tfrac{1}{2}mv^2$.

SET UP: Let A be the squid and B be the water it expels, so $m_A = 6.50 \text{ kg} - 1.75 \text{ kg} = 4.75 \text{ kg}$. Let $+x$ be the direction in which the water is expelled. $v_{A2x} = -2.50 \text{ m/s}$. Solve for v_{B2x}.

EXECUTE: **(a)** $P_{1x} = 0$. $P_{2x} = P_{1x}$, so $0 = m_A v_{A2x} + m_B v_{B2x}$.
$v_{B2x} = -\dfrac{m_A v_{A2x}}{m_B} = -\dfrac{(4.75 \text{ kg})(-2.50 \text{ m/s})}{1.75 \text{ kg}} = +6.79 \text{ m/s}$.

(b)
$K_2 = K_{A2} + K_{B2} = \tfrac{1}{2} m_A v_{A2}^2 + \tfrac{1}{2} m_B v_{B2}^2 = \tfrac{1}{2}(4.75 \text{ kg})(2.50 \text{ m/s})^2 + \tfrac{1}{2}(1.75 \text{ kg})(6.79 \text{ m/s})^2 = 55.2 \text{ J}$.
The initial kinetic energy is zero, so the kinetic energy produced is $K_2 = 55.2 \text{ J}$.

EVALUATE: The two objects end up with momenta that are equal in magnitude and opposite in direction, so the total momentum of the system remains zero. The kinetic energy is created by the work done by the squid as it expels the water.

8.21. **IDENTIFY:** Apply conservation of momentum to the system of the two pucks.

SET UP: Let $+x$ be to the right.

EXECUTE: **(a)** $P_{1x} = P_{2x}$ says $(0.250 \text{ kg})v_{A1} = (0.250 \text{ kg})(-0.120 \text{ m/s}) + (0.350 \text{ kg})(0.650 \text{ m/s})$ and $v_{A1} = 0.790$ m/s.

(b) $K_1 = \frac{1}{2}(0.250 \text{ kg})(0.790 \text{ m/s})^2 = 0.0780$ J.

$K_2 = \frac{1}{2}(0.250 \text{ kg})(0.120 \text{ m/s})^2 + \frac{1}{2}(0.350 \text{ kg})(0.650 \text{ m/s})^2 = 0.0757$ J and $\Delta K = K_2 - K_1 = -0.0023$ J.

EVALUATE: The total momentum of the system is conserved but the total kinetic energy decreases.

8.23. **IDENTIFY:** The momentum and the mechanical energy of the system are both conserved. The mechanical energy consists of the kinetic energy of the masses and the elastic potential energy of the spring. The potential energy stored in the spring is transformed into the kinetic energy of the two masses.

SET UP: Let the system be the two masses and the spring. The system is sketched in Figure 8.23, in its initial and final situations. Use coordinates where $+x$ is to the right. Call the masses A and B.

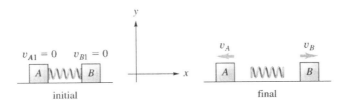

Figure 8.23

EXECUTE: $P_{1x} = P_{2x}$ so $0 = (0.900 \text{ kg})(-v_A) + (0.900 \text{ kg})(v_B)$ and, since the masses are equal, $v_A = v_B$. Energy conservation says the potential energy originally stored in the spring is all converted into kinetic energy of the masses, so $\frac{1}{2}kx_1^2 = \frac{1}{2}mv_A^2 + \frac{1}{2}mv_B^2$. Since $v_A = v_B$, this equation gives

$$v_A = x_1\sqrt{\frac{k}{2m}} = (0.200 \text{ m})\sqrt{\frac{175 \text{ N/m}}{2(0.900 \text{ kg})}} = 1.97 \text{ m/s}.$$

EVALUATE: If the objects have different masses they will end up with different speeds. The lighter one will have the greater speed, since they end up with equal magnitudes of momentum.

8.25. **IDENTIFY:** Since friction at the pond surface is neglected, there is no net external horizontal force, and the horizontal component of the momentum of the system of hunter plus bullet is conserved. Both objects are initially at rest, so the initial momentum of the system is zero. Gravity and the normal force exerted by the ice together produce a net vertical force while the rifle is firing, so the vertical component of momentum is not conserved.

SET UP: Let object A be the hunter and object B be the bullet. Let $+x$ be the direction of the horizontal component of velocity of the bullet. Solve for v_{A2x}.

EXECUTE: **(a)** $v_{B2x} = +965$ m/s. $P_{1x} = P_{2x} = 0$. $0 = m_A v_{A2x} + m_B v_{B2x}$ and

$$v_{A2x} = -\frac{m_B}{m_A}v_{B2x} = -\left(\frac{4.20\times10^{-3} \text{ kg}}{72.5 \text{ kg}}\right)(965 \text{ m/s}) = -0.0559 \text{ m/s}.$$

(b) $v_{B2x} = v_{B2}\cos\theta = (965 \text{ m/s})\cos 56.0° = 540 \text{ m/s}.$

$$v_{A2x} = -\left(\frac{4.20\times 10^{-3} \text{ kg}}{72.5 \text{ kg}}\right)(540 \text{ m/s}) = -0.0313 \text{ m/s}.$$

EVALUATE: The mass of the bullet is much less than the mass of the hunter, so the final mass of the hunter plus gun is still 72.5 kg, to three significant figures. Since the hunter has much larger mass, his final speed is much less than the speed of the bullet.

8.27. **IDENTIFY:** Each horizontal component of momentum is conserved. $K = \tfrac{1}{2}mv^2$.

SET UP: Let $+x$ be the direction of Rebecca's initial velocity and let the $+y$ axis make an angle of 36.9° with respect to the direction of her final velocity. $v_{D1x} = v_{D1y} = 0$. $v_{R1x} = 13.0$ m/s; $v_{R1y} = 0$. $v_{R2x} = (8.00 \text{ m/s})\cos 53.1° = 4.80$ m/s; $v_{R2y} = (8.00 \text{ m/s})\sin 53.1° = 6.40$ m/s. Solve for v_{D2x} and v_{D2y}.

EXECUTE: **(a)** $P_{1x} = P_{2x}$ gives $m_R v_{R1x} = m_R v_{R2x} + m_D v_{D2x}$.

$$v_{D2x} = \frac{m_R(v_{R1x} - v_{R2x})}{m_D} = \frac{(45.0 \text{ kg})(13.0 \text{ m/s} - 4.80 \text{ m/s})}{65.0 \text{ kg}} = 5.68 \text{ m/s}.$$

$P_{1y} = P_{2y}$ gives $0 = m_R v_{R2y} + m_D v_{D2y}$. $v_{D2y} = -\frac{m_R}{m_D} v_{R2y} = -\left(\frac{45.0 \text{ kg}}{65.0 \text{ kg}}\right)(6.40 \text{ m/s}) = -4.43 \text{ m/s}.$

The directions of $\vec{v}_{R1}$, $\vec{v}_{R2}$ and $\vec{v}_{D2}$ are sketched in Figure 8.27. $\tan\theta = \left|\frac{v_{D2y}}{v_{D2x}}\right| = \frac{4.43 \text{ m/s}}{5.68 \text{ m/s}}$ and $\theta = 38.0°$. $v_D = \sqrt{v_{D2x}^2 + v_{D2y}^2} = 7.20$ m/s.

(b) $K_1 = \tfrac{1}{2}m_R v_{R1}^2 = \tfrac{1}{2}(45.0 \text{ kg})(13.0 \text{ m/s})^2 = 3.80\times 10^3$ J.

$K_2 = \tfrac{1}{2}m_R v_{R2}^2 + \tfrac{1}{2}m_D v_{D2}^2 = \tfrac{1}{2}(45.0 \text{ kg})(8.00 \text{ m/s})^2 + \tfrac{1}{2}(65.0 \text{ kg})(7.20 \text{ m/s})^2 = 3.12\times 10^3$ J.

$\Delta K = K_2 - K_1 = -680$ J.

EVALUATE: Each component of momentum is separately conserved. The kinetic energy of the system decreases.

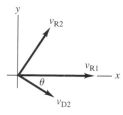

Figure 8.27

8.31. **IDENTIFY:** Momentum is conserved during the collision.

SET UP: $p_x = mv_x$. Call the $+x$-axis pointing northward. We want the speed v of the hockey players after they collide and become intertwined.

EXECUTE: The momentum before the collision is equal to the momentum after the collision.
$m_1 v_1 + m_2 v_2 = (m_1 + m_2)v \rightarrow (70 \text{ kg})(5.5 \text{ m/s}) + (110 \text{ kg})(-4.0 \text{ m/s}) = (180 \text{ kg})v \rightarrow v = -0.31$ m/s.
The minus sign tells that they are traveling toward the south.

EVALUATE: Even though the heavier player was traveling slower than the lighter player, his larger mass gave him greater momentum than that of the faster lighter player.

8.35. **IDENTIFY:** This problem involves a collision, so momentum is conserved.

SET UP: We use $p_x = mv_x$ and $K = \frac{1}{2}mv^2$. The total momentum P_x before the collision is equal to the total momentum after the collision, where $P_x = p_{1x} + p_{2x}$. Call the +x-axis eastward. We want to find the decrease in kinetic energy during the collision. Start by making a before-and-after sketch of the collision, as shown in Fig. 8.35.

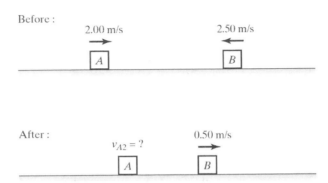

Figure 8.35

EXECUTE: We first need to find the velocity of A after the collision. Using $P_{1x} = P_{2x}$ gives $p_{A1} + p_{B1} = p_{A2} + p_{B2}$, so $m_A v_{A1} + m_B v_{B1} = m_A v_{A2} + m_B v_{B2}$. Putting in the numbers gives us
(4.00 kg)(2.00 m/s) – (6.00 kg)(2.50 m/s) = (4.00 kg)v_{A2} + (6.00 kg)(0.50 m/s), so v_{A2} = –2.50 m/s.
Now find the decrease in kinetic energy, which is the initial value minus the final value. Using v_{A2} = –2.50 m/s and the quantities given in the problem, we get

Decrease = $K_1 - K_2 = \frac{1}{2}m_A v_{A1}^2 + \frac{1}{2}m_B v_{B1}^2 - \left(\frac{1}{2}m_A v_{A2}^2 + \frac{1}{2}m_B v_{B2}^2\right) = 13.5$ J.

EVALUATE: We see that the momentum is conserved during this collision but the kinetic energy is not conserved. The final kinetic energy is 13.5 J less than the initial kinetic energy.

8.37. **IDENTIFY:** The forces the two players exert on each other during the collision are much larger than the horizontal forces exerted by the slippery ground and it is a good approximation to assume momentum conservation. Each component of momentum is separately conserved.

SET UP: Let +x be east and +y be north. After the collision the two players have velocity $\vec{v}_2$. Let the linebacker be object A and the halfback be object B, so $v_{A1x} = 0$, $v_{A1y} = 8.8$ m/s, $v_{B1x} = 7.2$ m/s and $v_{B1y} = 0$. Solve for v_{2x} and v_{2y}.

EXECUTE: $P_{1x} = P_{2x}$ gives $m_A v_{A1x} + m_B v_{B1x} = (m_A + m_B)v_{2x}$.

$$v_{2x} = \frac{m_A v_{A1x} + m_B v_{B1x}}{m_A + m_B} = \frac{(85 \text{ kg})(7.2 \text{ m/s})}{110 \text{ kg} + 85 \text{ kg}} = 3.14 \text{ m/s}.$$

$P_{1y} = P_{2y}$ gives $m_A v_{A1y} + m_B v_{B1y} = (m_A + m_B)v_{2y}$.

$$v_{2y} = \frac{m_A v_{A1y} + m_B v_{B1y}}{m_A + m_B} = \frac{(110 \text{ kg})(8.8 \text{ m/s})}{110 \text{ kg} + 85 \text{ kg}} = 4.96 \text{ m/s}.$$

$v = \sqrt{v_{2x}^2 + v_{2y}^2} = 5.9$ m/s.

$$\tan\theta = \frac{v_{2y}}{v_{2x}} = \frac{4.96 \text{ m/s}}{3.14 \text{ m/s}} \text{ and } \theta = 58°.$$

The players move with a speed of 5.9 m/s and in a direction 58° north of east.

EVALUATE: Each component of momentum is separately conserved.

8.41. IDENTIFY: Since friction forces from the road are ignored, the x- and y-components of momentum are conserved.

SET UP: Let object A be the subcompact and object B be the truck. After the collision the two objects move together with velocity $\vec{v}_2$. Use the x- and y-coordinates given in the problem.
$v_{A1y} = v_{B1x} = 0$. $v_{2x} = (16.0 \text{ m/s})\sin 24.0° = 6.5$ m/s; $v_{2y} = (16.0 \text{ m/s})\cos 24.0° = 14.6$ m/s.

EXECUTE: $P_{1x} = P_{2x}$ gives $m_A v_{A1x} = (m_A + m_B)v_{2x}$.

$$v_{A1x} = \left(\frac{m_A + m_B}{m_A}\right)v_{2x} = \left(\frac{950 \text{ kg} + 1900 \text{ kg}}{950 \text{ kg}}\right)(6.5 \text{ m/s}) = 19.5 \text{ m/s}.$$

$P_{1y} = P_{2y}$ gives $m_B v_{B1y} = (m_A + m_B)v_{2y}$.

$$v_{B1y} = \left(\frac{m_A + m_B}{m_B}\right)v_{2y} = \left(\frac{950 \text{ kg} + 1900 \text{ kg}}{1900 \text{ kg}}\right)(14.6 \text{ m/s}) = 21.9 \text{ m/s}.$$

Before the collision the subcompact car has speed 19.5 m/s and the truck has speed 21.9 m/s.

EVALUATE: Each component of momentum is independently conserved.

8.43. IDENTIFY: Apply conservation of momentum to the collision and conservation of energy to the motion after the collision. After the collision the kinetic energy of the combined object is converted to gravitational potential energy.

SET UP: Immediately after the collision the combined object has speed V. Let h be the vertical height through which the pendulum rises.

EXECUTE: **(a)** Conservation of momentum applied to the collision gives
$(12.0 \times 10^{-3} \text{ kg})(380 \text{ m/s}) = (6.00 \text{ kg} + 12.0 \times 10^{-3} \text{ kg})V$ and $V = 0.758$ m/s.

Conservation of energy applied to the motion after the collision gives $\frac{1}{2}m_{tot}V^2 = m_{tot}gh$ and

$$h = \frac{V^2}{2g} = \frac{(0.758 \text{ m/s})^2}{2(9.80 \text{ m/s}^2)} = 0.0293 \text{ m} = 2.93 \text{ cm}.$$

(b) $K = \frac{1}{2}m_b v_b^2 = \frac{1}{2}(12.0 \times 10^{-3} \text{ kg})(380 \text{ m/s})^2 = 866$ J.

(c) $K = \frac{1}{2}m_{tot}V^2 = \frac{1}{2}(6.00 \text{ kg} + 12.0 \times 10^{-3} \text{ kg})(0.758 \text{ m/s})^2 = 1.73$ J.

EVALUATE: Most of the initial kinetic energy of the bullet is dissipated in the collision.

8.45. IDENTIFY: The missile gives momentum to the ornament causing it to swing in a circular arc and thereby be accelerated toward the center of the circle.

SET UP: After the collision the ornament moves in an arc of a circle and has acceleration $a_{rad} = \frac{v^2}{r}$. During the collision, momentum is conserved, so $P_{1x} = P_{2x}$. The free-body diagram for the ornament plus missile is given in Figure 8.45. Take $+y$ to be upward, since that is the direction of the acceleration. Take the $+x$-direction to be the initial direction of motion of the missile.

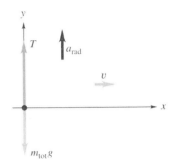

Figure 8.45

EXECUTE: Apply conservation of momentum to the collision. Using $P_{1x} = P_{2x}$, we get $(0.200 \text{ kg})(12.0 \text{ m/s}) = (1.00 \text{ kg})V$, which gives $V = 2.40$ m/s, the speed of the ornament immediately after the collision. Then $\Sigma F_y = ma_y$ gives $T - m_{\text{tot}} g = m_{\text{tot}} \dfrac{v^2}{r}$. Solving for T gives

$$T = m_{\text{tot}}\left(g + \dfrac{v^2}{r}\right) = (1.00 \text{ kg})\left(9.80 \text{ m/s}^2 + \dfrac{(2.40 \text{ m/s})^2}{1.50 \text{ m}}\right) = 13.6 \text{ N}.$$

EVALUATE: We cannot use energy conservation during the collision because it is an inelastic collision (the objects stick together).

8.47. **IDENTIFY:** When the spring is compressed the maximum amount the two blocks aren't moving relative to each other and have the same velocity $\vec{V}$ relative to the surface. Apply conservation of momentum to find V and conservation of energy to find the energy stored in the spring. Since the collision is elastic, $v_{A2x} = \left(\dfrac{m_A - m_B}{m_A + m_B}\right) v_{A1x}$ and $v_{B2x} = \left(\dfrac{2m_A}{m_A + m_B}\right) v_{A1x}$ give the final velocity of each block after the collision.

SET UP: Let $+x$ be the direction of the initial motion of A.
EXECUTE: **(a)** Momentum conservation gives $(2.00 \text{ kg})(2.00 \text{ m/s}) = (8.00 \text{ kg})V$ so $V = 0.500$ m/s. Both blocks are moving at 0.500 m/s, in the direction of the initial motion of block A. Conservation of energy says the initial kinetic energy of A equals the total kinetic energy at maximum compression plus the potential energy U_b stored in the bumpers:

$\tfrac{1}{2}(2.00 \text{ kg})(2.00 \text{ m/s})^2 = U_b + \tfrac{1}{2}(8.00 \text{ kg})(0.500 \text{ m/s})^2$ so $U_b = 3.00$ J.

(b) $v_{A2x} = \left(\dfrac{m_A - m_B}{m_A + m_B}\right) v_{A1x} = \left(\dfrac{2.00 \text{ kg} - 6.0 \text{ kg}}{8.00 \text{ kg}}\right)(2.00 \text{ m/s}) = -1.00$ m/s. Block A is moving in the $-x$-direction at 1.00 m/s.

$v_{B2x} = \left(\dfrac{2m_A}{m_A + m_B}\right) v_{A1x} = \dfrac{2(2.00 \text{ kg})}{8.00 \text{ kg}}(2.00 \text{ m/s}) = +1.00$ m/s. Block B is moving in the $+x$-direction at 1.00 m/s.

EVALUATE: When the spring is compressed the maximum amount, the system must still be moving in order to conserve momentum.

8.51. **IDENTIFY:** In any collision, momentum is conserved. But in this one, kinetic energy is conserved because it is an *elastic* collision.

SET UP: We use $p_x = mv_x$ and $K = \frac{1}{2}mv^2$. The total momentum is $P_x = p_{1x} + p_{2x}$. Call the $+x$-axis the original direction of object A. Start by making a before-and-after sketch of the collision, as shown in Fig. 8.51. Call v_0 the original speed of A. We want to find v_A after the collision.

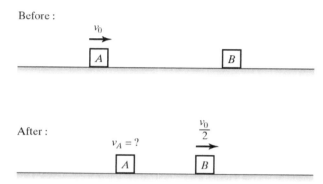

Figure 8.51

EXECUTE: **(a)** Which has greater mass, A or B? Apply momentum conservation and energy conservation to the collision. We know that $v_B = v_0/2$ after the collision. See the figure for the quantities used.

Momentum conservation: $m_A v_0 = m_A v_A + m_B \dfrac{v_0}{2}$ (Eq. 1)

Energy conservation: $\dfrac{1}{2} m_A v_0^2 = \dfrac{1}{2} m_A v_A^2 + \dfrac{1}{2} m_B \left(\dfrac{v_0}{2}\right)^2$ (Eq. 2)

Defining $R = m_B/m_A$, Eq. 1 becomes $v_A = v_0\left(1 - \dfrac{R}{2}\right)$. Using this result and simplifying, Eq. 2 becomes $1 = \left(1 - \dfrac{R}{2}\right)^2 + \dfrac{R}{4}$. Squaring and solving for R gives $R = 3$.

(b) Therefore $m_B/m_A = 3$, so B has 3 times the mass of A.

(c) From our result in part (a), $v_A = v_0\left(1 - \dfrac{R}{2}\right) = v_0\left(1 - \dfrac{3}{2}\right) = -\dfrac{v_0}{2} = -\dfrac{6.0 \text{ m/s}}{2} = -3.0$ m/s. The minus sign tells us that A is moving opposite to its original direction.

EVALUATE: Use our results to calculate the kinetic energy before and after the collision.

$K_1 = \dfrac{1}{2} m_A v_0^2$ and $K_2 = \dfrac{1}{2} m_A \left(\dfrac{v_0}{2}\right)^2 + \dfrac{1}{2}(3 m_A)\left(\dfrac{v_0}{2}\right)^2 = \dfrac{1}{2} m_A v_0^2 = K_1$. This agrees with the fact that it is an elastic collision.

8.53. **IDENTIFY:** The location of the center of mass is given by $x_{cm} = \dfrac{m_1 x_1 + m_2 x_2 + m_3 x_3 + \cdots}{m_1 + m_2 + m_3 + \cdots}$. The mass can be expressed in terms of the diameter. Each object can be replaced by a point mass at its center.

SET UP: Use coordinates with the origin at the center of Pluto and the $+x$-direction toward Charon, so $x_P = 0$, $x_C = 19{,}700$ km. $m = \rho V = \rho \frac{4}{3}\pi r^3 = \frac{1}{6}\rho \pi d^3$.

EXECUTE: $x_{cm} = \dfrac{m_P x_P + m_C x_C}{m_P + m_C} = \left(-\dfrac{m_C}{m_P + m_C}\right) x_C = \left(\dfrac{\frac{1}{6}\pi \rho d_C^3}{\frac{1}{6}\pi \rho d_P^3 + \frac{1}{6}\pi \rho d_C^3}\right) x_C = \left(\dfrac{d_C^3}{d_P^3 + d_C^3}\right) x_C.$

$x_{cm} = \left(\dfrac{[1250 \text{ km}]^3}{[2370 \text{ km}]^3 + [1250 \text{ km}]^3}\right)(19{,}700 \text{ km}) = 2.52 \times 10^3 \text{ km}.$

The center of mass of the system is 2.52×10^3 km from the center of Pluto.

EVALUATE: The center of mass is closer to Pluto because Pluto has more mass than Charon.

8.59. IDENTIFY: Apply $\sum \vec{F} = \dfrac{d\vec{P}}{dt}$ to the airplane.

SET UP: $\dfrac{d}{dt}(t^n) = nt^{n-1}$. $1 \text{ N} = 1 \text{ kg} \cdot \text{m/s}^2$

EXECUTE: $\dfrac{d\vec{P}}{dt} = [-(1.50 \text{ kg} \cdot \text{m/s}^3)t]\vec{i} + (0.25 \text{ kg} \cdot \text{m/s}^2)\vec{j}.$ $F_x = -(1.50 \text{ N/s})t$, $F_y = 0.25 \text{ N}$, $F_z = 0$.

EVALUATE: There is no momentum or change in momentum in the z-direction and there is no force component in this direction.

8.61. IDENTIFY: $a = -\dfrac{v_{ex}}{m}\dfrac{dm}{dt}$. Assume that dm/dt is constant over the 5.0 s interval, since m doesn't change much during that interval. The thrust is $F = -v_{ex}\dfrac{dm}{dt}$.

SET UP: Take m to have the constant value $110 \text{ kg} + 70 \text{ kg} = 180 \text{ kg}$. dm/dt is negative since the mass of the MMU decreases as gas is ejected.

EXECUTE: (a) $\dfrac{dm}{dt} = -\dfrac{m}{v_{ex}} a = -\left(\dfrac{180 \text{ kg}}{490 \text{ m/s}}\right)(0.029 \text{ m/s}^2) = -0.0106 \text{ kg/s}$. In 5.0 s the mass that is ejected is $(0.0106 \text{ kg/s})(5.0 \text{ s}) = 0.053 \text{ kg}$.

(b) $F = -v_{ex}\dfrac{dm}{dt} = -(490 \text{ m/s})(-0.0106 \text{ kg/s}) = 5.19 \text{ N}$.

EVALUATE: The mass change in the 5.0 s is a very small fraction of the total mass m, so it is accurate to take m to be constant.

8.65. IDENTIFY: The impulse, force, and change in velocity are related by $J_x = F_x \Delta t$.

SET UP: $m = w/g = 0.0571$ kg. Since the force is constant, $\vec{F} = \vec{F}_{av}$.

EXECUTE: (a) $J_x = F_x \Delta t = (-380 \text{ N})(3.00 \times 10^{-3} \text{ s}) = -1.14 \text{ N} \cdot \text{s}$.

$J_y = F_y \Delta t = (110 \text{ N})(3.00 \times 10^{-3} \text{ s}) = 0.330 \text{ N} \cdot \text{s}$.

(b) $v_{2x} = \dfrac{J_x}{m} + v_{1x} = \dfrac{-1.14 \text{ N} \cdot \text{s}}{0.0571 \text{ kg}} + 20.0 \text{ m/s} = 0.04 \text{ m/s}$.

$v_{2y} = \dfrac{J_y}{m} + v_{1y} = \dfrac{0.330 \text{ N} \cdot \text{s}}{0.0571 \text{ kg}} + (-4.0 \text{ m/s}) = +1.8 \text{ m/s}$.

EVALUATE: The change in velocity $\Delta \vec{v}$ is in the same direction as the force, so $\Delta \vec{v}$ has a negative x-component and a positive y-component.

8.67. **IDENTIFY AND SET UP:** When the spring is compressed the maximum amount the two blocks aren't moving relative to each other and have the same velocity V relative to the surface. Apply conservation of momentum to find V and conservation of energy to find the energy stored in the spring. Let $+x$ be the direction of the initial motion of A. The collision is elastic.

SET UP: $p = mv$, $K = \frac{1}{2} mv^2$, $v_{B2x} - v_{A2x} = -(v_{B1x} - v_{A1x})$ for an elastic collision.

EXECUTE: **(a)** The maximum energy stored in the spring is at maximum compression, at which time the blocks have the same velocity. Momentum conservation gives $m_A v_{A1} + m_B v_{B1} = (m_A + m_B)V$. Putting in the numbers we have $(2.00 \text{ kg})(2.00 \text{ m/s}) + (10.0 \text{ kg})(-0.500 \text{ m/s}) = (12.0 \text{ kg})V$, giving $V = -0.08333$ m/s. The energy U_{spring} stored in the spring is the loss of kinetic of the system. Therefore

$U_{\text{spring}} = K_1 - K_2 = \frac{1}{2} m_A v_{A1}^2 + \frac{1}{2} m_B v_{B1}^2 - \frac{1}{2}(m_A + m_B)V^2$. Putting in the same set of numbers as above, and using $V = -0.08333$ m/s, we get $U_{\text{spring}} = 5.21$ J. At this time, the blocks are both moving to the left, so their velocities are each -0.0833 m/s.

(b) Momentum conservation gives $m_A v_{A1} + m_B v_{B1} = m_A v_{A2} + m_B v_{B2}$. Putting in the numbers gives -1 m/s $= 2v_{A2} + 10v_{B2}$. Using $v_{B2x} - v_{A2x} = -(v_{B1x} - v_{A1x})$ we get $v_{B2x} - v_{A2x} = -(-0.500 \text{ m/s} - 2.00 \text{ m/s}) = +2.50$ m/s. Solving this equation and the momentum equation simultaneously gives $v_{A2x} = 2.17$ m/s and $v_{B2x} = 0.333$ m/s.

EVALUATE: The total kinetic energy before the collision is 5.25 J, and it is the same after, which is consistent with an elastic collision.

8.69. **IDENTIFY:** The x- and y-components of the momentum of the system are conserved.

SET UP: After the collision the combined object with mass $m_{\text{tot}} = 0.100$ kg moves with velocity $\vec{v}_2$. Solve for v_{Cx} and v_{Cy}.

EXECUTE: **(a)** $P_{1x} = P_{2x}$ gives $m_A v_{Ax} + m_B v_{Bx} + m_C v_{Cx} = m_{\text{tot}} v_{2x}$.

$$v_{Cx} = -\frac{m_A v_{Ax} + m_B v_{Bx} - m_{\text{tot}} v_{2x}}{m_C}$$

$$v_{Cx} = -\frac{(0.020 \text{ kg})(-1.50 \text{ m/s}) + (0.030 \text{ kg})(-0.50 \text{ m/s})\cos 60° - (0.100 \text{ kg})(0.50 \text{ m/s})}{0.050 \text{ kg}}.$$

$v_{Cx} = 1.75$ m/s.

$P_{1y} = P_{2y}$ gives $m_A v_{Ay} + m_B v_{By} + m_C v_{Cy} = m_{\text{tot}} v_{2y}$.

$$v_{Cy} = -\frac{m_A v_{Ay} + m_B v_{By} - m_{\text{tot}} v_{2y}}{m_C} = -\frac{(0.030 \text{ kg})(-0.50 \text{ m/s})\sin 60°}{0.050 \text{ kg}} = +0.260 \text{ m/s}.$$

(b) $v_C = \sqrt{v_{Cx}^2 + v_{Cy}^2} = 1.77$ m/s. $\Delta K = K_2 - K_1$.

$\Delta K = \frac{1}{2}(0.100 \text{ kg})(0.50 \text{ m/s})^2 - [\frac{1}{2}(0.020 \text{ kg})(1.50 \text{ m/s})^2 + \frac{1}{2}(0.030 \text{ kg})(0.50 \text{ m/s})^2 + \frac{1}{2}(0.050 \text{ kg})(1.77 \text{ m}$

$\Delta K = -0.092$ J.

EVALUATE: Since there is no horizontal external force the vector momentum of the system is conserved. The forces the spheres exert on each other do negative work during the collision and this reduces the kinetic energy of the system.

8.71. **IDENTIFY:** Momentum is conserved during the collision, and the wood (with the clay attached) is in free fall as it falls since only gravity acts on it.

SET UP: Apply conservation of momentum to the collision to find the velocity V of the combined object just after the collision. After the collision, the wood's downward acceleration is g and it has no horizontal acceleration, so we can use the standard kinematics equations: $y - y_0 = v_{0y}t + \frac{1}{2}a_y t^2$ and $x - x_0 = v_{0x}t + \frac{1}{2}a_x t^2$.

EXECUTE: Momentum conservation gives $(0.500 \text{ kg})(24.0 \text{ m/s}) = (8.50 \text{ kg})V$, so $V = 1.412$ m/s. Consider the projectile motion after the collision: $a_y = +9.8$ m/s^2, $v_{0y} = 0$, $y - y_0 = +2.20$ m, and t is unknown. $y - y_0 = v_{0y}t + \frac{1}{2}a_y t^2$ gives $t = \sqrt{\frac{2(y-y_0)}{a_y}} = \sqrt{\frac{2(2.20 \text{ m})}{9.8 \text{ m/s}^2}} = 0.6701$ s. The horizontal acceleration is zero so $x - x_0 = v_{0x}t + \frac{1}{2}a_x t^2 = (1.412 \text{ m/s})(0.6701 \text{ s}) = 0.946$ m.

EVALUATE: The momentum is *not* conserved after the collision because an external force (gravity) acts on the system. Mechanical energy is *not* conserved during the collision because the clay and block stick together, making it an inelastic collision.

8.73. **IDENTIFY:** During the collision, momentum is conserved, but after the collision mechanical energy is conserved. We cannot solve this problem in a single step because the collision and the motion after the collision involve different conservation laws.

SET UP: Use coordinates where $+x$ is to the right and $+y$ is upward. Momentum is conserved during the collision, so $P_{1x} = P_{2x}$. Energy is conserved after the collision, so $K_1 = U_2$, where $K = \frac{1}{2}mv^2$ and $U = mgh$.

EXECUTE: *Collision:* There is no external horizontal force during the collision so $P_{1x} = P_{2x}$. This gives $(5.00 \text{ kg})(12.0 \text{ m/s}) = (10.0 \text{ kg})v_2$ and $v_2 = 6.0$ m/s.

Motion after the collision: Only gravity does work and the initial kinetic energy of the combined chunks is converted entirely to gravitational potential energy when the chunk reaches its maximum height h above the valley floor. Conservation of energy gives $\frac{1}{2}m_{\text{tot}}v^2 = m_{\text{tot}}gh$ and

$$h = \frac{v^2}{2g} = \frac{(6.0 \text{ m/s})^2}{2(9.8 \text{ m/s}^2)} = 1.8 \text{ m}.$$

EVALUATE: After the collision the energy of the system is $\frac{1}{2}m_{\text{tot}}v^2 = \frac{1}{2}(10.0 \text{ kg})(6.0 \text{ m/s})^2 = 180$ J when it is all kinetic energy and the energy is $m_{\text{tot}}gh = (10.0 \text{ kg})(9.8 \text{ m/s}^2)(1.8 \text{ m}) = 180$ J when it is all gravitational potential energy. Mechanical energy is conserved during the motion after the collision. But before the collision the total energy of the system is $\frac{1}{2}(5.0 \text{ kg})(12.0 \text{ m/s})^2 = 360$ J; 50% of the mechanical energy is dissipated during the inelastic collision of the two chunks.

8.77. **IDENTIFY:** During the inelastic collision, momentum is conserved (in two dimensions), but after the collision we must use energy principles.

SET UP: The friction force is $\mu_k m_{\text{tot}} g$. Use energy considerations to find the velocity of the combined object immediately after the collision. Apply conservation of momentum to the collision. Use coordinates where $+x$ is west and $+y$ is south. For momentum conservation, we have $P_{1x} = P_{2x}$ and $P_{1y} = P_{2y}$.

EXECUTE: *Motion after collision:* The negative work done by friction takes away all the kinetic energy that the combined object has just after the collision. Calling ϕ the angle south of west at which the enmeshed cars slid, we have $\tan\phi = \dfrac{6.43 \text{ m}}{5.39 \text{ m}}$ and $\phi = 50.0°$. The wreckage slides 8.39 m in a direction 50.0° south of west. Energy conservation gives $\tfrac{1}{2}m_{tot}V^2 = \mu_k m_{tot} g d$, so $V = \sqrt{2\mu_k g d} = \sqrt{2(0.75)(9.80 \text{ m/s}^2)(8.39 \text{ m})} = 11.1$ m/s. The velocity components are $V_x = V\cos\phi = 7.13$ m/s; $V_y = V\sin\phi = 8.50$ m/s.

Collision: $P_{1x} = P_{2x}$ gives $(2200 \text{ kg})v_{SUV} = (1500 \text{ kg} + 2200 \text{ kg})V_x$ and $v_{SUV} = 12$ m/s. $P_{1y} = P_{2y}$ gives $(1500 \text{ kg})v_{sedan} = (1500 \text{ kg} + 2200 \text{ kg})V_y$ and $v_{sedan} = 21$ m/s.

EVALUATE: We cannot solve this problem in a single step because the collision and the motion after the collision involve different principles (momentum conservation and energy conservation).

8.79. IDENTIFY: Apply conservation of momentum to the collision and conservation of energy to the motion after the collision.

SET UP: Let $+x$ be to the right. The total mass is $m = m_{bullet} + m_{block} = 1.00$ kg. The spring has force constant $k = \dfrac{|F|}{|x|} = \dfrac{0.750 \text{ N}}{0.250\times 10^{-2} \text{ m}} = 300$ N/m. Let V be the velocity of the block just after impact.

EXECUTE: **(a)** Conservation of energy for the motion after the collision gives $K_1 = U_{el2}$. $\tfrac{1}{2}mV^2 = \tfrac{1}{2}kx^2$ and

$$V = x\sqrt{\dfrac{k}{m}} = (0.150 \text{ m})\sqrt{\dfrac{300 \text{ N/m}}{1.00 \text{ kg}}} = 2.60 \text{ m/s}.$$

(b) Conservation of momentum applied to the collision gives $m_{bullet}v_1 = mV$.

$$v_1 = \dfrac{mV}{m_{bullet}} = \dfrac{(1.00 \text{ kg})(2.60 \text{ m/s})}{8.00\times 10^{-3} \text{ kg}} = 325 \text{ m/s}.$$

EVALUATE: The initial kinetic energy of the bullet is 422 J. The energy stored in the spring at maximum compression is 3.38 J. Most of the initial mechanical energy of the bullet is dissipated in the collision.

8.83. IDENTIFY: This collision is elastic, so kinetic energy and momentum are both conserved.

SET UP: Use $v_A = \dfrac{m_A - m_B}{m_A + m_B}v_0$ (Eq. 8.24) and $v_B = \dfrac{2m_A}{m_A + m_B}v_0$ (Eq. 8.25), as well as $K = \dfrac{1}{2}mv^2$. Call the $+x$-axis the original direction that object A is moving. Where we let $v_{Ai} = v_0$ and we have simplified the notation of Eq. 8.24 and Eq. 8.25 somewhat. We have $m_A = \alpha m_B$.

EXECUTE: **(a)** Using Eq. 8.25 gives the final kinetic energy of B.

$K_{B,f} = \dfrac{1}{2}m_B v_B^2 = \dfrac{1}{2}m_B\left(\dfrac{2m_A v_0}{m_A + m_B}\right)^2 = \dfrac{1}{2}m_B\left(\dfrac{2m_A}{m_A + m_B}\right)^2 v_0^2$. This is equal to the initial kinetic energy of A, which is $K_{A,i} = \dfrac{1}{2}m_A v_0^2 = \dfrac{1}{2}\alpha m_B v_0^2$. Equating the two kinetic energies gives

$$\frac{1}{2}\alpha m_B v_0^2 = \frac{1}{2}m_B\left(\frac{2m_A}{m_A+m_B}\right)^2 v_0^2, \text{which simplifies to } \alpha = \left(\frac{2m_A}{m_A+m_B}\right)^2. \text{ Using } m_A = \alpha m_B \text{ gives}$$

$$\alpha = \left(\frac{2\alpha m_B}{\alpha m_B + m_B}\right)^2 = \left(\frac{2\alpha}{\alpha+1}\right)^2. \text{ Solving for } \alpha \text{ gives } \alpha = 1. \text{ This means that the masses are equal.}$$

(b) In this case, after the collision $K_A = K_B$, so $\frac{1}{2}m_A v_A^2 = \frac{1}{2}m_B v_B^2$. Using $m_A = \alpha m_B$ and simplifying gives $\alpha v_A^2 = v_B^2$. Now use Eq. 24 and Eq. 25 in the last equation, which gives

$$\alpha\left(\frac{m_A-m_B}{m_A+m_B}v_0\right)^2 = \left(\frac{2m_A v_0}{m_A+m_B}\right)^2. \text{ Using } m_A = \alpha m_B \text{ and simplifying gives } \alpha(\alpha-1)^2 = 4\alpha^2. \text{ The}$$

resulting quadratic equation has solutions $\alpha = 3 + 2\sqrt{2} \approx 5.83$ and $\alpha = 3 - 2\sqrt{2} \approx 0.172$.

EVALUATE: When $\alpha = 1$, $m_A = m_B$. Object A stops and object B moves ahead with the same speed that A had. Object A has lost all of its momentum and kinetic energy and object B has gained it all, so both momentum and kinetic energy are conserved.

8.85. **IDENTIFY:** Apply conservation of momentum to the collision between the bullet and the block and apply conservation of energy to the motion of the block after the collision.

(a) SET UP: For the collision between the bullet and the block, let object A be the bullet and object B be the block. Apply momentum conservation to find the speed v_{B2} of the block just after the collision (see Figure 8.85a).

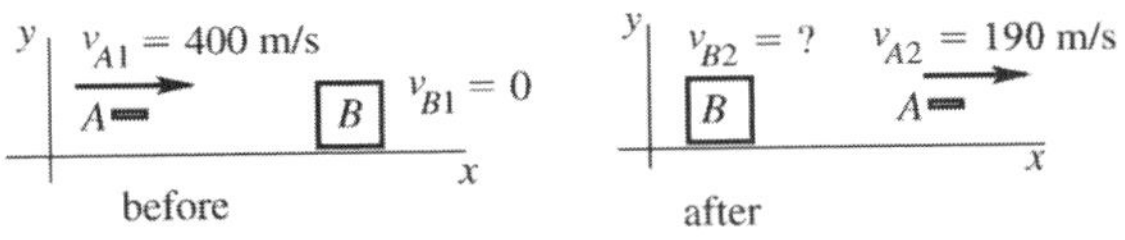

Figure 8.85a

EXECUTE: P_x is conserved so $m_A v_{A1x} + m_B v_{B1x} = m_A v_{A2x} + m_B v_{B2x}$. $m_A v_{A1} = m_A v_{A2} + m_B v_{B2x}$.

$$v_{B2x} = \frac{m_A(v_{A1}-v_{A2})}{m_B} = \frac{4.00 \times 10^{-3} \text{ kg}(400 \text{ m/s} - 190 \text{ m/s})}{0.800 \text{ kg}} = 1.05 \text{ m/s}.$$

SET UP: For the motion of the block after the collision, let point 1 in the motion be just after the collision, where the block has the speed 1.05 m/s calculated above, and let point 2 be where the block has come to rest (see Figure 8.85b).
$K_1 + U_1 + W_{\text{other}} = K_2 + U_2$.

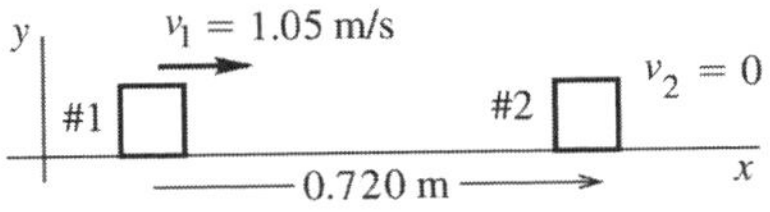

Figure 8.85b

EXECUTE: Work is done on the block by friction, so $W_{\text{other}} = W_f$.
$W_{\text{other}} = W_f = (f_k \cos\phi)s = -f_k s = -\mu_k mgs$, where $s = 0.720$ m. $U_1 = 0$, $U_2 = 0$,
$K_1 = \frac{1}{2}mv_1^2$, $K_2 = 0$ (the block has come to rest). Thus $\frac{1}{2}mv_1^2 - \mu_k mgs = 0$. Therefore

$$\mu_k = \frac{v_1^2}{2gs} = \frac{(1.05 \text{ m/s})^2}{2(9.80 \text{ m/s}^2)(0.720 \text{ m})} = 0.0781.$$

(b) For the bullet, $K_1 = \frac{1}{2}mv_1^2 = \frac{1}{2}(4.00\times10^{-3}\text{ kg})(400\text{ m/s})^2 = 320\text{ J}$ and $K_2 = \frac{1}{2}mv_2^2 = \frac{1}{2}(4.00\times10^{-3}\text{ kg})(190\text{ m/s})^2 = 72.2\text{ J}$. $\Delta K = K_2 - K_1 = 72.2\text{ J} - 320\text{ J} = -248\text{ J}$. The kinetic energy of the bullet decreases by 248 J.

(c) Immediately after the collision the speed of the block is 1.05 m/s, so its kinetic energy is $K = \frac{1}{2}mv^2 = \frac{1}{2}(0.800\text{ kg})(1.05\text{ m/s})^2 = 0.441\text{ J}$.

EVALUATE: The collision is highly inelastic. The bullet loses 248 J of kinetic energy but only 0.441 J is gained by the block. But momentum is conserved in the collision. All the momentum lost by the bullet is gained by the block.

8.87. IDENTIFY: Apply conservation of energy to the motion of the package before the collision and apply conservation of the horizontal component of momentum to the collision.

(a) SET UP: Apply conservation of energy to the motion of the package from point 1 as it leaves the chute to point 2 just before it lands in the cart. Take $y = 0$ at point 2, so $y_1 = 4.00$ m. Only gravity does work, so

$$K_1 + U_1 = K_2 + U_2.$$

EXECUTE: $\frac{1}{2}mv_1^2 + mgy_1 = \frac{1}{2}mv_2^2$.

$v_2 = \sqrt{v_1^2 + 2gy_1} = 9.35$ m/s.

(b) SET UP: In the collision between the package and the cart, momentum is conserved in the horizontal direction. (But not in the vertical direction, due to the vertical force the floor exerts on the cart.) Take $+x$ to be to the right. Let A be the package and B be the cart.

EXECUTE: P_x is constant gives $m_A v_{A1x} + m_B v_{B1x} = (m_A + m_B)v_{2x}$.

$v_{B1x} = -5.00$ m/s.

$v_{A1x} = (3.00\text{ m/s})\cos 37.0°$. (The horizontal velocity of the package is constant during its free fall.) Solving for v_{2x} gives $v_{2x} = -3.29$ m/s. The cart is moving to the left at 3.29 m/s after the package lands in it.

EVALUATE: The cart is slowed by its collision with the package, whose horizontal component of momentum is in the opposite direction to the motion of the cart.

8.91. IDENTIFY: No net external force acts on the Burt-Ernie-log system, so the center of mass of the system does not move.

SET UP: $x_{cm} = \dfrac{m_1 x_1 + m_2 x_2 + m_3 x_3}{m_1 + m_2 + m_3}$.

EXECUTE: Use coordinates where the origin is at Burt's end of the log and where $+x$ is toward Ernie, which makes $x_1 = 0$ for Burt initially. The initial coordinate of the center of mass is $x_{cm,1} = \dfrac{(20.0\text{ kg})(1.5\text{ m}) + (40.0\text{ kg})(3.0\text{ m})}{90.0\text{ kg}}$. Let d be the distance the log moves toward Ernie's original position. The final location of the center of mass is $x_{cm,2} = \dfrac{(30.0\text{ kg})d + (1.5\text{ kg} + d)(20.0\text{ kg}) + (40.0\text{ kg})d}{90.0\text{ kg}}$. The center of mass does not move, so $x_{cm,1} = x_{cm,2}$, which gives

$(20.0\text{ kg})(1.5\text{ m}) + (40.0\text{ kg})(3.0\text{ m}) = (30.0\text{ kg})d + (20.0\text{ kg})(1.5\text{ m} + d) + (40.0\text{ kg})d$. Solving for d gives $d = 1.33$ m.

EVALUATE: Burt, Ernie, and the log all move, but the center of mass of the system does not move.

8.93. **IDENTIFY:** This process involves a swing, a collision, and another swing. Energy is conserved during the two swings and momentum is conserved during the collision.

SET UP: We must break this problem into three parts: the first swing, the collision, and the second swing. We cannot solve it in a single step. We want to find h_{max}, the maximum height the combined spheres reach during the second swing after the collision.

EXECUTE: Swing of B: Energy conservation gives $mgH = \frac{1}{2}mv_B^2$, so $v_B = \sqrt{2gH}$.

Collision: Momentum conservation gives $mv_B = (3m)v_{AB}$, so $v_{AB} = \frac{v_B}{3} = \frac{1}{3}\sqrt{2gH}$.

Swing of $A + B$: Energy conservation gives $\frac{1}{2}(3m)v_{AB}^2 = 3mgh_{max}$. Using v_{AB}, this becomes

$\frac{1}{2}\left(\frac{1}{3}\sqrt{2gH}\right)^2 = gh_{max}$. Solving for h_{max} gives $h_{max} = \frac{H}{9}$.

EVALUATE: It is reasonable that we get $h_{max} < H$ because mechanical energy is lost during the inelastic collision.

8.95. **IDENTIFY:** The collision is inelastic since the blocks stick together. Momentum is conserved during the collision and energy is conserved before and after the collision.

SET UP: Hooke's law: $F = kx$. The elastic energy stored in a spring is $U_{spr} = \frac{1}{2}kx^2$. Energy conservation gives $U_1 + K_1 + W_{other} = U_2 + K_2$ and momentum is $p_x = mv_x$.

EXECUTE: **(a)** First find the distance the spring is compressed using Hooke's law. $mg = kx$, so

$x = \frac{mg}{k} = \frac{(0.500 \text{ kg})(9.80 \text{ m/s}^2)}{80.0 \text{ N/m}} = 0.06125 \text{ m}$. The energy stored with this compression is

$U_{spr} = \frac{1}{2}kx^2 = \frac{1}{2}(80.0 \text{ N})(0.06125 \text{ m})^2 = 0.150 \text{ J}$.

(b) After the collision, the two-block system has kinetic energy, which can be transferred to the spring. As the spring compresses, gravity also does work on the masses. Fig. 8.95 shows the system just after the collision.

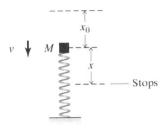

Figure 8.95

First we need to find v, the speed of the two-block system just after the collision. We do this in two steps: conservation of energy as the first block is dropped and reaches the block on the spring, followed by momentum conservation during the collision. Call v_0 the speed of the single block just before it hits the block on the spring. Energy conservation gives $mgh = \frac{1}{2}mv_0^2$, so $v_0 = \sqrt{2gh}$.

Momentum conservation during the collision gives $mv_0 = (2m)v$, which gives

$v = \frac{v_0}{2} = \frac{\sqrt{2gh}}{2} = \frac{\sqrt{2(9.80 \text{ m/s}^2)(4.00 \text{ m})}}{2} = 19.60$ m/s. Now we use energy conservation after the collision. $U_1 + K_1 + W_{other} = U_2 + K_2$ with $W_{other} = 0$. Choose point 1 to be the instant after the collision and point 2 to be when the spring has its maximum compression. At that point, the blocks stop, so $K_2 = 0$. Call x the maximum distance that the spring will compress after the collision (see Fig. 8.25). At point 1 the blocks have gravitational potential energy, which is $U_g = mgx$. At point 1 the system has two forms of potential, U_g and the elastic potential energy that is already stored in the spring. In part (a) we saw that this is 0.150 J. At point 2 the system has only elastic potential energy because the spring is now compressed a *total* distance of $x_0 + x$, where $x_0 = 0.06125$ m from part (a). Calling M the total mass of the system, $U_1 + K_1 = U_2 + K_2$ becomes

$Mgx + U_1 + \frac{1}{2}Mv^2 = \frac{1}{2}k(x_0 + x)^2$. Expanding the square, realizing that $U_1 = \frac{1}{2}kx_0^2$, and collecting terms, this equation becomes $-kx^2 + (2Mg - 2kx_0)x + Mv^2 = 0$. Using $k = 80.0$ N/m, $v = 19.60$ m/s, $M = 1.00$ kg, and $x_0 = 0.06125$ m, the quadratic formula gives two solutions. One of the solutions is negative, so we discard it as nonphysical. The other solution is $x = 0.560$ m. This is the maximum distance that the system compresses the spring *after* the collision. But the spring was already compressed by 0.06125 m before the collision, so the *total* distance that the spring is compressed at the instant the blocks stop moving is $x_{total} = 0.560$ m $+ 0.06125$ m. Therefore the maximum elastic energy stored in the spring is $U_{max} = \frac{1}{2}kx_{max}^2 = \frac{1}{2}k(x + x_0)^2 = \frac{1}{2}(80$ N/m$)(0.560$ m $+ 0.06125$ m$)^2 =$ 15.4 J.

(c) As shown in part (b), $x = 0.560$ m.

EVALUATE: We cannot overlook the gravitational potential energy in solving this problem because it could be significant compared to the elastic energy in the spring.

8.97. **IDENTIFY:** The rocket moves in projectile motion before the explosion and its fragments move in projectile motion after the explosion. Apply conservation of energy and conservation of momentum to the explosion.

(A) SET UP: Apply conservation of energy to the explosion. Just before the explosion the rocket is at its maximum height and has zero kinetic energy. Let A be the piece with mass 1.40 kg and B be the piece with mass 0.28 kg. Let v_A and v_B be the speeds of the two pieces immediately after the collision.

EXECUTE: $\frac{1}{2}m_A v_A^2 + \frac{1}{2}m_B v_B^2 = 860$ J

SET UP: Since the two fragments reach the ground at the same time, their velocities just after the explosion must be horizontal. The initial momentum of the rocket before the explosion is zero, so after the explosion the pieces must be moving in opposite horizontal directions and have equal magnitude of momentum: $m_A v_A = m_B v_B$.

EXECUTE: Use this to eliminate v_A in the first equation and solve for v_B:

$\frac{1}{2}m_B v_B^2(1 + m_B/m_A) = 860$ J and $v_B = 71.6$ m/s.

Then $v_A = (m_B/m_A)v_B = 14.3$ m/s.

(b) SET UP: Use the vertical motion from the maximum height to the ground to find the time it takes the pieces to fall to the ground after the explosion. Take $+y$ downward.

$v_{0y} = 0$, $a_y = +9.80$ m/s^2, $y - y_0 = 80.0$ m, $t = ?$

EXECUTE: $y - y_0 = v_{0y}t + \frac{1}{2}a_y t^2$ gives $t = 4.04$ s.

During this time the horizontal distance each piece moves is $x_A = v_A t = 57.8$ m and $x_B = v_B t = 289.1$ m. They move in opposite directions, so they are $x_A + x_B = 347$ m apart when they land.

EVALUATE: Fragment A has more mass so it is moving slower right after the collision, and it travels horizontally a smaller distance as it falls to the ground.

8.101. IDENTIFY: As the bullet strikes and embeds itself in the block, momentum is conserved. After that, we use $K_1 + U_1 + W_{other} = K_2 + U_2$, where W_{other} is due to kinetic friction.

SET UP: Momentum conservation during the collision gives $m_b v_b = (m_b + m)V$, where m is the mass of the block and m_b is the mass of the bullet. After the collision, $K_1 + U_1 + W_{other} = K_2 + U_2$ gives $\frac{1}{2}MV^2 - \mu_k Mgd = \frac{1}{2}kd^2$, where M is the mass of the block plus the bullet.

EXECUTE: **(a)** From the energy equation above, we can see that the greatest compression of the spring will occur for the greatest V (since $M \gg m_b$), and the greatest V will occur for the bullet with the greatest initial momentum. Using the data in the table with the problem, we get the following momenta expressed in units of grain · ft/s.

A: 1.334×10^5 grain · ft/s   B: 1.181×10^5 grain · ft/s   C: 2.042×10^5 grain · ft/s
D: 1.638×10^5 grain · ft/s   E: 1.869×10^5 grain · ft/s

From these results, it is clear that bullet C will produce the maximum compression of the spring and bullet B will produce the least compression.

(b) For bullet C, we use $p_b = m_b v_b = (m_b + m)V$. Converting mass (in grains) and speed to SI units gives $m_b = 0.01555$ kg and $v_b = 259.38$ m/s, we have
$(0.01555 \text{ kg})(259.38 \text{ m/s}) = (0.01555 \text{ kg} + 2.00 \text{ kg})V$, so $V = 2.001$ m/s.

Now use $\frac{1}{2}MV^2 - \mu_k Mgd = \frac{1}{2}kd^2$ and solve for k, giving

$k = (2.016 \text{ kg})[(2.001 \text{ m/s})^2 - 2(0.38)(9.80 \text{ m/s}^2)(0.25 \text{ m})]/(0.25 \text{ m})^2 = 69.1$ N/m, which rounds to 69 N/m.

(c) For bullet B, $m_b = 125$ grains $= 0.00810$ kg and $v_b = 945$ ft/s $= 288.0$ m/s. Momentum conservation gives
$V = (0.00810 \text{ kg})(288.0 \text{ m/s})/(2.00810 \text{ kg}) = 1.162$ m/s.

Using $\frac{1}{2}MV^2 - \mu_k Mgd = \frac{1}{2}kd^2$, the above numbers give $33.55d^2 + 7.478d - 1.356 = 0$. The quadratic formula, using the positive square root, gives $d = 0.118$ m, which rounds to 0.12 m.

EVALUATE: This method for measuring muzzle velocity involves a spring displacement of around 12 cm, which should be readily measurable.

STUDY GUIDE FOR ROTATION OF RIGID BODIES

Summary

In this chapter, we'll investigate the rotational motion of *rigid bodies*—objects that don't change size or shape as they move. We'll describe first the kinematics of rotation for a rigid body and then its rotational kinetic energy. We'll see how these quantities are analogous to linear kinematics and translational kinetic energy. Finally, we'll learn about the moment of inertia, how to calculate it, and how to use it to measure rotational inertia.

Objectives

After studying this chapter, you'll understand
- How to use radians in angular measurements.
- The definition and application of *angular displacement*, *velocity*, and *acceleration*.
- How to solve problems involving constant angular acceleration.
- How to define *moment of inertia* and apply it to systems of varying shapes.
- How to calculate the moment of inertia.
- How to solve conservation-of-energy problems that include rotational kinetic energy.
- How to draw analogies between translational and rotational motion and energy.

Concepts and Equations

Term	Description
Rigid Body	A rigid body is an object that maintains an unchanging size and shape. We neglect squeezing, stretching, and twisting in our analysis of a rigid body.
Radian	Angular displacements are usually measured in radians. A displacement θ measured in radians is the ratio of the arc length s to the radius r: $$\theta = \frac{s}{r}.$$ There are 2π radians in one revolution (360°).
Angular Velocity	The instantaneous angular velocity about the z-axis is the rate of change of angular displacement with respect to time: $$\omega_z = \lim_{\Delta t \to 0} \frac{\Delta \theta}{\Delta t} = \frac{d\theta}{dt}.$$ The term *angular velocity* refers to the instantaneous angular velocity. All pieces of a rigid object have the same angular velocity at any given instant. The direction of the angular velocity is given by the right-hand rule.
Angular Acceleration	The instantaneous angular acceleration about the z-axis is the rate of change of angular velocity with respect to time: $$\alpha_z = \lim_{\Delta t \to 0} \frac{\Delta \omega_z}{\Delta t} = \frac{d\omega_z}{dt} = \frac{d^2\theta}{dt^2}.$$ The term *angular acceleration* refers to the instantaneous angular acceleration. All pieces of a rigid object have the same angular acceleration at any given instant. The direction of the angular acceleration is the same as that of the angular speed when the object is speeding up and opposite that of the angular speed when the object is slowing down.
Rotation with Constant Angular Acceleration	When an object moves with constant angular acceleration, the angular displacement, velocity, acceleration, and time are related by the formulas $$\theta = \theta_0 + \omega_{0z} t + \tfrac{1}{2}\alpha_z t^2,$$ $$\omega_z = \omega_{0z} + \alpha_z t,$$ $$\omega_z^2 = \omega_{0z}^2 + 2\alpha_z(\theta - \theta_0),$$ $$\theta - \theta_0 = \tfrac{1}{2}(\omega_z + \omega_{0z})t,$$ where θ_0 and ω_{0z} are the initial values of the angular position and velocity, respectively.
Connecting Linear and Angular Quantities	The tangential speed v of a particle rotating in a rigid body at a distance r from the axis of rotation is $$v = r\omega.$$ The particle's acceleration $\vec{a}$ has a tangential component $$a_{\tan} = \frac{dv}{dt} = \frac{r\,d\omega}{dt} = r\alpha$$ and a radial component $$a_{\text{rad}} = \frac{v^2}{r} = \omega^2 r.$$

Moment of Inertia	The moment of inertia, I, of a body is a measure of its rotational inertia and depends on how the body's mass is distributed relative to the axis of rotation. The moment of inertia is given by $$I = m_1 r_1^2 + m_2 r_2^2 + m_3 r_3^2 + \ldots$$ $$= \sum_i m_i r_i^2.$$ For arbitrarily distributed masses, the moment of inertia is given by $$I = \int r^2 \, dm.$$ Moments of inertia for common shapes are given in Table 2 in the textbook. The parallel-axis theorem can be used to find the moment of inertia, I_P, about another parallel axis; that is, $$I_P = I_{cm} + Md^2,$$ where I_{cm} is the moment of inertia about the center of mass, M is the mass of the body, and d is the distance between the two axes.
Energy of Rotating Body	The kinetic energy of a rigid body rotating about a fixed axis is $$K = \tfrac{1}{2} I \omega^2.$$ This quantity is the sum of the kinetic energies of all of the particles that make up the rigid body.

Key Concept 1: To find the *average* angular velocity of a rotating rigid body, first find the body's angular displacement (final angular position minus initial angular position) during a time interval. Then divide the result by that time interval. To find the rigid body's *instantaneous* angular velocity, take the derivative of its angular position with respect to time.

Key Concept 2: To find the average angular acceleration of a rotating rigid body, first find the change in its angular velocity (final angular velocity minus initial angular velocity) during a time interval. Then divide the result by the time interval.

Key Concept 3: The relationships among angular position θ, angular velocity ω_z, and angular acceleration α_z for a rigid body rotating with constant angular acceleration are the same as the relationships among position x, velocity v_x, and acceleration a_x for an object moving in a straight line with constant linear acceleration.

Key Concept 4: Points on a rigid body have a *centripetal* (radial) acceleration component $a_{rad} = \omega^2 r$ whenever the rigid body is rotating; they have a *tangential* acceleration component $a_{tan} = r\alpha$ *only* if the angular speed ω is changing. These two acceleration components are perpendicular, so you can use the Pythagorean theorem to relate them to the magnitude of the acceleration.

Key Concept 5: If a rotating rigid body is also moving as a whole through space, use vector addition to find the velocity of a point on the rigid body.

Key Concept 6: The moment of inertia I of a rigid body about an axis depends on the position and orientation of the axis. The value of I for a given rigid body can be very different for different axes.

Key Concept 7: When using energy methods to solve problems about rotating rigid bodies, follow the same general steps as in Chapter 7, but include any rotational kinetic energy, $K = \tfrac{1}{2} I \omega^2$.

Key Concept 8: When a block is attached to a string that wraps around a cylinder or pulley of radius R, the speed v of the block is related to the angular speed ω of the cylinder or pulley by $v = R\omega$. You can use this to find the combined kinetic energy of the two objects.

Key Concept 9: You can use the parallel-axis theorem to relate the moment of inertia of a rigid body about any axis to the moment of inertia of the same rigid body through a parallel axis through its center of mass.

Key Concept 10: Use integration to calculate the moment of inertia of a rigid body that is a continuous distribution of mass. If the body is symmetrical, divide it into volume elements that make use of its symmetry.

Key Concept 11: When calculating a moment of inertia by integration, in general use geometry to determine the size and moment of inertia of each volume element.

Conceptual Questions

1: Rolling versus sliding down a hill

A ball travels down a hill. Will the ball reach the bottom of the hill faster if it rolls or if it slides without friction down the hill?

IDENTIFY, SET UP, AND EXECUTE No energy is lost as the ball travels down the hill, so we can use energy conservation to answer this question. At the top of the hill, the ball has gravitational potential energy. As the ball descends, the gravitational potential energy is transformed into kinetic energy. When the ball rolls down the hill, the kinetic energy is shared between translational kinetic energy and rotational kinetic energy. When the ball slides down the hill without friction, the gravitational potential energy transforms into translational kinetic energy alone. There is no rotational kinetic energy in this case. Therefore, more energy is transformed into translational kinetic energy if the ball slides without friction than if it rolls down the hill. Since it acquires more translational kinetic energy, its velocity is higher and it reaches the bottom faster when it slides without friction.

EVALUATE This question shows how we must include both translational and rotational kinetic energy in our energy analyses. We'll practice using rotational kinetic energy in the problem section.

2: Comparing moments of inertia

A light rod of length L has two lead weights of mass M attached at both ends of the rod. How does the system's moment of inertia compare when the rod is spun about an axis at its center as opposed to when it is spun around a point one-quarter along the length of the rod?

IDENTIFY, SET UP, AND EXECUTE The moment of inertia of an object is

$$I = m_1 r_1^2 + m_2 r_2^2 + g.$$

The moment of inertia when the axis is at the center of the rod is then

$$I_{\text{center axis}} = M\left(\frac{L}{2}\right)^2 + M\left(\frac{L}{2}\right)^2 = M\left(\frac{L^2}{2}\right) = \tfrac{1}{2} ML^2,$$

since each mass is positioned half of the length of the rod away from the axis. When the axis is one-quarter along the length of the rod, the moment of inertia is

$$I_{1/4 \text{ along rod}} = M\left(\frac{L}{4}\right)^2 + M\left(\frac{3L}{4}\right)^2 = M\left(\frac{L^2}{16}\right) + M\left(\frac{9L^2}{16}\right) = \tfrac{10}{16} ML^2,$$

since one mass is $\frac{1}{4}L$ from the axis and the other is $\frac{3}{4}L$ from the axis. The moment of inertia when the rod is spun one-quarter along its length is 25% larger than the moment of inertia when the rod is spun at the center.

KEY CONCEPT 6 **EVALUATE** The moment of inertia depends on both mass and the location of the mass. Since the location from the axis enters as the square of the distance, moving the axis changes the moment of inertia. Do you get the same result if you apply the parallel-axis theorem?

Practice Problem: What axis of rotation provides the largest moment of inertia? *Answer*: The axis located at one end of the rod gives the largest moment of inertia, $I = ML^2$.

3: Rolling up a ramp

A solid sphere and a thin-walled sphere roll without slipping along a horizontal surface. The two spheres roll with the same translational speed. The surface leads to a ramp. Which sphere rises to the greatest height on the ramp before stopping momentarily?

IDENTIFY AND SET UP We'll use conservation of energy and ignore air drag. Both spheres have initial translational and rotational kinetic energies that are transformed completely into gravitational potential energy when they stop momentarily on the ramp. The sphere with the greatest initial total kinetic energy will rise to the greatest height.

EXECUTE The initial kinetic energy of each sphere is

$$K_i = \tfrac{1}{2} m_{\text{sphere}} v_{\text{cm}}^2 + \tfrac{1}{2} I_{\text{sphere}} \omega^2.$$

The angular velocity is related to the velocity of the center of mass, since both spheres roll without slipping. Thus,

$$\omega = \frac{v_{\text{cm}}}{R},$$

where R is the radius of the sphere. The moment of inertia of the solid sphere is $I_{\text{solid}} = 2/5 \, m_{\text{solid}} R_{\text{solid}}^2$. The kinetic energy of the solid sphere is then

$$K_{\text{solid}} = \tfrac{1}{2} m_{\text{solid}} v_{\text{cm}}^2 + \tfrac{1}{2} I_{\text{solid}} \omega^2 = \tfrac{1}{2} m_{\text{solid}} v_{\text{cm}}^2 + \tfrac{1}{2} \tfrac{2}{5} m_{\text{solid}} R_{\text{solid}}^2 \left(\frac{v_{\text{cm}}}{R_{\text{solid}}}\right)^2 = \tfrac{7}{10} m_{\text{solid}} v_{\text{cm}}^2.$$

The moment of inertia of the thin-walled sphere (shell) is $I_{\text{shell}} = 2/3 \, m_{\text{shell}} R_{\text{shell}}^2$. The kinetic energy of the shell is then

$$K_{\text{shell}} = \tfrac{1}{2} m_{\text{shell}} v_{\text{cm}}^2 + \tfrac{1}{2} I_{\text{shell}} \omega^2 = \tfrac{1}{2} m_{\text{shell}} v_{\text{cm}}^2 + \tfrac{1}{2} \tfrac{2}{3} m_{\text{shell}} R_{\text{shell}}^2 \left(\frac{v_{\text{cm}}}{R_{\text{shell}}}\right)^2 = \tfrac{5}{6} m_{\text{shell}} v_{\text{cm}}^2.$$

Since the final gravitational potential energy depends on mass ($U = mgh$), the masses will cancel and we compare the leading fractions in the kinetic-energy terms to determine which sphere has the greatest initial kinetic energy. We see that the shell has more initial kinetic energy; therefore, the thin-walled sphere rises to the greatest height.

EVALUATE It is interesting to find that the results don't depend on either the mass or the radius of the two spheres. The results depend only on how the mass is distributed in the object.

4: Racing down a ramp

A thin-walled hollow cylinder, a solid cylinder, a solid sphere, and a thin-walled sphere start from rest at the same height and roll without slipping down a wide ramp. Rank the velocities of the four objects, from first to last.

IDENTIFY AND SET UP We'll use conservation of energy. All of the objects have gravitational potential energy that is transformed into translational and rotational kinetic energies. We'll find the velocity at the bottom of the ramp in terms of the height and other factors.

EXECUTE For any of the four objects, energy conservation applied to the starting point and the bottom of the ramp gives

$$mgh = \tfrac{1}{2}mv^2 + \tfrac{1}{2}I\omega^2.$$

The angular velocity is related to the velocity of the center of mass, since all of the objects roll without slipping; thus,

$$\omega = \frac{v}{R},$$

where R is the radius of the object. Replacing the angular velocity yields

$$mgh = \tfrac{1}{2}mv^2 + \tfrac{1}{2}I\left(\frac{v}{R}\right)^2.$$

We now solve for the velocity of each of the four objects. The thin-walled hollow cylinder (TWHC) gives

$$m_{TWHC}gh = \tfrac{1}{2}m_{TWHC}v_{TWHC}^2 + \tfrac{1}{2}\left(m_{TWHC}R^2\right)\left(\frac{v_{TWHC}}{R}\right)^2 = m_{TWHC}v_{TWHC}^2,$$

$$v_{TWHC} = \sqrt{gh}.$$

The solid cylinder (SC) results in

$$m_{SC}gh = \tfrac{1}{2}m_{SC}v_{SC}^2 + \tfrac{1}{2}\left(\tfrac{1}{2}m_{SC}R^2\right)\left(\frac{v_{SC}}{R}\right)^2 = \tfrac{3}{4}m_{SC}v_{SC}^2,$$

$$v_{SC} = \sqrt{\tfrac{4}{3}}\sqrt{gh}.$$

The solid sphere (SS) produces

$$m_{SS}gh = \tfrac{1}{2}m_{SS}v_{SS}^2 + \tfrac{1}{2}\left(\tfrac{2}{5}m_{SS}R^2\right)\left(\frac{v_{SS}}{R}\right)^2 = \tfrac{7}{10}m_{SS}v_{SS}^2,$$

$$v_{SS} = \sqrt{\tfrac{10}{7}}\sqrt{gh}.$$

The thin-walled hollow sphere (TWHS) gives

$$m_{TWHS}gh = \tfrac{1}{2}m_{TWHS}v_{TWHS}^2 + \tfrac{1}{2}\left(\tfrac{2}{3}m_{TWHS}R^2\right)\left(\frac{v_{TWHS}}{R}\right)^2 = \tfrac{5}{6}m_{TWHS}v_{TWHS}^2,$$

$$v_{TWHS} = \sqrt{\tfrac{6}{5}}\sqrt{gh}.$$

Comparing the leading factors, we see that the solid sphere has the largest velocity, followed by the solid cylinder, the thin-walled hollow sphere, and the thin-walled hollow cylinder.

KEY CONCEPT 7 **EVALUATE** The largest moment of inertias resulted in the smallest velocities at the bottom of the ramp, since energy went into rotational kinetic energy. We see that the radii and masses cancel in this problem: The velocity at the bottom is dependent only upon the distribution of mass in the object.

In what order do the four objects reach the bottom? Since all the velocities depend on the square root of the height, those with the fastest velocity reach the bottom first.

Problems

1: Constant angular acceleration in a pottery wheel

A pottery wheel is rotating with an initial angular velocity ω_0 when the wheel's drive motor is turned on. The wheel increases to a final angular velocity of 125 rpm while making 30.0 revolutions in 25.0 seconds. Find the initial angular velocity and angular acceleration, assuming that the latter is constant. A pottery wheel is essentially a cylinder rotating about a vertical axis driven by a motor.

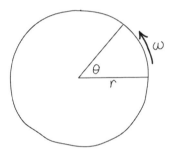

Figure 1 Problem 1 sketch.

IDENTIFY The wheel exhibits constant angular acceleration, so we use the equations for constant angular acceleration to solve the problem. The target variables are the initial angular velocity and the angular acceleration.

SET UP Figure 1 shows a sketch of the pottery wheel. For consistency, we'll use radians and seconds as our units and convert the given quantities.

We are given the final angular velocity, the angular displacement, and the time. We need to find the initial angular velocity and the angular acceleration. None of the equations for constant angular acceleration allow us to solve for both unknowns at once, so we'll solve for the initial angular velocity first and then use the results to solve for the angular acceleration.

EXECUTE The angular displacement can be written in terms of the average angular velocity and the time interval as

$$\theta - \theta_0 = \tfrac{1}{2}(\omega_z + \omega_{0z})t.$$

Solving for the initial angular velocity gives

$$\omega_{0z} = \frac{2(\theta - \theta_0)}{t} - \omega_z.$$

The quantity $(\theta - \theta_0)$ is our angular displacement, 30.0 revolutions. The quantity ω_z is the final angular velocity, 125 rpm. We convert the revolutions to radians by multiplying by

(2π rad/rev) and the rpm to radians/second by multiplying by (2π rad/rev) (1 min/60 s). Solving for ω_{0z} gives

$$\omega_{0z} = \frac{2(30.0 \text{ rev})(2\pi \text{ rad/rev})}{25.0 \text{ s}} - (125 \text{ rpm})(2\pi \text{ rad/rev})(1 \text{ min}/60 \text{ s}) = 1.99 \text{ rad/s}.$$

The initial angular velocity is 1.99 rad/s, or 19.0 rpm. To find the angular acceleration, we use the relationship between angular velocity, angular acceleration, and time:

$$\omega_z = \omega_{0z} + \alpha_z t.$$

Solving for the angular acceleration, we obtain

$$\alpha_z = \frac{\omega_z - \omega_{0z}}{t} = \frac{(125 \text{ rpm})(2\pi \text{ rad/rev})(1 \text{ min}/60 \text{ s}) - (1.99 \text{ rad/s})}{25.0 \text{ s}} = 0.444 \text{ rad/s}^2.$$

The angular acceleration is 0.444 rad/s^2.

KEY CONCEPT 3 **EVALUATE** This problem reminds us of the problems involving constant linear acceleration. The same problem-solving strategy applies: Draw a diagram, check for constant acceleration, find one or more equations that can be used to solve for the unknowns, and reflect upon the results. We may need to use more than one equation for the solution, and we must watch our units carefully.

Practice Problem: What would the angular acceleration be if the final angular velocity was 100 rpm? *Answer:* 0.237 rad/s^2

Extra Practice: If you keep the motor running for another 10 s, then what final angular velocity does the wheel have? *Answer:* 123 rpm

2: A slowing pottery wheel

A pottery wheel rotating at 30 revolutions per minute is shut off and slows uniformly, coming to a stop in 2 complete revolutions. Find the angular acceleration and the time it takes to come to a stop.

IDENTIFY The wheel exhibits constant angular acceleration as it slows uniformly, so we use the equations for constant angular acceleration to solve the problem. The target variables are the angular acceleration and the time required to stop.

SET UP We're given the initial angular velocity, the final angular velocity (zero), and the angular displacement for the wheel to come to a stop. We'll use two of the equations for constant angular acceleration to solve for the two unknowns.

EXECUTE The angular displacement, velocity, and acceleration are related by the formula

$$\omega_z^2 = \omega_{0z}^2 + 2\alpha_z(\theta - \theta_0).$$

Solving for the angular acceleration gives

$$\alpha_z = \frac{\omega_z^2 - \omega_{0z}^2}{2(\theta - \theta_0)} = \frac{(0)^2 - (30 \text{ rev/min})^2}{2(2 \text{ rev})} = -225 \text{ rev/min}^2 \left(\frac{2\pi \text{ rad}}{\text{rev}}\right)^2 \left(\frac{1 \text{ min}}{60 \text{ s}}\right)^2 = -0.393 \text{ rad/s}^2.$$

To solve for time, we use the acceleration–velocity relation:

$$\omega_z = \omega_{0z} + \alpha_z t.$$

Solving for the time yields

$$t = \frac{\omega_z - \omega_{0z}}{\alpha_z} = \frac{(0) - (30 \text{ rev/min})}{-225 \text{ rev/min}^2} = 0.133 \text{ min} = 8.0 \text{ s}.$$

The wheel slows to a stop at a rate of -0.393 rad/s^2, stopping in 8.0 s.

KEY CONCEPT 3 **EVALUATE** This problem reminds us of problems involving constant linear acceleration. You should be able to become proficient at these problems rather quickly. Just make sure that you watch your units!

Practice Problem: How long would it take to stop if the initial angular velocity was 40.0 rpm? *Answer:* 10.7s

Extra Practice: How many revolutions did the wheel turn in this time? *Answer:* 3.55 revolutions

3: Energy in a wheel–stone system

A thin light string is wrapped around the rim of a spoked wheel that can rotate without friction around its center axle. An 8.00 kg stone is attached to the end of the string as shown in Figure 2. If the stone is released from rest, how far does it travel before attaining a speed of 4.80 m/s? The spoked wheel is made of a central hub (a solid uniform cylinder of radius 7.50 cm and mass 22.0 kg) attached to a rim (a thin-walled hollow cylinder of radius 30.0 cm mass 12.0 kg) by spokes of negligible mass.

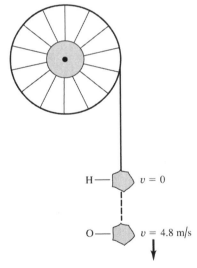

Figure 2 Problem 3.

IDENTIFY No work is done by external forces, so mechanical energy is conserved. The target variable is the height the stone falls.

SET UP Initially, there is only gravitational potential energy. When the stone is released, the gravitational potential energy is transformed into kinetic energy of the stone and wheel. We'll set the origin at the point where the stone reaches a speed of 4.80 m/s; the starting position is a distance H above the origin, as we see in Figure 2. We'll find the moment of inertia of the wheel by combining the moment of inertia of a solid cylinder with the moment of inertia of a thin-walled hollow cylinder.

EXECUTE Energy conservation relates the initial and final energies:

$$K_1 + U_1 = K_2 + U_2.$$

Initially, there is only U_{grav}. At the origin, the gravitational potential energy has been transformed into the kinetic energies of the stone and wheel:

$$U_{grav} = K_{stone} + K_{wheel}.$$

Replacing the energies yields

$$m_{stone}gH = \tfrac{1}{2}m_{stone}v^2 + \tfrac{1}{2}I_{wheel}\omega^2.$$

We need to find the moment of inertia of the wheel and its angular velocity when the stone reaches its final velocity. The moment of inertia of the wheel is the algebraic sum of the moment of inertia of the central hub plus the moment of inertia of the outer rim (the thin-walled cylinder.) Using Table 2 of the text, we find the total moment of inertia of the wheel:

$$I_{wheel} = I_{solid\ cylinder} + I_{thin\text{-walled cylinder}} = \tfrac{1}{2}M_{hub}R_{hub}^2 + M_{rim}R_{rim}^2.$$

The speed of the stone is the tangential speed of the wheel, so we can find the angular speed of the wheel from the equation

$$\omega = \frac{v}{R_{rim}}.$$

Substituting the two expressions we found into the energy relation gives

$$m_{stone}gH = \tfrac{1}{2}m_{stone}v^2 + \tfrac{1}{2}\left(\tfrac{1}{2}M_{hub}R_{hub}^2 + M_{rim}R_{rim}^2\right)\left(\frac{v}{R_{rim}}\right)^2.$$

Solving for H results in

$$H = \frac{1}{m_{stone}g}\mathsf{B}\tfrac{1}{2}m_{stone}v^2 + \tfrac{1}{2}\left(\tfrac{1}{2}M_{hub}R_{hub}^2 + M_{rim}R_{rim}^2\right)\left(\frac{v}{R_{rim}}\right)^2 \mathsf{R},$$

$$H = \frac{1}{(8.00\text{ kg})(9.8\text{ m/s}^2)}\mathsf{B}\tfrac{1}{2}(8.00\text{ kg})(4.80\text{ m/s})^2 + \tfrac{1}{2}\left(\tfrac{1}{2}(22.0\text{ kg})(0.0750\text{ m})^2 + (12.0\text{ kg})(0.300\text{ m})^2\right)\left(\frac{4.80\text{ m/s}}{0.300\text{ m}}\right)^2 \mathsf{R} = 3.04\text{ m}.$$

After the stone falls 3.04 m, its speed will be 4.80 m/s.

KEY CONCEPT 8 **EVALUATE** If the stone had fallen freely, it would have attained the final speed when $h = v/\sqrt{2g}$ (1.08 m). Why does it take almost three times this distance to reach the final speed? Looking at the energy conservation equation, we see that the energy is shared between the kinetic energies of the stone and wheel. Roughly two-thirds of the energy goes into the kinetic energy of the wheel.

Practice Problem: Repeat the problem with only the inner hub of the wheel. *Answer:* 2.78 m.

Extra Practice: How much rotational energy is there at this point? *Answer:* 126 J

4: Energy in a falling cylinder

A thin, light string is wrapped around a solid uniform cylinder of mass M and radius R as shown in Figure 3. The string is held stationary and the cylinder is released from rest. What is the cylinder's radius if it reaches an angular speed of 350.0 rpm after it falls 3.00 m?

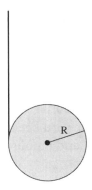

Figure 3 Problem 4.

IDENTIFY Gravity and tension are the only forces acting in this problem, so energy is conserved. The target variable is the cylinder's radius.

SET UP Initially, there is only gravitational potential energy. When the cylinder is released, the gravitational potential energy is transformed into rotational and translational kinetic energy. We'll set the origin 3.00 m below the initial position.

EXECUTE Energy conservation relates the initial and final energies:

$$K_1 + U_1 = K_2 + U_2.$$

Initially, there is only U_{grav}. At the origin, the gravitational potential energy has been transformed into the total kinetic energy of the cylinder:

$$U_{grav} = K_{translational} + K_{rotational}.$$

Replacing the energies, we obtain

$$Mgh = \tfrac{1}{2}Mv^2 + \tfrac{1}{2}I\omega^2.$$

Now, recall that the moment of inertia of a solid uniform cylinder is $\tfrac{1}{2}MR^2$. (See Table 2 in the text.) The speed of the cylinder is the tangential speed of the wheel, so we can find the angular speed of the wheel from the relation

$$v = \omega R.$$

Substituting the preceding expressions into the energy relation gives

$$Mgh = \tfrac{1}{2}Mv^2 + \tfrac{1}{2}I\omega^2 = \tfrac{1}{2}M(\omega R)^2 + \tfrac{1}{2}\left(\tfrac{1}{2}MR^2\right)\omega^2 = \tfrac{3}{4}M(\omega R)^2.$$

Solving for R, we get

$$R = \frac{\sqrt{\tfrac{4}{3}gh}}{\omega} = \frac{\sqrt{\tfrac{4}{3}(9.80 \text{ m/s}^2)(3.00 \text{ m})}}{(350 \text{ rpm})(2\pi \text{ rad/rev})(1 \text{ min}/60 \text{ s})} = 0.171 \text{ m}.$$

The cylinder's radius is 17.1 cm.

KEY CONCEPT 7 **EVALUATE** We see that the mass of the cylinder cancels in the conservation-of-energy equation; the results apply to a cylinder of any mass.

Practice Problem: Repeat the problem with a thin hoop replacing the cylinder. *Answer:* 14.8 cm.

Extra Practice: What is the radius of the thin hoop if it only takes 1.0 m to reach that speed? *Answer:* 8.54 cm

5: Cylinder rolling up a ramp

A uniform hollow cylinder rolls along a horizontal floor and then up a flat ramp without slipping. The ramp is inclined at 20.0°. How far along the ramp does the cylinder roll before stopping if its initial forward speed is 12.0 m/s? The hollow cylinder has a mass of 6.5 kg, an inner radius of 0.13 m, and an outer radius of 0.25 m.

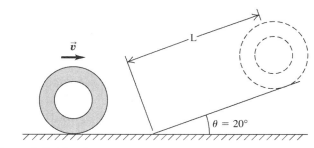

Figure 4 Problem 5.

IDENTIFY There are no nonconservative forces (ignoring air drag), so we use energy conservation. The target variable is the distance up the ramp the cylinder travels before stopping momentarily.

SET UP Figure 4 shows a sketch of the problem. Initially, the cylinder has translational and rotational kinetic energy. When the cylinder stops momentarily at the top of the ramp, the kinetic energy has been totally transformed to gravitational potential energy. We'll set the origin at the base of the ramp.

EXECUTE Energy conservation relates the initial and final energies:

$$K_1 + U_1 = K_2 + U_2.$$

Initially, the cylinder has only kinetic energy. At its highest point, the cylinder has only gravitational potential energy. Thus,

$$K_{\text{translational}} + K_{\text{rotational}} + 0 = 0 + U_{\text{grav}}.$$

Replacing the energies with their equivalent expressions yields

$$\tfrac{1}{2}Mv^2 + \tfrac{1}{2}I\omega^2 = Mgy.$$

Recall that the moment of inertia of a hollow uniform cylinder is $\tfrac{1}{2}M(R_1^2 + R_2^2)$. (See Table 2 in the text.) The speed of the cylinder is the tangential speed of the wheel at the outer radius, so we can replace the speed with

$$\omega = \frac{v}{R_2}.$$

Substituting these expressions into the energy relation gives

$$\tfrac{1}{2}Mv^2 + \tfrac{1}{2}I\omega^2 = \tfrac{1}{2}Mv^2 + \tfrac{1}{2}\left(\tfrac{1}{2}M\left(R_1^2 + R_2^2\right)\right)\left(\frac{v}{R_2}\right)^2 = Mgy.$$

Solving for y, the maximum vertical height, we obtain

$$y = \frac{v^2}{g}\text{B}\tfrac{1}{2} + \tfrac{1}{4}\left(\frac{R_1^2 + R_2^2}{R_2^2}\right)\text{R} = \frac{(12.0 \text{ m/s})^2}{(9.8 \text{ m/s}^2)}\text{B}\tfrac{1}{2} + \tfrac{1}{4}\left(\frac{(0.13 \text{ m})^2 + (0.25 \text{ m})^2}{(0.25 \text{ m})^2}\right)\text{R} = 12.0 \text{ m}.$$

The cylinder's maximum vertical height is 12.0 m. To find the distance along the ramp, we use the sine relation:

$$L = \frac{y}{\sin 20°} = \frac{(12.0 \text{ m})}{\sin 20°} = 35.1 \text{ m}.$$

The hollow cylinder rolls 35.1 m up along the ramp.

KEY CONCEPT 7 **EVALUATE** How far up the ramp would the cylinder travel without friction? It would move 21.5 m along the ramp without friction. This result shows how the added initial rotational kinetic energy results in a greater distance along the ramp.

Practice Problem: How far up the ramp would a solid cylinder of radius 0.25 m travel? *Answer:* 32.2 m

Extra Practice: What if the solid cylinder had the same mass but a radius of 0.3 m? *Answer:* 32.2 m

6: Moment of inertia of a rectangular sheet

Find the moment of inertia of a uniform thin rectangular sheet of metal with mass M, length L, and width W about the x-axis in Figure 5.

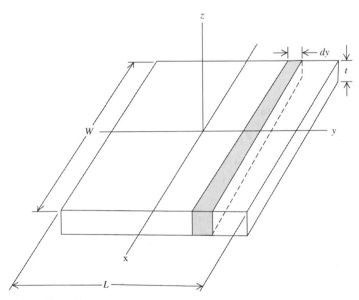

Figure 5 Problem 6.

IDENTIFY The sheet is a continuous distribution of mass, so we must integrate to find the moment of inertia. We break the sheet up into thin strips along the x-axis as shown. The target variable is the moment of inertia.

SET UP The mass density of the sheet is the total mass divided by the volume, or

$$\rho = \frac{M}{WLt},$$

where t is the thickness of the sheet. The volume of the thin strip along the x-axis is

$$dV = Wt\,dy.$$

The mass of the strip is the density multiplied by the volume. We'll integrate the product of the mass and r^2 to find the moment of inertia.

EXECUTE The moment of inertia is given by

$$I = \int r^2\,dm.$$

We'll integrate from $-L/2$ to $+L/2$, since the sheet is centered on the origin. Replacing dm and adding the limits of integration gives

$$I = \int_{-L/2}^{L/2} r^2 \rho\,dV$$

$$= \int_{-L/2}^{L/2} y^2 \frac{M}{WLt} Wt\,dy$$

$$= \frac{M}{L} \int_{-L/2}^{L/2} y^2\,dy$$

$$= \frac{M}{L} \frac{y^3}{3} \Big|_{-L/2}^{L/2} = \frac{M}{L}\left[\frac{(L/2)^3}{3} - \frac{(-L/2)^3}{3}\right]$$

$$= \frac{ML^2}{12}.$$

The moment of inertia of the sheet rotating about the x-axis is $ML^2/12$.

KEY CONCEPT 10 **EVALUATE** We see that the moment of inertia does not depend on the width or thickness of the sheet when it is rotated about the x-axis. If we look at other shapes in Table 2, we see that neither does the moment of inertia depend upon dimensions along the axis of rotation.

Practice Problem: Use the parallel-axis theorem to check that this result agrees with the moment of inertia of a thin rectangular plate rotated about the edge.

Extra Practice: What is the moment of inertia of a thin rectangular plate rotated about a line halfway between the edge and the x-axis? *Answer:* $7\,ML^2/48$

Try It Yourself!

1: Motion of flywheel

A flywheel is a disk-shaped mass that rotates about its central perpendicular axis. A flywheel 1.0 m in diameter rotates with an initial angular velocity of 500 rpm. It increases its speed to 1000 rpm in 20.0 s. Assuming constant acceleration, find the angular acceleration and the angular displacement of the flywheel as it increases its angular speed from 500 to 1000 rpm.

Solution Checkpoints

IDENTIFY AND SET UP Determine the target variables and identify the appropriate constant-angular-acceleration equations needed to find the target variables.

EXECUTE The angular acceleration is found from the relation

$$\omega_z = \omega_{0z} + \alpha_z t.$$

This gives an angular acceleration of 1500 rev/min^2. The angular displacement is found from the formula

$$\omega_z^2 = \omega_{0z}^2 + 2\alpha_z(\theta - \theta_0).$$

This gives an angular displacement of 250 rev.

KEY CONCEPT 2 **EVALUATE** We see that this problem closely parallels the linear kinematics problems. We'll continue to use angular kinematics as we investigate rotations.

Practice Problem: What is the angular displacement if it takes 30.0 s to increase the speed?
Answer: 375 revolutions

Extra Practice: What is the angular velocity 45 seconds later? *Answer:* 1750 rpm

2: Energy in a grinding wheel

How much energy is dissipated when a 2.0 kg grinding wheel of radius 0.1 m is brought to rest from an initial angular velocity of 3000 rpm? What is the average power dissipated if the wheel stops in 10 rev? Assume constant angular acceleration.

Solution Checkpoints

IDENTIFY AND SET UP The grinding wheel loses all of its energy as it stops, so you must find the initial kinetic energy. The moment of inertia of the wheel is that of a disk. To find the power, you must find the time it takes the wheel to stop and divide the energy by the time. The target variables are the energy and the power.

EXECUTE The initial kinetic energy of the grinding wheel is given by

$$K = \tfrac{1}{4} MR^2 \omega^2.$$

The final kinetic energy is zero, so the wheel loses 493 J.

We find the time it takes the wheel to stop by directly combining two kinematics equations:

Study Guide for Rotation of Rigid Bodies

$$\omega_z^2 = \omega_{0z}^2 + 2\alpha_z(\theta - \theta_0) \text{ and}$$
$$\omega_z = \omega_{0z} + \alpha_z t.$$

These equations result in a time of 0.40 s. The power is then the energy lost divided by the time, giving 1.23 kW.

EVALUATE This problem combines energy, kinematics, and power. How can we best check our results?

Practice Problem: How much time has passed after 5 revolutions? *Answer:* 0.117 s

Extra Practice: What is the rotational kinetic energy at this point? *Answer:* 246 J

Key Example Variation Problems

Solutions to these problems are in Chapter 9 of the Student's Solutions Manual.

Be sure to review EXAMPLE 9.3 (Section 9.2) before attempting these problems.

VP9.3.1 A machine part is initially rotating at 0.500 rad/s. Its rotation speeds up with constant angular acceleration 2.50 rad/s². Through what angle has the machine part rotated when its angular speed equals 3.25 rad/s? Give your answer in (a) radians, (b) degrees, and (c) revolutions.

VP9.3.2 The rotor of a helicopter is gaining angular speed with constant angular acceleration. At $t = 0$ it is rotating at 1.25 rad/s. From $t = 0$ to $t = 2.00$ s, the rotor rotates through 8.00 rad. (a) What is the angular acceleration of the rotor? (b) Through what angle (in radians) does the rotor rotate from $t = 0$ to $t = 4.00$ s?

VP9.3.3 A jeweler's grinding wheel slows down at a constant rate from 185 rad/s to 105 rad/s while it rotates through 16.0 revolutions. How much time does this take?

VP9.3.4 A disk rotates around an axis through its center that is perpendicular to the plane of the disk. The disk has a line drawn on it that extends from the axis of the disk to the rim. At $t = 0$ this line lies along the x-axis and the disk is rotating with positive angular velocity ω_{0z}. The disk has constant positive angular acceleration α_z. At what time after $t = 0$ has the line on the disk rotated through an angle θ?

Be sure to review EXAMPLES 9.4 and 9.5 (Section 9.3) before attempting these problems.

VP9.5.1 Shortly after a vinyl record (radius 0.152 m) starts rotating on a turntable, its angular velocity is 1.60 rad/s and increasing at a rate of 8.00 rad/s². At this instant, for a point at the rim of the record, what are (a) the tangential component of acceleration, (b) the centripetal component of acceleration, and (c) the magnitude of acceleration?

VP9.5.2 A superhero swings a magic hammer over her head in a horizontal plane. The end of the hammer moves around a circular path of radius 1.50 m at an angular speed of 6.00 rad/s. As the superhero swings the hammer, she then ascends vertically at a constant 2.00 m/s. (a) What is the speed of the end of the hammer relative to the ground? (b) What is the acceleration (magnitude and direction) of the end of the hammer?

VP9.5.3 If the magnitude of the acceleration of a propeller blade's tip exceeds a certain value a_{max}, the blade tip will fracture. If the propeller has radius r, is initially at rest, and has angular acceleration of magnitude α, at what angular speed ω will the blade tip fracture?

VP9.5.4 At a certain instant, a rotating turbine wheel of radius R has angular speed ω (measured in rad/s). (a) What must be the magnitude α of its angular acceleration (measured in rad/s^2) at this instant if the acceleration vector $\vec{a}$ of a point on the rim of the wheel makes an angle of exactly 30° with the velocity vector $\vec{v}$ of that point? (b) At this same instant, what is the angle between $\vec{a}$ and $\vec{v}$ for a point on the wheel halfway between the axis of rotation and the rim?

Be sure to review EXAMPLES 9.7 and 9.8 (Section 9.4) before attempting these problems.

VP9.8.1 A solid cylinder of mass 12.0 kg and radius 0.250 m is free to rotate without friction around its central axis. If you do 75.0 J of work on the cylinder to increase its angular speed, what will be its final angular speed if the cylinder (a) starts from rest; (b) is initially rotating at 12.0 rad/s?

VP9.8.2 A square plate has mass 0.600 kg and sides of length 0.150 m. It is free to rotate without friction around an axis through its center and perpendicular to the plane of the plate. How much work must you do on the plate to change its angular speed (a) from 0 to 40.0 rad/s and (b) from 40.0 rad/s to 80.0 rad/s?

VP9.8.3 A hollow cylinder of mass 2.00 kg, inner radius 0.100 m, and outer radius 0.200 m is free to rotate without friction around a horizontal shaft of radius 0.100 m along the axis of the cylinder. You wrap a light, nonstretching cable around the cylinder and tie the free end to a 0.500 kg block of cheese. You release the cheese from rest a distance h above the floor. (a) If the cheese is moving downward at 4.00 m/s just before it hits the ground, what is the value of h? (b) What is the angular speed of the cylinder just before the cheese hits the ground?

VP9.8.4 A pulley in the shape of a solid cylinder of mass 1.50 kg and radius 0.240 m is free to rotate around a horizontal shaft along the axis of the pulley. There is friction between the pulley and this shaft. A light, nonstretching cable is wrapped around the pulley, and the free end is tied to a 2.00 kg textbook. You release the textbook from rest a distance 0.900 m above the floor. Just before the textbook hits the floor, the angular speed of the pulley is 10.0 rad/s. (a) What is the speed of the textbook just before it hits the floor? (b) How much work was done on the pulley by the force of friction while the textbook was falling to the floor?

STUDENT'S SOLUTIONS MANUAL FOR ROTATION OF RIGID BODIES

VP9.3.1. **IDENTIFY:** We are dealing with angular motion having constant angular acceleration, so the constant angular acceleration formulas apply.

SET UP: A $\omega_z^2 = \omega_{0z}^2 + 2\alpha_z(\theta - \theta_0)$

EXECUTE: (a) $(3.25 \text{ rad/s})^2 = (0.500 \text{ rad/s})^2 + 2(2.50 \text{ rad/s}^2)(\theta - \theta_0) \rightarrow \theta - \theta_0 = 2.06$ rad.

(b) Convert using π rad = 180°: $2.06 \text{ rad} \left(\dfrac{180°}{\pi \text{ rad}} \right) = 118°$.

(c) Convert using 1 rev = 2π rad: $2.06 \text{ rad} \left(\dfrac{1 \text{ rev}}{2\pi \text{ rad}} \right) = 0.328$ rev.

EVALUATE: The conversion factors in (b) and (c) are very useful to remember.

VP9.3.2. **IDENTIFY:** The constant angular acceleration formulas apply.

SET UP: $\theta - \theta_0 = \omega_{0z}t + \dfrac{1}{2}\alpha_z t^2$ applies.

EXECUTE: (a) Use $\theta - \theta_0 = \omega_{0z}t + \dfrac{1}{2}\alpha_z t^2$ to find α_z.

$8.00 \text{ rad} = (1.25 \text{ rad/s})(2.00 \text{ s}) + \dfrac{1}{2}\alpha_z (2.00 \text{ s})^2 \rightarrow \alpha_z = 2.75 \text{ rad/s}^2$.

(b) Now use $\theta - \theta_0 = \omega_{0z}t + \dfrac{1}{2}\alpha_z t^2$ again to find $\theta - \theta_0$.

$\theta - \theta_0 = (1.25 \text{ rad/s})(4.00 \text{ s}) + \dfrac{1}{2}(2.75 \text{ rad/s}^2)(4.00 \text{ s})^2 = 27.0$ rad.

EVALUATE: During the first 2.00 s the rotor turned through 8.00 rad and during the first 4.00 s it turned through 27.0 rad. This means that during the *second* 2.00 s it turned through 19.0 rad, which is considerably more than 8.00 rad. The difference shows us that the rotor is speeding up due to angular acceleration, so it turns much farther during the second 2.00 s than during the first 2.00 s.

VP9.3.3. **IDENTIFY:** The constant angular acceleration formulas apply.

SET UP: We need α_z to find the time to slow down from 185 rad/s to 105 rad/s, so we use $\omega_z^2 = \omega_{0z}^2 + 2\alpha_z(\theta - \theta_0)$ and $\omega_z = \omega_{0z} + \alpha_z t$.

EXECUTE: First convert 16.0 rev to rad: 16.0 rev = (16.0)(2π) rad.

Now use $\omega_z^2 = \omega_{0z}^2 + 2\alpha_z(\theta - \theta_0)$ to find α_z.

$(105 \text{ rad/s})^2 = (185 \text{ rad/s})^2 + 2\alpha_z (16.0)(2\pi \text{ rad}) \rightarrow \alpha_z = -115.4 \text{ rad/s}^2$.

Now use $\omega_z = \omega_{0z} + \alpha_z t$ to find the time for the wheel to slow down.
105 rad/s = 185 rad/s + (−115.4 rad/s²)t → t = 0.693 s.
EVALUATE: The minus sign for α_z means that it is in the opposite angular direction from ω_z. This is reasonable because the wheel is slowing down. If α_z and ω_z were in the same direction, the wheel would be increasing its angular speed.

VP9.3.4. **IDENTIFY:** The constant angular acceleration formulas apply.
SET UP: We need the time to turn through a given angle θ, so we use $\theta - \theta_0 = \omega_{0z} t + \frac{1}{2}\alpha_z t^2$.
EXECUTE: The angular displacement is θ, so we use $\theta = \omega_{0z} t + \frac{1}{2}\alpha_z t^2$ and solve for t. Using the quadratic formula gives $t = \dfrac{-\omega_{0z} + \sqrt{\omega_{0z}^2 + 2\alpha_z \theta}}{\alpha_z}$.
EVALUATE: We used the positive square root because the time t should be positive.

VP9.5.1. **IDENTIFY:** We need to relate angular quantities to linear quantities.
SET UP: $a_{tan} = r\alpha$, $a_{rad} = \omega^2 r$, and $a = \sqrt{a_{tan}^2 + a_{rad}^2}$ all apply.
EXECUTE: **(a)** $a_{tan} = r\alpha = (0.152 \text{ m})(8.00 \text{ rad/s}) = 1.22 \text{ m/s}^2$.
(b) $a_{rad} = \omega^2 r = (1.60 \text{ rad/s})^2 (0.152 \text{ m}) = 0.389 \text{ m/s}^2$.
(c) $a = \sqrt{a_{tan}^2 + a_{rad}^2} = \sqrt{(0.122 \text{ m/s}^2)^2 + (0.399 \text{ m/s}^2)^2} = 1.28 \text{ m/s}^2$.
EVALUATE: Careful! In order to use the formulas $a_{tan} = r\alpha$ and $a_{rad} = \omega^2 r$, the angular quantities *must* be in *radian* measure. If they are in revolutions or degrees, you must convert them to radians before using these formulas.

VP9.5.2. **IDENTIFY:** We need to relate angular quantities to linear quantities.
SET UP: $v_{tan} = r\omega$, $a_{tan} = r\alpha$, and $a_{rad} = \omega^2 r$ apply. The magnitude A of a vector is $A = \sqrt{A_x^2 + A_y^2}$.
EXECUTE: **(a)** The end of the hammer has two components to its velocity: a horizontal component due to its rotation and the vertical component of 2.00 m/s. The horizontal component is its tangential speed: $v_{tan} = r\omega = (1.50 \text{ m})(6.00 \text{ rad/s}) = 9.00 \text{ m/s}$. Its speed v is
$v = \sqrt{v_{tan}^2 + v_y^2} = \sqrt{(9.00 \text{ m/s})^2 + (2.00 \text{ m/s})^2} = 9.22 \text{ m/s}$.
(b) $a_{tan} = 0$ and $a_y = 0$. $a_{rad} = \omega^2 r = (6.00 \text{ rad/s})^2 (1.50 \text{ m}) = 54.0 \text{ m/s}^2$. The direction is toward the center of the circular path of the end of the hammer.
EVALUATE: The end of the hammer has a very large acceleration, so it must take a large force to give it this acceleration.

VP9.5.3. **IDENTIFY:** We want to relate angular quantities to linear quantities.
SET UP: $a_{tan} = r\alpha$, $a_{rad} = \omega^2 r$, and $a = \sqrt{a_{tan}^2 + a_{rad}^2}$ apply.
EXECUTE: The maximum acceleration of the blade tip is a_{max}. We need to relate this to ω.
$a_{max} = \sqrt{a_{tan}^2 + a_{rad}^2} = \sqrt{(\omega^2 r)^2 + (r\alpha)^2} = r\sqrt{\omega^4 + \alpha^2}$. Squaring and solving for sa ω gives
$\omega = \left(\dfrac{a_{max}^2}{r^2} - \alpha^2\right)^{1/4}$.
EVALUATE: Our result says that the smaller a_{max}, the smaller ω can be. This is reasonable because it is the fast spin that causes a large acceleration of the tip.

VP9.5.4. **IDENTIFY:** We want to relate angular quantities to linear quantities.

SET UP: $a_{tan} = r\alpha$ and $a_{rad} = \omega^2 r$ apply. The angle between the acceleration vector and the velocity vector is 30°. The velocity vector is tangent to the circular path, so the angle between the acceleration vector and the tangent is also 30°. This tells us that $\tan 30° = a_{rad}/a_{tan}$. (See Fig. VP9.5.4.)

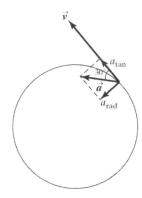

Figure VP9.5.4

EXECUTE: (a) $\tan 30° = \dfrac{a_{rad}}{a_{tan}} = \dfrac{\omega^2 R}{R\alpha} = \dfrac{\omega^2}{\alpha} \rightarrow \dfrac{1}{\sqrt{3}} = \dfrac{\omega^2}{\alpha} \rightarrow \alpha = \omega^2 \sqrt{3}$.

(b) This point is $R/2$ from the center of the circle. Using our work in (a) and using θ for the unknown angle, we see that $\tan\theta = \dfrac{a_{rad}}{a_{tan}} = \dfrac{\omega^2 R/2}{(R/2)\alpha} = \dfrac{\omega^2}{\alpha}$. The result is independent of R, so the angle is 30° as before.

EVALUATE: Careful! In our analysis we cannot say that $\tan 30° = a/v$ because a and v have different units. This would give $\tan 30°$ units of s^{-1}, which is not possible because the $\tan\theta$ is always dimensionless.

VP9.8.1. **IDENTIFY:** We are dealing with the rotation of a solid object having rotational kinetic energy. Work is done on it, so we can apply the work-energy theorem.

SET UP: $K = \dfrac{1}{2}I\omega^2$ for a rotating object, $I = \dfrac{1}{12}M(a^2 + b^2)$ for a rectangular plate, and the work-energy theorem is $W = K_2 - K_1$.

EXECUTE: (a) We want to find the work to change the angular speed.

$W = K_2 = \dfrac{1}{2}I\omega^2 = \dfrac{1}{2}\left[\dfrac{1}{12}M(a^2 + b^2)\right]\omega^2$. In this case, $a = b = 0.150$ m, so the equation reduces to

$W = \dfrac{1}{12}Ma^2\omega^2 = \dfrac{1}{12}(0.600\text{ kg})(0.150\text{ m})^2(40.0\text{ rad/s})^2 = 1.80$ J.

(b) $W = K_2 - K_1 = \dfrac{1}{12}Ma^2(\omega_2^2 - \omega_1^2) = \dfrac{1}{12}(0.600\text{ kg})(0.150\text{ m})^2[(80.0\text{ rad/s})^2 - (40.0\text{ rad/s})^2]$.

$W = 5.4$ J.

EVALUATE: Notice that in both cases, we changed the angular speed by 40.0 rad/s, yet the work required to do this was different. This is because the kinetic energy depends on the *square* of ω. Also note that in all of these formulas relating linear and rotational quantities, the angular measure must always be in terms of *radians*.

VP9.8.2. **IDENTIFY:** We are dealing with the rotation of a solid object having rotational kinetic energy. Work is done on it, so we can apply the work-energy theorem.

SET UP: $K = \frac{1}{2}I\omega^2$ for a rotating object, $I = \frac{1}{2}MR^2$ for a solid cylinder, and the work-energy theorem is $W = K_2 - K_1$.

EXECUTE: **(a)** We know the work and want to find the final angular velocity if $\omega_1 = 0$.

$W = K_2 - K_1 = \frac{1}{2}I\omega^2 = \frac{1}{2}\left[\frac{1}{2}MR^2\right]\omega^2 \rightarrow \omega = \sqrt{\frac{4(75.0 \text{ J})}{(12.0 \text{ kg})(0.250 \text{ m})^2}} = 20.0$ rad/s.

(b) We follow exactly the same procedure except that $K_1 = \frac{1}{2}I\omega_1^2 \neq 0$. This gives

$W = K_2 - K_1 = \frac{1}{2}I\left(\omega_2^2 - \omega_1^2\right) = \frac{1}{2}\left[\frac{1}{2}MR^2\right]\left(\omega_2^2 - \omega_1^2\right)$. Solving for ω_2 and using the same numbers as in (a) except $\omega_1 = 12.0$ rad/s, we get $\omega_2 = 23.2$ rad/s.

EVALUATE: In both cases, 75.0 J of work was done on the cylinder. When starting from rest, the angular velocity increased by 20.0 rad/s, but when starting from 12.0 rad/s, it increased by only 11.2 rad/s (23.2 rad/s – 12.0 rad/s). The kinetic energy increased by the same amount (75.0 J) in both cases, but the increase in the angular velocity was different.

VP9.8.3. **IDENTIFY:** We are dealing with the rotation of a cylinder that is connected to a piece of cheese. As the cheese falls, it causes the cylinder to turn. Work is done on the system by gravity, so we can apply energy conservation. The kinetic energy of the cheese is due to its linear motion and that of the cylinder is due to its rotation.

SET UP: $K = \frac{1}{2}I\omega^2$ for a rotating object and $K = \frac{1}{2}mv^2$ for a moving object. For a hollow cylinder $I = \frac{1}{2}M\left(R_1^2 + R_2^2\right)$ and energy conservation is $U_1 + K_1 + W_{\text{other}} = U_2 + K_2$. Measure y from the floor level, so $U_2 = 0$ and $U_1 = mgy_1 = mgh$. There is no friction, so $W_{\text{other}} = 0$, and $K_1 = 0$. The speed v of the cheese is $v = v_{\text{tan}} = r\omega$, so $\omega = v/R_2$. Call m the mass of the cheese and M the mass of the cylinder.

EXECUTE: **(a)** We want to find h. Energy conservation gives $mgh = \frac{1}{2}I\omega^2 + \frac{1}{2}mv^2$. Putting in the expressions for I and ω and solving for h, we get $h = \dfrac{\frac{M}{4}\left(R_1^2 + R_2^2\right)\left(\dfrac{v}{R_2}\right)^2 + \frac{1}{2}mv^2}{mg}$. Using m = 0.500 kg, M = 2.00 kg, R_1 = 0.100 m, R_2 = 0.200 m, and v = 4.00 m/s, we get h = 2.86 m.

(b) $\omega = v_{\text{tan}}/R_2 = v/R_2 = (4.00 \text{ m/s})/(0.200 \text{ m}) = 20.0$ rad/s.

EVALUATE: Checking for the proper units in the equation for h is a good way to spot algebraic errors in the solution.

VP9.8.4. **IDENTIFY:** We are dealing with the rotation of a cylinder that is connected to a textbook. As the book falls, it causes the cylinder to turn. Work is done on the system by gravity and by friction at the shaft of the cylinder, so we can apply energy conservation. The kinetic energy of the book is due to its linear motion and that of the cylinder is due to its rotation.

SET UP: $K = \frac{1}{2}I\omega^2$ for a rotating object and $K = \frac{1}{2}mv^2$ for a moving object. For a solid cylinder $I = \frac{1}{2}MR^2$ and energy conservation is $U_1 + K_1 + W_{\text{other}} = U_2 + K_2$, where $K_1 = 0$. Measure y from the floor level, so $U_2 = 0$ and $U_1 = mgh$. The speed v of the textbook is $v = v_{tan} = R\omega$. Call m the mass of the book and M the mass of the cylinder.

EXECUTE: **(a)** $v = v_{tan} = R\omega = (0.240 \text{ m})(10.0 \text{ rad/s}) = 2.40 \text{ m/s}$.

(b) From $U_1 + K_1 + W_{other} = U_2 + K_2$ we get $mgh + W_f = \frac{1}{2}I\omega^2 + \frac{1}{2}mv^2$. Solving for W_f using the expressions for v and I gives $W_f = \frac{1}{2}\left(m + \frac{M}{2}\right)(R\omega)^2 - mgh$. Putting in $m = 2.00$ kg, $M = 1.50$ kg, $R = 0.240$ m, and $\omega = 10.0$ rad/s gives $W_f = -9.72$ J.

EVALUATE: Friction does negative work, which is reasonable because it opposes the turning motion at the axle of the cylinder.

9.3. **IDENTIFY:** $\alpha_z(t) = \frac{d\omega_z}{dt}$. Using $\omega_z = d\theta/dt$ gives $\theta - \theta_0 = \int_{t_1}^{t_2} \omega_z dt$.

SET UP: $\frac{d}{dt} t^n = nt^{n-1}$ and $\int t^n dt = \frac{1}{n+1} t^{n+1}$

EXECUTE: **(a)** A must have units of rad/s and B must have units of rad/s³.

(b) $\alpha_z(t) = 2Bt = (3.00 \text{ rad/s}^3)t$. (i) For $t = 0$, $\alpha_z = 0$. (ii) For $t = 5.00$ s, $\alpha_z = 15.0 \text{ rad/s}^2$.

(c) $\theta_2 - \theta_1 = \int_{t_1}^{t_2}(A + Bt^2)dt = A(t_2 - t_1) + \frac{1}{3}B(t_2^3 - t_1^3)$. For $t_1 = 0$ and $t_2 = 2.00$ s,

$\theta_2 - \theta_1 = (2.75 \text{ rad/s})(2.00 \text{ s}) + \frac{1}{3}(1.50 \text{ rad/s}^3)(2.00 \text{ s})^3 = 9.50$ rad.

EVALUATE: Both α_z and ω_z are positive and the angular speed is increasing.

9.5. **IDENTIFY and SET UP:** Use $\omega_z = \frac{d\theta}{dt}$ to calculate the angular velocity and $\omega_{av-z} = \frac{\Delta\theta}{\Delta t} = \frac{\theta_2 - \theta_1}{t_2 - t_1}$ to calculate the average angular velocity for the specified time interval.

EXECUTE: $\theta = \gamma t + \beta t^3$; $\gamma = 0.400$ rad/s, $\beta = 0.0120$ rad/s³

(a) $\omega_z = \frac{d\theta}{dt} = \gamma + 3\beta t^2$

(b) At $t = 0$, $\omega_z = \gamma = 0.400$ rad/s

(c) At $t = 5.00$ s, $\omega_z = 0.400 \text{ rad/s} + 3(0.0120 \text{ rad/s}^3)(5.00 \text{ s})^2 = 1.30$ rad/s

$\omega_{av-z} = \frac{\Delta\theta}{\Delta t} = \frac{\theta_2 - \theta_1}{t_2 - t_1}$

For $t_1 = 0$, $\theta_1 = 0$.

For $t_2 = 5.00$ s, $\theta_2 = (0.400 \text{ rad/s})(5.00 \text{ s}) + (0.012 \text{ rad/s}^3)(5.00 \text{ s})^3 = 3.50$ rad

So $\omega_{av-z} = \frac{3.50 \text{ rad} - 0}{5.00 \text{ s} - 0} = 0.700$ rad/s.

EVALUATE: The average of the instantaneous angular velocities at the beginning and end of the time interval is $\frac{1}{2}(0.400 \text{ rad/s} + 1.30 \text{ rad/s}) = 0.850$ rad/s. This is larger than ω_{av-z}, because $\omega_z(t)$ is increasing faster than linearly.

9.11. **IDENTIFY:** Apply the constant angular acceleration equations to the motion. The target variables are t and $\theta - \theta_0$.

SET UP: **(a)** $\alpha_z = 1.50 \text{ rad/s}^2$; $\omega_{0z} = 0$ (starts from rest); $\omega_z = 36.0$ rad/s; $t = ?$

$\omega_z = \omega_{0z} + \alpha_z t$

EXECUTE: $t = \dfrac{\omega_z - \omega_{0z}}{\alpha_z} = \dfrac{36.0 \text{ rad/s} - 0}{1.50 \text{ rad/s}^2} = 24.0 \text{ s}$

(b) $\theta - \theta_0 = ?$

$\theta - \theta_0 = \omega_{0z} t + \tfrac{1}{2}\alpha_z t^2 = 0 + \tfrac{1}{2}(1.50 \text{ rad/s}^2)(24.0 \text{ s})^2 = 432 \text{ rad}$

$\theta - \theta_0 = 432 \text{ rad}(1 \text{ rev}/2\pi \text{ rad}) = 68.8 \text{ rev}$

EVALUATE: We could use $\theta - \theta_0 = \tfrac{1}{2}(\omega_z + \omega_{0z})t$ to calculate

$\theta - \theta_0 = \tfrac{1}{2}(0 + 36.0 \text{ rad/s})(24.0 \text{ s}) = 432 \text{ rad}$, which checks.

9.15. IDENTIFY: Apply constant angular acceleration equations.

SET UP: Let the direction the flywheel is rotating be positive.
$\theta - \theta_0 = 200 \text{ rev}$, $\omega_{0z} = 500 \text{ rev/min} = 8.333 \text{ rev/s}$, $t = 30.0 \text{ s}$.

EXECUTE: (a) $\theta - \theta_0 = \left(\dfrac{\omega_{0z} + \omega_z}{2}\right)t$ gives $\omega_z = 5.00 \text{ rev/s} = 300 \text{ rpm}$.

(b) Use the information in part (a) to find α_z: $\omega_z = \omega_{0z} + \alpha_z t$ gives $\alpha_z = -0.1111 \text{ rev/s}^2$. Then $\omega_z = 0$, $\alpha_z = -0.1111 \text{ rev/s}^2$, $\omega_{0z} = 8.333 \text{ rev/s}$ in $\omega_z = \omega_{0z} + \alpha_z t$ gives $t = 75.0 \text{ s}$ and

$\theta - \theta_0 = \left(\dfrac{\omega_{0z} + \omega_z}{2}\right)t$ gives $\theta - \theta_0 = 312 \text{ rev}$.

EVALUATE: The mass and diameter of the flywheel are not used in the calculation.

9.25. IDENTIFY: Use $a_{\text{rad}} = r\omega^2$ and solve for r.

SET UP: $a_{\text{rad}} = r\omega^2$ so $r = a_{\text{rad}}/\omega^2$, where ω must be in rad/s

EXECUTE: $a_{\text{rad}} = 3000g = 3000(9.80 \text{ m/s}^2) = 29{,}400 \text{ m/s}^2$

$\omega = (5000 \text{ rev/min})\left(\dfrac{1 \text{ min}}{60 \text{ s}}\right)\left(\dfrac{2\pi \text{ rad}}{1 \text{ rev}}\right) = 523.6 \text{ rad/s}$

Then $r = \dfrac{a_{\text{rad}}}{\omega^2} = \dfrac{29{,}400 \text{ m/s}^2}{(523.6 \text{ rad/s})^2} = 0.107 \text{ m}$.

EVALUATE: The diameter is then 0.214 m, which is larger than 0.127 m, so the claim is *not* realistic.

9.27. IDENTIFY: We relate angular speed to linear speed and use the constant-angular acceleration equations.

SET UP: Use $v_{\text{tan}} = r\omega$ and $\omega_z^2 = \omega_{0z}^2 + 2\alpha_z(\theta - \theta_0)$. We want the angular acceleration.

EXECUTE: First find the initial and final angular velocities. $v_{\text{tan}} = r\omega$ gives $\omega = \dfrac{v_{\text{tan}}}{r}$. Therefore $\omega_1 = \dfrac{3.00 \text{ m/s}}{0.300 \text{ m}} = 10.0 \text{ rad/s}$ and $\omega_2 = 20.0 \text{ rad/s}$. Now use $\omega_z^2 = \omega_{0z}^2 + 2\alpha_z(\theta - \theta_0)$ to find α_z. The wheel turns through 4 rev, which is 8π rad, so $(20.0 \text{ rad/s})^2 = (10.0 \text{ rad/s})^2 + 2\alpha_z(8\pi \text{ rad})$, so $\alpha_z = 5.97 \text{ rad/s}^2$.

EVALUATE: Notice that in many cases, the units of radians do not actually emerge, but must be inferred. For example in the calculation $\omega_1 = \dfrac{3.00 \text{ m/s}}{0.300 \text{ m}}$, the units are 1/s (or s^{-1}), so the radians must be put in. This is not a problem because a radian has no dimensions; it is a pure number since it is defined as the length of an arc of a circle divided by the radius of the circle: θ (in rad) $= s/r$.

9.29. **IDENTIFY:** We relate angular quantities to linear quantities and use the constant-angular accleration equations.

SET UP: Use $a_{\tan} = r\alpha$, $a_{\text{rad}} = \omega^2 r$, $a = \sqrt{a_{\tan}^2 + a_{\text{rad}}^2}$, and $\omega_z = \omega_{0z} + \alpha_z t$.

EXECUTE: First find ω at the end of 2.00 s.
$\omega_z = \omega_{0z} + \alpha_z t = 0 + (0.600 \text{ rad/s}^2)(2.00 \text{ s})^2 = 1.20$ rad/s.
$a_{\tan} = r\alpha = (0.300 \text{ m})(0.600 \text{ rad/s}^2) = 0.180 \text{ m/s}^2$.
$a_{\text{rad}} = \omega^2 r = (1.20 \text{ rad/s})^2 (0.300 \text{ m}) = 0.432 \text{ m/s}^2$.
$a = \sqrt{a_{\tan}^2 + a_{\text{rad}}^2} = \sqrt{(0.180 \text{ m/s}^2)^2 + (0.432 \text{ m/s}^2)^2} = 0.468 \text{ m/s}^2$.

EVALUATE: There is only a tangential acceleration when the object has an angular acceleration, but there is a radial acceleration even if there is no angular acceleration.

9.33. **IDENTIFY:** I for the object is the sum of the values of I for each part.

SET UP: For the bar, for an axis perpendicular to the bar, use the appropriate expression from Table 9.2. For a point mass, $I = mr^2$, where r is the distance of the mass from the axis.

EXECUTE: **(a)** $I = I_{\text{bar}} + I_{\text{balls}} = \frac{1}{12} M_{\text{bar}} L^2 + 2 m_{\text{balls}} \left(\frac{L}{2}\right)^2$.

$I = \frac{1}{12}(4.00 \text{ kg})(2.00 \text{ m})^2 + 2(0.300 \text{ kg})(1.00 \text{ m})^2 = 1.93 \text{ kg} \cdot \text{m}^2$

(b) $I = \frac{1}{3} m_{\text{bar}} L^2 + m_{\text{ball}} L^2 = \frac{1}{3}(4.00 \text{ kg})(2.00 \text{ m})^2 + (0.300 \text{ kg})(2.00 \text{ m})^2 = 6.53 \text{ kg} \cdot \text{m}^2$

(c) $I = 0$ because all masses are on the axis.

(d) All the mass is a distance $d = 0.500$ m from the axis and
$I = m_{\text{bar}} d^2 + 2 m_{\text{ball}} d^2 = M_{\text{Total}} d^2 = (4.60 \text{ kg})(0.500 \text{ m})^2 = 1.15 \text{ kg} \cdot \text{m}^2$.

EVALUATE: I for an object depends on the location and direction of the axis.

9.35. **IDENTIFY and SET UP:** $I = \sum m_i r_i^2$ implies $I = I_{\text{rim}} + I_{\text{spokes}}$

EXECUTE: $I_{\text{rim}} = MR^2 = (1.40 \text{ kg})(0.300 \text{ m})^2 = 0.126 \text{ kg} \cdot \text{m}^2$
Each spoke can be treated as a slender rod with the axis through one end, so
$I_{\text{spokes}} = 8\left(\frac{1}{3} ML^2\right) = \frac{8}{3}(0.280 \text{ kg})(0.300 \text{ m})^2 = 0.0672 \text{ kg} \cdot \text{m}^2$

$I = I_{\text{rim}} + I_{\text{spokes}} = 0.126 \text{ kg} \cdot \text{m}^2 + 0.0672 \text{ kg} \cdot \text{m}^2 = 0.193 \text{ kg} \cdot \text{m}^2$.

EVALUATE: Our result is smaller than $m_{\text{tot}} R^2 = (3.64 \text{ kg})(0.300 \text{ m})^2 = 0.328 \text{ kg} \cdot \text{m}^2$, since the mass of each spoke is distributed between $r = 0$ and $r = R$.

9.37. **IDENTIFY:** The flywheel has kinetic energy due to its rotation, but it is slowing down. We need to use rotational kinetic energy and constant-angular acceleration equations.

SET UP: $K = \frac{1}{2} I \omega^2$, $\omega_z = \omega_{0z} + \alpha_z t$. We want to find the time for the flywheel to lose half of its kinetic energy.

EXECUTE: First use kinetic energy to find the initial and final angular velocities. We know that $K_2 = \frac{1}{2}K_1$, so $\frac{1}{2}I\omega_2^2 = \frac{1}{2}\left(\frac{1}{2}I\omega_1^2\right)$, which gives $\omega_2 = \frac{\omega_1}{\sqrt{2}}$. Now use $K = \frac{1}{2}I\omega^2$ to find ω_1. $\frac{1}{2}I\omega_1^2 = K_1$ gives $\omega_1 = \sqrt{\frac{2K_1}{I}} = \sqrt{\frac{2(30.0 \text{ J})}{12.0 \text{ kg}\cdot\text{m}^2}} = 2.236$ rad/s. $\omega_2 = \frac{\omega_1}{\sqrt{2}} = \frac{2.236 \text{ rad/s}}{\sqrt{2}} = 1.581$ rad/s. Now use $\omega_z = \omega_{0z} + \alpha_z t$ to find t: 1.581 rad/s = 2.236 rad/s + $((-0.500)(2\pi)\text{ rad/s})t$, which gives $t = 0.208$ s.

EVALUATE: We must use radian measure in the formula $K = \frac{1}{2}I\omega^2$, but we could use any type of angular measure in the equation $\omega_z = \omega_{0z} + \alpha_z t$.

9.39. IDENTIFY: This problem requires moment of inertia calculations.

SET UP: The two spheres have the same mass, radius, and kinetic energy, but one is solid and the other is a hollow shell. $I_h = \frac{2}{3}MR^2$ and $I_s = \frac{2}{5}MR^2$. $K = \frac{1}{2}I\omega^2$. The solid sphere has angular speed ω_1 and we want to find the angular speed ω_h of the hollow sphere.

EXECUTE: Equate the kinetic energies of the two spheres: $\frac{2}{5}MR^2\omega_1^2 = \frac{2}{3}MR^2\omega_h^2$ which gives $\omega_h = \omega_1\sqrt{3/5}$.

EVALUATE: From our result, we see that the hollow sphere is spinning slower than the solid sphere. This is reasonable because the hollow sphere has a greater moment of inertia than the solid sphere because more of its mass is located farther from the axis of rotation.

9.41. IDENTIFY: Knowing the kinetic energy, mass and radius of the sphere, we can find its angular velocity. From this we can find the tangential velocity (the target variable) of a point on the rim.

SET UP: $K = \frac{1}{2}I\omega^2$ and $I = \frac{2}{5}MR^2$ for a solid uniform sphere. The tangential velocity is $v = r\omega$.

EXECUTE: $I = \frac{2}{5}MR^2 = \frac{2}{5}(28.0 \text{ kg})(0.380 \text{ m})^2 = 1.617 \text{ kg}\cdot\text{m}^2$. $K = \frac{1}{2}I\omega^2$ so
$\omega = \sqrt{\frac{2K}{I}} = \sqrt{\frac{2(236 \text{ J})}{1.617 \text{ kg}\cdot\text{m}^2}} = 17.085$ rad/s.
$v = r\omega = (0.380 \text{ m})(17.085 \text{ rad/s}) = 6.49$ m/s.

EVALUATE: This is the speed of a point on the surface of the sphere that is farthest from the axis of rotation (the "equator" of the sphere). Points off the "equator" would have smaller tangential velocity but the same angular velocity.

9.43. IDENTIFY: We need to use rotational kinetic energy.

SET UP: $K = \frac{1}{2}I\omega^2$. We know that $I_A = 3I_B$ and $\omega_B = 4\omega_A$, and we want to compare the kinetic energies of the wheels.

EXECUTE: (a) Using what we know gives $K_A = \frac{1}{2}I_A\omega_A^2$ and $K_B = \frac{1}{2}I_B\omega_B^2 = \frac{1}{2}\left(\frac{I_A}{3}\right)(4\omega_A)^2$. For K_B this becomes $K_B = \frac{1}{2}\left(\frac{I_A}{3}\right)(4\omega_A)^2 = \frac{16}{3}\left(\frac{1}{2}I_A\omega_A^2\right) = \frac{16}{3}K_A$, which tells us the B has more kinetic energy than A.

(b) $\dfrac{K_A}{K_B} = \dfrac{\frac{1}{2}I_A\omega_A^2}{\left(\dfrac{16}{3}\right)\left(\dfrac{1}{2}I_A\omega_A^2\right)} = \dfrac{3}{16}.$

EVALUATE: Wheel A has 3 times the moment of inertia of B but only ¼ the angular speed and therefore only 1/16 the *square* of the angular speed, so it has less kinetic energy than B.

9.45. **IDENTIFY and SET UP:** Combine $K = \frac{1}{2}I\omega^2$ and $a_{\text{rad}} = r\omega^2$ to solve for K. Use Table 9.2 to get I.

EXECUTE: $K = \frac{1}{2}I\omega^2$

$a_{\text{rad}} = R\omega^2$, so $\omega = \sqrt{a_{\text{rad}}/R} = \sqrt{(3500 \text{ m/s}^2)/1.20 \text{ m}} = 54.0$ rad/s

For a disk, $I = \frac{1}{2}MR^2 = \frac{1}{2}(70.0 \text{ kg})(1.20 \text{ m})^2 = 50.4 \text{ kg} \cdot \text{m}^2$

Thus $K = \frac{1}{2}I\omega^2 = \frac{1}{2}(50.4 \text{ kg} \cdot \text{m}^2)(54.0 \text{ rad/s})^2 = 7.35 \times 10^4$ J

EVALUATE: The limit on a_{rad} limits ω which in turn limits K.

9.47. **IDENTIFY:** Apply conservation of energy to the system of stone plus pulley. $v = r\omega$ relates the motion of the stone to the rotation of the pulley.

SET UP: For a uniform solid disk, $I = \frac{1}{2}MR^2$. Let point 1 be when the stone is at its initial position and point 2 be when it has descended the desired distance. Let $+y$ be upward and take $y = 0$ at the initial position of the stone, so $y_1 = 0$ and $y_2 = -h$, where h is the distance the stone descends.

EXECUTE: (a) $K_p = \frac{1}{2}I_p\omega^2$. $I_p = \frac{1}{2}M_pR^2 = \frac{1}{2}(2.50 \text{ kg})(0.200 \text{ m})^2 = 0.0500 \text{ kg} \cdot \text{m}^2$.

$\omega = \sqrt{\dfrac{2K_p}{I_p}} = \sqrt{\dfrac{2(4.50 \text{ J})}{0.0500 \text{ kg} \cdot \text{m}^2}} = 13.4$ rad/s. The stone has speed

$v = R\omega = (0.200 \text{ m})(13.4 \text{ rad/s}) = 2.68$ m/s. The stone has kinetic energy

$K_s = \frac{1}{2}mv^2 = \frac{1}{2}(1.50 \text{ kg})(2.68 \text{ m/s})^2 = 5.39$ J. $K_1 + U_1 = K_2 + U_2$ gives $0 = K_2 + U_2$.

$0 = 4.50 \text{ J} + 5.39 \text{ J} + mg(-h)$. $h = \dfrac{9.89 \text{ J}}{(1.50 \text{ kg})(9.80 \text{ m/s}^2)} = 0.673$ m.

(b) $K_{\text{tot}} = K_p + K_s = 9.89$ J. $I = 2MR^2$,

EVALUATE: The gravitational potential energy of the pulley doesn't change as it rotates. The tension in the wire does positive work on the pulley and negative work of the same magnitude on the stone, so no net work on the system.

9.51. **IDENTIFY:** Use the parallel-axis theorem to relate I for the wood sphere about the desired axis to I for an axis along a diameter.

SET UP: For a thin-walled hollow sphere, axis along a diameter, $I = \frac{2}{3}MR^2$.

For a solid sphere with mass M and radius R, $I_{\text{cm}} = \frac{2}{5}MR^2$, for an axis along a diameter.

EXECUTE: Find d such that $I_P = I_{\text{cm}} + Md^2$ with $I_P = \frac{2}{3}MR^2$:

$\frac{2}{3}MR^2 = \frac{2}{5}MR^2 + Md^2$

The factors of M divide out and the equation becomes $(\frac{2}{3} - \frac{2}{5})R^2 = d^2$

$d = \sqrt{(10-6)/15}\,R = 2R/\sqrt{15} = 0.516R$.

The axis is parallel to a diameter and is $0.516R$ from the center.

EVALUATE: $I_{cm}(\text{lead}) > I_{cm}(\text{wood})$ even though M and R are the same since for a hollow sphere all the mass is a distance R from the axis. The parallel-axis theorem says $I_P > I_{cm}$, so there must be a d where $I_P(\text{wood}) = I_{cm}(\text{lead})$.

9.53. IDENTIFY and **SET UP:** Use the parallel-axis theorem. The cm of the sheet is at its geometrical center. The object is sketched in Figure 9.53.

EXECUTE: $I_P = I_{cm} + Md^2$.

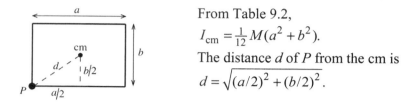

From Table 9.2,
$I_{cm} = \frac{1}{12}M(a^2+b^2)$.
The distance d of P from the cm is
$d = \sqrt{(a/2)^2 + (b/2)^2}$.

Figure 9.53

Thus $I_P = I_{cm} + Md^2 = \frac{1}{12}M(a^2+b^2) + M(\frac{1}{4}a^2 + \frac{1}{4}b^2) = (\frac{1}{12}+\frac{1}{4})M(a^2+b^2) = \frac{1}{3}M(a^2+b^2)$

EVALUATE: $I_P = 4I_{cm}$. For an axis through P mass is farther from the axis.

9.57. IDENTIFY: Apply $I = \int r^2\, dm$ and $M = \int dm$.

SET UP: For this case, $dm = \gamma x\, dx$.

EXECUTE: (a) $M = \int dm = \int_0^L \gamma x\, dx = \gamma \frac{x^2}{2}\Big|_0^L = \frac{\gamma L^2}{2}$

(b) $I = \int_0^L x^2(\gamma x)dx = \gamma \frac{x^4}{4}\Big|_0^L = \frac{\gamma L^4}{4} = \frac{M}{2}L^2$. This is larger than the moment of inertia of a uniform rod of the same mass and length, since the mass density is greater farther away from the axis than nearer the axis.

(c) $I = \int_0^L (L-x)^2 \gamma x\, dx = \gamma \int_0^L (L^2 x - 2Lx^2 + x^3)dx = \gamma\left(L^2\frac{x^2}{2} - 2L\frac{x^3}{3} + \frac{x^4}{4}\right)\Big|_0^L = \gamma\frac{L^4}{12} = \frac{M}{6}L^2$.

This is a third of the result of part (b), reflecting the fact that more of the mass is concentrated at the right end.

EVALUATE: For a uniform rod with an axis at one end, $I = \frac{1}{3}ML^2$. The result in (b) is larger than this and the result in (c) is smaller than this.

9.59. IDENTIFY: The target variable is the horizontal distance the piece travels before hitting the floor. Using the angular acceleration of the blade, we can find its angular velocity when the piece breaks off. This will give us the linear horizontal speed of the piece. It is then in free fall, so we can use the linear kinematics equations.

SET UP: $\omega_z^2 = \omega_{0z}^2 + 2\alpha_z(\theta - \theta_0)$ for the blade, and $v = r\omega$ is the horizontal velocity of the piece. $y - y_0 = v_{0y}t + \frac{1}{2}a_y t^2$ for the falling piece.

EXECUTE: Find the initial horizontal velocity of the piece just after it breaks off.
$\theta - \theta_0 = (155 \text{ rev})(2\pi \text{ rad}/1 \text{ rev}) = 973.9 \text{ rad}$.

$\alpha_z = (2.00 \text{ rev/s}^2)(2\pi \text{ rad}/1 \text{ rev}) = 12.566 \text{ rad/s}^2$. $\omega_z^2 = \omega_{0z}^2 + 2\alpha_z(\theta - \theta_0)$.

$\omega_z = \sqrt{2\alpha_z(\theta-\theta_0)} = \sqrt{2(12.566 \text{ rad/s}^2)(973.9 \text{ rad})} = 156.45 \text{ rad/s}$. The horizontal velocity of the piece is $v = r\omega = (0.120 \text{ m})(156.45 \text{ rad/s}) = 18.774 \text{ m/s}$. Now consider the projectile motion of the piece. Take $+y$ downward and use the vertical motion to find t. Solving $y - y_0 = v_{0y}t + \frac{1}{2}a_y t^2$ for t

gives $t = \sqrt{\dfrac{2(y-y_0)}{a_y}} = \sqrt{\dfrac{2(0.820 \text{ m})}{9.8 \text{ m/s}^2}} = 0.4091 \text{ s}$. Then

$x - x_0 = v_{0x}t + \frac{1}{2}a_x t^2 = (18.774 \text{ m/s})(0.4091 \text{ s}) = 7.68 \text{ m}$.

EVALUATE: Once the piece is free of the blade, the only force acting on it is gravity so its acceleration is g downward.

9.61. IDENTIFY: The angular acceleration α of the disk is related to the linear acceleration a of the ball by $a = R\alpha$. Since the acceleration is not constant, use $\omega_z - \omega_{0z} = \int_0^t \alpha_z dt$ and $\theta - \theta_0 = \int_0^t \omega_z dt$ to relate θ, ω_z, α_z, and t for the disk. $\omega_{0z} = 0$.

SET UP: $\int t^n dt = \dfrac{1}{n+1} t^{n+1}$. In $a = R\alpha$, α is in rad/s^2.

EXECUTE: (a) $A = \dfrac{a}{t} = \dfrac{1.80 \text{ m/s}^2}{3.00 \text{ s}} = 0.600 \text{ m/s}^3$

(b) $\alpha = \dfrac{a}{R} = \dfrac{(0.600 \text{ m/s}^3)t}{0.250 \text{ m}} = (2.40 \text{ rad/s}^3)t$

(c) $\omega_z = \int_0^t (2.40 \text{ rad/s}^3)t\, dt = (1.20 \text{ rad/s}^3)t^2$. $\omega_z = 15.0$ rad/s for $t = \sqrt{\dfrac{15.0 \text{ rad/s}}{1.20 \text{ rad/s}^3}} = 3.54 \text{ s}$.

(d) $\theta - \theta_0 = \int_0^t \omega_z dt = \int_0^t (1.20 \text{ rad/s}^3)t^2 dt = (0.400 \text{ rad/s}^3)t^3$. For $t = 3.54$ s, $\theta - \theta_0 = 17.7$ rad.

EVALUATE: If the disk had turned at a constant angular velocity of 15.0 rad/s for 3.54 s it would have turned through an angle of 53.1 rad in 3.54 s. It actually turns through less than half this because the angular velocity is increasing in time and is less than 15.0 rad/s at all but the end of the interval.

9.63. IDENTIFY: We use energy conservation for this problem. The accelerations are constant, so we can use the constant-acceleration equations.

SET UP: We can use $U_1 + K_1 + W_{\text{other}} = U_2 + K_2$, $K_{\text{rot}} = \dfrac{1}{2}I\omega^2$, $K_{\text{tr}} = \dfrac{1}{2}mv^2$, $U_g = mgy$, $\omega_z = \omega_{0z} + \alpha_z t$, $I = \dfrac{1}{2}MR^2$, and $\theta - \theta_0 = \omega_{0z}t + \dfrac{1}{2}\alpha_z t^2$. The system starts from rest, and the wheel turns through 8.00 rev in the first 5.00 s. We want to find the mass of the block.

EXECUTE: First use $\theta - \theta_0 = \omega_{0z}t + \dfrac{1}{2}\alpha_z t^2$ to find α. 16π rad $= 0 + \dfrac{1}{2}\alpha(5.00 \text{ s})^2$, which gives $\alpha = 4.021$ rad/s^2. Now use $\omega_z = \omega_{0z} + \alpha_z t$ to find the angular speed of the wheel at the end of 5.00 s. $\omega = 0 + (4.021 \text{ rad/s}^2)(5.00 \text{ s}) = 20.11$ rad/s. Now use energy conservation for the system, $U_1 + K_1 + W_{\text{other}} = U_2 + K_2$. $W_{\text{other}} = 0$ and $K_1 = 0$. For gravitational potential energy, call $y = 0$ the level of the block at the end of the first 5.00 s. This makes $U_2 = 0$ and $U_1 = mgy$, where y is the length of rope that has come off the rim of the wheel as it turns through 8.00 rev (which is 16.00π rad). This length is $R(16.00\pi \text{ rad}) = (0.800 \text{ m})(16.00\pi \text{ rad}) = 40.21$ m; therefore $y = 40.21$ m. Putting this into the energy equation gives $m_B g y = \frac{1}{2} m_B v^2 + \frac{1}{2} I \omega^2$. Using $I = \frac{1}{2} M R^2$ and $v = R\omega$, the energy equation becomes $m_B g y = \frac{1}{2} m_B (R\omega)^2 + \frac{1}{2}\left(\frac{1}{2} m_{\text{wheel}} R^2\right)\omega^2$. Solving for m_B gives

$m_B = \dfrac{\frac{1}{4} m_{\text{wheel}} R^2 \omega^2}{gy - \frac{1}{2} R^2 \omega^2}$. Putting in $m_{\text{wheel}} = 5.00$ kg, $R = 0.800$ m, $\omega = 20.11$ rad/s, and $y = 40.21$ m,

we get $m_B = 1.22$ kg.

EVALUATE: As the block falls, its gravitational potential energy is converted to kinetic energy, but part of that kinetic energy goes to the wheel and part goes to the block.

9.65. **IDENTIFY:** We use energy conservation. The kinetic energy is shared between the two blocks and the pulley.

SET UP: We can use $U_1 + K_1 + W_{\text{other}} = U_2 + K_2$, $K_{\text{rot}} = \frac{1}{2} I \omega^2$, $K_{\text{tr}} = \frac{1}{2} m v^2$, and $U_g = mgy$. The system starts from rest, and the pulley is turning at 8.00 rad/s after block B has descended 1.20 m. We want to find the mass of block A.

EXECUTE: First use $v_{\tan} = r\omega$ to find the speed of the block when $\omega = 8.00$ rad/s. $v_B = v_{\tan} = R\omega = (0.200 \text{ m})(8.00 \text{ rad/s}) = 1.60$ m/s. Now use energy conservation, $U_1 + K_1 + W_{\text{other}} = U_2 + K_2$. $K_1 = 0$ and $W_{\text{other}} = 0$. $K_2 = K_A + K_B + K_{\text{pulley}}$. Choose $y = 0$ at 1.20 m below the starting point of B, which makes $U_2 = 0$ and $U_1 = mgy$, where $y = 1.20$ m. We also know that $v_A = v_B = v = 1.60$ m/s. Therefore we have $m_B g y = \frac{1}{2} m_A v^2 + \frac{1}{2} m_B v^2 + \frac{1}{2} I \omega^2$. Solving for m_A gives $m_A = \dfrac{2 m_B g y - m_B v^2 - I \omega^2}{v^2}$. Using $m_B = 5.00$ kg, $y = 1.20$ m, $I = 1.30 \text{ kg} \cdot \text{m}^2$, $v = 1.60$ m/s, and $\omega = 8.00$ rad/s, we get $m_A = 8.44$ kg.

EVALUATE: As block B falls, its gravitational potential energy gets converted to kinetic energy which is shared between itself, block A, and the pulley. This is a case where the pulley is *not* massless!

9.67. **IDENTIFY:** Apply $v = r\omega$.

SET UP: Points on the chain all move at the same speed, so $r_r \omega_r = r_f \omega_f$.

EXECUTE: The angular velocity of the rear wheel is $\omega_r = \dfrac{v_r}{r_r} = \dfrac{5.00 \text{ m/s}}{0.330 \text{ m}} = 15.15$ rad/s.

The angular velocity of the front wheel is $\omega_f = 0.600$ rev/s $= 3.77$ rad/s. $r_r = r_f (\omega_f / \omega_r) = 2.99$ cm.

EVALUATE: The rear sprocket and wheel have the same angular velocity and the front sprocket and wheel have the same angular velocity. α is the same for both, so the rear sprocket has a smaller radius since it has a larger angular velocity. The speed of a point on the chain is $v = r_r \omega_r = (2.99 \times 10^{-2} \text{ m})(15.15 \text{ rad/s}) = 0.453$ m/s. The linear speed of the bicycle is 5.00 m/s.

9.69. **IDENTIFY:** This problem involves energy conservation and Newton's second law.

SET UP: Only conservative forces act on the system, so $U_1 + K_1 = U_2 + K_2$. The kinetic energy is due to the translation of the block and the rotation of the wheel, and $K_1 = 0$. Call point 2 the location of the block after it has fallen for 2.00 s. This makes $U_2 = 0$ and $U_1 = mgy$. We also use

$v_y = v_{0y} + a_y t$, $y = y_0 + v_{0y}t + \frac{1}{2}a_y t^2$, and $\Sigma F_y = ma_y$. We want to find the kinetic energy of the wheel 2.00 s after motion has begun.

EXECUTE: $U_1 + K_1 = U_2 + K_2$ gives $mgy = \frac{1}{2}mv^2 + K_{\text{wheel}}$. We need to find v, so apply $K = \frac{1}{2}I\omega^2 + \frac{1}{2}m(\omega R)^2 = \frac{1}{2}(I + mR^2)\omega^2$. to the block, taking downward positive because that is the direction of its acceleration. This gives $mg - T = ma$ so Using $I = \frac{1}{2}mR^2$ and solving for ω, = $9.80 \text{ m/s}^2 - \frac{9.00 \text{ N}}{1.50 \text{ kg}} = 3.80 \text{ m/s}^2$. Now use $v_y = v_{0y} + a_y t$ to find v, which gives $v = 0 + (3.80 \text{ m/s}^2)(2.00 \text{ s}) = 7.60$ m/s. Find y using $y = y_0 + v_{0y}t + \frac{1}{2}a_y t^2$, from which we get $y = \frac{1}{2}at^2 = \frac{1}{2}(3.80 \text{ m/s}^2)(2.00 \text{ s})^2 = 7.60$ m. Now return to energy conservation, where we found that $mgy = \frac{1}{2}mv^2 + K_{\text{wheel}}$. Solve for K_{wheel} to get

$$K_{\text{wheel}} = m\left(gy - \frac{v^2}{2}\right) = (1.50 \text{ kg})\left[(9.80 \text{ m/s}^2)(7.60 \text{ m}) - \frac{(7.60 \text{ m/s})^2}{2}\right] = 68.4 \text{ J}.$$

EVALUATE: The kinetic energy of the block is $K_{\text{block}} = \frac{1}{2}mv^2 = \frac{1}{2}(1.50 \text{ kg})(7.60 \text{ m/s})^2 = 43.3$ J, which is around 2/3 the kinetic energy of the wheel. So the wheel has a large effect on the motion.

9.71. IDENTIFY: The falling wood accelerates downward as the wheel undergoes angular acceleration. Newton's second law applies to the wood and the wheel, and the linear kinematics formulas apply to the wood because it has constant acceleration.

SET UP: $\Sigma \vec{F} = m\vec{a}$, $\tau = I\alpha$, $a_{\text{tan}} = R\alpha$, $y - y_0 = v_{0y}t + \frac{1}{2}a_y t^2$.

EXECUTE: First use $y - y_0 = v_{0y}t + \frac{1}{2}a_y t^2$ to find the downward acceleration of the wood. With $v_0 = 0$, we have $a_y = 2(y - y_0)/t^2 = 2(12.0 \text{ m})/(4.00 \text{ s})^2 = 1.50 \text{ m/s}^2$. Now apply Newton's second to the wood to find the tension in the rope. $\Sigma \vec{F} = m\vec{a}$ gives $mg - T = ma$, $T = m(g - a)$, which gives $T = (8.20 \text{ kg})(9.80 \text{ m/s}^2 - 1.50 \text{ m/s}^2) = 68.06$ N. Now use $a_{\text{tan}} = R\alpha$ and apply Newton's second law (in its rotational form) to the wheel. $\tau = I\alpha$ gives $TR = I\alpha$, $I = TR/\alpha = TR/(a/R) = TR^2/a$ $I = (68.06 \text{ N})(0.320 \text{ m})^2/(1.50 \text{ m/s}^2) = 4.65 \text{ kg} \cdot \text{m}^2$.

EVALUATE: The tension in the rope affects the acceleration of the wood and causes the angular acceleration of the wheel.

9.75. IDENTIFY: Mechanical energy is conserved since there is no friction.

SET UP: $K_1 + U_1 = K_2 + U_2$, $K = \frac{1}{2}I\omega^2$ (for rotational motion), $K = \frac{1}{2}mv^2$ (for linear motion), $I = \frac{1}{12}ML^2$ for a slender rod.

EXECUTE: Take the initial position with the rod horizontal, and the final position with the rod vertical. The heavier sphere will be at the bottom and the lighter one at the top. Call the gravitational potential energy zero with the rod horizontal, which makes the initial potential energy zero. The initial kinetic energy is also zero. Applying $K_1 + U_1 = K_2 + U_2$ and calling A and B the spheres gives

$0 = K_A + K_B + K_{rod} + U_A + U_B + U_{rod}$. $U_{rod} = 0$ in the final position since its center of mass has not moved. Therefore $0 = \frac{1}{2} m_A v_A^2 + \frac{1}{2} m_B v_B^2 + \frac{1}{2} I \omega^2 + m_A g \frac{L}{2} - m_B g \frac{L}{2}$. We also know that $v_A = v_B = (L/2) \omega$. Calling v the speed of the spheres, we get
$0 = \frac{1}{2} m_A v^2 + \frac{1}{2} m_B v^2 + \frac{1}{2} (\frac{1}{12})(ML^2)(2v/L)^2 + m_A g \frac{L}{2} - m_B g \frac{L}{2}$ Putting in $m_A = 0.0200$ kg, $m_B = 0.0500$ kg, $M = 0.120$ kg, and $L = 800$ m, we get $v = 1.46$ m/s.
EVALUATE: As the rod turns, the heavier sphere loses potential energy but the lighter one gains potential energy.

9.79. IDENTIFY: $I = I_1 + I_2$. Apply conservation of energy to the system. The calculation is similar to Example 9.8.

SET UP: $\omega = \frac{v}{R_1}$ for part (b) and $\omega = \frac{v}{R_2}$ for part (c).

EXECUTE: (a) $I = \frac{1}{2} M_1 R_1^2 + \frac{1}{2} M_2 R_2^2 = \frac{1}{2}((0.80 \text{ kg})(2.50 \times 10^{-2} \text{ m})^2 + (1.60 \text{ kg})(5.00 \times 10^{-2} \text{ m})^2)$
$I = 2.25 \times 10^{-3}$ kg·m^2.

(b) The method of Example 9.8 yields $v = \sqrt{\frac{2gh}{1 + (I/mR_1^2)}}$.

$v = \sqrt{\frac{2(9.80 \text{ m/s}^2)(2.00 \text{ m})}{(1 + ((2.25 \times 10^{-3} \text{ kg} \cdot \text{m}^2)/(1.50 \text{ kg})(0.025 \text{ m})^2))}} = 3.40$ m/s.

(c) The same calculation, with R_2 instead of R_1 gives $v = 4.95$ m/s.
EVALUATE: The final speed of the block is greater when the string is wrapped around the larger disk. $v = R\omega$, so when $R = R_2$ the factor that relates v to ω is larger. For $R = R_2$ a larger fraction of the total kinetic energy resides with the block. The total kinetic energy is the same in both cases (equal to mgh), so when $R = R_2$ the kinetic energy and speed of the block are greater.

9.85. IDENTIFY: The density depends on the distance from the center of the sphere, so it is a function of r. We need to integrate to find the mass and the moment of inertia.

SET UP: $M = \int dm = \int \rho dV$ and $I = \int dI$.

EXECUTE: (a) Divide the sphere into thin spherical shells of radius r and thickness dr. The volume of each shell is $dV = 4\pi r^2 dr$. $\rho(r) = a - br$, with $a = 3.00 \times 10^3$ kg/m^3 and $b = 9.00 \times 10^3$ kg/m^4. Integrating gives $M = \int dm = \int \rho dV = \int_0^R (a - br) 4\pi r^2 dr = \frac{4}{3} \pi R^3 \left(a - \frac{3}{4} bR \right)$.

$M = \frac{4}{3} \pi (0.200 \text{ m})^3 \left(3.00 \times 10^3 \text{ kg/m}^3 - \frac{3}{4}(9.00 \times 10^3 \text{ kg/m}^4)(0.200 \text{ m}) \right) = 55.3$ kg.

(b) The moment of inertia of each thin spherical shell is
$dI = \frac{2}{3} r^2 dm = \frac{2}{3} r^2 \rho dV = \frac{2}{3} r^2 (a - br) 4\pi r^2 dr = \frac{8\pi}{3} r^4 (a - br) dr$.

$I = \int_0^R dI = \frac{8\pi}{3} \int_0^R r^4 (a - br) dr = \frac{8\pi}{15} R^5 \left(a - \frac{5b}{6} R \right)$.

$I = \frac{8\pi}{15} (0.200 \text{ m})^5 \left(3.00 \times 10^3 \text{ kg/m}^3 - \frac{5}{6}(9.00 \times 10^3 \text{ kg/m}^4)(0.200 \text{ m}) \right) = 0.804$ kg·m^2.

EVALUATE: We cannot use the formulas $M = \rho V$ and $I = \frac{1}{2}MR^2$ because this sphere is not uniform throughout. Its density increases toward the surface. For a uniform sphere with density 3.00×10^3 kg/m^3, the mass is $\frac{4}{3}\pi R^3 \rho = 100.5$ kg. The mass of the sphere in this problem is less than this. For a uniform sphere with mass 55.3 kg and $R = 0.200$ m, $I = \frac{2}{5}MR^2 = 0.885$ kg·m^2. The moment of inertia for the sphere in this problem is less than this, since the density decreases with distance from the center of the sphere.

9.87. **IDENTIFY:** The graph with the problem in the text shows that the angular acceleration increases linearly with time and is therefore not constant.

SET UP: $\omega_z = d\theta/dt$, $\alpha_z = d\omega_z/dt$.

EXECUTE: **(a)** Since the angular acceleration is not constant, Eq. (9.11) cannot be used, so we must use $\alpha_z = d\omega_z/dt$ and $\omega_z = d\theta/dt$ and integrate to find the angle. The graph passes through the origin and has a constant positive slope of 6/5 rad/s^3, so the equation for α_z is $\alpha_z = (1.2 \text{ rad/s}^3)t$.

Using $\alpha_z = d\omega_z/dt$ gives $\omega_z = \omega_{0z} + \int_0^t \alpha_z dt = 0 + \int_0^t (1.2 \text{ rad/s}^3)t\,dt = (0.60 \text{ rad/s}^3)t^2$. Now we must use $\omega_z = d\theta/dt$ and integrate again to get the angle.

$\theta_2 - \theta_1 = \int_0^t \omega_z dt = \int_0^t (0.60 \text{ rad/s}^3)t^2 dt = (0.20 \text{ rad/s}^3)t^3 = (0.20 \text{ rad/s}^3)(5.0 \text{ s})^3 = 25$ rad.

(b) The result of our first integration gives $\omega_z = (0.60 \text{ rad/s}^3)(5.0 \text{ s})^2 = 15$ rad/s.

(c) The result of our second integration gives 4π rad = $(0.20 \text{ rad/s}^3)t^3$, so $t = 3.98$ s. Therefore $\omega_z = (0.60 \text{ rad/s}^3)(3.98 \text{ s})^2 = 9.48$ rad/s.

EVALUATE: When the constant-acceleration angular kinematics formulas do not apply, we must go back to basic definitions.

9.89. **IDENTIFY and SET UP:** The equation of the graph in the text is $d = (165 \text{ cm/s}^2)t^2$. For constant acceleration, the second time derivative of the position (d in this case) is a constant.

EXECUTE: **(a)** $\dfrac{d(d)}{dt} = (330 \text{ cm/s}^2)t$ and $\dfrac{d^2(d)}{dt^2} = 330$ cm/s^2, which is a constant. Therefore the acceleration of the metal block is a constant 330 cm/s^2 = 3.30 m/s^2.

(b) $v = \dfrac{d(d)}{dt} = (330 \text{ cm/s}^2)t$. When $d = 1.50$ m = 150 cm, we have 150 cm = $(165 \text{ cm/s}^2)t^2$, which gives $t = 0.9535$ s. Thus $v = 330$ cm/s^2(0.9535 s) = 315 cm/s = 3.15 m/s.

(c) Energy conservation $K_1 + U_1 = K_2 + U_2$ gives $mgd = \frac{1}{2}I\omega^2 + \frac{1}{2}mv^2$. Using $\omega = v/r$, solving for I and putting in the numbers $m = 5.60$ kg, $d = 1.50$ m, $r = 0.178$ m, $v = 3.15$ m/s, we get $I = 0.348$ kg·m^2.

(d) Newton's second law gives $mg - T = ma$, $T = m(g - a) = (5.60 \text{ kg})(9.80 \text{ m/s}^2 - 3.30 \text{ m/s}^2) = 36.4$ N.

EVALUATE: When dealing with non-uniform objects, such as this flywheel, we cannot use the standard moment of inertia formulas and must resort to other ways.

STUDY GUIDE FOR DYNAMICS OF ROTATIONAL MOTION

Summary

In this chapter, we'll investigate the dynamics of rotational motion to learn what gives an object angular acceleration. We'll define *torque*—the turning or twisting effort of a force—and learn how to apply it to both equilibrium and nonequilibrium situations. Work and power for rotating systems will also be investigated. Angular momentum will be introduced and become the basis of an important new conservation law that will lead to an analysis of a spinning gyroscope and the motion called precession. The linear dynamics foundation we have developed throughout the text will help build our intuition about rotational dynamics.

Objectives

After studying this chapter, you'll understand

- The definition and meaning of *torque*.
- How to identify torques acting on a rigid body.
- The equation of motion for rotational systems and how to apply it to problems.
- How to apply work and power to rotational dynamics problems.
- The definition of *angular momentum* and how it changes with time.
- How to apply conservation of angular momentum to problems.
- The concept of precession as it applies to gyroscopes.

Concepts and Equations

Term	Description
Torque	Torque is the tendency of a force to cause or change rotational motion about a chosen axis. The magnitude of torque is the magnitude of the force (F) times the moment arm (l), which is the perpendicular distance between the axis and the line of force: $$\tau = Fl.$$ For a force $\vec{F}$ applied at point O and a vector $\vec{r}$ from the chosen axis to point O, the torque is given by $$\vec{\tau} = \vec{r} \times \vec{F}.$$ The SI unit of torque is the newton-meter (Nm).
Combined Translation and Rotation	The motion of a rigid body, moving through space and rotating, can be regarded as translational motion of the center of mass plus rotational motion about the center of mass. The kinetic energy, net external force, and net external torque are given respectively, by $$K = \tfrac{1}{2} M v_{cm}^2 + \tfrac{1}{2} I_{cm} \omega^2,$$ $$\sum \vec{F}_{ext} = M\vec{a}_{cm},$$ $$\sum \vec{\tau}_z = I_{cm} \alpha_z.$$ In the case of rolling without slipping, the motion of the center of mass is related to the angular velocity by $$v_{cm} = R\omega.$$
Work Done by a Torque	The work done by a torque is given by $$W = \int_{\theta_1}^{\theta_2} \tau_z \, d\theta.$$ For a constant torque, $$W = \tau_z \Delta\theta.$$ The power provided by a torque is the product of the torque and the angular velocity of the object: $$P = \tau_z \omega_z.$$
Angular Momentum	The angular momentum L of a particle with respect to point O is the vector product of the position vector and momentum of the particle: $$\vec{L} = \vec{r} \times \vec{p}.$$ The angular momentum of a symmetrical object rotating about a stationary axis of symmetry is $$\vec{L} = I\vec{\omega}.$$ Our convention is that counterclockwise rotations have positive L and clockwise rotations have negative L.
Rotational Dynamics and Angular Momentum	The net external torque on a system is equal to the rate of change of the angular momentum of the system: $$\sum \vec{\tau} = \frac{d\vec{L}}{dt}.$$ If the net external torque acting on a system is zero, the total angular momentum is conserved and remains constant.

Key Concept 1: You can determine the magnitude of the torque due to a force $\vec{F}$ in any of three ways: (i) from the magnitude of $\vec{F}$ and the lever arm; (ii) from the magnitude of $\vec{F}$, the magnitude of the vector $\vec{r}$ from the origin to where $\vec{F}$ acts, and the angle between $\vec{r}$ and $\vec{F}$; or (iii) from the magnitude of $\vec{r}$ and the tangential component of $\vec{F}$. Find the direction of the torque using the right-hand rule.

Key Concept 2: For any problem involving torques on a rigid body, first draw a free-body diagram to identify where on the rigid body each external force acts with respect to the axis of rotation. Then apply the rotational analog of Newton's second law, $\sum \tau_z = I\alpha_z$.

Key Concept 3: When an object is connected to a string that wraps around a rotating pulley of radius R, the linear acceleration a_y of the object is related to the angular acceleration α_z of the pulley by $a_y = R\alpha_z$.

Key Concept 4: If a rigid body is both translating (moving as a whole through space) and rotating, its total kinetic energy is the sum of the kinetic energy of translation of the center of mass and the kinetic energy of rotation around an axis through the center of mass.

Key Concept 5: For a rigid body that rolls without slipping, has a given mass and radius, and moves with a given center-of-mass speed, the kinetic energy of rotation depends on the shape of the rigid body.

Key Concept 6: To analyze the motion of a rigid body that is both translating and rotating, use Newton's second law for the translational motion of the center of mass and the rotational analog of Newton's second law for the rotation around the center of mass.

Key Concept 7: If an object is rolling without slipping on an incline, a friction force must act on it. The direction of this friction force is always such as to prevent slipping.

Key Concept 8: If a torque acts on a rigid body, the work done equals torque times angular displacement and the power equals torque times angular velocity.

Key Concept 9: The angular momentum vector of a rotating rigid body points along the rigid body's rotation axis. The rate of change of angular momentum equals the net external torque on the rigid body.

Key Concept 10: If there is zero net external torque on a rigid body, its angular momentum is conserved. If the body changes shape so that its moment of inertia changes, its angular velocity changes to keep the angular momentum constant.

Key Concept 11: In processes that conserve angular momentum, the kinetic energy can change if nonconservative forces act.

Key Concept 12: The total angular momentum of a system that includes a rigid body and a particle is the sum of the angular momenta for the rigid body and for the particle. You can find the magnitude of the angular momentum of a particle about a rotation axis by multiplying the magnitude of its linear momentum by the perpendicular distance from the axis to the line of the particle's velocity.

Key Concept 13: A spinning rigid body will precess if the net external torque on the rigid body is perpendicular to the body's angular momentum vector.

Conceptual Questions

1: Ranking torques

In the following diagram, each rod pivots about the indicated axis with the indicated force:

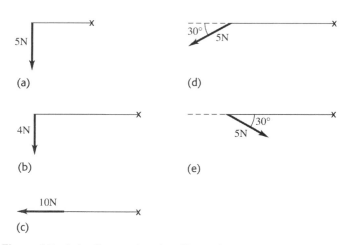

Figure 1 Rank the diagrams in order of increasing torque.

IDENTIFY, SET UP, AND EXECUTE Torque is given by $\tau = rF\sin\phi$, so we must examine the magnitude of the force, the location at which the force is applied, and the direction in which the force is applied. Comparing (a) and (b), we see that both forces act in the same direction, the force is larger in (a), and the moment arm is larger in (b). We estimate the moment arm in (a) as half the moment arm of (b). Since the force in (a) is less than double the force in (b), the torque in (b) is greater than that in (a). No torque is generated in (c), because the force acts along the moment arm. The force and moment arm are the same in (d) and (e), but the directions are different. However, since both forces act 30° from the horizontal, the components of the forces perpendicular to the moment arm are the same for both and the torque in (d) and (e) are the same. The vertical component of the force in (d) and (e) is 2.5 N, which is less than the force in (b) and less than half of the force in (a). Diagrams (d) and (e) both show less torque than diagrams (a) and (b).

Ranked in order of increasing torque, the diagrams are therefore (c), (d) = (e), (a), (b).

KEY CONCEPT 1   **EVALUATE** Torque is the most complicated quantity. It depends on the magnitude and direction of the force responsible for it, as well as where that force acts relative to the rotation axis. Gaining intuition about torque will help guide you through problems.

2: Massless versus massive pulleys

Why were massless pulleys used?

IDENTIFY, SET UP, AND EXECUTE Consider the net external torque acting on a pulley with a rope resting on the pulley. The left segment of rope has a tension T_L, and the right segment of rope has a tension T_R. The net external torque is then

$$\sum \tau = T_R R - T_L R = I\alpha.$$

Each tension creates a torque TR, since the rope will be perpendicular to the radius and the two torques are opposite in direction. Massless pulleys have no moment of inertia (since their mass is zero); therefore, the net external torque acting on the pulley is zero. If the net external torque is zero, the torques in each segment must be equal and the tensions are equal in each rope.

When we include the pulley's mass, the pulley has a moment of inertia. If the pulley accelerates, there must be a difference in the left and right torques and the tensions must also be different. Only when the angular acceleration is zero are the two tensions equal.

Using massless pulleys simplified our analysis and let us focus on learning about forces.

KEY CONCEPT 2 **EVALUATE** From now on, we must assume that the tensions in segments of rope may vary and we must identify each segment's tension separately.

How does the acceleration of each segment of rope compare when we include a pulley's mass? There is no change: Objects connected by the rope are constrained to have the same magnitude of acceleration.

3: Spinning on a roundabout

You are standing at the center of a rotating playground roundabout (a round, horizontal plate that spins about its center axis). As you move to the edge, will your angular speed increase, decrease, or stay the same?

IDENTIFY, SET UP, AND EXECUTE To simplify our analysis, we ignore friction in the roundabout. Without friction, there is no torque to slow the roundabout, so angular momentum is conserved. As you move to the edge, the moment of inertia of the system increases. (Your mass moves to a greater radius.) For angular momentum to be conserved, the angular speed must be reduced.

EVALUATE Like linear momentum and energy, angular momentum is a conserved quantity. We could solve this problem numerically by picking initial and final angular momenta before and after you moved and setting them equal to each other.

4: Standing on a turntable

You are standing on a small frictionless turntable with your arms outstretched and spinning about the axis of the turntable. As you pull your arms in, you spin faster. Does your rotational kinetic energy increase, decrease, or stay the same?

IDENTIFY AND SET UP There are no external torques acting on you or the turntable, so angular momentum is conserved. You spin faster, since you decrease your moment of inertia as you bring your arms in, leading to a greater angular velocity. Rotational kinetic energy is given by

$$K = \tfrac{1}{2} I \omega^2.$$

Since kinetic energy depends on both moment of inertia and angular velocity, and since both change, we'll have to consider more information. Angular momentum is conserved, so $I\omega$ is constant.

EXECUTE We include angular momentum in the kinetic energy:

$$K = \tfrac{1}{2} L \omega.$$

Since L is constant and ω increases as you bring your arms in, the kinetic energy increases.

KEY CONCEPT 10 **EVALUATE** Where does the increased energy come from? You must do work to pull your arms in; doing work on the system increases its kinetic energy.

This example also applies to a figure skater spinning on ice.

Problems

1: Balancing a food tray

A waiter balances a tray of food on his hand. On the tray is a 0.40 kg drink and a 2.0 kg lobster dinner. The drink is placed 6.5 cm from one edge of the tray, and the lobster dinner is placed 8.0 cm from the opposite edge. The tray has a mass of 1.2 kg and a diameter of 42 cm. Where should the waiter hold the tray so that it doesn't tip over?

IDENTIFY Figure 2 shows a sketch, and a free-body diagram, of the food tray. The tray should be held so that it is in equilibrium and the net external torque on it is zero. The target variable is the location at which the waiter's hand holds the tray.

SET UP The forces on the food tray include the force of the waiter's hand holding the tray, the weight of the tray, and the normal forces due to the lobster dinner and drink. To find the location of the hand, we need to consider the torque acting on the tray. When the tray is in equilibrium, the net torque must be zero about any axis. We'll take the axis to be the left edge, marked by an X.

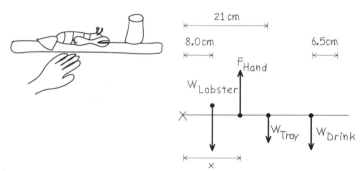

Figure 2 Problem 1 sketch and free-body diagram.

When we choose the left edge as the axis, we note that four torques act on the tray, corresponding to the four forces on the tray. Each of the four forces is applied perpendicular to the moment arm (the plane of the tray); each torque is the magnitude of the force times the distance from the axis. We'll take counterclockwise torques to be positive.

EXECUTE Since the tray is in equilibrium, the net external torque is zero:

$$\sum \tau = \tau_{tray} + \tau_{lobster} + \tau_{drink} + \tau_{hand} = 0.$$

Writing the four torques explicitly, we have

$$\sum \tau = \tau_{tray} + \tau_{lobster} + \tau_{drink} + \tau_{hand} = -m_{tray}\, g x_{tray} - n_{lobster}\, x_{lobster} - n_{drink}\, x_{drink} + n_{hand}\, x_{hand} = 0.$$

The first term is the torque due to the weight of the tray; the moment arm is half the tray diameter (at the center of mass). The second and third terms are the torques due to the normal forces of the lobster dinner and drink, respectively. The normal force is equal to the weight of the objects, and the moment arm is the distance from the left edge of the tray. The first three terms are negative, since they are all in the clockwise direction. The last term is the torque due to the normal force of the waiter's hand and is the only positive term, because it is counterclockwise. We need to find this normal force of his hand to solve the problem. We find the force by using Newton's first law:

$$\sum F = 0.$$

In the vertical direction, four forces act on the tray. We have
$$\Sigma F = -m_{tray} g - n_{lobster} - n_{drink} + n_{hand} = 0.$$
Solving yields
$$n_{hand} = (m_{tray} + m_{lobster} + m_{drink})g = ((1.2 \text{ kg}) + (2.0 \text{ kg}) + (0.40 \text{ kg}))(9.8 \text{ m/s}^2) = 35.3 \text{ N}.$$
We can now solve for the location of the waiter's hand:
$$x_{hand} = \frac{m_{tray} g x_{tray} + m_{lobster} g x_{lobster} + m_{drink} g x_{drink}}{n_{hand}}$$
$$= \frac{((1.2 \text{ kg})(0.21 \text{ m}) + (2.0 \text{ kg})(0.080 \text{ m}) + (0.40 \text{ kg})(0.42 \text{ m} - 0.065 \text{ m}))(9.8 \text{ m/s}^2)}{(35.3 \text{ N})}$$
$$= 0.15 \text{ m}.$$
The waiter should hold the tray 15 cm from the left edge (closest to the lobster dinner).

KEY CONCEPT 2 **EVALUATE** This equilibrium problem required us to apply both the equilibrium torque and the equilibrium force conditions to solve the problem. You should be familiar with the equilibrium force condition and should need to gain expertise only in torque to solve similar problems.

Experience shows that we may be able to simplify similar problems by picking an axis that coincides with the location at which a force is applied. This choice will reduce the number of torques in the problem. In the current problem, for example, we could have chosen the axis to be at the location of the lobster dinner, thus removing the torque due to the normal force of the lobster dinner.

Practice Problem: What force must the waiter's hand provide if there is no drink? *Answer:* 31.4 N

Extra Practice: Now where does his hand need to be? *Answer:* 12.9 cm

2: Tension in string attached to a falling cylinder

A thin, light string is wrapped around the outer rim of a uniform hollow cylinder of mass 12.0 kg, inner radius 15.0 cm, and outer radius 30.0 cm. The cylinder is released from rest. What is the tension in the string as the cylinder falls?

IDENTIFY We'll apply Newton's second law and its rotational analog to solve for the tension, the target variable.

SET UP Figure 3 shows a sketch, and a free-body diagram, of the situation. As the cylinder falls, it'll accelerate downward and rotate about its central axis. In falling, the cylinder will rotate faster and undergo angular acceleration. The cylinder has both a net external force and a net external torque acting on it.

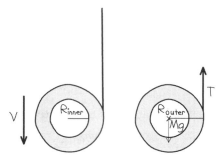

Figure 3 Problem 2 sketch and free-body diagram.

EXECUTE We first apply Newton's second law to the translational motion of the center of mass in the vertical direction. The only forces acting in the vertical direction are gravity and tension. We have

$$\sum F_y = Mg - T = Ma_{cm,y}.$$

The moment of inertia of the hollow cylinder is $I = \frac{1}{2}M(R_{inner}^2 + R_{outer}^2)$. The one torque acting on the cylinder as it rotates about its central axis is due to the tension force. Gravity acts on the center of mass, but creates no torque about the central axis. The torque acts perpendicular to the outer radius. Thus,

$$\sum \tau_z = TR_{outer} = I\alpha_z = \frac{1}{2}M(R_{inner}^2 + R_{outer}^2)\alpha_z.$$

We can relate the two accelerations, since the cylinder falls without slipping:

$$a_y = R_{outer}\alpha_z.$$

Solving for the acceleration of the center of mass in the first equation yields

$$a_y = \frac{Mg - T}{M}.$$

Substituting the last two equations into the torque result gives

$$TR_{outer} = \frac{1}{2}M(R_{inner}^2 + R_{outer}^2)\frac{a_y}{R_{outer}} = \frac{1}{2}M(R_{inner}^2 + R_{outer}^2)\frac{Mg - T}{MR_{outer}},$$

$$T = \frac{(R_{inner}^2 + R_{outer}^2)Mg}{R_{inner}^2 + 3R_{outer}^2} = \frac{((15.0 \text{ cm})^2 + (30.0 \text{ cm})^2)(12.0 \text{ kg})(9.8 \text{ m/s}^2)}{(15.0 \text{ cm})^2 + 3(30.0 \text{ cm})^2} = 45.2 \text{ N}.$$

The tension in the string is 45.2 N.

KEY CONCEPT 3 **EVALUATE** This problem resembles Problem 1, but with the addition of torque. Once torque is included, the problem becomes a relatively straightforward algebraic one. We also see that the tension in the string is less than that for a stationary cylinder. If the cylinder were stationary, the tension would have been 118 N.

Practice Problem: At what rate does the cylinder accelerate? *Answer:* 6.03 m/s^2

Extra Practice: What is the angular acceleration? *Answer:* 20.1 rad/s^2

3: Acceleration and tension of two blocks connected by a pulley

Two blocks are connected to each other by a light cord passing over a pulley as shown in Figure 4. Block *A* has a mass of 5.00 kg and block *B* has a mass of 4.00 kg. The pulley has a mass of 8.00 kg and a radius of 4.00 cm. Find the acceleration of the blocks and the tensions in the horizontal and vertical segments of the cord. Assume that the pulley is a solid, uniform disk and there is no friction between block A and the table.

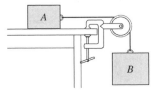

Figure 4 Problem 3.

IDENTIFY We'll apply the net-force and net-torque equations to solve the problem. The accelerations of both blocks are the same. Our target variables are the two tensions and the acceleration of the blocks.

SET UP Figure 5 shows the free-body diagram of the two blocks and the pulley. The forces on the blocks include tension, gravity, and the normal force (block A). We assume that the tensions of the two segments are not equal and label them T_A and T_B. The two tension forces lead to two torques acting on the pulley. (The axis of rotation is the center of the pulley.)

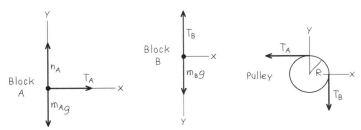

Figure 5 Problem 3 free-body diagrams.

EXECUTE We first apply Newton's second law to each block. Block A (with mass m_A) accelerates in the x- direction due to tension T_A, so

$$\sum F_x = T_A = m_A a.$$

Gravity and tension T_B act on block B (with mass m_B) and that block accelerates at the same rate as block A in the y- direction, so

$$\sum F_y = m_B g + (-T_B) = m_B a.$$

As block B falls, the pulley's rotational speed increases. The net torque on the pulley is

$$\sum \tau_z = \tau_A - \tau_B = I \alpha_z,$$

where we have taken the counterclockwise torque as positive and clockwise torque as negative. The moment of inertia of a uniform cylinder is $I = \frac{1}{2} M R^2$. We assume that the cord doesn't slip on the pulley, so we relate the angular acceleration of the pulley to the tangential acceleration of the cord (a):

$$\alpha_z = -\frac{a}{R}.$$

We included a minus sign in the equation because the pulley rotates clockwise (negative, according to our convention). The tension forces act perpendicular to the moment arm, so the torques are simply TR. Rewriting the net external torque, we have

$$\sum \tau_z = \tau_A - \tau_B = T_A R - T_B R = I\alpha_z = \tfrac{1}{2} MR^2 \left(-\frac{a}{R}\right).$$

Simplifying yields

$$T_B - T_A = \tfrac{1}{2} Ma.$$

Our second-law equations are used to replace the tensions:

$$m_B g - m_B a - m_A a = \tfrac{1}{2} Ma.$$

Solving for the acceleration gives

$$a = \frac{m_B g}{m_B + m_A + \tfrac{1}{2} M} = \frac{(4.00 \text{ kg})(9.8 \text{ m/s}^2)}{(4.00 \text{ kg}) + (5.00 \text{ kg}) + \tfrac{1}{2}(8.00 \text{ kg})} = 3.02 \text{ m/s}^2.$$

Using this result to find the two tensions, we get

$$T_A = m_A a = (5.00 \text{ kg})(3.02 \text{ m/s}^2) = 15.1 \text{ N},$$

$$T_B = m_B g - m_B a = (4.00 \text{ kg})(9.80 \text{ m/s}^2) - (4.00 \text{ kg})(3.02 \text{ m/s}^2) = 27.1 \text{ N}.$$

The blocks accelerate at 3.02 m/s², the tension in the horizontal segment of the cord is 15.1 N, and the tension in the vertical segment of the cord is 27.1 N.

KEY CONCEPT 3 **EVALUATE** The tensions in the two segments of the cord differ by almost a factor of two. The tension was constant in both segments of the cord. What causes this difference? The current problem includes the pulley's mass, resulting in some energy spent on increasing the pulley's angular velocity, leaving less energy available for the blocks. This problem also illustrates why we let the first pulleys we encountered be massless.

Practice Problem: What is the angular acceleration of the pulley? *Answer:* -75.5 rad/s²

Extra Practice: What is the acceleration of the blocks if they have the same mass as the pulley? *Answer:* 3.92 m/s²

4: Solid cylinder rolling down ramp

A solid cylinder rolls without slipping down an incline of 40°. Find the acceleration and minimum coefficient of friction needed to prevent slipping.

IDENTIFY We'll apply translational and rotational dynamics to the cylinder. Since the cylinder doesn't slip, we'll use the relationship between the linear and angular acceleration of the cylinder. The target variables are the acceleration and coefficient of friction.

SET UP Figure 6 shows a sketch, and a free-body diagram, of the cylinder on the incline. Gravity, the normal force, and the frictional force act on the cylinder. If we set the axis of rotation at the center of the cylinder, a torque due to friction acts on the cylinder. We'll apply the net-force and net-torque equations to solve the problem.

A rotated coordinate system is included in the free-body diagram to simplify our analysis.

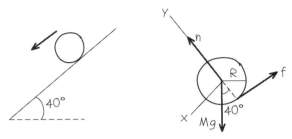

Figure 6 Problem 4 sketch and free-body diagram.

EXECUTE We first apply Newton's second law to the translational motion along the x-axis:

$$\Sigma F_x = Mg\sin\theta - f_s = Ma_x.$$

The equation of motion for the rotation about the axis is

$$\Sigma \tau_z = f_s R = I_{cm}\alpha_z = \tfrac{1}{2}MR^2\alpha_z,$$

where we included the moment of inertia ($I = \tfrac{1}{2}MR^2$). The translational and rotational accelerations are related by $a_{cm} = \alpha_z R$, since the cylinder rolls without slipping. Combining and writing the second equation in terms of f_s yields

$$f_s = \tfrac{1}{2}Ma_x.$$

We use this result in the first equation and solve for the acceleration:

$$Mg\sin\theta - \tfrac{1}{2}Ma_x = Ma_x,$$

$$a_x = \tfrac{2}{3}g\sin\theta = \tfrac{2}{3}(9.80 \text{ m/s}^2)\sin 40° = 4.20 \text{ m/s}^2.$$

To find the minimum coefficient of static friction, we use the equilibrium equation along the y-axis:

$$\Sigma F_y = n - Mg\cos\theta = 0.$$

The friction force is then

$$f_s = \mu_s n = \mu_s Mg\cos\theta = \tfrac{1}{2}Ma_x.$$

Solving for μ_s gives

$$\mu_s = \frac{\tfrac{1}{2}a_x}{g\cos\theta} = \frac{\tfrac{1}{2}(4.20 \text{ m/s}^2)}{(9.8 \text{ m/s}^2)\cos 40°} = 0.280.$$

The cylinder accelerates at 4.20 m/s^2, and the minimum coefficient of static friction that will prevent slipping is 0.280.

KEY CONCEPT 7 **EVALUATE** This solution is a straightforward application of net-force and net-torque problem-solving techniques. We see that neither the mass nor the radius of the cylinder affect the results, which are valid for *any* cylinder rolling down a 40° incline.

Practice Problem: What coefficient is needed if it is a 30° ramp? *Answer:* 0.192

Extra Practice: What if it is a 60° ramp? *Answer:* 0.577

5: Pivoting rod

A rod of mass 1.0 kg and length 0.50 m is connected to the ceiling by a frictionless hinge. It is released from rest when it is in the horizontal position. What is its angular velocity when it is vertical?

IDENTIFY There are no dissipative forces, so we'll use conservation of energy. The rod begins with only gravitational potential energy, which is subsequently transformed to kinetic energy. The target variable is the rod's angular velocity when the rod is vertical.

SET UP Figure 7 shows a diagram of the rod as it falls. The center of mass of the rod is at the center of the rod; the gravitational potential energy depends on the position of the center of mass. As the rod falls, its rotational kinetic energy increases.

We set the origin of the coordinate axis at the center of the rod when it is in the vertical position. The initial height of the rod in this coordinate system is $h = L/2$.

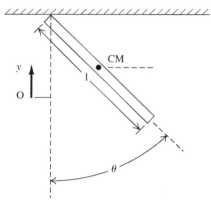

Figure 7 Problem 5 sketch and free-body diagram.

EXECUTE Conservation of energy gives
$$K_1 + U_1 = K_2 + U_2.$$
Initially, there is only gravitational potential energy ($MgL/2$). When the rod is in the vertical position, there is only kinetic energy. The kinetic energy of the rod is
$$K = \tfrac{1}{2}I\omega_z^2 = \tfrac{1}{2}(\tfrac{1}{3}ML^2)\omega_z^2.$$
Combining these energies, we find that
$$U_1 = K_2,$$
$$Mg\frac{L}{2} = \tfrac{1}{6}ML^2\omega_z^2.$$
Solving for angular velocity gives
$$\omega_z = \sqrt{\frac{3g}{L}} = \sqrt{\frac{3(9.8 \text{ m/s}^2)}{(0.5 \text{ m})}} = 7.67 \text{ rad/s}.$$
The angular velocity of the rod when it is vertical is 7.67 rad/s.

EVALUATE Could we have obtained this result by using rotational dynamics? The forces and torques change with position; hence, the solution would have been more challenging.

Practice Problem: What is the angular acceleration initially? *Answer:* 29.4 rad/s²

Extra Practice: What is the torque initially? *Answer:* 2.45 N m

Study Guide for Dynamics of Rotational Motion

6: Baggage carousel

A baggage carousel has a mass of 500.0 kg and can be approximated as a disk of radius 2.0 m. It is rotating freely at an angular velocity of 1.0 rad/s when 10 pieces of baggage, each with a mass of 20.0 kg, are dropped on the edge of the carousel. Assuming that no external torques act on the system, what is the final angular velocity of the system?

IDENTIFY Since there are no external torques, the total angular momentum of the system remains constant. We'll find the initial angular momentum and set it equal to the final angular momentum. The target variable is the carousel's angular velocity after the baggage is added.

SET UP The initial angular momentum is that of the carousel. As the bags are added, they share the angular momentum, resulting in a slower final angular velocity.

EXECUTE The initial angular momentum is

$$L_1 = I_{carousel}\, \omega_1 = (\tfrac{1}{2} M_{carousel}\, R_{carousel}^2)\, \omega_1.$$

The final angular momentum will be the angular momentum of the carousel plus the angular momentum of the 10 bags:

$$L_2 = I_{carousel}\, \omega_2 + 10 m_{bag}\, r_{bag}^2\, \omega_2 = (\tfrac{1}{2} M_{carousel}\, R_{carousel}^2)\, \omega_2 + 10 m_{bag}\, r_{bag}^2\, \omega_2.$$

Equating the initial and final angular momenta gives

$$(\tfrac{1}{2} M_{carousel}\, R_{carousel}^2)\, \omega_1 = (\tfrac{1}{2} M_{carousel}\, R_{carousel}^2)\, \omega_2 + 10 m_{bag}\, r_{bag}^2\, \omega_2.$$

All of the radii are 2.0 m, so they cancel. Solving for the final angular velocity gives

$$\omega_2 = \frac{(\tfrac{1}{2} M_{carousel})\, \omega_1}{(\tfrac{1}{2} M_{carousel}) + 10 m_{bag}} = \frac{(\tfrac{1}{2}(500.0\ \text{kg}))(1.0\ \text{rad/s})}{(\tfrac{1}{2}(500.0\ \text{kg})) + 10(20.0\ \text{kg})} = 0.556\ \text{rad/s}.$$

The final angular velocity of the system is 0.556 rad/s.

KEY CONCEPT 12 EVALUATE We see how to apply conservation of angular momentum in this problem. Was energy conserved? Even though there were no external torques, there must have been external forces, since energy was not conserved.

Practice Problem: How much energy was lost as the baggage was added? *Answer:* 218 J

Extra Practice: How much energy is lost if the initial angular velocity was 2.0 rad/s? *Answer:* 887 J

7: Rotating mass on a string

A 0.10 kg block of mass is attached to a cord that passes through a hole in a horizontal frictionless surface. The block initially is rotating in a circle of radius 0.20 m at an angular velocity of 7.0 rad/s. A force is applied to the cord, shortening it to 0.10 m. What is the new angular velocity of the block?

IDENTIFY AND SET UP There is no external torque, as the force exerts no torque at the hole. Therefore, the total angular momentum of the system remains constant. The target variable is the block's final angular velocity.

EXECUTE The initial angular momentum is

$$L_1 = I_1 \omega_1 = m r_1^2 \omega_1.$$

The final angular momentum is

$$L_2 = I_2\omega_2 = mr_2^2\omega_2.$$

Setting these equal to each other gives

$$mr_1^2\omega_1 = mr_2^2\omega_2.$$

Solving results in

$$\omega_2 = \frac{r_1^2}{r_2^2}\omega_1 = 4(7.0 \text{ rad/s}) = 28 \text{ rad/s}.$$

The final angular velocity of the block is 28 rad/s.

KEY CONCEPT 11 **EVALUATE** We see that the final angular velocity is greater than the initial angular velocity, since the radius decreased. The final result does not depend on the mass of the block. We can go on to find the amount of work done by the force by comparing the initial and final kinetic energies.

Practice Problem: How much work did the force do when shortening the cord? *Answer:* 0.29 J

Extra Practice: How fast does the block move if the cord is suddenly cut? *Answer:* 2.8 m/s

Try It Yourself!

1: Torque in a grinding wheel

How much torque is required to bring a 2.0 kg grinding wheel of radius 0.1 m to rest from an initial velocity of 3000 rpm? The grinding wheel stops in 10 rev. How much work is done by the torque to bring the grinding wheel to a halt? Assume constant angular acceleration.

IDENTIFY AND SET UP We begin by finding the angular acceleration and use that in combination with the moment of inertia to find the torque. The work is the torque times the angular displacement.

EXECUTE The angular acceleration is -4.5×10^5 rev/min^2, which, when combined with the moment of inertia, gives a torque of -7.85 Nm.
The work done by the torque is -493 J.

KEY CONCEPT 8 **EVALUATE** Why are the angular acceleration, torque, and work done all negative values? The negative sign indicates that the grinding wheel is slowing.

Practice Problem: What is the initial angular momentum? *Answer:* 3.14 kg m^2/s

Extra Practice: How long does it take the wheel to stop? *Answer:* 0.4 s

2: Mass on a flywheel

A cord is wrapped around the rim of a uniform flywheel of radius 0.20 m and mass 10.0 kg. A 10.0 kg mass is suspended from the cord 10.0 m above the floor. How much time does it take the mass to hit the floor? What is the tension in the rope as it falls?

IDENTIFY AND SET UP Apply the net-force and net-torque equations to solve the problem. The length of cord pulled from the flywheel is equal to the arc length $r\theta$ at the wheel.

EXECUTE Combining the torque and force equations yields the relation

$$a = \frac{g}{1 + \frac{m_{\text{flywheel}}}{2 m_{\text{mass}}}}$$

for the acceleration. This result can be combined with linear kinematics to find the time to fall, 1.74 s. The tension in the rope is found from the net-force equation:

$$m_{\text{mass}} g - T = m_{\text{mass}} a.$$

The tension is 32.7 N.

KEY CONCEPT 3 **EVALUATE** Can energy conservation be used to check the results?

Practice Problem: What is the final angular speed of the flywheel? *Answer:* 546 rpm

Extra Practice: How much angular momentum does the flywheel have when the block hits the ground? *Answer:* 11.4 kg m²/s

3: Man on a turntable

A turntable with moment of inertia of 2000 kg m² makes one revolution every 5.0 s. A man of mass 100 kg standing at the center of the turntable runs out along a radius fixed on the turntable. What is the angular velocity of the turntable when the man is 3.0 m from the center?

IDENTIFY AND SET UP There are no external torques acting on the system, so angular momentum is conserved. As the man runs out, he changes the angular velocity of the system.

EXECUTE Conservation of angular momentum gives

$$I_{\text{turntable}} \omega_1 = I_{\text{turntable}} \omega_2 + MR^2 \omega_2.$$

This results in a final angular velocity of 0.14 rev/s, or one revolution every 7.25 s.

KEY CONCEPT 10 **EVALUATE** Does the man do positive or negative work on the system? Work is done on the man as he runs out.

Practice Problem: How much work is done? *Answer:* –457 J

Extra Practice: What torque must be provided to stop the turntable in 15 seconds? *Answer:* 170 N m

Problem Summary

Our problem-solving methodology continues to encompass the following techniques:

- Identifying the general procedure to find the solution.
- Sketching the situation when no figure is provided.
- Identifying the forces and torques acting on the objects.
- Identifying the forms of energy included in the problem.
- Drawing free-body diagrams of the bodies.
- Applying appropriate coordinate systems to the diagrams.
- Applying the equations of motion to find relations among the forces, masses, and accelerations.

- Applying conservation of energy and conservation of momentum when appropriate.
- Solving the equations through algebra and substitutions.
- Reflecting on the results and checking for inconsistencies.

Expert problem solvers use this foundation at all levels of physics investigations, from introductory courses through cutting-edge research projects.

Key Example Variation Problems

Solutions to these problems are in Chapter 10 of the Student's Solutions Manual.

Be sure to review EXAMPLES 10.2 and 10.3 (Section 10.2) before attempting these problems.

VP10.3.1 In the cylinder and cable apparatus of Example 10.2, you apply a force to the cable so that a point on the horizontal part of the cable accelerates to the left at 0.60 m/s². What are the magnitudes of (a) the angular acceleration of the cylinder, (b) the torque that the cable exerts on the cylinder, and (c) the force that you exert on the cable?

VP10.3.2 In the cylinder, cable, and block apparatus of Example 10.3, you replace the solid cylinder with a thin-walled, hollow cylinder of mass M and radius R. Find (a) the acceleration of the falling block and (b) the tension in the cable as the block falls.

VP10.3.3 A bucket of mass m is hanging from the free end of a rope whose other end is wrapped around a drum (radius R, mass M) that can rotate with negligible friction about a stationary horizontal axis. The drum is not a uniform cylinder and has unknown moment of inertia.

When you release the bucket from rest, you find that it has a downward acceleration of magnitude a. What are (a) the tension in the cable between the drum and the bucket and (b) the moment of inertia of the drum about its rotation axis?

VP10.3.4 In the cylinder, cable, and block apparatus of Example 10.3, you attach an electric motor to the axis of the cylinder of mass M and radius R and turn the motor on. As a result the block of mass m moves upward with an upward acceleration of magnitude a. What are (a) the tension in the cable between the cylinder and the block, (b) the magnitude of the torque that the cable exerts on the cylinder, and (c) the magnitude of the torque that the motor exerts on the cylinder?

Be sure to review EXAMPLES 10.6 and 10.7 (Section 10.3) before attempting these problems.

VP10.7.1 In the primitive yo-yo apparatus of Example 10.6, you replace the solid cylinder with a hollow cylinder of mass M, outer radius R, and inner radius $R/2$. Find (a) the downward acceleration of the hollow cylinder and (b) the tension in the string.

VP10.7.2 A thin-walled, hollow sphere of mass M rolls without slipping down a ramp that is inclined at an angle b to the horizontal. Find (a) the acceleration of the sphere, (b) the magnitude of the friction force that the ramp exerts on the sphere, and (c) the magnitude of the torque that this force exerts on the sphere.

VP10.7.3 You redo the primitive yo-yo experiment of Example 10.6, but instead of holding the free end of the string stationary, you move your hand vertically so that the tension in the string equals $2Mg/3$. (a) What is the magnitude of the vertical acceleration of the yo-yo's center of mass? Does it accelerate upward or downward? (b) What is the angular acceleration of the yo-yo around its axis?

VP10.7.4 You place a solid cylinder of mass M on a ramp that is inclined at an angle β to the horizontal. The coefficient of static friction for the cylinder on the ramp is μ_s. (a) If the cylinder rolls downhill without slipping, what is the magnitude of the friction force that the ramp exerts on the cylinder? (b) You find by varying the angle of the ramp that the cylinder rolls without slipping if β is less than a certain critical value but the cylinder slips if β is greater than this critical value. What is this critical value of β?

Be sure to review EXAMPLES 10.11 and 10.12 (Section 10.6) before attempting these problems.

VP10.12.1 In the situation shown in Example 10.11, suppose disk A has moment of inertia I_A and initial angular speed ω_A, while disk B has moment of inertia $I_A/4$ and initial angular speed $\omega_A/2$. Initially disks A and B are rotating in the *same* direction. (a) What is the final common angular speed of the two disks? (b) What fraction of the initial rotational kinetic energy remains as rotational kinetic energy after the disks have come to their final common angular speed?

VP10.12.2 In the situation shown in Example 10.11, suppose disk A has moment of inertia I_A and initial angular speed ω_A, while disk B has moment of inertia $I_A/4$ and initial angular speed $\omega_A/2$. Initially disks A and B are rotating in *opposite* directions. (a) What is the final common angular speed of the two disks? (b) What fraction of the initial rotational kinetic energy remains as rotational kinetic energy after the disks have come to their final common angular speed?

VP10.12.3 Suppose that instead of hitting the center of the door, the bullet in Example 10.12 strikes the door at the edge farthest away from the hinge and embeds itself there. (a) What is the angular speed of the door just after the bullet embeds itself? (b) What fraction of the initial kinetic energy of the bullet remains as kinetic energy after the collision?

VP10.12.4 A thin-walled, hollow sphere of mass M and radius R is free to rotate around a vertical shaft that passes through the center of the sphere. Initially the sphere is at rest. A small ball of clay of the same mass M moving horizontally at speed v grazes the surface of the sphere at its equator. After grazing the surface, the ball of clay is moving at speed $v/2$. (a) What is the angular momentum of the ball of clay about the shaft before it grazes the surface? After it grazes the surface?
(b) What is the angular speed of the sphere after being grazed by the ball of clay?
(c) What fraction of the ball of clay's initial kinetic energy remains as the combined kinetic energy of the sphere and the ball of clay?

STUDENT'S SOLUTIONS MANUAL FOR DYNAMICS OF ROTATIONAL MOTION

VP10.3.1. **IDENTIFY:** The force of the cable produces a torque on the cylinder, giving it an angular acceleration. We apply the rotational analog of Newton's second law.

SET UP: $\sum \tau_z = I\alpha_z$, $I = \frac{1}{2}MR^2$ for a solid cylinder, $a_{\tan} = r\alpha_z$ apply in this case.

EXECUTE: **(a)** $a_{\tan} = r\alpha_z$ gives $\alpha_z = a_{\tan}/R = (0.60 \text{ m/s}^2)/(0.060 \text{ m}) = 10 \text{ rad/s}^2$.

(b) Apply $\sum \tau_z = I\alpha_z$ to the cylinder. There is only one torque, so $\tau_z = I\alpha_z = \frac{1}{2}MR^2\alpha_z$.

$\tau_z = \frac{1}{2}(50 \text{ kg})(0.060 \text{ m})^2(10 \text{ rad/s}^2) = 0.90 \text{ N}\cdot\text{m}$.

(c) $\tau_z = FR \rightarrow F = \tau_z/R = (0.90 \text{ N}\cdot\text{m})/(0.060 \text{ m}) = 15 \text{ N}$.

EVALUATE: When using $\sum \tau_z = I\alpha_z$, α_z *must* be in radian measure.

VP10.3.2. **IDENTIFY:** Gravity pulls the block, causing tension in the cable. This tension produces a torque on the cylinder, giving it an angular acceleration. We apply the rotational analog of Newton's second law to the wheel and the linear form to the falling block.

SET UP: $\sum \tau_z = I\alpha_z$, $I = MR^2$ for a hollow cylinder, $a_{\tan} = r\alpha_z$, $\sum F_y = ma_y$ apply in this case. Call m the mass of the block and M the mass of the wheel.

EXECUTE: **(a)** Apply $\sum F_y = ma_y$ to the block. Since the block accelerates downward, it is best to call the y-axis positive downward. This gives $mg - T = ma_y$ (Eq. 1)
Now apply $\sum \tau_z = I\alpha_z$ to the wheel. $TR = MR^2 \alpha_z$. Using $a_y = a_{\tan} = R\alpha_z$, the second equation becomes $T = MR(a_y/R) = Ma_y$ (Eq. 2)
Combining Eq. 1 and Eq. 2 gives $mg - Ma_y = ma_y$. Solving for a_y gives
$a_y = \frac{mg}{m+M} = \frac{g}{1 + M/m}$.

(b) Eq. 2 gives $T = Ma_y = \frac{Mg}{1 + M/m}$, which can also be written as $T = \frac{mg}{1 + m/M}$.

EVALUATE: Careful! The torque on the wheel is *not* equal to mgR! It is the *tension* that turns the wheel, and $T \neq mg$. Look at our answers in some limiting cases. If $m \gg M$, the ratio M/m is very small, so the acceleration approaches g. This is reasonable because the wheel has very little effect on the block, so the block is essentially in freefall. The tension approaches zero because M approaches zero. This is reasonable because the block is essentially in freefall. If $M \gg m$, the acceleration approaches zero since the small m produces very little acceleration of the much heavier wheel. The tension approaches mg because $m/M \ll 1$, so the block is essentially just hanging with almost no acceleration.

VP10.3.3. **IDENTIFY:** Gravity pulls the block, causing tension in the cable. The torque on the cylinder is due to the tension in the cable. We apply the rotational analog of Newton's second law to the cylinder and the linear form to the block. We do not know the moment of inertia of the drum.
SET UP: $\sum \tau_z = I\alpha_z$, $a_{\tan} = r\alpha_z$, $\sum F_y = ma_y$ apply in this case. Call m the mass of the block.
EXECUTE: (a) Apply $\sum F_y = ma_y$ to the block. Since the block accelerates downward, it is best to call the y-axis positive downward. This gives $mg - T = ma$. Solving for T gives
$T = m(g - a)$ (Eq. 1)
(b) Applying $\sum \tau_z = I\alpha_z$ to the drum gives $TR = I\alpha_z$ (Eq. 2)
We also have $a_y = a_{\tan} = R\alpha_z$, which gives $\alpha_z = a_y/R$ (Eq. 3)
Combining Equations 1, 2, and 3 and solving for I gives $I = mR^2(g/a - 1)$.
EVALUATE: Check our result for $a \to g$, which gives $I \to 0$. This is reasonable because the block is then essentially in freefall, meaning that the drum had almost no effect on it. This would be the case of the drum's moment of inertia was very very small. You could also interpret our result for $a > g$ to understand its meaning in that case.

VP10.3.4. **IDENTIFY:** Gravity pulls the block, causing tension in the cable. The torque on the cylinder is due to the tension in the cable and the torque produced by the motor. We apply the rotational analog of Newton's second law to the cylinder and the linear form to the block.
SET UP: $\sum \tau_z = I\alpha_z$, $a_{\tan} = r\alpha_z$, $\sum F_y = ma_y$ apply in this case. Call m the mass of the block and M the mass of the cylinder. $I = \frac{1}{2}MR^2$ for a solid uniform cylinder.
EXECUTE: (a) Apply $\sum F_y = ma_y$ to the block. Since the block accelerates upward, it is best to call the y-axis positive upward. This gives $T - mg = ma$. Solving for T gives $T = m(g + a)$.
(b) $\tau_z = TR = mR(g + a)$.
(c) Apply $\sum \tau_z = I\alpha_z$ to the cylinder. The torque due to the motor must be opposite in direction to that of the tension and it must have a greater magnitude.
$\tau_{\text{motor}} + \tau_{\text{tension}} = I\alpha_z \to \tau_{\text{motor}} - TR = \left(\frac{1}{2}MR^2\right)\left(\frac{a}{R}\right)$. Using the tension from part (a) and solving for τ_{motor}, we get $\tau_{\text{motor}} = mR(g + a) + fc\, MRa$.
EVALUATE: Check if our result is reasonable. If either m or M are large, the torque must be large to produce the given acceleration.

VP10.7.1. **IDENTIFY:** The cylinder is rotating and translating at the same time. The rotational analog of Newton's second law applies to its rotational motion, and the linear form applies to its center of mass motion.
SET UP: $\sum F_y = ma_y$ applies to the vertical motion and $\sum \tau_z = I\alpha_z$ applies to the rotational motion about the center of mass. For a solid hollow cylinder, $I = \frac{1}{2}M\left(R_1^2 + R_2^2\right)$ and $a = r\alpha$.
EXECUTE: (a) We want the acceleration, so we apply $\sum F_y = ma_y$: $Mg - T = Ma_y$ (Eq. 1)
Now apply $\sum \tau_z = I\alpha_z$. For this cylinder, $R_1 = R$ and $R_2 = R/2$, so $I = 5MR^2/4$. Using this result and $\alpha = a_y/R$, we have $TR = \frac{1}{2}\left(\frac{5R^2}{4}\right)\left(\frac{a_y}{R}\right)M$, so $T = \frac{5}{8}Ma_y$ (Eq. 2)

Combining Eq. 1 and Eq. 2 and solving for a_y gives $a_y = \frac{8}{13}g$.

(b) From Eq. 2, we have $T = \frac{5}{8} Ma_y = \frac{5}{8} M \left(\frac{8}{13} g\right) = \frac{5}{13} Mg$.

EVALUATE: Note the difference between the answer to this problem and Ex. 10.6. In both cases, the yo-yo has the same mass but a different distribution of that mass. The acceleration is less in this problem than in Example 10.6 because the moment of inertia is greater than in the example.

VP10.7.2. **IDENTIFY:** As the solid sphere rolls down a ramp its linear velocity and its angular velocity both increase. Newton's second law in both its linear and rotational forms applies.
SET UP: $\sum F_x = ma_x$ applies to the linear motion and $\sum \tau_z = I\alpha_z$ applies to the rotational motion about the center of mass. For a solid sphere $I = \frac{2}{3} MR^2$, and $a_x = R\alpha$ because the sphere does not slip. Take the x-axis along the surface of the ramp, pointing downward.
EXECUTE: Apply $\sum F_x = ma_x$. The friction force is up the ramp, so we get
$Mg \sin \beta - f = Ma_x$ (Eq. 1)
Apply $\sum \tau_z = I\alpha_z$, which gives $fR = \frac{2}{3} MR^2 \alpha_z = \frac{2}{3} MR^2 (a_x/R)$, which simplifies to
$f = \frac{2}{3} Ma_x$ (Eq. 2)
Combining Eq. 1 and Eq. 2 and solving for a_x gives $a_x = \frac{3}{5} g \sin \beta$.

(b) From Eq. 2 and the result from (a), we have $f = \frac{2}{3} Ma_x = \frac{2}{3} M \left(\frac{3}{5} g \sin \beta\right) = \frac{2}{5} Mg \sin \beta$.

(c) $\tau_z = fR = \frac{2}{5} MgR \sin \beta$.

EVALUATE: If there were no friction and the ball just slid down the incline, its acceleration would be $g \sin \beta$. But with rolling, friction is *up* the ramp, so it opposes the component of gravity down the ramp. Therefore the acceleration is *less* than $g \sin \beta$.

VP10.7.3. **IDENTIFY:** The yo-yo is rotating and translating at the same time. The rotational analog of Newton's second law applies to its rotational motion, and the linear form applies to its center of mass motion.
SET UP: $\sum F_y = ma_y$ applies to the vertical motion and $\sum \tau_z = I\alpha_z$ applies to the rotational motion about the center of mass. For a solid cylinder, $I = \frac{1}{2} MR^2$. We know that the tension in the string is 2/3 the weight of the yo-yo, and we want to find the acceleration of its center of mass and the angular acceleration about its center of mass. The only torque about the center of mass is due to the tension in the string.
EXECUTE: **(a)** We want the acceleration of the center of mass, so we apply $\sum F_y = ma_y$.
$Mg - T = Ma_y \rightarrow Mg - 2Mg/3 = Ma_y \rightarrow a_y = -g/3$. The magnitude is $g/3$ and the direction is downward.
(b) We want the angular acceleration about the center of mass, so we apply $\sum \tau_z = I\alpha_z$.
$TR = \frac{1}{2} MR^2 \alpha_{cm} \rightarrow \alpha_{cm} = \frac{4g}{3R}$.
EVALUATE: Notice that a_{cm} is *not* equal to $R\alpha_{cm}$. The cylinder is turning and moving, but it is not rolling.

VP10.7.4. **IDENTIFY:** The cylinder is rotating and translating at the same time and is rolling without slipping. The rotational analog of Newton's second law applies to its rotational motion, and the linear form applies to its center of mass motion.

SET UP: $\sum F_x = ma_x$ applies to the center of mass motion and $\sum \tau_z = I\alpha_z$ applies to the rotational motion about the center of mass. For a solid cylinder, $I = \frac{1}{2}MR^2$, and for rolling without slipping $a_{cm} = r\alpha_{cm}$. We want to find the friction force on the cylinder and the maximum angle of the ramp for which slipping will not occur. Call the x-axis along the surface of the ramp pointing downward.

EXECUTE: **(a)** We want the acceleration of the center of mass, so we apply $\sum F_x = ma_x$. The friction force f acts up the ramp to prevent sliding, so $Mg \sin\beta - f = Ma_x$ (Eq. 1)

Applying $\sum \tau_z = I\alpha_z$ about the center of mass of the cylinder gives $fR = \frac{1}{2}MR^2 \alpha_{cm}$. For rolling we also have $a_{cm} = r\alpha_{cm}$, so this becomes $fR = \frac{1}{2}MR^2(a_{cm}/R)$, which gives $f = \frac{1}{2}Ma_{cm}$ (Eq. 2)

Putting Eq. 2 into Eq. 1 gives $f = \frac{Mg}{3}\sin\beta$.

(b) When the ramp is at the maximum angle β, the cylinder is just ready to slip, so f_s is at its maximum value of $\mu_s n$. Applying $\sum F_y = 0$ to the cylinder gives $n = Mg\cos\beta$. Combining this result with the answer to (a) gives $\frac{Mg}{3}\sin\beta = \mu_s Mg \cos\beta \rightarrow \tan\beta = 3\mu_s$, so $\beta = \arctan(3\mu_s)$.

EVALUATE: Check a special case: If $\mu_s = 0$ (perfectly smooth ramp), then $\beta = 0$, which means that any elevation at all will cause slipping. This is a reasonable result for a perfectly smooth ramp.

VP10.12.1. **IDENTIFY:** The two-disk system does not experience any external torque, so its angular momentum is conserved.

SET UP: Angular momentum is $L = I\omega$, $I = \frac{1}{2}MR^2$ for a solid disk, and rotational kinetic energy is $K = \frac{1}{2}I\omega^2$.

EXECUTE: **(a)** Conservation of angular momentum tells us that $L_1 = L_2$, so $I_A\omega_A + I_B\omega_B = I_{A+B}\omega_2$. We know that $I_B = I_A/4$ and $\omega_B = \omega_A/2$, so this gives $I_A\omega_A + \frac{I_A}{4} \cdot \frac{\omega_A}{2} = \left(I_A + \frac{I_A}{4}\right)\omega_2$. Solving for ω_2 gives $\omega_2 = \frac{9}{10}\omega_A$.

(b) We want K_2/K_1. $K_1 = \frac{1}{2}I_A\omega_A^2 + \frac{1}{2}I_B\omega_B^2 = \frac{1}{2}I_A\omega_A^2 + \frac{1}{2}\left(\frac{I_A}{4}\right)\left(\frac{\omega_A}{2}\right)^2 = \frac{17}{32}I_A\omega_A^2$.

Using the result from (a) gives $K_2 = \frac{1}{2}I_{A+B}\omega^2 = \frac{1}{2}\left(I_A + \frac{I_A}{4}\right)\left(\frac{9}{10}\omega_A\right)^2 = \frac{405}{800}I_A\omega_A^2$.

The fraction of the initial rotational kinetic energy that remains is $\frac{K_2}{K_1} = \frac{\frac{405}{800}I_A\omega_A^2}{\frac{17}{32}I_A\omega_A^2} = \frac{81}{85} = 0.953$.

EVALUATE: The disks stick together so kinetic energy is lost, just as when objects collide in an inelastic collision. In this case, 95.3% of the kinetic energy remains, so only 4.7% is lost.

VP10.12.2. IDENTIFY: We follow exactly the same procedure as in VP10.12.1 *except* that the initial angular velocities are in *opposite* directions.

SET UP: The set up is the same as in VP10.12.1 except that $\omega_B = -\omega_A/2$

EXECUTE: (a) Conservation of angular momentum tells us that $L_1 = L_2$, so $I_A\omega_A + I_B\omega_B = I_{A+B}\omega_2$. We know that $I_B = I_A/4$ and $\omega_B = -\omega_A/2$, so this gives $I_A\omega_A - \frac{I_A}{4}\cdot\frac{\omega_A}{2} = \left(I_A + \frac{I_A}{4}\right)\omega_2$. Solving for ω_2 gives $\omega_2 = \frac{7}{10}\omega_A$.

(b) We want K_2/K_1. $K_1 = \frac{1}{2}I_A\omega_A^2 + \frac{1}{2}I_B\omega_B^2 = \frac{1}{2}I_A\omega_A^2 + \frac{1}{2}\left(\frac{I_A}{4}\right)\left(\frac{\omega_A}{2}\right)^2 = \frac{17}{32}I_A\omega_A^2$.

Using the result from (a) gives $K_2 = \frac{1}{2}I_{A+B}\omega^2 = \frac{1}{2}\left(I_A + \frac{I_A}{4}\right)\left(\frac{7}{10}\omega_A\right)^2 = \frac{49}{160}I_A\omega_A^2$.

The fraction of the initial rotational kinetic energy that remains is $\dfrac{K_2}{K_1} = \dfrac{\frac{49}{160}I_A\omega_A^2}{\frac{17}{32}I_A\omega_A^2} = \dfrac{49}{85} = 0.576$.

EVALUATE: The disks stick together so kinetic energy is lost, just as when objects collide in an inelastic collision. In this case, only 57.6% of the kinetic energy remains, so 42.4% is lost.

VP10.12.3. IDENTIFY: During the collision, the hinge exerts a force on the door. But if we look at the angular momentum about the hinge, the hinge exerts no torque, so the angular momentum of the bullet-door system is conserved about the hinge.

SET UP: About the hinge $L_{\text{bullet}} = mvd$, $L_{\text{door}} = I_{\text{door}}\omega$, and $I_{\text{door}} = \frac{1}{3}Md^2$. $K = \frac{1}{2}I\omega^2$ for rotation and $K = \frac{1}{2}mv^2$ for linear motion. Figure VP10.12.3 shows before and after sketches.

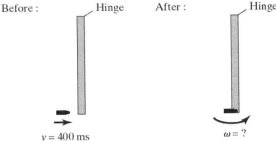

Figure VP10.12.3

EXECUTE: (a) We want the angular velocity ω just after the bullet hits the door, so use conservation of angular momentum about the hinge. $L_{\text{bullet}} + L_{\text{door}} = L_{\text{door + bullet}}$.

$mvd = (I_{\text{door}} + I_{\text{bullet}})\omega = \left(\frac{1}{3}Md^2 + md^2\right)\omega$

$\omega = \dfrac{mv}{d\left(\dfrac{M}{3}+m\right)} = \dfrac{(0.0100\text{ kg})(400\text{ m/s})}{(1.00\text{ m})\left(\dfrac{15\text{ kg}}{3}+0.0100\text{ kg}\right)} = 0.80$ rad/s.

(b) $K_1 = K_{\text{bullet}} = \frac{1}{2}mv^2 = \frac{1}{2}(0.0100\text{ kg})(400\text{ m/s})^2 = 800$ J

$K_2 = K_{\text{bullet}} + K_{\text{door}} = \frac{1}{2}\left(md^2\omega^2 + \frac{1}{3}Md^2\omega^2\right) = \left(m+\frac{M}{3}\right)\frac{(d\omega)^2}{2}$

$$K_2 = \left(0.010 \text{ kg} + \frac{15 \text{ kg}}{3}\right)\frac{[(1.00 \text{ m})(0.80 \text{ rad/s})]^2}{2} = 1.6 \text{ J}$$

$K_2/K_1 = (1.6 \text{ J})/(800 \text{ J}) = 0.0020 = 1/500$.

EVALUATE: This is a *very* inelastic collision. Only 0.20% of the original kinetic energy remained, so 99.8% was lost.

VP10.12.4. IDENTIFY: During the collision, the angular momentum of the clay-sphere system is conserved about the vertical shaft.

SET UP: About the shaft $L_{clay} = MvR$, $L_{sphere} = I_{sphere}\omega$, and $I_{sphere} = \frac{2}{3}MR^2$. $K = \frac{1}{2}I\omega^2$ for rotation and $K = \frac{1}{2}mv^2$ for linear motion.

EXECUTE: (a) Before: $L = MvR$. After: $L = M(v/2)R = \frac{1}{2}MvR$.

(b) The clay clay lost half of its angular momentum, so the sphere must have gained that amount by conservation of angular momentum. Therefore $L_{sphere} = \frac{1}{2}MvR = I_{sphere}\omega$. This gives $\frac{2}{3}MR^2\omega = \frac{1}{2}MvR$, so $\omega = \frac{3v}{4R}$.

(c) $K_1 = K_{clay} = \frac{1}{2}Mv^2$

$$K_2 = K_{clay} + K_{sphere} = \frac{1}{2}M(v/2)^2 + \frac{1}{2}I\omega^2 = \frac{1}{2}M\left(\frac{v}{2}\right)^2 + \frac{1}{2}\left(\frac{2}{3}MR^2\right)\left(\frac{3v}{4R}\right)^2 = \frac{5}{16}Mv^2$$

$$\frac{K_2}{K_1} = \frac{\frac{5}{16}Mv^2}{\frac{1}{2}Mv^2} = \frac{5}{8}.$$

EVALUATE: The system lost 3/8 of its kinetic energy during this collision, so it was *not* an elastic collision.

10.3. IDENTIFY and SET UP: Use $\tau = Fl$ to calculate the magnitude of each torque and use the right-hand rule (Figure 10.4 in the textbook) to determine the direction. Consider Figure 10.3.

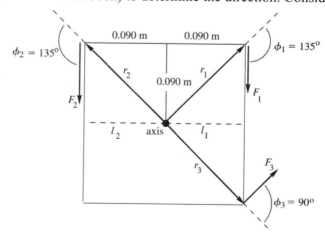

Figure 10.3

Let counterclockwise be the positive sense of rotation.

EXECUTE: $r_1 = r_2 = r_3 = \sqrt{(0.090 \text{ m})^2 + (0.090 \text{ m})^2} = 0.1273 \text{ m}$

$\tau_1 = -F_1 l_1$

$l_1 = r_1 \sin\phi_1 = (0.1273 \text{ m})\sin 135° = 0.0900 \text{ m}$

$\tau_1 = -(18.0 \text{ N})(0.0900 \text{ m}) = -1.62 \text{ N} \cdot \text{m}$

$\vec{\tau}_1$ is directed into paper

$\tau_2 = +F_2 l_2$

$l_2 = r_2 \sin\phi_2 = (0.1273 \text{ m})\sin 135° = 0.0900 \text{ m}$

$\tau_2 = +(26.0 \text{ N})(0.0900 \text{ m}) = +2.34 \text{ N} \cdot \text{m}$

$\vec{\tau}_2$ is directed out of paper

$\tau_3 = +F_3 l_3$

$l_3 = r_3 \sin\phi_3 = (0.1273 \text{ m})\sin 90° = 0.1273 \text{ m}$

$\tau_3 = +(14.0 \text{ N})(0.1273 \text{ m}) = +1.78 \text{ N} \cdot \text{m}$

$\vec{\tau}_3$ is directed out of paper

$\Sigma \tau = \tau_1 + \tau_2 + \tau_3 = -1.62 \text{ N} \cdot \text{m} + 2.34 \text{ N} \cdot \text{m} + 1.78 \text{ N} \cdot \text{m} = 2.50 \text{ N} \cdot \text{m}$

EVALUATE: The net torque is positive, which means it tends to produce a counterclockwise rotation; the vector torque is directed out of the plane of the paper. In summing the torques it is important to include + or − signs to show direction.

10.5. IDENTIFY and **SET UP:** Calculate the torque using Eq. (10.3) and also determine the direction of the torque using the right-hand rule.

(a) $\vec{r} = (-0.450 \text{ m})\hat{i} + (0.150 \text{ m})\hat{j}$; $\vec{F} = (-5.00 \text{ N})\hat{i} + (4.00 \text{ N})\hat{j}$. The sketch is given in Figure 10.5.

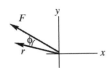

Figure 10.5

EXECUTE: (b) When the fingers of your right hand curl from the direction of $\vec{r}$ into the direction of $\vec{F}$ (through the smaller of the two angles, angle ϕ) your thumb points into the page (the direction of $\vec{\tau}$, the $-z$-direction).

(c) $\vec{\tau} = \vec{r} \times \vec{F} = [(-0.450 \text{ m})\hat{i} + (0.150 \text{ m})\hat{j}] \times [(-5.00 \text{ N})\hat{i} + (4.00 \text{ N})\hat{j}]$

$\vec{\tau} = +(2.25 \text{ N} \cdot \text{m})\hat{i} \times \hat{i} - (1.80 \text{ N} \cdot \text{m})\hat{i} \times \hat{j} - (0.750 \text{ N} \cdot \text{m})\hat{j} \times \hat{i} + (0.600 \text{ N} \cdot \text{m})\hat{j} \times \hat{j}$

$\hat{i} \times \hat{i} = \hat{j} \times \hat{j} = 0$

$\hat{i} \times \hat{j} = \hat{k}, \quad \hat{j} \times \hat{i} = -\hat{k}$

Thus $\vec{\tau} = -(1.80 \text{ N} \cdot \text{m})\hat{k} - (0.750 \text{ N} \cdot \text{m})(-\hat{k}) = (-1.05 \text{ N} \cdot \text{m})\hat{k}$.

EVALUATE: The calculation gives that $\vec{\tau}$ is in the $-z$-direction. This agrees with what we got from the right-hand rule.

10.9. IDENTIFY: Apply $\Sigma \tau_z = I\alpha_z$.

SET UP: $\omega_{0z} = 0$. $\omega_z = (400 \text{ rev/min})\left(\dfrac{2\pi \text{ rad/rev}}{60 \text{ s/min}}\right) = 41.9 \text{ rad/s}$

EXECUTE: $\tau_z = I\alpha_z = I\dfrac{\omega_z - \omega_{0z}}{t} = (1.60 \text{ kg} \cdot \text{m}^2)\dfrac{41.9 \text{ rad/s}}{8.00 \text{ s}} = 8.38 \text{ N} \cdot \text{m}.$

EVALUATE: In $\tau_z = I\alpha_z$, α_z must be in rad/s^2.

10.11. IDENTIFY: Use $\Sigma\tau_z = I\alpha_z$ to calculate α. Use a constant angular acceleration kinematic equation to relate α_z, ω_z, and t.

SET UP: For a solid uniform sphere and an axis through its center, $I = \tfrac{2}{5}MR^2$. Let the direction the sphere is spinning be the positive sense of rotation. The moment arm for the friction force is $l = 0.0150$ m and the torque due to this force is negative.

EXECUTE: (a) $\alpha_z = \dfrac{\tau_z}{I} = \dfrac{-(0.0200 \text{ N})(0.0150 \text{ m})}{\tfrac{2}{5}(0.225 \text{ kg})(0.0150 \text{ m})^2} = -14.8 \text{ rad/s}^2$

(b) $\omega_z - \omega_{0z} = -22.5$ rad/s. $\omega_z = \omega_{0z} + \alpha_z t$ gives $t = \dfrac{\omega_z - \omega_{0z}}{\alpha_z} = \dfrac{-22.5 \text{ rad/s}}{-14.8 \text{ rad/s}^2} = 1.52$ s.

EVALUATE: The fact that α_z is negative means its direction is opposite to the direction of spin. The negative α_z causes ω_z to decrease.

10.13. IDENTIFY: Apply $\Sigma\vec{F} = m\vec{a}$ to each book and apply $\Sigma\tau_z = I\alpha_z$ to the pulley. Use a constant acceleration equation to find the common acceleration of the books.

SET UP: $m_1 = 2.00$ kg, $m_2 = 3.00$ kg. Let T_1 be the tension in the part of the cord attached to m_1 and T_2 be the tension in the part of the cord attached to m_2. Let the $+x$-direction be in the direction of the acceleration of each book. $a = R\alpha$.

EXECUTE: (a) $x - x_0 = v_{0x}t + \tfrac{1}{2}a_x t^2$ gives $a_x = \dfrac{2(x - x_0)}{t^2} = \dfrac{2(1.20 \text{ m})}{(0.800 \text{ s})^2} = 3.75 \text{ m/s}^2.$

$a_1 = 3.75$ m/s^2 so $T_1 = m_1 a_1 = 7.50$ N and $T_2 = m_2(g - a_1) = 18.2$ N.

(b) The torque on the pulley is $(T_2 - T_1)R = 0.803$ N·m, and the angular acceleration is $\alpha = a_1/R = 50$ rad/s^2, so $I = \tau/\alpha = 0.016$ kg·m^2.

EVALUATE: The tensions in the two parts of the cord must be different, so there will be a net torque on the pulley.

10.17. IDENTIFY: The rotational form of Newton's second law applies to the cylinder. Interpretation of graphical data is necessary.

SET UP: Since $\theta - \theta_0$ is proportional to t^2, the equation $\theta - \theta_0 = \omega_{0z}t + \dfrac{1}{2}\alpha_z t^2$ applies to the rotational motion. Since the rotation starts from rest, $\omega_{0z} = 0$, so we have $\theta - \theta_0 = \dfrac{1}{2}\alpha_z t^2$. Therefore a graph of $\theta - \theta_0$ versus t^2 should be a straight line having slope $\dfrac{1}{2}\alpha_z$. Our target variable is the moment of inertia of the cylinder. Once we know α_z, we can apply $\Sigma\tau_z = I\alpha_z$ to find I.

EXECUTE: Use the slope of the graph to find α_z. As we discussed above, slope = $\frac{1}{2}\alpha_z$, so α_z = 2(slope) = 2(16.0 rad/s²) = 32.0 rad/s². Now use $\Sigma\tau_z = I\alpha_z$ to find I. $FR = I\alpha_z$, so $I = \dfrac{FR}{\alpha_z} = \dfrac{(3.00\text{ N})(0.140\text{ m})}{32.0\text{ rad/s}^2} = 0.0131\text{ kg}\cdot\text{m}^2$.

EVALUATE: Using the slope to find α_z is more accurate than using individual data points because individual measurements vary, but finding the slope essentially "averages out" the data points.

10.21. IDENTIFY: Apply $K = K_{cm} + K_{rot}$.

SET UP: For an object that is rolling without slipping, $v_{cm} = R\omega$.

EXECUTE: The fraction of the total kinetic energy that is rotational is

$$\frac{(1/2)I_{cm}\omega^2}{(1/2)Mv_{cm}^2 + (1/2)I_{cm}\omega^2} = \frac{1}{1+(M/I_{cm})v_{cm}^2/\omega^2} = \frac{1}{1+(MR^2/I_{cm})}$$

(a) $I_{cm} = (1/2)MR^2$, so the above ratio is 1/3.

(b) $I_{cm} = (2/5)MR^2$ so the above ratio is 2/7.

(c) $I_{cm} = (2/3)MR^2$ so the ratio is 2/5.

(d) $I_{cm} = (5/8)MR^2$ so the ratio is 5/13.

EVALUATE: The moment of inertia of each object takes the form $I = \beta MR^2$. The ratio of rotational kinetic energy to total kinetic energy can be written as $\dfrac{1}{1+1/\beta} = \dfrac{\beta}{1+\beta}$. The ratio increases as β increases.

10.23. IDENTIFY: Apply $\Sigma\vec{F}_{ext} = m\vec{a}_{cm}$ and $\Sigma\tau_z = I_{cm}\alpha_z$ to the motion of the ball.

(a) SET UP: The free-body diagram is given in Figure 10.23a.

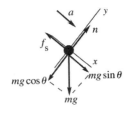

EXECUTE: $\Sigma F_y = ma_y$
$n = mg\cos\theta$ and $f_s = \mu_s mg\cos\theta$
$\Sigma F_x = ma_x$
$mg\sin\theta - \mu_s mg\cos\theta = ma$
$g(\sin\theta - \mu_s \cos\theta) = a$ (Eq. 1)

Figure 10.23a

SET UP: Consider Figure 10.23b.

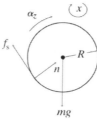

The normal force n is directed through the center of the ball and mg acts at the center of the ball, so neither of them produces a torque about the center.

Figure 10.23b

EXECUTE: $\Sigma \tau = \tau_f = \mu_s mg (\cos\theta) R$; $I = \frac{2}{5}mR^2$

$\Sigma \tau_z = I_{cm}\alpha_z$ gives $\mu_s mg (\cos\theta) R = \frac{2}{5}mR^2 \alpha$

No slipping means $\alpha = a/R$, so $\mu_s g \cos\theta = \frac{2}{5}a$ (Eq. 2)

We have two equations in the two unknowns a and μ_s. Solving gives $a = \frac{5}{7}g \sin\theta$ and
$\mu_s = \frac{2}{7}\tan\theta = \frac{2}{7}\tan 65.0° = 0.613$.

(b) Repeat the calculation of part (a), but now $I = \frac{2}{3}mR^2$. $a = \frac{3}{5}g \sin\theta$ and
$\mu_s = \frac{2}{5}\tan\theta = \frac{2}{5}\tan 65.0° = 0.858$

The value of μ_s calculated in part (a) is not large enough to prevent slipping for the hollow ball.

(c) EVALUATE: There is no slipping at the point of contact. More friction is required for a hollow ball since for a given m and R it has a larger I and more torque is needed to provide the same α. Note that the required μ_s is independent of the mass or radius of the ball and only depends on how that mass is distributed.

10.25. IDENTIFY: Apply conservation of energy to the motion of the wheel.

SET UP: The wheel at points 1 and 2 of its motion is shown in Figure 10.25.

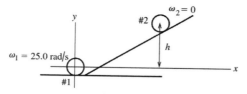

Take $y = 0$ at the center of the wheel when it is at the bottom of the hill.

Figure 10.25

The wheel has both translational and rotational motion so its kinetic energy is
$K = \frac{1}{2}I_{cm}\omega^2 + \frac{1}{2}Mv_{cm}^2$.

EXECUTE: $K_1 + U_1 + W_{other} = K_2 + U_2$

$W_{other} = W_{fric} = -2600$ J (the friction work is negative)

$K_1 = \frac{1}{2}I\omega_1^2 + \frac{1}{2}Mv_1^2$; $v = R\omega$ and $I = 0.800MR^2$ so

$K_1 = \frac{1}{2}(0.800)MR^2\omega_1^2 + \frac{1}{2}MR^2\omega_1^2 = 0.900MR^2\omega_1^2$

$K_2 = 0$, $U_1 = 0$, $U_2 = Mgh$

Thus $0.900MR^2\omega_1^2 + W_{fric} = Mgh$

$M = w/g = 392$ N$/(9.80$ m/s$^2) = 40.0$ kg

$$h = \frac{0.900 MR^2 \omega_1^2 + W_{fric}}{Mg}$$

$$h = \frac{(0.900)(40.0 \text{ kg})(0.600 \text{ m})^2 (25.0 \text{ rad/s})^2 - 2600 \text{ J}}{(40.0 \text{ kg})(9.80 \text{ m/s}^2)} = 14.0 \text{ m}.$$

EVALUATE: Friction does negative work and reduces h.

10.29. IDENTIFY: As the cylinder falls, its potential energy is transformed into both translational and rotational kinetic energy. Its mechanical energy is conserved.

SET UP: The hollow cylinder has $I = \frac{1}{2}m(R_a^2 + R_b^2)$, where $R_a = 0.200$ m and $R_b = 0.350$ m. Use coordinates where $+y$ is upward and $y = 0$ at the initial position of the cylinder. Then $y_1 = 0$ and $y_2 = -d$, where d is the distance it has fallen. $v_{cm} = R\omega$. $K_{cm} = \frac{1}{2}Mv_{cm}^2$ and $\omega_{0z} = 10.47$ rad/s.

EXECUTE: **(a)** Conservation of energy gives $K_1 + U_1 = K_2 + U_2$. $K_1 = 0$, $U_1 = 0$. $0 = U_2 + K_2$ and $0 = -mgd + \frac{1}{2}mv_{cm}^2 + \frac{1}{2}I_{cm}\omega^2$. $\frac{1}{2}I\omega^2 = \frac{1}{2}\left[\frac{1}{2}m(R_a^2 + R_b^2)\right](v_{cm}/R_b)^2 = \frac{1}{4}m\left[1 + (R_a/R_b)^2\right]v_{cm}^2$,

so $\frac{1}{2}\{1 + \frac{1}{2}[1 + (R_a/R_b)^2]\}v_{cm}^2 = gd$ and

$$d = \frac{\{1 + \frac{1}{2}[1 + (R_a/R_b)^2]\}v_{cm}^2}{2g} = \frac{(1 + 0.663)(6.66 \text{ m/s})^2}{2(9.80 \text{ m/s}^2)} = 3.76 \text{ m}.$$

(b) $K_2 = \frac{1}{2}mv_{cm}^2$ since there is no rotation. So $mgd = \frac{1}{2}mv_{cm}^2$ which gives

$v_{cm} = \sqrt{2gd} = \sqrt{2(9.80 \text{ m/s}^2)(3.76 \text{ m})} = 8.58$ m/s.

(c) In part (a) the cylinder has rotational as well as translational kinetic energy and therefore less translational speed at a given kinetic energy. The kinetic energy comes from a decrease in gravitational potential energy and that is the same, so in (a) the translational speed is less.

EVALUATE: If part (a) were repeated for a solid cylinder, $R_a = 0$ and $d = 3.39$ m. For a thin-walled hollow cylinder, $R_a = R_b$ and $d = 4.52$ cm. Note that all of these answers are independent of the mass m of the cylinder.

10.31. IDENTIFY: As the ball rolls up the hill, its kinetic energy (translational and rotational) is transformed into gravitational potential energy. Since there is no slipping, its mechanical energy is conserved.

SET UP: The ball has moment of inertia $I_{cm} = \frac{2}{3}mR^2$. Rolling without slipping means $v_{cm} = R\omega$. Use coordinates where $+y$ is upward and $y = 0$ at the bottom of the hill, so $y_1 = 0$ and $y_2 = h = 5.00$ m. The ball's kinetic energy is $K = \frac{1}{2}mv_{cm}^2 + \frac{1}{2}I_{cm}\omega^2$ and its potential energy is $U = mgh$.

EXECUTE: **(a)** Conservation of energy gives $K_1 + U_1 = K_2 + U_2$. $U_1 = 0$, $K_2 = 0$ (the ball stops). Therefore $K_1 = U_2$ and $\frac{1}{2}mv_{cm}^2 + \frac{1}{2}I_{cm}\omega^2 = mgh$. $\frac{1}{2}I_{cm}\omega^2 = \frac{1}{2}(\frac{2}{3}mR^2)\left(\frac{v_{cm}}{R}\right)^2 = \frac{1}{3}mv_{cm}^2$, so

$\frac{5}{6}mv_{cm}^2 = mgh$. Therefore $v_{cm} = \sqrt{\frac{6gh}{5}} = \sqrt{\frac{6(9.80 \text{ m/s}^2)(5.00 \text{ m})}{5}} = 7.67$ m/s and

$\omega = \frac{v_{cm}}{R} = \frac{7.67 \text{ m/s}}{0.113 \text{ m}} = 67.9$ rad/s.

(b) $K_{rot} = \frac{1}{2}I\omega^2 = \frac{1}{3}mv_{cm}^2 = \frac{1}{3}(0.426 \text{ kg})(7.67 \text{ m/s})^2 = 8.35 \text{ J}.$

EVALUATE: Its translational kinetic energy at the base of the hill is $\frac{1}{2}mv_{cm}^2 = \frac{3}{2}K_{rot} = 12.52 \text{ J}$. Its total kinetic energy is 20.9 J, which equals its final potential energy:

$mgh = (0.426 \text{ kg})(9.80 \text{ m/s}^2)(5.00 \text{ m}) = 20.9 \text{ J}.$

10.33. **(a) IDENTIFY:** Use $\Sigma\tau_z = I\alpha_z$ to find α_z and then use a constant angular acceleration equation to find ω_z.

SET UP: The free-body diagram is given in Figure 10.33.

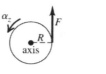

EXECUTE: Apply $\Sigma\tau_z = I\alpha_z$ to find the angular acceleration:

$FR = I\alpha_z$

$\alpha_z = \dfrac{FR}{I} = \dfrac{(18.0 \text{ N})(2.40 \text{ m})}{2100 \text{ kg}\cdot\text{m}^2} = 0.02057 \text{ rad/s}^2$

Figure 10.33

SET UP: Use the constant α_z kinematic equations to find ω_z.

$\omega_z = ?$; ω_{0z} (initially at rest); $\alpha_z = 0.02057 \text{ rad/s}^2$; $t = 15.0 \text{ s}$

EXECUTE: $\omega_z = \omega_{0z} + \alpha_z t = 0 + (0.02057 \text{ rad/s}^2)(15.0 \text{ s}) = 0.309 \text{ rad/s}$

(b) IDENTIFY and SET UP: Calculate the work from $W = \tau_z\Delta\theta$, using a constant angular acceleration equation to calculate $\theta - \theta_0$, or use the work-energy theorem. We will do it both ways.

EXECUTE: (1) $W = \tau_z\Delta\theta$

$\Delta\theta = \theta - \theta_0 = \omega_{0z}t + \frac{1}{2}\alpha_z t^2 = 0 + \frac{1}{2}(0.02057 \text{ rad/s}^2)(15.0 \text{ s})^2 = 2.314 \text{ rad}$

$\tau_z = FR = (18.0 \text{ N})(2.40 \text{ m}) = 43.2 \text{ N}\cdot\text{m}$

Then $W = \tau_z\Delta\theta = (43.2 \text{ N}\cdot\text{m})(2.314 \text{ rad}) = 100 \text{ J}.$

or

(2) $W_{tot} = K_2 - K_1$

$W_{tot} = W$, the work done by the child

$K_1 = 0$; $K_2 = \frac{1}{2}I\omega^2 = \frac{1}{2}(2100 \text{ kg}\cdot\text{m}^2)(0.309 \text{ rad/s})^2 = 100 \text{ J}$

Thus $W = 100$ J, the same as before.

EVALUATE: Either method yields the same result for W.

(c) IDENTIFY and SET UP: Use $P_{av} = \dfrac{\Delta W}{\Delta t}$ to calculate P_{av}.

EXECUTE: $P_{av} = \dfrac{\Delta W}{\Delta t} = \dfrac{100 \text{ J}}{15.0 \text{ s}} = 6.67 \text{ W}.$

EVALUATE: Work is in joules, power is in watts.

10.41. **IDENTIFY:** $\omega_z = d\theta/dt$. $L_z = I\omega_z$ and $\tau_z = dL_z/dt$.

SET UP: For a hollow, thin-walled sphere rolling about an axis through its center, $I = \frac{2}{3}MR^2$. $R = 0.240$ m.

EXECUTE: **(a)** $A = 1.50 \text{ rad/s}^2$ and $B = 1.10 \text{ rad/s}^4$, so that $\theta(t)$ will have units of radians.

(b) (i) $\omega_z = \dfrac{d\theta}{dt} = 2At + 4Bt^3$. At $t = 3.00$ s,

$\omega_z = 2(1.50 \text{ rad/s}^2)(3.00 \text{ s}) + 4(1.10 \text{ rad/s}^4)(3.00 \text{ s})^3 = 128$ rad/s.

$L_z = (\tfrac{2}{3}MR^2)\omega_z = \tfrac{2}{3}(12.0 \text{ kg})(0.240 \text{ m})^2(128 \text{ rad/s}) = 59.0 \text{ kg}\cdot\text{m}^2/\text{s}$.

(ii) $\tau_z = \dfrac{dL_z}{dt} = I\dfrac{d\omega_z}{dt} = I(2A + 12Bt^2)$ and

$\tau_z = \tfrac{2}{3}(12.0 \text{ kg})(0.240 \text{ m})^2\left[2(1.50 \text{ rad/s}^2) + 12(1.10 \text{ rad/s}^4)(3.00 \text{ s})^2\right] = 56.1 \text{ N}\cdot\text{m}$.

EVALUATE: The angular speed of rotation is increasing. This increase is due to an acceleration α_z that is produced by the torque on the sphere. When I is constant, as it is here, $\tau_z = dL_z/dt = Id\omega_z/dt = I\alpha_z$.

10.45. IDENTIFY: Apply conservation of angular momentum to the motion of the skater.

SET UP: For a thin-walled hollow cylinder $I = mR^2$. For a slender rod rotating about an axis through its center, $I = \tfrac{1}{12}ml^2$.

EXECUTE: $L_i = L_f$ so $I_i\omega_i = I_f\omega_f$.

$I_i = 0.40 \text{ kg}\cdot\text{m}^2 + \tfrac{1}{12}(8.0 \text{ kg})(1.8 \text{ m})^2 = 2.56 \text{ kg}\cdot\text{m}^2$.

$I_f = 0.40 \text{ kg}\cdot\text{m}^2 + (8.0 \text{ kg})(0.25 \text{ m})^2 = 0.90 \text{ kg}\cdot\text{m}^2$.

$\omega_f = \left(\dfrac{I_i}{I_f}\right)\omega_i = \left(\dfrac{2.56 \text{ kg}\cdot\text{m}^2}{0.90 \text{ kg}\cdot\text{m}^2}\right)(0.40 \text{ rev/s}) = 1.14 \text{ rev/s}$.

EVALUATE: $K = \tfrac{1}{2}I\omega^2 = \tfrac{1}{2}L\omega$. ω increases and L is constant, so K increases. The increase in kinetic energy comes from the work done by the skater when he pulls in his hands.

10.47. IDENTIFY and SET UP: There is no net external torque about the rotation axis so the angular momentum $L = I\omega$ is conserved.

EXECUTE: **(a)** $L_1 = L_2$ gives $I_1\omega_1 = I_2\omega_2$, so $\omega_2 = (I_1/I_2)\omega_1$

$I_1 = I_{tt} = \tfrac{1}{2}MR^2 = \tfrac{1}{2}(120 \text{ kg})(2.00 \text{ m})^2 = 240 \text{ kg}\cdot\text{m}^2$

$I_2 = I_{tt} + I_p = 240 \text{ kg}\cdot\text{m}^2 + mR^2 = 240 \text{ kg}\cdot\text{m}^2 + (70 \text{ kg})(2.00 \text{ m})^2 = 520 \text{ kg}\cdot\text{m}^2$

$\omega_2 = (I_1/I_2)\omega_1 = (240 \text{ kg}\cdot\text{m}^2/520 \text{ kg}\cdot\text{m}^2)(3.00 \text{ rad/s}) = 1.38 \text{ rad/s}$

(b) $K_1 = \tfrac{1}{2}I_1\omega_1^2 = \tfrac{1}{2}(240 \text{ kg}\cdot\text{m}^2)(3.00 \text{ rad/s})^2 = 1080$ J

$K_2 = \tfrac{1}{2}I_2\omega_2^2 = \tfrac{1}{2}(520 \text{ kg}\cdot\text{m}^2)(1.38 \text{ rad/s})^2 = 495$ J

EVALUATE: The kinetic energy decreases because of the negative work done on the turntable and the parachutist by the friction force between these two objects.
The angular speed decreases because I increases when the parachutist is added to the system.

10.49. (a) IDENTIFY and SET UP: Apply conservation of angular momentum $\vec{L}$, with the axis at the nail. Let object A be the bug and object B be the bar. Initially, all objects are at rest and $L_1 = 0$. Just after the bug jumps, it has angular momentum in one direction of rotation and the bar is rotating with angular velocity ω_B in the opposite direction.

EXECUTE: $L_2 = m_A v_A r - I_B \omega_B$ where $r = 1.00$ m and $I_B = \tfrac{1}{3}m_B r^2$

$L_1 = L_2$ gives $m_A v_A r = \frac{1}{3} m_B r^2 \omega_B$

$\omega_B = \dfrac{3 m_A v_A}{m_B r} = 0.120$ rad/s

(b) $K_1 = 0$;

$K_2 = \frac{1}{2} m_A v_A^2 + \frac{1}{2} I_B \omega_B^2 = \frac{1}{2}(0.0100 \text{ kg})(0.200 \text{ m/s})^2 + \frac{1}{2}(\frac{1}{3}(0.0500 \text{ kg})(1.00 \text{ m})^2)(0.120 \text{ rad/s})^2$

$= 3.2 \times 10^{-4}$ J.

(c) The increase in kinetic energy comes from work done by the bug when it pushes against the bar in order to jump.

EVALUATE: There is no external torque applied to the system and the total angular momentum of the system is constant. There are internal forces, forces the bug and bar exert on each other. The forces exert torques and change the angular momentum of the bug and the bar, but these changes are equal in magnitude and opposite in direction. These internal forces do positive work on the two objects and the kinetic energy of each object and of the system increases.

10.51. IDENTIFY: Energy is conserved as the cylinder rolls down the incline without slipping.

SET UP: The total energy of the cylinder is $K_{tot} = \frac{1}{2} m v_{cm}^2 + \frac{1}{2} I \omega^2$, $I_{cyl} = \frac{1}{2} MR^2$. As the cylinder rolls down the incline, gravitational potential energy is transformed into kinetic energy, so we use $U_1 + K_1 + W_{other} = U_2 + K_2$. Our target variable is the acceleration due to gravity on the planet.

EXECUTE: Energy conservation gives $mgh = \frac{1}{2} m v_{cm}^2 + \frac{1}{2} I \omega^2$, which, for a solid cylinder, gives

$mgh = \frac{1}{2} m v_{cm}^2 + \frac{1}{2} \left(\frac{1}{2} MR^2\right) \left(\frac{v_{cm}}{R}\right)^2$. Solving for v_{cm}^2 gives $v_{cm}^2 = \left(\frac{4}{3} g\right) h$. The graph of v_{cm}^2 versus h should be a straight line having slope equal to $4g/3$. Thus $g = \frac{3}{4}(\text{slope}) = \frac{3}{4}(6.42 \text{ m/s}^2) = 4.82$ m/s².

EVALUATE: On this planet, g is about half of what it is on the earth.

10.53. IDENTIFY: When the teenager throws the rock, it causes the wooden disk to spin. Because no external torques act on the system due to the throw, the angular momentum of the system is conserved.

SET UP: $L = I\omega$ for a rotating object, and $L = mvr$ for a small object. $I_{disk} = \frac{1}{2} MR^2$. The target variable is the angular speed of the disk just after the rock is thrown.

EXECUTE: The initial angular momentum is zero since nothing is turning. Conservation of angular momentum tells us that $L_1 = L_2$, so $0 = L_{teen} + L_{disk} + L_{rock}$. The rock's motion is opposite to that of the teen and disk. This gives us $0 = mR^2 \omega + \frac{1}{2} MR^2 \omega - m_{rock} vR$. Solving for ω gives

$\omega = \dfrac{m_{rock} v}{R\left(m + \dfrac{M}{2}\right)}$.

EVALUATE: Check some special cases. If $m_{rock} \to 0$, then $\omega \to 0$, which means that throwing the very light rock had no effect on the disk. If m or M are very large, $\omega \to 0$, which means that the teen or the disk were too massive to be moved by the light rock. If the teen and the rock are both much more massive than the disk, then $\omega \to v/R$, which means that the teen and rock have the same speed but in opposite directions.

10.55. **IDENTIFY:** An external torque will cause precession of the telescope.

SET UP: $I = MR^2$, with $R = 2.5 \times 10^{-2}$ m. 1.0×10^{-6} degree $= 1.745 \times 10^{-8}$ rad. $\omega = 19{,}200$ rpm $= 2.01 \times 10^3$ rad/s. $t = 5.0$ h $= 1.8 \times 10^4$ s.

EXECUTE: $\Omega = \dfrac{\Delta\phi}{\Delta t} = \dfrac{1.745 \times 10^{-8} \text{ rad}}{1.8 \times 10^4 \text{ s}} = 9.694 \times 10^{-13}$ rad/s. $\Omega = \dfrac{\tau}{I\omega}$ so $\tau = \Omega I \omega = \Omega M R^2 \omega$.

Putting in the numbers gives
$\tau = (9.694 \times 10^{-13} \text{ rad/s})(2.0 \text{ kg})(2.5 \times 10^{-2} \text{ m})^2 (2.01 \times 10^3 \text{ rad/s}) = 2.4 \times 10^{-12}$ N·m.

EVALUATE: The external torque must be very small for this degree of stability.

10.59. **IDENTIFY:** Use the kinematic information to solve for the angular acceleration of the grindstone. Assume that the grindstone is rotating counterclockwise and let that be the positive sense of rotation. Then apply $\Sigma\tau_z = I\alpha_z$ to calculate the friction force and use $f_k = \mu_k n$ to calculate μ_k.

SET UP: $\omega_{0z} = 850$ rev/min$(2\pi$ rad/1 rev$)(1$ min/60 s$) = 89.0$ rad/s
$t = 7.50$ s; $\omega_z = 0$ (comes to rest); $\alpha_z = ?$

EXECUTE: $\omega_z = \omega_{0z} + \alpha_z t$

$\alpha_z = \dfrac{0 - 89.0 \text{ rad/s}}{7.50 \text{ s}} = -11.9 \text{ rad/s}^2$

SET UP: Apply $\Sigma\tau_z = I\alpha_z$ to the grindstone. The free-body diagram is given in Figure 10.59.

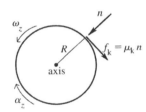

Figure 10.59

The normal force has zero moment arm for rotation about an axis at the center of the grindstone, and therefore zero torque. The only torque on the grindstone is that due to the friction force f_k exerted by the ax; for this force the moment arm is $l = R$ and the torque is negative.

EXECUTE: $\Sigma\tau_z = -f_k R = -\mu_k n R$

$I = \tfrac{1}{2}MR^2$ (solid disk, axis through center)

Thus $\Sigma\tau_z = I\alpha_z$ gives $-\mu_k n R = (\tfrac{1}{2}MR^2)\alpha_z$

$\mu_k = -\dfrac{MR\alpha_z}{2n} = -\dfrac{(50.0 \text{ kg})(0.260 \text{ m})(-11.9 \text{ rad/s}^2)}{2(160 \text{ N})} = 0.483$

EVALUATE: The friction torque is clockwise and slows down the counterclockwise rotation of the grindstone.

10.61. **IDENTIFY:** Use $\sum \tau_z = I\alpha_z$ to find the angular acceleration just after the ball falls off and use conservation of energy to find the angular velocity of the bar as it swings through the vertical position.

SET UP: The axis of rotation is at the axle. For this axis the bar has $I = \frac{1}{12} m_{bar} L^2$, where $m_{bar} = 3.80$ kg and $L = 0.800$ m. Energy conservation gives $K_1 + U_1 = K_2 + U_2$. The gravitational potential energy of the bar doesn't change. Let $mg(\sin 36.9° - \mu_k \cos 36.9°) - T = ma$ so $y_2 = -L/2$.

EXECUTE: **(a)** $\tau_z = m_{ball} g(L/2)$ and $I = I_{ball} + I_{bar} = \frac{1}{12} m_{bar} L^2 + m_{ball}(L/2)^2$. $\sum \tau_z = I\alpha_z$ gives

$$\alpha_z = \frac{m_{ball} g(L/2)}{\frac{1}{12} m_{bar} L^2 + m_{ball}(L/2)^2} = \frac{2g}{L}\left(\frac{m_{ball}}{m_{ball} + m_{bar}/3}\right)$$ and

$$\alpha_z = \frac{2(9.80 \text{ m/s}^2)}{0.800 \text{ m}}\left(\frac{2.50 \text{ kg}}{2.50 \text{ kg} + [3.80 \text{ kg}]/3}\right) = 16.3 \text{ rad/s}^2.$$

(b) As the bar rotates, the moment arm for the weight of the ball decreases and the angular acceleration of the bar decreases.

(c) $K_1 + U_1 = K_2 + U_2$. $0 = K_2 + U_2$. $\frac{1}{2}(I_{bar} + I_{ball})\omega^2 = -m_{ball} g(-L/2)$.

$$\omega = \sqrt{\frac{m_{ball} gL}{m_{ball} L^2/4 + m_{bar} L^2/12}} = \sqrt{\frac{g}{L}\left(\frac{4 m_{ball}}{m_{ball} + m_{bar}/3}\right)} = \sqrt{\frac{9.80 \text{ m/s}^2}{0.800 \text{ m}}\left(\frac{4(2.50 \text{ kg})}{2.50 \text{ kg} + (3.80 \text{ kg})/3}\right)}$$

$\omega = 5.70$ rad/s.

EVALUATE: As the bar swings through the vertical, the linear speed of the ball that is still attached to the bar is $v = (0.400 \text{ m})(5.70 \text{ rad/s}) = 2.28$ m/s. A point mass in free-fall acquires a speed of 2.80 m/s after falling 0.400 m; the ball on the bar acquires a speed less than this.

10.63. **IDENTIFY:** Blocks A and B have linear acceleration and therefore obey the linear form of Newton's second law $\sum F_y = ma_y$. The wheel C has angular acceleration, so it obeys the rotational form of Newton's second law $\sum \tau_z = I\alpha_z$.

SET UP: A accelerates downward, B accelerates upward and the wheel turns clockwise. Apply $\sum F_y = ma_y$ to blocks A and B. Let $+y$ be downward for A and $+y$ be upward for B. Apply $\sum \tau_z = I\alpha_z$ to the wheel, with the clockwise sense of rotation positive. Each block has the same magnitude of acceleration, a, and $a = R\alpha$. Call the T_A the tension in the cord between C and A and T_B the tension between C and B.

EXECUTE: For A, $\sum F_y = ma_y$ gives $m_A g - T_A = m_A a$. For B, $\sum F_y = ma_y$ gives $T_B - m_B g = m_B a$. For the wheel, $\sum \tau_z = I\alpha_z$ gives $T_A R - T_B R = I\alpha = I(a/R)w$ and $T_A - T_B = \left(\frac{I}{R^2}\right)a$. Adding these three equations gives $(m_A - m_B)g = \left(m_A + m_B + \frac{I}{R^2}\right)a$. Solving for a, we have

$$a = \left(\frac{m_A - m_B}{m_A + m_B + I/R^2}\right)g = \left(\frac{4.00 \text{ kg} - 2.00 \text{ kg}}{4.00 \text{ kg} + 2.00 \text{ kg} + (0.220 \text{ kg} \cdot \text{m}^2)/(0.120 \text{ m})^2}\right)(9.80 \text{ m/s}^2)$$

$= 0.921 \text{ m/s}^2$.

$$\alpha = \frac{a}{R} = \frac{0.921 \text{ m/s}^2}{0.120 \text{ m}} = 7.68 \text{ rad/s}^2.$$

$T_A = m_A(g - a) = (4.00 \text{ kg})(9.80 \text{ m/s}^2 - 0.921 \text{ m/s}^2) = 35.5 \text{ N}.$

$T_B = m_B(g + a) = (2.00 \text{ kg})(9.80 \text{ m/s}^2 + 0.921 \text{ m/s}^2) = 21.4 \text{ N}.$

EVALUATE: The tensions must be different in order to produce a torque that accelerates the wheel when the blocks accelerate.

10.65. IDENTIFY: A hollow sphere and a solid sphere roll up a ramp without slipping starting with the same speed at the base. Energy conservation applies to both of them.

SET UP: We use $U_1 + K_1 + W_{other} = U_2 + K_2$, with $K_2 = 0$, $U_1 = 0$, $U_2 = mgh$, and $W_{other} = 0$. The total kinetic energy is $K_{tot} = \frac{1}{2}mv_{cm}^2 + \frac{1}{2}I\omega^2$, where $I_{solid} = \frac{2}{5}MR^2$ and $I_{hollow} = \frac{2}{3}MR^2$. We want to know which sphere reaches the greater height on the ramp.

EXECUTE: Apply energy conservation to each sphere to find h in each case.

Solid sphere: $mgh_s = \frac{1}{2}mv_{cm}^2 + \frac{1}{2}\left(\frac{2}{5}mR^2\right)\left(\frac{v_{cm}}{R}\right)^2$, so $h_s = \frac{7}{10}\frac{v_{cm}^2}{g} = 0.700\frac{v_{cm}^2}{g}$.

Hollow sphere: $mgh_H = \frac{1}{2}mv_{cm}^2 + \frac{1}{2}\left(\frac{2}{3}mR^2\right)\left(\frac{v_{cm}}{R}\right)^2$, so $h_H = \frac{5}{6}\frac{v_{cm}^2}{g} = 0.833\frac{v_{cm}^2}{g}$.

The hollow sphere reaches a greater height than the solid sphere.

EVALUATE: Even though the two spheres have the same size, mass, linear speed, and angular speed at the bottom of the ramp, they do not have the same kinetic energy because the hollow sphere has a greater moment of inertia. Therefore the hollow sphere goes higher up the ramp.

10.67. IDENTIFY: A force produces a torque on a wheel, giving it an angular acceleration. But the force is not constant, so the angular acceleration is not constant.

SET UP: The force is $F = kt$, where $k = 5.00$ N/s. Our target variable is the magnitude of the force at the instant the wheel has turned through 8.00 rev (which is 16.0π rad). We can apply $\Sigma\tau_z = I\alpha_z$ to the wheel but we cannot use the constant-acceleration equations. We must return to the basic definitions $\omega_z = d\theta/dt$ and $\alpha_z = d\omega_z/dt$ and integrate.

EXECUTE: First apply $\Sigma\tau_z = I\alpha_z$ to get α_z, calling R the wheel radius. $FR = ktR = I\alpha_z$, so $\alpha_z = Rkt/I = Bt$, where $B = Rk/I$. Since α_z is a function of time, we must integrate to get the angle turned through $\Delta\theta$. $\omega_z = \int \alpha_z dt = \int Bt \, dt = \frac{Bt^2}{2}$, where we have used $\omega_z = 0$ at $t = 0$. Now integrate ω_z to find $\Delta\theta$. $\Delta\theta = \int \omega dt = \int \frac{Bt^2}{2} dt = \frac{Bt^3}{6}$, so $t = \left(\frac{6\Delta\theta}{B}\right)^{1/3}$. The force at this time is

$F = kt = k\left(\frac{6\Delta\theta}{B}\right)^{1/3} = k\left(\frac{96\pi I}{kR}\right)^{1/3} = (5.00 \text{ N/s})\left[\frac{96\pi(2.50 \text{ kg}\cdot\text{m}^2)}{(5.00 \text{ N/s})(0.0600 \text{ m})}\right]^{1/3} = 68.0 \text{ N}.$

EVALUATE: When the acceleration (linear or angular) is not constant, we have little choice but to return to basic definitions and use calculus.

10.69. **IDENTIFY:** Apply $\sum \vec{F} = m\vec{a}$ to each object and apply $\sum \tau_z = I\alpha_z$ to the pulley.

SET UP: Call the 75.0 N weight A and the 125 N weight B. Let T_A and T_B be the tensions in the cord to the left and to the right of the pulley. For the pulley, $I = \frac{1}{2}MR^2$, where $Mg = 80.0$ N and $R = 0.300$ m. The 125 N weight accelerates downward with acceleration a, the 75.0 N weight accelerates upward with acceleration a and the pulley rotates clockwise with angular acceleration α, where $a = R\alpha$.

EXECUTE: $\sum \vec{F} = m\vec{a}$ applied to the 75.0 N weight gives $T_A - w_A = m_A a$. $\sum \vec{F} = m\vec{a}$ applied to the 125.0 N weight gives $w_B - T_B = m_B a$. $\sum \tau_z = I\alpha_z$ applied to the pulley gives $(T_B - T_A)R = (\frac{1}{2}MR^2)\alpha_z$ and $T_B - T_A = \frac{1}{2}Ma$. Combining these three equations gives $w_B - w_A = (m_A + m_B + M/2)a$ and

$$a = \left(\frac{w_B - w_A}{w_A + w_B + w_{\text{pulley}}/2}\right)g = \left(\frac{125\text{ N} - 75.0\text{ N}}{75.0\text{ N} + 125\text{ N} + 40.0\text{ N}}\right)g = 0.2083g.$$

$T_A = w_A(1 + a/g) = 1.2083w_A = 90.62$ N. $T_B = w_B(1 - a/g) = 0.792w_B = 98.96$ N. $\sum \vec{F} = m\vec{a}$ applied to the pulley gives that the force F applied by the hook to the pulley is $F = T_A + T_B + w_{\text{pulley}} = 270$ N. The force the ceiling applies to the hook is 270 N.

EVALUATE: The force the hook exerts on the pulley is less than the total weight of the system, since the net effect of the motion of the system is a downward acceleration of mass.

10.71. **IDENTIFY:** Apply $\sum \vec{F}_{\text{ext}} = m\vec{a}_{\text{cm}}$ to the motion of the center of mass and apply $\sum \tau_z = I_{\text{cm}}\alpha_z$ to the rotation about the center of mass.

SET UP: $I = 2(\frac{1}{2}mR^2) = mR^2$. The moment arm for T is b.

EXECUTE: The tension is related to the acceleration of the yo-yo by $(2m)g - T = (2m)a$, and to the angular acceleration by $Tb = I\alpha = I\frac{a}{b}$. Dividing the second equation by b and adding to the first to eliminate T yields $a = g\frac{2m}{(2m + I/b^2)} = g\frac{2}{2 + (R/b)^2}$, $\alpha = g\frac{2}{2b + R^2/b}$. The tension is found by substitution into either of the two equations:

$$T = (2m)(g - a) = (2mg)\left(1 - \frac{2}{2 + (R/b)^2}\right) = 2mg\frac{(R/b)^2}{2 + (R/b)^2} = \frac{2mg}{(2(b/R)^2 + 1)}.$$

EVALUATE: $a \to 0$ when $b \to 0$. As $b \to R$, $a \to 2g/3$.

10.73. **IDENTIFY:** As it rolls down the rough slope, the basketball gains rotational kinetic energy as well as translational kinetic energy. But as it moves up the smooth slope, its rotational kinetic energy does not change since there is no friction.

SET UP: $I_{\text{cm}} = \frac{2}{3}mR^2$. When it rolls without slipping, $v_{\text{cm}} = R\omega$. When there is no friction the angular speed of rotation is constant. Take $+y$ upward and let $y = 0$ in the valley.

EXECUTE: (a) Find the speed v_{cm} in the level valley: $K_1 + U_1 = K_2 + U_2$. $y_1 = H_0$, $y_2 = 0$. $K_1 = 0$, $U_2 = 0$. Therefore, $U_1 = K_2$. $mgH_0 = \frac{1}{2}mv_{\text{cm}}^2 + \frac{1}{2}I_{\text{cm}}\omega^2$.

$\frac{1}{2}I_{\text{cm}}\omega^2 = \frac{1}{2}(\frac{2}{3}mR^2)\left(\frac{v_{\text{cm}}}{R}\right)^2 = \frac{1}{3}mv_{\text{cm}}^2$, so $mgH_0 = \frac{5}{6}mv_{\text{cm}}^2$ and $v_{\text{cm}}^2 = \frac{6gH_0}{5}$. Find the height H it goes up the other side. Its rotational kinetic energy stays constant as it rolls on the frictionless surface. $\frac{1}{2}mv_{cm}^2 + \frac{1}{2}I_{cm}\omega^2 = \frac{1}{2}I_{cm}\omega^2 + mgH$. $H = \dfrac{v_{cm}^2}{2g} = \frac{3}{5}H_0$.

(b) Some of the initial potential energy has been converted into rotational kinetic energy so there is less potential energy at the second height H than at the first height H_0.

EVALUATE: Mechanical energy is conserved throughout this motion. But the initial gravitational potential energy on the rough slope is not all transformed into potential energy on the smooth slope because some of that energy remains as rotational kinetic energy at the highest point on the smooth slope.

10.75. IDENTIFY: Apply conservation of energy to the motion of the boulder.

SET UP: $K = \frac{1}{2}mv^2 + \frac{1}{2}I\omega^2$ and $v = R\omega$ when there is rolling without slipping. $I = \frac{2}{5}mR^2$.

EXECUTE: Break into two parts, the rough and smooth sections.

Rough: $mgh_1 = \frac{1}{2}mv^2 + \frac{1}{2}I\omega^2$. $mgh_1 = \frac{1}{2}mv^2 + \frac{1}{2}\left(\frac{2}{5}mR^2\right)\left(\dfrac{v}{R}\right)^2$. $v^2 = \dfrac{10}{7}gh_1$.

Smooth: Rotational kinetic energy does not change. $mgh_2 + \frac{1}{2}mv^2 + K_{rot} = \frac{1}{2}mv_{Bottom}^2 + K_{rot}$.

$gh_2 + \dfrac{1}{2}\left(\dfrac{10}{7}gh_1\right) = \dfrac{1}{2}v_{Bottom}^2$. $v_{Bottom} = \sqrt{\dfrac{10}{7}gh_1 + 2gh_2}$

$= \sqrt{\dfrac{10}{7}(9.80 \text{ m/s}^2)(25 \text{ m}) + 2(9.80 \text{ m/s}^2)(25 \text{ m})} = 29.0$ m/s.

EVALUATE: If all the hill was rough enough to cause rolling without slipping,

$v_{Bottom} = \sqrt{\dfrac{10}{7}g(50 \text{ m})} = 26.5$ m/s. A smaller fraction of the initial gravitational potential energy goes into translational kinetic energy of the center of mass than if part of the hill is smooth. If the entire hill is smooth and the boulder slides without slipping, $v_{Bottom} = \sqrt{2g(50 \text{ m})} = 31.3$ m/s. In this case all the initial gravitational potential energy goes into the kinetic energy of the translational motion.

10.77. IDENTIFY: Apply conservation of energy to the motion of the wheel.

SET UP: $K = \frac{1}{2}mv^2 + \frac{1}{2}I\omega^2$. No slipping means that $\omega = v/R$. Uniform density means $m_r = \lambda 2\pi R$ and $m_s = \lambda R$, where m_r is the mass of the rim and m_s is the mass of each spoke. For the wheel, $I = I_{rim} + I_{spokes}$. For each spoke, $I = \frac{1}{3}m_sR^2$.

EXECUTE: (a) $mgh = \frac{1}{2}mv^2 + \frac{1}{2}I\omega^2$. $I = I_{rim} + I_{spokes} = m_rR^2 + 6(\frac{1}{3}m_sR^2)$

Also, $m = m_r + m_s = 2\pi R\lambda + 6R\lambda = 2R\lambda(\pi + 3)$. Substituting into the conservation of energy equation gives $2R\lambda(\pi + 3)gh = \frac{1}{2}(2R\lambda)(\pi + 3)(R\omega)^2 + \frac{1}{2}\left[2\pi R\lambda R^2 + 6(\frac{1}{3}\lambda RR^2)\right]\omega^2$.

$\omega = \sqrt{\dfrac{(\pi + 3)gh}{R^2(\pi + 2)}} = \sqrt{\dfrac{(\pi + 3)(9.80 \text{ m/s}^2)(58.0 \text{ m})}{(0.210 \text{ m})^2(\pi + 2)}} = 124$ rad/s and $v = R\omega = 26.0$ m/s

(b) Doubling the density would have no effect because it does not appear in the answer. ω is inversely proportional to R so doubling the diameter would double the radius which would reduce ω by half, but $v = R\omega$ would be unchanged.

EVALUATE: Changing the masses of the rim and spokes by different amounts would alter the speed v at the bottom of the hill.

10.79. **IDENTIFY:** Use conservation of energy to relate the speed of the block to the distance it has descended. Then use a constant acceleration equation to relate these quantities to the acceleration.

SET UP: For the cylinder, $I = \frac{1}{2}M(2R)^2$, and for the pulley, $I = \frac{1}{2}MR^2$.

EXECUTE: Doing this problem using kinematics involves four unknowns (six, counting the two angular accelerations), while using energy considerations simplifies the calculations greatly. If the block and the cylinder both have speed v, the pulley has angular velocity v/R and the cylinder has angular velocity $v/2R$, the total kinetic energy is

$$K = \frac{1}{2}\left[Mv^2 + \frac{M(2R)^2}{2}(v/2R)^2 + \frac{MR^2}{2}(v/R)^2 + Mv^2\right] = \frac{3}{2}Mv^2.$$

This kinetic energy must be the work done by gravity; if the hanging mass descends a distance y, $K = Mgy$, or $v^2 = (2/3)gy$. For constant acceleration, $v^2 = 2ay$, and comparison of the two expressions gives $a = g/3$.

EVALUATE: If the pulley were massless and the cylinder slid without rolling, $Mg = 2Ma$ and $a = g/2$. The rotation of the objects reduces the acceleration of the block.

10.81. **IDENTIFY:** Apply conservation of angular momentum to the collision. Linear momentum is not conserved because of the force applied to the rod at the axis. But since this external force acts at the axis, it produces no torque and angular momentum is conserved.

SET UP: The system before and after the collision is sketched in Figure 10.81.

EXECUTE: **(a)** $m_b = \frac{1}{4}m_{rod}$

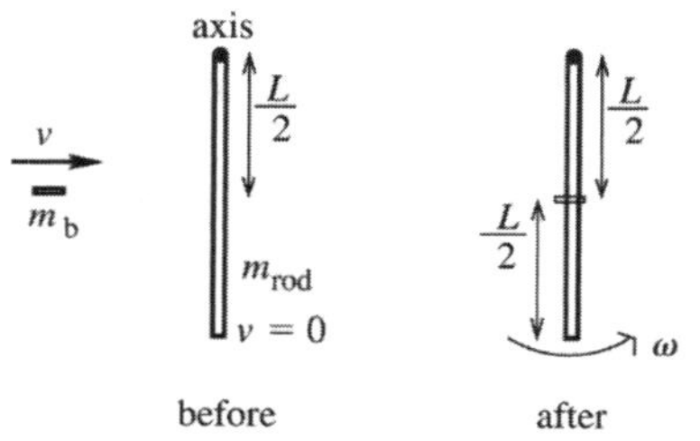

EXECUTE: $L_1 = m_b v r = \frac{1}{4}m_{rod}v(L/2)$

$L_1 = \frac{1}{8}m_{rod}vL$

$L_2 = (I_{rod} + I_b)\omega$

$I_{rod} = \frac{1}{3}m_{rod}L^2$

$I_b = m_b r^2 = \frac{1}{4}m_{rod}(L/2)^2$

$I_b = \frac{1}{16}m_{rod}L^2$

Figure 10.81

Thus $L_1 = L_2$ gives $\frac{1}{8}m_{rod}vL = (\frac{1}{3}m_{rod}L^2 + \frac{1}{16}m_{rod}L^2)\omega$

$\frac{1}{8}v = \frac{19}{48}L\omega$

$\omega = \frac{6}{19}v/L$

(b) $K_1 = \frac{1}{2}mv^2 = \frac{1}{8}m_{rod}v^2$

$K_2 = \frac{1}{2}I\omega^2 = \frac{1}{2}(I_{rod} + I_b)\omega^2 = \frac{1}{2}(\frac{1}{3}m_{rod}L^2 + \frac{1}{16}m_{rod}L^2)(6v/19L)^2$

$K_2 = \frac{1}{2}(\frac{19}{48})(\frac{6}{19})^2 m_{rod}v^2 = \frac{3}{152}m_{rod}v^2$

Then $\dfrac{K_2}{K_1} = \dfrac{\frac{3}{152}m_{rod}v^2}{\frac{1}{8}m_{rod}v^2} = 3/19.$

EVALUATE: The collision is inelastic and $K_2 < K_1$.

10.85. **IDENTIFY:** Apply conservation of angular momentum to the collision between the bird and the bar and apply conservation of energy to the motion of the bar after the collision.

SET UP: For conservation of angular momentum take the axis at the hinge. For this axis the initial angular momentum of the bird is $m_{\text{bird}}(0.500 \text{ m})v$, where $m_{\text{bird}} = 0.500$ kg and $v = 2.25$ m/s. For this axis the moment of inertia is $I = \frac{1}{3}m_{\text{bar}}L^2 = \frac{1}{3}(1.50 \text{ kg})(0.750 \text{ m})^2 = 0.281 \text{ kg} \cdot \text{m}^2$. For conservation of energy, the gravitational potential energy of the bar is $U = m_{\text{bar}}gy_{\text{cm}}$, where y_{cm} is the height of the center of the bar. Take $y_{\text{cm},1} = 0$, so $y_{\text{cm},2} = -0.375$ m.

EXECUTE: **(a)** $L_1 = L_2$ gives $m_{\text{bird}}(0.500 \text{ m})v = (\frac{1}{3}m_{\text{bar}}L^2)\omega$.

$$\omega = \frac{3m_{\text{bird}}(0.500 \text{ m})v}{m_{\text{bar}}L^2} = \frac{3(0.500 \text{ kg})(0.500 \text{ m})(2.25 \text{ m/s})}{(1.50 \text{ kg})(0.750 \text{ m})^2} = 2.00 \text{ rad/s}.$$

(b) $U_1 + K_1 = U_2 + K_2$ applied to the motion of the bar after the collision gives

$$\frac{1}{2}I\omega_1^2 = m_{\text{bar}}g(-0.375 \text{ m}) + \frac{1}{2}I\omega_2^2. \quad \omega_2 = \sqrt{\omega_1^2 + \frac{2}{I}m_{\text{bar}}g(0.375 \text{ m})}.$$

$$\omega_2 = \sqrt{(2.00 \text{ rad/s})^2 + \frac{2}{0.281 \text{ kg} \cdot \text{m}^2}(1.50 \text{ kg})(9.80 \text{ m/s}^2)(0.375 \text{ m})} = 6.58 \text{ rad/s}.$$

EVALUATE: Mechanical energy is not conserved in the collision. The kinetic energy of the bar just after the collision is less than the kinetic energy of the bird just before the collision.

Study Guide for Equilibrium and Elasticity

Summary

We'll explore equilibrium and elasticity in this chapter. We'll focus on extended objects in equilibrium: objects having no net external force or torque acting on them. Objects deform when forces act on them, so we'll examine deformations that describe the stretching, twisting, and compressing of an object. We'll learn about stress, strain, and elastic modulus, and we'll further clarify Hooke's law.

Objectives

After studying this chapter, you'll understand:

- The conditions required for an object to be in equilibrium.
- The definition of *center of gravity* and how to apply it to a problem.
- How to solve problems when objects are in equilibrium.
- How to analyze problems involving the deformation of objects.
- Stress and strain with respect to tension, compression, and shear forces.
- How to use Young's, bulk, and shear moduli to predict the changes due to stress.
- The limits of stress and strain.

Concepts and Equations

Term	Description
Equilibrium of a Rigid Body	No net external force and no net external torque acts on a rigid body in equilibrium: $$\sum \vec{F} = 0 \text{ and } \sum \vec{\tau} = 0.$$
Center of Mass	The torque due to the weight of an object is found by assuming that the entire weight of the object is located at the center of gravity, given by $$\vec{r}_{cm} = \frac{m_1 \vec{r}_1 + m_2 \vec{r}_2 + m_3 \vec{r}_3 + \cdots}{m_1 + m_2 + m_3 + \cdots}$$ The center of gravity is equivalent to the center of mass when gravity is constant.
Stress and Strain	Stress characterizes the strength of a force that stretches, squeezes, or twists an object. Strain is the resulting deformation. Stress and strain are often directly proportional, with the proportionality—the elastic modulus—given by Hooke's law: $$\text{elastic modulus} = \frac{\text{stress}}{\text{stain}}.$$
Tensile and Compressive Stress	Tensile stress is the ratio of the perpendicular component of a force to the cross-sectional area where the force is applied: $$\text{Tensile stress } = \frac{F_\perp}{A}.$$ The SI unit of stress is the pascal (Pa), equal to 1 newton per meter squared. Tensile strain is the ratio of the change in an object's length under stress to its original length: $$\text{Tensile Strain } = \frac{\Delta l}{l_0}.$$ Young's modulus Y is the elastic modulus: $$Y = \frac{\text{Tensile stress}}{\text{Tensile strain}} = \frac{F_\perp / A}{\Delta l / l_0}.$$ Compressive stress and strain are defined in the same manner.
Bulk Stress	The pressure in a fluid is the force per unit area of the fluid: $$p = \frac{F_\perp}{A}.$$ Bulk stress is the change in pressure, and bulk strain is the fractional change in volume, of the fluid. The bulk modulus is the elastic modulus: $$B = \frac{\text{Bulk stress}}{\text{Bulk Strain}} = -\frac{\Delta p}{\Delta V / V_0}.$$
Shear Stress	Shear stress is the force tangent to an object's surface, divided by the area on which the force acts. The shear modulus (S) is the ratio of shear stress to strain: $$S = \frac{\text{Shear stress}}{\text{Shear strain}} = \frac{F_\parallel / A}{x / h}.$$
Limits of Hooke's law	There is a maximum stress for which stress and strain are proportional, beyond which Hooke's law is not valid. The elastic limit is the stress

	beyond which irreversible deformation occurs.

Key Concept 1: If an extended object supported at two or more points is to be in equilibrium, its center of gravity must be somewhere within the area bounded by the supports. If the object is supported at only one point, its center of gravity must be above that point.

Key Concept 2: For an extended object to be in equilibrium, both the net external *force* and the net external *torque* on the object must be zero. The weight of the object acts at its center of gravity.

Key Concept 3: In an equilibrium problem, you can calculate torques around any point you choose. The torque equation will not include any force whose line of action goes through your chosen point, so choosing the point wisely can simplify your calculations.

Key Concept 4: In a two-dimensional equilibrium problem, you have *three* equilibrium equations: two from the condition that the net external force is zero and one from the condition that the net external torque is zero. Use these equations to solve for the unknowns.

Key Concept 5: For an object under tensile or compressive stress, the stress equals the force exerted on either end of the object divided by the cross-sectional area of either end. The strain equals the fractional change in length. The ratio of stress to strain equals Young's modulus for the material of which the object is made.

Key Concept 6: For an object under bulk stress, the stress equals the additional pressure applied to all sides of the object. The strain equals the fractional change in volume. The ratio of stress to strain equals the bulk modulus for the material of which the object is made.

Key Concept 7: For an object under shear stress, the stress equals the force exerted tangent to either of two opposite surfaces of the object divided by the area of that surface. The strain equals the deformation divided by the distance between the two surfaces. The ratio of stress to strain equals the shear modulus for the material of which the object is made.

Conceptual Questions

1: Object in equilibrium

An object is acted upon by no net external force and no net external torque. Is it at rest?

IDENTIFY, SET UP, AND EXECUTE An object that is moving with constant velocity is in equilibrium. An object could exhibit translational motion with a constant velocity or rotate with a constant angular velocity (or both) and remain in equilibrium.

EVALUATE Just as we saw with forces, constant velocity is a state of equilibrium.

2: Ladder on a frictionless surface

A ladder is placed against a wall. The wall is rough, but the floor is frictionless. Can the ladder be in equilibrium?

IDENTIFY, SET UP, AND EXECUTE Four forces act on the ladder: the normal force due to the wall, the normal force due to the floor, gravity, and friction due to the wall. Three of the forces act in the vertical direction: gravity (acting downward), friction due to the wall (acting upward), and the normal force due to the floor (acting upward). These forces can sum to zero, since they act in different directions. One force acts in the horizontal direction: the normal force due to the wall. Since there is only one force, the net external force on the ladder cannot be zero and therefore the ladder cannot be in equilibrium.

EVALUATE Equilibrium is needed in order for us to use the ladder. How can we achieve equilibrium? Friction with the floor is needed to establish equilibrium. Can the ladder be in equilibrium if placed against a frictionless wall on a rough floor? Yes, the forces and torques can be in equilibrium in this case.

Problems

1: Forces on a diving board

A 4.0-m-long diving board with a uniform mass of 150.0 kg is mounted as shown in Figure 1. Find the forces holding the board in place when a 100.0 kg man is standing on the end of the board.

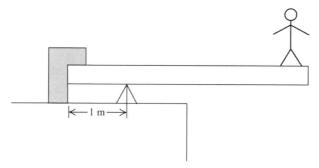

Figure 1 Problem 1.

IDENTIFY We'll use the conditions of equilibrium to solve the problem. The board has two forces holding it in place: a downward force at the left end and an upward force at the pivot. The target variables are the forces acting on the board.

SET UP Figure 2 shows the free-body diagram of the board. Forces A and B hold the board in place, the weight of the board acts at the board's center, and the weight of the man acts at the end. We'll take counterclockwise torques as positive.

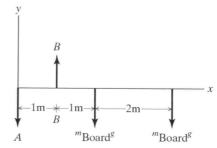

Figure 2 Problem 1 free-body diagram.

EXECUTE Newton's first law applied to the board gives

$$\Sigma F_y = 0 = -F_A + F_B - m_{board}g - m_{man}g.$$

We have two unknowns in this equation, so we need to use the net external torque equation. The net external torque about the left end is zero:

$$\Sigma \tau = 0 = F_B(1.0 \text{ m}) - m_{board}g(2.0 \text{ m}) - m_{man}g(4.0 \text{ m}).$$

Solving for the force at B gives

$$F_B = m_{board}g(2.0) + m_{man}g(4.0) = 6860 \text{ N}.$$

Substituting and solving for the force at A gives
$$F_A = F_B - m_{board}g - m_{man}g = 4410 \text{ N}.$$
The force at the left end is 4410 N downward, and the force at the pivot point is 6860 N upward.

KEY CONCEPT 3 **EVALUATE** To simplify our analysis, we chose the axis for the net external torque such that the torque due to force A was zero. We can double check the result by calculating the torque about the pivot point. If we do, we find the same result.

Practice Problem: What are the forces when no one stands on the board? *Answer:* $F_A = -470$ N, $F_B = +2940$ N

Extra Practice: Where does the man need to stand so that F_B is twice as big as F_A? *Answer:* 2.0 m

2: Force on a support strut

Find the force exerted by the wall on the uniform strut shown in Figure 3 if the strut weighs 100.0 N.

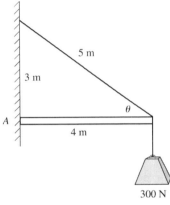

Figure 3 Problem 2.

IDENTIFY We'll use equilibrium conditions to solve the problem. The strut has four forces acting on it; both the net external force and the net external torque are zero. The target variable is the force acting on the strut due to the wall.

SET UP Figure 4 shows the free-body diagram of the strut. The weights of the strut and hanging mass, tension, and the force of the wall act on the strut. We assume that the force of the wall on the strut acts to the right and upward. The forces act in two directions, so we'll need to include net external forces in those directions. We'll take counterclockwise torques as positive.

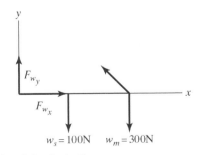

Figure 4 Problem 2 free-body diagram.

EXECUTE Newton's first law applied in the x- direction gives
$$\Sigma F_x = 0 = F_{Wx} - T\cos\theta.$$
Newton's first law applied in the y- direction gives
$$\Sigma F_y = 0 = F_{Wy} + T\sin\theta - w_S - w_M.$$

We have three unknowns in these equations, so we need to use the net external torque equation. The net external torque about the left end of the strut is zero:
$$\Sigma\tau = 0 = -w_S(2.0\text{ m}) - w_M(4.0\text{ m}) + T\sin\theta(4.0\text{ m}).$$

Inspecting the figure, we see that the sine and cosine of the angle are 3/5 and 4/5, respectively. We solve the torque equation for tension, giving
$$T = \frac{w_S(2.0\text{ m}) + w_M(4.0\text{ m})}{\sin\theta(4.0\text{ m})} = 583\text{ N}.$$

With the tension, we can solve for the components of the force due to the wall:
$$F_{Wx} = T\cos\theta = 467\text{ N},$$
$$F_{Wy} = w_S + w_M - T\sin\theta = 50\text{ N}.$$

The magnitude of the force is
$$F_W = \sqrt{F_{Wx}^2 + F_{Wy}^2} = 470\text{ N},$$
and it acts at an angle
$$\phi = \tan^{-1}\left(\frac{F_{Wy}}{F_{Wx}}\right) = 6.1°$$
above the positive x-axis.

KEY CONCEPT 4 **EVALUATE** We see that we chose the correct directions for the force of the wall on the strut. If we hadn't guessed correctly, we would've found negative results for one or both of the force components.

You can double check the result by calculating the torque about any other point. Do you find the same result when you do?

CAUTION Pick pivots carefully! Carefully choosing your pivot point simplifies the net external torque equation, as we have seen in the previous two examples. Note how the pivot point is chosen in the next two problems.

Practice Problem: What is the force if the hanging mass is doubled? *Answer:* 868 N

Extra Practice: What is the force if the hanging mass is removed? *Answer:* 83.6 N

3: Boom in equilibrium

A horizontal wire supports a boom of length L. The boom supports a 200.0 N weight as shown in Figure 5. The boom weighs 200.0 N. Find the tension in the wire and the force exerted by the ground on the boom.

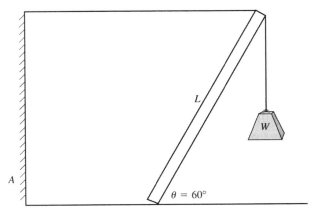

Figure 5 Problem 3.

IDENTIFY We'll use equilibrium conditions to solve the problem. The boom has four forces acting on it; both the net external force and the net external torque are zero. The target variables are the tension and the force acting on the boom due to the ground.

SET UP Figure 6 shows the free-body diagram of the boom. The weights of the boom and the hanging mass, tension, and the force of the floor act on the boom. We assume that the force of the ground on the boom acts to the right and upward. The forces act in two directions, so we'll need to include net external forces in those directions. We'll take counterclockwise torques as positive.

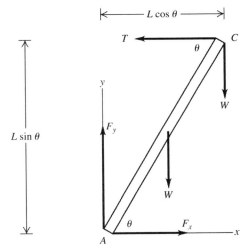

Figure 6 Problem 3 free-body diagram.

EXECUTE Newton's first law applied in the x- direction gives
$$\sum F_x = 0 = F_x - T.$$
Newton's first law applied in the y- direction gives
$$\sum F_y = 0 = F_y - w - w.$$

We have three unknowns in these equations, so we need to use the net external torque equation. The net external torque about the bottom end of the boom is zero:
$$\sum \tau = 0 = LT \sin\theta - Lw\cos\theta - \tfrac{1}{2} Lw\cos\theta.$$

We solve the torque equation for tension, giving

$$T = \frac{3}{2}\frac{w\cos\theta}{\sin\theta} = \frac{3}{2}\frac{(200.0\text{ N})\cos 60°}{\sin 60°} = 173\text{ N}.$$

With the tension, we can solve for the components of the force due to the ground:
$$F_x = T = 173\text{ N},$$
$$F_y = 2w = 400.0\text{ N}.$$

The magnitude of the force is
$$F = \sqrt{F_x^2 + F_y^2} = 436\text{ N},$$

and it acts at an angle
$$\phi = \tan^{-1}\left(\frac{F_y}{F_x}\right) = 66.6°$$

above the positive x-axis.

KEY CONCEPTS 3, 4 **EVALUATE** We see that we chose the correct directions for the force of the ground on the boom. If we hadn't guessed correctly, we would've found negative results for one or both of the force components.

You can double check the result by calculating the torque about any other point. Do you find the same result when you do?

Practice Problem: What is the force if $\theta = 30°$? *Answer:* 656 N

Extra Practice: What if you now double the hanging weight? *Answer:* 1054 N

4: Coefficient of friction for a strut

Find the minimum coefficient of friction between the weightless horizontal strut and the wall in the system shown in Figure 7.

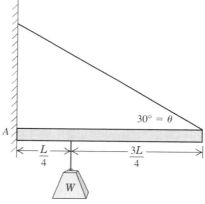

Figure 7 Problem 4.

IDENTIFY We'll use equilibrium conditions to solve the problem. The strut has four forces acting on it; both the net external force and the net external torque are zero. The target variable is the coefficient of friction at the wall.

SET UP Figure 8 shows the free-body diagram of the strut. The weights of the hanging mass, tension, friction, and the normal force of the wall act on the strut. The forces act in two directions, so we'll need to include net external forces in those directions. We'll take counterclockwise torques as positive.

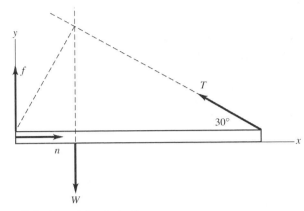

Figure 8 Problem 4 free-body diagram.

EXECUTE Newton's first law applied in the x- direction gives
$$\sum F_x = 0 = n - T\cos 30°.$$
Newton's first law applied in the y- direction gives
$$\sum F_y = 0 = f + T\sin 30° - w.$$
We have three unknowns in these equations, so we need to use the net torque equation. The net external torque about the left end of the strut is zero:
$$\sum \tau = 0 = -w\tfrac{1}{4}L + T\sin 30° \, L.$$
We solve the torque equation for tension, giving
$$T = \frac{w}{4\sin 30°}.$$
With the tension, we can solve for the normal and friction forces:
$$n = T\cos 30° = \frac{w\cos 30°}{4\sin 30°},$$
$$f = w - T\sin 30° = w - \tfrac{1}{4}w = \tfrac{3}{4}w.$$
For there to be no slipping,
$$f < \mu n.$$
Solving for the coefficient of friction gives
$$\mu = \frac{f}{n} = \frac{\tfrac{3}{4}w}{\frac{w}{4\tan 30°}} = 3\tan 30° = 1.73.$$
The minimum coefficient of friction is 1.73.

KEY CONCEPTS 3, 4

EVALUATE How can you double-check the result? Do you find the same result when you do?

Practice Problem: What is the tension if the weight is 300 N? *Answer:* 150 N

Extra Practice: What minimum coefficient is required if the strut weighs w/3? *Answer:* 1.27

5: Strain in a steel cable

A 10.0 kg weight is hung from a steel wire having an unstretched length of 1.0 m and a diameter of 2.0 mm. How much does the wire stretch?

IDENTIFY The force acting on the cable is the weight of the mass. Young's modulus will be used to find the change in length.

SET UP We look up Young's modulus for steel and find that it is 2×10^{11} Pa.

EXECUTE Young's modulus is

$$Y = \frac{F/A}{\Delta L/L_0}.$$

Rearranging terms to find the change in length gives

$$\Delta L = \frac{FL_0}{YA} = \frac{mgL_0}{Y\pi(d/2)^2} = \frac{(10.0 \text{ kg})(9.8 \text{ m/s}^2)(1.0 \text{ m})}{(2 \times 10^{11} \text{ N/m}^2)\pi(0.002 \text{ mm}/2)^2} = 1.56 \times 10^{-4} \text{ m} = 0.156 \text{ mm}.$$

KEY CONCEPT 5 **EVALUATE** We see that the stretch is very small for the wire. This agrees with experience: Steel is a difficult material to stretch.

Practice Problem: How much does it stretch if it is made out of aluminum? *Answer:* 0.446 mm

Extra Practice: How much does the aluminum wire stretch if you double the length? *Answer:* 0.892 mm

6: Strain on an elevator cable

A steel elevator cable can support a maximum stress of 9.0×10^7 Pa. If the maximum mass of the fully loaded elevator is 2100 kg and the maximum upward acceleration is 3.0 m/s², what should the diameter of the cable be? By how much does the cable stretch when the elevator is accelerating upward at 3.0 m/s² and 120 m of cable has been released? (Young's modulus for steel is 2×10^{11} Pa.)

IDENTIFY We'll use Newton's second law to find the tension in the elevator cable and then use the maximum stress to find the cable diameter. To solve the second part, we'll use Young's modulus. The target variables are the cable diameter and the elongation of the cable.

SET UP Figure 9 shows the free-body diagram of the elevator. Gravity and tension act on the elevator.

Figure 9 Problem 6 free-body diagram.

EXECUTE We'll apply Newton's second law to find the maximum tension in the cable:

$$\Sigma F_y = T - mg = ma_y.$$

Study Guide for Equilibrium and Elasticity

The tension is then
$$T = m(g + a_x) = (2100 \text{ kg})((9.8 \text{ m/s}^2) + (3.0 \text{ m/s}^2)) = 26{,}900 \text{ N}.$$
Stress is the force per unit area, or
$$S = \frac{F}{A} = \frac{T}{A}.$$
The area can be written in terms of the diameter as
$$A = \frac{\pi d^2}{4}.$$
The diameter is then
$$d = 2\sqrt{\frac{T}{S\pi}} = 2\sqrt{\frac{(26{,}900 \text{ N})}{(9.0\times 10^7)\pi}} = 0.020 \text{ m} = 2.0 \text{ cm},$$
where we have replaced the stress with the maximum stress. Young's modulus then leads to the amount of cable stretch:
$$Y = \frac{l_0 F}{A \Delta l} = \frac{4 l_0 T}{\pi d^2 \Delta l}.$$
Rearranging terms to find the cable stretch gives
$$\Delta l = \frac{4 l_0 T}{\pi d^2 Y} = \frac{4(120 \text{ m})(26{,}900 \text{ N})}{\pi (0.020 \text{ m})^2 (2.0\times 10^{11} \text{ Pa})} = 0.051 \text{ m} = 5.1 \text{ cm}.$$
The cable must have a 2.0 cm diameter and stretch 5.1 cm when 120 m of cable has been released.

KEY CONCEPT 5 **EVALUATE** We see that the 2-cm-thick cable stretches over 5 cm. This may appear to be a significant elongation, but it represents only 0.04% of the cable length.

Practice Problem: How much would the same cable stretch if the elevator is accelerating downward at 3.0 m/s²? *Answer:* 2.71 cm

Extra Practice: How much would it stretch if the elevator is stationary? *Answer:* 3.90 cm

7: Compressibility of oil

Find the compressibility of a 0.1 m³ sample of oil whose volume decreases 2.04×10^{-4} m³ when subjected to an increase in pressure of 1.02×10^7 Pa.

IDENTIFY AND SET UP We'll use the bulk modulus to find the compressibility. With the given information, the compressibility follows directly from the definition of the bulk modulus.

EXECUTE The bulk modulus is given by
$$B = -\frac{\Delta p}{\Delta V / V_0}.$$
Solving, we have
$$B = -\frac{\Delta p V_0}{\Delta V} = -\frac{(1.02\times 10^7 \text{ Pa})(0.1 \text{ m}^3)}{(-2.04\times 10^{-4} \text{ m}^3)} = 5.0\times 10^9 \text{ Pa} = 4.9\times 10^4 \text{ atm}.$$
The compressibility is the inverse of the bulk modulus, or

$$k = \frac{1}{B} = \frac{1}{4.9 \times 10^4 \text{ atm}} = 2.0 \times 10^{2\,5}/\text{atm}.$$

KEY CONCEPT 6 EVALUATE We see that the oil does not compress when subjected to pressure. It requires an increase of 500 atmospheres of pressure to change the volume by 1%.

Practice Problem: How much does the volume decrease if you double the pressure increase?
Answer: 4.08×10^{-4} m

Extra Practice: How much pressure increase is required to reduce the volume by 0.5%?
Answer: 2.5×10^7 Pa

Try It Yourself!

1: Tension in support cable

Find the tension in the supporting cable and the force acting on the strut due to the wall for the weightless horizontal strut shown in Figure 10.

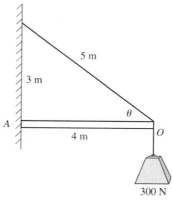

Figure 10 Try it yourself 1.

Solution Checkpoints

IDENTIFY AND SET UP The strut is in equilibrium, so apply equilibrium conditions to solve the problem. Start with a free-body diagram, and include the force due to the weight hanging off the end of the strut, tension, and the force of the wall. Set the net external forces and net external torques equal to zero.

EXECUTE Newton's first law applied in the *x*- and *y*- directions gives

$$\Sigma F_x = 0 = F_{\text{wall }x} - T\cos\theta,$$
$$\Sigma F_y = 0 = F_{\text{wall }y} + T\sin\theta - w.$$

The net torque about the right end of the strut is zero:

$$\Sigma \tau = 0 = -F_{\text{wall }y}(4.0 \text{ m}).$$

Determine the sine and cosine of the angle and solve.

The tension is 500.0 N. The force due to the wall is 400 N, directed perpendicular to the wall.

KEY CONCEPTS **EVALUATE** Choosing a good pivot point simplifies calculations. By calculating the torques
3, 4 about the right end of the strut, we immediately learned that there is no y- component of force due to the wall. Do you get the same result if you set the pivot on the left end of the strut?

Practice Problem: What is the tension if the strut weighs 100 N? *Answer:* 583 N

Extra Practice: What force does the wall exert in this case? *Answer:* 469 N

2: Force acting on a ladder

A ladder of mass 25.0 kg rests against a frictionless wall as shown in Figure 11. Find all the forces acting on the ladder.

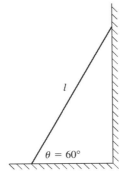

Figure 11 Try it yourself 2.

Solution Checkpoints

IDENTIFY AND SET UP The ladder is in equilibrium, so apply equilibrium conditions to solve the problem. Start with a free-body diagram, and include the forces due to the wall, the ground, and gravity. Set the net external forces and net external torques equal to zero.

EXECUTE Newton's first law applied in the x- and y- directions gives

$$\Sigma F_x = 0 = F_{\text{ground}\, x} - n,$$
$$\Sigma F_y = 0 = F_{\text{ground}\, y} - w.$$

The net torque about the bottom end of the ladder is zero:

$$\Sigma \tau = 0 = -\frac{L}{2} w \cos\theta + L n \sin\theta.$$

The normal force due to the wall is 71 N. The force due to the ground is 71 N in the positive x- direction and 245 N upward.

KEY CONCEPTS **EVALUATE** Does the wall exert a force in the vertical direction? How can you check your
3, 4 results?

Practice Problem: What is the normal force when the angle is 30°? *Answer:* 212 N

Extra Practice: What is the total force from the ground in this case? *Answer:* 324 N

3: Friction force acting on a ladder

A ladder of mass 25.0 kg rests against a frictionless wall as shown in Figure 11. What is the minimum coefficient of friction between the ladder and the ground that allows the ladder to stand without slipping?

Solution Checkpoints

IDENTIFY AND SET UP The x- component of the force due to the ground is the friction force in the previous problem. Use the definition of friction to solve for the coefficient.

EXECUTE The static friction force is
$$F_{\text{ground }x} = f \leq \mu_s n.$$
The coefficient of static friction is 0.29.

EVALUATE How did you find the normal force?

Practice Problem: What coefficient is required when the angle is 30°? *Answer:* 0.87

Extra Practice: What coefficient is required if you now double the mass of the ladder? *Answer:* 0.87

4: Stress in wire

A copper wire of cross-sectional area 0.050 cm^2 and length 5.0 m is attached end to end to a steel wire of length 3.0 m and cross-sectional area 0.020 cm^2. The wires are stretched under a tension of 200.0 N. Find the stress in each wire and the total change in length for the combination.

Solution Checkpoints

IDENTIFY Tensile stress is the force per area. Young's modulus will be used to find the change in length. You can find Young's modulus for steel and copper from Table 1 of the text. Then add the changes in lengths to find the total change in length.

EXECUTE The stresses are
$$\text{Stress}_{\text{Copper}} = \frac{F}{A} = 4.0 \times 10^7 \text{ Pa},$$
$$\text{Stress}_{\text{Steel}} = \frac{F}{A} = 1.0 \times 10^8 \text{ Pa}.$$
The change in length of the copper is
$$\Delta L = \frac{FL_0}{YA} = 1.8 \times 10^{-3} \text{ m}.$$
For steel, the change in length is 1.5×10^{-3} m. The total length is 3.3×10^{-3} m, or 3.3 mm.

KEY CONCEPT 5 **EVALUATE** We see that the stretch is very small for the combined length of wire.

Practice Problem: What is the total change if the copper wire is replaced with an aluminum wire having the same dimensions? *Answer:* 4.36 mm

Extra Practice: What is the total change if the entire wire is steel ($L = 8$ m and $A = 0.020 \text{ cm}^2$)? *Answer:* 4.00 mm

Key Example Variation Problems

Solutions to these problems are in Chapter 11 of the Student's Solutions Manual.

Be sure to review EXAMPLE 11.1 (Section 11.2) before attempting these problems.

VP11.1.1 A uniform plank 8.00 m in length with mass 40.0 kg is supported at two points located 1.00 m and 5.00 m, respectively, from the left-hand end. What is the maximum additional mass you could place on the right-hand end of the plank and have the plank still be at rest?

VP11.1.2 A bowling ball (which we can regard as a uniform sphere) has a mass of 7.26 kg and a radius of 0.216 m. A baseball has a mass of 0.145 kg. If you connect these two balls with a lightweight rod, what must be the distance between the center of the bowling ball and the center of the baseball so that the system of the two balls and the rod will balance at the point where the rod touches the surface of the bowling ball?

VP11.1.3 Three small objects are arranged along a uniform rod of mass m and length L: one of mass m at the left end, one of mass m at the center, and one of mass $2m$ at the right end. How far to the left or right of the rod's center should you place a support so that the rod with the attached objects will balance there?

VP11.1.4 A small airplane with full fuel tanks, but no occupants or baggage, has a mass of 1.17×10^3 kg and a center of gravity that is 2.58 m behind the nose of the airplane. The pilot's seat is 2.67 m behind the nose, and the baggage compartment is 4.30 m behind the nose. A 75.0 kg pilot boards the plane and is the only occupant. If the center of gravity of the airplane with pilot can be no more than 2.76 m behind the nose for in-flight stability, what is the maximum mass that the baggage compartment can hold?

Be sure to review EXAMPLES 11.2, 11.3, and 11.4 (Section 11.3) before attempting these problems.

VP11.4.1 The rear wheels of a truck support 57.0% of the weight of the truck, while the front wheels support 43.0% of the weight. The center of gravity of the truck is 2.52 m in front of the rear wheels. What is the wheelbase of the truck (the distance between the front and rear wheels)?

VP11.4.2 A small airplane is sitting at rest on the ground. Its center of gravity is 2.58 m behind the nose of the airplane, the front wheel (nose wheel) is 0.800 m behind the nose, and the main wheels are 3.02 m behind the nose. What percentage of the airplane's weight is supported by the nose wheel, and what percentage is supported by the main wheels?

VP11.4.3 Figure 11.11 shows a metal advertising sign of weight w suspended from the end of a horizontal rod of negligible mass and length L. The end of the rod with the sign is supported by a cable at an angle θ from the horizontal, and the other end is supported by a hinge at point P. (a) Using the idea that there is zero net torque about the end of the rod with the attached sign, find the vertical component of the force $\vec{F}_{\text{hinge}}$ exerted by the hinge. (b) Using the idea that there is zero net vertical force on the rod with the attached sign, find the tension in the cable. (c) Using the idea that there is zero net horizontal force on the rod with the attached sign, find the horizontal component of the force exerted by the hinge.

VP11.4.4 Suppose that in Figure 11.11 the horizontal rod has weight w (the same as the hanging sign). (a) Using the idea that there is zero net torque about the end of the rod with the attached sign, find the vertical component of the force $\vec{F}_{\text{hinge}}$ exerted by the hinge. (b) Using the idea that there is zero net vertical force on the rod with the attached sign, find the tension in the cable. (c) Using the idea that there is zero net horizontal force on the rod with the attached sign, find the horizontal component of the force exerted by the hinge.

Be sure to review EXAMPLES 11.5, 11.6, and 11.7 (Section 11.4) before attempting these problems.

VP11.7.1 A copper wire has a radius of 4.5 mm. When forces of a certain equal magnitude but opposite directions are applied to the ends of the wire, the wire stretches by 5.0×10^{-3} of its original length. (a) What is the tensile stress on the wire? (b) What is the magnitude of the force on either end?

VP11.7.2 An aluminum cylinder with a radius of 2.5 cm and a height of 82 cm is used as one leg of a workbench. The workbench pushes down on the cylinder with a force of 3.2×10^4 N. (a) What is the compressive strain of the cylinder? (b) By what distance does the cylinder's height decrease as a result of the forces on it?

VP11.7.3 The pressure on the surface of a sphere of radius 1.2 cm is increased by 2.5×10^7 Pa 2.5. Calculate the resulting decrease in volume of the sphere if it is made (a) of lead and (b) of mercury.

VP11.7.4 You apply forces of magnitude 4.2×10^4 N to the top and bottom surfaces of a brass cube. The forces are tangent to each surface and parallel to the sides of each surface. If the cube is 2.5 cm on a side, what is the resulting shear displacement?

STUDENT'S SOLUTIONS MANUAL FOR EQUILIBRIUM AND ELASTICITY

VP11.1.1. **IDENTIFY:** We are dealing with center of gravity. If the center of gravity of the mass-plank system moves beyond the right-hand support point, the plank will tip over.

SET UP: $x_{cg} = \dfrac{m_1 x_1 + m_2 x_2 + m_3 x_3 + \cdots}{m_1 + m_2 + m_3 + \cdots}$. Call the origin the right-hand support point. The center of mass of the system is at that point, so $x_{cm} = 0$. Call m the unknown mass.

EXECUTE: Applying the center of gravity formula $x_{cg} = \dfrac{m_1 x_1 + m_2 x_2}{m_1 + m_2}$ gives

$0 = \dfrac{(40.0 \text{ kg})(-1.00 \text{ m}) + (3.00 \text{ m}) m}{40.0 \text{ kg} + m} \rightarrow m = 13.3 \text{ kg}$.

EVALUATE: The added mass of 13.3 kg is much less than the 40.0-kg mass of the plank because this added mass is much farther from the pivot point than the center of mass is.

VP11.1.2. **IDENTIFY:** We are dealing with center of gravity. If the center of gravity of the two-ball system is at the bowling ball, the system will balance.

SET UP: $x_{cg} = \dfrac{m_1 x_1 + m_2 x_2}{m_1 + m_2}$. Call the origin the center of the bowling ball and call d the distance between the centers of the two balls.

EXECUTE: With our choice of origin, $x_{cm} = 0.216$ m, so we have

$0.216 \text{ m} = \dfrac{(7.26 \text{ kg})(0) + (0.145 \text{ kg})d}{7.26 \text{ kg} + 0.145 \text{ kg}} \rightarrow d = 11.0 \text{ m}$.

EVALUATE: To check, we can balance torques about the surface of the bowling ball with the rod horizontal. Call M the mass of the bowling ball and R its radius and m the mass of the baseball. This gives $MgR - mg(d - R) = 0 \rightarrow d = R(M + m)/m = 11.0$ m, which agrees with our result.

VP11.1.3. **IDENTIFY:** For balance, the center of gravity of the system of three objects and the rod must be at the support point.

SET UP: Call the center of the rod the origin and call x the distance of the support point from the center of the rod; this is also the distance of the center of gravity from the center. We treat the bar as a point-mass of m located at its center.

EXECUTE: $x_{cg} = \dfrac{m_1 x_1 + m_2 x_2 + m_3 x_3 + \cdots}{m_1 + m_2 + m_3 + \cdots} = \dfrac{m(-L/2) + 2m(0) + 2m(L/2)}{5m} = \dfrac{L}{10}$. The support point should be a distance $L/10$ to the right of the center of the bar, which is the location of the center of gravity of the system.

EVALUATE: To check, substitute the calculated answer back into the center-of-mass formula but using a different origin and use it to calculate the location of the center of mass. For example, using the left end of the bar as the origin, we have $x_{cg} = \frac{2m(L/2) + 2m(L)}{5m} = 6L/10$, which is $L/10$ to the right of the center of the bar, as we just found.

VP11.1.4. IDENTIFY: We know where the center of gravity of the loaded plane should be, and we want to find out how much mass we can have in the baggage compartment.

SET UP: Use $x_{cg} = \frac{m_1 x_1 + m_2 x_2 + m_3 x_3 + \cdots}{m_1 + m_2 + m_3 + \cdots}$. Let the x-axis point horizontally to the right with the nose of the plane on the left end and the origin at the center of gravity of the *loaded* plane. With this choice, $x_{cm} = 0$.

EXECUTE: Let M be the maximum mass in the baggage compartment.

$$0 = \frac{-(1170 \text{ kg})(2.76 \text{ m} - 2.58 \text{ m}) - (75.0 \text{ kg})(2.76 \text{ m} - 2.67 \text{ m}) + M(4.30 \text{ m} - 2.76 \text{ m})}{m_{\text{total}}}$$

$M = 141$ kg.

EVALUATE: Choosing the origin at the center of gravity of the loaded plane makes the algebra rather simple compared to other choices because we do not have to deal with m_{total} in the denominator. Since m_{total} includes M, we would have more work (and more chances for error) with other choices of the origin. It helps to plan ahead!

VP11.4.1. IDENTIFY: The truck is in equilibrium, so the torques on it must balance.

SET UP: Apply $\Sigma \tau_z = 0$. Take torques about the rear wheel. Call d the wheelbase and w the weight of the truck. See Fig. VP11.4.1.

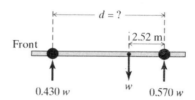

Figure VP11.4.1

EXECUTE: $\Sigma \tau_z = 0$ gives $(0.430 \, w)d = (2.52 \text{ m})w \rightarrow d = 5.86$ m.

EVALUATE: To check, calculate the center of gravity of the truck. This gives $(0.570 \, w)(2.52 \text{ m}) - (0.430 \, w)(5.86 \text{ m} - 2.52 \text{ m}) = 0$, as it should since the truck is in equilibrium.

VP11.4.2. IDENTIFY: The plane is in equilibrium, so the torques on it must balance.

SET UP: Apply $\Sigma \tau_z = 0$. Take torques about the nose wheel. Call F the force due to the main wheel and w the weight of the plane.

EXECUTE: $\Sigma \tau_z = 0$ gives $-w(2.58 \text{ m} - 0.800 \text{ m}) + F(3.02 \text{ m} - 0.800 \text{ m}) = 0$, so $F = 0.802w$. The main wheel supports 80.2% of the weight of the plane, so the nose wheels must support 19.8% of the weight.

EVALUATE: To check, calculate torques about another point.

VP11.4.3. IDENTIFY: The sign is in equilibrium, so the forces and torques on the rod must balance.

SET UP: Apply $\Sigma \tau_z = 0$, $\Sigma F_x = 0$, and $\Sigma F_y = 0$.

EXECUTE: **(a)** Apply $\Sigma \tau_z = 0$ about the right end of the rod, giving
$F_{\text{hinge-}y} L + F_{\text{hinge-}x}(0) + T(0) + w(0) = 0 \rightarrow F_{\text{hinge-}y} = 0$.
(b) $\Sigma F_y = 0$ gives $F_{\text{hinge-}y} + T \sin \theta - w = 0 \rightarrow 0 + T \sin \theta - w = 0 \rightarrow T = w/\sin \theta$.
(c) $\Sigma F_x = 0$ gives $F_{\text{hinge-}x} - T \cos \theta = 0$. Using T from part (b) gives

$F_{\text{hinge-}x} = \left(\dfrac{w}{\sin\theta}\right)\cos\theta = w/\tan\theta.$

EVALUATE: Our result in (c) says that if θ is small, the hinge exerts a large horizontal force on the rod. This is reasonable because the cable pulls nearly horizontally against the rod which causes it to push very hard against the hinge.

VP11.4.4. **IDENTIFY:** The sign is in equilibrium, so the forces and torques on the rod must balance.

SET UP: Apply $\Sigma\tau_z = 0$, $\Sigma F_x = 0$, and $\Sigma F_y = 0$. The rod now also has weight w.

EXECUTE: **(a)** Apply $\Sigma\tau_z = 0$ about the right end of the rod, giving
$F_{\text{hinge-}y}L + F_{\text{hinge-}x}(0) + wL/2 + T(0) + w(0) = 0 \rightarrow F_{\text{hinge-}y} = w/2$.

(b) $\Sigma F_y = 0$ gives $F_{\text{hinge-}y} + T\sin\theta - w/2 - w = 0 \rightarrow T\sin\theta - 3w/2 = 0$

so $T = \dfrac{3w}{2\sin\theta}$.

(c) $\Sigma F_x = 0$ gives $F_{\text{hinge-}x} - T\cos\theta = 0$. Using T from part (b) gives

$F_{\text{hinge-}x} = \left(\dfrac{3w}{2\sin\theta}\right)\cos\theta = \dfrac{3w}{2\tan\theta}$.

EVALUATE: Our result in (c) says that if θ is small, the hinge exerts a large horizontal force on the rod. This is reasonable because the cable pulls nearly horizontally against the rod, which causes it to push very hard against the hinge. We also find that if the rod weighs as much as the sign, the tension and hinge force each are 1.5 times as great as before.

VP11.7.1. **IDENTIFY:** This problem involves elasticity, tensile stress, tensile stress, and Young's modulus.

SET UP: Young's modulus is $Y = \dfrac{\text{Tensile stress}}{\text{Tensile strain}}$, tensile strain $= \dfrac{\Delta\ell}{\ell_0}$, tensile stress $= \dfrac{F_\perp}{A}$.

EXECUTE: **(a)** Tensile stress = Y(tensile strain). $\dfrac{\Delta\ell}{\ell_0} = \dfrac{5.0\times 10^{-3}\ell_0}{\ell_0} = 5.0\times 10^{-3}$, so

tensile stress = $(11\times 10^{10}\text{ Pa})(5.0\times 10^{-3}) = 5.5\times 10^8$ Pa.

(b) Tensile stress = $\dfrac{F_\perp}{A}$, where $A = \pi r^2$. Therefore $F_\perp$ = (tensile strain)(πr^2) so

$F_\perp = (5.5\times 10^8\text{ Pa})(\pi)(4.5\times 10^{-3}\text{ m})^2 = 3.5\times 10^4$ N.

EVALUATE: The pressure is much less than atmospheric pressure, but the force is about 7600 lb, which is very large. But a wire 4.5 mm in radius is quite thick compared to ordinary electrical copper wires.

VP11.7.2. **IDENTIFY:** This problem involves elasticity, compressive stress and compressive strain, and Young's modulus.

SET UP: Young's modulus is $Y = \dfrac{\text{Compressive stress}}{\text{Compressive strain}}$, compressive strain $= \dfrac{\Delta\ell}{\ell_0}$, compressive stress $= \dfrac{F_\perp}{A}$. From Table 11.1, $Y = 7.0\times 10^{10}$ Pa for aluminum.

EXECUTE: **(a)** We want the compressive strain. $Y = \dfrac{\text{Compressive stress}}{\text{Compressive strain}}$ tells us that compressive strain $= \dfrac{\text{Compressive stress}}{Y} = \dfrac{F_\perp}{AY} = \dfrac{3.2\times 10^4\text{ N}}{\pi(0.025\text{ m})^2(7.0\times 10^{10}\text{ Pa})} = 2.3\times 10^{-4}$.

(b) Compressive strain $= \dfrac{\Delta\ell}{\ell_0} \rightarrow \Delta\ell = \ell_0 \times$ compressive strain, which gives

$\Delta \ell = (82 \text{ cm})(2.3 \times 10^{-4}) = 1.9 \times 10^{-2} \text{ cm} = 0.19 \text{ mm}.$

EVALUATE: Note that an 82-cm cylinder compresses only 0.19 mm under a force of 32,000 N, which is about 7200 lb. So fractional changes in length are normally quite small.

VP11.7.3. **IDENTIFY:** The increase in pressure compresses the sphere. Using the bulk modulus we can find the change in volume.

SET UP: Bulk modulus is $B = -\dfrac{\Delta p}{\Delta V / V_0}$. To find the *decrease* in volume, we can neglect the minus sign. From Table 11.1, $B = 4.1 \times 10^{10}$ Pa for lead, and from Table 11.2, $B = 1/k = 2.7 \times 10^{10}$ Pa for mercury.

EXECUTE: The decrease in volume is $\Delta V = \dfrac{V_0 \Delta p}{B}$.

(a) Lead: $\Delta V = \dfrac{V_0 \Delta p}{B} = \dfrac{\frac{4}{3}\pi (0.012 \text{ m})^3 (2.5 \times 10^7 \text{ Pa})}{4.1 \times 10^{10} \text{ Pa}} = 4.4 \times 10^{-9} \text{ m}^3.$

(b) Mercury: Table 11.2 gives compressibility (k), so we use $B = 1/k$ to find $B = 2.7 \times 10^{10}$ Pa. Using the same formula as in (a), but with a different B, gives $\Delta V = 6.7 \times 10^{-9}$ m³.

EVALUATE: If we divide the equations for ΔV, we get $\Delta V_M = \Delta V_L (B_L/B_M)$
$\Delta V_M = (4.4 \times 10^{-9} \text{ m}^3)(4.1/2.7) = 6.7 \times 10^{-9}$ m³, which agrees with our result. Also note that since $B_L \approx 1.5 B_M$, we would expect ΔV_M to be about 1.5 times as great as ΔV_L (since ΔV is *inversely* proportional to B), which is what we find.

VP11.7.4. **IDENTIFY:** The shear forces distort the cube by a small distance x.

SET UP: The shear modulus is $S = \dfrac{F_\parallel h}{Ax}$, so $x = \dfrac{F_\parallel h}{AS}$. For brass $S = 3.5 \times 10^{10}$ Pa from Table 11.1.

EXECUTE: $x = \dfrac{(4.2 \times 10^4 \text{ N})(0.025 \text{ m})}{(0.025 \text{ m})^2 (3.5 \times 10^{10} \text{ Pa})} = 4.8 \times 10^{-5} \text{ m}.$

EVALUATE: The length of a side of this cube is 2.5 cm, but the shear displacement is only 4.8×10^{-3} cm, which is 4.8 thousandths of a centimeter. Shear displacements are typically very small.

11.3. **IDENTIFY:** Treat the rod and clamp as point masses. The center of gravity of the rod is at its midpoint, and we know the location of the center of gravity of the rod-clamp system.

SET UP: $x_{cm} = \dfrac{m_1 x_1 + m_2 x_2}{m_1 + m_2}$.

EXECUTE: $1.20 \text{ m} = \dfrac{(1.80 \text{ kg})(1.00 \text{ m}) + (2.40 \text{ kg})x_2}{1.80 \text{ kg} + 2.40 \text{ kg}}$.

$x_2 = \dfrac{(1.20 \text{ m})(1.80 \text{ kg} + 2.40 \text{ kg}) - (1.80 \text{ kg})(1.00 \text{ m})}{2.40 \text{ kg}} = 1.35 \text{ m}$

EVALUATE: The clamp is to the right of the center of gravity of the system, so the center of gravity of the system lies between that of the rod and the clamp, which is reasonable.

11.5. **IDENTIFY:** We need to calculate the center of gravity of a compound object consisting of two spheres and a steel rod. The center of gravity of each of them is at their midpoint since they are all uniform.

SET UP: Use $x_{cg} = \dfrac{m_1 x_1 + m_2 x_2 + m_3 x_3}{m_1 + m_2 + m_3}$ with the origin at the center of the rod. Let the 0.900-kg sphere be on the left end of the rod and the 0.380-kg sphere on the right end.

EXECUTE: $x_{cg} = \dfrac{m_1 x_1 + m_2 x_2 + m_3 x_3}{m_1 + m_2 + m_3} =$

$\dfrac{(0.900 \text{ kg})(-28.0 \text{ m}) + (0.300 \text{ kg})(0 \text{ m}) + (0.380 \text{ kg})(26.0 \text{ cm})}{0.900 \text{ kg} + 0.300 \text{ kg} + 0.380 \text{ kg}} = -9.70$ cm. So the center of gravity is 9.70 cm from the center of the rod toward the 0.900-kg sphere.

EVALUATE: The center of gravity is toward the heavier sphere, which is reasonable.

11.7. IDENTIFY: Apply $\Sigma \tau_z = 0$ to the ladder.

SET UP: Take the axis to be at point A. The free-body diagram for the ladder is given in Figure 11.7. The torque due to F must balance the torque due to the weight of the ladder.

EXECUTE: $F(8.0 \text{ m})\sin 40° = (3400 \text{ N})(10.0 \text{ m})$, so $F = 6.6$ kN.

EVALUATE: The force required is greater than the weight of the ladder, because the moment arm for F is less than the moment arm for w.

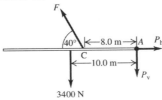

Figure 11.7

11.9. IDENTIFY: Apply $\Sigma F_y = 0$ and $\Sigma \tau_z = 0$ to the board.

SET UP: Let $+y$ be upward. Let x be the distance of the center of gravity of the motor from the end of the board where the 400 N force is applied.

EXECUTE: **(a)** If the board is taken to be massless, the weight of the motor is the sum of the applied forces, 1000 N. The motor is a distance $\dfrac{(2.00 \text{ m})(600 \text{ N})}{(1000 \text{ N})} = 1.20$ m from the end where the 400 N force is applied, and so is 0.800 m from the end where the 600 N force is applied.

(b) The weight of the motor is $400 \text{ N} + 600 \text{ N} - 200 \text{ N} = 800$ N. Applying $\Sigma \tau_z = 0$ with the axis at the end of the board where the 400 N acts gives $(600 \text{ N})(2.00 \text{ m}) = (200 \text{ N})(1.00 \text{ m}) + (800 \text{ N})x$ and $x = 1.25$ m. The center of gravity of the motor is 0.75 m from the end of the board where the 600 N force is applied.

EVALUATE: The motor is closest to the end of the board where the larger force is applied.

11.13. IDENTIFY: The system of the person and diving board is at rest so the two conditions of equilibrium apply.

(a) SET UP: The free-body diagram for the diving board is given in Figure 11.13. Take the origin of coordinates at the left-hand end of the board (point A).

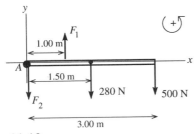

$\vec{F}_1$ is the force applied at the support point and $\vec{F}_2$ is the force at the end that is held down.

Figure 11.13

EXECUTE: $\sum \tau_A = 0$ gives $+F_1(1.0 \text{ m}) - (500 \text{ N})(3.00 \text{ m}) - (280 \text{ N})(1.50 \text{ m}) = 0$

$$F_1 = \frac{(500 \text{ N})(3.00 \text{ m}) + (280 \text{ N})(1.50 \text{ m})}{1.00 \text{ m}} = 1920 \text{ N}$$

(b) $\sum F_y = ma_y$

$F_1 - F_2 - 280 \text{ N} - 500 \text{ N} = 0$

$F_2 = F_1 - 280 \text{ N} - 500 \text{ N} = 1920 \text{ N} - 280 \text{ N} - 500 \text{ N} = 1140 \text{ N}$

EVALUATE: We can check our answers by calculating the net torque about some point and checking that $\sum \tau_z = 0$ for that point also. Net torque about the right-hand end of the board:

$(1140 \text{ N})(3.00 \text{ m}) + (280 \text{ N})(1.50 \text{ m}) - (1920 \text{ N})(2.00 \text{ m}) = 3420 \text{ N} \cdot \text{m} + 420 \text{ N} \cdot \text{m} - 3840 \text{ N} \cdot \text{m} = 0$, which checks.

11.15. IDENTIFY: Apply the first and second conditions of equilibrium to the strut.

(a) SET UP: The free-body diagram for the strut is given in Figure 11.15a. Take the origin of coordinates at the hinge (point A) and $+y$ upward. Let F_h and F_v be the horizontal and vertical components of the force $\vec{F}$ exerted on the strut by the pivot. The tension in the vertical cable is the weight w of the suspended object. The weight w of the strut can be taken to act at the center of the strut. Let L be the length of the strut.

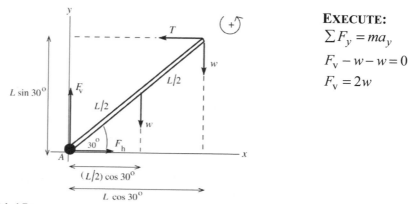

EXECUTE:
$\sum F_y = ma_y$
$F_v - w - w = 0$
$F_v = 2w$

Figure 11.15a

Sum torques about point A. The pivot force has zero moment arm for this axis and so doesn't enter into the torque equation.

$\tau_A = 0$

$TL \sin 30.0° - w((L/2)\cos 30.0°) - w(L \cos 30.0°) = 0$

$T \sin 30.0° - (3w/2)\cos 30.0° = 0$

$$T = \frac{3w \cos 30.0°}{2 \sin 30.0°} = 2.60w$$

Then $\sum F_x = ma_x$ implies $T - F_h = 0$ and $F_h = 2.60w$.

We now have the components of $\vec{F}$ so can find its magnitude and direction (Figure 11.15b).

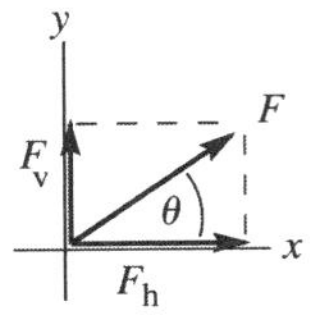

$$F = \sqrt{F_h^2 + F_v^2}$$
$$F = \sqrt{(2.60w)^2 + (2.00w)^2}$$
$$F = 3.28w$$
$$\tan\theta = \frac{F_v}{F_h} = \frac{2.00w}{2.60w}$$
$$\theta = 37.6°$$

Figure 11.15b

(b) SET UP: The free-body diagram for the strut is given in Figure 11.15c.

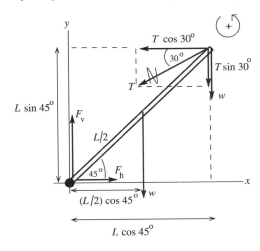

Figure 11.15c

The tension T has been replaced by its x and y components. The torque due to T equals the sum of the torques of its components, and the latter are easier to calculate.

EXECUTE: $\sum \tau_A = 0 + (T\cos 30.0°)(L\sin 45.0°) - (T\sin 30.0°)(L\cos 45.0°) -$
$$w[(L/2)\cos 45.0°] - w(L\cos 45.0°) = 0$$

The length L divides out of the equation. The equation can also be simplified by noting that $\sin 45.0° = \cos 45.0°$.
Then $T(\cos 30.0° - \sin 30.0°) = 3w/2$.

$$T = \frac{3w}{2(\cos 30.0° - \sin 30.0°)} = 4.10w$$

$\sum F_x = ma_x$
$F_h - T\cos 30.0° = 0$
$F_h = T\cos 30.0° = (4.10w)(\cos 30.0°) = 3.55w$
$\sum F_y = ma_y$
$F_v - w - w - T\sin 30.0° = 0$
$F_v = 2w + (4.10w)\sin 30.0° = 4.05w$

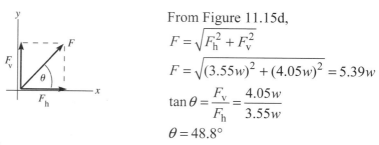

From Figure 11.15d,
$$F = \sqrt{F_h^2 + F_v^2}$$
$$F = \sqrt{(3.55w)^2 + (4.05w)^2} = 5.39w$$
$$\tan\theta = \frac{F_v}{F_h} = \frac{4.05w}{3.55w}$$
$$\theta = 48.8°$$

Figure 11.15d

EVALUATE: In each case the force exerted by the pivot does not act along the strut. Consider the net torque about the upper end of the strut. If the pivot force acted along the strut, it would have zero torque about this point. The two forces acting at this point also have zero torque and there would be one nonzero torque, due to the weight of the strut. The net torque about this point would then not be zero, violating the second condition of equilibrium.

11.17. **IDENTIFY:** The boom is at rest, so the forces and torques on it must each balance.

SET UP: $\Sigma\tau = 0$, $\Sigma F_x = 0$, $\Sigma F_y = 0$. The free-body is shown in Figure 11.17. Call L the length of the boom.

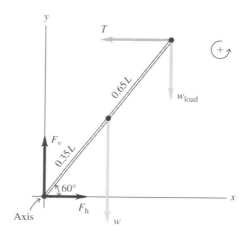

Figure 11.17

EXECUTE: **(a)** $\Sigma\tau = 0$ gives $T(L\sin 60.0°) - w_{load}(L\cos 60.0°) - w(0.35L\cos 60.0°) = 0$ and

$$T = \frac{w_{load}\cos 60.0° + w(0.35\cos 60.0°)}{\sin 60.0°} = \frac{(5000\text{ N})\cos 60.0° + (2600\text{ N})(0.35\cos 60.0°)}{\sin 60.0°} = 3.41\times 10^3\text{ N}.$$

(b) $\Sigma F_x = 0$ gives $F_h - T = 0$ and $F_h = 3410$ N.

$\Sigma F_y = 0$ gives $F_v - w - w_{load} = 0$ and $F_v = 5000\text{ N} + 2600\text{ N} = 7600\text{ N}$

EVALUATE: The bottom of the boom is the best point about which to take torques because only one unknown (the tension) appears in our equation. Using the top (or the center of mass) would give a torque equation with two (or three) unknowns.

11.19. **IDENTIFY:** The beam is at rest so the forces and torques on it must each balance.

SET UP: $\Sigma\tau = 0$, $\Sigma F_x = 0$, $\Sigma F_y = 0$. The distance along the beam from the hinge to where the cable is attached is 3.0 m. The angle ϕ that the cable makes with the beam is given by

$\sin\phi = \dfrac{4.0 \text{ m}}{5.0 \text{ m}}$, so $\phi = 53.1°$. The center of gravity of the beam is 4.5 m from the hinge. Use coordinates with $+y$ upward and $+x$ to the right. Take the pivot at the hinge and let counterclockwise torque be positive. Express the hinge force as components H_v and H_h. Assume H_v is downward and that H_h is to the right. If one of these components is actually in the opposite direction we will get a negative value for it. Set the tension in the cable equal to its maximum possible value, $T = 1.00$ kN.

EXECUTE: (a) The free-body diagram is shown in Figure 11.19, with $\vec{T}$ resolved into its x- and y-components.

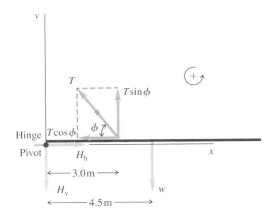

Figure 11.19

(b) $\Sigma\tau = 0$ gives $(T\sin\phi)(3.0 \text{ m}) - w(4.5 \text{ m}) = 0$

$$w = \dfrac{(T\sin\phi)(3.00 \text{ m})}{4.50 \text{ m}} = \dfrac{(1000 \text{ N})(\sin 53.1°)(3.00 \text{ m})}{4.50 \text{ m}} = 533 \text{ N}$$

(c) $\Sigma F_x = 0$ gives $H_h - T\cos\phi = 0$ and $H_h = (1.00 \text{ kN})(\cos 53.1°) = 600$ N

$\Sigma F_y = 0$ gives $T\sin\phi - H_v - w = 0$ and $H_v = (1.00 \text{ kN})(\sin 53.1°) - 533 \text{ N} = 267$ N.

EVALUATE: $T\cos\phi$, H_v and H_h all have zero moment arms for a pivot at the hinge and therefore produce zero torque. If we consider a pivot at the point where the cable is attached we can see that H_v must be downward to produce a torque that opposes the torque due to w.

11.21. IDENTIFY: Apply the first and second conditions of equilibrium to the rod.

SET UP: The force diagram for the rod is given in Figure 11.21.

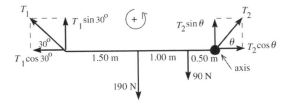

Figure 11.21

EXECUTE: $\Sigma\tau_z = 0$, axis at right end of rod, counterclockwise torque is positive

$(190 \text{ N})(1.50 \text{ m}) + (90 \text{ N})(0.50 \text{ m}) - (T_1 \sin 30.0°)(3.00 \text{ m}) = 0$

$T_1 = \dfrac{285 \text{ N} \cdot \text{m} + 45 \text{ N} \cdot \text{m}}{1.50 \text{ m}} = 220$ N

$\Sigma F_x = ma_x$

$T_2 \cos\theta - T_1 \cos 30° = 0$ and $T_2 \cos\theta = (220 \text{ N})(\cos 30°) = 190.5$ N

$\Sigma F_y = ma_y$

$T_1 \sin 30° + T_2 \sin\theta - 190 \text{ N} - 90 \text{ N} = 0$

$T_2 \sin\theta = 280 \text{ N} - (220 \text{ N})\sin 30° = 170$ N

Then $\dfrac{T_2 \sin\theta}{T_2 \cos\theta} = \dfrac{170 \text{ N}}{190.5 \text{ N}}$ gives $\tan\theta = 0.89239$ and $\theta = 41.7°$

And $T_2 = \dfrac{170 \text{ N}}{\sin 41.7°} = 255$ N.

EVALUATE: The monkey is closer to the right rope than to the left one, so the tension is larger in the right rope. The horizontal components of the tensions must be equal in magnitude and opposite in direction. Since $T_2 > T_1$, the rope on the right must be at a greater angle above the horizontal to have the same horizontal component as the tension in the other rope.

11.25. IDENTIFY: This problem involves the use of torques and graphical interpretation.

SET UP: In order to interpret the graph, apply $\Sigma \tau_z = 0$ to the rod and solve for the tension T in terms of the angle θ. Take torques about the hinge and call L the length of the rod. The mass of the rod is the target variable.

EXECUTE: $\Sigma \tau_z = 0$: $TL\sin\theta - mg\dfrac{L}{2}\cos\theta$, which gives $T = \dfrac{mg}{2}\dfrac{\cos\theta}{\sin\theta} = \left(\dfrac{mg}{2}\right)\cot\theta$. From this we see that a graph of T versus $\cot\theta$ should be a straight line having slope $mg/2$. Using the given slope we have $m = \dfrac{2(\text{slope})}{g} = \dfrac{2(30.0 \text{ N})}{9.80 \text{ m/s}^2} = 6.12$ kg.

EVALUATE: A situation like this might occur if you had a very heavy rod that could not easily be removed to weigh it, but could be pivoted about the hinge. The tension could be measured using a strain gauge.

11.27. IDENTIFY and SET UP: Apply $Y = \dfrac{l_0 F_\perp}{A\Delta l}$ and solve for A and then use $A = \pi r^2$ to get the radius and $d = 2r$ to calculate the diameter.

EXECUTE: $Y = \dfrac{l_0 F_\perp}{A\Delta l}$ so $A = \dfrac{l_0 F_\perp}{Y\Delta l}$ (A is the cross-section area of the wire)

For steel, $Y = 2.0\times 10^{11}$ Pa (Table 11.1)

Thus $A = \dfrac{(2.00 \text{ m})(700 \text{ N})}{(2.0\times 10^{11} \text{ Pa})(0.25\times 10^{-2} \text{ m})} = 2.8\times 10^{-6}$ m^2.

$A = \pi r^2$, so $r = \sqrt{A/\pi} = \sqrt{2.8\times 10^{-6} \text{ m}^2/\pi} = 9.44\times 10^{-4}$ m

$d = 2r = 1.9\times 10^{-3}$ m $= 1.9$ mm.

EVALUATE: Steel wire of this diameter doesn't stretch much; $\Delta l/l_0 = 0.12\%$.

11.31. IDENTIFY: The increased pressure compresses the lead sphere, so we are dealing with bulk stress and strain.

SET UP: The bulk modulus is $B = -\dfrac{\Delta p}{\Delta V/V_0}$, and for lead it is $B = 4.1\times 10^{10}$ Pa. The volume compresses by 0.50%, so $\Delta V = -0.0050 V_0$. The target variable is the pressure increase that causes this amount of compression.

EXECUTE: Solve $-\dfrac{\Delta p}{\Delta V / V_0}$ for Δp: $\Delta p = -B\left(\dfrac{\Delta V}{V_0}\right) = -(4.1 \times 10^{10} \text{ Pa})\left(\dfrac{-0.0050 V_0}{V_0}\right) = 2.05 \times 10^8$ Pa $= 2.0 \times 10^3$ atm. The pressure is 2000 atmospheres above atmospheric pressure.

EVALUATE: This is a very large pressure, but it would take a large pressure to compress a lead sphere.

11.33. IDENTIFY: The amount of compression depends on the bulk modulus of the bone.

SET UP: $\dfrac{\Delta V}{V_0} = -\dfrac{\Delta p}{B}$ and 1 atm $= 1.01 \times 10^5$ Pa.

EXECUTE: **(a)** $\Delta p = -B \dfrac{\Delta V}{V_0} = -(15 \times 10^9 \text{ Pa})(-0.0010) = 1.5 \times 10^7$ Pa $= 150$ atm.

(b) The depth for a pressure increase of 1.5×10^7 Pa is 1.5 km.

EVALUATE: An extremely large pressure increase is needed for just a 0.10% bone compression, so pressure changes do not appreciably affect the bones. Unprotected dives do not approach a depth of 1.5 km, so bone compression is not a concern for divers.

11.37. IDENTIFY: Apply $S = \dfrac{F_\parallel}{A} \dfrac{h}{x}$.

SET UP: $F_\parallel = 9.0 \times 10^5$ N. $A = (0.100 \text{ m})(0.500 \times 10^{-2} \text{ m})$. $h = 0.100$ m. From Table 11.1, $S = 7.5 \times 10^{10}$ Pa for steel.

EXECUTE: **(a)** Shear strain $= \dfrac{F_\parallel}{AS} = \dfrac{(9 \times 10^5 \text{ N})}{[(0.100 \text{ m})(0.500 \times 10^{-2} \text{ m})][7.5 \times 10^{10} \text{ Pa}]} = 2.4 \times 10^{-2}$.

(b) Since shear strain $= x/h$, $x = $ (Shear strain) $\cdot h = (0.024)(0.100 \text{ m}) = 2.4 \times 10^{-3}$ m.

EVALUATE: This very large force produces a small displacement; $x/h = 2.4\%$.

11.39. IDENTIFY: The problem involves the stretching of metal wires, so it makes use of tensile stress and strain and Young's modulus.

SET UP: The steel and aluminum wires have the same fractional change in length. We use $Y = \dfrac{F_\perp}{A} \dfrac{\ell_0}{\Delta \ell}$. We know that $\dfrac{\Delta \ell}{\ell_0}$ is the same for both wires, $r_{al} = 2 r_{st}$, and $A = \pi r^2$. Table 11.1 gives us the values of Y for both metals. Our target variable is the tension T_{al} in the aluminum wire in terms of the tension T_{st} in the steel wire. Since $\dfrac{\Delta \ell}{\ell_0}$ is the same for both wires, we can equate this expression for both wires and solve for T_{al}.

EXECUTE: Steel: $\dfrac{\Delta \ell}{\ell_0} = \dfrac{T_{st}}{Y_{st} \pi r_{st}^2}$. Aluminum: $\dfrac{T_{al}}{Y_{al} \pi r_{al}^2}$. Equating gives $\dfrac{T_{st}}{Y_{st} \pi r_{st}^2} = \dfrac{T_{al}}{Y_{al} \pi r_{al}^2}$. Using the fact that $r_{al} = 2 r_{st}$ and $A = \pi r^2$, we have $T_{al} = \dfrac{Y_{al}}{Y_{st}} \left(\dfrac{r_{al}}{r_{st}}\right)^2 T_{st} = \dfrac{7.0 \times 10^{10} \text{ Pa}}{20 \times 10^{10} \text{ Pa}} \left(\dfrac{2 r_{st}}{r_{st}}\right)^2 T_{st} = 1.4 T_{st}$.

EVALUATE: Although aluminum has a smaller Young's modulus than does steel, the aluminum wire is twice as thick and hence has 4 times the area as the steel wire, so it can support more tension than the steel for the same fractional stretch.

11.43. **IDENTIFY** and **SET UP:** Use stress $= \dfrac{F_\perp}{A}$.

EXECUTE: Tensile stress $= \dfrac{F_\perp}{A} = \dfrac{F_\perp}{\pi r^2} = \dfrac{90.8 \text{ N}}{\pi (0.92 \times 10^{-3} \text{ m})^2} = 3.41 \times 10^7$ Pa

EVALUATE: A modest force produces a very large stress because the cross-sectional area is small.

11.47. **IDENTIFY:** The center of gravity of the combined object must be at the fulcrum. Use $x_{cm} = \dfrac{m_1 x_1 + m_2 x_2 + m_3 x_3 + ...}{m_1 + m_2 + m_3 + ...}$ to calculate x_{cm}.

SET UP: The center of gravity of the sand is at the middle of the box. Use coordinates with the origin at the fulcrum and $+x$ to the right. Let $m_1 = 25.0$ kg, so $x_1 = 0.500$ m. Let $m_2 = m_{sand}$, so $x_2 = -0.625$ m. $x_{cm} = 0$.

EXECUTE: $x_{cm} = \dfrac{m_1 x_1 + m_2 x_2}{m_1 + m_2} = 0$ and $m_2 = -m_1 \dfrac{x_1}{x_2} = -(25.0 \text{ kg})\left(\dfrac{0.500 \text{ m}}{-0.625 \text{ m}}\right) = 20.0$ kg.

EVALUATE: The mass of sand required is less than the mass of the plank since the center of the box is farther from the fulcrum than the center of gravity of the plank is.

11.49. **IDENTIFY:** Apply the conditions of equilibrium to the climber. For the minimum coefficient of friction the static friction force has the value $f_s = \mu_s n$.

SET UP: The free-body diagram for the climber is given in Figure 11.49. f_s and n are the vertical and horizontal components of the force exerted by the cliff face on the climber. The moment arm for the force T is $(1.4 \text{ m})\cos 10°$.

EXECUTE: **(a)** $\Sigma \tau_z = 0$ gives $T(1.4 \text{ m})\cos 10° - w(1.1 \text{ m})\cos 35.0° = 0$.

$T = \dfrac{(1.1 \text{ m})\cos 35.0°}{(1.4 \text{ m})\cos 10°} (82.0 \text{ kg})(9.80 \text{ m/s}^2) = 525$ N

(b) $\Sigma F_x = 0$ gives $n = T \sin 25.0° = 222$ N. $\Sigma F_y = 0$ gives $f_s + T\cos 25° - w = 0$ and $f_s = (82.0 \text{ kg})(9.80 \text{ m/s}^2) - (525 \text{ N})\cos 25° = 328$ N.

(c) $\mu_s = \dfrac{f_s}{n} = \dfrac{328 \text{ N}}{222 \text{ N}} = 1.48$

EVALUATE: To achieve this large value of μ_s the climber must wear special rough-soled shoes.

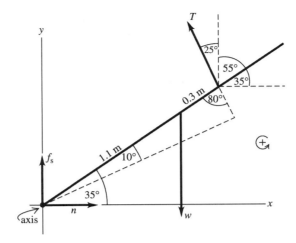

Figure 11.49

11.53. **IDENTIFY:** Apply the conditions of equilibrium to the horizontal beam. Since the two wires are symmetrically placed on either side of the middle of the sign, their tensions are equal and are each equal to $T_w = mg/2 = 137$ N.

SET UP: The free-body diagram for the beam is given in Figure 11.53. F_v and F_h are the vertical and horizontal forces exerted by the hinge on the beam. Since the cable is 2.00 m long and the beam is 1.50 m long, $\cos\theta = \dfrac{1.50 \text{ m}}{2.00 \text{ m}}$ and $\theta = 41.4°$. The tension T_c in the cable has been replaced by its horizontal and vertical components.

EXECUTE: (a) $\Sigma \tau_z = 0$ gives

$T_c (\sin 41.4°)(1.50 \text{ m}) - w_{\text{beam}} (0.750 \text{ m}) - T_w (1.50 \text{ m}) - T_w (0.60 \text{ m}) = 0.$

$T_c = \dfrac{(16.0 \text{ kg})(9.80 \text{ m/s}^2)(0.750 \text{ m}) + (137 \text{ N})(1.50 \text{ m} + 0.60 \text{ m})}{(1.50 \text{ m})(\sin 41.4°)} = 408.6$ N, which rounds to 409 N.

(b) $\Sigma F_y = 0$ gives $F_v + T_c \sin 41.4° - w_{\text{beam}} - 2T_w = 0$ and

$F_v = 2T_w + w_{\text{beam}} - T_c \sin 41.4° = 2(137 \text{ N}) + (16.0 \text{ kg})(9.80 \text{ m/s}^2) - (408.6 \text{ N})(\sin 41.4°) = 161$ N. The hinge must be able to supply a vertical force of 161 N.

EVALUATE: The force from the two wires could be replaced by the weight of the sign acting at a point 0.60 m to the left of the right-hand edge of the sign.

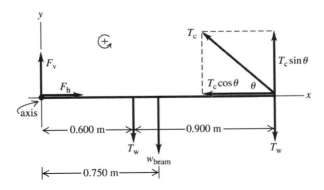

Figure 11.53

11.55. IDENTIFY: We want to locate the center of mass of the leg-cast system. We can treat each segment of the leg and cast as a point-mass located at its center of mass.

SET UP: The force diagram for the leg is given in Figure 11.55. The weight of each piece acts at the center of mass of that piece. The mass of the upper leg is $m_{ul} = (0.215)(37 \text{ kg}) = 7.955$ kg. The mass of the lower leg is $m_{ll} = (0.140)(37 \text{ kg}) = 5.18$ kg. Use the coordinates shown, with the origin at the hip and the x-axis along the leg, and use $x_{cm} = \dfrac{x_{ul}m_{ul} + x_{ll}m_{ll} + x_{cast}m_{cast}}{m_{ul} + m_{ll} + m_{cast}}$.

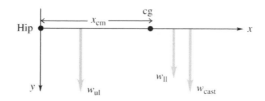

Figure 11.55

EXECUTE: Using $x_{cm} = \dfrac{x_{ul}m_{ul} + x_{ll}m_{ll} + x_{cast}m_{cast}}{m_{ul} + m_{ll} + m_{cast}}$, we have

$$x_{cm} = \dfrac{(18.0 \text{ cm})(7.955 \text{ kg}) + (69.0 \text{ cm})(5.18 \text{ kg}) + (78.0 \text{ cm})(5.50 \text{ kg})}{7.955 \text{ kg} + 5.18 \text{ kg} + 5.50 \text{ kg}} = 49.9 \text{ cm}$$

EVALUATE: The strap is attached to the left of the center of mass of the cast, but it is still supported by the rigid cast since the cast extends beyond its center of mass.

11.57. IDENTIFY: The rod is suspended at rest, so the forces and torques on it must balance. Once the wire breaks, it rotates downward about the hinge, so we can use energy conservation.

SET UP: While the rod is at rest, we use $\sum \tau_z = 0$. After the wire breaks, we apply energy conservation $U_1 + K_1 + W_{other} = U_2 + K_2$ with $I = \dfrac{1}{3}ML^2$, $K_1 = 0$, $U_2 = 0$, and $W_{other} = 0$ because the hinge is frictionless. Our target variables are the angle the wire makes with the horizontal and its angular speed after the wire breaks. Begin with a free-body diagram of the rod, as in Fig. 11.57.

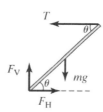

Figure 11.57

EXECUTE: (a) $\sum \tau_z = 0$: $TL \sin\theta - mg\dfrac{L}{2}\cos\theta = 0$, which gives $\theta = \arctan\left(\dfrac{mg}{2T}\right)$.

(b) Calling $y = 0$ at the level when the rod is horizontal, $U_1 + K_1 + W_{other} = U_2 + K_2$ gives $mg\dfrac{L}{2}\sin\theta = \dfrac{1}{2}\left(\dfrac{1}{3}mL^2\right)\omega^2$, from which we get $\omega = \sqrt{\dfrac{3g\sin\theta}{L}}$.

EVALUATE: Check is some special cases: If θ is large, ω is large, which is reasonable because a large θ means that the center of gravity of the rod would be higher than for a small θ. If $\theta = 0$, $\omega = 0$, which is reasonable since the rod started from rest horizontally.

11.59. **IDENTIFY:** The rod is held in position, so the forces and torques on it must balance.

SET UP: Start with a free-body diagram of the rod, as in Fig. 11.59. Apply $\sum \tau_z = 0$, $\sum F_x = 0$, and $\sum F_y = 0$. The target variable is the angle β in the figure.

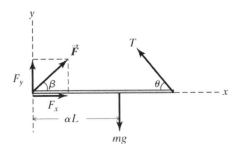

Figure 11.59

EXECUTE: **(a)** If we can find the components F_x and F_y of the hinge force $\vec{F}$, we can use them to find β. Taking torques about the right end of the rod gives $F_y L = mg(L - \alpha L) = mgL(1 - \alpha)$, which gives $F_y = mg(1-\alpha)$ (Eq. 1)

$\sum F_x = 0$: $F_x = T\cos\theta$ (Eq. 2)

$\sum \tau_z = 0$ about the hinge: $TL\sin\theta = mg\alpha L$ (Eq. 3)

Dividing Eq. 3 by Eq. 2 gives $\dfrac{TL\sin\theta}{T\cos\theta} = \dfrac{mg\alpha L}{F_x}$, which gives $F_x = \dfrac{mg\alpha}{\tan\theta}$. Now use this result and Eq. 1 to find β. $\tan\beta = \dfrac{F_y}{F_x} = \dfrac{mg(1-\alpha)}{\dfrac{mg\alpha}{\tan\theta}} = \left(\dfrac{1}{\alpha} - 1\right)\tan\theta$, so $\beta = \arctan\left(\dfrac{1}{\alpha} - 1\right)$.

(b) If $\beta = \theta$ we get $\tan\beta = \left(\dfrac{1}{\alpha} - 1\right)\tan\theta = \tan\theta$, so $\alpha = \frac{1}{2}$.

(c) If $\alpha = 1$, we get $\tan\beta = \left(\dfrac{1}{\alpha} - 1\right)\tan\theta = 0$, so $\beta = 0$.

EVALUATE: In part (c), if $\beta = 0$ then $F_y = 0$, so all the weight of m is supported by T_y. Taking torques about the hinge gives $TL\sin\theta - mgL = 0$, so $T\sin\theta = mg$, which agrees with our answer with $\beta = 0$.

11.63. **IDENTIFY:** The amount the tendon stretches depends on Young's modulus for the tendon material. The foot is in rotational equilibrium, so the torques on it balance.

SET UP: $Y = \dfrac{F_T / A}{\Delta l / l_0}$. The foot is in rotational equilibrium, so $\sum \tau_z = 0$.

EXECUTE: **(a)** The free-body diagram for the foot is given in Figure 11.63. T is the tension in the tendon and A is the force exerted on the foot by the ankle. $n = (75 \text{ kg})g$, the weight of the person.

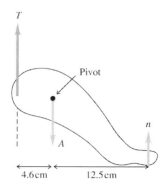

Figure 11.63

(b) Apply $\Sigma \tau_z = 0$, letting counterclockwise torques be positive and with the pivot at the ankle: $T(4.6 \text{ cm}) - n(12.5 \text{ cm}) = 0$. $T = \left(\dfrac{12.5 \text{ cm}}{4.6 \text{ cm}}\right)(75 \text{ kg})(9.80 \text{ m/s}^2) = 2000$ N, which is 2.72 times his weight.

(c) The foot pulls downward on the tendon with a force of 2000 N.

$$\Delta l = \left(\dfrac{F_T}{YA}\right)l_0 = \dfrac{2000 \text{ N}}{(1470 \times 10^6 \text{ Pa})(78 \times 10^{-6} \text{ m}^2)}(25 \text{ cm}) = 4.4 \text{ mm}.$$

EVALUATE: The tension is quite large, but the Achilles tendon stretches about 4.4 mm, which is only about 1/6 of an inch, so it must be a strong tendon.

11.65. IDENTIFY: The rod is held in place, so the torques on it must balance. The added weight of the object causes the wire to stretch slightly, so we need to use tensile stress and strain.

SET UP: We use $\Sigma \tau_z = 0$ and $Y = \dfrac{F_\perp}{A}\dfrac{l_0}{\Delta l}$. The target variable is the distance the aluminum wire stretches due to the added weight.

EXECUTE: (a) A little trigonometry gives $\cos 30.0° = (1.20 \text{ m})/L_{\text{wire}}$, so $L_{\text{wire}} = 1.39$ m.

(b) Fig. 11.65 shows a free-body diagram of the rod with the object attached. The stretching of the wire is due to the *increase* in tension due to the addition of the 90.0-kg object. Therefore in Fig. 11.65 we do not use the weight *mg* of the rod when computing the torque about the hinge.

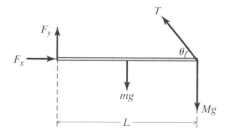

Figure 11.65

$\Sigma \tau_z = 0: TL \sin \theta = MgL \rightarrow T = \dfrac{mg}{\sin \theta} = \dfrac{(90.0 \text{ kg})(9.80 \text{ m/s}^2)}{\sin 30.0°} = 1764$ N. Now use Young's modulus for aluminum (from Table 11.1) to find the increase in the length of the wire. Solving $Y = \dfrac{F_\perp}{A}\dfrac{l_0}{\Delta l}$ for Δl gives $\Delta l = \dfrac{F_\perp l_0}{AY} = \dfrac{F_\perp l_0}{\pi r^2 Y} = \dfrac{(1764 \text{ N})(1.39 \text{ m})}{\pi (0.00250 \text{ m})^2 (7.0 \times 10^{10} \text{ Pa})} = 1.8 \times 10^{-3}$ m = 1.8 mm.

EVALUATE: Since $\Delta \ell \ll \ell_0$ we are justified in treating the rod as being horizontal after object is added. Our result is reasonable since most materials stretch very little under ordinary circumstances.

11.67. IDENTIFY: The torques must balance since the person is not rotating.

SET UP: Figure 11.67a shows the distances and angles. $\theta + \phi = 90°$. $\theta = 56.3°$ and $\phi = 33.7°$. The distances x_1 and x_2 are $x_1 = (90 \text{ cm})\cos\theta = 50.0 \text{ cm}$ and $x_2 = (135 \text{ cm})\cos\phi = 112 \text{ cm}$. The free-body diagram for the person is given in Figure 11.67b. $w_l = 277$ N is the weight of his feet and legs, and $w_t = 473$ N is the weight of his trunk. n_f and f_f are the total normal and friction forces exerted on his feet and n_h and f_h are those forces on his hands. The free-body diagram for his legs is given in Figure 11.67c. F is the force exerted on his legs by his hip joints. For balance, $\Sigma \tau_z = 0$.

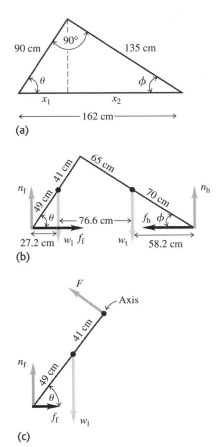

Figure 11.67

EXECUTE: **(a)** Consider the force diagram of Figure 11.67b. $\Sigma \tau_z = 0$ with the pivot at his feet and counterclockwise torques positive gives $n_h(162 \text{ cm}) - (277 \text{ N})(27.2 \text{ cm}) - (473 \text{ N})(103.8 \text{ cm}) = 0$. $n_h = 350$ N, so there is a normal force of 175 N at each hand. $n_f + n_h - w_l - w_t = 0$ so $n_f = w_l + w_t - n_h = 750 \text{ N} - 350 \text{ N} = 400 \text{ N}$, so there is a normal force of 200 N at each foot.

(b) Consider the force diagram of Figure 11.67c. $\Sigma \tau_z = 0$ with the pivot at his hips and counterclockwise torques positive gives $f_f(74.9 \text{ cm}) + w_l(22.8 \text{ cm}) - n_f(50.0 \text{ cm}) = 0$.

$$f_f = \frac{(400 \text{ N})(50.0 \text{ cm}) - (277 \text{ N})(22.8 \text{ cm})}{74.9 \text{ cm}} = 182.7 \text{ N}.$$ There is a friction force of 91 N at each foot.

$\Sigma F_x = 0$ in Figure 11.67b gives $f_h = f_f$, so there is a friction force of 91 N at each hand.

EVALUATE: In this position the normal forces at his feet and at his hands don't differ very much.

11.69. **IDENTIFY:** Apply the equilibrium conditions to the crate. When the crate is on the verge of tipping it touches the floor only at its lower left-hand corner and the normal force acts at this point. The minimum coefficient of static friction is given by the equation $f_s = \mu_s n$.

SET UP: The free-body diagram for the crate when it is ready to tip is given in Figure 11.69.

EXECUTE: (a) $\Sigma \tau_z = 0$ gives $P(1.50 \text{ m}) \sin 53.0° - w(1.10 \text{ m}) = 0$.

$P = w\left(\dfrac{1.10 \text{ m}}{[1.50 \text{ m}][\sin 53.0°]}\right) = 1.15 \times 10^3$ N

(b) $\Sigma F_y = 0$ gives $n - w - P \cos 53.0° = 0$.

$n = w + P \cos 53.0° = 1250 \text{ N} + (1.15 \times 10^3 \text{ N}) \cos 53° = 1.94 \times 10^3$ N

(c) $\Sigma F_y = 0$ gives $f_s = P \sin 53.0° = (1.15 \times 10^3 \text{ N}) \sin 53.0° = 918$ N.

(d) $\mu_s = \dfrac{f_s}{n} = \dfrac{918 \text{ N}}{1.94 \times 10^3 \text{ N}} = 0.473$

EVALUATE: The normal force is greater than the weight because P has a downward component.

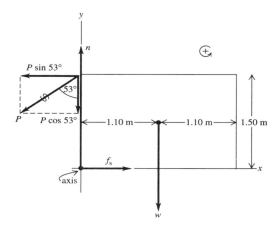

Figure 11.69

11.71. **IDENTIFY:** Apply the first and second conditions of equilibrium to the crate.

SET UP: The free-body diagram for the crate is given in Figure 11.71.

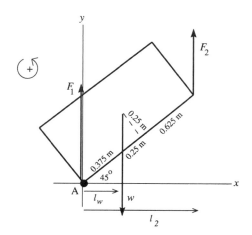

$l_w = (0.375 \text{ m}) \cos 45°$

$l_2 = (1.25 \text{ m}) \cos 45°$

Let $\vec{F}_1$ and $\vec{F}_2$ be the vertical forces exerted by you and your friend. Take the origin at the lower left-hand corner of the crate (point A).

Figure 11.71

EXECUTE: $\sum F_y = ma_y$ gives $F_1 + F_2 - w = 0$

$F_1 + F_2 = w = (200 \text{ kg})(9.80 \text{ m/s}^2) = 1960 \text{ N}$

$\sum \tau_A = 0$ gives $F_2 l_2 - w l_w = 0$

$F_2 = w\left(\dfrac{l_w}{l_2}\right) = 1960 \text{ N}\left(\dfrac{0.375 \text{ m}\cos 45°}{1.25 \text{ m}\cos 45°}\right) = 590 \text{ N}$

Then $F_1 = w - F_2 = 1960 \text{ N} - 590 \text{ N} = 1370 \text{ N}$.

EVALUATE: The person below (you) applies a force of 1370 N. The person above (your friend) applies a force of 590 N. It is better to be the person above. As the sketch shows, the moment arm for $\vec{F}_1$ is less than for $\vec{F}_2$, so must have $F_1 > F_2$ to compensate.

11.73. IDENTIFY: Apply $\sum \tau_z = 0$ to the forearm.

SET UP: The free-body diagram for the forearm is given in Figure 11.10 in the textbook.
EXECUTE: (a) $\sum \tau_z = 0$, axis at elbow gives

$wL - (T \sin\theta)D = 0.$ $\sin\theta = \dfrac{h}{\sqrt{h^2 + D^2}}$ so $w = T\dfrac{hD}{L\sqrt{h^2 + D^2}}$.

$w_{max} = T_{max}\dfrac{hD}{L\sqrt{h^2 + D^2}}$.

(b) $\dfrac{dw_{max}}{dD} = \dfrac{T_{max} h}{L\sqrt{h^2 + D^2}}\left(1 - \dfrac{D^2}{h^2 + D^2}\right)$; the derivative is positive.

EVALUATE: (c) The result of part (b) shows that w_{max} increases when D increases, since the derivative is positive. w_{max} is larger for a chimp since D is larger.

11.77. IDENTIFY: Apply the first and second conditions of equilibrium, first to both marbles considered as a composite object and then to the bottom marble.

(a) SET UP: The forces on each marble are shown in Figure 11.77.

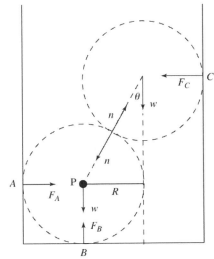

EXECUTE:
$F_B = 2w = 1.47 \text{ N}$
$\sin\theta = R/2R$ so $\theta = 30°$
$\sum \tau_z = 0$, axis at P
$F_C(2R\cos\theta) - wR = 0$

$F_C = \dfrac{mg}{2\cos 30°} = 0.424 \text{ N}$

$F_A = F_C = 0.424 \text{ N}$

Figure 11.77

(b) Consider the forces on the bottom marble. The horizontal forces must sum to zero, so
$F_A = n\sin\theta$.

$$n = \frac{F_A}{\sin 30°} = 0.848 \text{ N}$$

Could use instead that the vertical forces sum to zero

$$F_B - mg - n\cos\theta = 0$$

$$n = \frac{F_B - mg}{\cos 30°} = 0.848 \text{ N, which checks.}$$

EVALUATE: If we consider each marble separately, the line of action of every force passes through the center of the marble so there is clearly no torque about that point for each marble. We can use the results we obtained to show that $\Sigma F_x = 0$ and $\Sigma F_y = 0$ for the top marble.

11.83. **IDENTIFY:** Use the second condition of equilibrium to relate the tension in the two wires to the distance w is from the left end. Use $\text{stress} = \frac{F_\perp}{A}$ and $Y = \frac{l_0 F_\perp}{A\Delta l}$ to relate the tension in each wire to its stress and strain.

(a) SET UP: $\text{stress} = F_\perp/A$, so equal stress implies T/A same for each wire.

$T_A/2.00 \text{ mm}^2 = T_B/4.00 \text{ mm}^2$ so $T_B = 2.00 T_A$

The question is where along the rod to hang the weight in order to produce this relation between the tensions in the two wires. Let the weight be suspended at point C, a distance x to the right of wire A. The free-body diagram for the rod is given in Figure 11.83.

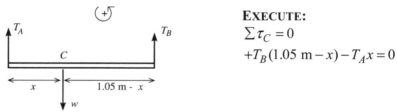

EXECUTE:
$\Sigma \tau_C = 0$
$+T_B(1.05 \text{ m} - x) - T_A x = 0$

Figure 11.83

But $T_B = 2.00 T_A$ so $2.00 T_A(1.05 \text{ m} - x) - T_A x = 0$
$2.10 \text{ m} - 2.00x = x$ and $x = 2.10 \text{ m}/3.00 = 0.70 \text{ m}$ (measured from A).

(b) SET UP: $Y = \text{stress/strain}$ gives that $\text{strain} = \text{stress}/Y = F_\perp/AY$.

EXECUTE: Equal strain thus implies

$$\frac{T_A}{(2.00 \text{ mm}^2)(1.80 \times 10^{11} \text{ Pa})} = \frac{T_B}{(4.00 \text{ mm}^2)(1.20 \times 10^{11} \text{ Pa})}$$

$$T_B = \left(\frac{4.00}{2.00}\right)\left(\frac{1.20}{1.80}\right) T_A = 1.333 T_A.$$

The $\Sigma \tau_C = 0$ equation still gives $T_B(1.05 \text{ m} - x) - T_A x = 0$.
But now $T_B = 1.333 T_A$ so $(1.333 T_A)(1.05 \text{ m} - x) - T_A x = 0$.
$1.40 \text{ m} = 2.33x$ and $x = 1.40 \text{ m}/2.33 = 0.60 \text{ m}$ (measured from A).

EVALUATE: Wire B has twice the diameter so it takes twice the tension to produce the same stress. For equal stress the moment arm for T_B (0.35 m) is half that for T_A (0.70 m), since the torques must be equal. The smaller Y for B partially compensates for the larger area in determining the strain and for equal strain the moment arms are closer to being equal.

11.87. **IDENTIFY:** The bar is at rest, so the forces and torques on it must all balance.

SET UP: $\Sigma F_y = 0$, $\Sigma \tau_z = 0$.

EXECUTE: **(A)** The free-body diagram is shown in Figure 11.87a, where F_p is the force due to the knife-edge pivot.

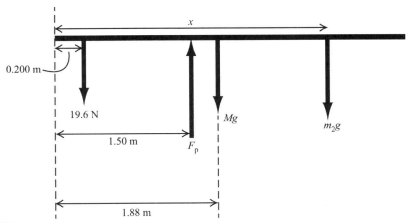

Figure 11.87a

(b) $\sum \tau_z = 0$, with torques taken about the location of the knife-edge pivot, gives
$(2.00 \text{ kg})g(1.30 \text{ m}) - Mg(0.38 \text{ m}) - m_2 g(x - 1.50 \text{ m}) = 0$
Solving for x gives
$x = [(2.00 \text{ kg})(1.30 \text{ m}) - M(0.38 \text{ m})](1/m_2) + 1.50 \text{ m}$
The graph of this equation (x versus $1/m_2$) is a straight line of slope $[(2.00 \text{ kg})(1.30 \text{ m}) - M(0.38 \text{ m})]$.
(c) The plot of x versus $1/m_2$ is shown in Figure 11.87b. The equation of the best-fit line is
$x = (1.9955 \text{ m} \cdot \text{kg})/m_2 + 1.504 \text{ m}$. The slope of the best-fit line is $1.9955 \text{ m} \cdot \text{kg}$, so
$[(2.00 \text{ kg})(1.30 \text{ m}) - M(0.38 \text{ m})] = 1.9955 \text{ m} \cdot \text{kg}$, which gives $M = 1.59$ kg.

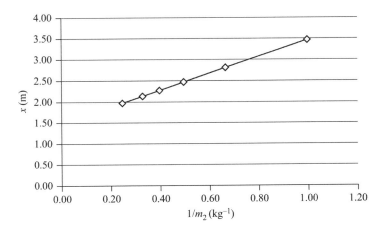

Figure 11.87b

(d) The y-intercept of the best-fit line is 1.50 m. This is plausible. As the graph approaches the y-axis, $1/m_2$ approaches zero, which means that m_2 is getting extremely large. In that case, it would be much larger than any other masses involved, so to balance the system, m_2 would have to be at the knife-point pivot, which is at $x = 1.50$ m.
EVALUATE: The fact that the graph gave a physically plausible result in part (d) suggests that this graphical analysis is reasonable.

STUDY GUIDE FOR FLUID MECHANICS

Summary

We interact with fluids on a continual basis, from walking through air to swimming in the ocean. This chapter examines fluids, or substances that can flow, including liquids and gases. We'll begin with fluid statics and use Newton's laws to describe the behavior of fluids at rest. *Density, pressure, buoyancy,* and *surface tension,* concepts needed for our investigation, will be defined. We'll also delve into fluid dynamics and see how to analyze fluids in motion. Conservation of energy and Newton's laws will guide us in this examination. Although fluid dynamics can be quite complex, several examples will give us insight into the subject.

Objectives

After studying this chapter, you'll understand
- The definition of a *material's density*.
- The definition of *pressure* in a fluid and its measurement.
- How to analyze fluids in equilibrium and find the pressure at varying depths.
- Buoyancy and how to calculate the buoyancy acting on an object.
- How to compare and contrast laminar and turbulent fluid flow.
- How to apply Bernoulli's equation to fluid dynamics problems.

Concepts and Equations

Term	Description
Density	Density is the mass per unit volume of a material. For a homogeneous material with mass m and volume V, the density is $$\rho = \frac{m}{V}.$$ The SI unit of density is the kilogram per cubic meter (1 kg/m^3). The cgs unit is used to express density in grams per cubic centimeter ($1 \text{ gm/cm}^3 = 1000 \text{ kg/m}^3$).
Pressure	The pressure p in a fluid is the normal force per unit area: $$p = \frac{dF_\perp}{dA}.$$ The SI unit of pressure is the pascal (Pa); $1 \text{ Pa} = 1 \text{ N/m}^2$. Also common are the bar (10^5 Pa) and millibar (10^2 bar).
Pressure in a Fluid	The pressure difference between two points in a fluid with uniform density ρ is proportional to the difference in elevations between the two points: $$p_2 - p_1 = -\rho g(y_2 - y_1).$$ Pascal's law states that the pressure applied to a fluid is transmitted through the fluid and depends only on depth.
Buoyant Force	Archimedes' principle states that when an object is immersed in a fluid, the fluid exerts an upward buoyant force on the object equal in magnitude to the weight of the fluid displaced by the object.
Fluid Flow	An ideal fluid is incompressible and has no viscosity. Conservation of mass requires that the amount of fluid flowing through a cross section of a tube per unit time be the same for all cross sections: $$\frac{\Delta V}{\Delta t} = A_1 v_1 = A_2 v_2.$$
Bernoulli's Equation	Bernoulli's equation relates the pressure p, flow speed v, and elevation y of an ideal fluid at any two points: $$p_1 + \rho g y_1 + \frac{1}{2}\rho v_1^2 = p_2 + \rho g y_2 + \frac{1}{2}\rho v_2^2 = \text{constant}.$$

Key Concept 1: To find the density of a uniform substance, divide the mass of the substance by the volume that it occupies.

Key Concept 2: To find the force exerted by a fluid perpendicular to a surface, multiply the pressure of the fluid by the surface's area. This relationship comes from the definition of pressure as the normal force per unit area within the fluid.

Key Concept 3: Absolute pressure is the total pressure at a given point in a fluid. Gauge pressure is the difference between absolute pressure and atmospheric pressure.

Key Concept 4: The pressure is the same at all points *at the same level* in a fluid at rest. This is true even if the fluid contains different substances with different densities.

Key Concept 5: The buoyant force on an object immersed in a fluid is equal to the weight of the fluid that the object displaces (Archimedes's principle). The greater the density of the fluid, the greater the buoyant force that the fluid exerts.

Key Concept 6: The continuity equation states that as an incompressible fluid moves through a flow tube, the volume flow rate (the flow tube's cross-sectional area times the flow speed) is the same *at all points* along the flow tube. The narrower the flow tube, the faster the flow speed.

Key Concept 7: Bernoulli's equation allows you to relate the flow speeds at two different points in a fluid to the pressures and heights at those two points.

Key Concept 8: When solving problems about the flow of an incompressible fluid with negligible internal friction, you can use both Bernoulli's equation and the continuity equation.

Key Concept 9: When an incompressible fluid with negligible internal friction flows through a pipe of varying size, the pressure and flow speed both change. Where the cross-sectional area is small, the pressure is low and the speed is high; where the cross-sectional area is large, the pressure is high and the speed is low.

Key Concept 10: The pressure in a flowing incompressible fluid with negligible internal friction is low at points where the flow lines are crowded together, such as above the upper surface of an airplane wing.

Key Concept 11: Even in a fluid with low viscosity (that is, little internal friction) such as air, the effects of viscosity can be important for determining how that fluid flows around objects.

Conceptual Questions

1: Ice in a glass

Two glasses are filled with water to the same level. In one glass, ice cubes float on the top. If the two glasses are made of the same material and have the same shape, how do their total weights compare?

IDENTIFY, SET UP, AND EXECUTE The glasses must have the same weight, since they are made of the same material and have the same shape. We solve the problem by comparing the mass of the water alone to the mass of the water-plus-ice mix.
Archimedes' principle states that an object will displace its own weight in a fluid. The volume of water displaced by the ice has the same weight as the ice; therefore, the water in one glass weighs the same as the water plus ice in the second glass.

EVALUATE The volume of the glass with the ice is greater than the volume of the glass without ice, but weight depends on *both* density *and* volume. As with any new physical principle, we need to develop our skills carefully and not jump to conclusions.

2: Energy in a hydraulic lift

A hydraulic lift is used to lift a car. The piston supporting the car has a cross-sectional area 100 times larger than the cross-sectional area of the piston driving the lift. The drive piston will therefore require a force 100 times smaller than the weight of the car to lift the car. Does this mean that energy conservation is violated?

IDENTIFY, SET UP, AND EXECUTE Pascal's law states that the pressure is the same at both pistons. The drive piston requires a small force to create the pressure that will lift the car. A large displacement in the drive piston creates a small displacement in the lift piston, due to the differences in areas. The amount of work done in moving the drive piston a long distance is equal to the work done by the lift piston moving a small distance (ignoring friction). The amounts of work are equivalent; energy conservation is not violated.

EVALUATE If you have ever operated a hydraulic jack to lift your car or a house, you should recall that you had to pump the jack several times to move a small distance. The work produced by your small force applied over a long distance was equivalent to the work done in lifting the object.

3: Race car spoilers

Why do race cars have spoilers, or wings, on their bodies?

IDENTIFY, SET UP, AND EXECUTE Spoilers are essentially inverted airplane wings. We've seen that airplane wings produce lift by reducing the pressure above the plane's wing. The inverted wing produces a downward force to help hold the race car on the pavement and maintain contact between the wheels and the road. The spoiler also helps stabilize the car as it moves around the track.

EVALUATE Spoiler design for race cars is critical: There is a careful balance between enough downward force to keep the car on the track and too much force that causes lost fuel economy and premature tire wear. Some race cars have downward forces of up to three times the force of gravity (i.e., they could operate on an upside-down track and not fall off). Many cars have spoilers; most serve only an aesthetic purpose and have no effect on the car's performance.

Problems

1: How much seawater in a tank

Seawater is stored under pressure in a tank of horizontal cross-sectional area 4.5 m^2. The pressure above the seawater in the tank is 7.2×10^5 Pa, and the pressure at the bottom of the tank is 1.2×10^6 Pa. What is the mass of the seawater in the tank?

IDENTIFY We'll use the relations among pressure, density, and height to solve the problem. The target variable is the mass of the seawater.

SET UP The tank is sketched in Figure 1. We can find the mass by first finding the volume of the seawater and then multiplying by the density. To find the volume, we multiply the height by the cross-sectional area of the tank. To find the height, we use the variation in the pressures due to the height of seawater. We assume that the seawater is incompressible.

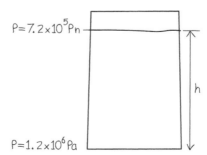

Figure 1 Problem 1 sketch.

EXECUTE We start by finding the height of the seawater in the container. The difference in pressure is related to the height by
$$p_2 - p_1 = \rho g(y_2 - y_1),$$
where we set p_1 as the pressure at the bottom of the tank and p_2 as the pressure at the top of the tank. Solving for the height, we get
$$h = y_2 - y_1 = \frac{p - p_0}{\rho g} = \frac{(1.2 \times 10^6 \text{ Pa}) - (7.2 \times 10^5 \text{ Pa})}{(1.03 \times 10^3 \text{ kg/m}^3)(9.8 \text{ m/s}^2)} = 47.6 \text{ m},$$
where we used the density of seawater given in Table 1 in the text $(1.03 \times 10^3 \text{ kg/m}^3)$. We now find the volume of seawater in the tank. The volume is the height times the cross-sectional area:
$$V = hA = (47.6 \text{ m})(4.5 \text{ m}^2) = 214 \text{ m}^3.$$
The mass is the volume times the density we found from the table:
$$m = \rho V = (1.03 \times 10^3 \text{ kg/m}^3)(214 \text{ m}^3) = 221{,}000 \text{ kg}.$$
The tank holds 221,000 kg of seawater.

EVALUATE This problem shows how to determine height from change in pressure. Altimeters find changes in altitude by monitoring the change in pressure of air.

Practice Problem: What would the pressure at the bottom be if the area of the tank was 9.0 m²?
Answer: 9.5×10^5 Pa

Extra Practice: What would the mass be if the tank was filled with regular water?
Answer: 214,000 kg

2: Water pressure in a town

Water pressure in a town is maintained by a water tower 35.0 m high, open to the atmosphere at the top. (a) What is the gauge pressure at ground level? (b) A 1.75-cm-diameter garden hose at the bottom of the tower is open and spilling water. How much force is needed at the end of the hose to seal the end?

IDENTIFY We'll use the relations among pressure, density, and height to solve the problem. The target variables are the gauge pressure at the ground and the force needed to seal the hose.

SET UP The gauge pressure is found by multiplying the height by the density by g, the acceleration due to gravity. The force is found by finding the pressure differential at the hose and multiplying by the area of the opening. We assume that the water is incompressible.

EXECUTE We start by finding the pressure at the surface. The gauge pressure is given by
$$p_g = p - p_a = \rho g(y_2 - y_1)$$
$$= (10^3 \text{ kg/m}^3)(9.8 \text{ m/s}^2)(35.0 \text{ m})$$
$$= 3.43 \times 10^5 \text{ Pa}$$
$$= 3.40 \text{ atm}.$$

The pressure difference between the inside and the outside of the hose is the gauge pressure, equivalent to the force per unit area needed to seal the hose. The force needed is then

$$F = (p - p_a)A = (3.43 \times 10^5 \text{ Pa})(\pi(0.0175/2 \text{ m})^2) = 83 \text{ N}.$$

The gauge pressure at the ground level is 3.40 atm, and the force required to seal the hose is 83 N.

KEY CONCEPT 3 **EVALUATE** This problem illustrates how force, pressure, and height are related in a fluid.

Practice Problem: What force is required to seal the end if the water tower is 45.0 m high? *Answer:* 107 N

Extra Practice: What force is required if we increase the hose diameter to 4.0 cm (with the 45.0 m tower)? *Answer:* 554 N

3: Velocity of water exiting a fire hose

Water enters a round fire hose of diameter 3.5 cm and exits from a round, 0.60-cm-diameter nozzle. If the water enters the hose at 2.0 m/s, what is the velocity of the exiting water? What is the maximum horizontal range of the water leaving the hose?

IDENTIFY The continuity equation relates the velocities and cross-sectional areas of incompressible fluids in a tube. The target variables are the velocity of the water at the outlet of the nozzle and the maximum range of the water.

SET UP We use the continuity equation to find the velocity of the water exiting the nozzle. To find the range, we employ projectile motion. We treat the water as incompressible.

EXECUTE The amount of fluid flowing through a tube per unit time is constant. The flow through the hose is equal to the flow through the nozzle:

$$A_{\text{hose}} v_{\text{hose}} = A_{\text{nozzle}} v_{\text{nozzle}}.$$

The area of the hose or nozzle is π times the square of half the diameter. Solving for the velocity of the nozzle gives

$$v_{\text{nozzle}} = \frac{A_{\text{hose}} v_{\text{hose}}}{A_{\text{nozzle}}} = \frac{\pi (D_{\text{hose}}/2)^2 v_{\text{hose}}}{\pi (D_{\text{nozzle}}/2)^2} = \frac{(3.5 \text{ cm})^2 (2.0 \text{ m/s})}{(0.6 \text{ cm})^2} = 68 \text{ m/s},$$

where the factors π and 2 cancel. The water molecules leaving the hose have a velocity of 68 m/s and undergo acceleration due to gravity. We use the kinematic relations for two-dimensional motion to find the range. Recall that the horizontal range of a projectile in terms of the launch angle θ_0 and the initial velocity v_0 is

$$R = \frac{v_0^2 \sin 2\theta_0}{g}.$$

The maximum range occurs when the launch angle is 45°. Substituting our values, we obtain

$$R = \frac{v_0^2 \sin 2(45°)}{g} = \frac{(68 \text{ m/s})^2 (1)}{(9.8 \text{ m/s}^2)} = 470 \text{ m}.$$

The maximum horizontal range of the water leaving the nozzle at 68 m/s is 470 m.

KEY CONCEPT 6 **EVALUATE** This problem illustrates why nozzles are placed at the ends of hoses. The reduced diameter of the nozzle increases the exit velocity and thereby increases the range of the water. Next time you wash your car, compare the velocity and range of the water leaving the hose with and without the nozzle attached.

Practice Problem: How far does the water travel if a 1.20 cm nozzle is used? *Answer:* 29.5 m

Extra Practice: How much time does it take to travel that distance? *Answer:* 2.45 s

4: Ice cube in glycerine

What fraction of an ice cube is submerged when floating in glycerine?

IDENTIFY We'll use Newton's law and the definition of *buoyancy* to solve the problem. The target variable is the fraction of the ice cube submerged.

SET UP The buoyant force acts upward and gravity acts downward. The ice cube is in equilibrium, so the forces sum to zero.

EXECUTE The weight of the ice cube is

$$w_w = \rho_w g V_w.$$

The buoyant force is equal to the amount of glycerine displaced by the ice cube, which is equal to the weight of the ice cube. Combining terms yields

$$F_B = \rho_g g V_g = \rho_w g V_w.$$

The fraction of the ice cube that is submerged is the volume of glycerine displaced divided by the volume of the ice cube. Rearranging terms in the previous equation yields the fraction

$$\frac{V_g}{V_w} = \frac{\rho_w}{\rho_g} = \left(\frac{0.92 \text{ g/cm}^2}{1.26 \text{ g/cm}^2} \right) = 0.73.$$

Thus, 73% of the ice cube is submerged when floating in glycerine. Note that the densities of ice and glycerine were taken from Table 1 in the text.

KEY CONCEPT 5 **EVALUATE** We see how we can determine the fraction of ice located below the surface when the ice is placed in a liquid. You can use the same procedure to find out how much of an iceberg is below the surface in the ocean.

Practice Problem: Draw the free-body diagram of the problem.

Extra Practice: What fraction do you get if the ice cube is floating in seawater? *Answer:* 0.89

5: Examining the buoyant force

A 4.5-cm-radius sphere of wood is held in fresh water below the surface by a spring. If the spring's force constant is 55 N/m, by how much is the spring stretched from its equilibrium position? Take the density of wood to be 700 kg/m³.

IDENTIFY We'll use Newton's law and the definition of buoyancy to solve for the stretch of the spring, the target variable. The wood is in equilibrium.

SET UP The free-body diagram of the block of wood is shown in Figure 2. The buoyant force is directed upward, and both gravity and tension due to the spring are directed downward. The block of wood is in equilibrium, so the forces sum to zero. The size and density of the wood determine its volume and mass, needed for the buoyant force and gravity. Hooke's law will be used to determine the amount of stretch.

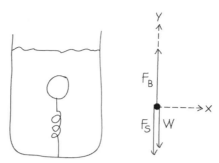

Figure 2 Problem 5 sketch and free-body diagram.

EXECUTE We sum the three forces acting on the block of wood. The forces act only in the vertical direction and add to zero:

$$\Sigma F_y = F_B - mg - F_s = 0.$$

The buoyant force is equal to the amount of water displaced by the wooden sphere. The volume of a sphere is $4/3\pi r^2$. Combining terms yields

$$F_B = \rho_{water} V_{sphere}\, g = \rho_{water}\left(\frac{4}{3}\pi r_{sphere}^3\right)g.$$

The spring force is equal to the spring constant times its displacement, kx. The mass of the sphere is its volume times its density. Inserting these expressions into the equilibrium equation gives

$$\Sigma F_y = F_B - mg - F_s = \rho_{water}\left(\frac{4}{3}\pi r_{sphere}^3\right)g - \rho_{wood}\left(\frac{4}{3}\pi r_{sphere}^3\right)g - kx = 0.$$

Rearranging terms, we solve for x:

$$x = \frac{\rho_{water}\left(\frac{4}{3}\pi r_{sphere}^3\right)g - \rho_{wood}\left(\frac{4}{3}\pi r_{sphere}^3\right)g}{k} = \frac{(\rho_{water} - \rho_{wood})\left(\frac{4}{3}\pi r_{sphere}^3\right)g}{k},$$

$$x = \frac{((1\times 10^3\ \text{kg/m}^3) - (700\ \text{kg/m}^3))\left(\frac{4}{3}\pi(0.045\ \text{m})^3\right)(9.8\ \text{m/s}^2)}{(55\ \text{N/m})} = 0.020\ \text{m}.$$

The spring is stretched 0.020 m, or 2.0 cm, from its equilibrium position.

KEY CONCEPT 5 **EVALUATE** This problem illustrates how to incorporate the buoyant force with previously encountered forces to solve equilibrium problems. We've used the same procedure as in the past: starting with a free-body diagram and setting the net external force equal to zero. Working with fluids requires conversions among volume, mass, and density.

Practice Problem: How much does the spring stretch if the density of wood is 800 kg/m³?
Answer: 13.6 mm

Extra Practice: How much is the spring compressed if the density of wood is 1100 kg/m³?
Answer: 6.80 mm

6: Water from a tank

A 10.0-m-high cylindrical tank of cross-sectional area 0.75 m^2 is filled with water. (a) Find the velocity of discharge as a function of the height of water remaining in the tank when a hole of area 0.40 m^2 is opened at the bottom of the tank. (b) Find the initial discharge velocity. (c) Find the initial volume rate of discharge.

IDENTIFY Bernoulli's equation and the continuity equation will be used to relate the pressure difference, height, and velocity of the flowing water. The target variables are the discharge velocity and volume rate of discharge.

SET UP Bernoulli's equation will be used to relate the velocities at the top and bottom of the tank to the change in height of the water. The continuity equation also relates the two velocities. Combining both equations will yield the velocity at the bottom of the tank, from which we can determine the solutions to parts (b) and (c).

EXECUTE Bernoulli's equation applied to the top and bottom of the cylinder gives

$$p_{top} + \rho g y_{top} + \frac{1}{2}\rho v_{top}^2 = p_{bottom} + \rho g y_{bottom} + \frac{1}{2}\rho v_{bottom}^2.$$

Both sides are open to atmospheric pressure, so the pressures are the same. We derive an expression for the velocities:

$$v_{bottom}^2 - v_{top}^2 = 2g(y_{top} - y_{bottom}) = 2gh.$$

The continuity equation yields another relation between the two velocities:

$$v_{bottom} A_{bottom} = v_{top} A_{top}$$

or

$$v_{top} = \frac{A_{bottom}}{A_{top}} v_{bottom}.$$

Placing the right-hand side of the latter equation into the previous equation and solving for the velocity at the bottom gives

$$v_{bottom}^2 - \left(\frac{A_{bottom}}{A_{top}} v_{bottom}\right)^2 = 2gh,$$

or

$$v_{bottom} = \sqrt{2gh}\left(1 - \left(\frac{A_{bottom}}{A_{top}}\right)^2\right)^{1/2}.$$

The discharge velocity as a function of the height of the water remaining in the tank is

$$v_{bottom} = 1.20\sqrt{gh}.$$

The initial discharge velocity is when the tank begins to drain ($h = 10.0$ m) and is equal to 11.8 m/s. The discharge rate when the tank begins to drain is

$$\text{volume discharge rate} = A_{bottom}\, v_{bottom} = 4.73 \text{ m}^3/\text{s}.$$

KEY CONCEPT 8 **EVALUATE** We must carefully check the units in this problem. Do they cancel correctly?

Practice Problem: What is the volume rate of discharge when half of the water is gone?
Answer: 3.36 m³/s

Extra Practice: At what height is the discharge velocity equal to half of the initial discharge velocity? *Answer:* 2.50 m

7: Lift on a car on a highway

As a car travels down the highway, the speed of the air flowing over the top of the car is higher than the speed of the air flowing under the car, thus creating lift. Estimate the lift on a car as it travels at 100 kph.

Take the density of air to be 1.20 kg/m³, the car's area to be 6 m², and the height of the car to be 1.0 m. Assume that air travels under the car at 100 kph and over the top of the car at 140 kph.

IDENTIFY Bernoulli's equation gives the pressure difference between the top and bottom of the car. We'll use that equation to find the target variable, the lift on the car.

SET UP We find the lift force acting on the car from the definition of *pressure* as force per unit surface area.

EXECUTE The car is moving through a fluid (air), so Bernoulli's equation can be applied. We'll compare the pressures below and above the car to find the pressure difference. Bernoulli's equation is

$$p_{above} + \rho g y_{above} + \frac{1}{2}\rho v_{above}^2 = p_{below} + \rho g y_{below} + \frac{1}{2}\rho v_{below}^2.$$

The pressure difference is then

$$\Delta p = \rho g y_{above} + \frac{1}{2}\rho v_{above}^2 - \frac{1}{2}\rho v_{below}^2,$$

where we have set the origin below the car ($y_{below} = 0$). The pressure difference is

$$\Delta p = \rho \left(g y_{above} + \frac{1}{2}(v_{above}^2 - v_{below}^2) \right)$$

$$= (1.20 \text{ kg/m}^3)\left((9.8 \text{ m/s}^2)(1.0 \text{ m}) + \frac{1}{2}\left((38.9 \text{ m/s})^2 - (27.7 \text{ m/s})^2\right) \right)$$

$$= 459 \text{ Pa},$$

where we replaced 100 kpm with 27.7 m/s and 140 kph with 38.9 m/s. We find the force by multiplying the pressure by the area of the car:

$$F = PA = (459 \text{ Pa})(6.0 \text{ m}^2) = 2750 \text{ N}.$$

The lift on the car is 2750 N.

KEY CONCEPT 7 **EVALUATE** We see that the lift is significant in this case—roughly equivalent to a weight of 280 kg. It is not enough to lift the car off the highway, since most cars weigh over 1000 kg. We assumed that the flow of air around the car was smooth and that air is incompressible. Neither are valid assumptions and should be modified in a careful examination. Our results show the maximum lift of the car.

Practice Problem: What would the pressure difference be if the car had a height of 0.75 m?
Answer: 456 Pa

Extra Practice: What is the pressure difference if both the over and under air speeds increase by 10 kph? *Answer:* 494 Pa

8: Pressure in a water system

Water is discharged from a closed system, reaching a maximum height of 10.0 m. What is the gauge pressure of the water system at the hose nozzle?

IDENTIFY Bernoulli's equation gives the pressure difference between the tank and the top of water in flight. The target variable is the gauge pressure in the tank.

SET UP Kinematics is used to find the initial velocity of the water, taking into account the maximum height. Bernoulli's equation is then used to find the pressure in the tank.

EXECUTE From kinematics, for the water to reach a height h, it must have an initial velocity of

$$v_{noz}^2 = 2gh.$$

Bernoulli's equation is applied to two points, one inside the tank and one at the nozzle, both at the same height, to find the pressure:

$$p_{in} + \rho gy + \frac{1}{2}\rho v_{in}^2 = p_{noz} + \rho gy + \frac{1}{2}\rho v_{noz}^2.$$

The outside pressure is atmospheric pressure, so the difference between the two pressures is the gauge pressure, our target variable. The velocity inside the tank is zero, giving

$$p_{gauge} = p_{in} - p_a = \tfrac{1}{2}\rho v_{noz}^2.$$

Combining the results produces

$$p_{gauge} = \frac{1}{2}\rho(2gh) = (10^3 \text{ kg/m}^3)(9.8 \text{ m/s}^2)(10.0 \text{ m}) = 9.8\times10^4 \text{ Pa}.$$

The tank is at a gauge pressure of 9.8×10^4 Pa.

KEY CONCEPT 7 **EVALUATE** We've combined kinematics and fluid dynamics in this problem. We'll continue building our physics models and call upon older material to help us as we move forward.

Practice Problem: What is the velocity of the water leaving the nozzle? *Answer:* 14 m/s

Extra Practice: How much water is leaving the nozzle if it has a diameter of 0.80 cm? *Answer:* 7.04×10^{-4} m^3/s

Try It Yourself!

1: Leak in a submarine

A submarine descends 35.0 m into the ocean and springs a leak. The hole out of which water is leaking has a diameter of 2.5 cm. What force must be used to plug the hole?

Solution Checkpoints

IDENTIFY AND SET UP We'll use the relations among pressure, density, and depth to solve the problem. The target variable is the force needed to seal the hole. We must find the pressure at depth first. The density of seawater is 1.03×10^3 kg/m^3.

EXECUTE The pressure difference between the water and the submarine (assuming that the latter is at a pressure of 1 atmosphere inside) is given by

$$p - p_a = \rho g (y_2 - y_1).$$

This equation yields a pressure difference of 3.50 atm. The force needed to plug the hole is

$$F = (p - p_a) A.$$

The force required to seal the hole is 173 N.

KEY CONCEPT 2 **EVALUATE** Do you think it is possible for someone to push with a force of 173 N and seal the hole?

Practice Problem: What is the diameter of a hole that only requires 20 N of force to plug? *Answer:* 8.47 mm

Extra Practice: What force is required if the hole has a diameter of 5.0 cm? *Answer:* 694 N

2: Floating dumpling

A dumpling floats two-thirds submerged in water. What is its density?

Solution Checkpoints

IDENTIFY AND SET UP The dumpling is in equilibrium. What forces are acting on it? How do you quantify the buoyant force?

EXECUTE The fraction of the dumpling that is submerged is given by

$$\frac{V_D}{V_w} = \frac{\rho_w}{\rho_D}.$$

This equation can be rearranged to solve for the density of the dumpling. Doing so yields a density of 0.66 g/cm^3.

KEY CONCEPT 5 **EVALUATE** This problem illustrates how we can find the density of an object by examining how it floats.

Practice Problem: How much of the dumpling is submerged if it is floating in ethanol? *Answer:* 82%

Extra Practice: What is the density of an object that is 28% submerged in seawater? *Answer:* 0.288 g/cm^3

3: Weighing a sphere under water

A sphere of volume 10 cm^3 displaces a spring scale and is found to weigh 0.49 N when submerged in water. What are the sphere's mass and density?

Solution Checkpoints

IDENTIFY AND SET UP The sphere is in equilibrium. What three forces are acting on it?

EXECUTE The net external force acting on the sphere is
$$\sum F_y = F_s + F_B - mg = 0.$$

The weight and buoyant force are written in terms of density, volume, and g, the acceleration due to gravity. These quantities can be rearranged to yield
$$\rho = \frac{F_s}{Vg} + \rho_w.$$

From this equation, the density of the sphere is 6.0 g/cm³ and the mass of the sphere is 60.0 g.

KEY CONCEPT 5 **EVALUATE** How did you decide the direction that the spring force acted in?

Practice Problem: What is the spring constant if the spring is displaced 0.03 m? *Answer:* 16.3 N/m

Extra Practice: What would the sphere appear to weigh if it were submerged in glycerin? *Answer:* 0.46 N

4: Water from a tank

Water inside an enclosed tank is subjected to a pressure of two atmospheres at the top of the tank. What is the velocity of discharge from a small hole 3.0 m below the surface of the water?

Solution Checkpoints

IDENTIFY AND SET UP Use Bernoulli's equation and the continuity equation to solve this problem. The target variable is the discharge velocity. The small hole indicates that the ratio of the hole to the surface area is small.

EXECUTE Bernoulli's equation applied to the top of the water and hole gives
$$p_{top} + \rho g y_{top} + \frac{1}{2}\rho v_{top}^2 = p_{hole} + \rho g y_{hole} + \frac{1}{2}\rho v_{hole}^2.$$

What is the pressure outside and at the top? Can you assume that the velocity at the top surface is small? The continuity equation indicates that
$$v_{top} = \frac{A_{hole}}{A_{top}} v_{hole} \approx 0.$$

Solving for the velocity at the hole gives
$$v_{hole}^2 = \frac{2}{\rho}(p_a + \rho g h),$$

or a velocity of 16 m/s.

KEY CONCEPT 8 **EVALUATE** How can you check these results?

Practice Problem: What velocity do you get if the tank is open at the top? *Answer:* 7.67 m/s

Extra Practice: What velocity do you get if the open tank is filled with ethanol? *Answer:* 7.67 m/s

5: Water from a rocket

A toy rocket of diameter 2.0 in consists of water under the pressure of compressed air pumped into the nose chamber. When the gauge air pressure is 60 lb/in^2, the water is ejected through a hole of diameter 0.2 in. Find the propelling force, or thrust, of the rocket.

Solution Checkpoints

IDENTIFY AND SET UP Use Bernoulli's equation and the continuity equation to solve this problem. The target variable is the thrust.

EXECUTE Bernoulli's equation applied to points inside and outside of the rocket, at the same height, gives

$$p_{in} + \rho g y + \frac{1}{2}\rho v_{in}^2 = p_{out} + \rho g y + \frac{1}{2}\rho v_{out}^2.$$

What is the pressure p_a outside the rocket? What is the velocity of water inside the rocket? The thrust is given by

$$F = v_{out}\frac{dm_{out}}{dt} = v_{out}\rho\frac{dV_{out}}{dt} = v_{out}\rho A_{out} v_{out}.$$

Solving for the force, we find that it is 3.8 lb.

KEY CONCEPT 7 **EVALUATE** Did you check units?

Practice Problem: What is the velocity of the water leaving the rocket? *Answer:* 28.8 m/s

Extra Practice: What thrust do you get when the water is ejected through a 0.1-inch-diameter hole? *Answer:* 0.94 lb

Key Example Variation Problems

Solutions to these problems are in Chapter 12 of the Student's Solutions Manual.

Be sure to review EXAMPLES 12.3 and 12.4 (Section 12.2) before attempting these problems.

VP12.4.1 Liquefied natural gas (LNG) in a vertical storage tank has a density of 455 kg/m^3. The depth of LNG in the tank is 2.00 m, and the absolute pressure of the air inside the tank above the upper surface of the LNG is 1.22×10^5 Pa. What is the absolute pressure at the bottom of the tank?

VP12.4.2 You fill a vertical glass tube to a depth of 15.0 cm with freshwater. You then pour on top of the water an additional 15.0 cm of gasoline (density 7.40×10^2 kg/m^3), which does not mix with water. The upper surface of the gasoline is exposed to the air. Find the gauge pressure (a) at the interface between the gasoline and water and (b) at the bottom of the tube.

VP12.4.3 In an open-tube manometer (see Fig. 12.8a), the absolute pressure in the container of gas on the left is 2.10×10^5 Pa. If atmospheric pressure is 1.01×10^5 Pa and the liquid in the manometer is mercury (density 1.36×10^4 kg/m^3), what will be the difference in the heights of the left and right columns of liquid?

VP12.4.4 In the manometer tube shown in Fig. 12.10, the oil in the right-hand arm is olive oil of density 916 kg/m³. (a) If the top of the oil is 25.0 cm above the bottom of the tube, what is the height of the top of the water above the bottom of the tube? (b) What is the gauge pressure 15.0 cm beneath the surface of the water? (c) At what depth below the surface of the oil is the gauge pressure the same as in part (b)?

Be sure to review EXAMPLE 12.5 (Section 12.3) before attempting these problems.

VP12.5.1 An object of volume 7.50×10^{-4} m³ and density 1.15×10^3 kg/m³ is completely submerged in a fluid. (a) Calculate the weight of this object. (b) Calculate the buoyant force on this object if the fluid is (i) air (density 1.20 kg/m³), (ii) water (density 1.00×10^3 kg/m³), and (iii) glycerin (density 1.26×10^3 kg/m³). In each case state whether the object will rise or sink if released while submerged in the fluid.

VP12.5.2 A sphere of volume 1.20×10^{-3} m³ hangs from a cable. When the sphere is completely submerged in water, the tension in the cable is 29.4 N. (a) What is the buoyant force on the submerged sphere? (b) What is the weight of the sphere? (c) What is the density of the sphere?

VP12.5.3 A cube of volume 5.50×10^{-3} m³ and density 7.50×10^3 kg/m³ hangs from a cable. When the cube has the lower half of its volume submerged in an unknown liquid, the tension in the cable is 375 N. What is the density of the liquid? (Ignore the small buoyant force exerted by the air on the upper half of the cube.)

VP12.5.4 A wooden cylinder of length L and cross-sectional area A is partially submerged in a liquid with the axis of the cylinder oriented straight up and down. The density of the liquid is ρ_L. (a) If the length of the cylinder that is below the surface of the liquid is d, what is the buoyant force that the liquid exerts on the cylinder? (b) If the cylinder floats in the position described in part (a), what is the density of the cylinder? (Ignore the small buoyant force exerted by the air on the part of the cylinder above the surface of the liquid.)

Be sure to review EXAMPLES 12.7, 12.8, and 12.9 (Section 12.5) before attempting these problems.

VP12.9.1 A pipe leads from a storage tank on the roof of a building to the ground floor. The absolute pressure of the water in the storage tank where it connects to the pipe is 3.0×10^5 Pa, the pipe has a radius of 1.0 cm where it connects to the storage tank, and the speed of flow in this pipe is 1.6 m/s. The pipe on the ground floor has a radius of 0.50 cm and is 9.0 m below the storage tank. Find (a) the speed of flow and (b) the pressure in the pipe on the ground floor.

VP12.9.2 The storage tank in Fig. 12.25 contains ethanol (density 8.1×10^2 kg/m³). The tank is 4.0 m in radius, the short pipe at the bottom of the tank is 1.0 cm in radius, and the height of ethanol in the tank is 3.2 m. The volume flow rate of ethanol from the short pipe is 4.4×10^{-3} m³/s. (a) What is the speed at which ethanol flows out of the short pipe? (b) What is the gauge pressure of the air inside the tank in the space above the ethanol?

VP12.9.3 In a Venturi meter (see Fig. 12.26) that uses freshwater, the pressure difference between points 1 and 2 is 8.1×10^2 Pa, and the wide and narrow parts of the pipe have radii 2.5 cm and 1.2 cm, respectively. Find (a) the difference in the heights of the liquid levels in the two vertical tubes and (b) the volume flow rate through the horizontal pipe.

VP12.9.4 In the storage tank shown in Fig. 12.25, suppose area A_1 is *not* very much larger than area A_2. In this case we cannot treat the speed v_1 of the upper surface of the gasoline as zero. Derive an expression for the flow speed in the pipe in this situation in terms of p_0, p_{atm}, g, h, A_1, A_2 and ρ.

STUDENT'S SOLUTIONS MANUAL FOR FLUID MECHANICS

VP12.4.1. **IDENTIFY:** We want the pressure at a depth in a fluid.
SET UP: $p = p_0 + \rho g h$ gives the absolute pressure.
EXECUTE: $p = p_0 + \rho g h = 1.22 \times 10^5$ Pa $+ (455$ kg/m$^3)(9.80$ m/s$^2)(2.00$ m$) = 1.31 \times 10^5$ Pa.
EVALUATE: This pressure is about 30% above atmospheric pressure.

VP12.4.2. **IDENTIFY:** We are dealing with the gauge pressure at a depth in a fluid.
SET UP: $p = \rho g h$ gives the gauge pressure which is the pressure above atmospheric pressure.
EXECUTE: **(a)** The gauge pressure is due only to the gasoline.
$p = \rho g h = (740$ kg/m$^3)(9.80$ m/s$^2)(0.150$ m$) = 1.09 \times 10^3$ Pa.
(b) The gauge pressure at the top of the water is 1.09×10^3 Pa, so the pressure at the bottom is
$p = p_0 + \rho g h = 1.09 \times 10^3$ Pa $+ (1000$ kg/m$^3)(9.80$ m/s$^2)(0.150$ m$) = 2.56 \times 10^3$ Pa.
EVALUATE: To find the *absolute* pressure in each case we would have to add 1.01×10^5 Pa to our answers. A pressure gauge would read the gauge pressure.

VP12.4.3. **IDENTIFY:** We are dealing with the pressure at a depth in a fluid in a manometer.
SET UP: $p = p_0 + \rho g h$ gives the absolute pressure. Using Fig. 12.8a in the text, and looking at the right-hand column, the pressure in that column at a level y_1 is p. This is also the pressure due to the column from y_1 to the top which is a depth h in the right-hand column. Therefore the difference in heights h is $p - p_{atm} = \rho g h$.
EXECUTE: Solve for h: $h = \dfrac{p - p_{atm}}{\rho g} = \dfrac{2.1 \times 10^5 \text{ Pa} - 1.01 \times 10^5 \text{ Pa}}{(1.36 \times 10^4 \text{ kg/m}^3)(9.80 \text{ m/s}^2)} = 0.818$ m $= 81.8$ cm.
EVALUATE: According to our results, if the pressure p in the container were greater, h would be greater. If the density of the fluid in the manometer were less, h would also be greater. Both cases are physically reasonable, which suggests our solution is correct.

VP12.4.4. **IDENTIFY:** We are dealing with the pressure at a depth in a fluid in a manometer.
SET UP: $p = \rho g h$ gives the gauge pressure. The gauge pressure at the bottom of the oil is the same as the gauge pressure at the bottom of the water since both tubes are open to the air.
EXECUTE: **(a)** Equating the gauge pressures gives $\rho_{oil} g h_{oil} = \rho_{water} g h_{water}$.
$h_{water} = \dfrac{\rho_{oil}}{\rho_{water}} h_{oil} = \dfrac{916 \text{ kg/m}^3}{1000 \text{ kg/m}^3} \cdot 25.0$ cm $= 22.9$ cm.
(b) $p_{gauge} = \rho g h = (1000$ kg/m$^3)(9.80$ m/s$^2)(0.150$ m$) = 1.47 \times 10^3$ Pa.
(c) Solve $p_{gauge} = \rho g h$ for h, giving
$h = p_{gauge}/\rho g = (1.47 \times 10^3$ Pa$)/[(916$ kg/m$^3)(9.80$ m/s$^2)] = 0.164$ m $= 16.4$ cm.

EVALUATE: From (b) and (c), we see that we would have to be 15.0 cm deep in water to have the same pressure as at 16.4 cm in oil. This is reasonable since the density of oil is less than that of water.

VP12.5.1. **IDENTIFY:** This problem involves, density, buoyancy, and Archimedes's principle.
SET UP: Density is $\rho = m/V$. Archimedes's principle says that the buoyant force B on an immersed object is equal to the weight of the fluid displaced.
EXECUTE: (a) $m = \rho V$, so $w = \rho Vg = (1150 \text{ kg/m}^3)(7.50\times10^{-4} \text{ m}^3)(9.80 \text{ m/s}^2) = 8.45$ N.
(b) B equals the weight of the displaced fluid or gas, so $B = \rho Vg$.
(i) $B_{\text{air}} = \rho_{\text{air}} Vg = (1.20 \text{ kg/m}^3)(7.50\times10^{-4} \text{ m}^3)(9.80 \text{ m/s}^2) = 8.82\times10^{-3}$ N. The object would sink since the buoyant force on it is less than its weight.
(ii) $B_{\text{water}} = \rho_{\text{water}} Vg = (1000 \text{ kg/m}^3)(7.50\times10^{-4} \text{ m}^3)(9.80 \text{ m/s}^2) = 7.35$ N. The object would sink since the buoyant force on it is less than its weight.
(iii) $B_{\text{glycerine}} = \rho_{\text{glycerine}} Vg = (1260 \text{ kg/m}^3)(7.50\times10^{-4} \text{ m}^3)(9.80 \text{ m/s}^2) = 9.26$ N. The object would rise since the buoyant force on it is greater than its weight.
EVALUATE: Notice that in each case in (b), the object sinks if its density is greater than the fluid and rises if its density is less than that of the fluid. This is a useful point to keep in mind.

VP12.5.2. **IDENTIFY:** This problem involves density, buoyancy, and Archimedes's principle.
SET UP: Density is $\rho = m/V$. Archimedes's principle says that the buoyant force B on an immersed object is equal to the weight of the fluid displaced. $\Sigma F_y = 0$ for the sphere.
EXECUTE: (a) $B = \rho Vg = (1000 \text{ kg/m}^3)(9.80 \text{ m/s}^2)(1.20\times10^{-3} \text{ m}^3) = 11.8$ N.
(b) $\Sigma F_y = 0$ gives $w = T + B = 29.4 \text{ N} + 11.8 \text{ N} = 41.2$ N.
(c) $\rho = m/V = (W/g)/V = (41.2 \text{ N})/[(9.80 \text{ m/s}^2)(1.20\times10^{-3} \text{ m}^3)] = 3.50\times10^3$ kg/m^3.
EVALUATE: The sphere is denser than water, so the buoyant force is less than its weight, so there must be a tension in the cable. This agrees with our results.

VP12.5.3. **IDENTIFY:** This problem involves density, buoyancy, and Archimedes's principle.
SET UP: Density is $\rho = m/V$. Archimedes's principle says that the buoyant force B on an immersed object is equal to the weight of the fluid displaced. $\Sigma F_y = 0$ for the cube. Call V the volume of the cube and ρ_c its density. Let ρ be the density of the fluid.
EXECUTE: Archimedes's principle gives $B = \rho g(V/2)$. $\Sigma F_y = 0$ for the cube gives

$$T + B - w_{\text{cube}} = 0 \rightarrow T + \rho gV/2 - \rho_c gV = 0 \rightarrow \rho = \frac{2(\rho_c gV - T)}{gV}$$

$$\rho = \frac{2\left[(7.50\times10^3 \text{kg/m}^3)(9.80 \text{ m/s}^2)(5.50\times10^{-3}\text{m}^3) - 375 \text{ N}\right]}{(9.80 \text{ m/s}^2)(5.50\times10^{-3}\text{m}^3)} = 1.1\times10^3 \text{ kg/m}^3.$$

EVALUATE: The density of this fluid is slightly greater than that of water.

VP12.5.4. **IDENTIFY:** This problem involves density, buoyancy, and Archimedes's principle.
SET UP: Density is $\rho = m/V$. Archimedes's principle says that the buoyant force B on an immersed object is equal to the weight of the fluid displaced.
EXECUTE: (a) $B = \rho_L Vg$ and $V = Ad$, so $B = \rho_L Adg$.
(b) For floating $B = w_{\text{cyl}} \rightarrow \rho_L Vg = \rho_{\text{cyl}} ALg \rightarrow \rho_{\text{cyl}} = \frac{d}{L}\rho_L.$

EVALUATE: If $d = L$, our result gives $\rho_{cyl} = \rho_L$. In that case the cylinder would be fully submerged. Since the cylinder and liquid have the same densities, the volumes of the cylinder and the displaced liquid would have to be equal for the cylinder to float.

VP12.9.1. **IDENTIFY:** This is a problem in fluid flow involving Bernoulli's equation and the continuity equation.

SET UP: Bernoulli's equation is $p_1 + \rho g y_1 + \frac{1}{2}\rho v_1^2 = p_2 + \rho g y_2 + \frac{1}{2}\rho v_2^2$ and the continuity equation is $A_1 v_1 = A_2 v_2$.

EXECUTE: (a) Use $A_1 v_1 = A_2 v_2$ to find the speed of flow at ground level.
$A_1 v_1 = A_2 v_2 \to \pi r_1^2 v_1 = \pi r_2^2 v_2 \to v_2 = (r_1/r_2)^2 v_1$
$v_2 = [(1.0 \text{ cm})/(0.50 \text{ cm})]^2 (1.6 \text{ m/s}) = 6.4 \text{ m/s}$.
(b) Use Bernoulli's equation with point 2 to be floor level. This makes $v_1 = 1.6$ m/s, $v_2 = 6.4$ m/s, $y_1 = 9.0$ m, $y_2 = 0$, $p_1 = 3.0 \times 10^5$ Pa, and $\rho = 1000$ kg/m³. We want p_2 at ground level. Solving Bernoulli's equation gives $p_2 = 3.7 \times 10^5$ Pa.

EVALUATE: The pressure in the pipe is about 3.7 times atmospheric pressure. It would be even greater if the pipe were longer than 9.0 m.

VP12.9.2. **IDENTIFY:** This problem is about fluid flow. It involves Bernoulli's equation, volume flow rate, and the continuity equation.

SET UP: Bernoulli's equation is $p_1 + \rho g y_1 + \frac{1}{2}\rho v_1^2 = p_2 + \rho g y_2 + \frac{1}{2}\rho v_2^2$, the volume flow rate is $dV/dt = Av$, and the continuity equation is $A_1 v_1 = A_2 v_2$.

EXECUTE: (a) We know the volume flow rate, so use $dV/dt = Av$ to find v_2.
$$v = \frac{dV/dt}{A_2} = \frac{dV/dt}{\pi r_2^2} = \frac{4.4 \times 10^{-3} \text{ m}^3/\text{s}}{\pi (0.010 \text{ m})^2} = 14 \text{ m/s}.$$
(b) Apply Bernoulli's equation. Call point 1 the top of the ethanol and point 2 the level of the outcoming ethanol at the bottom. We want $p_1 - p_{atm}$ (the *gauge* pressure) and we know that $p_2 = p_{atm}$, $y_1 = 3.2$ m, $y_2 = 0$, $\rho = 810$ kg/m³, $v_2 = 14$ m/s, and $v_1 \approx 0$ (because $A_1 \gg A_2$). Solve for $p_1 - p_{atm}$ gives $p_{gauge} = p_1 - p_{atm} = 5.4 \times 10^4$ Pa.

EVALUATE: The tank is pressurized because $p_1 > p_{atm}$.

VP12.9.3. **IDENTIFY:** This problem deals with a Venturi meter.

SET UP: From Example 12.9, we have $p_1 - p_2 = \frac{1}{2}\rho v_1^2 \left[\left(\frac{A_1}{A_2}\right)^2 - 1\right]$. We want the difference in height h of the liquid levels in the two tubes. We also know from Example 12.9 that
$$v = \sqrt{\frac{2gh}{(A_1/A_2)^2 - 1}}.$$
First solve for v_1 and then use that result to solve for h. The volume flow rate is $dV/dt = Av$.

EXECUTE: (a) First use $p_1 - p_2 = \frac{1}{2}\rho v_1^2 \left[\left(\frac{A_1}{A_2}\right)^2 - 1\right]$ to find v_1.

$810 \text{ Pa} = \frac{1}{2}(1000 \text{ kg/m}^3) v_1^2 \left[\left(\frac{\pi(2.5 \text{ cm})^2}{\pi(1.2 \text{ cm})^2}\right)^2 - 1\right] \to v_1 = 0.3014$ m/s.

Now use $v_1 = \sqrt{\dfrac{2gh}{(A_1/A_2)^2 - 1}}$ to solve for h. Use $v_1 = 0.3014$ m/s and the same areas as above. This gives $h = 5.9 \times 10^{-4}$ m³/s.

(b) $dV/dt = A_1 v_1 = \pi r_1^2 v_1 = \pi(0.025 \text{ m})^2(0.3014 \text{ m/s}) = 5.9 \times 10^{-4}$ m³/s.

EVALUATE: The volume flow rate is the same throughout the pipe, so we calculated it at point 1. But it would also be true at point 2 or any other point in the pipe.

VP12.9.4. **IDENTIFY:** This problem is about fluid flow. It involves Bernoulli's equation, volume flow rate, and the continuity equation.

SET UP: Bernoulli's equation is $p_1 + \rho g y_1 + \frac{1}{2}\rho v_1^2 = p_2 + \rho g y_2 + \frac{1}{2}\rho v_2^2$, the volume flow rate is $dV/dt = Av$, and the continuity equation is $A_1 v_1 = A_2 v_2$. We follow exactly the same procedure as in problem VP12.9.2 *except* that v_1 cannot be neglected.

EXECUTE: The continuity equation gives v_1: $A_1 v_1 = A_2 v_2 \to v_1 = v_2(A_2/A_1)$. Now use Bernoulli's equation with $p_1 = p_0$, $p_2 = p_{\text{atm}}$, $y_1 = h$, $y_2 = 0$, $v_1 = v_2(A_2/A_1)$. This equation becomes

$$p_0 - p_{\text{atm}} + \rho g h = \frac{1}{2}\rho v_2^2 - \frac{1}{2}\rho v_2^2 \left(\frac{A_2}{A_1}\right)^2. \text{ Solving for } v_2 \text{ gives } v_2 = \sqrt{\dfrac{2\left(\dfrac{p_0 - p_{\text{atm}}}{\rho}\right) + 2gh}{1 - (A_2/A_1)^2}}.$$

EVALUATE: To check our result, consider the case of $A_1 \gg A_2$. In that case the denominator approaches 1 and $v_2 \to \sqrt{2\left(\dfrac{p_0 - p_{\text{atm}}}{\rho}\right) + 2gh}$. This is the result obtained in Example 2.8 in the text, so our result looks reasonable. Also if $A_1 \gg A_2$ and the tank is open at the top, $p_0 = p_{\text{atm}}$ and $v_2 \to \sqrt{2gh}$, which is the as for an object in freefall, as we would expect.

12.5. **IDENTIFY:** Apply $\rho = m/V$ to relate the densities and volumes for the two spheres.

SET UP: For a sphere, $V = \frac{4}{3}\pi r^3$. For lead, $\rho_l = 11.3 \times 10^3$ kg/m³ and for aluminum, $\rho_a = 2.7 \times 10^3$ kg/m³.

EXECUTE: $m = \rho V = \frac{4}{3}\pi r^3 \rho$. Same mass means $r_a^3 \rho_a = r_l^3 \rho_l$.

$\dfrac{r_a}{r_l} = \left(\dfrac{\rho_l}{\rho_a}\right)^{1/3} = \left(\dfrac{11.3 \times 10^3}{2.7 \times 10^3}\right)^{1/3} = 1.6.$

EVALUATE: The aluminum sphere is larger, since its density is less.

12.7. **IDENTIFY:** The gauge pressure $p - p_0$ at depth h is $p - p_0 = \rho g h$.

SET UP: Freshwater has density 1.00×10^3 kg/m³ and seawater has density 1.03×10^3 kg/m³.

EXECUTE: **(a)** $p - p_0 = (1.00 \times 10^3 \text{ kg/m}^3)(3.71 \text{ m/s}^2)(500 \text{ m}) = 1.86 \times 10^6$ Pa.

(b) $h = \dfrac{p - p_0}{\rho g} = \dfrac{1.86 \times 10^6 \text{ Pa}}{(1.03 \times 10^3 \text{ kg/m}^3)(9.80 \text{ m/s}^2)} = 184$ m

EVALUATE: The pressure at a given depth is greater on earth because a cylinder of water of that height weighs more on earth than on Mars.

12.9. **IDENTIFY:** Apply $p = p_0 + \rho g h$.

SET UP: Gauge pressure is $p - p_{\text{air}}$.

EXECUTE: The pressure difference between the top and bottom of the tube must be at least 5980 Pa in order to force fluid into the vein: $\rho g h = 5980$ Pa and

$$h = \frac{5980 \text{ Pa}}{\rho g} = \frac{5980 \text{ N/m}^2}{(1050 \text{ kg/m}^3)(9.80 \text{ m/s}^2)} = 0.581 \text{ m}.$$

EVALUATE: The bag of fluid is typically hung from a vertical pole to achieve this height above the patient's arm.

12.11. **IDENTIFY:** The pressure due to the glycerin balances the pressure due to the unknown liquid, so we are dealing with the pressure at a depth in a liquid.

SET UP: Both sides of the U-shaped tube are open to the air, so it is the gauge pressures that balance. Therefore we use $p_g = \rho g h$. At the level of the bottom of the column of the unknown, the pressure in the right side is equal to the pressure in the left side. (See Fig. 12.11.) On the right side, we have a 35.0-cm column of the unknown, and on the left side we have a column of glycerin whose top is 12.0 cm below that of the unknown, so its height is 35.0 cm − 12.0 cm = 23.0 cm above the bottom of the bottom of the column of the unknown liquid. The target variable is the density of the unknown liquid.

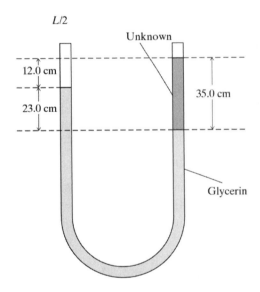

Figure 12.11

EXECUTE: We use $p_g = \rho g h$ to find the gauge pressure at the bottom of each column.

Right-hand column (the unknown): $p = \rho_x g (35.0 \text{ cm})$

Left-hand column (glycerin): $p = \rho_g g (35.0 \text{ cm} - 12.0 \text{ cm}) = \rho_g g (23.0 \text{ cm})$

Equating these pressures and solving for ρ_x gives $\rho_x = \frac{23 \text{ cm}}{35 \text{ cm}} \rho_g$. Using $\rho_g = 1260$ kg/m³ from Table 12.1, we have $\rho_x = \left(\frac{23 \text{ cm}}{35 \text{ cm}}\right)(1260 \text{ kg/m}^3) = 828 \text{ kg/m}^3$.

EVALUATE: Our result shows that the unknown is less dense than glycerin. This is reasonable because it takes only a 23-cm column of glycerin to balance a 35-cm column of the unknown.

12.15. IDENTIFY: Apply $p = p_0 + \rho g h$.

SET UP: For water, $\rho = 1.00 \times 10^3$ kg/m^3.

EXECUTE: $p - p_{\text{air}} = \rho g h = (1.00 \times 10^3 \text{ kg/m}^3)(9.80 \text{ m/s}^2)(6.1 \text{ m}) = 6.0 \times 10^4$ Pa.

EVALUATE: The pressure difference increases linearly with depth.

12.17. IDENTIFY: $p = p_0 + \rho g h$. $F = pA$.

SET UP: For seawater, $\rho = 1.03 \times 10^3$ kg/m^3.

EXECUTE: The force F that must be applied is the difference between the upward force of the water and the downward forces of the air and the weight of the hatch. The difference between the pressure inside and out is the gauge pressure, so

$F = (\rho g h) A - w = (1.03 \times 10^3 \text{ kg/m}^3)(9.80 \text{ m/s}^2)(30 \text{ m})(0.75 \text{ m}^2) - 300 \text{ N} = 2.27 \times 10^5$ N.

EVALUATE: The force due to the gauge pressure of the water is much larger than the weight of the hatch and would be impossible for the crew to apply it just by pushing.

12.19. IDENTIFY: The gauge pressure at the top of the oil column must produce a force on the disk that is equal to its weight.

SET UP: The area of the bottom of the disk is $A = \pi r^2 = \pi (0.150 \text{ m})^2 = 0.0707$ m^2.

EXECUTE: **(a)** $p - p_0 = \dfrac{w}{A} = \dfrac{45.0 \text{ N}}{0.0707 \text{ m}^2} = 636$ Pa.

(b) The increase in pressure produces a force on the disk equal to the increase in weight. By Pascal's law the increase in pressure is transmitted to all points in the oil.

(i) $\Delta p = \dfrac{83.0 \text{ N}}{0.0707 \text{ m}^2} = 1170$ Pa. (ii) 1170 Pa

EVALUATE: The absolute pressure at the top of the oil produces an upward force on the disk but this force is partially balanced by the force due to the air pressure at the top of the disk.

12.23. IDENTIFY: We are dealing with the pressure at a depth in a fluid.

SET UP: The pressure with the wine is the same as the pressure with mercury, except that the heights of the fluids are different because they have different densities. $p = p_0 + \rho g h$ gives the pressure in each case, with p_0 the same for both cases. The target variable is the height of the wine column.

EXECUTE: $p_0 + \rho_w g h_w = p_0 + \rho_m g h_m$, so $h_w = h_m \dfrac{\rho_m}{\rho_w} = (0.750 \text{ m}) \dfrac{13.6 \times 10^3 \text{ kg/m}^3}{990 \text{ kg/m}^3} = 10.3$ m.

EVALUATE: A wine barometer over 10 m high is highly unwieldy compared to one around 1 m high for mercury. Besides, there are better uses for wine!

12.25. IDENTIFY: A buoyant force acts on the athlete, so Archimedes's principle applies. He doesn't sink, so the forces on him must balance.

SET UP: Apply $\Sigma F_y = 0$ to the athlete. The volume of water he displaces is equal to his volume since he is totally submerged, so $V_w = V_{\text{ath}}$. The buoyant force it exerts is equal to the weight of that volume of water. We want to find the volume and average density of the athlete. $\rho_{\text{av}} = m/V$.

EXECUTE: Using $\Sigma F_y = 0$ gives $B + 20 \text{ N} - 900 \text{ N} = 0$, so $B = 880$ N. The buoyant force is equal to the weight of the water he displaces, and $V_w = V_{\text{ath}}$, so $B = \rho_w g V_w = \rho_w g V_{\text{ath}}$. Solving for his volume gives $V_{ath} = \dfrac{B}{\rho_w g} = \dfrac{880 \text{ N}}{(1000 \text{ kg/m}^3)(9.80 \text{ m/s}^2)} = 8.98 \times 10^{-2} \text{ m}^3$. His average density

is $\rho_{av} = \dfrac{m}{V} = \dfrac{W/g}{V} = \dfrac{(900 \text{ N})/(9.80 \text{ m/s}^2)}{8.98 \times 10^{-2} \text{ m}^3} = 1020 \text{ kg/m}^3$.

EVALUATE: The athlete is slightly denser than pure water. This is reasonable because he probably has little fat, which is less dense than muscle.

12.29. IDENTIFY: Apply $\Sigma F_y = ma_y$ to the sample, with $+y$ upward. $B = \rho_{water} V_{obj} g$.

SET UP: $w = mg = 17.50$ N and $m = 1.79$ kg.

EXECUTE: $T + B - mg = 0$. $a_x = 0$

$V_{obj} = \dfrac{B}{\rho_{water} g} = \dfrac{6.30 \text{ N}}{(1.00 \times 10^3 \text{ kg/m}^3)(9.80 \text{ m/s}^2)} = 6.43 \times 10^{-4} \text{ m}^3$.

$\rho = \dfrac{m}{V} = \dfrac{1.79 \text{ kg}}{6.43 \times 10^{-4} \text{ m}^3} = 2.78 \times 10^3 \text{ kg/m}^3$.

EVALUATE: The density of the sample is greater than that of water and it doesn't float.

12.33. IDENTIFY and SET UP: Use $p = p_0 + \rho g h$ to calculate the gauge pressure at the two depths.

(a) The distances are shown in Figure 12.33a.

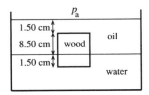

EXECUTE: $p - p_0 = \rho g h$

The upper face is 1.50 cm below the top of the oil, so

$p - p_0 = (790 \text{ kg/m}^3)(9.80 \text{ m/s}^2)(0$

$p - p_0 = 116$ Pa

Figure 12.33a

(b) The pressure at the interface is $p_{interface} = p_a + \rho_{oil} g(0.100 \text{ m})$. The lower face of the block is 1.50 cm below the interface, so the pressure there is $p = p_{interface} + \rho_{water} g(0.0150 \text{ m})$. Combining these two equations gives

$p - p_a = \rho_{oil} g(0.100 \text{ m}) + \rho_{water} g(0.0150 \text{ m})$

$p - p_a = [(790 \text{ kg/m}^3)(0.100 \text{ m}) + (1000 \text{ kg/m}^3)(0.0150 \text{ m})](9.80 \text{ m/s}^2)$

$p - p_a = 921$ Pa

(c) IDENTIFY and SET UP: Consider the forces on the block. The area of each face of the block is $A = (0.100 \text{ m})^2 = 0.0100 \text{ m}^2$. Let the absolute pressure at the top face be p_t and the pressure at the bottom face be p_b. In $p = \dfrac{F_\perp}{A}$, use these pressures to calculate the force exerted by the fluids at the top and bottom of the block. The free-body diagram for the block is given in Figure 12.33b.

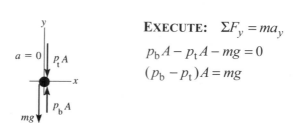

EXECUTE: $\Sigma F_y = ma_y$

$p_b A - p_t A - mg = 0$

$(p_b - p_t)A = mg$

Figure 12.33b

Note that $(p_b - p_t) = (p_b - p_a) - (p_t - p_a) = 921 \text{ Pa} - 116 \text{ Pa} = 805 \text{ Pa}$; the difference in absolute pressures equals the difference in gauge pressures.

$$m = \frac{(p_b - p_t)A}{g} = \frac{(805 \text{ Pa})(0.0100 \text{ m}^2)}{9.80 \text{ m/s}^2} = 0.821 \text{ kg}.$$

And then $\rho = m/V = 0.821 \text{ kg}/(0.100 \text{ m})^3 = 821 \text{ kg/m}^3$.

EVALUATE: We can calculate the buoyant force as $B = (\rho_{\text{oil}}V_{\text{oil}} + \rho_{\text{water}}V_{\text{water}})g$ where $V_{\text{oil}} = (0.0100 \text{ m}^2)(0.0850 \text{ m}) = 8.50 \times 10^{-4} \text{ m}^3$ is the volume of oil displaced by the block and $V_{\text{water}} = (0.0100 \text{ m}^2)(0.0150 \text{ m}) = 1.50 \times 10^{-4} \text{ m}^3$ is the volume of water displaced by the block. This gives $B = (0.821 \text{ kg})g$. The mass of water displaced equals the mass of the block.

12.35. **IDENTIFY:** The vertical forces on the rock sum to zero. The buoyant force equals the weight of liquid displaced by the rock. $V = \frac{4}{3}\pi R^3$.

SET UP: The density of water is $1.00 \times 10^3 \text{ kg/m}^3$.

EXECUTE: The rock displaces a volume of water whose weight is $39.2 \text{ N} - 28.4 \text{ N} = 10.8 \text{ N}$. The mass of this much water is thus $10.8 \text{ N}/(9.80 \text{ m/s}^2) = 1.102 \text{ kg}$ and its volume, equal to the rock's volume, is $\frac{1.102 \text{ kg}}{1.00 \times 10^3 \text{ kg/m}^3} = 1.102 \times 10^{-3} \text{ m}^3$. The weight of unknown liquid displaced is $39.2 \text{ N} - 21.5 \text{ N} = 17.7 \text{ N}$, and its mass is $(17.7 \text{ N})/(9.80 \text{ m/s}^2) = 1.806 \text{ kg}$. The liquid's density is thus $(1.806 \text{ kg})/(1.102 \times 10^{-3} \text{ m}^3) = 1.64 \times 10^3 \text{ kg/m}^3$.

EVALUATE: The density of the unknown liquid is a little more than 1.5 times the density of water.

12.37. **IDENTIFY:** The cylinder is partially submerged in water, so we apply Archimedes's principle. The vertical forces on it must balance.

SET UP: The density of this cylinder is $370/1000 = 37\%$ that of water, so it would float with 37% of its volume under the water and 63% above water. But it is partially submerged with 70.0% under water. Therefore the buoyant force B on it must be greater than its weight, and this would force the cylinder upward. The tension in the cable prevents this from happening. The target variable is the tension T in the cable. We apply $\Sigma F_y = 0$ and $w = \rho g V$.

EXECUTE: Start with $\Sigma F_y = 0$: $B - T - w_c = 0$. We see that we first need to find B. Call V the volume of the cylinder and ρ_c its density. Therefore $w_c = \rho_c g V$. By Archimedes's principle, the buoyant force B is equal to the weight of the water displaced by the cylinder, and we know that the cylinder has 70% of its volume below the water, which is $0.700V$. Therefore $B = \rho_w g(0.700V)$. Di-

viding B by w_c gives $\dfrac{B}{w_c} = \dfrac{\rho_w g(0.700V)}{\rho_c gV} = \dfrac{\rho_w}{\rho_c} = \dfrac{1000 \text{ kg/m}^3}{370 \text{ kg/m}^3} = 2.7027$, so $B = 2.7027 w_c$. Putting this result into $\Sigma F_y = 0$ gives $2.7027 w_c - T - w_c = 0$. Solving for T gives
$T = 1.7027 w_c = (1.7027)(30.0 \text{ kg})(9.80 \text{ m/s}^2) = 501$ N.

EVALUATE: If the cable were to break, the net upward force at that instant on the cylinder would be $B - w_c = 2.7027 w_c - w_c = 1.70 w_c$, so the cylinder would accelerate upward and eventually float with 63% of its volume above water.

12.41. IDENTIFY and SET UP: Apply the continuity equation, $v_1 A_1 = v_2 A_2$. In part (a) the target variable is V. In part (b) solve for A and then from that get the radius of the pipe.

EXECUTE: (a) $vA = 1.20 \text{ m}^3/\text{s}$

$v = \dfrac{1.20 \text{ m}^3/\text{s}}{A} = \dfrac{1.20 \text{ m}^3/\text{s}}{\pi r^2} = \dfrac{1.20 \text{ m}^3/\text{s}}{\pi (0.150 \text{ m})^2} = 17.0$ m/s

(b) $vA = 1.20 \text{ m}^3/\text{s}$

$v\pi r^2 = 1.20 \text{ m}^3/\text{s}$

$r = \sqrt{\dfrac{1.20 \text{ m}^3/\text{s}}{v\pi}} = \sqrt{\dfrac{1.20 \text{ m}^3/\text{s}}{(3.80 \text{ m/s})\pi}} = 0.317$ m

EVALUATE: The speed is greater where the area and radius are smaller.

12.43. IDENTIFY: Water is flowing out of the tank and collecting in a bucket, so we use Bernoulli's equation and the volume flow rate.

SET UP: $p_1 + \rho g y_1 + \dfrac{1}{2}\rho v_1^2 = p_2 + \rho g y_2 + \dfrac{1}{2}\rho v_2^2$, $dV/dt = Av$. Choose $y = 0$ at the base of the tank, so $y_2 = 0$ and $y_1 = h$. The top of the tank and the end of the pipe are open to the atmosphere, so $p_1 = p_2$. The target variable is the time it takes to collect a gallon of water. Call R the radius of the tank and r the radius of the small hole at the bottom.

EXECUTE: (a) As 1 gal flows out of the tank the change in volume in the tank is 1 gal = 3.788×10^{-3} m^3. The change in volume ΔV of water in the tank due to a height change Δh is $\Delta V = \pi R^2 \Delta h$, so $\Delta h = \Delta V/(\pi R^2) = (3.788 \times 10^{-3} \text{ m}^3)/[\pi(1.50 \text{ m})^2] = 5.36 \times 10^{-4}$ m = 0.536 mm.

(b) Now use $p_1 + \rho g y_1 + \dfrac{1}{2}\rho v_1^2 = p_2 + \rho g y_2 + \dfrac{1}{2}\rho v_2^2$. Based upon the result in part (a), we can treat the height of water in the tank as constant while a gallon flows out and treat the speed v_1 of the water at the top of the tank as being essentially zero. The top of the tank and the end of the pipe are open to the atmosphere, so $p_1 = p_2$. Take $y = 0$ at the bottom of the tank. We need to find the speed v_2 of the water as it comes out of the small hole at the bottom of the tank. Bernoulli's equation becomes $\rho g h = \dfrac{1}{2}\rho v_2^2$, which gives $v_2 = \sqrt{2gh}$. Now use the volume flow rate at the bottom.

$dV/dt = Av_2 = \pi r^2 \sqrt{2gh}$. This gives $\Delta V = \pi r^2 \sqrt{2gh}\, \Delta t$, so $\Delta t = \dfrac{\Delta V}{\pi r^2 \sqrt{2gh}}$. Using $\Delta V = 3.788 \times 10^{-3}$ m^3, $r = 0.250$ cm $= 0.0025$ m, and $h = 2.00$ m gives $\Delta t = 30.8$ s.

EVALUATE: It is reasonable to neglect the change in depth of the water in the tank as one gallon flows out because $\Delta h \ll h$: 0.536 mm $\ll$ 2.00 m.

12.45. IDENTIFY and SET UP:

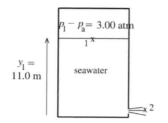

Figure 12.45

Apply Bernoulli's equation with points 1 and 2 chosen as shown in Figure 12.45. Let $y = 0$ at the bottom of the tank so $y_1 = 11.0$ m and $y_2 = 0$. The target variable is v_2.

$p_1 + \rho g y_1 + \frac{1}{2}\rho v_1^2 = p_2 + \rho g y_2 + \frac{1}{2}\rho v_2^2$

$A_1 v_1 = A_2 v_2$, so $v_1 = (A_2/A_1)v_2$. But the cross-sectional area of the tank (A_1) is much larger than the cross-sectional area of the hole (A_2), so $v_1 \ll v_2$ and the $\frac{1}{2}\rho v_1^2$ term can be neglected.

EXECUTE: This gives $\frac{1}{2}\rho v_2^2 = (p_1 - p_2) + \rho g y_1$.

Use $p_2 = p_a$ and solve for v_2:

$v_2 = \sqrt{2(p_1 - p_a)/\rho + 2g y_1} = \sqrt{\dfrac{2(3.039 \times 10^5 \text{ Pa})}{1030 \text{ kg/m}^3} + 2(9.80 \text{ m/s}^2)(11.0 \text{ m})}$

$v_2 = 28.4$ m/s

EVALUATE: If the pressure at the top surface of the water were air pressure, then Torricelli's theorem (Example: 12.8) gives $v_2 = \sqrt{2g(y_1 - y_2)} = 14.7$ m/s. The actual afflux speed is much larger than this due to the excess pressure at the top of the tank.

12.47. IDENTIFY and SET UP:

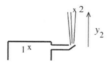

Figure 12.47

Apply Bernoulli's equation to points 1 and 2 as shown in Figure 12.47. Point 1 is in the mains and point 2 is at the maximum height reached by the stream, so $v_2 = 0$.

Solve for p_1 and then convert this absolute pressure to gauge pressure.

EXECUTE: $p_1 + \rho g y_1 + \frac{1}{2}\rho v_1^2 = p_2 + \rho g y_2 + \frac{1}{2}\rho v_2^2$

Let $y_1 = 0$, $y_2 = 15.0$ m. The mains have large diameter, so $v_1 \approx 0$.

Thus $p_1 = p_2 + \rho g y_2$.

But $p_2 = p_a$, so $p_1 - p_a = \rho g y_2 = (1000 \text{ kg/m}^3)(9.80 \text{ m/s}^2)(15.0 \text{ m}) = 1.47 \times 10^5$ Pa.

EVALUATE: This is the gauge pressure at the bottom of a column of water 15.0 m high.

12.51. IDENTIFY and SET UP: Let point 1 be where $r_1 = 4.00$ cm and point 2 be where $r_2 = 2.00$ cm. The volume flow rate vA has the value 7200 cm^3/s at all points in the pipe. Apply $v_1 A_1 = v_2 A_2$ to find the fluid speed at points 1 and 2 and then use Bernoulli's equation for these two points to find p_2.

EXECUTE: $v_1 A_1 = v_1 \pi r_1^2 = 7200$ cm^3, so $v_1 = 1.43$ m/s

$v_2 A_2 = v_2 \pi r_2^2 = 7200$ cm^3/s, so $v_2 = 5.73$ m/s

$$p_1 + \rho g y_1 + \tfrac{1}{2}\rho v_1^2 = p_2 + \rho g y_2 + \tfrac{1}{2}\rho v_2^2$$

$y_1 = y_2$ and $p_1 = 2.40 \times 10^5$ Pa, so $p_2 = p_1 + \tfrac{1}{2}\rho(v_1^2 - v_2^2) = 2.25 \times 10^5$ Pa.

EVALUATE: Where the area decreases the speed increases and the pressure decreases.

12.55. IDENTIFY: In part (a), the force is the weight of the water. In part (b), the pressure due to the water at a depth h is $\rho g h$. $F = pA$ and $m = \rho V$.

SET UP: The density of water is 1.00×10^3 kg/m^3.

EXECUTE: (a) The weight of the water is
$\rho g V = (1.00 \times 10^3 \text{ kg/m}^3)(9.80 \text{ m/s}^2)((5.00 \text{ m})(4.0 \text{ m})(3.0 \text{ m})) = 5.9 \times 10^5$ N.

(b) Integration gives the expected result that the force is what it would be if the pressure were uniform and equal to the pressure at the midpoint. If d is the depth of the pool and A is the area of one end of the pool, then

$$F = \rho g A \frac{d}{2} = (1.00 \times 10^3 \text{ kg/m}^3)(9.80 \text{ m/s}^2)((4.0 \text{ m})(3.0 \text{ m}))(1.50 \text{ m}) = 1.76 \times 10^5 \text{ N}.$$

EVALUATE: The answer to part (a) can be obtained as $F = pA$, where $p = \rho g d$ is the gauge pressure at the bottom of the pool and $A = (5.0 \text{ m})(4.0 \text{ m})$ is the area of the bottom of the pool.

12.57. IDENTIFY: Use $p = p_0 + \rho g h$ to find the gauge pressure versus depth, use $p = \dfrac{F_\perp}{A}$ to relate the pressure to the force on a strip of the gate, calculate the torque as force times moment arm, and follow the procedure outlined in the hint to calculate the total torque.

SET UP: The gate is sketched in Figure 12.57a.

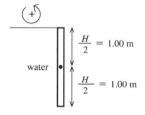

Let τ_u be the torque due to the net force of the water on the upper half of the gate, and τ_l be the torque due to the force on the lower half.

Figure 12.57a

With the indicated sign convention, τ_l is positive and τ_u is negative, so the net torque about the hinge is $\tau = \tau_l - \tau_u$. Let H be the height of the gate.

Upper half of gate:
Calculate the torque due to the force on a narrow strip of height dy located a distance y below the top of the gate, as shown in Figure 12.57b. Then integrate to get the total torque.

The net force on the strip is $dF = p(y)\,dA$, where $p(y) = \rho g y$ is the pressure at this depth and $dA = W\,dy$ with $W = 4.00$ m.
$dF = \rho g y W\,dy$

Figure 12.57b

The moment arm is $(H/2 - y)$, so $d\tau = \rho g W (H/2 - y) y\,dy$.

$$\tau_u = \int_0^{H/2} d\tau = \rho g W \int_0^{H/2} (H/2 - y) y\, dy = \rho g W ((H/4)y^2 - y^3/3)\Big|_0^{H/2}$$

$$\tau_u = \rho g W (H^3/16 - H^3/24) = \rho g W (H^3/48)$$

$$\tau_u = (1000 \text{ kg/m}^3)(9.80 \text{ m/s}^2)(4.00 \text{ m})(2.00 \text{ m})^3/48 = 6.533 \times 10^3 \text{ N} \cdot \text{m}$$

Lower half of gate:

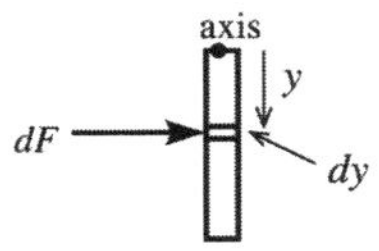

Consider the narrow strip shown in Figure 12.57c.
The depth of the strip is $(H/2 + y)$ so the force dF is
$$dF = p(y)\, dA = \rho g (H/2 + y) W\, dy.$$

Figure 12.57c

The moment arm is y, so $d\tau = \rho g W (H/2 + y) y\, dy$.

$$\tau_l = \int_0^{H/2} d\tau = \rho g W \int_0^{H/2} (H/2 + y) y\, dy = \rho g W ((H/4)y^2 + y^3/3)\Big|_0^{H/2}$$

$$\tau_l = \rho g W (H^3/16 + H^3/24) = \rho g W (5H^3/48)$$

$$\tau_l = (1000 \text{ kg/m}^3)(9.80 \text{ m/s}^2)(4.00 \text{ m})5(2.00 \text{ m})^3/48 = 3.267 \times 10^4 \text{ N} \cdot \text{m}$$

Then $\tau = \tau_l - \tau_u = 3.267 \times 10^4 \text{ N} \cdot \text{m} - 6.533 \times 10^3 \text{ N} \cdot \text{m} = 2.61 \times 10^4 \text{ N} \cdot \text{m}$.

EVALUATE: The forces and torques on the upper and lower halves of the gate are in opposite directions so find the net value by subtracting the magnitudes. The torque on the lower half is larger than the torque on the upper half since pressure increases with depth.

12.59. **IDENTIFY:** This problem deals with the pressure at a depth in a fluid and requires graphical interpretation.

SET UP: The gauge pressure on the bottom of the block supports the weight of the block and the coins. The graph is a plot of the depth h of the bottom of the block versus the mass m of the coins, so we need to look for a relationship between these quantities. The target variable is the mass M of the block. The gauge pressure is $p = \rho g h$ and the force on an area is $F = pA$.

EXECUTE: The net force on the block is $p_g A$, and that force must be equal to the weight of the box plus the coins. This gives $p_g A = mg + Mg$. Using $p_g = \rho g h$, this becomes $\rho g h A = mg + Mg$ so

$h = \dfrac{1}{\rho A} m + \dfrac{M}{\rho A}$. Therefore a graph of h versus m should have a slope of $\dfrac{1}{\rho A}$ and a y-intercept of $\dfrac{M}{\rho A}$. From the slope we get $A = \dfrac{1}{\rho(\text{slope})}$. The y-intercept gives $y\text{-int} = \dfrac{M}{\rho A}$, so $M = (y\text{-int})\rho A$.

Putting the result for A into the equation for M gives $M = (y\text{-int})\rho A = (y\text{-int})\rho \left(\dfrac{1}{\rho(\text{slope})}\right) = \dfrac{y\text{-int}}{\text{slope}}$

$= \dfrac{0.0312 \text{ m}}{0.0390 \text{ m/kg}} = 0.800 \text{ kg}.$

EVALUATE: Careful graphing is important because the answer depends on *both* the slope and y-intercept of the graph.

12.61. IDENTIFY: The buoyant force on an object in a liquid is equal to the weight of the liquid it displaces.

SET UP: $V = \dfrac{m}{\rho}$.

EXECUTE: When it is floating, the ice displaces an amount of glycerin equal to its weight. From Table 12.1, the density of glycerin is 1260 kg/m^3. The volume of this amount of glycerin is $V = \dfrac{m}{\rho} = \dfrac{0.180 \text{ kg}}{1260 \text{ kg/m}^3} = 1.429 \times 10^{-4}$ m^3. The ice cube produces 0.180 kg of water. The volume of this mass of water is $V = \dfrac{m}{\rho} = \dfrac{0.180 \text{ kg}}{1000 \text{ kg/m}^3} = 1.80 \times 10^{-4}$ m^3. The volume of water from the melted ice is greater than the volume of glycerin displaced by the floating cube and the level of liquid in the cylinder rises. The distance the level rises is

$$\dfrac{1.80 \times 10^{-4} \text{ m}^3 - 1.429 \times 10^{-4} \text{ m}^3}{\pi (0.0350 \text{ m})^2} = 9.64 \times 10^{-3} \text{ m} = 0.964 \text{ cm}.$$

EVALUATE: The melted ice has the same mass as the solid ice, but a different density.

12.65. IDENTIFY: Archimedes's principle applies.

SET UP: $\rho = m/V$, the buoyant force B is equal to the weight of the liquid displaced. Call m the mass of the block.

EXECUTE: **(a)** The volume of water displaced by the block is 80.0% of the volume of the block. Using $B = mg$: $\rho_w V_w g = mg$ gives $\rho_w(0.800 V_{block}) = m$. Therefore $V_{block} = m/(0.800\, \rho_w)$, so $V_{block} = (40.0 \text{ kg})/[(0.800)(1000 \text{ kg/m}^3)] = 0.0500$ m^3.

(b) With the maximum amount of bricks added on, the block is completely submerged but the bricks are not under water. Therefore $m_{bricks}g + mg = \rho_w V_{block}g$. Solving for m_{bricks} and putting in the numbers gives

$m_{bricks} = \rho_w V_{block} - m = (1000 \text{ kg/m}^3)(0.0500 \text{ m}^3) - 40.0 \text{ kg} = 10.0$ kg.

EVALUATE: If the bricks were to go under water, the buoyant force would increase because a greater volume of water would be displaced.

12.71. IDENTIFY: After the water leaves the hose the only force on it is gravity. Use conservation of energy to relate the initial speed to the height the water reaches. The volume flow rate is Av.

SET UP: $A = \pi D^2/4$

EXECUTE: **(a)** $\tfrac{1}{2}mv^2 = mgh$ gives $v = \sqrt{2gh} = \sqrt{2(9.80 \text{ m/s}^2)(28.0 \text{ m})} = 23.4$ m/s.

$(\pi D^2/4)v = 0.500$ m/s^3. $D = \sqrt{\dfrac{4(0.500 \text{ m/s}^3)}{\pi v}} = \sqrt{\dfrac{4(0.500 \text{ m/s}^3)}{\pi(23.4 \text{ m/s})}} = 0.165$ m $= 16.5$ cm.

(b) D^2v is constant so if D is twice as great, then v is decreased by a factor of 4. h is proportional to v^2, so h is decreased by a factor of 16. $h = \dfrac{28.0 \text{ m}}{16} = 1.75$ m.

EVALUATE: The larger the diameter of the nozzle the smaller the speed with which the water leaves the hose and the smaller the maximum height.

12.73. IDENTIFY: As water flows from the tank, the water level changes. This affects the speed with which the water flows out of the tank and the pressure at the bottom of the tank.

SET UP: Bernoulli's equation, $p_1 + \rho g y_1 + \frac{1}{2}\rho v_1^2 = p_2 + \rho g y_2 + \frac{1}{2}\rho v_2^2$, and the continuity equation, $A_1 v_1 = A_2 v_2$, both apply.

EXECUTE: (a) Let point 1 be at the surface of the water in the tank and let point 2 be in the stream of water that is emerging from the tank. $p_1 + \rho g y_1 + \frac{1}{2}\rho v_1^2 = p_2 + \rho g y_2 + \frac{1}{2}\rho v_2^2$. $v_1 = \dfrac{\pi d_2^2}{\pi d_1^2} v_2$, with $d_2 = 0.0200$ m and $d_1 = 2.00$ m. $v_1 \ll v_2$ so the $\frac{1}{2}\rho v_1^2$ term can be neglected. $v_2 = \sqrt{\dfrac{2 p_0}{\rho} + 2gh}$, where $h = y_1 - y_2$ and $p_0 = p_1 - p_2 = 5.00 \times 10^3$ Pa. Initially $h = h_0 = 0.800$ m and when the tank has drained $h = 0$. At $t = 0$,

$$v_2 = \sqrt{\dfrac{2(5.00 \times 10^3 \text{ Pa})}{1000 \text{ kg/m}^3} + 2(9.8 \text{ m/s}^2)(0.800 \text{ m})} = \sqrt{10 + 15.68} \text{ m/s} = 5.07 \text{ m/s}.$$

If the tank is open to the air, $p_0 = 0$ and $v_2 = 3.96$ m/s. The ratio is 1.28.

(b) $v_1 = -\dfrac{dh}{dt} = \dfrac{A_2}{A_1} v_2 = \left(\dfrac{d_2}{d_1}\right)^2 \sqrt{\dfrac{2 p_0}{\rho} + 2gh} = \left(\dfrac{d_2}{d_1}\right)^2 \sqrt{2g} \sqrt{\dfrac{p_0}{g\rho} + h}$. Separating variables gives

$\dfrac{dh}{\sqrt{\dfrac{p_0}{g\rho} + h}} = -\left(\dfrac{d_2}{d_1}\right)^2 \sqrt{2g}\, dt$. We now must integrate $\displaystyle\int_{h_0}^{0} \dfrac{dh'}{\sqrt{\dfrac{p_0}{g\rho} + h'}} = -\left(\dfrac{d_2}{d_1}\right)^2 \sqrt{2g} \int_0^t dt'$. To do the left-hand side integral, make the substitution $u = \dfrac{p_0}{g\rho} + h'$, which makes $du = dh'$. The integral is then of the form $\displaystyle\int \dfrac{du}{u^{1/2}}$, which can be readily integrated using $\displaystyle\int u^n\, du = \dfrac{u^{n+1}}{n+1}$. The result is

$2\left(\sqrt{\dfrac{p_0}{g\rho}} - \sqrt{\dfrac{p_0}{g\rho} + h_0}\right) = -\left(\dfrac{d_2}{d_1}\right)^2 \sqrt{2g}\, t$. Solving for t gives $t = \left(\dfrac{d_1}{d_2}\right)^2 \sqrt{\dfrac{2}{g}}\left(\sqrt{\dfrac{p_0}{g\rho} + h_0} - \sqrt{\dfrac{p_0}{g\rho}}\right)$.

Since $\dfrac{p_0}{g\rho} = \dfrac{5.00 \times 10^3 \text{ Pa}}{(9.8 \text{ m/s}^2)(1000 \text{ kg/m}^3)} = 0.5102$ m, we get

$t = \left(\dfrac{2.00}{0.0200}\right)^2 \sqrt{\dfrac{2}{9.8 \text{ m/s}^2}} \left(\sqrt{0.5102 \text{ m} + 0.800 \text{ m}} - \sqrt{0.5102 \text{ m}}\right) = 1.944 \times 10^3$ s = 32.4 min. When $p_0 = 0$, $t = \left(\dfrac{2.00}{0.0200}\right)^2 \sqrt{\dfrac{2}{9.8 \text{ m/s}^2}} \left(\sqrt{0.800 \text{ m}}\right) = 4.04 \times 10^3$ s = 67.3 min. The ratio is 2.08.

EVALUATE: Both ratios are greater than one because a surface pressure greater than atmospheric pressure causes the water to drain with a greater speed and in a shorter time than if the surface were open to the atmosphere with a pressure of one atmosphere.

12.77. IDENTIFY: After leaving the tank, the water is in free fall, with $a_x = 0$ and $a_y = +g$.

SET UP: The speed of efflux is $\sqrt{2gh}$.

EXECUTE: (a) The time it takes any portion of the water to reach the ground is $t = \sqrt{\dfrac{2(H-h)}{g}}$, in which time the water travels a horizontal distance $R = vt = 2\sqrt{h(H-h)}$.

(b) Note that if $h' = H - h$, $h'(H - h') = (H - h)h$, and so $h' = H - h$ gives the same range. A hole $H - h$ below the water surface is a distance h above the bottom of the tank.

EVALUATE: For the special case of $h = H/2$, $h = h'$ and the two points coincide. For the upper hole the speed of efflux is less but the time in the air during the free fall is greater.

12.79. **IDENTIFY:** As you constrict the hose, you decrease its area, but the equation of continuity applies to the water.

SET UP: $A_1 v_1 = A_2 v_2$. The distance traveled by a projectile that is fired from a height h with an initial horizontal velocity v is $x = vt$ where $t = \sqrt{\dfrac{2h}{g}}$.

EXECUTE: Since h is fixed, t does not change as we constrict the nozzle. Looking at the ratio of distances we obtain $\dfrac{x_1}{x_2} = \dfrac{v_1 t}{v_2 t} = \dfrac{A_2}{A_1} = \dfrac{\pi r_2^2}{\pi r_1^2}$, which gives

$$x_1 = x_2 \left(\dfrac{r_2}{r_1}\right)^2 = (0.950 \text{ m})\left(\dfrac{1.80 \text{ cm}}{0.750 \text{ cm}}\right)^2 = 5.47 \text{ m}.$$

EVALUATE: A smaller constriction results in a higher exit velocity, which results in a greater range, so our result is plausible.

12.81. **IDENTIFY:** Apply Bernoulli's equation and the equation of continuity.

SET UP: The speed of efflux is $\sqrt{2gh}$, where h is the distance of the hole below the surface of the fluid.

EXECUTE: (a) $v_3 A_3 = \sqrt{2g(y_1 - y_3)} A_3 = \sqrt{2(9.80 \text{ m/s}^2)(8.00 \text{ m})}(0.0160 \text{ m}^2) = 0.200 \text{ m}^3/\text{s}$.

(b) Since p_3 is atmospheric pressure, the gauge pressure at point 2 is

$$p_2 = \dfrac{1}{2}\rho(v_3^2 - v_2^2) = \dfrac{1}{2}\rho v_3^2 \left(1 - \left(\dfrac{A_3}{A_2}\right)^2\right) = \dfrac{8}{9}\rho g(y_1 - y_3),$$ using the expression for v_3 found above.

Substitution of numerical values gives $p_2 = 6.97 \times 10^4$ Pa.

EVALUATE: We could also calculate p_2 by applying Bernoulli's equation to points 1 and 2.

12.83. **IDENTIFY:** Apply Bernoulli's equation and the equation of continuity.

SET UP: The speed of efflux at point D is $\sqrt{2gh_1}$.

EXECUTE: Applying the equation of continuity to points at C and D gives that the fluid speed is $\sqrt{8gh_1}$ at C. Applying Bernoulli's equation to points A and C gives that the gauge pressure at C is $\rho g h_1 - 4\rho g h_1 = -3\rho g h_1$, and this is the gauge pressure at the surface of the fluid at E. The height of the fluid in the column is $h_2 = 3h_1$.

EVALUATE: The gauge pressure at C is less than the gauge pressure $\rho g h_1$ at the bottom of tank A because of the speed of the fluid at C.

12.85. **IDENTIFY AND SET UP:** We are given the densities of elements in the table and look up their atomic masses in Appendix D.

EXECUTE: For example, for aluminum, the density is 2.7 g/cm^3 and the atomic mass is 26.98 g/mol.
(a) Figure 12.85 shows the graph of density versus atomic mass.

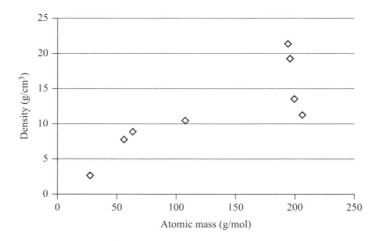

Figure 12.85

(b) From the graph, we see that there is no obvious mathematical relation between the two variables. No straight line or simple curve can be fitted to the data points.
(c) Density depends not only on atomic mass, but also on how tightly atoms are packed together. This packing is determined by the electrical interactions between atoms.
EVALUATE: Not all data can be reduced to straight-line graphs!

12.87. **IDENTIFY:** Bernoulli's equation applies. We have free-fall projectile motion after the liquid leaves the tank. The pressure at the hole where the liquid exits is atmospheric pressure p_0. The absolute pressure at the top of the liquid is $p_g + p_0$.

SET UP: $p_1 + \frac{1}{2}\rho v_1^2 + \rho g y_1 = p_2 + \frac{1}{2}\rho v_2^2 + \rho g y_2.$

EXECUTE: **(a)** The graph of R^2 versus p_g is shown in Figure 12.87.

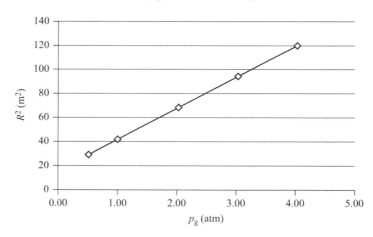

Figure 12.87

Applying Bernoulli's equation between the top and bottom of the liquid in the tank gives

$p_0 + p_g + \rho g h = \tfrac{1}{2}\rho v^2 + p_0$, which simplifies to $p_g + \rho g h = \tfrac{1}{2}\rho v^2$.

The free-fall motion after leaving the tank gives $vt = R$ and $y = \tfrac{1}{2}gt^2$, where $y = 50.0$ cm. Eliminating t between these two equations gives $v^2 = (\rho g/4y)R^2$. Putting this into the result from Bernoulli's equation gives $p_g + \rho g h = (\rho g/4y)R^2$. Solving for R^2 in terms of h gives $R^2 = (4y/\rho g)\,p_g + 4yh$.

This is the equation of a straight line of slope $4y/\rho g$, which gives $\rho = 4y/[g(\text{slope})]$ and y-intercept $4yh$. The best-fit equation is $R^2 = (25.679\ \text{m}^2/\text{atm})\,p_g + 16.385\ \text{m}^2$. The y-intercept gives us h:

$4yh = y$-intercept, so $h = (y\text{-intercept})/(4y) = (16.385\ \text{m}^2)/[4(0.500\ \text{m})] = 8.2$ m. And the density is $\rho = 4y/[g(\text{slope})] = 4(0.500\ \text{m})/[(9.80\ \text{m/s}^2)(25.679\ \text{m}^2/\text{atm})(1\ \text{atm}/1.01 \times 10^5\ \text{Pa})] = 803\ \text{kg/m}^3$.

EVALUATE: The liquid is about 80% as dense as water, and $h = 8.2$ m which is about 25 ft, so this is a rather large tank.

STUDY GUIDE FOR GRAVITATION

Summary

In this chapter, we'll delve into the gravitational interaction by learning that the gravity we experience on earth also applies to planets and celestial objects and is responsible for their motion. We'll see how to apply Newton's law of gravitation and gain a better understanding of the concept of weight. We'll use this knowledge to explain the orbits of satellites and planets. We'll also examine an extreme case of gravity: black holes.

Objectives

After studying this chapter, you'll understand

- How to apply Newton's law of gravitation to pairs of masses.
- The general definition of *weight*.
- How to use the generalized expression for gravitational potential energy.
- How satellites orbit astronomical bodies.
- How to predict the motion of satellites.
- Kepler's three laws of planetary motion.
- The definition of a *black hole* and the properties of black holes.

Concepts and Equations

Term	Description
Newton's Law of Gravitation	Newton's law of gravitation states that the magnitude of the force between two objects with masses m_1 and m_2, separated by a distance r, is given by $$F_g = G\frac{m_1 m_2}{r^2},$$ where G denotes the gravitational constant and is equal to $6.67 \times 10^{-11} \, \text{N} \cdot \text{m}^2/\text{kg}^2$. The gravitational force is always attractive and is directed along the line that separates the objects.
Weight	The weight of an object is the total gravitational force exerted on the object by all other objects in the universe. Near the surface of the earth, an object's weight is very nearly equal to the gravitational force of the earth on the object alone.
Gravitational Potential Energy	The gravitational potential energy of two objects with masses m and m_E, separated by a distance r, is given by $$U = -\frac{G m_E m}{r}.$$ The potential energy is never positive and is zero only when the two objects are infinitely far apart.
Orbits of Satellites	For a satellite moving in a circular orbit, the gravitational attraction between the satellite and the astronomical body provides the centripetal acceleration. The velocity v and period T of a satellite orbiting at a radius r are given, respectively, by $$v = \sqrt{\frac{GM}{r}},$$ $$T = \frac{2\pi r^{3/2}}{\sqrt{GM}},$$ where M is the mass of the astronomical body.
Kepler's Laws	Kepler's three laws describe the motion of a planet or satellite around the sun or another planet. They describe the elliptical motion and the area swept out per unit time in the orbit, as well as relate the period of the planet or satellite to the major axis of its orbit.
Black Holes	A black hole is a nonrotating spherical mass distribution with total mass M contained within a radius R_S, the Schwarzschild radius, given by $$R_S = \frac{2GM}{c^2}.$$ Gravity prevents matter and light from escaping from within a sphere with radius R_S.

Key Concept 1: Any two objects exert attractive gravitational forces on each other that are proportional to the product of the masses of the two objects and inversely proportional to the square of the distance between their centers (Newton's law of gravitation).

Key Concept 2: The gravitational forces that two objects exert on each other are always equal in magnitude. The accelerations caused by these forces, however, will be different in magnitude for the two objects if they have different masses.

Key Concept 3: When there are two objects that exert gravitational forces on a third object, use vector addition to find the magnitude and direction of the net gravitational force on the third object.

Key Concept 4: An object's weight at the surface of a planet equals the gravitational force exerted on the object by that planet. The distance between the planet and the object equals the planet's radius. The acceleration due to gravity at a planet's surface is proportional to the planet's mass and inversely proportional to the square of its radius.

Key Concept 5: When an object moves far away from a planet's surface, use Eq. (13.11) for the gravitational potential energy. The potential energy for the system of planet and object increases as the object moves farther away; it reaches its maximum value of zero when the object is infinitely far away.

Key Concept 6: The larger the radius of a satellite's orbit, the greater (less negative) the total mechanical energy, the greater (less negative) the potential energy, and the smaller the kinetic energy (the satellite moves more slowly in a larger orbit). The radius of a satellite's orbit is always measured from the center of the earth.

Key Concept 7: Total mechanical energy remains constant for a planet in an elliptical orbit. The planet's kinetic energy (and speed) therefore increases as it moves closer to the sun and the gravitational potential energy decreases (becomes more negative).

Key Concept 8: Kepler's third law allows you to relate the period of an orbit, the semi-major axis of that orbit, and the mass of the object being orbited.

Key Concept 9: Kepler's third law applies to orbits of any eccentricity e between 0 and 1. You can determine e from the semi-major axis of the orbit and the distance at perihelion from the orbiting object to the object being orbited.

Key Concept 10: The gravitational force on a particle at a point a radius r from the center of a spherically symmetric mass distribution is caused solely by the mass *inside* a sphere of radius r. The mass outside radius r has no effect on the particle.

Key Concept 11: The event horizon of a black hole has a radius of R_S, the Schwarzschild radius, which is proportional to the mass of the black hole. Nothing can escape from inside the event horizon.

Conceptual Questions

1: Is the earth falling?

There is a net gravitational force between the earth and the sun, so why doesn't the earth fall into the sun?

IDENTIFY, SET UP, AND EXECUTE The earth is constantly falling toward the sun, but the earth doesn't get closer to the sun, since the sun's surface curves away beneath the earth. If we launch an object parallel to the ground, it follows a parabolic path. If we give the object a larger initial velocity, then the object moves farther away from the launch site as it falls. The earth is round, so as the object moves farther away, the object will have a larger distance to fall. At a sufficiently high launch velocity, the object will make a complete revolution and not land on the ground. This is the same physical situation as the earth revolving around the sun. If the earth had a lower velocity, it would fall into the sun.

EVALUATE This result may seem a bit odd, but is indeed accurate. The result also shows how our understanding of one physical phenomenon helps us understand other phenomena: Our experience with projectile motion helped us interpret the motion of the earth around the sun.

2: Does the earth maintain a constant speed?

In the previous question, we saw that the earth is constantly falling. Does it maintain a constant speed?

IDENTIFY, SET UP, AND EXECUTE There is a net gravitational force acting on the earth due to the sun. The direction of the net external force is toward the sun; however, the earth's velocity is perpendicular to the direction of force. The gravitational force can change only the direction of the earth's velocity around the sun and not the magnitude of the velocity. Thus, the earth maintains a constant speed as it orbits the sun. The earth does *not* maintain a constant velocity, since its direction is always changing.

EVALUATE We've come to know that a net external force causes acceleration—a change in velocity. The magnitude of an object's velocity changed when the object was acted upon by a net force. This chapter examines additional consequences of the influences of forces.

3: Two satellites and the earth

The moon and the international space station are located on opposite sides of the earth. How does the presence of the earth influence the gravitational force between the moon and the space station?

IDENTIFY, SET UP, AND EXECUTE The gravitational force between two objects depends only on the mass of the objects and their separation, according to Newton's law of gravitation. The force between the moon and the space station is not affected by the presence of the earth. There are forces between the earth and the two orbiting satellites, but those forces do not affect the force between the satellites.

EVALUATE Forces are between two objects. The net external force on a single object may include forces due to many objects, but each force acts between two objects.

Problems

1: Gravitational force due to three masses

Three masses are arranged as shown in Figure 1. Find the net external force acting on the top mass (*A*). Each mass is 5.00 kg.

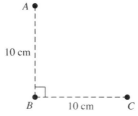

Figure 1 Problem 1.

IDENTIFY The net external force on *A* is found by adding the force on *A* due to *B* and the force on *A* due to *C*. The target variable is the force on *A*.

SETUP A free-body diagram illustrating the two forces on *A* is shown in Figure 2. We'll need to add the two forces by using components. Newton's law of gravitation gives the magnitude of the forces. We'll use the coordinate axes provided in the figure.

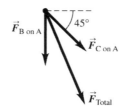

Figure 2 Problem 1 free-body diagram.

EXECUTE To apply Newton's law of gravity, we need the distances between the masses. The distance between A and B is 10.0 cm. By summing the squares of the sides of the triangle and taking the square root, we find that the distance between A and C is 14.1 cm. The force of B on A is

$$F_{B\text{ on }A} = \frac{Gm_B m_A}{r_{BA}^2} = \frac{(6.67\times10^{-11}\text{ N}\cdot\text{m}^2/\text{kg}^2)(5.00\text{ kg})(5.00\text{ kg})}{(0.100\text{ cm})^2} = 1.67\times10^{-7}\text{ N}.$$

The force of C on A is

$$F_{C\text{ on }A} = \frac{Gm_C m_A}{r_{CA}^2} = \frac{(6.67\times10^{-11}\text{ N}\cdot\text{m}^2/\text{kg}^2)(5.00\text{ kg})(5.00\text{ kg})}{(0.141\text{ cm})^2} = 8.39\times10^{-8}\text{ N}.$$

With the magnitudes of the forces determined, we simply add the two vectors together, using components. The force of C on A has the only x- component:

$$\Sigma F_x = F_{C\text{ on }A}\cos 45° = 5.93\times10^{-8}\text{ N}.$$

The 45° angle results from the masses arranged as an isosceles triangle. Both forces have y- components:

$$\Sigma F_y = -F_{B\text{ on }A} - F_{C\text{ on }A}\sin 45° = -2.26\times10^{-7}\text{ N}.$$

The negative result indicates that the y- component points downward. We find the magnitude of the net force by combining the components:

$$F = \sqrt{F_x^2 + F_y^2} = 2.34\times10^{-7}\text{ N}.$$

The direction of the net external force is found by using the tangent. We want to specify the angle ϕ with respect to the x-axis:

$$\phi = \tan^{-1}\frac{F_y}{F_x} = \tan^{-1}\frac{(-2.26\times10^{-7}\text{ N})}{(5.93\times10^{-8}\text{ N})} = -75.3°.$$

The net external force on A has magnitude 2.34×10^{-7} N and points 75.3° below the positive x-axis.

KEY CONCEPT 3 **EVALUATE** We see that the gravitational force between the masses is very small. To have an appreciable gravitational force, we need at least one large mass, such as the earth. Also, we can see that Newton's third law is valid: Reversing indices in the first two equations would result in a force of the same magnitude, but opposite in direction.

Practice Problem: What is the magnitude of the net external force on A if mass B is doubled? *Answer:* 3.98×10^{-7} N

Extra Practice: What is the angle of this force? *Answer:* 81.4° below the positive x-axis

2: Orbit of a weather satellite

Imagine you are designing a new weather satellite. The goal is to have the satellite orbit the earth in a circular orbit every 6 hours. At what distance above the earth's surface should the satellite be placed to obtain the correct period?

IDENTIFY The force acting on the satellite is the force of gravitation between the satellite and the earth. The satellite follows a circular orbit and so has a radial acceleration toward the center of the earth. The target variable is the height of the satellite's orbit.

SET UP Newton's law of gravitation gives the force on the satellite due to the earth. The acceleration of the satellite is centripetal. Combining the two equations will lead to the velocity of the satellite and the period of rotation. We solve for the distance by setting the period to 6 hours.

EXECUTE Newton's law of gravitation gives the force on the satellite, namely,

$$F_g = \frac{Gmm_E}{r^2},$$

where m is the mass of the satellite, m_E is the mass of the earth, and r is the distance from the center of the earth to the satellite. Newton's second law gives the net external force on the satellite (the acceleration is v^2/r):

$$\Sigma F = ma_{\text{rad}}, \Sigma F = F_g = \frac{Gmm_E}{r^2} = m\frac{v^2}{r}.$$

Solving for v, we find that

$$v = \sqrt{\frac{Gm_E}{r}}.$$

We can also write the velocity in terms of the distance the satellite travels $(2\pi r)$ in one period (T):

$$v = \frac{2\pi r}{T}.$$

To find the radius, we equate the last two equations and solve for r, obtaining

$$r = \sqrt[3]{\frac{Gm_E T^2}{4\pi^2}} = \sqrt[3]{\frac{(6.67 \times 10^{2\,11}\ \text{N} \cdot \text{m}^2/\text{kg}^2)(5.98 \times 10^{24}\ \text{kg})(2.16 \times 10^4\ \text{s})^2}{4\pi^2}} = 1.68 \times 10^7\ \text{m},$$

where we replaced the 6 hour period with the equivalent 21,600 s. The satellite should be placed in an orbit of radius 16,800 km. Subtracting the radius of the earth (6380 km) from the radius of the satellite's orbit, we find that the satellite should be placed 10,400 km above the earth's surface to achieve a 6 hour orbital period.

KEY CONCEPT 6 **EVALUATE** We could've avoided our derivation and used Eq. (13.12) in the chapter to arrive at the solution directly. However, this review helps remind us how to find the solution without searching the text.

Practice Problem: What does the free-body diagram of the satellite look like?
Answer: A single vector

Extra Practice: What distance do you need if you want the orbit to take 12 hours?
Answer: 20,200 km

3: Velocity of rocket

What velocity must a rocket have at the surface of the earth if it is to rise to a height equal to the earth's radius before it begins to descend? Ignore air resistance.

IDENTIFY We'll use energy conservation to solve the problem. The target variable is the initial velocity of the rocket.

SET UP The rocket is given an initial kinetic energy for lift-off and has initial gravitational potential energy. At the top of the flight, its kinetic energy drops to zero, leaving only gravitational potential energy. We'll set these equal to each other to solve for the initial velocity.

EXECUTE Energy conservation gives
$$K_1 + U_1 = K_2 + U_2.$$
Replacing both sides with the expressions for the two forms of energy gives
$$\frac{1}{2}mv^2 - \frac{Gmm_E}{r_E} = -\frac{Gmm_E}{2r_E},$$
where r_E is the radius of the earth and $K_2 = 0$. Solving for the initial velocity, we obtain
$$v = \sqrt{\frac{Gm_E}{r_E}} = \sqrt{\frac{r_{earth}^2 g}{r_E}} = \sqrt{gr_E} = \sqrt{(9.8 \text{ m/s}^2)(6.36 \times 10^6 \text{ m})} = 7.9 \times 10^3 \text{ m/s}.$$

The initial velocity of the rocket must be 7.9×10^3 m/s.

KEY CONCEPT 5 **EVALUATE** By examining the solution to the problem, we see why rockets must have large initial velocities to be propelled into space. The actual value is higher due to air resistance.

Practice Problem: What velocity do you need to have if you want to triple that height?
Answer: 9700 m/s

Extra Practice: How far above the surface of the earth do you get if your initial velocity is only 4.0×10^3 m/s? *Answer:* 1280 km

4: What if the sun were a black hole?

What would the sun's radius need to be in order for the surface escape velocity to be c?

IDENTIFY AND SET UP A radius corresponding to an escape velocity of c is the Schwarzschild radius. We can use the expression for the Schwarzschild radius to solve the problem.

EXECUTE The Schwarzschild radius is given by
$$R_S = \frac{2GM}{c^2}.$$
Substituting gives
$$R_S = \frac{2GM}{c^2} = \frac{2(6.67 \times 10^{-11} \text{ Nm}^2/\text{kg}^2)(1.99 \times 10^{30} \text{ kg})}{(3.0 \times 10^8 \text{ m/s})^2} = 2960 \text{ m}.$$

The radius would be 2960 m.

KEY CONCEPT 11 **EVALUATE** The sun would have to be compressed by a factor of over 200,000 to become a black hole.

Practice Problem: What is the Schwarzschild radius for the earth? *Answer:* 8.8 mm

Extra Practice: What is the Schwarzschild radius for a 60 kg student? *Answer:* 8.9×10^{-26} m

Try It Yourself!

1: Sun's gravity on earth

The sun is a distance of 1.48×10^{11} m from earth and has a mass of 1.99×10^{30} kg. Find the ratio of the sun's gravitational force to the earth's gravitational force on an object on the earth's surface.

Solution Checkpoints

IDENTIFY AND SET UP Newton's law of gravitation is used to find the force of gravity due to the earth and the force of gravity due to the sun. Taking their ratio solves the problem.

EXECUTE Substituting the values given results in a ratio of 6.03×10^{-4}.

KEY CONCEPT 1 **EVALUATE** Do you need to include the force of gravity due to the sun in physics problems on earth?

Practice Problem: Repeat the problem with the moon replacing the sun? *Answer:* 3.37×10^{-6}

Extra Practice: Repeat the problem for when Jupiter is as close as it can get to the earth? *Answer:* 3.26×10^{-8}

2: Three masses positioned in a triangle

Three 1000.0 kg masses are at the vertices of an equilateral triangle with sides of length 1.0 m. Find the force due to any two masses on the third.

Solution Checkpoints

IDENTIFY AND SET UP Newton's law of gravitation is used to find the force of gravitation due to the other masses. The two forces must be added as vectors. A free-body diagram should be used.

EXECUTE The force between two masses is

$$F_g = \frac{Gm_1 m_2}{r^2} = 6.67 \times 10^{-5} \text{ N}.$$

The net external force is 1.16×10^{-4} N, directed toward the line separating the other two masses.

KEY CONCEPT 3 **EVALUATE** Did you use symmetry arguments to determine that the force has no component parallel to the line separating the two masses?

Practice Problem: What would the force be if all masses were doubled? *Answer:* 4.62×10^{-4} N

Extra Practice: What is the total force on the mass at the top of the triangle if it is the only mass that is doubled? *Answer:* 2.31×10^{-4} N

3: Orbit of a communications satellite

Communications satellites revolve in orbits over the earth's equator, adjusted so that their period of rotation is the same as the period of rotation of the earth about its axis. This speed and a period of rotation together cause the satellite to remain in a fixed position in the sky. Find the height of these satellites' orbit above the earth.

Solution Checkpoints

IDENTIFY AND SET UP You can find the height by setting the period equal to 24 hours. The solution can be found by solving Newton's second law.

EXECUTE The period is given by

$$T = \frac{2\pi r^{3/2}}{\sqrt{gR^2}}.$$

The terms in this equation can be rearranged to solve for r. The height above earth is r minus the radius of earth. The satellite must be placed 36,000 km above the surface of the earth to remain in geosynchronous orbit.

KEY CONCEPT 6   **EVALUATE** Such heights require additional power for radio signals to reach the satellites and for the signal to be redirected back to earth.

Practice Problem: What is the speed of one of these satellites? *Answer:* 3080 m/s

Extra Practice: How long would it take a radio signal to travel to the satellite and then return (note: radio signals travel at the speed of light)? *Answer:* 0.24 seconds

4: Escape from the sun

What is the escape velocity of a particle on the surface of the sun?

Solution Checkpoints

IDENTIFY AND SET UP The escape velocity can be found from energy conservation or the equation given in the text.

EXECUTE The escape velocity is given by

$$v = \sqrt{\frac{2GM}{r}}.$$

After substituting, we find the escape velocity to be 6.48×10^5 m/s.

KEY CONCEPT 11   **EVALUATE** Is the resulting velocity greater or less than the escape velocity of a particle on the surface of the earth? Why?

Practice Problem: What is the escape velocity of a particle on the earth? *Answer:* 11,200 m/s

Extra Practice: What is the escape velocity of a particle on the moon? *Answer:* 2370 m/s

Key Example Variation Problems

Solutions to these problems are in Chapter 13 of the Student's Solutions Manual.

Be sure to review EXAMPLE 13.5 (Section 13.3) before attempting these problems.

VP13.5.1 The *New Horizons* spacecraft (mass 478 kg) was launched from earth in 2006 and flew past Pluto in 2015. What minimum amount of work has to be done on a spacecraft of this mass to send it from the earth's surface to infinitely far away from the earth? Neglect air resistance and the gravitational effects of the sun, moon, and other planets.

VP13.5.2 A piece of spacecraft debris initially at rest falls to the earth's surface from a height above the earth equal to one-half of the earth's radius. Find the speed at which the piece of debris hits the surface. Neglect air resistance and the gravitational pull of the moon.

VP13.5.3 An astronaut on Mars (mass 6.42×10^{23} kg, radius 3.9×10^6 m) launches a probe straight upward from the surface at 3.00×10^3 m/s. What is the maximum height above the surface that the probe reaches? Neglect air resistance and the gravitational pull of the two small moons of Mars.

VP13.5.4 In order to rendezvous with an asteroid passing close to the earth, a spacecraft must be moving at 8.50×10^3 m/s relative to the earth at a distance of 2.50×10^8 m from the center of the earth. At what speed must this spacecraft be launched from the earth's surface? Neglect air resistance and the gravitational pull of the moon.

Be sure to review EXAMPLE 13.6 (Section 13.4) before attempting these problems.

VP13.6.1 You wish to place a spacecraft in a circular orbit around the earth so that its orbital speed will be 4.00×10^3 m/s. What are this orbit's (a) radius, (b) altitude above the earth's surface, and (c) period (in hours)?

VP13.6.2 You are designing a spacecraft intended to monitor a human expedition to Mars (mass 6.42×10^{23} kg, radius 3.39×10^6 m). This spacecraft will orbit around the Martian equator with an orbital period of 24.66 h, the same as the rotation period of Mars, so that it will always be above the same point on the equator. (a) What must be the radius of the orbit? (b) What will be the speed of the spacecraft in its orbit?

VP13.6.3 A spacecraft of mass 1.00×10^3 kg orbits the sun (mass 1.99×10^{30} kg) in a circular orbit of radius 1.50×10^{11} m (equal to the average distance from the sun to the earth). You wish to move the spacecraft into a smaller circular orbit around the sun of radius 1.08×10^{11} m (equal to the average distance from the sun to Venus). In doing this, what will be the changes in (a) the spacecraft's kinetic energy, (b) the gravitational potential energy of the sun–spacecraft system, and (c) the total mechanical energy of the sun–spacecraft system? Neglect the gravitational pulls of the planets on the spacecraft.

VP13.6.4 A satellite of mass 1.50×10^3 kg is at point P, at an altitude of 3.50×10^6 m above the earth's surface, and traveling at 7.50×10^3 m/s. (a) If you wanted to put the satellite into a circular orbit at this altitude above the earth's surface, how much work would you have to do on it at point P? (b) If instead you wanted to make the satellite escape the earth, how much work would you have to do on it at point P?

Be sure to review EXAMPLES 13.7, 13.8, and 13.9 (Section 13.5) before attempting these problems.

VP13.9.1 Some comets are in highly elongated orbits that come very close to the sun at perihelion. The distance from one such comet to the center of the sun is 6.00×10^9 m at perihelion and 3.00×10^{12} m at aphelion. For this comet's orbit, find (a) the semi-major axis, (b) the eccentricity, and (c) the period (in years).

VP13.9.2 The orbit of a certain asteroid around the sun has period 7.85 y and eccentricity 0.250. Find (a) the semi-major axis and the distances from the sun to the asteroid at (b) perihelion and (c) aphelion.

VP13.9.3 The minor planet (33340) 1998 VG44 makes exactly two orbits of the sun during the time that Neptune makes three orbits. The eccentricity of the orbit of this minor planet is 0.25, and the orbit radius of Neptune is 4.5×10^{12} m. (a) Find the semi-major axis of this minor planet. (b) Find the distance between the sun and the minor planet at perihelion. At perihelion, is this minor planet inside or outside the orbit of Neptune?

VP13.9.4 In 2017 astronomers discovered a planet orbiting the star HATS-43. The orbit of the planet around HATS-43 has semi-major axis 7.41×10^9 m, eccentricity 0.173, and period 4.39 days. Find (a) the distance between HATS-43 and the planet at its closest approach and (b) the mass of HATS-43. (For comparison, the distance between the sun and Mercury at its closest approach is 4.60×10^{10} m and the mass of the sun is 1.99×10^{30} kg.)

STUDENT'S SOLUTIONS MANUAL FOR GRAVITATION

VP13.5.1. **IDENTIFY:** This problem involves gravitational potential energy.
SET UP: The minimum amount of work is just the magnitude of the spacecraft's initial potential energy at the earth's surface. $U_{grav} = -Gm_1m_2/r$.
EXECUTE: $U_{grav} = -Gm_1m_2/r = -Gm_s m_E/R_E$. The magnitude of this energy is equal to the work, so
$$W = \frac{(6.67 \times 10^{-11} \text{ N} \cdot \text{m}^2/\text{kg}^2)(478 \text{ kg})(5.97 \times 10^{24} \text{ kg})}{6.37 \times 10^6 \text{ m}} = 2.99 \times 10^{10} \text{ J.}$$
EVALUATE: Technically this is the energy given to the earth-satellite system, but it essentially all goes to the satellite because the earth is so massive that it does not change its kinetic energy.

VP13.5.2. **IDENTIFY:** This problem involves gravitational potential energy. Energy conservation applies.
SET UP: No air resistance, so $K_1 + U_1 = K_2 + U_2$, $U_{grav} = -Gm_1m_2/r$, $K = \frac{1}{2}mv^2$. Let m be the mass of the debris, R the radius of the earth, and m_E its mass. Call point 1 the original location of the debris and point 2 the surface of the earth.
EXECUTE: $K_1 = 0$, so $-Gmm_E/(3R/2) = -Gmm_E/R + \frac{1}{2}mv^2$, which gives $v = \sqrt{\frac{2Gm_E}{3R}}$. Using $R = 6.37 \times 10^6$ m and $m_E = 5.97 \times 10^{24}$ kg gives $v = 6.46 \times 10^3$ m/s.
EVALUATE: The equation $\sqrt{\frac{2Gm_E}{3R}}$ does not contain the mass of the satellite, so *any* object dropped from the same height would reach the earth's surface with the same speed.

VP13.5.3. **IDENTIFY:** This problem involves gravitational potential energy. Energy conservation applies.
SET UP: No air resistance, so $K_1 + U_1 = K_2 + U_2$, $U_{grav} = -Gm_1m_2/r$, $K = \frac{1}{2}mv^2$. Let m be the mass of the probe, R the radius of Mars, and M its mass. Call point 1 at the surface of Mars and point 2 at the maximum height of the probe. At that point, the probe's distance from the center of Mars is h and its speed is zero. Call v the speed at the surface.
EXECUTE: $K_1 + U_1 = K_2 + U_2$ gives $\frac{1}{2}mv^2 - \frac{GmM}{R} = -\frac{GmM}{h}$. Using $M = 6.42 \times 10^{23}$ kg, $R = 3.39 \times 10^6$ m, and $v = 3.00 \times 10^3$ m/s, we find $h = 5.27 \times 10^6$ m from the center of Mars. The height H above the surface is $H = h - R = 5.27 \times 10^6$ m $- 3.39 \times 10^6$ m $= 1.88 \times 10^6$ m.
EVALUATE: The answer is independent of the mass of the probe since m cancels from the equations, so *any* probe would reach that height if it started at 3000 m/s.

VP13.5.4. **IDENTIFY:** This problem involves gravitational potential energy. Energy conservation applies.

SET UP: No air resistance, so $K_1 + U_1 = K_2 + U_2$, $U_{grav} = -Gm_1m_2/r$, $K = \frac{1}{2}mv^2$. Let m be the mass of the spacecraft, R the radius of the earth and M its mass. Call point 1 at the surface of the earth and point 2 at the height of the spacecraft when it has reached the desired speed. Call v_1 its initial speed at the surface and v_2 the desired speed when it has reached the desired height r from the earth's center. We want to find v_1.

EXECUTE: Energy conservation gives $\frac{1}{2}mv_1^2 - \frac{GmM}{R} = \frac{1}{2}mv_2^2 - \frac{GmM}{r}$. Using $v_2 = 8.50 \times 10^3$ m/s, $r = 2.50 \times 10^8$ m, $M = 5.97 \times 10^{24}$ kg, and $R = 6.37 \times 10^6$ m, we get $v_1 = 1.39 \times 10^4$ m/s = 13.9 km/s.

EVALUATE: Any spacecraft, regardless of mass, would need the same launch speed since the mass m cancels from the equations.

VP13.6.1. **IDENTIFY:** This problem deals with satellite orbits. The force of gravity applies, as well as Newton's second law for circular motion.

SET UP: $F = Gm_1m_2/r^2$. $\Sigma F = m\frac{v^2}{r}$ for circular motion. Let m be the satellite mass, v its speed, and M the earth's mass. $M = 5.97 \times 10^{24}$ kg and $v = 4.00 \times 10^3$ m/s.

EXECUTE: (a) $\Sigma F = m\frac{v^2}{r}$ gives $\frac{GmM}{r^2} = \frac{mv^2}{r} \rightarrow r = \frac{GM}{v^2} = 2.49 \times 10^7$ m.

(b) The height H above the surface is $H = r - R = 2.49 \times 10^7$ m $- 6.37 \times 10^6$ m $= 1.85 \times 10^7$ m.
(c) $v = 2\pi r/T$, so $T = 2\pi r/v = 2\pi(2.49 \times 10^7$ m$)/(4.00 \times 10^3$ m/s$) = 3.91 \times 10^4$ s $= 10.9$ h.

EVALUATE: The results are independent of the mass of the satellite.

VP13.6.2. **IDENTIFY:** This problem deals with satellite orbits. The force of gravity applies, as well as Newton's second law for circular motion.

SET UP: $F = Gm_1m_2/r^2$. $\Sigma F = m\frac{v^2}{r}$ for circular motion. Let m be the satellite mass, v its speed, T its orbital period, and M the mass of Mars.

EXECUTE: (a) We want the radius r of the orbit. $\Sigma F = m\frac{v^2}{r}$ gives $\frac{GmM}{r^2} = \frac{m\left(\frac{2\pi r}{T}\right)}{r}$. Solving for r gives $r = \left(\frac{GMT^2}{4\pi^2}\right)^{1/3}$. Using $M = 6.42 \times 10^{23}$ kg and $T = 24.66$ h $= 8.8776 \times 10^4$ s gives $r = 2.04 \times 10^7$ m.

(b) Using the $r = 2.04 \times 10^7$ m and $R = 3.39 \times 10^6$ m gives $v = 2\pi r/T = 1.45 \times 10^3$ m/s.

EVALUATE: The orbital radius does not depend on the mass of the satellite, but it *does* depend on the mass of the planet. On the earth the radius would be larger since $r \propto M^{1/3}$.

VP13.6.3. **IDENTIFY:** This problem deals with orbits around the sun and involves gravitational potential energy and kinetic energy. Newton's second law also applies.

SET UP: $U_{grav} = -Gm_1m_2/r$, $K = \frac{1}{2}mv^2$. Let m be the mass of the spacecraft, M the mass of the sun, r_1 the radius of the outer orbit and r_2 that of the inner orbit. Apply $\Sigma F = m\frac{v^2}{r}$ to the spacecraft.

EXECUTE: (a) First find the kinetic energy in both orbits. $\Sigma F = m\frac{v^2}{r}$ to the spacecraft gives

$\dfrac{GmM}{r^2} = \dfrac{mv^2}{r} = \dfrac{2}{r}\left(\dfrac{1}{2}mv^2\right) = \dfrac{2K}{r}$, so $K = \dfrac{GmM}{2r}$. Therefore change in kinetic energy is

$\Delta K = K_2 - K_1 = \dfrac{GmM}{2}\left(\dfrac{1}{r_2} - \dfrac{1}{r_1}\right)$. Using $m = 1000$ kg, $M = 1.99 \times 10^{30}$ kg, $r_1 = 1.50 \times 10^{11}$ m, and $r_2 = 1.08 \times 10^{11}$ m, we get $\Delta K = 1.72 \times 10^{11}$ J.

(b) $U_2 - U_1 = -\dfrac{GmM}{r_2} - \left(-\dfrac{GmM}{r_1}\right) = GmM\left(\dfrac{1}{r_1} - \dfrac{1}{r_2}\right) = -3.44 \times 10^{11}$ J.

(c) $E = K + U$, so $\Delta E = \Delta K + \Delta U = 1.72 \times 10^{11}$ J $+ (-3.44 \times 10^{11}$ J$) = -1.72 \times 10^{11}$ J.

EVALUATE: The spacecraft lost twice as much potential energy as it gained in kinetic energy, so it had a net loss equal to the kinetic energy gain.

VP13.6.4 **IDENTIFY:** This problem deals with orbits around the sun and involves gravitational potential energy and kinetic energy. Newton's second law also applies.

SET UP: $U_{grav} = -Gm_1m_2/r$, $K = \dfrac{1}{2}mv^2$. Let m be the mass of the spacecraft and M the mass of the earth. Apply $\sum F = m\dfrac{v^2}{r}$ to the spacecraft.

EXECUTE: **(a)** The work you would need to do is equal to the change in energy of the satellite. The potential energy does not change because r is the same, so the change in energy is the change in the kinetic energy of the satellite. Apply $\sum F = m\dfrac{v^2}{r}$ for a circular orbit.

6.37×10^6 m $= \dfrac{2}{r}\left(\dfrac{1}{2}mv^2\right) = \dfrac{2K}{r}$, which gives $K = \dfrac{GmM}{2r}$.

$\Delta K = K_2 - K_1 = \dfrac{GmM}{2r} - \dfrac{1}{2}mv_1^2$, where v_1 is the initial speed. Using $m = 1500$ kg, $M = 5.97 \times 10^{24}$ kg, $v_1 = 7500$ m/s, and $r = 3.50 \times 10^6$ m $+ 6.37 \times 10^6$ m $= 9.87 \times 10^6$ m, we find that $\Delta E = \Delta K = -1.19 \times 10^{10}$ J.

(b) The minimum energy you would have to give the satellite for it to escape earth would be its total energy, which is $E_1 = K_1 + U_1 = \dfrac{1}{2}mv_1^2 - \dfrac{GmM}{r}$. Using the numbers from part (a), this energy is $E_1 = -1.83 \times 10^{10}$ J. The work you would have to would be $+1.83 \times 10^{10}$ J.

EVALUATE: Orbital parameters such as period, speed, and radius do not depend on the mass of the satellite, but energies do depend on that mass.

VP13.9.1. **IDENTIFY:** We are dealing with a comet in noncircular orbit.

SET UP: The semi-major axis a is $2a = r_p + r_a$, where r_p is the perihelion distance and r_a is the aphelion distance. $T = \dfrac{2\pi a^{3/2}}{\sqrt{Gm_S}}$, where m_S is the mass of the sun (or central star for another solar system). $a = ea + r_p$, where e is the eccentricity of the orbit. Refer to Fig. 13.18 in the textbook.

EXECUTE: **(a)** We want a, so we use $2a = = r_p + r_a = 6.00 \times 10^9$ m $+ 3.00 \times 10^{12}$ m, so $a = 1.50 \times 10^{12}$ m.

(b) $a = ea + r_p$, so $e = (a - r_p)/a = 1 - r_p/a = 1 - (6.00 \times 10^9$ m$)/(1.50 \times 10^{12}$ m$) = 0.996$.

(c) Use $T = \dfrac{2\pi a^{3/2}}{\sqrt{Gm_S}}$ with a from (a) and $m_S = 1.99 \times 10^{30}$ kg gives $T = 1.00 \times 10^9$ s $= 31.8$ y.

EVALUATE: This orbital period is roughly the same as that of Saturn.

VP13.9.2. **IDENTIFY:** We are dealing with an asteroid in a noncircular orbit.
SET UP: The semi-major axis a is $2a = r_p + r_a$, where r_p is the perihelion distance and r_a is the aphelion distance. $T = \dfrac{2\pi a^{3/2}}{\sqrt{Gm_S}}$, where m_S is the mass of the sun (or central star for another solar system). $a = ea + r_p$, where e is the eccentricity of the orbit. Refer to Fig. 13.18 in the textbook. $T = 7.85$ y $= 2.4775 \times 10^8$ s, and $e = 0.250$.

EXECUTE: **(a)** Use $T = \dfrac{2\pi a^{3/2}}{\sqrt{Gm_S}}$ to find a. Solve for a: $a = \left[\dfrac{T^2 Gm_S}{4\pi^2}\right]^{1/3}$. Using T from above and $m_S = 1.99 \times 10^{30}$ kg gives $a = 5.91 \times 10^{11}$ m.
(b) $a = ea + r_p \rightarrow r_p = a(1 - e) = (5.91 \times 10^{11}\text{ m})(1 - 0.250) = 4.43 \times 10^{11}$ m.
(c) $r_a = 2a - r_p = 2(5.91 \times 10^{11}\text{ m}) - 4.43 \times 10^{11}\text{ m} = 7.39 \times 10^{11}$ m.
EVALUATE: This orbit is considerably more eccentric than those of the major planets. Its aphelion distance is about 1.7 times greater than its perihelion distance.

VP13.9.3. **IDENTIFY:** We are dealing with a satellite in a noncircular orbit.
SET UP: The semi-major axis a is $2a = r_p + r_a$, where r_p is the perihelion distance and r_a is the aphelion distance. $T = \dfrac{2\pi a^{3/2}}{\sqrt{Gm_S}}$, where m_S is the mass of the sun (or central star for another solar system). $a = ea + r_p$, where e is the eccentricity of the orbit. Refer to Fig. 13.18 in the textbook. The minor planet makes 2 orbits in the time it takes Neptune to make 3 orbits, so the minor planet must have a longer period than Neptune: $T_{mp} = 3/2\ T_N$, and its eccentricity is $e = 0.330$.

EXECUTE: **(a)** We want to find the semi-major axis a. If we take the ratio of the period of Neptune and minor planet, the factors in common will cancel.

$$\dfrac{T_{mp}}{T_N} = \dfrac{\dfrac{2\pi a^{3/2}}{\sqrt{Gm_S}}}{\dfrac{2\pi r_N^{3/2}}{\sqrt{Gm_S}}} = \left(\dfrac{a}{r_N}\right)^{3/2} = \dfrac{3}{2} \rightarrow (a/r_N)^{3/2} = 3/2 \rightarrow a = r_N(3/2)^{2/3}$$

$a = (4.50 \times 10^{12}\text{ m})(3/2)^{2/3} = 5.90 \times 10^{12}$ m.
(b) $a = ea + r_p \rightarrow r_p = a(1 - e) = (5.90 \times 10^{12}\text{ m})(1 - 0.330) = 3.95 \times 10^{12}$ m.
EVALUATE: At perihelion, this comet is *inside* Neptune's orbit. The eccentricity of its orbit is much greater than that of Neptune's orbit.

VP13.9.4. **IDENTIFY:** We are dealing with a planet in a noncircular orbit around another star.
SET UP: The semi-major axis a is $2a = r_p + r_a$, where r_p is the perihelion distance and r_a is the aphelion distance. $T = \dfrac{2\pi a^{3/2}}{\sqrt{Gm_S}}$, where m_S is the mass of the central star. $a = ea + r_p$, where e is the eccentricity of the orbit. In this case, $T = 4.39$ d $= 3.7930 \times 10^5$ s. Refer to Fig. 13.18 in the textbook.

EXECUTE: **(a)** We want the distance of the planet from its star at perigee (point of closest approach). $a = ea + r_p \rightarrow r_p = a(1 - e) = (7.41 \times 10^9\text{ m})(1 - 0.173) = 6.13 \times 10^9$ m.
(b) We want the mass of HATS-43, so we use $T = \dfrac{2\pi a^{3/2}}{\sqrt{Gm_S}}$. Solving for m_S gives $m_S = \dfrac{4\pi^2 a^3}{GT^2}$.
Using $a = 7.41 \times 10^9$ m and $T = 3.793 \times 10^5$ s, we get $m_S = 1.67 \times 10^{30}$ kg.
EVALUATE: This planet is considerably closer to its star than Mercury is to our sun. Comparing the mass of this star to that of our sun gives $m_{HATS}/m_{sun} = 1.67/1.99 = 0.839$, so the mass of this star is about 84% the mass of our sun.

13.5. **IDENTIFY:** Use $F_g = \dfrac{Gm_1 m_2}{r^2}$ to find the force exerted by each large sphere. Add these forces as vectors to get the net force and then use Newton's second law to calculate the acceleration.

SET UP: The forces are shown in Figure 13.5.

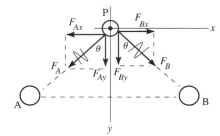

$\sin\theta = 0.80$
$\cos\theta = 0.60$
Take the origin of coordinate at point P.

Figure 13.5

EXECUTE: $F_A = G\dfrac{m_A m}{r^2} = G\dfrac{(0.26 \text{ kg})(0.010 \text{ kg})}{(0.100 \text{ m})^2} = 1.735\times 10^{-11}$ N

$F_B = G\dfrac{m_B m}{r^2} = 1.735\times 10^{-11}$ N

$F_{Ax} = -F_A \sin\theta = -(1.735\times 10^{-11} \text{ N})(0.80) = -1.39\times 10^{-11}$ N

$F_{Ay} = +F_A \cos\theta = +(1.735\times 10^{-11} \text{ N})(0.60) = +1.04\times 10^{-11}$ N

$F_{Bx} = +F_B \sin\theta = +1.39\times 10^{-11}$ N

$F_{By} = +F_B \cos\theta = +1.04\times 10^{-11}$ N

$\Sigma F_x = ma_x$ gives $F_{Ax} + F_{Bx} = ma_x$

$0 = ma_x$ so $a_x = 0$

$\Sigma F_y = ma_y$ gives $F_{Ay} + F_{By} = ma_y$

$2(1.04\times 10^{-11} \text{ N}) = (0.010 \text{ kg})a_y$

$a_y = 2.1\times 10^{-9}$ m/s^2, directed downward midway between A and B

EVALUATE: For ordinary size objects the gravitational force is very small, so the initial acceleration is very small. By symmetry there is no x-component of net force and the y-component is in the direction of the two large spheres, since they attract the small sphere.

13.9. IDENTIFY: Use $F_g = Gm_1 m_2/r^2$ to calculate the gravitational force each particle exerts on the third mass. The equilibrium is stable when for a displacement from equilibrium the net force is directed toward the equilibrium position and it is unstable when the net force is directed away from the equilibrium position.

SET UP: For the net force to be zero, the two forces on M must be in opposite directions. This is the case only when M is on the line connecting the two particles and between them. The free-body diagram for M is given in Figure 13.9. $m_1 = 3m$ and $m_2 = m$. If M is a distance x from m_1, it is a distance 1.00 m $- x$ from m_2.

EXECUTE: (a) $F_x = F_{1x} + F_{2x} = -G\dfrac{3mM}{x^2} + G\dfrac{mM}{(1.00 \text{ m} - x)^2} = 0$. Cancelling and simplifying gives $3(1.00 \text{ m} - x)^2 = x^2$. Taking square roots gives $1.00 \text{ m} - x = \pm x/\sqrt{3}$. Since M is between the two particles, x must be less than 1.00 m and $x = \dfrac{1.00 \text{ m}}{1 + 1/\sqrt{3}} = 0.634$ m. M must be placed at a point that is 0.634 m from the particle of mass $3m$ and 0.366 m from the particle of mass m.

(b) (i) If M is displaced slightly to the right in Figure 13.9, the attractive force from m is larger than the force from $3m$ and the net force is to the right. If M is displaced slightly to the left in Figure 13.9, the attractive force from $3m$ is larger than the force from m and the net force is to the left. In each case the net force is away from equilibrium and the equilibrium is unstable.

(ii) If M is displaced a very small distance along the y-axis in Figure 13.9, the net force is directed opposite to the direction of the displacement and therefore the equilibrium is stable.
EVALUATE: The point where the net force on M is zero is closer to the smaller mass.

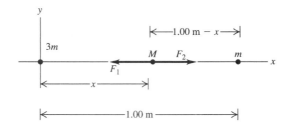

Figure 13.9

13.11. IDENTIFY: $F_g = G\dfrac{mm_E}{r^2} = mg$, so ssc where r is the distance of the object from the center of the earth.

SET UP: $r = h + R_E$, where h is the distance of the object above the surface of the earth and $R_E = 6.37 \times 10^6$ m is the radius of the earth.

EXECUTE: To decrease the acceleration due to gravity by one-tenth, the distance from the center of the earth must be increased by a factor of $\sqrt{10}$, and so the distance above the surface of the earth is $(\sqrt{10}-1)R_E = 1.38 \times 10^7$ m.

EVALUATE: This height is about twice the radius of the earth.

13.15. IDENTIFY: We are dealing with the acceleration due to gravity on another planet.

SET UP: We want to find g at the surface of a planet. At the surface, $g = GM/R^2$. Since we know the density ρ of the planet, we should put g in terms of ρ using $\rho = m/V$.

EXECUTE: (a) We want g at the surface of this planet. Using $m = \rho V = \rho(4/3\,\pi R^3)$, we have

$$g = \dfrac{Gm}{R^2} = \dfrac{G\rho\left(\dfrac{4}{3}\pi R^3\right)}{R^2} = \dfrac{4}{3}\pi G\rho R.$$ Now take the ratio of g_{planet}/g_{earth} using the same density for both planets but $R = 1.25 R_{earth}$. We get $\dfrac{g_p}{g_e} = \dfrac{\tfrac{4}{3}\pi G\rho R_p}{\tfrac{4}{3}\pi G\rho R_e} = \dfrac{R_p}{R_e} = \dfrac{1.25 R_e}{R_e} = 1.25$, from which we get $g_p = 1.25\,g_e = (1.25)(9.80\text{ m/s}^2) = 12.3\text{ m/s}^2$.

(b) We now want to change the density of this planet (a neat trick if you can do it!) so that g at its surface is the same as on earth. Use the result we found in part (a) for g: $g = \dfrac{4}{3}\pi G\rho R$. Since $g_p = g_e$, we equate the two expressions for g, giving $\dfrac{4}{3}\pi G\rho_p R_p = \dfrac{4}{3}\pi G\rho_e R_e$. Solving for ρ_p gives $\rho_p = \dfrac{R_e}{R_p}\rho_e = \dfrac{R_e}{1.25 R_e}\rho_e = 0.800\rho_e$. The planet should have 80.0% the density of earth.

EVALUATE: The planet is less dense than earth but it has more mass since it is larger, so g at its surface is the same as it is on earth.

13.19. IDENTIFY: Mechanical energy is conserved. At the escape speed, the object has no kinetic energy when it is very far away from the planet.

SET UP: Call m the mass of the object, M the mass of the planet, and r its radius. $K_1 + U_1 = K_2 + U_2$,
$K = \tfrac{1}{2}mv^2$, $U = -GmM/r$, $g = GM/r^2$.

EXECUTE: Energy conservation gives ½ $mv^2 - GmM/r = 0 + 0$. $M = rv^2/2G$. Putting this into g

$= GM/r^2$ gives $g = \dfrac{G\left(\dfrac{rv^2}{2G}\right)}{r^2} = \dfrac{v^2}{2r}$. Putting in the numbers gives

$g = (7.65 \times 10^3 \text{ m/s})^2/[2(3.24 \times 10^6 \text{ m})] = 9.03 \text{ m/s}^2$.

EVALUATE: This result is not very different from g on earth, so it is physically reasonable for a planet.

13.21. IDENTIFY: This problem involves the gravitational force and gravitational potential energy.

SET UP: $F_g = \dfrac{Gm_1 m_2}{r^2}$ and $U_g = -\dfrac{Gm_p m}{r}$. We know U_g and want to find the force F_g on you at the surface of the planet when $r = R$.

EXECUTE: Relate F_g and U_g: $F_g = \dfrac{Gmm_p}{R^2} = \left(\dfrac{Gmm_p}{R}\right)\dfrac{1}{R} = -\dfrac{U_g}{R} = -\dfrac{-1.20 \times 10^9 \text{ J}}{5.00 \times 10^6 \text{ m}} = 240$ N.

EVALUATE: Knowing your mass, we could find g at the surface.

13.27. IDENTIFY: We know orbital data (speed and orbital radius) for one satellite and want to use it to find the orbital speed of another satellite having a known orbital radius. Newton's second law and the law of universal gravitation apply to both satellites.

SET UP: For circular motion, $F_{\text{net}} = ma = mv^2/r$, which in this case is $G\dfrac{mm_p}{r^2} = m\dfrac{v^2}{r}$.

EXECUTE: Using $G\dfrac{mm_p}{r^2} = m\dfrac{v^2}{r}$, we get $Gm_p = rv^2 =$ constant. $r_1 v_1^2 = r_2 v_2^2$.

$v_2 = v_1 \sqrt{\dfrac{r_1}{r_2}} = (4800 \text{ m/s})\sqrt{\dfrac{7.00 \times 10^7 \text{ m}}{3.00 \times 10^7 \text{ m}}} = 7330$ m/s.

EVALUATE: The more distant satellite moves slower than the closer satellite, which is reasonable since the planet's gravity decreases with distance. The masses of the satellites do not affect their orbits.

13.29. IDENTIFY: Apply $\Sigma \vec{F} = m\vec{a}$ to the motion of the baseball. $v = \dfrac{2\pi r}{T}$.

SET UP: $r_D = 6 \times 10^3$ m.

EXECUTE: **(a)** $F_g = ma_{\text{rad}}$ gives $G\dfrac{m_D m}{r_D^2} = m\dfrac{v^2}{r_D}$.

$v = \sqrt{\dfrac{Gm_D}{r_D}} = \sqrt{\dfrac{(6.673 \times 10^{-11} \text{ N} \cdot \text{m}^2/\text{kg}^2)(1.5 \times 10^{15} \text{ kg})}{6 \times 10^3 \text{ m}}} = 4.07$ m/s which rounds to 4.1 m/s.

4.1 m/s = 9.1 mph, which is easy to achieve.

(b) $T = \dfrac{2\pi r}{v} = \dfrac{2\pi(6 \times 10^3 \text{ m})}{4.07 \text{ m/s}} = 9263 \text{ s} = 154.4 \text{ min} = 2.6$ h. The game would last a very long time indeed!

EVALUATE: The speed v is relative to the center of Deimos. The baseball would already have some speed before we throw it because of the rotational motion of Deimos.

13.31. IDENTIFY: The orbital speed is given by $v = \sqrt{Gm/r}$, where m is the mass of the star. The orbital period is given by $T = \dfrac{2\pi r}{v}$.

SET UP: The sun has mass $m_S = 1.99 \times 10^{30}$ kg. The orbit radius of the earth is 1.50×10^{11} m.

EXECUTE: **(a)** $v = \sqrt{Gm/r}$.

$v = \sqrt{(6.673 \times 10^{-11} \text{ N} \cdot \text{m}^2/\text{kg}^2)(0.85 \times 1.99 \times 10^{30} \text{ kg})/((1.50 \times 10^{11} \text{ m})(0.11))} = 8.27 \times 10^4$ m/s.

(b) $2\pi r/v = 1.25 \times 10^6$ s $= 14.5$ days (about two weeks).

EVALUATE: The orbital period is less than the 88-day orbital period of Mercury; this planet is orbiting very close to its star, compared to the orbital radius of Mercury.

13.33. IDENTIFY: Kepler's third law applies.

SET UP: $T = \dfrac{2\pi a^{3/2}}{\sqrt{Gm_S}}$, $d_{min} = a(1-e)$, $d_{max} = a(1+e)$.

EXECUTE: **(a)** Kepler's third law gives

$T = \dfrac{2\pi a^{3/2}}{\sqrt{Gm_S}} = \dfrac{2\pi (5.91 \times 10^{12} \text{ m})^{3/2}}{\sqrt{(6.67 \times 10^{-11} \text{ N} \cdot \text{m}^2/\text{kg}^2)(1.99 \times 10^{30} \text{ kg})}} = 7.84 \times 10^9$ s [(1 y)/(3.156$\times 10^7$ s)] = 248 y.

(b) $d_{min} = a(1-e) = (5.91 \times 10^{12} \text{ m})(1 - 0.249) = 4.44 \times 10^{12}$ m; $d_{max} = a(1+e) = 7.38 \times 10^{12}$ m.

EVALUATE: $d_{max} = 1.66 d_{min}$, which is *much* greater than for the earth's orbit since the earth moves in a much more circular orbit than Pluto.

13.35. IDENTIFY: We are dealing with the gravitational force due to spherical shells.

SET UP: Outside a uniform spherical shell, $F_g = \dfrac{Gm_1 m_2}{r^2}$, and inside of it $F_g = 0$. Our target variable is the net force F at various distances from the center, where $F = F_A + F_B$.

EXECUTE: **(a)** At $r = 2.00$ m, the point mass is inside both shells, so $F = 0$.

(b) At $r = 5.00$ m, the mass is between the shells, so $F_B = 0$, so $F = F_A$. This gives

$F = F_A = \dfrac{Gmm_A}{r^2} = \dfrac{G(0.0200 \text{ kg})(20.0 \text{ kg})}{(5.00 \text{ m})^2} = 1.07 \times 10^{-12}$ N.

(c) At $r = 8.00$ m, the mass is outside of both shells. We treat each one as a point mass at its center. $F = \dfrac{G(m_A + m_B)}{r^2} = \dfrac{G(0.0200 \text{ kg})(60.0 \text{ kg})}{(8.00 \text{ m})^2} = 1.25 \times 10^{-12}$ N.

EVALUATE: If a shell is not uniform, we cannot treat it as a point mass at its center. And the force on a mass inside of it is not necessarily zero.

13.37. IDENTIFY: Section 13.6 states that for a point mass outside a uniform sphere the gravitational force is the same as if all the mass of the sphere were concentrated at its center. It also states that for a point mass a distance r from the center of a uniform sphere, where r is less than the radius of the sphere, the gravitational force on the point mass is the same as though we removed all the mass at points farther than r from the center and concentrated all the remaining mass at the center.

SET UP: The density of the sphere is $\rho = \dfrac{M}{\frac{4}{3}\pi R^3}$, where M is the mass of the sphere and R is its radius. The mass inside a volume of radius $r < R$ is $M_r = \rho V_r = \left(\dfrac{M}{\frac{4}{3}\pi R^3}\right)\left(\frac{4}{3}\pi r^3\right) = M\left(\dfrac{r}{R}\right)^3$.

$r = 5.01$ m is outside the sphere and $r = 2.50$ m is inside the sphere. $F_g = \dfrac{Gm_1 m_2}{r^2}$.

EXECUTE: (a) (i) $F_g = \dfrac{GMm}{r^2} = (6.67 \times 10^{-11}\ \text{N} \cdot \text{m}^2/\text{kg}^2)\dfrac{(1000.0\ \text{kg})(2.00\ \text{kg})}{(5.01\ \text{m})^2} = 5.31 \times 10^{-9}$ N.

(ii) $F_g = \dfrac{GM'm}{r^2}$. $M' = M\left(\dfrac{r}{R}\right)^3 = (1000.0\ \text{kg})\left(\dfrac{2.50\ \text{m}}{5.00\ \text{m}}\right)^3 = 125$ kg.

$F_g = (6.67 \times 10^{-11}\ \text{N} \cdot \text{m}^2/\text{kg}^2)\dfrac{(125\ \text{kg})(2.00\ \text{kg})}{(2.50\ \text{m})^2} = 2.67 \times 10^{-9}$ N.

(b) $F_g = \dfrac{GM(r/R)^3 m}{r^2} = \left(\dfrac{GMm}{R^3}\right) r$ for $r < R$ and $F_g = \dfrac{GMm}{r^2}$ for $r > R$. The graph of F_g versus r is sketched in Figure 13.37.

EVALUATE: At points outside the sphere the force on a point mass is the same as for a shell of the same mass and radius. For $r < R$ the force is different in the two cases of uniform sphere versus hollow shell.

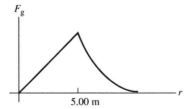

Figure 13.37

13.39. IDENTIFY: Find the potential due to a small segment of the ring and integrate over the entire ring to find the total U.

(a) SET UP:

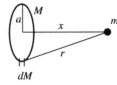

Divide the ring up into small segments dM, as indicated in Figure 13.39.

Figure 13.39

EXECUTE: The gravitational potential energy of dM and m is $dU = -Gm\,dM/r$.

The total gravitational potential energy of the ring and particle is $U = \int dU = -Gm \int dM/r$.

But $r = \sqrt{x^2 + a^2}$ is the same for all segments of the ring, so

$$U = -\frac{Gm}{r}\int dM = -\frac{GmM}{r} = -\frac{GmM}{\sqrt{x^2+a^2}}.$$

(b) EVALUATE: When $x \gg a$, $\sqrt{x^2+a^2} \to \sqrt{x^2} = x$ and $U = -GmM/x$. This is the gravitational potential energy of two point masses separated by a distance x. This is the expected result.

(c) IDENTIFY and **SET UP:** Use $F_x = -dU/dx$ with $U(x)$ from part (a) to calculate F_x.

EXECUTE: $F_x = -\dfrac{dU}{dx} = -\dfrac{d}{dx}\left(-\dfrac{GmM}{\sqrt{x^2+a^2}}\right)$

$F_x = +GmM\dfrac{d}{dx}(x^2+a^2)^{-1/2} = GmM\left(-\dfrac{1}{2}(2x)(x^2+a^2)^{-3/2}\right)$

$F_x = -GmMx/(x^2+a^2)^{3/2}$; the minus sign means the force is attractive.

EVALUATE: **(d)** For $x \gg a$, $(x^2+a^2)^{3/2} \to (x^2)^{3/2} = x^3$

Then $F_x = -GmMx/x^3 = -GmM/x^2$. This is the force between two point masses separated by a distance x and is the expected result.

(e) For $x=0$, $U = -GMm/a$. Each small segment of the ring is the same distance from the center and the potential is the same as that due to a point charge of mass M located at a distance a. For $x=0$, $F_x = 0$. When the particle is at the center of the ring, symmetrically placed segments of the ring exert equal and opposite forces and the total force exerted by the ring is zero.

13.41. IDENTIFY and **SET UP:** At the north pole, $F_g = w_0 = mg_0$, where g_0 is given by $g = G\dfrac{m_E}{r^2}$ applied to Neptune. At the equator, the apparent weight is given by $w = w_0 - mv^2/R$. The orbital speed v is obtained from the rotational period using $v = 2\pi R/T$.

EXECUTE: (a)
$g_0 = Gm/R^2 = (6.673\times10^{-11}\ \text{N}\cdot\text{m}^2/\text{kg}^2)(1.02\times10^{26}\ \text{kg})/(2.46\times10^7\ \text{m})^2 = 11.25\ \text{m/s}^2$. This agrees with the value of g given in the problem.

$F = w_0 = mg_0 = (3.00\ \text{kg})(11.25\ \text{m/s}^2) = 33.74\ \text{N}$, which rounds to 33.7 N. This is the true weight of the object.

(b) We have $w = w_0 - mv^2/R$

$T = \dfrac{2\pi r}{v}$ gives $v = \dfrac{2\pi r}{T} = \dfrac{2\pi(2.46\times10^7\ \text{m})}{(16\ \text{h})(3600\ \text{s}/1\ \text{h})} = 2.683\times10^3\ \text{m/s}$

$v^2/R = (2.683\times10^3\ \text{m/s})^2/(2.46\times10^7\ \text{m}) = 0.2927\ \text{m/s}^2$

Then $w = 33.74\ \text{N} - (3.00\ \text{kg})(0.2927\ \text{m/s}^2) = 32.9\ \text{N}$.

EVALUATE: The apparent weight is less than the true weight. This effect is larger on Neptune than on earth.

13.43. IDENTIFY: The orbital speed for an object a distance r from an object of mass M is $v = \sqrt{\dfrac{GM}{r}}$.

The mass M of a black hole and its Schwarzschild radius R_S are related by $R_S = \dfrac{2GM}{c^2}$.

SET UP: $c = 3.00\times10^8$ m/s. 1 ly $= 9.461\times10^{15}$ m.

EXECUTE:

(a) $M = \dfrac{rv^2}{G} = \dfrac{(7.5\ \text{ly})(9.461\times10^{15}\ \text{m/ly})(200\times10^3\ \text{m/s})^2}{(6.673\times10^{-11}\ \text{N}\cdot\text{m}^2/\text{kg}^2)} = 4.3\times10^{37}\ \text{kg} = 2.1\times10^7\ M_S$.

(b) No, the object has a mass very much greater than 50 solar masses.

(c) $R_S = \dfrac{2GM}{c^2} = \dfrac{2v^2 r}{c^2} = 6.32 \times 10^{10}$ m, which does fit.

EVALUATE: The Schwarzschild radius of a black hole is approximately the same as the radius of Mercury's orbit around the sun.

13.45. IDENTIFY: Use $F_g = Gm_1 m_2 / r^2$ to find each gravitational force. Each force is attractive. In part (b) apply conservation of energy.

SET UP: For a pair of masses m_1 and m_2 with separation r, $U = -G\dfrac{m_1 m_2}{r}$.

EXECUTE: **(a)** From symmetry, the net gravitational force will be in the direction 45° from the x-axis (bisecting the x- and y-axes), with magnitude

$$F = (6.673 \times 10^{-11}~\text{N} \cdot \text{m}^2 / \text{kg}^2)(0.0150~\text{kg}) \left[\dfrac{(2.0~\text{kg})}{(2(0.50~\text{m})^2)} + 2\dfrac{(1.0~\text{kg})}{(0.50~\text{m})^2} \sin 45° \right] = 9.67 \times 10^{-12}~\text{N}$$

(b) The initial displacement is so large that the initial potential energy may be taken to be zero. From the work-energy theorem, $\dfrac{1}{2}mv^2 = Gm\left[\dfrac{(2.0~\text{kg})}{\sqrt{2}~(0.50~\text{m})} + 2\dfrac{(1.0~\text{kg})}{(0.50~\text{m})}\right]$. Cancelling the factor of m, solving for v, and using the numerical values gives $v = 3.02 \times 10^{-5}$ m/s.

EVALUATE: The result in part (b) is independent of the mass of the particle. It would take the particle a long time to reach point P.

13.47. IDENTIFY: Apply conservation of energy and conservation of linear momentum to the motion of the two spheres.

SET UP: Denote the 50.0-kg sphere by a subscript 1 and the 100-kg sphere by a subscript 2.

EXECUTE: **(a)** Linear momentum is conserved because we are ignoring all other forces, that is, the net external force on the system is zero. Hence, $m_1 v_1 = m_2 v_2$.

(b) (i) From the work-energy theorem in the form $K_i + U_i = K_f + U_f$, with the initial kinetic energy $K_i = 0$ and $U = -G\dfrac{m_1 m_2}{r}$, $Gm_1 m_2 \left[\dfrac{1}{r_f} - \dfrac{1}{r_i}\right] = \dfrac{1}{2}(m_1 v_1^2 + m_2 v_2^2)$. Using the conservation of momentum relation $m_1 v_1 = m_2 v_2$ to eliminate v_2 in favor of v_1 and simplifying yields

$v_1^2 = \dfrac{2Gm_2^2}{m_1 + m_2}\left[\dfrac{1}{r_f} - \dfrac{1}{r_i}\right]$, with a similar expression for v_2. Substitution of numerical values gives

$v_1 = 1.49 \times 10^{-5}$ m/s, $v_2 = 7.46 \times 10^{-6}$ m/s. (ii) The magnitude of the relative velocity is the sum of the speeds, 2.24×10^{-5} m/s.

(c) The distance the centers of the spheres travel (x_1 and x_2) is proportional to their acceleration, and $\dfrac{x_1}{x_2} = \dfrac{a_1}{a_2} = \dfrac{m_2}{m_1}$, or $x_1 = 2x_2$. When the spheres finally make contact, their centers will be a distance of $2r$ apart, or $x_1 + x_2 + 2r = 40$ m, or $2x_2 + x_2 + 2r = 40$ m. Thus, $x_2 = 40/3$ m $- 2r/3$, and $x_1 = 80/3$ m $- 4r/3$. The point of contact of the surfaces is $80/3$ m $- r/3 = 26.6$ m from the initial position of the center of the 50.0-kg sphere.

EVALUATE: The result $x_1 / x_2 = 2$ can also be obtained from the conservation of momentum result that $\dfrac{v_1}{v_2} = \dfrac{m_2}{m_1}$, at every point in the motion.

EVALUATE: The work done by the attractive gravity forces is negative. The work you do is positive.

13.49. **IDENTIFY** and **SET UP:** (a) To stay above the same point on the surface of the earth the orbital period of the satellite must equal the orbital period of the earth:

$T = 1\text{ d}(24\text{ h}/1\text{ d})(3600\text{ s}/1\text{ h}) = 8.64 \times 10^4$ s. The equation $T = \dfrac{2\pi r^{3/2}}{\sqrt{Gm_E}}$ gives the relation between the orbit radius and the period.

EXECUTE: $T = \dfrac{2\pi r^{3/2}}{\sqrt{Gm_E}}$ gives $T^2 = \dfrac{4\pi^2 r^3}{Gm_E}$. Solving for r gives

$r = \left(\dfrac{T^2 Gm_E}{4\pi^2}\right)^{1/3} = \left(\dfrac{(8.64 \times 10^4\text{ s})^2 (6.673 \times 10^{-11}\text{ N}\cdot\text{m}^2/\text{kg}^2)(5.97 \times 10^{24}\text{ kg})}{4\pi^2}\right)^{1/3} = 4.23 \times 10^7$ m.

This is the radius of the orbit; it is related to the height h above the earth's surface and the radius R_E of the earth by $r = h + R_E$. Thus $h = r - R_E = 4.23 \times 10^7$ m $- 6.37 \times 10^6$ m $= 3.59 \times 10^7$ m.

EVALUATE: The orbital speed of the geosynchronous satellite is $2\pi r/T = 3080$ m/s. The altitude is much larger and the speed is much less than for the satellite in Example 13.6.

(b) Consider Figure 13.49.

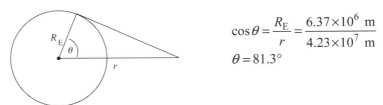

$\cos\theta = \dfrac{R_E}{r} = \dfrac{6.37 \times 10^6\text{ m}}{4.23 \times 10^7\text{ m}}$

$\theta = 81.3°$

Figure 13.49

A line from the satellite is tangent to a point on the earth that is at an angle of 81.3° above the equator. The sketch shows that points at higher latitudes are blocked by the earth from viewing the satellite.

13.51. **IDENTIFY:** From Example 13.5, the escape speed is $v = \sqrt{\dfrac{2GM}{R}}$. Use $\rho = M/V$ to write this expression in terms of ρ.

SET UP: For a sphere $V = \tfrac{4}{3}\pi R^3$.

EXECUTE: In terms of the density ρ, the ratio M/R is $(4\pi/3)\rho R^2$, and so the escape speed is

$v = \sqrt{(8\pi/3)(6.673 \times 10^{-11}\text{ N}\cdot\text{m}^2/\text{kg}^2)(2500\text{ kg/m}^3)(150 \times 10^3\text{ m})^2} = 177$ m/s.

EVALUATE: This is much less than the escape speed for the earth, 11,200 m/s.

13.53. IDENTIFY: Apply the law of gravitation to the astronaut at the north pole to calculate the mass of the planet. Then apply $\Sigma \vec{F} = m\vec{a}$ to the astronaut, with $a_{rad} = \dfrac{4\pi^2 R}{T^2}$, toward the center of the planet, to calculate the period T. Apply $T = \dfrac{2\pi r^{3/2}}{\sqrt{Gm_E}}$ to the satellite in order to calculate its orbital period.

SET UP: Get radius of X: $\tfrac{1}{4}(2\pi R) = 18{,}850$ km and $R = 1.20 \times 10^7$ m. Astronaut mass: $m = \dfrac{w}{g} = \dfrac{943 \text{ N}}{9.80 \text{ m/s}^2} = 96.2$ kg.

EXECUTE: $\dfrac{GmM_X}{R^2} = w$, where $w = 915.0$ N.

$M_X = \dfrac{mg_X R^2}{Gm} = \dfrac{(915 \text{ N})(1.20 \times 10^7 \text{ m})^2}{(6.67 \times 10^{-11} \text{ N}\cdot\text{m}^2/\text{kg}^2)(96.2 \text{ kg})} = 2.05 \times 10^{25}$ kg

Apply Newton's second law to the astronaut on a scale at the equator of X.

$F_{grav} - F_{scale} = ma_{rad}$, so $F_{grav} - F_{scale} = \dfrac{4\pi^2 mR}{T^2}$.

$915.0 \text{ N} - 850.0 \text{ N} = \dfrac{4\pi^2 (96.2 \text{ kg})(1.20 \times 10^7 \text{ m})}{T^2}$ and $T = 2.65 \times 10^4 \text{ s}\left(\dfrac{1 \text{ h}}{3600 \text{ s}}\right) = 7.36$ h.

(b) For the satellite,

$T = \sqrt{\dfrac{4\pi^2 r^3}{Gm_X}} = \sqrt{\dfrac{4\pi^2(1.20 \times 10^7 \text{ m} + 2.0 \times 10^6 \text{ m})^3}{(6.67 \times 10^{-11} \text{ N}\cdot\text{m}^2/\text{kg}^2)(2.05 \times 10^{25} \text{ kg})}} = 8.90 \times 10^3 \text{ s} = 2.47$ hours.

EVALUATE: The acceleration of gravity at the surface of the planet is $g_X = \dfrac{915.0 \text{ N}}{96.2 \text{ kg}} = 9.51 \text{ m/s}^2$, similar to the value on earth. The radius of the planet is about twice that of earth. The planet rotates more rapidly than earth and the length of a day is about one-third what it is on earth.

13.57. IDENTIFY: Use the orbital speed and altitude to find the mass of the planet. Use this mass and the planet's radius to find g at the surface. Use projectile motion to find the horizontal range x, where $x = \dfrac{v_0^2 \sin(2\alpha)}{g}$.

SET UP: For an object in a circular orbit, $v = \sqrt{GM/r}$. $g = GM/r^2$. Call r the orbital radius and R the radius of the planet.

EXECUTE: $v = \sqrt{GM/r}$ gives $M = rv^2/G$. Using this to find g gives
$g = GM/R^2 = G(rv^2/G)/R^2 = v^2 r/R^2 = (4900 \text{ m/s})^2(4.48 \times 10^6 \text{ m} + 6.30 \times 10^5 \text{ m})/(4.48 \times 10^6 \text{ m})^2 = 6.113 \text{ m/s}^2$. Now use this acceleration to find the horizontal range.

$x = \dfrac{v_0^2 \sin(2\alpha)}{g} = (12.6 \text{ m/s})^2 \sin[2(30.8°)]/(6.113 \text{ m/s}^2) = 22.8$ m.

EVALUATE: On this planet, $g = 0.624 g_E$, so the range is about 1.6 times what it would be on earth.

13.63. **IDENTIFY:** Use $F_g = Gm_1m_2/r^2$ to calculate F_g. Apply Newton's second law to circular motion of each star to find the orbital speed and period. Apply the conservation of energy to calculate the energy input (work) required to separate the two stars to infinity.

(a) SET UP: The cm is midway between the two stars since they have equal masses. Let R be the orbit radius for each star, as sketched in Figure 13.63.

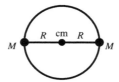

The two stars are separated by a distance $2R$, so
$$F_g = GM^2/(2R)^2 = GM^2/4R^2$$

Figure 13.63

(b) EXECUTE: $F_g = ma_{rad}$

$GM^2/4R^2 = M(v^2/R)$ so $v = \sqrt{GM/4R}$

And $T = 2\pi R/v = 2\pi R\sqrt{4R/GM} = 4\pi\sqrt{R^3/GM}$

(c) SET UP: Apply $K_1 + U_1 + W_{other} = K_2 + U_2$ to the system of the two stars. Separate to infinity implies $K_2 = 0$ and $U_2 = 0$.

EXECUTE: $K_1 = \frac{1}{2}Mv^2 + \frac{1}{2}Mv^2 = 2\left(\frac{1}{2}M\right)(GM/4R) = GM^2/4R$

$U_1 = -GM^2/2R$

Thus the energy required is $W_{other} = -(K_1 + U_1) = -(GM^2/4R - GM^2/2R) = GM^2/4R$.

EVALUATE: The closer the stars are and the greater their mass, the larger their orbital speed, the shorter their orbital period and the greater the energy required to separate them.

13.65. **IDENTIFY** and **SET UP:** Use conservation of energy, $K_1 + U_1 + W_{other} = K_2 + U_2$. The gravity force exerted by the sun is the only force that does work on the comet, so $W_{other} = 0$.

EXECUTE: $K_1 = \frac{1}{2}mv_1^2$, $v_1 = 2.0 \times 10^4$ m/s

$U_1 = -Gm_Sm/r_1$, where $r_1 = 2.5 \times 10^{11}$ m

$K_2 = \frac{1}{2}mv_2^2$

$U_2 = -Gm_Sm/r_2$, $r_2 = 5.0 \times 10^{10}$ m

$\frac{1}{2}mv_1^2 - Gm_Sm/r_1 = \frac{1}{2}mv_2^2 - Gm_Sm/r_2$

$v_2^2 = v_1^2 + 2Gm_S\left(\frac{1}{r_2} - \frac{1}{r_1}\right) = v_1^2 + 2Gm_S\left(\frac{r_1 - r_2}{r_1 r_2}\right)$

$v_2 = 6.8 \times 10^4$ m/s

EVALUATE: The comet has greater speed when it is closer to the sun.

13.67. **(a) IDENTIFY** and **SET UP:** Use $T = \dfrac{2\pi a^{3/2}}{\sqrt{Gm_S}}$, applied to the satellites orbiting the earth rather than the sun.

EXECUTE: Find the value of a for the elliptical orbit:

$2a = r_a + r_p = R_E + h_a + R_E + h_p$, where h_a and h_p are the heights at apogee and perigee, respectively.

$a = R_E + (h_a + h_p)/2$

$a = 6.37\times 10^6$ m $+ (400\times 10^3$ m $+ 4000\times 10^3$ m$)/2 = 8.57\times 10^6$ m

$$T = \frac{2\pi a^{3/2}}{\sqrt{GM_E}} = \frac{2\pi(8.57\times 10^6 \text{ m})^{3/2}}{\sqrt{(6.67\times 10^{-11} \text{ N}\cdot\text{m}^2/\text{kg}^2)(5.97\times 10^{24} \text{ kg})}} = 7.90\times 10^3 \text{ s}$$

(b) Conservation of angular momentum gives $r_a v_a = r_p v_p$

$$\frac{v_p}{v_a} = \frac{r_a}{r_p} = \frac{6.37\times 10^6 \text{ m} + 4.00\times 10^6 \text{ m}}{6.37\times 10^6 \text{ m} + 4.00\times 10^5 \text{ m}} = 1.53.$$

(c) Conservation of energy applied to apogee and perigee gives $K_a + U_a = K_p + U_p$

$\frac{1}{2}mv_a^2 - Gm_E m/r_a = \frac{1}{2}mv_p^2 - Gm_E m/r_p$

$v_p^2 - v_a^2 = 2Gm_E(1/r_p - 1/r_a) = 2Gm_E(r_a - r_p)/r_a r_p$

But $v_p = 1.532 v_a$, so $1.347 v_a^2 = 2Gm_E(r_a - r_p)/r_a r_p$

$v_a = 5.51\times 10^3$ m/s, $v_p = 8.43\times 10^3$ m/s

(d) Need v so that $E = 0$, where $E = K + U$.

<u>at perigee:</u> $\frac{1}{2}mv_p^2 - Gm_E m/r_p = 0$

$v_p = \sqrt{2Gm_E/r_p} = \sqrt{2(6.67\times 10^{-11} \text{ N}\cdot\text{m}^2/\text{kg}^2)(5.97\times 10^{24} \text{ kg})/(6.77\times 10^6 \text{ m})} = 1.085\times 10^4$ m/s

This means an increase of 1.085×10^4 m/s $- 8.43\times 10^3$ m/s $= 2.42\times 10^3$ m/s.

<u>at apogee:</u>

$v_a = \sqrt{2Gm_E/r_a} = \sqrt{2(6.67\times 10^{-11} \text{ N}\cdot\text{m}^2/\text{kg}^2)(5.97\times 10^{24} \text{ kg})/(1.037\times 10^7 \text{ m})} = 8.763\times 10^3$ m/s

This means an increase of 8.763×10^3 m/s $- 5.51\times 10^3$ m/s $= 3.25\times 10^3$ m/s.

EVALUATE: Perigee is more efficient. At this point r is smaller so v is larger and the satellite has more kinetic energy and more total energy.

13.69. **IDENTIFY** and **SET UP:** Apply conservation of energy, $K_1 + U_1 + W_{other} = K_2 + U_2$, and solve for W_{other}. Only $r = h + R_E$ is given, so use $v = \sqrt{GM/r}$ to relate r and v.

EXECUTE: $K_1 + U_1 + W_{other} = K_2 + U_2$

$U_1 = -Gm_M m/r_1$, where m_M is the mass of Mars and $r_1 = R_M + h$, where R_M is the radius of Mars and $h = 2000\times 10^3$ m.

$U_1 = -(6.67\times 10^{-11} \text{ N}\cdot\text{m}^2/\text{kg}^2)\dfrac{(6.42\times 10^{23} \text{ kg})(5000 \text{ kg})}{3.39\times 10^6 \text{ m} + 2000\times 10^3 \text{ m}} = -3.97230\times 10^{10}$ J

$U_2 = -Gm_M m/r_2$, where r_2 is the new orbit radius.

$U_2 = -(6.67\times 10^{-11} \text{ N}\cdot\text{m}^2/\text{kg}^2)\dfrac{(6.42\times 10^{23} \text{ kg})(5000 \text{ kg})}{3.39\times 10^6 \text{ m} + 4000\times 10^3 \text{ m}} = -2.89725\times 10^{10}$ J

For a circular orbit $v = \sqrt{Gm_M/r}$, with the mass of Mars rather than the mass of the earth.

Using this gives $K = \frac{1}{2}mv^2 = \frac{1}{2}m(Gm_M/r) = \frac{1}{2}Gm_M m/r$, so $K = -\frac{1}{2}U$.

$K_1 = -\frac{1}{2}U_1 = +1.98615\times 10^{10}$ J and $K_2 = -\frac{1}{2}U_2 = +1.44863\times 10^{10}$ J

Then $K_1 + U_1 + W_{other} = K_2 + U_2$ gives

$W_{other} = (K_2 - K_1) + (U_2 - U_1)$

$W_{other} = (1.44863\times 10^{10} \text{ J} - 1.98615\times 10^{10} \text{ J}) + (+3.97230\times 10^{10} \text{ J} - 2.89725\times 10^{10} \text{ J})$

$W_{other} = 5.38\times 10^9$ J.

EVALUATE: When the orbit radius increases the kinetic energy decreases and the gravitational potential energy increases. $K = -U/2$ so $E = K + U = -U/2$ and the total energy also increases (becomes less negative). Positive work must be done to increase the total energy of the satellite.

13.71. **IDENTIFY:** Integrate $dm = \rho dV$ to find the mass of the planet. Outside the planet, the planet behaves like a point mass, so at the surface $g = GM/R^2$.

SET UP: A thin spherical shell with thickness dr has volume $dV = 4\pi r^2 dr$. The earth has radius $R_E = 6.37 \times 10^6$ m.

EXECUTE: Get M: $M = \int dm = \int \rho dV = \int \rho 4\pi r^2 dr$. The density is $\rho = \rho_0 - br$, where $\rho_0 = 15.0 \times 10^3$ kg/m^3 at the center and at the surface, $\rho_S = 2.0 \times 10^3$ kg/m^3, so $b = \dfrac{\rho_0 - \rho_s}{R}$.

$$M = \int_0^R (\rho_0 - br) 4\pi r^2 dr = \dfrac{4\pi}{3}\rho_0 R^3 - \pi b R^4 = \dfrac{4}{3}\pi R^3 \rho_0 - \pi R^4 \left(\dfrac{\rho_0 - \rho_s}{R}\right) = \pi R^3 \left(\dfrac{1}{3}\rho_0 + \rho_s\right)$$

and $M = 5.71 \times 10^{24}$ kg. Then $g = \dfrac{GM}{R^2} = \dfrac{G\pi R^3 (\frac{1}{3}\rho_0 + \rho_s)}{R^2} = \pi RG\left(\dfrac{1}{3}\rho_0 + \rho_s\right)$.

$$g = \pi(6.37\times10^6 \text{m})(6.67\times10^{-11} \text{ N}\cdot\text{m}^2/\text{kg}^2)\left(\dfrac{15.0\times10^3 \text{ kg/m}^3}{3} + 2.0\times10^3 \text{ kg/m}^3\right).$$

$g = 9.34$ m/s^2.

EVALUATE: The average density of the planet is

$\rho_{av} = \dfrac{M}{V} = \dfrac{M}{\frac{4}{3}\pi R^3} = \dfrac{3(5.71\times10^{24} \text{ kg})}{4\pi(6.37\times10^6 \text{ m})^3} = 5.27\times10^3$ kg/m^3. Note that this is not $(\rho_0 + \rho_s)/2$.

13.75. **IDENTIFY:** Compare F_E to Hooke's law.

SET UP: The earth has mass $m_E = 5.97\times10^{24}$ kg and radius $R_E = 6.37\times10^6$ m.

EXECUTE: **(A)** For $F_x = -kx$, $U = \frac{1}{2}kx^2$. The force here is in the same form, so by analogy $U(r) = \dfrac{Gm_E m}{2R_E^3}r^2$. This is also given by the integral of F_g from 0 to r with respect to distance.

(b) From part (a), the initial gravitational potential energy is $\dfrac{Gm_E m}{2R_E}$. Equating initial potential energy and final kinetic energy (initial kinetic energy and final potential energy are both zero) gives $v^2 = \dfrac{Gm_E}{R_E}$, so $v = 7.91\times10^3$ m/s.

EVALUATE: When $r = 0$, $U(r) = 0$, as specified in the problem.

13.77. **IDENTIFY and SET UP:** At the surface of a planet, $g = \dfrac{GM}{R^2}$, and average density is $\rho = m/V$, where $V = 4/3\ \pi R^3$ for a sphere.

EXECUTE: We have expressions for g and M: $g = \dfrac{GM}{R^2}$ and $M = \rho V = \rho\left(\dfrac{4}{3}\pi R^3\right)$. Combining them we get $g = \dfrac{G\rho\left(\dfrac{4}{3}\pi R^3\right)}{R^2} = \dfrac{4\pi G\rho R}{3}$. Using $R = D/2$ gives $g = \dfrac{2\pi G\rho D}{3}$.

(a) A graph of g versus D is shown in Figure 13.77. As this graph shows, the densities vary considerably and show no apparent pattern.

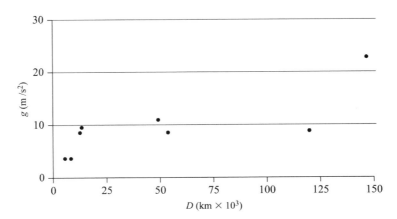

Figure 13.77

(b) Using the equation we just derived, $g = \dfrac{2\pi G \rho D}{3}$, we solve for ρ and use the values from the table given in the problem. For example, for Mercury we have

$\rho = \dfrac{3g}{2\pi DG} = \dfrac{3(3.7 \text{ m/s}^2)}{2\pi(4.879\times 10^6 \text{ m})(6.67\times 10^{-11} \text{ N}\cdot\text{m}^2/\text{kg}^2)} = 5400 \text{ kg/m}^3$. Continuing the calculations and putting the results in order of decreasing density, we get the following results.

Earth: 5500 kg/m^3
Mercury: 5400 kg/m^3
Venus: 5300 kg/m^3
Mars: 3900 kg/m^3
Neptune: 1600 kg/m^3
Uranus: 1200 kg/m^3
Jupiter: 1200 kg/m^3
Saturn: 534 kg/m^3

(c) For several reasons, it is reasonable that the other planets would be denser toward their centers. Gravity is stronger at close distances, so it would compress matter near the center. In addition, during the formation of planets, heavy elements would tend to sink toward the center and displace light elements, much as a rock sinks in water. This variation in density would have no effect on our analysis however, since the planets are still spherically symmetric.

(d) $g = \dfrac{2\pi G \rho D}{3} = \dfrac{2\pi(6.67\times 10^{-11} \text{ N}\cdot\text{m}^2/\text{kg}^2)(1.20536\times 10^8 \text{ m})(5500 \text{ kg/m}^3)}{3} = 93 \text{ m/s}^2$.

EVALUATE: Saturn is less dense than water, so it would float if we could throw it into our ocean (which of course is impossible since it is much larger than the earth). This low density is the reason that g at its "surface" is less than g at the earth's surface, even though the mass of Saturn is much greater than that of the earth. Also note in our results in (b) that the inner four planets are much denser than the outer four (the gas giants), with the earth being the densest of all.

STUDY GUIDE FOR PERIODIC MOTION

Summary

We'll examine periodic motion, or oscillation, in this chapter. Many systems exhibit periodic motion, such as a swinging pendulum, a ball on a spring, or the membrane of a drum. We'll describe the motion of oscillating objects, characterized by amplitude, period, frequency, and angular frequency. We'll use force equations and energy concepts to analyze their motions. We'll look at several models of periodic motion that can be used to represent the motion of many oscillators. Periodic motion plays a vital role in many areas of physics, and this chapter will lay the foundation for further studies.

Objectives

After studying this chapter, you'll understand
- Periodic motion and the terminology used to describe oscillations.
- How to identify and analyze simple harmonic motion.
- Energy and motion as a function of time for a particle in simple harmonic motion.
- The simple pendulum, the physical pendulum, damped and forced oscillations, and resonance.

Concepts and Equations

Term	Description
Periodic Motion	Periodic motion is motion that repeats in a definite cycle. Periodic motion occurs when an object is displaced from its equilibrium position and a restoring force exists that tends to return the object to equilibrium. The amplitude is the maximum magnitude of displacement from equilibrium. A cycle is one complete round-trip. The period is the time taken to complete one cycle. Frequency (f) is the number of cycles per unit time. Angular frequency (ω) is 2π times the frequency. Period, frequency, and angular frequency are related: $$T = \frac{1}{f}, \qquad f = \frac{1}{T}, \qquad \omega = 2\pi f = \frac{2\pi}{T}.$$
Simple Harmonic Motion	Simple harmonic motion (SHM) is periodic motion in which the restoring force is directly proportional to the object's displacement. Often, SHM occurs when the displacement is small. The equation of motion is $$x = A\cos(\omega t + \phi).$$ A system attached to a spring with spring constant k and having mass m will oscillate at a frequency of $$\omega = 2\pi f \frac{\omega}{2\pi} = \sqrt{\frac{k}{m}}$$ and a period of $$T = \frac{1}{f} = 2\pi\sqrt{\frac{m}{k}}.$$ The total mechanical energy remains constant in SHM and can be expressed in terms of its amplitude: $$E = \tfrac{1}{2}mv^2 + \tfrac{1}{2}kx^2 = \tfrac{1}{2}kA^2.$$ For angular simple harmonic motion, the frequency is related to the moment of inertia and the torsion constant by $$\omega = \sqrt{\frac{\kappa}{I}}.$$
Simple Pendulum	A simple pendulum is a model of a point mass suspended by a massless string in a gravitational field. For small displacements, a pendulum of length L has frequency $$\omega = 2\pi f = \sqrt{\frac{g}{L}}.$$
Physical Pendulum	A physical pendulum is any object suspended from an axis of rotation. The angular frequency for small-amplitude oscillations is given by $$\omega = \sqrt{\frac{mgd}{I}}.$$
Damped Oscillations	A simple harmonic oscillator impelled by a force that is proportional to velocity exhibits *damped oscillations*. The angular frequency becomes $$\omega' = \sqrt{\frac{k}{m} - \frac{b^2}{4m^2}}.$$

	Systems are called critically damped, overdamped, and underdamped according to how they return to equilibrium.
Forced Oscillations	Periodic motion in a system with a sinusoidally varying driving force is called *forced oscillation* or *driven oscillation*. Resonance occurs when the driving angular frequency is near the natural-oscillation angular frequency, increasing the amplitude of the motion. The amplitude is given by $$A = \frac{F_{\max}}{\sqrt{(k - m\omega_d^2)^2 + b\omega_d^2}}.$$

Key Concept 1: The period of an oscillation is the reciprocal of the oscillation frequency. The angular frequency equals the frequency multiplied by 2π.

Key Concept 2: The magnitude of the force exerted by an ideal spring equals the spring constant k times the distance the spring is stretched or compressed. An object of mass m attached to an ideal spring oscillates in simple harmonic motion, with its period, frequency, and angular frequency determined by the values of k and m.

Key Concept 3: You can determine the amplitude and phase angle of a simple harmonic oscillation from the initial position, initial velocity, and angular frequency of the motion.

Key Concept 4: In simple harmonic motion, the sum of kinetic energy and elastic potential energy is conserved. You can use this to find the velocity of an object in SHM as a function of its displacement.

Key Concept 5: You can use conservation of total mechanical energy in simple harmonic motion to help solve problems that involve motion at a given position, not at a given time.

Key Concept 6: The same equations apply to both horizontal and vertical simple harmonic motion. The only effect of gravity on vertical SHM is to change the equilibrium position of the oscillating object.

Key Concept 7: Molecular vibrations, as well as the vibrations of many other systems, are approximately simple harmonic if the oscillation amplitude is sufficiently small.

Key Concept 8: The period of a simple pendulum depends only on the length of the pendulum and the value of g, not on the mass of the pendulum bob.

Key Concept 9: The period of a physical pendulum depends on the value of g, the distance from the pivot to the physical pendulum's center of gravity, and how the mass is distributed within the physical pendulum.

Key Concept 10: You can apply the idea of a physical pendulum to many oscillating systems, including the swinging limbs of humans and animals.

Conceptual Questions

1: Glider in simple harmonic motion

A glider attached to a spring and set on a horizontal air track is allowed to oscillate with a 5.0 cm amplitude. How far does the glider travel in one period?

IDENTIFY, SET UP, AND EXECUTE We answer the question by considering how the glider moves during one period. The period is the time an object takes to move from any position through one complete periodic cycle and return to the starting position. Imagine that the glider starts from an equilibrium position, moves to the right, and momentarily stops at a displacement equal to the amplitude. It has traveled a distance of one amplitude, or 5.0 cm. The glider then returns to its equilibrium position, traveling a second distance equal to the amplitude, or a total of 10.0 cm. The glider continues moving to the left until it reaches its maximum displacement on the left side, thus traveling a third distance equal to the amplitude

(15.0 cm total). The glider then moves to the right and returns to the starting position, traveling a fourth distance equal to the amplitude (20.0 cm total).

In one period, the glider travels a distance equal to four amplitudes, or 20.0 cm.

EVALUATE You must distinguish between amplitude and total distance traveled. Comprehending this difference helps build an understanding of simple harmonic motion.

2: Gravity on the moon

You are asked to estimate the moon's gravitational acceleration by watching a video of the early lunar explorations. How could you estimate the acceleration due to gravity on the moon?

IDENTIFY, SET UP, AND EXECUTE We've seen that, for small oscillations, the period of a simple pendulum is related to the gravitational constant and the length of the pendulum. If you can find an object that can be approximated by a simple pendulum, then you can determine the gravitational acceleration from the object's motion. One approach would be to look for a dangling object during a moonwalk. You can estimate the length of the pendulum by comparing it with the size of the astronaut on the walk and measure the time with a stopwatch or by counting video frames.

EVALUATE The moon's gravitational acceleration was estimated by physics students around the world watching the early moonwalks. The technique can also be used to estimate the sizes of objects in videos by taking the known gravitational acceleration value and combining it with the period to find the length of the pendulum.

Problems

1: Mass on a spring

A spring stretches 4.7 cm from its equilibrium position when a 1.2 kg mass is hung from it. If the mass is now stretched 6.5 cm from the equilibrium position and released, find (a) the period of the motion, (b) the maximum velocity of the mass, and (c) the maximum acceleration of the mass.

IDENTIFY Since the net external force acting on the block is proportional to the displacement of the block, the motion is simple harmonic motion. The target variables are the period, maximum velocity, and maximum acceleration.

SET UP We'll use the equations of simple harmonic motion to find the solutions to the problem. We first find the spring constant, using the preliminary information.

EXECUTE We find the spring constant from Hooke's law. When the mass is initially attached to the spring, it hangs in equilibrium, so the spring force is equal to the product of the mass and the acceleration due to gravity:

$$F_s = k\,x = mg.$$

The spring constant is

$$k = \frac{mg}{x} = \frac{(1.2 \text{ kg})(9.8 \text{ m/s}^2)}{0.047 \text{ m}} = 250 \text{ N/m}.$$

With the spring constant, we can directly find the period:

$$T = 2\pi\sqrt{\frac{m}{k}} = 2\pi\sqrt{\frac{(1.2 \text{ kg})}{(250 \text{ N/m})}} = 0.44 \text{ s}.$$

The maximum velocity is

$$v_{max} = \sqrt{\frac{k}{m}}A = \sqrt{\frac{(250 \text{ N/m})}{(1.2 \text{ kg})}}(0.065 \text{ m}) = 0.94 \text{ m/s}.$$

The maximum (positive) acceleration occurs when the mass is at its most negative position, so

$$a_{max} = -\frac{k}{m}x = -\frac{(250 \text{ N/m})}{(1.2 \text{ kg})}(-0.065 \text{ m}) = 13.5 \text{ m/s}^2.$$

The mass oscillates with a period of 0.44 s and has a maximum velocity of 0.94 m/s and a maximum acceleration of 13.5 m/s^2, upward.

KEY CONCEPT 2 **EVALUATE** Simple harmonic motion is the most complicated motion we have studied to date. However, our previous experiences led to straightforward relationships from which we can easily extract useful information.

We could also have found the motion as a function of time, taken derivatives to find the velocity and acceleration as a function of time, and then determined the maxima—the amplitudes of the velocity and acceleration functions.

Practice Problem: How far does the mass travel in one complete oscillation? *Answer:* 0.26 m

Extra Practice: What is the maximum acceleration if you double the mass? *Answer:* 6.78 m/s^2

2: Object in SHM

A 200.0 g mass vibrates in SHM with a total energy of 25.0 J and a frequency of 5.0 Hz. Find the time it takes to move from 25.0 cm below to 25.0 cm above the equilibrium position.

IDENTIFY We'll use the simple harmonic motion relations to find the solution. The target variable is the time needed to move the specified distance.

SET UP We'll use the equation of simple harmonic motion to find the solutions to the problem. We'll find the times the mass is −0.25 cm and +25.0 cm from the equilibrium position. We'll need to find the spring constant and amplitude from the information given.

EXECUTE The spring constant can be found from the frequency equation:

$$f = \frac{1}{2\pi}\sqrt{\frac{k}{m}}.$$

Solving for k gives

$$k = m(2\pi f)^2 = (0.200 \text{ kg})(2\pi(5.0 \text{ Hz}))^2 = 197.4 \text{ N/m}.$$

The total energy is given by

$$E = \tfrac{1}{2}mv^2 + \tfrac{1}{2}kx^2.$$

The amplitude is the maximum displacement. At the maximum displacement, the velocity is zero. Solving for the amplitude results in

$$E = \tfrac{1}{2} k A^2,$$

$$A = \sqrt{2E/k} = \sqrt{2(24.7 \text{ J})/(197 \text{ N/m})} = 0.500 \text{ m}.$$

The position as a function of time is given by

$$x = A \sin \omega t = A \sin 2\pi f t.$$

We need to solve for the time when $x = -0.25$ m and $x = 0.25$ m. Solving gives

$$t_{-0.25 \text{ m}} = \frac{1}{2\pi f} \sin^{-1}\left(\frac{x}{A}\right) = \frac{1}{2\pi(5.0 \text{ Hz})} \sin^{-1}\left(\frac{-0.25 \text{ m}}{0.50 \text{ m}}\right) = -0.0167 \text{ s},$$

$$t_{+0.25 \text{ m}} = \frac{1}{2\pi f} \sin^{-1}\left(\frac{x}{A}\right) = \frac{1}{2\pi(5.0 \text{ Hz})} \sin^{-1}\left(\frac{0.25 \text{ m}}{0.50 \text{ m}}\right) = 0.0167 \text{ s}.$$

The time the mass takes to move from 25.0 cm below to 25.0 cm above the equilibrium position is 0.0333 s.

KEY CONCEPT 5 **EVALUATE** We see that the time the mass takes to move from half the amplitude below to half the amplitude above the equilibrium position is about 15% of the period. Does this make sense? Yes, it makes sense, since the velocity near the equilibrium point is maximal.

Practice Problem: How long would it take the mass to move from the equilibrium point to the maximum amplitude? *Answer:* 0.05 s, or one-fourth the period.

CAUTION Use radians! When you work with trigonometric functions, the arguments are in radians. You must either set your calculator to radian mode or convert degrees to radians after taking the inverse of the trigonometric function.

Extra Practice: How fast does the mass move at the equilibrium point? *Answer:* 15.7 m/s

3: Period of a simple pendulum

A simple pendulum reaches a maximum angle of 7.2° after swinging through the bottom of its path with a maximum speed of 0.35 m/s. What is the period of the pendulum's oscillation?

IDENTIFY The period of a simple pendulum depends on its length and the gravitational constant. We can find the target variable—the length—from the velocity and maximum angle.

SET UP The maximum angle, arc length, and length are related to each other. The amplitude and the maximum speed will be used to find the length of the pendulum, and the period will be derived from the length.

EXECUTE The amplitude of a simple pendulum is the maximum arc length, which is related to the maximum angle by the length:

$$S_{max} = L\theta_{max}.$$

The maximum velocity is

$$v_{max} = 2\pi f A = 2\pi f L\theta_{max}.$$

For a simple pendulum, the frequency is found from the length:

$$f = \frac{1}{2\pi}\sqrt{\frac{g}{L}}.$$

Combining these equations, we obtain the length:

$$v_{max} = 2\pi\left(\frac{1}{2\pi}\sqrt{\frac{g}{L}}\right)L\theta_{max} = \sqrt{gL}\,\theta_{max},$$

$$L = \frac{v_{max}^2}{g\theta_{max}^2} = \frac{(0.35 \text{ m/s})^2}{(9.8 \text{ m/s}^2)(0.126)^2} = 0.79 \text{ m}.$$

Note that we replaced the maximum angle of 7.2° with the equivalent 0.126 radian. The period is then

$$T = 2\pi\sqrt{\frac{L}{g}} = 2\pi\sqrt{\frac{(0.79 \text{ m})}{(9.8 \text{ m/s}^2)}} = 1.8 \text{ s}.$$

The pendulum has a length of 79 cm and a period of 1.8 s.

KEY CONCEPT 8 **EVALUATE** The period of a simple pendulum depends only on the length of the pendulum and the gravitational constant. The maximum angle and velocity provided enough information to solve the problem.

Practice Problem: What is the angular frequency of this pendulum? *Answer:* 3.5 rad/s

Extra Practice: What is the length of a pendulum having a period of 1.0 s? *Answer:* 0.25 m

4: Period of a physical pendulum

A thin, uniform rod is pivoted at a point one-quarter of its length from one end and is then pivoted at a point at its end. Find the ratio of the two periods.

IDENTIFY The period of a physical pendulum depends on its moment of inertia, its mass, the distance to its center of mass, and the gravitational constant. The target variable is the ratio of the periods for the two pivot points.

SET UP We'll calculate the moment of inertia for each of the two pivot points and then combine the two moments to form the ratio.

EXECUTE The moment of inertia of a rod about its end point is

$$I_{end} = \frac{1}{3}ML^2.$$

When the rod is pivoted at a point one-quarter along its length, the moment of inertia is found by the parallel-axis theorem:

$$I_{1/4} = I_{end} + mx^2 = \frac{1}{3}ML^2 + M\left(\frac{L}{4}\right)^2 = \frac{7}{48}ML^2.$$

The period of a physical pendulum is given by

$$T = 2\pi\sqrt{\frac{I}{mgd}}.$$

The ratio of the two periods is then

$$\frac{T_{1/4}}{T_{end}} = \frac{2\pi\sqrt{\frac{I_{1/4}}{mgd_{1/4}}}}{2\pi\sqrt{\frac{I_{end}}{mgd_{end}}}} = \sqrt{\frac{I_{1/4}d_{end}}{I_{end}d_{1/4}}} = \sqrt{\frac{\frac{7}{48}ML^2 \frac{L}{2}}{\frac{1}{3}ML^2 \frac{L}{4}}} = \sqrt{\frac{(7)(3)(4)}{(2)(48)}} = 0.94.$$

The ratio of the period when the rod is pivoted at a point one-quarter of its length from one end to the period when the rod is pivoted at a point at its end is 0.94.

KEY CONCEPT 9 **EVALUATE** We see that the period does not change substantially when the pivot point moves between the two positions.

Practice Problem: What is the period of a 1.0 m rod with a mass of 0.25 kg pivoting at one end? *Answer:* 1.64 s

Extra Practice: What is the period of a 1.0 m simple pendulum? *Answer:* 2.01 s

5: Oscillating blocks

Two blocks shown in Figure 1 oscillate on a frictionless surface with a frequency of 0.30 Hz. The top block has a mass of 2.0 kg and the bottom block has a mass of 4.5 kg. If the amplitude is increased to 25 cm, the top block begins to slide. What is the coefficient of static friction?

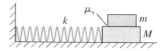

Figure 1 Problem 5.

IDENTIFY When the top block just begins to slide, the force applied must be equal to the maximum static frictional force. The maximum applied force occurs at the maximum displacement (equal to the amplitude). We will determine the force at the turning point and use that to solve for the coefficient of friction, the target variable.

SET UP We'll need the spring constant, which we extract from the initial frequency. The motion is simple harmonic motion, as the only horizontal force acting on the blocks is the spring force, a restoring force that is directly proportional to the displacement.

EXECUTE The frequency in simple harmonic motion depends on the spring constant according to the formula

$$f = \frac{1}{2\pi}\sqrt{\frac{k}{m+M}},$$

where we include the combined mass of the oscillating blocks. Solving for k gives

$$k = (2\pi f)^2(m+M) = (2\pi(0.30 \text{ Hz}))^2((2.0 \text{ kg}) + (4.5 \text{ kg})) = 23.1 \text{ N/m}.$$

Recall that the maximum static frictional force is $f_s = \mu_s n$. For the top block, the normal force is mg. The maximum force applied by the spring is kA. Equating these forces gives

$$f_s = \mu_s n = \mu_s mg = kA.$$

Rearranging terms to find the coefficient of static friction yields

$$\mu_s = \frac{kA}{mg} = \frac{(23.1 \text{ N/m})(0.25 \text{ m})}{(2.0 \text{ kg})(9.8 \text{ m/s}^2)} = 0.29.$$

The coefficient of static friction between the blocks is 0.29.

KEY CONCEPT 2 **EVALUATE** This problem brings together topics from several areas we have studied throughout the text, including the normal force, frictional forces, the spring force, and simple harmonic motion. Combining our knowledge helps us understand complex phenomena.

Practice Problem: At what amplitude does the top block start to slide if the spring constant is 50.0 N/m? *Answer:* 0.11 m

Extra Practice: How much time does it take for one complete oscillation? *Answer:* 3.33 s

6: Damped oscillation

An object with mass 0.30 kg hangs by a spring with force constant 50.0 N/m. By what factor is the frequency of oscillation reduced if the oscillation is damped and reaches $1/e$ of its original amplitude in 100 oscillations?

IDENTIFY The amplitude in damped SHM diminishes by a factor of $e^{-bt/2m}$ as a function of time. We'll set this quantity to $1/e$ to solve. The target variable is the fractional frequency shift, which is the change in frequency divided by the undamped frequency.

SET UP We'll write the fractional frequency shift in terms of the damped and undamped frequencies. We'll need to expand the frequency equation to simplify the shift and substitute the exponential decay information to solve.

EXECUTE The undamped frequency is given by

$$\omega = \sqrt{\frac{k}{m}}.$$

The damped frequency is given by

$$\omega' = \sqrt{\frac{k}{m} - \frac{b^2}{4m^2}}.$$

Combining these two equations to find the fractional frequency shift produces

$$\frac{\Delta\omega}{\omega} = \frac{\omega' - \omega}{\omega} = \frac{\sqrt{\frac{k}{m} - \frac{b^2}{4m^2}} - \sqrt{\frac{k}{m}}}{\sqrt{\frac{k}{m}}}$$

$$= \sqrt{\left[1 - \left(\frac{m}{k}\right)\frac{b^2}{4m^2}\right]} - 1.$$

For small damping, the second term in the square root is small, so we can simplify by using the approximation

$$\sqrt{1-x} \approx 1 - \frac{1}{2}x.$$

This gives

$$\frac{\Delta \omega}{\omega} \approx 1 - \frac{1}{2}\left(\frac{m}{k}\right)\frac{b^2}{4m^2} - 1 = \frac{1}{2}\left(\frac{m}{k}\right)\frac{b^2}{4m^2}.$$

We need to eliminate the damping term. We use the exponential decay information. The amplitude in damped oscillations changes as

$$A(t) = A_0 e^{-(b/2m)t}.$$

We know that after 100 oscillations the amplitude drops to $1/e$. This gives

$$e^{-(b/2m)t} = e^{-1}.$$

Taking the logarithm of each side yields an expression for $b/2m$:

$$-(b/2m)[100T] = -1$$

$$\frac{b}{2m} = \frac{1}{100}\frac{1}{T} = \frac{1}{100}\frac{1}{2\pi}\sqrt{\frac{k}{m}}.$$

We now solve for the shift:

$$\frac{\Delta \omega}{\omega} = \frac{1}{2}\left(\frac{m}{k}\right)\frac{b^2}{4m^2} = \frac{1}{2}\left(\frac{m}{k}\right)\left(\frac{1}{100}\frac{1}{2\pi}\sqrt{\frac{k}{m}}\right)^2 = \frac{1}{2}\left(\frac{1}{100}\frac{1}{2\pi}\right)^2 = 1.27 \times 10^{-6}.$$

The frequency shift is 1.27×10^{-6}.

EVALUATE We see that the frequency shift for this problem is very small.

Practice Problem: What is the frequency shift if the spring constant is doubled? *Answer:* 1.27×10^{-6}

Extra Practice: What is the frequency shift if it only takes 10 oscillations to reach $1/e$? *Answer:* 1.27×10^{-4}

Try It Yourself!

1: SHM practice

An object of mass 0.5 kg is attached to a spring with spring constant 100.0 N/m and is allowed to oscillate on a horizontal frictionless surface. It is given an initial velocity, at $x = 0$, of 5.0 m/s. Find (a) the total energy of the object, (b) the amplitude of oscillation, (c) the velocity when the displacement is half of the amplitude, (d) the displacement when the velocity is half of its initial value, (e) the displacement when the kinetic and potential energies are equal, and (f) the frequency and period of the motion.

Solution Checkpoints

IDENTIFY AND SET UP The object is in simple harmonic motion. Use the energy equations for SHM to solve for the many target variables. Remember that energy is conserved in SHM.

EXECUTE Energy conservation is used to solve (a) through (e) by substituting the appropriate knowns to find the unknowns. Energy conservation for the system states that

$$E = \tfrac{1}{2}mv^2 + \tfrac{1}{2}kx^2 = \tfrac{1}{2}kA^2.$$

This equation can be used to find (a) the total energy of the body (6.25 J), (b) the amplitude of oscillation (0.35 m), (c) the velocity when the displacement is half of the amplitude (± 4.34 m/s,) (d) the displacement when the velocity is half of its initial value (±0.31 m), and (e) the displacement when the kinetic and potential energies are equal (0.25 m).
The period and frequency are found from the relationships

$$f = \frac{1}{2\pi}\sqrt{\frac{k}{m}}, \quad T = \frac{1}{f}.$$

The period is 0.44 s and the frequency is 2.25 Hz.

KEY CONCEPT 4 **EVALUATE** This problem illustrates how to apply simple harmonic energy relations to find displacements, velocities, and the amplitude, period, and frequency of the motion.

Practice Problem: If the initial velocity is 2.5 m/s, then what is the period? *Answer:* 0.44 s

Extra Practice: What is the total energy in this case? *Answer:* 1.56 J

2: SHM practice

An object in SHM with angular frequency 0.5 rad/s is initially 10.0 cm from its equilibrium position and is moving back toward equilibrium with a velocity of 5.0 cm/s, as shown in Figure 2. How long does it take for the object to return to its equilibrium position?

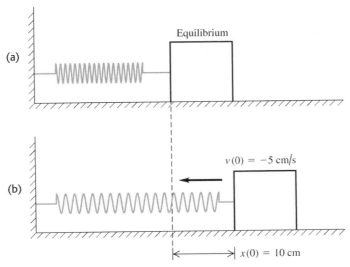

Figure 2 Try It Yourself 2.

Solution Checkpoints

IDENTIFY AND SET UP The object is in simple harmonic motion. Find the motion equation in terms of amplitude, angular frequency, and phase angle. Then solve for the time needed to move back to equilibrium, the target variable.

EXECUTE The general forms of the position and velocity equations are

$$x = A\cos(\omega t + \theta_0),$$
$$v = -\omega A \sin(\omega t + \theta_0).$$

To find the two constants, we use the initial conditions. At time zero, the position and velocity are given. This yields

$$A = 14.1 \text{ cm},$$
$$\theta_0 = 0.79 \text{ rad}.$$

The position is solved for time when $x = 0$, giving a time of 1.56 s.

KEY CONCEPT 3 **EVALUATE** This problem shows how to use initial conditions to solve for the equations of motion.

CAUTION Watch **f** and **ω**! Be careful to distinguish the frequency f from the angular frequency $\omega = 2\pi f$. Some problems involve both quantities.

Practice Problem: How much time would it take if the initial velocity was *away* from the equilibrium position? *Answer:* 4.72 s

Extra Practice: What is the spring constant if the mass is 3.5 kg? *Answer:* 0.88 N/m

3: Simple pendulum

A clock pendulum with mass 5.0 kg is set to swing with a 2.0 s period. How long should the pendulum be made if you approximate it as a simple pendulum?

Solution Checkpoints

IDENTIFY AND SET UP The period of a simple pendulum is given in terms of the length of the pendulum and the gravitational constant.

EXECUTE The period of a simple pendulum is given by

$$T = 2\pi\sqrt{\frac{L}{g}}.$$

Rearranging terms and solving yields a length of 0.99 m.

KEY CONCEPT 8 **EVALUATE** Mass does not affect the results for a simple pendulum.

Practice Problem: What is the angular frequency of this pendulum? *Answer:* 3.14 rad/s

Extra Practice: What would the length be if the period was doubled? *Answer:* 3.97 m

4: Physical pendulum

An object of mass 2.0 kg is suspended at a point 3.0 cm from its center of mass and observed to oscillate with a 2.0 s period. Find its moment of inertia.

Solution Checkpoints

IDENTIFY AND SET UP Is this a physical pendulum?

EXECUTE The period of a physical pendulum is given by

$$T = 2\pi \sqrt{\frac{I}{mgd}}.$$

Rearranging terms and solving yields a moment of inertia of 5.96×10^{22} kg m².

KEY CONCEPT 9 **EVALUATE** This problem shows another method of determining the moment of inertia: Set the object in oscillation and measure the period.

Practice Problem: How many oscillations occur in 30 seconds? *Answer:* 15

Extra Practice: What is the period when the same object is suspended 1.5 cm from its center of mass? *Answer:* 2.83 s

Key Example Variation Problems

Solutions to these problems are in Chapter 14 of the Student's Solutions Manual.

Be sure to review EXAMPLES 14.2 and 14.3 (Section 14.2) before attempting these problems.

VP14.3.1 A glider of mass 0.400 kg is placed on a frictionless, horizontal air track. One end of a horizontal spring is attached to the glider, and the other end is attached to the end of the track. When released, the glider oscillates in SHM with frequency 4.15 Hz. (a) Find the period and angular frequency of the motion. (b) Find the force constant k of the spring. (c) Find the magnitude of the force that the spring exerts on the glider when the spring is stretched by 0.0200 m.

VP14.3.2 A hockey puck attached to a horizontal spring oscillates on a frictionless, horizontal surface. The spring has force constant 4.50 N/m and the oscillation period is 1.20 s. (a) What is the mass of the puck? (b) During an oscillation, the acceleration of the puck has maximum magnitude 1.20 m/s². What is the amplitude of the oscillation?

VP14.3.3 The piston of a gasoline engine oscillates in SHM with frequency 50.0 Hz. At one point in the cycle the piston is 0.0300 m from equilibrium and moving at 12.4 m/s. (a) What is the amplitude of the motion? (b) What is the maximum speed the piston attains during its oscillation?

VP14.3.4 A cat is sleeping on a platform that oscillates from side to side in SHM. The combined mass of the cat and platform is 5.00 kg, and the force constant of the horizontal spring attached to the platform that makes it oscillate is 185 N/m. Ignore friction. (a) What is the frequency of the oscillation? (b) The cat will wake up if the acceleration of the platform is greater than 1.52 m/s². What is the maximum amplitude of oscillation that will allow the cat to stay asleep?

Be sure to review EXAMPLE 14.4 (Section 14.3) before attempting these problems.

VP14.4.1 A hockey puck oscillates on a frictionless, horizontal track while attached to a horizontal spring. The puck has mass 0.150 kg and the spring has force constant 8.00 N/m. The maximum speed of the puck during its oscillation is 0.350 m/s. (a) What is the amplitude of the oscillation? (b) What is the total mechanical energy of the oscillation? (c) What are the potential energy and the kinetic energy of the puck when the displacement of the glider is 0.0300 m?

VP14.4.2 A block of mass 0.300 kg attached to a horizontal spring oscillates on a frictionless surface. The oscillation has amplitude 0.0440 m, and total mechanical energy $E = 6.00 \times 10^{-2}$ J Find (a) the force constant of the spring and (b) the block's speed when the potential energy equals exactly $E/2$.

VP14.4.3 A glider attached to a horizontal spring oscillates on a horizontal air track. The total mechanical energy of the oscillation is 4.00×10^{-3} J, the amplitude of the oscillation is 0.0300 m, and the maximum speed of the glider is 0.125 m/s. (a) What are the force constant of the spring and the mass of the glider? (b) What is the maximum acceleration of the glider? (c) What is the magnitude of the glider's acceleration when the potential energy equals 3.00×10^{-3} J?

VP14.4.4 An object is undergoing SHM with amplitude A. For what values of the displacement is the kinetic energy equal to (a) 13 of the total mechanical energy; (b) 45 of the total mechanical energy?

Be sure to review EXAMPLES 14.8 and 14.9 (Sections 14.5 and 14.6) before attempting these problems.

VP14.9.1 On an alien planet, a simple pendulum of length 0.500 m has oscillation frequency 0.609 Hz. Find (a) the period of the pendulum and (b) the acceleration due to gravity on this planet's surface.

VP14.9.2 What must be the length of a simple pendulum if its oscillation frequency is to be equal to that of an air-track glider of mass 0.350 kg attached to a spring of force constant 8.75 N/m?

VP14.9.3 At a bicycle repair shop, a bicycle tire of mass M and radius R is suspended from a peg on the wall. The moment of inertia of the tire around the peg is $2MR^2$. If the tire is displaced from equilibrium and starts swinging back and forth, what will be its frequency of oscillation?

VP14.9.4 A rod has length 0.900 m and mass 0.600 kg and is pivoted at one end. The rod is *not* uniform; the center of mass of the rod is not at its center but is 0.500 m from the pivot. The period of the rod's motion as a pendulum is 1.59 s. What is the moment of inertia of the rod around the pivot?

STUDENT'S SOLUTIONS MANUAL FOR PERIODIC MOTION

VP14.3.1. **IDENTIFY:** The glider undergoes SHM on the spring.
SET UP: For SHM, $T = 1/f$, $\omega = 2\pi f$, $f = \frac{1}{2\pi}\sqrt{k/m}$, and $F = kx$ for an ideal spring.
EXECUTE: **(a)** $\omega = 2\pi f = 2\pi(4.15 \text{ Hz}) = 26.1 \text{ rad/s}$. $T = 1/f = 1/(4.15 \text{ Hz}) = 0.241 \text{ s}$.
(b) Use $f = \frac{1}{2\pi}\sqrt{k/m}$ to solve for k: $k = 4\pi^2 f^2 m = 4\pi^2 (4.15 \text{ Hz})^2 (0.400 \text{ kg}) = 272 \text{ N/m}$.
(c) $F = kx = (272 \text{ N/m})(0.0200 \text{ m}) = 5.44 \text{ N}$.
EVALUATE: The force in part (c) is around a pound.

VP14.3.2. **IDENTIFY:** The puck is executing SHM on the spring.
SET UP: For SHM $T = 1/f$, $T = 2\pi\sqrt{m/k}$, and $a_{max} = A\omega^2$. Our target variables are the mass of the puck and the amplitude of the oscillations.
EXECUTE: **(a)** Solve $T = 2\pi\sqrt{m/k}$ for m: $m = \frac{T^2 k}{4\pi^2} = \frac{(1.20 \text{ s})^2 (4.50 \text{ N/m})}{4\pi^2} = 0.164 \text{ kg}$.
(b) Solve $a_{max} = A\omega^2$ for A: $A = \frac{a_{max}}{\omega^2} = \frac{a_{max}}{(2\pi/T)^2} = \frac{1.20 \text{ m/s}^2}{[2\pi/(1.20 \text{ s})]^2} = 0.0438 \text{ m} = 4.38 \text{ cm}$.
EVALUATE: The maximum acceleration occurs at the instants that the puck has stopped moving, which are at the extremes of its motion when $x = \pm A$.

VP14.3.3. **IDENTIFY:** The piston is moving in SHM.
SET UP: For SHM, $x(t) = A\cos(\omega t + \phi)$ and $v_{max} = A\omega$. We know that the frequency is 50.0 Hz and at a certain instant $x = 0.0300$ m and $v = 12.5$ m/s; but we don't know if either x or v are positive or negative. The target variables are the amplitude of the motion and the maximum speed of the piston.

EXECUTE: (a) With $x = A\cos(\omega t + \phi)$, we have $v = dx/dt = -A\omega\sin(\omega t + \phi)$. Since we don't know the sign of x or v at the instant in question, so we can neglect the minus sign for v. Taking the ration of v/x gives $\dfrac{v}{x} = \dfrac{A\omega\sin(\omega t + \phi)}{A\cos(\omega t + \phi)} = \omega\tan(\omega t + \phi)$. We know this ratio at the instant in question, so we can use this result to find $(\omega t + \phi)$. Knowing this, we can use the known value of x to find the amplitude A. Using the known values, first find $(\omega t + \phi)$. Solving $\dfrac{v}{x} = \omega\tan(\omega t + \phi)$ for $(\omega t + \phi)$ gives

$(\omega t + \phi) = \arctan\left(\dfrac{v}{x\omega}\right) = \arctan\left(\dfrac{v}{2\pi f x}\right)$. Using the known values we have $(\omega t + \phi) =$ $\arctan\left(\dfrac{12.5 \text{ m/s}}{2\pi(50.0 \text{ Hz})(0.0300 \text{ m})}\right) = 52.984°$. Now use $x = A\cos(\omega t + \phi)$ to find A, giving 0.0300 m $= A\cos(52.984°)$, so $A = 0.0498$ m.

(b) $v_{max} = A\omega = 2\pi f A = 2\pi(50.0 \text{ Hz})(0.0498 \text{ m}) = 15.7$ m/s.

EVALUATE: From the information give, we don't know the exact position or velocity of the piston. It could be to the right or left of the origin moving either right or left. But none of this affects the amplitude or maximum speed.

VP14.3.4. **IDENTIFY:** The cat and platform oscillate together in SHM.

SET UP: We use $f = \dfrac{1}{2\pi}\sqrt{k/m}$ and $a_{max} = \omega^2 A$ with $\omega = 2\pi f$. The target variables are the frequency of vibration and the amplitude of the motion so that the acceleration that will not disturb the sleeping cat.

EXECUTE: (a) $f = \dfrac{1}{2\pi}\sqrt{k/m} = \dfrac{1}{2\pi}\sqrt{\dfrac{185 \text{ N/m}}{5.00 \text{ kg}}} = 0.968$ Hz. .

(b) Use $a_{max} = \omega^2 A = (2\pi f)^2 A$ to solve for A: $A = \dfrac{a_{max}}{(2\pi f)^2} = \dfrac{1.52 \text{ m/s}^2}{[2\pi(0.968 \text{ Hz})]^2} = 0.0411$ m.

EVALUATE: This motion makes about one vibration per second with an amplitude of about 4 cm, so it is not particularly fast.

VP14.4.1. **IDENTIFY:** This problem deals with the energy of a glider attached to a spring and oscillating with SHM.

SET UP: We use $v_{max} = A\omega$, $\omega = \sqrt{k/m}$, $K = \dfrac{1}{2}mv^2$, and $U = \dfrac{1}{2}kx^2$. The target variables are the amplitude A of the motion, the total mechanical energy E of the system, and the potential energy and kinetic energy at a certain point.

EXECUTE: (a) Use $v_{max} = A\omega$ and $\omega = \sqrt{k/m}$ to solve for A. We get $A = \dfrac{v_{max}}{\omega}$, which gives

$A = v_{max}\sqrt{\dfrac{m}{k}} = (0.350 \text{ m/s})\sqrt{\dfrac{0.150 \text{ kg}}{8.00 \text{ N/m}}} = 0.0479$ m.

(b) $E = K_{max} = \dfrac{1}{2}mv_{max}^2 = \dfrac{1}{2}(0.150 \text{ kg})(0.350 \text{ m/s})^2 = 9.19 \times 10^{-3}$ J.

(c) $U = \dfrac{1}{2}kx^2 = \dfrac{1}{2}(8.00 \text{ N/m})(0.0300 \text{ m})^2 = 3.60 \times 10^{-3}$ J.

$K = E - U = 9.19 \times 10^{-3}$ J $- 3.60 \times 10^{-3}$ J $= 5.59 \times 10^{-3}$ J.

EVALUATE: Another way to find the amplitude is to realize that $K_{max} = U_{max}$. Therefore $\frac{1}{2}mv_{max}^2 = \frac{1}{2}kA^2$, which gives $A = \sqrt{\frac{mv_{max}^2}{k}} = \sqrt{\frac{(0.150 \text{ kg})(0.350 \text{ m/s})^2}{8.00 \text{ N/m}}} = 0.479$ m, which agrees with your result.

VP14.4.2. **IDENTIFY:** The block oscillates in SHM on the spring.

SET UP: The total mechanical energy is $E = K + U$, where $K = \frac{1}{2}mv^2$. We also know that $\omega = \sqrt{k/m}$ and $v_{max} = A\omega$. Our target variables are the force constant of the spring and the speed of the block when the potential energy equals one-half the total mechanical energy.

EXECUTE: **(a)** When $x = 0$, $U = 0$ so $K = K_{max}$. Therefore $v = v_{max} = A\omega$. This tells us that

$E = \frac{1}{2}mv_{max}^2 = \frac{1}{2}m(A\omega)^2$. Using $\omega = \sqrt{k/m}$, this becomes $E = \frac{1}{2}m\left(A\sqrt{\frac{k}{m}}\right)^2$, which gives

$k = \frac{2E}{A^2} = \frac{2(6.00 \times 10^{-2} \text{ J})}{(0.0440 \text{ m})^2} = 62.0$ N/m.

(b) If $U = E/2$, then K must also equal $E/2$, so $\frac{1}{2}mv^2 = \frac{1}{2}E$. This gives $v = \sqrt{\frac{E}{m}} = \sqrt{\frac{6.00 \times 10^{-2} \text{ J}}{0.300 \text{ kg}}} = 0.447$ m/s.

EVALUATE: Note that we *cannot* say that $v = v_{max}/2$ when $K = K_{max}/2$ because K depends on the *square* of v.

VP14.4.3. **IDENTIFY:** The glider oscillates in SHM on the spring.

SET UP: The total mechanical energy is $E = K + U$, where $K = \frac{1}{2}mv^2$ and $U = \frac{1}{2}kx^2$. We know the total mechanical energy E of the system, the amplitude of the oscillations, and the maximum speed of the glider. We also know that $a_{max} = \omega^2 A$ and $\omega = \sqrt{k/m}$, $x(t) = A\cos(\omega t + \phi)$, and $a(t) = -\omega^2 A\cos(\omega t + \phi)$. Our target variables are the force constant of the spring, the mass m of the glider, maximum acceleration of the glider, and its acceleration when the potential energy is 3.00×10^{-3} J.

EXECUTE: **(a)** When $x = 0$, $U = 0$ so $K = K_{max}$ and $E = K_{max}$. So $\frac{1}{2}mv_{max}^2 = E$, which gives

$m = \frac{2E}{v_{max}^2} = \frac{2(4.00 \times 10^{-3} \text{ J})}{(0.125 \text{ m/s})^2} = 0.512$ kg.

When $x = A$, $K = 0$ so $U = U_{max}$, so $E = U_{max} = \frac{1}{2}kA^2$ which gives $k = \frac{2E}{A^2} = \frac{2(4.00 \times 10^{-3} \text{ J})}{(0.0300 \text{ m})^2} = 8.89$ N/m.

(b) $a_{max} = \omega^2 A = \frac{k}{m}A = \left(\frac{8.89 \text{ N/m}}{0.512 \text{ kg}}\right)(0.0300 \text{ m}) = 0.521$ m/s².

(c) We can use if we can find the value of $\cos(\omega t + \phi)$. We know that $U = 3.00 \times 10^{-3}$ J $= \frac{1}{2}kx^2$ and that $x(t) = A\cos(\omega t + \phi)$. Therefore $U = \frac{1}{2}kx^2 = \frac{1}{2}kA^2\cos^2(\omega t + \phi)$, which gives

$\cos(\omega t + \phi) = \sqrt{\dfrac{2U}{kA^2}}$. Now find the acceleration for this value of $\cos(\omega t + \phi)$. We can drop the minus sign since we only want the magnitude of the acceleration. $a(t) = \omega^2 A \cos(\omega t + \phi) = \omega^2 A \sqrt{\dfrac{2U}{kA^2}} = \dfrac{k}{m}\sqrt{\dfrac{2U}{k}}$. Using $k = 8.89$ N/m, $m = 0.512$ kg, and $U = 3.00 \times 10^{-3}$ J, we have $a = 0.451$ m/s^2.

EVALUATE: In part (c) the acceleration is less than the maximum of 0.521 m/s^2 from part (b), so our result is reasonable.

VP14.4.4. **IDENTIFY:** An object is in SHM, and the energy of the system is conserved.

SET UP: The total energy is $E = K + U$ and is constant, where $U = \dfrac{1}{2}kx^2$. We want to know the values of x when the kinetic energy is equal to 1/3 of the total mechanical energy and to 1/5 of the total mechanical energy.

EXECUTE: **(a)** If $K = 1/3\ E$, then $U = 2/3\ E$, and $E = \dfrac{1}{2}kA^2$. This gives $\dfrac{1}{2}kx^2 = \dfrac{2}{3}\left(\dfrac{1}{2}kA^2\right)$, so $x = \pm A\sqrt{\dfrac{2}{3}}$.

(b) Proceed as in part (a). If $K = 4/5\ E$, then $U = 1/5\ E$, which gives $\dfrac{1}{2}kx^2 = \dfrac{1}{5}\left(\dfrac{1}{2}kA^2\right)$, so $x = \pm \dfrac{A}{\sqrt{5}}$.

EVALUATE: We get square roots in our answers because U depends on the *square* of x and K depends on the *square* of v.

VP14.9.1. **IDENTIFY:** We are dealing with the small oscillations of a simple pendulum.

SET UP: $T = 1/f$ and $T = 2\pi\sqrt{L/g}$ for small oscillations. We want to find g on the alien planet, but first we need the period T.

EXECUTE: **(a)** $T = 1/f = 1/(0.609$ Hz$) = 1.64$ s.

(b) Solve $T = 2\pi\sqrt{L/g}$ for g, giving $g = \dfrac{4\pi^2 L}{T^2} = \dfrac{4\pi^2 (0.500\text{ m})}{(1.64\text{ s})^2} = 7.32$ m/s^2.

EVALUATE: This would be a very simple way to determine g since measurements of L and T are quite easy to make.

VP14.9.2. **IDENTIFY:** We are comparing the oscillation frequencies of a simple pendulum and an object attached to a spring.

SET UP: For the glider $\omega = \sqrt{k/m}$, and for a simple pendulum $\omega = \sqrt{g/L}$. The target variable is the length L of the pendulum so that it oscillates with the same frequency as the object attached to the spring.

EXECUTE: If the oscillation frequencies are the same, the angular frequencies must also be the same. So $\sqrt{k/m} = \sqrt{g/L}$, which gives $L = \dfrac{mg}{k} = \dfrac{(0.350\text{ kg})(9.80\text{ m/s}^2)}{8.75\text{ N/m}} = 0.392$ m.

EVALUATE: Our result is only accurate if the pendulum makes small oscillations.

VP14.9.3. IDENTIFY: The bicycle tire oscillates, but it is a *physical* pendulum, not a simple pendulum.

SET UP: The angular frequency is $\omega = \sqrt{\dfrac{Mgd}{I}}$ and $\omega = 2\pi f$.

EXECUTE: Using the given moment of inertia, we have $2\pi f = \sqrt{\dfrac{Mgd}{I}} = \sqrt{\dfrac{MgR}{2MR^2}} = \sqrt{\dfrac{g}{2R}}$. The oscillation frequency is $f = \dfrac{1}{2\pi}\sqrt{\dfrac{g}{2R}}$.

EVALUATE: If R is large, f is small, meaning that the wheel oscillates with a long period. This is analogous to a long simple pendulum, which oscillates slowly. If g were large, the frequency would be large since gravity could pull it back quickly from its extremes. Both cases suggest that our result is physically plausible.

VP14.9.4. IDENTIFY: The rod swings back and forth, but it is not a *simple* pendulum because its mass is spread out. Instead it is a *physical* pendulum.

SET UP: The period of swing is $T = 2\pi\sqrt{\dfrac{I}{mgd}}$. We want to find the moment of inertia of the rod.

EXECUTE: Use $T = 2\pi\sqrt{\dfrac{I}{mgd}}$ and solve for I, giving $I = mgd\left(\dfrac{T}{2\pi}\right)^2$. Therefore we get

$I = (0.600 \text{ kg})(9.80 \text{ m/s}^2)(0.500 \text{ m})\left(\dfrac{1.59 \text{ s}}{2\pi}\right)^2 = 0.188 \text{ kg}\cdot\text{m}^2$.

EVALUATE: If this rod were uniform, its moment of inertia about the pivot would be $I = \dfrac{1}{3}ML^2 = \dfrac{1}{3}(0.600 \text{ kg})(0.900 \text{ m})^2 = 0.162 \text{ kg}\cdot\text{m}^2$, which is *less than* we found. This is reasonable because the center of mass of this pendulum is *below* the midpoint of the rod, so it will have a larger moment of inertia about the pivot point than if it were uniform.

14.5. IDENTIFY: This displacement is $\tfrac{1}{4}$ of a period.

SET UP: $T = 1/f = 0.250$ s.

EXECUTE: $t = 0.0625$ s

EVALUATE: The time is the same for $x = A$ to $x = 0$, for $x = 0$ to $x = -A$, for $x = -A$ to $x = 0$ and for $x = 0$ to $x = A$.

14.9. IDENTIFY and SET UP: Use $T = 1/f$ to calculate T, $\omega = 2\pi f$ to calculate ω, and $\omega = \sqrt{k/m}$ for m.

EXECUTE: (a) $T = 1/f = 1/6.00 \text{ Hz} = 0.167$ s.
(b) $\omega = 2\pi f = 2\pi(6.00 \text{ Hz}) = 37.7$ rad/s.
(c) $\omega = \sqrt{k/m}$ implies $m = k/\omega^2 = (120 \text{ N/m})/(37.7 \text{ rad/s})^2 = 0.0844$ kg.

EVALUATE: We can verify that k/ω^2 has units of mass.

14.11. IDENTIFY: For SHM the motion is sinusoidal.

SET UP: $x(t) = A\cos(\omega t)$.

EXECUTE: $x(t) = A\cos(\omega t)$, where $A = 0.320$ m and $\omega = \dfrac{2\pi}{T} = \dfrac{2\pi}{0.900 \text{ s}} = 6.981$ rad/s.

(a) $x = 0.320$ m at $t_1 = 0$. Let t_2 be the instant when $x = 0.160$ m. Then we have
0.160 m $= (0.320$ m$) \cos(\omega t_2)$. $\cos(\omega t_2) = 0.500$. $\omega t_2 = 1.047$ rad. $t_2 = \dfrac{1.047 \text{ rad}}{6.981 \text{ rad/s}} = 0.150$ s. It takes $t_2 - t_1 = 0.150$ s.

(b) Let t_3 be when $x = 0$. Then we have $\cos(\omega t_3) = 0$ and $\omega t_3 = 1.571$ rad.
$t_3 = \dfrac{1.571 \text{ rad}}{6.981 \text{ rad/s}} = 0.225$ s. It takes $t_3 - t_2 = 0.225$ s $- 0.150$ s $= 0.0750$ s.

EVALUATE: Note that it takes twice as long to go from $x = 0.320$ m to $x = 0.160$ m than to go from $x = 0.160$ m to $x = 0$, even though the two distances are the same, because the speeds are different over the two distances.

14.13. IDENTIFY: Use $A = \sqrt{x_0^2 + \dfrac{v_{0x}^2}{\omega^2}}$ to calculate A. The initial position and velocity of the block determine ϕ. $x(t)$ is given by $x = A\cos(\omega t + \phi)$.

SET UP: $\cos\theta$ is zero when $\theta = \pm\pi/2$ and $\sin(\pi/2) = 1$.

EXECUTE: (a) From $A = \sqrt{x_0^2 + \dfrac{v_{0x}^2}{\omega^2}}$, $A = \left|\dfrac{v_0}{\omega}\right| = \left|\dfrac{v_0}{\sqrt{k/m}}\right| = 0.98$ m.

(b) Since $x(0) = 0$, $x = A\cos(\omega t + \phi)$ requires $\phi = \pm\dfrac{\pi}{2}$. Since the block is initially moving to the left, $v_{0x} < 0$ and $v_{0x} = -\omega A\sin\phi$ requires that $\sin\phi > 0$, so $\phi = +\dfrac{\pi}{2}$.

(c) $\cos(\omega t + \pi/2) = -\sin\omega t$, so $x = (-0.98$ m$) \sin[(12.2$ rad/s$)t]$.

EVALUATE: The $x(t)$ result in part (c) does give $x = 0$ at $t = 0$ and $x < 0$ for t slightly greater than zero.

14.15. IDENTIFY: The block oscillates in SHM.

SET UP: For the given initial conditions $x(t) = A\cos\omega t$ where $\omega = \sqrt{k/m} = 2\pi/T$ and $T = 2\pi\sqrt{m/k}$.

EXECUTE: (a) We want to find the time when $x(t) = A/2$ for the first time.
$x(t) = A\cos\omega t = A/2$, so $\omega t = 60° = \pi/3$ rad, which gives $t = \pi/3\omega$. Since $\omega = 2\pi/T$, we have
$t = \dfrac{\pi}{3\left(\dfrac{2\pi}{T}\right)} = T/6$.

(b) We want to find the time when the block first $v_{\max}/2$. The velocity is $v = dx/dt$, which gives
$\dfrac{d(A\cos\omega t)}{dt} = -A\omega\sin\omega t = -v_{\max}\sin\omega t$. We can drop the minus sign because we are interested only in the speed. Therefore $v = \dfrac{1}{2}v_{\max} = v_{\max}\sin\omega t$, so $\sin\omega t = \dfrac{1}{2}$, which means that $\omega t = 30° = \pi/6$.
Using $\omega = 2\pi/T$, we get $t = \dfrac{\pi/6}{\omega} = \dfrac{\pi}{6\left(\dfrac{2\pi}{T}\right)} = T/12$.

(c) The answer is no, the block does not reach $v_{\max}/2$ when $x = A/2$.

EVALUATE: We can check our result using energy conservation. The total mechanical energy E is
$E = \dfrac{1}{2}mv^2 + \dfrac{1}{2}kx^2 = \dfrac{1}{2}mv_{\max}^2 = \dfrac{1}{2}kA^2$. We know that $v_{\max} = A\omega = A\sqrt{k/m}$. Use energy to find x

when $v = v_{max}/2$, which gives $\frac{1}{2}m\left(\frac{v_{max}}{2}\right)^2 + \frac{1}{2}kx^2 = \frac{1}{2}kA^2$. Using $v_{max} = A\omega = A\sqrt{k/m}$, this becomes $\frac{1}{2}m\left(\frac{A\sqrt{k/m}}{2}\right)^2 + \frac{1}{2}kx^2 = \frac{1}{2}kA^2$. Squaring and solving for x gives $x = \frac{\sqrt{3}}{2}A$, which is *not* equal to $A/2$. This agrees with our result in part (c).

14.19. IDENTIFY: $T = 2\pi\sqrt{\frac{m}{k}}$. $a_x = -\frac{k}{m}x$ so $a_{max} = \frac{k}{m}A$. $F = -kx$.

SET UP: a_x is proportional to x so a_x goes through one cycle when the displacement goes through one cycle. From the graph, one cycle of a_x extends from $t = 0.10$ s to $t = 0.30$ s, so the period is $T = 0.20$ s. $k = 2.50$ N/cm $= 250$ N/m. From the graph the maximum acceleration is 12.0 m/s^2.

EXECUTE: (a) $T = 2\pi\sqrt{\frac{m}{k}}$ gives $m = k\left(\frac{T}{2\pi}\right)^2 = (250 \text{ N/m})\left(\frac{0.20 \text{ s}}{2\pi}\right)^2 = 0.253$ kg

(b) $A = \frac{ma_{max}}{k} = \frac{(0.253 \text{ kg})(12.0 \text{ m/s}^2)}{250 \text{ N/m}} = 0.0121$ m $= 1.21$ cm

(c) $F_{max} = kA = (250 \text{ N/m})(0.0121 \text{ m}) = 3.03$ N.

EVALUATE: We can also calculate the maximum force from the maximum acceleration: $F_{max} = ma_{max} = (0.253 \text{ kg})(12.0 \text{ m/s}^2) = 3.04$ N, which agrees with our previous results.

14.21. IDENTIFY: Compare the specific $x(t)$ given in the problem to the general form $x = A\cos(\omega t + \phi)$.

SET UP: $A = 7.40$ cm, $\omega = 4.16$ rad/s, and $\phi = -2.42$ rad.

EXECUTE: (a) $T = \frac{2\pi}{\omega} = \frac{2\pi}{4.16 \text{ rad/s}} = 1.51$ s.

(b) $\omega = \sqrt{\frac{k}{m}}$ so $k = m\omega^2 = (1.50 \text{ kg})(4.16 \text{ rad/s})^2 = 26.0$ N/m

(c) $v_{max} = \omega A = (4.16 \text{ rad/s})(7.40 \text{ cm}) = 30.8$ cm/s

(d) $F_x = -kx$ so $F_{max} = kA = (26.0 \text{ N/m})(0.0740 \text{ m}) = 1.92$ N.

(e) $x(t)$ evaluated at $t = 1.00$ s gives $x = -0.0125$ m. $v_x = -\omega A \sin(\omega t + \phi) = 30.4$ cm/s. $a_x = -kx/m = -\omega^2 x = +0.216$ m/s^2.

(f) $F_x = -kx = -(26.0 \text{ N/m})(-0.0125 \text{ m}) = +0.325$ N

EVALUATE: The maximum speed occurs when $x = 0$ and the maximum force is when $x = \pm A$.

14.25. IDENTIFY: The mechanical energy of the system is conserved. The maximum acceleration occurs at the maximum displacement and the motion is SHM.

SET UP: Energy conservation gives $\frac{1}{2}mv_{max}^2 = \frac{1}{2}kA^2$ and $a_{max} = \frac{kA}{m}$.

EXECUTE: $A = 0.165$ m. $\frac{1}{2}mv_{max}^2 = \frac{1}{2}kA^2$ gives $\frac{k}{m} = \left(\frac{v_{max}}{A}\right)^2 = \left(\frac{3.90 \text{ m/s}}{0.165 \text{ m}}\right)^2 = 558.7$ s^{-2}.

$a_{max} = \frac{kA}{m} = (558.7 \text{ s}^{-2})(0.165 \text{ m}) = 92.2$ m/s^2.

EVALUATE: The acceleration is much greater than g.

14.29. **IDENTIFY:** Velocity and position are related by $E = \frac{1}{2}kA^2 = \frac{1}{2}mv_x^2 + \frac{1}{2}kx^2$. Acceleration and position are related by $-kx = ma_x$.

SET UP: The maximum speed is at $x = 0$ and the maximum magnitude of acceleration is at $x = \pm A$.

EXECUTE: **(a)** For $x = 0$, $\frac{1}{2}mv_{max}^2 = \frac{1}{2}kA^2$ and $v_{max} = A\sqrt{\frac{k}{m}} = (0.040 \text{ m})\sqrt{\frac{450 \text{ N/m}}{0.500 \text{ kg}}} = 1.20 \text{ m/s}$

(b) $v_x = \pm\sqrt{\frac{k}{m}}\sqrt{A^2 - x^2} = \pm\sqrt{\frac{450 \text{ N/m}}{0.500 \text{ kg}}}\sqrt{(0.040 \text{ m})^2 - (0.015 \text{ m})^2} = \pm 1.11 \text{ m/s}$.

The speed is $v = 1.11$ m/s.

(c) For $x = \pm A$, $a_{max} = \frac{k}{m}A = \left(\frac{450 \text{ N/m}}{0.500 \text{ kg}}\right)(0.040 \text{ m}) = 36 \text{ m/s}^2$

(d) $a_x = -\frac{kx}{m} = -\frac{(450 \text{ N/m})(-0.015 \text{ m})}{0.500 \text{ kg}} = +13.5 \text{ m/s}^2$

(e) $E = \frac{1}{2}kA^2 = \frac{1}{2}(450 \text{ N/m})(0.040 \text{ m})^2 = 0.360$ J

EVALUATE: The speed and acceleration at $x = -0.015$ m are less than their maximum values.

14.33. **IDENTIFY:** Conservation of energy says $\frac{1}{2}mv^2 + \frac{1}{2}kx^2 = \frac{1}{2}kA^2$ and Newton's second law says $-kx = ma_x$.

SET UP: Let $+x$ be to the right. Let the mass of the object be m.

EXECUTE: $k = -\frac{ma_x}{x} = -m\left(\frac{-8.40 \text{ m/s}^2}{0.600 \text{ m}}\right) = (14.0 \text{ s}^{-2})m$.

$A = \sqrt{x^2 + (m/k)v^2} = \sqrt{(0.600 \text{ m})^2 + \left(\frac{m}{(14.0 \text{ s}^{-2})m}\right)(2.20 \text{ m/s})^2} = 0.840$ m. The object will therefore travel $0.840 \text{ m} - 0.600 \text{ m} = 0.240$ ma to the right before stopping at its maximum amplitude.

EVALUATE: The acceleration is not constant and we cannot use the constant acceleration kinematic equations.

14.39. **IDENTIFY:** $K = \frac{1}{2}mv^2$, $U_{grav} = mgy$ and $U_{el} = \frac{1}{2}kx^2$.

SET UP: At the lowest point of the motion, the spring is stretched an amount $2A$.

EXECUTE: **(a)** At the top of the motion, the spring is unstretched and so has no potential energy, the cat is not moving and so has no kinetic energy, and the gravitational potential energy relative to the bottom is $2mgA = 2(4.00 \text{ kg})(9.80 \text{ m/s}^2)(0.050 \text{ m}) = 3.92$ J. This is the total energy, and is the same total for each part.

(b) $U_{grav} = 0$, $K = 0$, so $U_{spring} = 3.92$ J.

(c) At equilibrium the spring is stretched half as much as it was for part (a), and so $U_{spring} = \frac{1}{4}(3.92 \text{ J}) = 0.98$ J, $U_{grav} = \frac{1}{2}(3.92 \text{ J}) = 1.96$ J, and so $K = 0.98$ J.

EVALUATE: During the motion, work done by the forces transfers energy among the forms kinetic energy, gravitational potential energy and elastic potential energy.

14.41. IDENTIFY and SET UP: The number of ticks per second tells us the period and therefore the frequency. We can use a formula from Table 9.2 to calculate I. Then $f = \dfrac{1}{2\pi}\sqrt{\dfrac{\kappa}{I}}$ allows us to calculate the torsion constant κ.

EXECUTE: Ticks four times each second implies 0.25 s per tick. Each tick is half a period, so $T = 0.50$ s and $f = 1/T = 1/0.50$ s $= 2.00$ Hz.

(a) Thin rim implies $I = MR^2$ (from Table 9.2).
$I = (0.900 \times 10^{-3}\text{ kg})(0.55 \times 10^{-2}\text{ m})^2 = 2.7 \times 10^{-8}\text{ kg}\cdot\text{m}^2$

(b) $T = 2\pi\sqrt{I/\kappa}$ so $\kappa = I(2\pi/T)^2 = (2.7 \times 10^{-8}\text{ kg}\cdot\text{m}^2)(2\pi/0.50\text{ s})^2 = 4.3 \times 10^{-6}\text{ N}\cdot\text{m/rad}$

EVALUATE: Both I and κ are small numbers.

14.47. IDENTIFY: Apply $T = 2\pi\sqrt{L/g}$

SET UP: The period of the pendulum is $T = (136\text{ s})/100 = 1.36$ s.

EXECUTE: $g = \dfrac{4\pi^2 L}{T^2} = \dfrac{4\pi^2(0.500\text{ m})}{(1.36\text{ s})^2} = 10.7\text{ m/s}^2$.

EVALUATE: The same pendulum on earth, where g is smaller, would have a larger period.

14.49. IDENTIFY: $a_{\text{tan}} = L\alpha$, $a_{\text{rad}} = L\omega^2$ and $a = \sqrt{a_{\text{tan}}^2 + a_{\text{rad}}^2}$. Use energy conservation in parts (b) and (c).

SET UP: Just after the sphere is released, $\omega = 0$ and $a_{\text{rad}} = 0$. When the rod is vertical, $a_{\text{tan}} = 0$.

EXECUTE: (a) The forces and acceleration are shown in Figure 14.49(a). $a_{\text{rad}} = 0$ so $a = a_{\text{tan}} = g\sin\theta$.

(b) The forces and acceleration are shown in Figure 14.49(b). In this case, the sphere has radial and tangential acceleration, so we need to use $a = \sqrt{a_{\text{tan}}^2 + a_{\text{rad}}^2}$. Use energy conservation, calling point 1 the instant that the sphere is released from rest at angle θ and point 2 the instant the rod makes an angle ϕ with the vertical. This gives $U_1 = U_2 + K_2$, so $mgL(1 - \cos\theta) = mgL(1 - \cos\phi) + \dfrac{1}{2}mv^2$.

Solving for v^2 gives $v^2 = 2gL(\cos\phi - \cos\theta)$. Therefore $a_{\text{rad}} = \dfrac{v^2}{L} = 2g(\cos\phi - \cos\theta)$. To find a_{tan}, apply $\Sigma\vec{\tau} = I\vec{\alpha}$: $mgL\sin\phi = mL^2\alpha = mL(L\alpha) = mL\tan\alpha$, which gives $a_{\text{tan}} = g\sin\phi$. The magnitude of the acceleration is $a = \sqrt{a_{\text{tan}}^2 + a_{\text{rad}}^2} = \sqrt{(g\sin\phi)^2 + [2g(\cos\phi - \cos\theta)]^2}$, which simplifies to $a = g\sqrt{\sin^2\phi + 4(\cos\phi - \cos\theta)^2}$. In this case $\phi = \theta/2$, so the acceleration is
$a = g\sqrt{\sin^2(\theta/2) + 4[\cos(\theta/2) - \cos\theta]^2}$.

(c) The forces and acceleration are shown in Figure 14.49(c). Calling point 2 the lowest part of the swing, $U_1 = K_2$ gives $mgL(1 - \cos\theta) = \dfrac{1}{2}mv^2$ and $v = \sqrt{2gL(1 - \cos\theta)}$. Using the formula derived in part (b) with $\phi = 0°$, the acceleration is $a = g\sqrt{\sin^2 0° + 4(\cos 0° - \cos\theta)^2} = 2g(1 - \cos\theta)$.

EVALUATE: As the rod moves toward the vertical, v increases, a_{rad} increases and a_{tan} decreases. The result in (c) agrees with the fact that $a_{tan} = 0$ when the rod is vertical because in that case $a = a_r$.

$$a_{rad} = \frac{v^2}{L} = \frac{\left(\sqrt{2gL(1-\cos\theta)}\right)^2}{L} = 2g(1-\cos\theta),$$ which is what we found in part (c).

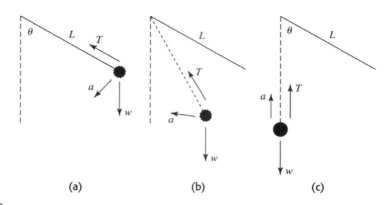

Figure 14.49

14.51. IDENTIFY: Pendulum A can be treated as a simple pendulum. Pendulum B is a physical pendulum.

SET UP: For pendulum B the distance d from the axis to the center of gravity is $3L/4$.
$I = \frac{1}{3}(m/2)L^2$ for a bar of mass $m/2$ and the axis at one end. For a small ball of mass $m/2$ at a distance L from the axis, $I_{ball} = (m/2)L^2$.

EXECUTE: Pendulum A: $T_A = 2\pi\sqrt{\frac{L}{g}}$.

Pendulum B: $I = I_{bar} + I_{ball} = \frac{1}{3}(m/2)L^2 + (m/2)L^2 = \frac{2}{3}mL^2$.

$T_B = 2\pi\sqrt{\frac{I}{mgd}} = 2\pi\sqrt{\frac{\frac{2}{3}mL^2}{mg(3L/4)}} = 2\pi\sqrt{\frac{L}{g}}\sqrt{\frac{2}{3}\cdot\frac{4}{3}} = \sqrt{\frac{8}{9}}\left(2\pi\sqrt{\frac{L}{g}}\right) = 0.943T_A$. The period is longer for pendulum A.

EVALUATE: Example 14.9 shows that for the bar alone, $T = \sqrt{\frac{2}{3}}T_A = 0.816T_A$. Adding the ball of equal mass to the end of the rod increases the period compared to that for the rod alone.

14.53. IDENTIFY: Pendulum A can be treated as a simple pendulum. Pendulum B is a physical pendulum. Use the parallel-axis theorem to find the moment of inertia of the ball in B for an axis at the top of the string.

SET UP: For pendulum B the center of gravity is at the center of the ball, so $d = L$. For a solid sphere with an axis through its center, $I_{cm} = \frac{2}{5}MR^2$. $R = L/2$ and $I_{cm} = \frac{1}{10}ML^2$.

EXECUTE: Pendulum A: $T_A = 2\pi\sqrt{\frac{L}{g}}$.

Pendulum B: The parallel-axis theorem says $I = I_{cm} + ML^2 = \tfrac{11}{10}ML^2$.

$$T = 2\pi\sqrt{\frac{I}{mgd}} = 2\pi\sqrt{\frac{11ML^2}{10MgL}} = \sqrt{\frac{11}{10}}\left(2\pi\sqrt{\frac{L}{g}}\right) = \sqrt{\frac{11}{10}}T_A = 1.05T_A.$$ It takes pendulum B longer to completea swing.

EVALUATE: The center of the ball is the same distance from the top of the string for both pendulums, but the mass is distributed differently and I is larger for pendulum B, even though the masses are the same.

14.57. **IDENTIFY and SET UP:** Use $\omega' = \sqrt{(k/m)-(b^2/4m^2)}$ to calculate ω', and then $f' = \omega'/2\pi$.

(a) EXECUTE: $\omega' = \sqrt{(k/m)-(b^2/4m^2)} = \sqrt{\dfrac{2.50\text{ N/m}}{0.300\text{ kg}} - \dfrac{(0.900\text{ kg/s})^2}{4(0.300\text{ kg})^2}} = 2.47$ rad/s

$f' = \omega'/2\pi = (2.47\text{ rad/s})/2\pi = 0.393$ Hz

(b) IDENTIFY and SET UP: The condition for critical damping is $b = 2\sqrt{km}$.

EXECUTE: $b = 2\sqrt{(2.50\text{ N/m})(0.300\text{ kg})} = 1.73$ kg/s

EVALUATE: The value of b in part (a) is less than the critical damping value found in part (b). With no damping, the frequency is $f = 0.459$ Hz; the damping reduces the oscillation frequency.

14.59. **IDENTIFY:** Apply $A = \dfrac{F_{max}}{\sqrt{\left(k - m\omega_d^2\right)^2 + b^2\omega_d^2}}$.

SET UP: $\omega_d = \sqrt{k/m}$ corresponds to resonance, and in this case $A = \dfrac{F_{max}}{\sqrt{\left(k - m\omega_d^2\right)^2 + b^2\omega_d^2}}$ reduces to $A = F_{max}/b\omega_d$.

EXECUTE: **(a)** $A_1/3$

(b) $2A_1$

EVALUATE: Note that the resonance frequency is independent of the value of b. (See Figure 14.28 in the textbook).

14.61. **IDENTIFY:** Two objects are in SHM on different springs, and we want to compare their maximum speed and maximum acceleration.

SET UP: We know that $v_{max} = A\omega = A\sqrt{k/m}$ and $a_{max} = \omega^2 A = (k/m)A$.

EXECUTE: **(a)** $\dfrac{v_{max,A}}{v_{max,B}} = \dfrac{\sqrt{\dfrac{k_A}{m_A}}A_A}{\sqrt{\dfrac{k_B}{m_B}}A_B} = \sqrt{\left(\dfrac{k_A}{k_B}\right)\left(\dfrac{m_B}{m_A}\right)}\left(\dfrac{A_A}{A_B}\right) = \sqrt{\left(\dfrac{9k_B}{k_B}\right)\left(\dfrac{4m_A}{m_A}\right)}\left(\dfrac{2A_B}{A_B}\right) = 12.$

(b) $\dfrac{a_{max,A}}{a_{max,B}} = \dfrac{\dfrac{k_A}{m_A}A_A}{\dfrac{k_B}{m_B}A_B} = \left(\dfrac{k_A}{k_B}\right)\left(\dfrac{m_B}{m_A}\right)\left(\dfrac{A_A}{A_B}\right) = \left(\dfrac{9k_B}{k_B}\right)\left(\dfrac{4m_A}{m_A}\right)\left(\dfrac{2A_B}{A_B}\right) = 72.$

EVALUATE: The ratio of the accelerations is considerably greater than that of the speeds because the acceleration depends on k/m while the speed depends on $\sqrt{k/m}$.

14.65. **IDENTIFY:** An object is executing SHM on a spring.

SET UP: We want to change the spring so that the amplitude A_2 is half the original amplitude A_1 and the mechanical energy E_2 is 4 times its original value E_1. Our target variables are the new force constant k_2 and the new maximum speed $v_{2,\text{max}}$ of the object. We know that $U_{\text{max}} = \frac{1}{2}kA^2$, $E = K + U = K_{\text{max}} = E_{\text{max}}$, and $v_{\text{max}} = A\omega = A\sqrt{k/m}$.

EXECUTE: **(a)** We want to relate k_2 to k_1. The original and final mechanical energies are

$$E_1 = \frac{1}{2}k_1 A_1^2 \text{ and } E_2 = \frac{1}{2}k_2 A_2^2 = 4E_1 = 4\left(\frac{1}{2}k_1 A_1^2\right), \text{ so } k_2 = 4\left(\frac{A_1}{A_2}\right)^2 k_1 = 4\left(\frac{2A_2}{A_2}\right)^2 k_1 = 16k_1.$$

(b) Using $v_{\text{max}} = A\omega = A\sqrt{k/m}$ and taking the ratio of the maximum speeds gives

$$\frac{v_{2,\text{max}}}{v_{1,\text{max}}} = \frac{\sqrt{\frac{k_2}{m}}A_2}{\sqrt{\frac{k_1}{m}}A_1} = \sqrt{\frac{k_2}{k_1}}\left(\frac{A_2}{A_1}\right) = \sqrt{\frac{16k_1}{k_1}}\left(\frac{A_2}{2A_2}\right) = 2, \text{ so } v_{2,\text{max}} = 2v_{1,\text{max}}.$$

EVALUATE: $E = \frac{1}{2}mv_{\text{max}}^2$, so if we increase E by a factor of 4, v_{max}^2 must increase by a factor of 4, so v_{max} must increase by a factor of 2, which is what we found in part (b).

14.67. **IDENTIFY:** A block is moving with SHM on a spring. Using measurements of its maximum speed and mass, we will use graphical interpretation.

SET UP: $v_{\text{max}} = A\omega = A\sqrt{k/m}$. Our target variable is the force constant of the spring. The data is plotted as v_{max}^2 versus $1/m$, so we need to find a relation between these quantities to interpret the graph.

EXECUTE: Solving $v_{\text{max}} = A\omega = A\sqrt{k/m}$ gives $v_{\text{max}}^2 = (A^2 k)(1/m)$, so a graph of v_{max}^2 versus $1/m$ should be a straight line having slope equal to $A^2 k$. Therefore $A^2 k =$ slope $= 8.62$ N·m, which gives $k = (8.62 \text{ N}\cdot\text{m})/(0.120 \text{ m})^2 = 599$ N/m.

EVALUATE: Converting 599 N/m to lb/in gives $k = 3.42$ lb/in. which is reasonable for a rather stiff spring.

14.69. **IDENTIFY:** The largest downward acceleration the ball can have is g whereas the downward acceleration of the tray depends on the spring force. When the downward acceleration of the tray is greater than g, then the ball leaves the tray. $y(t) = A\cos(\omega t + \phi)$.

SET UP: The downward force exerted by the spring is $F = kd$, where d is the distance of the object above the equilibrium point. The downward acceleration of the tray has magnitude $\frac{F}{m} = \frac{kd}{m}$, where m is the total mass of the ball and tray. $x = A$ at $t = 0$, so the phase angle ϕ is zero and $+x$ is downward.

EXECUTE: **(a)** $\frac{kd}{m} = g$ gives $d = \frac{mg}{k} = \frac{(1.775 \text{ kg})(9.80 \text{ m/s}^2)}{185 \text{ N/m}} = 9.40$ cm. This point is 9.40 cm above the equilibrium point so is $9.40 \text{ cm} + 15.0 \text{ cm} = 24.4 \text{ cm}$ above point A.

(b) $\omega = \sqrt{\dfrac{k}{m}} = \sqrt{\dfrac{185 \text{ N/m}}{1.775 \text{ kg}}} = 10.2$ rad/s. The point in (a) is above the equilibrium point so

$x = -9.40$ cm. $x = A\cos(\omega t)$ gives $\omega t = \arccos\left(\dfrac{x}{A}\right) = \arccos\left(\dfrac{-9.40 \text{ cm}}{15.0 \text{ cm}}\right) = 2.25$ rad.

$t = \dfrac{2.25 \text{ rad}}{10.2 \text{ rad/s}} = 0.221$ s.

(c) $\tfrac{1}{2}kx^2 + \tfrac{1}{2}mv^2 = \tfrac{1}{2}kA^2$ gives

$v = \sqrt{\dfrac{k}{m}(A^2 - x^2)} = \sqrt{\dfrac{185 \text{ N/m}}{1.775 \text{ kg}}([0.150 \text{ m}]^2 - [-0.0940 \text{ m}]^2)} = 1.19$ m/s.

EVALUATE: The period is $T = 2\pi\sqrt{\dfrac{m}{k}} = 0.615$ s. To go from the lowest point to the highest point takes time $T/2 = 0.308$ s. The time in (b) is less than this, as it should be.

14.71. IDENTIFY and SET UP: The bounce frequency is given by $f = \dfrac{1}{2\pi}\sqrt{\dfrac{k}{m}}$ and the pendulum frequency by $f = \dfrac{1}{2\pi}\sqrt{\dfrac{g}{L}}$. Use the relation between these two frequencies that is specified in the problem to calculate the equilibrium length L of the spring, when the apple hangs at rest on the end of the spring.

EXECUTE: Vertical SHM: $f_b = \dfrac{1}{2\pi}\sqrt{\dfrac{k}{m}}$

Pendulum motion (small amplitude): $f_p = \dfrac{1}{2\pi}\sqrt{\dfrac{g}{L}}$

The problem specifies that $f_p = \tfrac{1}{2}f_b$, so $\dfrac{1}{2\pi}\sqrt{\dfrac{g}{L}} = \dfrac{1}{2}\dfrac{1}{2\pi}\sqrt{\dfrac{k}{m}}$. Thus $g/L = k/4m$, which gives $L = 4gm/k = 4w/k = 4(1.00 \text{ N})/1.50 \text{ N/m} = 2.67$ m.

EVALUATE: This is the *stretched* length of the spring, its length when the apple is hanging from it. (Note: Small angle of swing means v is small as the apple passes through the lowest point, so a_{rad} is small and the component of mg perpendicular to the spring is small. Thus the amount the spring is stretched changes very little as the apple swings back and forth.)

IDENTIFY: Use Newton's second law to calculate the distance the spring is stretched from its unstretched length when the apple hangs from it.

SET UP: The free-body diagram for the apple hanging at rest on the end of the spring is given in Figure 14.71.

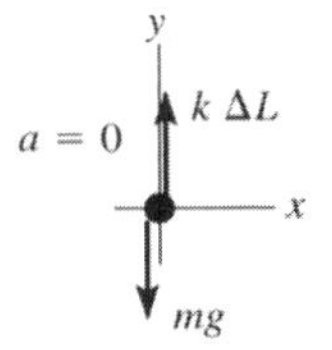

EXECUTE: $\Sigma F_y = ma_y$

$k\Delta L - mg = 0$

$\Delta L = mg/k = w/k = 1.00 \text{ N}/1.50 \text{ N/m} = 0.667$ m.

Thus the unstretched length of the spring is

$2.67 \text{ m} - 0.67 \text{ m} = 2.00$ m.

Figure 14.71

EVALUATE: The spring shortens to its unstretched length when the apple is removed.

14.73. IDENTIFY: The object oscillates as a physical pendulum, so $f = \dfrac{1}{2\pi}\sqrt{\dfrac{m_{object}gd}{I}}$. Use the parallel-axis theorem, $I = I_{cm} + Md^2$, to find the moment of inertia of each stick about an axis at the hook.

SET UP: The center of mass of the square object is at its geometrical center, so its distance from the hook is $L\cos 45° = L/\sqrt{2}$. The center of mass of each stick is at its geometrical center. For each stick, $I_{cm} = \tfrac{1}{12}mL^2$.

EXECUTE: The parallel-axis theorem gives I for each stick for an axis at the center of the square to be $\tfrac{1}{12}mL^2 + m(L/2)^2 = \tfrac{1}{3}mL^2$ and the total I for this axis is $\tfrac{4}{3}mL^2$. For the entire object and an axis at the hook, applying the parallel-axis theorem again to the object of mass $4m$ gives

$I = \tfrac{4}{3}mL^2 + 4m(L/\sqrt{2})^2 = \tfrac{10}{3}mL^2$.

$f = \dfrac{1}{2\pi}\sqrt{\dfrac{m_{object}gd}{I}} = \dfrac{1}{2\pi}\sqrt{\dfrac{4m_{object}gL/\sqrt{2}}{\tfrac{10}{3}m_{object}L^2}} = \sqrt{\dfrac{6}{5\sqrt{2}}}\left(\dfrac{1}{2\pi}\sqrt{\dfrac{g}{L}}\right) = 0.921\left(\dfrac{1}{2\pi}\sqrt{\dfrac{g}{L}}\right)$.

EVALUATE: Just as for a simple pendulum, the frequency is independent of the mass. A simple pendulum of length L has frequency $f = \dfrac{1}{2\pi}\sqrt{\dfrac{g}{L}}$ and this object has a frequency that is slightly less than this.

14.75. IDENTIFY: $T = 2\pi\sqrt{\dfrac{m}{k}}$ so the period changes because the mass changes.

SET UP: $\dfrac{dm}{dt} = -2.00\times 10^{-3}$ kg/s. The rate of change of the period is $\dfrac{dT}{dt}$.

EXECUTE: **(a)** When the bucket is half full, $m = 7.00$ kg. $T = 2\pi\sqrt{\dfrac{7.00\ \text{kg}}{450\ \text{N/m}}} = 0.784$ s.

(b) $\dfrac{dT}{dt} = \dfrac{2\pi}{\sqrt{k}}\dfrac{d}{dt}(m^{1/2}) = \dfrac{2\pi}{\sqrt{k}}\dfrac{1}{2}m^{-1/2}\dfrac{dm}{dt} = \dfrac{\pi}{\sqrt{mk}}\dfrac{dm}{dt}$.

$\dfrac{dT}{dt} = \dfrac{\pi}{\sqrt{(7.00\ \text{kg})(450\ \text{N/m})}}(-2.00\times 10^{-3}\ \text{kg/s}) = -1.12\times 10^{-4}$ s per s. $\dfrac{dT}{dt}$ is negative, so the period is getting shorter.

(c) The shortest period is when all the water has leaked out and $m = 2.00$ kg. In that case, $T = 2\pi\sqrt{m/k} = 0.419$ s.

EVALUATE: The rate at which the period changes is not constant but instead increases in time, even though the rate at which the water flows out is constant.

14.79. IDENTIFY: Apply conservation of linear momentum to the collision between the steak and the pan. Then apply conservation of energy to the motion after the collision to find the amplitude of the subsequent SHM. Use $T = 2\pi\sqrt{\dfrac{m}{k}}$ to calculate the period.

(a) SET UP: First find the speed of the steak just before it strikes the pan. Use a coordinate system with $+y$ downward.

$v_{0y} = 0$ (released from the rest); $y - y_0 = 0.40$ m; $a_y = +9.80$ m/s^2; $v_y = ?$

$v_y^2 = v_{0y}^2 + 2a_y(y - y_0)$

EXECUTE: $v_y = +\sqrt{2a_y(y-y_0)} = +\sqrt{2(9.80 \text{ m/s}^2)(0.40 \text{ m})} = +2.80 \text{ m/s}$

SET UP: Apply conservation of momentum to the collision between the steak and the pan. After the collision the steak and the pan are moving together with common velocity v_2. Let A be the steak and B be the pan. The system before and after the collision is shown in Figure 14.79.

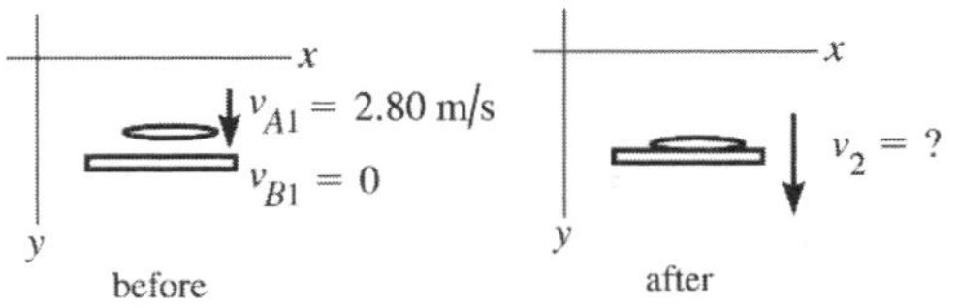

Figure 14.79

EXECUTE: P_y conserved: $m_A v_{A1y} + m_B v_{B1y} = (m_A + m_B) v_{2y}$

$m_A v_{A1} = (m_A + m_B) v_2$

$v_2 = \left(\dfrac{m_A}{m_A + m_B}\right) v_{A1} = \left(\dfrac{2.2 \text{ kg}}{2.2 \text{ kg} + 0.20 \text{ kg}}\right)(2.80 \text{ m/s}) = 2.57 \text{ m/s}$

(b) SET UP: Conservation of energy applied to the SHM gives: $\frac{1}{2} m v_0^2 + \frac{1}{2} k x_0^2 = \frac{1}{2} k A^2$ where v_0 and x_0 are the initial speed and displacement of the object and where the displacement is measured from the equilibrium position of the object.

EXECUTE: The weight of the steak will stretch the spring an additional distance d given by

$kd = mg$ so $d = \dfrac{mg}{k} = \dfrac{(2.2 \text{ kg})(9.80 \text{ m/s}^2)}{400 \text{ N/m}} = 0.0539 \text{ m}$. So just after the steak hits the pan, before the pan has had time to move, the steak plus pan is 0.0539 m above the equilibrium position of the combined object. Thus $x_0 = 0.0539$ m. From part (a) $v_0 = 2.57$ m/s, the speed of the combined object just after the collision. Then $\frac{1}{2} m v_0^2 + \frac{1}{2} k x_0^2 = \frac{1}{2} k A^2$ gives

$A = \sqrt{\dfrac{m v_0^2 + k x_0^2}{k}} = \sqrt{\dfrac{2.4 \text{ kg}(2.57 \text{ m/s})^2 + (400 \text{ N/m})(0.0539 \text{ m})^2}{400 \text{ N/m}}} = 0.21 \text{ m}$

(c) $T = 2\pi\sqrt{m/k} = 2\pi\sqrt{\dfrac{2.4 \text{ kg}}{400 \text{ N/m}}} = 0.49 \text{ s}$

EVALUATE: The amplitude is less than the initial height of the steak above the pan because mechanical energy is lost in the inelastic collision.

14.81. IDENTIFY: Use $x = A\cos(\omega t + \phi)$ to relate x and t. $T = 3.5$ s.

SET UP: The motion of the raft is sketched in Figure 14.81.

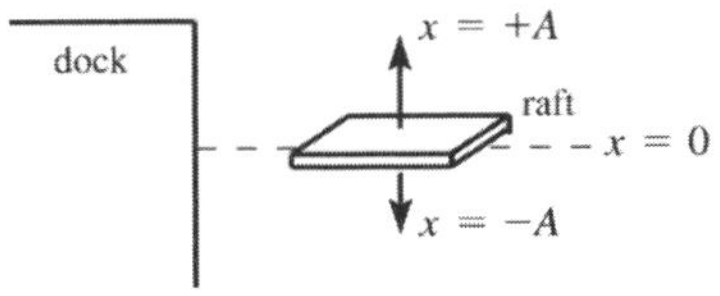

Let the raft be at $x = +A$ when $t = 0$.
Then $\phi = 0$ and $x(t) = A\cos\omega t$.

Figure 14.81

EXECUTE: Calculate the time it takes the raft to move from $x = +A = +0.200$ m to $x = A - 0.100$ m = 0.100 m.
Write the equation for $x(t)$ in terms of T rather than ω. $\omega = 2\pi/T$ gives that $x(t) = A\cos(2\pi t/T)$

$x = A$ at $t = 0$

$x = 0.100$ m implies 0.100 m $= (0.200$ m$) \cos(2\pi t/T)$

$\cos(2\pi t/T) = 0.500$ so $2\pi t/T = \arccos(0.500) = 1.047$ rad

$t = (T/2\pi)(1.047$ rad$) = (3.5$ s$/2\pi)(1.047$ rad$) = 0.583$ s

This is the time for the raft to move down from $x = 0.200$ m to $x = 0.100$ m. But people can also get off while the raft is moving up from $x = 0.100$ m to $x = 0.200$ m, so during each period of the motion the time the people have to get off is $2t = 2(0.583$ s$) = 1.17$ s.

EVALUATE: The time to go from $x = 0$ to $x = A$ and return is $T/2 = 1.75$ s. The time to go from $x = A/2$ to A and return is less than this.

14.83. **IDENTIFY:** During the collision, linear momentum is conserved. After the collision, mechanical energy is conserved and the motion is SHM.

SET UP: The linear momentum is $p_x = mv_x$, the kinetic energy is $\frac{1}{2}mv^2$, and the potential energy is $\frac{1}{2}kx^2$. The period is $T = 2\pi\sqrt{\frac{m}{k}}$, which is the target variable.

EXECUTE: Apply conservation of linear momentum to the collision:

$(8.00 \times 10^{-3}$ kg$)(280$ m/s$) = (1.00$ kg$)v$. $v = 2.24$ m/s. This is v_{max} for the SHM. $A = 0.150$ m (given). So $\frac{1}{2}mv_{max}^2 = \frac{1}{2}kA^2$. $k = \left(\frac{v_{max}}{A}\right)^2 m = \left(\frac{2.24 \text{ m/s}}{0.150 \text{ m}}\right)^2 (1.00$ kg$) = 223.0$ N/m.

$T = 2\pi\sqrt{\frac{m}{k}} = 2\pi\sqrt{\frac{1.00 \text{ kg}}{223.0 \text{ N/m}}} = 0.421$ s.

EVALUATE: This block would weigh about 2 pounds, which is rather heavy, but the spring constant is large enough to keep the period within an easily observable range.

14.87. **IDENTIFY:** The motion is simple harmonic if the equation of motion for the angular oscillations is of the form $\frac{d^2\theta}{dt^2} = -\frac{\kappa}{I}\theta$, and in this case the period is $T = 2\pi\sqrt{I/\kappa}$.

SET UP: For a slender rod pivoted about its center, $I = \frac{1}{12}ML^2$.

EXECUTE: The torque on the rod about the pivot is $\tau = -\left(k\frac{L}{2}\theta\right)\frac{L}{2}$. $\tau = I\alpha = I\frac{d^2\theta}{dt^2}$ gives

$\frac{d^2\theta}{dt^2} = -k\frac{L^2/4}{I}\theta = -\frac{3k}{M}\theta$. $\frac{d^2\theta}{dt^2}$ is proportional to θ and the motion is angular SHM. $\frac{\kappa}{I} = \frac{3k}{M}$,

$T = 2\pi\sqrt{\frac{M}{3k}}$.

EVALUATE: The expression we used for the torque, $\tau = -\left(k\frac{L}{2}\theta\right)\frac{L}{2}$, is valid only when θ is small enough for $\sin\theta \approx \theta$ and $\cos\theta \approx 1$.

14.89. **IDENTIFY:** The velocity is a sinusoidal function. From the graph we can read off the period and use it to calculate the other quantities.

SET UP: The period is the time for 1 cycle; after time T the motion repeats. The graph shows that $T = 1.60$ s and $v_{max} = 20.0$ cm/s. Mechanical energy is conserved, so $\frac{1}{2}mv_x^2 + \frac{1}{2}kx^2 = \frac{1}{2}kA^2$, and Newton's second law applies to the mass.

EXECUTE: (a) $T = 1.60$ s (from the graph with the problem).

(b) $f = \dfrac{1}{T} = 0.625$ Hz.

(c) $\omega = 2\pi f = 3.93$ rad/s.

(d) $v_x = v_{max}$ when $x = 0$ so $\frac{1}{2}kA^2 = \frac{1}{2}mv_{max}^2$. $A = v_{max}\sqrt{\dfrac{m}{k}}$. $f = \dfrac{1}{2\pi}\sqrt{\dfrac{k}{m}}$ so $A = v_{max}/(2\pi f)$.

From the graph in the problem, $v_{max} = 0.20$ m/s, so $A = \dfrac{0.20 \text{ m/s}}{2\pi(0.625 \text{ Hz})} = 0.051$ m $= 5.1$ cm. The mass is at $x = \pm A$ when $v_x = 0$, and this occurs at $t = 0.4$ s, 1.2 s, and 1.8 s.

(e) Newton's second law gives $-kx = ma_x$, so

$a_{max} = \dfrac{kA}{m} = (2\pi f)^2 A = (4\pi^2)(0.625 \text{ Hz})^2(0.051 \text{ m}) = 0.79$ m/s^2 = 79 cm/s^2. The acceleration is maximum when $x = \pm A$ and this occurs at the times given in (d).

(f) $T = 2\pi\sqrt{\dfrac{m}{k}}$ so $m = k\left(\dfrac{T}{2\pi}\right)^2 = (75 \text{ N/m})\left(\dfrac{1.60 \text{ s}}{2\pi}\right)^2 = 4.9$ kg.

EVALUATE: The speed is maximum at $x = 0$, when $a_x = 0$. The magnitude of the acceleration is maximum at $x = \pm A$, where $v_x = 0$.

14.91. **IDENTIFY AND SET UP:** For small-amplitude oscillations, the period of a simple pendulum is $T = 2\pi\sqrt{L/g}$.

EXECUTE: (a) The graph of T^2 versus L is shown in Figure 14.91a. Using $T = 2\pi\sqrt{L/g}$, we solve for T^2 in terms of L, which gives $T^2 = \left(\dfrac{4\pi^2}{g}\right)L$. The graph of T^2 versus L should be a straight line having slope $4\pi^2/g$. The best-fit line for our data has the equation $T^2 = (3.9795 \text{ s}^2/\text{m})L + 0.6674 \text{ s}^2$.

The quantity $4\pi^2/g = 4\pi^2/(9.80 \text{ m/s}^2) = 4.03$ s^2/m. Our line has slope 3.98 s^2/m, which is in very close agreement with the expected slope.

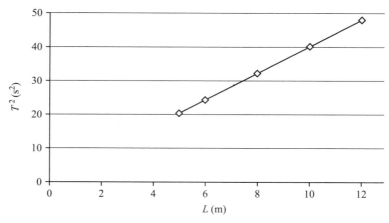

Figure 14.91a

(b) As L decreases, the angle the string makes with the vertical increases because the metal sphere is always released when it is touching the vertical wall. The formula $T = 2\pi\sqrt{L/g}$ is valid only for small angles. Figure 14.91b shows the graph of T/T_0 versus L.

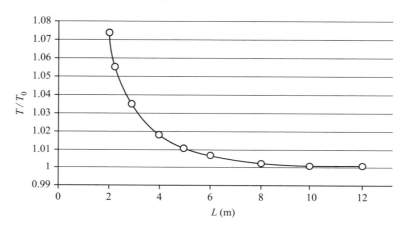

Figure 14.91b

(c) Since $T > T_0$, if T_0 is in error by 5%, $T/T_0 = 1.05$. From the graph in Figure 14.91b, that occurs for $L \approx 2.5$ m. In that case, $\sin\theta = (2.0 \text{ m})/(2.5 \text{ m}) = 0.80$, which gives $\theta = 53°$.

EVALUATE: Even for an angular amplitude of 53°, the error in using the formula $T = 2\pi\sqrt{L/g}$ is only 5%, so this formula is very useful in most situations. But for very large angular amplitudes it is not reliable.

STUDY GUIDE FOR MECHANICAL WAVES

Summary

In this chapter, we expand the concept of the periodic motion of an object to the periodic motion of many particles connected together as a medium. The periodic motion of a medium is a mechanical wave. Waves occur in many forms, including ocean waves, sound, light, earthquakes, and television transmission. This chapter will form the foundation for studying a variety of waves. We'll begin with the description of transverse and longitudinal waves and their amplitudes, periods, frequencies, and wavelengths. We'll see how waves move, interact, transmit energy, reflect, and combine in a variety of ways and how to describe their frequencies.

Objectives

After studying this chapter, you'll understand

- How to identify longitudinal and transverse waves and their media.
- The relations among the period, velocity, frequency, and wavelength of a wave.
- How a wave function that satisfies the wave equation describes a wave.
- The concepts of superposition, standing waves, nodes, and antinodes.
- The allowed frequencies for standing waves.
- How waves interact and interfere.

Concepts and Equations

Term	Description
Mechanical Wave	A mechanical wave is a disturbance from equilibrium that propagates from one region of space to another through a medium. In a transverse wave, the particles in the medium are displaced perpendicular to the direction of travel. In a longitudinal wave, the particles in the medium are displaced parallel to the direction of travel.
Periodic Mechanical Waves	In a periodic wave, particles in the medium exhibit periodic motion. The speed, wavelength, period, and frequency of a periodic wave are related by $$v = \lambda f = \frac{\lambda}{T}.$$ The speed of a transverse wave in a string under tension is given by $$v = \sqrt{\frac{F_T}{\mu}},$$ where F_T is the tension in the rope and μ is the mass per unit length.
Wave functions	The wave function $y(x, t)$ describes the displacements of individual particles in the medium. For a sinusoidal wave traveling in the $+x$- direction, the equation $$y(x,t) = A\cos\left[\omega\left(\frac{x}{v} - t\right)\right]$$ $$= A\cos(kx - \omega t)$$ describes the wave. The wave functions are solutions of the wave equation $$\frac{\partial^2 y(x,t)}{\partial x^2} = \frac{1}{v^2}\frac{\partial^2 y(x,t)}{\partial t^2}.$$
Wave Power	Waves convey energy from one region to another. The average power of a sinusoidal wave is proportional to the squares of the wave frequency and amplitude: $$P_{av} = \frac{1}{2}\sqrt{\mu F}\,\omega^2 A^2.$$ As a wave spreads out into three dimensions, the intensity drop is inversely proportional to the distance from the source: $$\frac{I_1}{I_2} = \frac{r_2^2}{r_1^2}.$$
Principle of Superposition	The principle of superposition states that when two waves overlap, the net displacement at any point at any time is found by taking the sum of the displacements of the individual waves: $$y(x,t) = y_1(x,t) + y_2(x,t).$$
Standing Waves	A standing wave is that combination of sinusoidal waves which produces a stationary sinusoidal pattern. Nodes are points where the standing wave pattern does not change with time. Antinodes are positions halfway between nodes; the amplitude is maximum at an antinode. The distance between successive nodes or antinodes is one-half of the wavelength. A string of length L held stationary at both ends can have standing waves only with frequencies such that $$f_n = n\frac{v}{2L} = nf_1 \quad (n = 1, 2, 3, \ldots).$$

	The fundamental frequency is given by $$f_1 = \frac{1}{2L}\sqrt{\frac{F}{\mu}}.$$ The multiples of f_1 are the harmonics. Each frequency and its associated vibration pattern is called a normal mode.

Key Concept 1: The product of a wave's wavelength and frequency has the same value no matter what the frequency is. This product equals the wave speed.

Key Concept 2: The wave function of a wave describes the displacement from equilibrium of the wave medium. It gives this displacement at any position in the medium and at any time.

Key Concept 3: The speed of a transverse wave on a string is determined by the tension in the string and the string's linear mass density (mass per unit length). For a given string, the wave speed is independent of frequency or wavelength.

Key Concept 4: The power (energy per second) carried by a wave on a string depends on the string tension and the linear mass density, as well as the frequency and amplitude of the wave. If the wave is sinusoidal, the average power equals one-half of the maximum instantaneous power.

Key Concept 5: If waves emitted from a source spread out equally in all directions, the wave intensity is inversely proportional to the square of the distance from the source.

Key Concept 6: A sinusoidal standing wave is the superposition of two sinusoidal waves of the same amplitude and frequency traveling in opposite directions. The wave nodes, which are spaced one half-wavelength apart, are points where the displacement in the standing wave is always zero.

Key Concept 7: A string with its ends fixed vibrates in a type of standing wave called a normal mode when all of its particles move sinusoidally with the same frequency. The frequencies of oscillation of the normal modes are integer multiples of a minimum normal-mode frequency, called the fundamental frequency.

Key Concept 8: A string vibrating at a certain frequency produces sound waves of the same frequency in the surrounding air. The standing wave on the string and the sound wave will have different wavelengths, however, if the speed of waves on the string does not equal the speed of sound.

Conceptual Questions

1: Waves in a jump rope

Your younger sister is playing with a jump rope. She ties one end to a fence post and moves the other end up and down, observing waves in the rope. She sees waves moving down the rope and becomes confused. She asks you why the rope doesn't move toward the fence, since that is how the waves move. How do you answer her question?

IDENTIFY, SET UP, AND EXECUTE Having just learned about mechanical waves, you explain that the rope is made up of lots of little pieces (particles) and the pieces at the end she holds move up and down as she moves her hand up and down. The pieces of rope next to her end are connected to the pieces she is moving and so are pulled up and down at the same time. These pieces, which take a little bit longer to move than the ones she holds, are connected to other pieces, which also move up and down. Each successive piece takes a bit longer to start moving than the piece before it, which is why there is a wave pattern. This wave pattern is what your sister observes moving down the rope. All of the individual pieces of rope move only up and down. Since they are connected to each other, their moving creates a disturbance in the rope that appears to move down the rope. The rope doesn't move toward the fence because none of the pieces of the rope move toward the fence.

EVALUATE Remember that when you pluck a string, you pull the string to the side, so you give the string a velocity to the side of, or perpendicular to, the string. During the pluck, you impart a force *perpendicular* to the string, not along the string.

2: Velocity of a wave

A taut string is plucked and a wave travels down the string at speed v. How can you double the speed of the wave?

IDENTIFY, SET UP, AND EXECUTE The speed of a transverse wave on a string is proportional to the square root of the tension on the string and inversely proportional to the square root of the mass per unit length. To double the speed, you can quadruple the tension in the string or decrease the mass per unit length by a factor of 4.

EVALUATE This problem illustrates the dependence of the speed of a wave in a string on the mass per length and the tension in the string.

3: Heavy rope

A heavy rope is suspended vertically and is stretched taut by a 10.0-kg mass attached to the bottom of the rope. The top end of the rope is plucked, creating a wave. Does the speed of the wave change as it propagates down the rope? If so, how?

IDENTIFY, SET UP, AND EXECUTE The speed of a transverse wave on a string depends on the tension and the mass per unit length in the rope. We can assume that the mass per unit length is constant. Because the rope is heavy, the top end must have greater tension than the bottom end, since the top end supports both the mass at the end and the rope itself. The tension, therefore, decreases toward the bottom of the rope. The decreasing tension will slow the speed of the wave as it travels down the rope.

The wave speed is not constant and slows as it approaches the bottom of the rope.

EVALUATE We'll normally not encounter the varying tension and speed found in this problem. We'll focus on light strings to understand wave propagation better.

Problems

1: An unusual scale

Your strange physics professor builds an unusual scale by hanging an object from a 3.0-m-long wire attached to the ceiling. She plucks the string just above the object and finds that the pulse takes 0.50 s to propagate up and down the wire. What is the mass of the object? The mass of the wire is 0.50 kg.

IDENTIFY AND SET UP The speed of a wave in a wire under tension is related to the tension in the wire. By finding the speed of the wave in the wire, we'll determine the mass of the object.

EXECUTE The speed of the wave in the wire is given by

$$v = \sqrt{\frac{F}{\mu}}.$$

The tension at the bottom of the wire is equal to the gravitational force on the object, since the object is in equilibrium. To calculate the tension, we first need the velocity and the mass per unit length. The velocity is found by noting that the wave takes 0.50 s to travel 6.0 m (up and down the wire):

$$v = \frac{\Delta d}{\Delta t} = \frac{6.0 \text{ m}}{0.5 \text{ s}} = 12 \text{ m/s}.$$

The mass per unit length is

$$\mu = \frac{m}{L} = \frac{0.50 \text{ kg}}{3.0 \text{ m}} = 0.167 \text{ kg/m}.$$

Substituting to find the tension, we obtain

$$F = v^2 \mu = (12 \text{ m/s})^2 (0.167 \text{ kg/m}) = 24 \text{ N}.$$

The tension force is equal to the weight, so the mass of the object is

$$m = \frac{F}{g} = \frac{24 \text{ N}}{9.8 \text{ m/s}^2} = 2.4 \text{ kg}.$$

The object's mass is 2.4 kg.

KEY CONCEPT 3 **EVALUATE** This unusual scale illustrates how we can use mechanical waves to measure mass, but it is impractical for several reasons. First, the scale requires a high ceiling and a method of accurately measuring the speed of waves in the wire. Also, we have omitted the mass of the wire, which is roughly 15% higher at the ceiling, in our calculation of the tension.

Practice Problem: How long would the pulse take to propagate up and down if a 6.0 kg object was hanging from the wire? *Answer:* 0.32 s

Extra Practice: What does the mass of the object need to be so that the waves propagate at a speed of 25.0 m/s? *Answer:* 10.7 kg

2: Write a wave equation

Write the wave equation for a traveling transverse wave that propagates in the $+x$- direction, has a maximum disturbance from equilibrium of 1.0 cm, and has a wavelength of 2.0 m and a period 0.02 s. At $x = 0.5$ m and $t = 0$, the instantaneous particle velocity is $\pi/2$ m/s downward.

IDENTIFY AND SET UP We'll begin with the general form of the wave function and use the given conditions to determine the constants.

EXECUTE The general form of a wave equation is

$$y(x, t) = A \sin(\omega t - kx + \phi).$$

Here, we used a minus sign in front of the wave number to ensure that the wave propagates toward positive x, and we included a phase angle. The frequency is found from the period:

$$\omega = 2\pi \frac{1}{T} = 314 \text{ rad/s}.$$

The wave number is related to the wavelength:

$$k = \frac{2\pi}{\lambda} = 3.14 \text{ /m}.$$

The amplitude is the maximum displacement from zero, so $A = 0.01$ m. To find the phase angle, we'll have to use the given velocity at the specified point. The velocity is the first derivative:

$$v_y = \frac{\partial y}{\partial t} = A\omega\cos(\omega t - kx + \phi).$$

At $x = 0.5$ m and $t = 0$, the instantaneous particle velocity is $\pi/2$ m/s downward, or negative. This gives

$$v_y(0.5 \text{ m}, 0) = -\pi/2 \text{ m/s}$$

$$(0.01 \text{ m})(314 \text{ rad/s})\cos((314 \text{ rad/s})(0) - (3.14 \text{ /m})(0.5 \text{ m}) + \phi) = -\pi/2 \text{ m/s}$$

$$(\pi)\cos(-\pi/2 + \phi) = -\pi/2$$

$$\sin(\phi) = -1/2.$$

The phase angle must be $-30°$, or $-\pi/6$. The complete wave function is

$$y(x,t) = (0.01 \text{ m})\sin((314 \text{ rad/s})t - (3.14 \text{ /m})x - \pi/6).$$

KEY CONCEPT 2   **EVALUATE** This problem illustrates how to construct a wave function, given the properties of the wave.

Practice Problem: How fast is the particle moving when $x = 0.25$ m and $t = 1.0$ s?
Answer: 3.03 m/s

Extra Practice: What is the initial phase angle if at $x = 0.5$ m and $t = 0$ s the instantaneous particle velocity is $\pi/2$ m/s *upward*? *Answer:* $+\pi/6$

3: Wave function check

Does $y(x,t) = Ae^{-kx}\sin\omega t$ satisfy the wave function equation?

IDENTIFY AND SET UP We'll take the second derivatives of the function with respect to both time and position. We'll then substitute the results into the wave equation and see if it is satisfied.

EXECUTE The wave equation is given by

$$\frac{\partial^2 y(x,t)}{\partial x^2} = \frac{1}{v^2}\frac{\partial^2 y(x,t)}{\partial t^2}.$$

We start by taking the derivative of the given function with respect to position. We have

$$\frac{\partial y(x,t)}{\partial x} = \frac{\partial Ae^{-kx}\sin\omega t}{\partial x} = A(-k)e^{-kx}\sin\omega t.$$

Next, we take the second derivative with respect to position:

$$\frac{\partial^2 y(x,t)}{\partial x^2} = \frac{\partial}{\partial x}(A(-k)e^{-kx}\sin\omega t) = Ak^2 e^{-kx}\sin\omega t.$$

We now switch to the time derivatives. The first derivative with respect to time is

$$\frac{\partial y(x,t)}{\partial t} = \frac{\partial Ae^{-kx}\sin\omega t}{\partial t} = A(\omega)e^{-kx}\cos\omega t.$$

The second derivative is

$$\frac{\partial^2 y(x,t)}{\partial t^2} = \frac{\partial^2 Ae^{-kx}\sin\omega t}{\partial t^2} = A(-\omega^2)e^{-kx}\sin\omega t.$$

Combining the results gives

$$A(-\omega^2)e^{-kx}\sin\omega t = A(k^2)e^{-kx}\sin\omega t,$$

$$Ae^{-kx}\sin\omega t = -\frac{1}{v^2}Ae^{-kx}\sin\omega t.$$

The latter formula would satisfy the wave equation were it not for the minus sign. The function does not satisfy the wave equation and is not a valid wave function.

EVALUATE Constructing valid wave functions takes practice and experience. We'll see that there are several common types of equations that describe most waves and satisfy the wave equation.

4: Combining strings

Two strings of mass per unit length μ_1 and μ_2 are joined together at their ends. The tension in the two strings is the same. If the wavelength in the first string with $\mu_1 = 5.0$ g/m is 3.0 cm, what is the mass per unit length in the second string if the wavelength in that string is 5.0 cm? Assume that the wave frequencies are the same in the two strings.

IDENTIFY We'll use the fact that the tension is the same in both strings to determine the target variable, the mass per unit length in the second string.

SET UP The speed of propagation depends on the mass per unit length and the tension. The tension and frequency are the same in both strings. We'll set these equal to each other to solve.

EXECUTE The speed is given by

$$v = \sqrt{\frac{F}{\mu}}.$$

Squaring both sides of this equation and solving for the tension gives

$$v^2\mu = F.$$

Since both tensions are the same, we have

$$v_1^2\mu_1 = v_2^2\mu_2.$$

The speeds may not be the same, but we know the frequencies and wavelengths in both strings. The frequencies are the same in both strings. Rewriting the speed in terms of frequency and wavelength gives

$$\lambda_1^2 f^2 \mu_1 = \lambda_2^2 f^2 \mu_2.$$

Solving for the mass per unit length yields

$$\mu_2 = \frac{\lambda_1^2 \mu_1}{\lambda_2^2} = \frac{(3.0 \text{ cm})^2 (5.0 \text{ g/m})}{(5.0 \text{ cm})^2} = 1.8 \text{ g/m}.$$

The second string has a mass per unit length of 1.8 g/m.

KEY CONCEPT 3 **EVALUATE** Do we expect a smaller mass per unit length in the second string? Yes, since the wavelength is larger in the second string, the mass per length must be less in order for the tensions to be the same.

Practice Problem: How do the speeds of the waves compare in the two strings? *Answer:* The speed of the wave in string 2 must be $\frac{5}{3}$ the speed in string 1.

Extra Practice: What is the wave frequency if the tension is 20.0 N? *Answer:* 2.11×10^3 Hz

5: Modes in a string

A uniform string of length 0.50 m is fixed at both ends. Find the wavelength of the fundamental mode of vibration. If the wave speed is 300 m/s, find the frequency of the fundamental and next possible modes.

IDENTIFY We'll use the properties and definitions of modes for a standing wave on a string to solve the problem.

SET UP The fundamental mode has a wavelength twice the length of the string. The frequency of the fundamental mode is the velocity divided by the wavelength. Higher frequencies are integer multiples of the fundamental frequency.

EXECUTE The wavelength of the fundamental mode is twice the length of the string. The wavelength is 1.0 m. For a wave speed of 300 m/s, the frequency of the fundamental mode is

$$f_1 = \frac{v}{2L} = 300 \text{ Hz}.$$

The next possible mode will have half of the fundamental mode's wavelength, or a wavelength of 0.5 m. Its frequency will be

$$f_2 = 2f_1 = \frac{v}{\lambda_2} = 600 \text{ Hz}.$$

KEY CONCEPT 7 **EVALUATE** This problem gives us practice understanding the properties of standing waves.

Practice Problem: What is the frequency of the third overtone? *Answer:* 1200 Hz

Extra Practice: What is the fundamental frequency if the length of the string is increased to 2.0 m? *Answer:* 75 Hz

Try It Yourself!

1: Wave function check

Does $y(x,t) = A(x - vt)^n$ satisfy the wave function equation? A is a constant and $n > 1$.

Solution Checkpoints

IDENTIFY AND SET UP Take the second derivatives of the function with respect to both time and position. Then substitute the results into the wave equation and see if it is satisfied.

EXECUTE The second derivative with respect to position is

$$\frac{\partial^2 y(x,t)}{\partial x^2} = n(n-1)A(x-vt)^{n-2}.$$

The second derivative with respect to time is

$$\frac{\partial^2 y(x,t)}{\partial t^2} = n(n-1)Av^2(x-vt)^{n-2}.$$

Combining the results, we see that the function does satisfy the wave equation and is a valid wave function.

EVALUATE What type of wave does the function represent? Is it periodic?

2: Changing the diameter

Waves propagate through a rope under tension, provided by a hanging mass of 20.0 kg, with a speed of 30.0 m/s. The rope is replaced by ropes made of the same material but different diameters. For the velocity to remain the same, what mass should be hung from the end of the rope if the replacement rope has (a) half the diameter and (b) twice the diameter of the original rope?

Solution Checkpoints

IDENTIFY AND SET UP The speed of propagation depends on the tension and the mass per unit length.

EXECUTE How must the mass per unit length and the tension relate if the speed is to remain constant? How does the mass change with (a) half the diameter rope and (b) twice the diameter rope?

With half the diameter, the volume decreases by a factor of four, so the mass should be 5.0 kg. With twice the diameter, the volume increases by a factor of four, so the mass should be 80.0 kg.

KEY CONCEPT 3 **EVALUATE** How do you confirm these results?

Practice Problem: What is the mass per unit length of the original rope? *Answer:* 0.218 kg/m

Extra Practice: At what speed does the wave propagate if the "half diameter" rope still has a hanging mass of 20.0 kg? *Answer:* 60 m/s

3: Modes in a free string

A uniform string of length 0.50 m is fixed at one end and free at the other end. Find the wavelength of the fundamental mode of vibration. If the wave speed is 300 m/s, find the frequency of the fundamental and next possible modes.

Solution Checkpoints

IDENTIFY Use the properties and definitions of modes for a standing wave on a string to solve the problem.

SET UP How do the fundamental mode's frequency and wavelength for a standing wave on a string with only one end fixed compare with the fundamental mode's frequency and wavelength for a string with both ends fixed?

EXECUTE The wavelength of the fundamental mode is four times the length, or 2.0 m. The frequency of the fundamental mode is then 150 Hz.

The next mode has a wavelength of $\frac{4}{3}L$, or 0.67 m, and a frequency of 450 Hz.

KEY CONCEPT 6 **EVALUATE:** Sketch the standing waves on the string for the fundamental and next possible modes. Does your sketch agree with the results you just calculated?

Practice Problem: As the rope vibrates in the second harmonic, how far along the rope do you have to go before reaching the first antinode? *Answer:* L/3

Extra Practice: Where is the second node located in that case? *Answer:* 2L/3

Key Example Variation Problems

Solutions to these problems are in Chapter 15 of the Student's Solutions Manual.

Be sure to review EXAMPLES 15.1, 15.2, and 15.3 (Sections 15.2, 15.3, and 15.4) before attempting these problems.

VP15.3.1 A boat is at anchor outside a harbor. A steady sinusoidal ocean wave makes the boat bob up and down with a period of 5.10 s and an amplitude of 1.00 m. The wave has wavelength 30.5 m. For this wave, what are (a) the frequency, (b) the wave speed, (c) the angular frequency, and (d) the wave number?

VP15.3.2 Sound waves in the thin Martian atmosphere travel at 245 m/s. (a) What are the period and wavelength of a 125 Hz sound wave in the Martian atmosphere?
(b) What are the frequency and angular frequency of a sound wave in the Martian atmosphere that has wavelength 3.00 m?

VP15.3.3 You are testing a mountain climbing rope that has a linear mass density of 0.0650 kg/m. The rope is held horizontal and is under a tension of 8.00×10^2 N to simulate the stress of supporting a mountain climber's weight. (a) What is the speed of transverse waves on this rope? (b) You oscillate one end of the rope up and down in SHM with frequency 25.0 Hz and amplitude 5.00 mm. What is the wavelength of the resulting waves on the rope? (c) At $t = 0$ the end you are oscillating is at its maximum positive displacement and is instantaneously at rest. Write an equation for the displacement as a function of time at a point 2.50 m from that end. Assume that no wave bounces back from the other end.

VP15.3.4 The tension in a long string is 25.0 N. You oscillate one end of the string up and down with frequency 45.0 Hz. When this end is at its maximum upward displacement, the nearest point that is at its maximum negative displacement is 0.400 m down the string. What are (a) the speed of waves on the string and (b) the linear mass density of the string?

Be sure to review EXAMPLES 15.4 and 15.5 (Section 15.5) before attempting these problems.

VP15.5.1 An athlete exerts a tension of 6.00×10^2 N on one end of a horizontal rope that has length 50.0 m and mass 2.50 kg. The other end is tied to a post. If she wiggles the rope with period 0.575 s and amplitude 3.00 cm, what are (a) the angular frequency of the oscillation and (b) the average rate at which energy is transferred along the rope?

VP15.5.2 A length of piano wire (mass density 5.55×10^{-4} kg/m) is under 185 N of tension. A sinusoidal wave of frequency 256 Hz carries a maximum power of 5.20 W along the wire. What is the amplitude of this wave?

VP15.5.3 A portable audio speaker has a power output of 8.00 W. (a) If the speaker emits sound equally in all directions, what is the sound intensity at a distance of 2.00 m from the speaker? (b) At what distance from the speaker is the intensity equal to 0.045 W/m²?

VP15.5.4 The "ears" of a frog are two circular membranes located behind the frog's eyes. In one species of frog each membrane is 0.500 cm in radius. If a source of sound has a power output of 2.50×10^{-6} W, emits sound equally in all directions, and is located 1.50 m from the frog, how much sound energy arrives at one of the membranes each second?

Be sure to review EXAMPLES 15.6, 15.7, and 15.8 (Sections 15.7 and 15.8) before attempting these problems.

VP15.8.1 For a standing wave on a string, the distance between nodes is 0.125 m, the frequency is 256 Hz, and the amplitude is 1.40×10^{-3} m. What are (a) the speed of waves on this string, (b) the maximum transverse velocity at an antinode, and (c) the maximum transverse acceleration at an antinode?

VP15.8.2 The G string of a guitar has a fundamental frequency of 196 Hz. The linear mass density of the string is 2.29×10^{-3} kg/m, and the length of string that is free to vibrate (between the nut and bridge of the guitar) is 0.641 m. What are (a) the speed of waves on the G string and (b) the tension in this string?

VP15.8.3 A cable is stretched between two posts 3.00 m apart. The speed of waves on this cable is 96.0 m/s, and the tension in the cable is 175 N. If a standing wave on this cable has five antinodes, what are (a) the wavelength of the standing wave, (b) the frequency of the standing wave, and (c) the linear mass density of the cable?

VP15.8.4 One of the strings on a musical instrument is 0.500 m in length and has linear mass density 1.17×10^{-3} kg/m. The second harmonic on this string has frequency 512 Hz. (a) What is the tension in the string? (b) The speed of sound in air at 20°C is 344 m/s. If the string is vibrating at its fundamental frequency, what is the wavelength of the sound wave that the string produces in air?

15

STUDENT'S SOLUTIONS MANUAL FOR MECHANICAL WAVES

VP15.3.1. IDENTIFY: We are dealing with general characteristics of waves.
SET UP: We know that $f = 1/T$, $v = f\lambda$, $\omega = 2\pi f$, and $k = 2\pi/\lambda$.
EXECUTE: (a) $f = 1/T = 1/(5.10 \text{ s}) = 0.196$ Hz.
(b) $v = f\lambda = (0.196 \text{ Hz})(30.5 \text{ m}) = 5.98$ m/s.
(c) $\omega = 2\pi f = 2\pi(0.196 \text{ Hz}) = 1.23$ rad/s.
(d) $k = 2\pi/\lambda = 2\pi/(0.206 \text{ m}) = 0.206$ m^{-1}.
EVALUATE: Caution! The wave number k is *not* the number of waves. It is simply defined as $k = 2\pi/\lambda$.

VP15.3.2. IDENTIFY: This problem involves the characteristics of a sound wave on Mars.
SET UP: $T = 1/f$, $\omega = 2\pi f$, and $v = f\lambda$.
EXECUTE: (a) $T = 1/f = 1/(125 \text{ Hz}) = 8.00 \times 10^{-3}$ s.
Solve $v = f\lambda$ for λ: $\lambda = v/f = (245 \text{ m/s})/(125 \text{ Hz}) = 1.96$ m.
(b) Solve $v = f\lambda$ for f: $f = v/\lambda = (245 \text{ m/s})/(3.00 \text{ m}) = 81.7$ Hz.
$\omega = 2\pi f = 2\pi(81.7 \text{ Hz}) = 513$ rad/s.
EVALUATE: On Earth the wavelength would be almost twice as long as on Mars for the same frequency because the speed of sound is about twice as great as on Mars.

VP15.3.3. IDENTIFY: We are dealing with a traveling wave on a string.
SET UP: $y(x,t) = A\cos(kx - \omega t)$, $\omega = 2\pi f$, $v = f\lambda$, $k = 2\pi/\lambda$, and $v = \sqrt{\dfrac{F}{\mu}}$. Call the $+x$-axis the direction in which the wave is traveling.
EXECUTE: (a) $v = \sqrt{\dfrac{F}{\mu}} = \sqrt{\dfrac{810 \text{ N}}{0.0650 \text{ kg/m}}} = 112$ m/s.
(b) Solve $v = f\lambda$ for λ: $\lambda = v/f = (112 \text{ m/s})/(25.0 \text{ Hz}) = 4.47$ m.
(c) Use $y(x,t) = A\cos(kx - \omega t)$ with $x = 2.50$ m, $k = 2\pi/\lambda = 2\pi/(4.47 \text{ m}) = 1.41$ m^{-1}, $A = 5.00$ mm, and $\omega = 2\pi f = 2\pi(25.0 \text{ Hz}) = 157$ rad/s. Using these numbers gives
$y(2.50 \text{ m}, t) = (5.00 \text{ mm})\cos[3.52 - (157 \text{ rad/s})t]$.
EVALUATE: This wave is traveling in the $+x$ direction. The equation for reflected waves having the same amplitude would be $y(2.50 \text{ m}, t) = (5.00 \text{ mm})\cos[3.52 + (157 \text{ rad/s})t]$.

VP15.3.4. **IDENTIFY:** We are dealing with a traveling wave on a string.

SET UP: $v = f\lambda$, $v = \sqrt{\dfrac{F}{\mu}}$. The 0.400 m is the distance from a crest to the adjacent trough, which is one-half of a complete wave. Therefore $\lambda = 2(0.400 \text{ m}) = 0.800$ m.

EXECUTE: (a) $v = f\lambda = (45.0 \text{ Hz})(0.800 \text{ m}) = 36.0$ m/s.

(b) Solve $v = \sqrt{\dfrac{F}{\mu}}$ for μ: $v = \sqrt{\dfrac{F}{\mu}} = 0.193$ kg/m.

EVALUATE: A linear mass density of 0.193 kg/m is a bit large but not unreasonable for a string.

VP15.5.1. **IDENTIFY:** We are dealing with a traveling wave on a rope and the average power it transfers.

SET UP: $P_{av} = \dfrac{1}{2}\sqrt{\mu F}\,\omega^2 A^2$, $\omega = 2\pi f$, and $f = 1/T$. The target variable is the average power delivered by the wave.

EXECUTE: (a) $\omega = 2\pi f = 2\pi/T = 2\pi/(0.575 \text{ s}) = 10.9$ rad/s.

(b) $P_{av} = \dfrac{1}{2}\sqrt{\mu F}\,\omega^2 A^2 = \dfrac{1}{2}\sqrt{\dfrac{m}{L}F}\,\omega^2 A^2 = \dfrac{1}{2}\sqrt{\left(\dfrac{2.50 \text{ kg}}{50.0 \text{ m}}\right)(600 \text{ N})}\,(10.9 \text{ rad/s})^2 (0.0300 \text{ m})^2 = 0.294$ W.

EVALUATE: Note that the power depends on the *square* of the frequency and the amplitude. This power is much less than that of an ordinary 60 W light bulb.

VP15.5.2. **IDENTIFY:** We are dealing with the maximum power carried by a traveling wave on a piano wire.

SET UP: $P_{max} = \sqrt{\mu F}\,\omega^2 A^2$, $\omega = 2\pi f$. We know all the quantities except the amplitude, and our target variable is the amplitude of the wave.

EXECUTE: Solve $P_{max} = \sqrt{\mu F}\,\omega^2 A^2$ for A: $A = \sqrt{\dfrac{P_{max}}{\omega^2 \sqrt{\mu F}}}$. Using $P_{max} = 5.20$ W, $F = 185$ N, $\mu = 5.55 \times 10^{-4}$ kg/m, and $\omega = 2\pi f = 2\pi(256 \text{ Hz}) = 512\pi$ Hz, we get $A = 2.50 \times 10^{-3}$ m = 2.50 mm.

EVALUATE: An amplitude of 2.50 mm on a piano string is large enough to see readily.

VP15.5.3. **IDENTIFY:** We are investigating the power output of a sound speaker.

SET UP: $I = P/A$ where $A = 4\pi r^2$.

EXECUTE: (a) $I = P/A = (8.00 \text{ W})/[4\pi(2.00 \text{ m})^2] = 0.159$ W/m^2.

(b) Solve $I = P/(4\pi r^2)$ for r: $r = \sqrt{\dfrac{P}{4\pi I}} = \sqrt{\dfrac{8.00 \text{ W}}{4\pi(0.045 \text{ W/m}^2)}} = 3.76$ m.

EVALUATE: The distance from the speaker in part (b) is nearly twice as great as in part (a), which is reasonable because the intensity in (b) is considerably less than in (a).

VP15.5.4. **IDENTIFY:** We are dealing with the sonic power received by a frog's "ears."

SET UP: $I = P/4\pi r^2$ and the energy E received by an area A during a time t is $E = IAt$. Our target variable is the amount of energy the frog's membrane receives in one second.

EXECUTE: $I = P/4\pi r^2$ where r is the distance from the source of sound and P is the power the source is emitting. The area A is $A = \pi R^2$, where R is the radius of the membrane. Combining these quantities gives $E = IAt = \left(\dfrac{P}{4\pi r^2}\right)(\pi R^2)t$. Using $P = 2.00 \times 10^{-6}$ W, $r = 1.50$ m, $t = 1.00$ s, and $R = 5.00 \times 10^{-3}$ m, we get $E = 5.56 \times 10^{-12}$ J.

EVALUATE: This is a very small amount of energy, so the frog must have very sensitive "ears"!

VP15.8.1. **IDENTIFY:** We are dealing with a standing wave on a string.

SET UP: The equation for the standing wave is $y(x,t) = A_{SW} \sin k_n x \cos \omega_n t$, where $\omega = 2\pi f$. The speed of the wave is $v = f\lambda$; it is a constant and directed along the string. The transverse velocity is $v_y(x,t) = \dfrac{\partial y}{\partial t}$; it is different for different points on the string and is directed perpendicular to the string.

EXECUTE: **(a)** The distance of 0.125 m between nodes is one-half a wavelength, so $\lambda = 0.250$ m. Therefore $v = f\lambda = (256 \text{ Hz})(0.250 \text{ m}) = 64.0$ m/s.

(b) $v_y = \dfrac{\partial y}{\partial t} = \dfrac{\partial((A_{SW} \sin k_n x) \sin \omega_n t)}{\partial t} = A_{SW} \omega_n \sin k_n x \cos \omega_n t$. The maximum speed is $A_{SW}\omega_n$, so

$v_{y\text{-max}} = (1.40 \times 10^{-3} \text{ m})[2\pi(256 \text{ Hz})] = 2.25$ m/s.

(c) $a_y = \dfrac{\partial v_y}{\partial t} = -A_{SW}\omega_n^2 \sin k_n x \sin \omega_n t$. The $= 2L$ maximum acceleration is $A_{SW}\omega_n^2$, so

$a_{y\text{-max}} = (1.40 \times 10^{-3} \text{ m})[2\pi(256 \text{ Hz})]^2 = 3620$ m/s^2.

EVALUATE: For a very taut string or wire, the maximum acceleration can be extremely large.

VP15.8.2. **IDENTIFY:** This problem involves a standing wave on the G string of a guitar string.

SET UP: $v = \sqrt{\dfrac{F}{\mu}}$, $v = f\lambda$, $\lambda_n = \dfrac{2L}{n}$, and for the fundamental mode $n = 1$ so $\lambda_1 = \dfrac{2L}{1} = 2L$.

EXECUTE: **(a)** We want the wave speed. $v = f_1 \lambda_1 = f_1(2L) = (196 \text{ Hz})(2)(0.641 \text{ m}) = 251$ m/s.

(b) We want the tension in the string, so solve $v = \sqrt{\dfrac{F}{\mu}}$ for F, giving $F = \mu v^2$. Therefore

$F = (2.29 \times 10^{-3} \text{ kg/m})(251 \text{ m/s})^2 = 145$ N.

EVALUATE: This is a very fast wave since the tension is large.

VP15.8.3. **IDENTIFY:** Standing waves form on the cable that is fixed at both ends.

SET UP: $\lambda_n = \dfrac{2L}{n}$, $v = f\lambda$, $v = \sqrt{\dfrac{F}{\mu}}$. Since there are 5 antinodes on the cable, it is vibrating in its 5th harmonic, so $n = 5$. We want to know the wavelength and frequency of the waves and the linear mass density of the cable.

EXECUTE: **(a)** $\lambda_n = \dfrac{2L}{n}$ so $\lambda_5 = \dfrac{2L}{5} = \dfrac{6.00 \text{ m}}{5} = 1.20$ m.

(b) Solve $v = f\lambda$ for f, giving $f = \dfrac{v}{\lambda} = \dfrac{96.0 \text{ m/s}}{1.20 \text{ m}} = 80.0$ Hz.

(c) Solve $v = \sqrt{\dfrac{F}{\mu}}$ for μ giving $\mu = \dfrac{F}{v^2} = \dfrac{175 \text{ N}}{(96.0 \text{ m/s})^2} = 1.90 \times 10^{-2}$ kg/m.

EVALUATE: This density is 19 kg/m, which is not unreasonable for a thin cable.

VP15.8.4. **IDENTIFY:** A string fixed at both ends is vibrating in its second harmonic of a standing wave pattern. This vibration produces a sound wave.

SET UP: $v = \sqrt{\dfrac{F}{\mu}}$, $\lambda_n = \dfrac{2L}{n}$, $f_n = nf_1$, $v = f\lambda$. For the second harmonic, $n = 2$. The target variables are the tension in the string and the wavelength of the sound wave the vibrating string produces.

EXECUTE: **(a)** We want the tension in the string. First find the wavelength and use it to find the speed of the wave. Then use $v = \sqrt{\frac{F}{\mu}}$ to find the tension. For the string in the second harmonic, $\lambda_2 = 2L/2 = L = 0.500$ m. The wave speed is $v = f\lambda = (512\text{ Hz})(0.500\text{ m}) = 256$ m/s. Now solve $v = \sqrt{\frac{F}{\mu}}$ for F giving $F = \mu v^2 = (1.17\times 10^{-3}\text{ kg/m})(256\text{ m/s})^2 = 76.7$ N.

(b) We want the wavelength of the sound wave this string produces when vibrating in its fundamental harmonic. Each cycle of the string produces a cycle of sound waves, so the frequency of the sound will be the same as the frequency of the string. The string is now vibrating in its *fundamental* frequency, not its second harmonic. Using $f_n = nf_1$, we have $f_2 = 2f_1 = 512$ Hz, so its frequency is $f_1 = (512\text{ Hz})/2 = 256$ Hz. Now solve $v_s = f\lambda$ for λ, where v_s is the speed of sound. This gives

$$\lambda = \frac{v_s}{f} = \frac{344\text{ m/s}}{256\text{ Hz}} = 1.34\text{ m}.$$

EVALUATE: The wavelength of the sound wave is very different from the wavelength of the waves on the string, even though both have the same frequency. This difference is due to the difference in the speeds of the two waves.

15.7. **IDENTIFY:** Use $v = f\lambda$ to calculate v. $T = 1/f$ and k is defined by $k = 2\pi/\lambda$. The general form of the wave function is given by $y(x,t) = A\cos 2\pi(x/\lambda + t/T)$, which is the equation for the transverse displacement.

SET UP: $v = 8.00$ m/s, $A = 0.0700$ m, $\lambda = 0.320$ m
EXECUTE: **(a)** $v = f\lambda$ so $f = v/\lambda = (8.00\text{ m/s})/(0.320\text{ m}) = 25.0$ Hz
$T = 1/f = 1/25.0\text{ Hz} = 0.0400$ s
$k = 2\pi/\lambda = 2\pi\text{ rad}/0.320\text{ m} = 19.6$ rad/m
(b) For a wave traveling in the $-x$-direction,
$y(x,t) = A\cos 2\pi(x/\lambda + t/T)$
At $x = 0$, $y(0,t) = A\cos 2\pi(t/T)$, so $y = A$ at $t = 0$. This equation describes the wave specified in the problem.
Substitute in numerical values:
$y(x,t) = (0.0700\text{ m})\cos\left[2\pi\left(x/(0.320\text{ m}) + t/(0.0400\text{ s})\right)\right]$.
Or, $y(x,t) = (0.0700\text{ m})\cos\left[(19.6\text{ m}^{-1})x + (157\text{ rad/s})t\right]$.
(c) From part (b), $y = (0.0700\text{ m})\cos[2\pi(x/0.320\text{ m} + t/0.0400\text{ s})]$.
Plug in $x = 0.360$ m and $t = 0.150$ s:
$y = (0.0700\text{ m})\cos[2\pi(0.360\text{ m}/0.320\text{ m} + 0.150\text{ s}/0.0400\text{ s})]$
$y = (0.0700\text{ m})\cos[2\pi(4.875\text{ rad})] = +0.0495\text{ m} = +4.95$ cm
(d) In part (c) $t = 0.150$ s.
$y = A$ means $\cos[2\pi(x/\lambda + t/T)] = 1$
$\cos\theta = 1$ for $\theta = 0, 2\pi, 4\pi, \ldots = n(2\pi)$ or $n = 0, 1, 2, \ldots$
So $y = A$ when $2\pi(x/\lambda + t/T) = n(2\pi)$ or $x/\lambda + t/T = n$
$t = T(n - x/\lambda) = (0.0400\text{ s})(n - 0.360\text{ m}/0.320\text{ m}) = (0.0400\text{ s})(n - 1.125)$
For $n = 4$, $t = 0.1150$ s (before the instant in part (c))
For $n = 5$, $t = 0.1550$ s (the first occurrence of $y = A$ after the instant in part (c)). Thus the elapsed time is $0.1550\text{ s} - 0.1500\text{ s} = 0.0050$ s.

EVALUATE: Part (d) says $y = A$ at 0.115 s and next at 0.155 s; the difference between these two times is 0.040 s, which is the period. At $t = 0.150$ s the particle at $x = 0.360$ m is at $y = 4.95$ cm and traveling upward. It takes $T/4 = 0.0100$ s for it to travel from $y = 0$ to $y = A$, so our answer of 0.0050 s is reasonable.

15.9. **IDENTIFY:** Evaluate the partial derivatives and see if the wave equation $\dfrac{\partial^2 y(x,t)}{\partial x^2} = \dfrac{1}{v^2} \dfrac{\partial^2 y(x,t)}{\partial t^2}$ is satisfied.

SET UP: $\dfrac{\partial}{\partial x} \cos(kx + \omega t) = -k \sin(kx + \omega t)$. $\dfrac{\partial}{\partial t} \cos(kx + \omega t) = -\omega \sin(kx + \omega t)$.

$\dfrac{\partial}{\partial x} \sin(kx + \omega t) = k \cos(kx + \omega t)$. $\dfrac{\partial}{\partial t} \sin(kx + \omega t) = \omega \cos(kx + \omega t)$.

EXECUTE: (a) $\dfrac{\partial^2 y}{\partial x^2} = -Ak^2 \cos(kx + \omega t)$. $\dfrac{\partial^2 y}{\partial t^2} = -A\omega^2 \cos(kx + \omega t)$. The wave equation is satisfied, if $v = \omega/k$.

(b) $\dfrac{\partial^2 y}{\partial x^2} = -Ak^2 \sin(kx + \omega t)$. $\dfrac{\partial^2 y}{\partial t^2} = -A\omega^2 \sin(kx + \omega t)$. The wave equation is satisfied, if $v = \omega/k$.

(c) $\dfrac{\partial y}{\partial x} = -kA \sin(kx)$. $\dfrac{\partial^2 y}{\partial x^2} = -k^2 A \cos(kx)$. $\dfrac{\partial y}{\partial t} = -\omega A \sin(\omega t)$. $\dfrac{\partial^2 y}{\partial t^2} = -\omega^2 A \cos(\omega t)$. The wave equation is not satisfied.

(d) $v_y = \dfrac{\partial y}{\partial t} = \omega A \cos(kx + \omega t)$. $a_y = \dfrac{\partial^2 y}{\partial t^2} = -A\omega^2 \sin(kx + \omega t)$

EVALUATE: The functions $\cos(kx + \omega t)$ and $\sin(kx + \omega t)$ differ only in phase.

15.11. **IDENTIFY and SET UP:** Read A and T from the graph. Apply $y(x,t) = A \cos 2\pi \left(\dfrac{x}{\lambda} - \dfrac{t}{T} \right)$ to determine λ and then use $v = f\lambda$ to calculate v.

EXECUTE: (a) The maximum y is 4 mm (read from graph).
(b) For either x the time for one full cycle is 0.040 s; this is the period.
(c) Since $y = 0$ for $x = 0$ and $t = 0$ and since the wave is traveling in the $+x$-direction then $y(x, t) = A \sin[2\pi(t/T - x/\lambda)]$. (The phase is different from the wave described by $y(x,t) = A \cos 2\pi \left(\dfrac{x}{\lambda} - \dfrac{t}{T} \right)$; for that wave $y = A$ for $x = 0$, $t = 0$.) From the graph, if the wave is traveling in the $+x$-direction and if $x = 0$ and $x = 0.090$ m are within one wavelength the peak at $t = 0.01$ s for $x = 0$ moves so that it occurs at $t = 0.035$ s (read from graph so is approximate) for $x = 0.090$ m. The peak for $x = 0$ is the first peak past $t = 0$ so corresponds to the first maximum in $\sin[2\pi(t/T - x/\lambda)]$ and hence occurs at $2\pi(t/T - x/\lambda) = \pi/2$. If this same peak moves to $t_1 = 0.035$ s at $x_1 = 0.090$ m, then
$2\pi(t/T - x/\lambda) = \pi/2$.
Solve for λ: $t_1/T - x_1/\lambda = 1/4$
$x_1/\lambda = t_1/T - 1/4 = 0.035$ s/0.040 s $- 0.25 = 0.625$
$\lambda = x_1/0.625 = 0.090$ m/0.625 $= 0.14$ m.
Then $v = f\lambda = \lambda/T = 0.14$ m/0.040 s $= 3.5$ m/s.

(d) If the wave is traveling in the $-x$-direction, then $y(x, t) = A\sin(2\pi(t/T + x/\lambda))$ and the peak at $t = 0.050$ s for $x = 0$ corresponds to the peak at $t_1 = 0.035$ s for $x_1 = 0.090$ m. This peak at $x = 0$ is the second peak past the origin so corresponds to $2\pi(t/T + x/\lambda) = 5\pi/2$. If this same peak moves to $t_1 = 0.035$ s for $x_1 = 0.090$ m, then $2\pi(t_1/T + x_1/\lambda) = 5\pi/2$.

$t_1/T + x_1/\lambda = 5/4$

$x_1/\lambda = 5/4 - t_1/T = 5/4 - 0.035 \text{ s}/0.040 \text{ s} = 0.375$

$\lambda = x_1/0.375 = 0.090 \text{ m}/0.375 = 0.24$ m.

Then $v = f\lambda = \lambda/T = 0.24 \text{ m}/0.040 \text{ s} = 6.0$ m/s.

EVALUATE: **(e)** No. Wouldn't know which point in the wave at $x = 0$ moved to which point at $x = 0.090$ m.

15.17. IDENTIFY: The speed of the wave depends on the tension in the wire and its mass density. The target variable is the mass of the wire of known length.

SET UP: $v = \sqrt{\dfrac{F}{\mu}}$ and $\mu = m/L$.

EXECUTE: First find the speed of the wave: $v = \dfrac{3.80 \text{ m}}{0.0492 \text{ s}} = 77.24$ m/s. $v = \sqrt{\dfrac{F}{\mu}}$. $\mu = \dfrac{F}{v^2} = \dfrac{(54.0 \text{ kg})(9.8 \text{ m/s}^2)}{(77.24 \text{ m/s})^2} = 0.08870$ kg/m. The mass of the wire is

$m = \mu L = (0.08870 \text{ kg/m})(3.80 \text{ m}) = 0.337$ kg.

EVALUATE: This mass is 337 g, which is a bit large for a wire 3.80 m long. It must be fairly thick.

15.19. IDENTIFY: For transverse waves on a string, $v = \sqrt{F/\mu}$. $v = f\lambda$.

SET UP: The wire has $\mu = m/L = (0.0165 \text{ kg})/(0.750 \text{ m}) = 0.0220$ kg/m.

EXECUTE: **(a)** $v = f\lambda = (625 \text{ Hz})(3.33 \times 10^{-2} \text{ m}) = 20.813$ m/s. The tension is

$F = \mu v^2 = (0.0220 \text{ kg/m})(20.813 \text{ m/s})^2 = 9.53$ N.

(b) $v = 20.8$ m/s

EVALUATE: If λ is kept fixed, the wave speed and the frequency increase when the tension is increased.

15.25. IDENTIFY: For a point source, $I = \dfrac{P}{4\pi r^2}$ and $\dfrac{I_1}{I_2} = \dfrac{r_2^2}{r_1^2}$.

SET UP: $1 \mu\text{W} = 10^{-6}$ W

EXECUTE: **(a)** $r_2 = r_1\sqrt{\dfrac{I_1}{I_2}} = (30.0 \text{ m})\sqrt{\dfrac{10.0 \text{ W/m}^2}{1\times 10^{-6} \text{ W/m}^2}} = 95$ km

(b) $\dfrac{I_2}{I_3} = \dfrac{r_3^2}{r_2^2}$, with $I_2 = 1.0 \mu\text{W/m}^2$ and $r_3 = 2r_2$. $I_3 = I_2\left(\dfrac{r_2}{r_3}\right)^2 = I_2/4 = 0.25 \mu\text{W/m}^2$.

(c) $P = I(4\pi r^2) = (10.0 \text{ W/m}^2)(4\pi)(30.0 \text{ m})^2 = 1.1\times 10^5$ W

EVALUATE: These are approximate calculations, that assume the sound is emitted uniformly in all directions and that ignore the effects of reflection, for example reflections from the ground.

15.29. IDENTIFY: The distance the wave shape travels in time t is vt. The wave pulse reflects at the end of the string, at point O.

SET UP: The reflected pulse is inverted when O is a fixed end and is not inverted when O is a free end.

EXECUTE: **(a)** The wave form for the given times, respectively, is shown in Figure 15.29(a).
(b) The wave form for the given times, respectively, is shown in Figure 15.29(b).
EVALUATE: For the fixed end the result of the reflection is an inverted pulse traveling to the right and for the free end the result is an upright pulse traveling to the right.

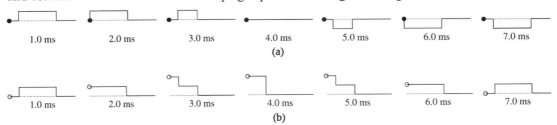

Figure 15.29

15.31. IDENTIFY: Apply the principle of superposition.

SET UP: The net displacement is the algebraic sum of the displacements due to each pulse.
EXECUTE: The shape of the string at each specified time is shown in Figure 15.31.
EVALUATE: The pulses interfere when they overlap but resume their original shape after they have completely passed through each other.

15.37. IDENTIFY: Use $v = f\lambda$ for v and $v = \sqrt{F/\mu}$ for the tension F. $v_y = \partial y/\partial t$ and $a_y = \partial v_y/\partial t$.

(a) SET UP: The fundamental standing wave is sketched in Figure 15.37.

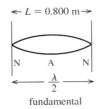

$f = 60.0$ Hz
From the sketch,
$\lambda/2 = L$ so
$\lambda = 2L = 1.60$ m

Figure 15.37

EXECUTE: $v = f\lambda = (60.0 \text{ Hz})(1.60 \text{ m}) = 96.0$ m/s

(b) The tension is related to the wave speed by $v = \sqrt{F/\mu}$:

$v = \sqrt{F/\mu}$ so $F = \mu v^2$.

$\mu = m/L = 0.0400$ kg/0.800 m $= 0.0500$ kg/m

$F = \mu v^2 = (0.0500 \text{ kg/m})(96.0 \text{ m/s})^2 = 461$ N.

(c) $\omega = 2\pi f = 377$ rad/s and $y(x,t) = A_{SW} \sin kx \sin \omega t$

$v_y = \omega A_{SW} \sin kx \cos \omega t$; $a_y = -\omega^2 A_{SW} \sin kx \sin \omega t$

$(v_y)_{max} = \omega A_{SW} = (377 \text{ rad/s})(0.300 \text{ cm}) = 1.13$ m/s.

$(a_y)_{max} = \omega^2 A_{SW} = (377 \text{ rad/s})^2(0.300 \text{ cm}) = 426$ m/s^2.

EVALUATE: The transverse velocity is different from the wave velocity. The wave velocity and tension are similar in magnitude to the values in the examples in the text. Note that the transverse acceleration is quite large.

15.39. **IDENTIFY:** Compare $y(x,t)$ given in the problem to $y(x,t) = (A_{SW} \sin kx)\sin \omega t$. From the frequency and wavelength for the third harmonic find these values for the eighth harmonic.

(a) SET UP: The third harmonic standing wave pattern is sketched in Figure 15.39.

Figure 15.39

EXECUTE: (b) Use the general equation for a standing wave on a string:
$y(x,t) = (A_{SW} \sin kx)\sin \omega t$
$A_{SW} = 2A$, so $A = A_{SW}/2 = (5.60 \text{ cm})/2 = 2.80$ cm
(c) The sketch in part (a) shows that $L = 3(\lambda/2)$. $k = 2\pi/\lambda$, $\lambda = 2\pi/k$
Comparison of $y(x,t)$ given in the problem to $y(x,t) = (A_{SW} \sin kx)\sin \omega t$ gives
$k = 0.0340$ rad/cm. So, $\lambda = 2\pi/(0.0340 \text{ rad/cm}) = 184.8$ cm
$L = 3(\lambda/2) = 277$ cm
(d) $\lambda = 185$ cm, from part (c)
$\omega = 50.0$ rad/s so $f = \omega/2\pi = 7.96$ Hz
period $T = 1/f = 0.126$ s $v = f\lambda = 1470$ cm/s
(e) $v_y = \partial y/\partial t = \omega A_{SW} \sin kx \cos \omega t$
$v_{y,\text{max}} = \omega A_{SW} = (50.0 \text{ rad/s})(5.60 \text{ cm}) = 280$ cm/s
(f) $f_3 = 7.96$ Hz $= 3f_1$, so $f_1 = 2.65$ Hz is the fundamental
$f_8 = 8f_1 = 21.2$ Hz; $\omega_8 = 2\pi f_8 = 133$ rad/s
$\lambda = v/f = (1470 \text{ cm/s})/(21.2 \text{ Hz}) = 69.3$ cm and $k = 2\pi/\lambda = 0.0906$ rad/cm
$y(x,t) = (5.60 \text{ cm})\sin[(0.0906 \text{ rad/cm})x]\sin[(133 \text{ rad/s})t]$.
EVALUATE: The wavelength and frequency of the standing wave equals the wavelength and frequency of the two traveling waves that combine to form the standing wave. In the eighth harmonic the frequency and wave number are larger than in the third harmonic.

15.41. **IDENTIFY:** Standing waves are formed on a string fixed at both ends.

SET UP: The distance between adjacent nodes is one-half of a wavelength. The second overtone is the third harmonic ($n = 3$). $\lambda_n = 2L/n$.

EXECUTE: (a) We want the length of the string. In the third harmonic $\lambda_3 = 2L/3$, so $L = 3\lambda_3/2$. The distance between adjacent nodes is $\lambda/2 = 8.00$ cm, so $\lambda = 16.0$ cm. Therefore $L = 3\lambda_3/2 = 24.0$ cm $= 0.240$ m.
(b) For the 4th harmonic, $n = 4$, so $\lambda_4 = 2L/4 = (48.0 \text{ cm})/4 = 12.0$ cm $= 0.120$ m. The node-to-node distance is $\lambda/2 = (12.0 \text{ cm})/2 = 6.00$ cm $= 0.0600$ m.
EVALUATE: The node-to-node distance decreases with higher harmonics because the wavelength decreases, which is consistent with your results.

15.43. **IDENTIFY** and **SET UP:** Use the information given about the A_4 note to find the wave speed that depends on the linear mass density of the string and the tension. The wave speed isn't affected by the placement of the fingers on the bridge. Then find the wavelength for the D_5 note and relate this to the length of the vibrating portion of the string.

EXECUTE: **(a)** $f = 440$ Hz when a length $L = 0.600$ m vibrates; use this information to calculate the speed v of waves on the string. For the fundamental $\lambda/2 = L$ so $\lambda = 2L = 2(0.600 \text{ m}) = 1.20$ m. Then $v = f\lambda = (440 \text{ Hz})(1.20 \text{ m}) = 528$ m/s. Now find the length $L = x$ of the string that makes $f = 587$ Hz.

$$\lambda = \frac{v}{f} = \frac{528 \text{ m/s}}{587 \text{ Hz}} = 0.900 \text{ m}$$

$L = \lambda/2 = 0.450$ m, so $x = 0.450$ m $= 45.0$ cm.

(b) No retuning means same wave speed as in part (a). Find the length of vibrating string needed to produce $f = 392$ Hz.

$$\lambda = \frac{v}{f} = \frac{528 \text{ m/s}}{392 \text{ Hz}} = 1.35 \text{ m}$$

$L = \lambda/2 = 0.675$ m; string is shorter than this. No, not possible.

EVALUATE: Shortening the length of this vibrating string increases the frequency of the fundamental.

15.45. **IDENTIFY:** We are dealing with a traveling sinusoidal wave on a string.

SET UP: The equation $y(x,t) = A\cos(kx - \omega t)$ describes this wave. We know the maximum transverse speed and acceleration. We use $v_{y\text{-max}} = A\omega$, $a_{y\text{-max}} = A\omega^2$, and $v = f\lambda$. The target variables are the speed v of the wave and its amplitude A.

EXECUTE: Take the ratio of the maximum acceleration to the maximum speed, giving

$$\frac{a_{y\text{-max}}}{v_{y\text{-max}}} = \frac{A\omega^2}{A\omega} = \omega.$$ Using the known values gives $\omega = \frac{8.50 \times 10^4 \text{ m/s}^2}{3.00 \text{ m/s}} = 2.8333 \times 10^4$ rad/s. Now

use $v = f\lambda = (\omega/2\pi)\lambda = \left(\frac{2.8333 \times 10^4 \text{ rad/s}}{2\pi}\right)(0.400 \text{ m}) = 1.80 \times 10^3$ m/s.

From the velocity equation $v_{y\text{-max}} = A\omega$ we have $A = \frac{v_{y\text{-max}}}{\omega}$, which gives

$$A = \frac{3.00 \text{ m/s}}{2.8333 \times 10^4 \text{ rad/s}} = 1.06 \times 10^{-4} \text{ m}.$$

EVALUATE: The propagation speed v is constant. But the transverse speed $v_y = \frac{\partial y}{\partial t}$ is *not* constant; it varies with time and from place to place on the string.

15.47. **IDENTIFY** and **SET UP:** Calculate v, ω, and k from $v = f\lambda$, $\omega = vk$, $k = 2\pi/\lambda$. Then apply $y(x,t) = A\cos(kx - \omega t)$ to obtain $y(x,t)$.

$A = 2.50 \times 10^{-3}$ m, $\lambda = 1.80$ m, $v = 36.0$ m/s

EXECUTE: **(a)** $v = f\lambda$ so $f = v/\lambda = (36.0 \text{ m/s})/1.80$ m $= 20.0$ Hz

$\omega = 2\pi f = 2\pi(20.0 \text{ Hz}) = 126$ rad/s

$k = 2\pi/\lambda = 2\pi$ rad/1.80 m $= 3.49$ rad/m

(b) For a wave traveling to the right, $y(x,t) = A\cos(kx - \omega t)$. This equation gives that the $x=0$ end of the string has maximum upward displacement at $t=0$.

Put in the numbers: $y(x,t) = (2.50\times 10^{-3} \text{ m})\cos[(3.49 \text{ rad/m})x - (126 \text{ rad/s})t]$.

(c) The left-hand end is located at $x = 0$. Put this value into the equation of part (b):

$y(0,t) = +(2.50\times 10^{-3} \text{ m})\cos((126 \text{ rad/s})t)$.

(d) Put $x = 1.35$ m into the equation of part (b):

$y(1.35 \text{ m}, t) = (2.50\times 10^{-3} \text{ m})\cos((3.49 \text{ rad/m})(1.35 \text{ m}) - (126 \text{ rad/s})t)$.

$y(1.35 \text{ m}, t) = (2.50\times 10^{-3} \text{ m})\cos(4.71 \text{ rad} - (126 \text{ rad/s})t)$

$4.71 \text{ rad} = 3\pi/2$ and $\cos(\theta) = \cos(-\theta)$, so

$y(1.35 \text{ m}, t) = (2.50\times 10^{-3} \text{ m})\cos((126 \text{ rad/s})t - 3\pi/2 \text{ rad})$

(e) $y = A\cos(kx - \omega t)$ (part (b))

The transverse velocity is given by $v_y = \dfrac{\partial y}{\partial t} = A\dfrac{\partial}{\partial t}\cos(kx - \omega t) = +A\omega\sin(kx - \omega t)$.

The maximum v_y is $A\omega = (2.50\times 10^{-3} \text{ m})(126 \text{ rad/s}) = 0.315$ m/s.

(f) $y(x,t) = (2.50\times 10^{-3} \text{ m})\cos((3.49 \text{ rad/m})x - (126 \text{ rad/s})t)$

$t = 0.0625$ s and $x = 1.35$ m gives

$y = (2.50\times 10^{-3} \text{ m})\cos((3.49 \text{ rad/m})(1.35 \text{ m}) - (126 \text{ rad/s})(0.0625 \text{ s})) = -2.50\times 10^{-3}$ m.

$v_y = +A\omega\sin(kx - \omega t) = +(0.315 \text{ m/s})\sin((3.49 \text{ rad/m})x - (126 \text{ rad/s})t)$

$t = 0.0625$ s and $x = 1.35$ m gives

$v_y = (0.315 \text{ m/s})\sin((3.49 \text{ rad/m})(1.35 \text{ m}) - (126 \text{ rad/s})(0.0625 \text{ s})) = 0.0$

EVALUATE: The results of part (f) illustrate that $v_y = 0$ when $y = \pm A$, as we saw from SHM in Chapter 14.

15.49. IDENTIFY: The block causes tension in the wire supporting the rod. A standing wave pattern exists on the wire, and the torques on the rod balance.

SET UP: The graph is a plot of the square of the frequency f versus the mass m hanging from the rod. The target variable is the mass of the wire. To interpret the graph, we need to find a relationship between f^2 and m. We apply $\Sigma \tau_z = 0$ about the hinge and use $v = \sqrt{\dfrac{F}{\mu}}$ and $v = f\lambda$.

EXECUTE: $\Sigma \tau_z = 0$: $TL\sin\theta - mg\dfrac{L}{2} = 0$ gives $T = \dfrac{mg}{2\sin\theta}$. Now use $v = \sqrt{\dfrac{F}{\mu}}$ and $v = f\lambda$. Equating these velocity equations gives $f\lambda = \sqrt{\dfrac{F}{\mu}}$, so $f^2 = \dfrac{1}{\lambda^2}\left(\dfrac{F}{\mu}\right)$. Using the result for the tension gives $f^2 = \dfrac{1}{\mu\lambda^2}\left(\dfrac{mg}{2\sin\theta}\right) = \left(\dfrac{g}{2\mu\lambda^2 \sin\theta}\right)m$. The wire is vibrating in its fundamental mode so $\lambda = 2L$. The linear mass density of the wire is $\mu = \dfrac{m_\text{w}}{L}$. Therefore the final equation becomes

$f^2 = \left(\dfrac{g}{8m_\text{w} L\sin\theta}\right)m$. From this we see that a graph of f^2 versus m should be a straight line with slope equal to $\dfrac{g}{8m_w L \sin\theta}$. Thus $m_w = \dfrac{g}{(\text{slope})8L\sin\theta}$ which gives

$$m_w = \dfrac{9.80 \text{ m/s}^2}{(20.4 \text{ kg}^{-1}\cdot \text{s}^{-2})(8)(2.00 \text{ m})(\sin 30.0°)} = 0.0600 \text{ kg} = 60.0 \text{ g}.$$

EVALUATE: A mass of 60 g for a 2.00-m wire is not unreasonable.

15.51. IDENTIFY: Calculate the speed of the wave and use that to find the length of the wire since we know how long it takes the wave to travel the length of the wire.

SET UP: $v = \sqrt{F/\mu}$, $x = v_x t$, and $\mu = m/L$.

EXECUTE: **(a)** $\mu = m/L = (14.5\times 10^{-9} \text{ kg})/(0.0200 \text{ m}) = 7.25\times 10^{-7}$ kg/m. Now combine

$v = \sqrt{F/\mu}$ and $x = v_x t$: $vt = L$, so $L = t\sqrt{F/\mu} = (26.7\times 10^{-3} \text{ s})\sqrt{\dfrac{(0.400 \text{ kg})(9.80 \text{ m/s}^2)}{7.25\times 10^{-7} \text{ kg/m}}} = 62.1$ m.

(b) The mass of the wire is $m = \mu L = (7.25\times 10^{-7} \text{ kg/m})(62.1 \text{ m}) = 4.50\times 10^{-5}$ kg $= 0.0450$ g.

EVALUATE: The mass of the wire is negligible compared to the 0.400-kg object hanging from the wire.

15.53. IDENTIFY: Apply $\Sigma\tau_z = 0$ to one post and calculate the tension in the wire. $v = \sqrt{F/\mu}$ for waves on the wire. $v = f\lambda$. The standing wave on the wire and the sound it produces have the same frequency. For standing waves on the wire, $\lambda_n = \dfrac{2L}{n}$.

SET UP: For the fifth overtone, $n = 6$. The wire has $\mu = m/L = (0.732 \text{ kg})/(5.00 \text{ m}) = 0.146$ kg/m. The free-body diagram for one of the posts is given in Figure 15.53. Forces at the pivot aren't shown. We take the rotation axis to be at the pivot, so forces at the pivot produce no torque.

EXECUTE: $\Sigma\tau_z = 0$ gives $w\left(\dfrac{L}{2}\cos 57.0°\right) - T(L\sin 57.0°) = 0$.

$T = \dfrac{w}{2\tan 57.0°} = \dfrac{235 \text{ N}}{2\tan 57.0°} = 76.3$ N. For waves on the wire, $v = \sqrt{\dfrac{F}{\mu}} = \sqrt{\dfrac{76.3 \text{ N}}{0.146 \text{ kg/m}}} = 22.9$ m/s.

For the fifth overtone standing wave on the wire, $\lambda = \dfrac{2L}{6} = \dfrac{2(5.00 \text{ m})}{6} = 1.67$ m.

$f = \dfrac{v}{\lambda} = \dfrac{22.9 \text{ m/s}}{1.67 \text{ m}} = 13.7$ Hz. The sound waves have frequency 13.7 Hz and wavelength

$\lambda = \dfrac{344 \text{ m/s}}{13.7 \text{ Hz}} = 25.0$ m.

EVALUATE: The frequency of the sound wave is just below the lower limit of audible frequencies. The wavelength of the standing wave on the wire is much less than the wavelength of the sound waves, because the speed of the waves on the wire is much less than the speed of sound in air.

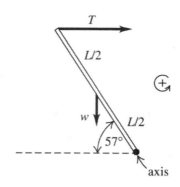

Figure 15.53

15.55. IDENTIFY: The wavelengths of standing waves depend on the length of the string (the target variable), which in turn determine the frequencies of the waves.

SET UP: $f_n = nf_1$ where $f_1 = \dfrac{v}{2L}$.

EXECUTE: $f_n = nf_1$ and $f_{n+1} = (n+1)f_1$. We know the wavelengths of two adjacent modes, so $f_1 = f_{n+1} - f_n = 630 \text{ Hz} - 525 \text{ Hz} = 105 \text{ Hz}$. Solving $f_1 = \dfrac{v}{2L}$ for L gives

$$L = \dfrac{v_1}{2f} = \dfrac{384 \text{ m/s}}{2(105 \text{ Hz})} = 1.83 \text{ m}.$$

EVALUATE: The observed frequencies are both audible which is reasonable for a string that is about a half meter long.

15.57. IDENTIFY: The tension in the wires along with their lengths determine the fundamental frequency in each one (the target variables). These frequencies are different because the wires have different linear mass densities. The bar is in equilibrium, so the forces and torques on it balance.

SET UP: $T_a + T_c = w$, $\Sigma \tau_z = 0$, $v = \sqrt{\dfrac{F}{\mu}}$, $f_1 = v/2L$ and $\mu = \dfrac{m}{L}$, where $m = \rho V = \rho \pi r^2 L$. The densities of copper and aluminum are given in a table in the text.

EXECUTE: Using the subscript "a" for aluminum and "c" for copper, we have $T_a + T_c = w = 638 \text{ N}$. $\Sigma \tau_z = 0$, with the axis at left-hand end of bar, gives $T_c (1.40 \text{ m}) = w(0.90 \text{ m})$, so $T_c = 410.1 \text{ N}$.

$T_a = 638 \text{ N} - 410.1 \text{ N} = 227.9 \text{ N}$. $f_1 = \dfrac{v}{2L}$. $\mu = \dfrac{m}{L} = \dfrac{\rho \pi r^2 L}{L} = \rho \pi r^2$.

<u>For the copper wire:</u> $F = 410.1 \text{ N}$ and
$\mu = (8.90 \times 10^3 \text{ kg/m}^3)\pi(0.280 \times 10^{-3} \text{ m})^2 = 2.19 \times 10^{-3} \text{ kg/m}$, so
$v = \sqrt{\dfrac{F}{\mu}} = \sqrt{\dfrac{410.1 \text{ N}}{2.19 \times 10^{-3} \text{ kg/m}}} = 432.7 \text{ m/s}$. $f_1 = \dfrac{v}{2L} = \dfrac{432.7 \text{ m/s}}{2(0.600 \text{ m})} = 361 \text{ Hz}$.

<u>For the aluminum wire:</u> $F = 227.9 \text{ N}$ and
$\mu = (2.70 \times 10^3 \text{ kg/m}^3)\pi(0.280 \times 10^{-3} \text{ m})^2 = 6.65 \times 10^{-4} \text{ kg/m}$, so
$v = \sqrt{\dfrac{F}{\mu}} = \sqrt{\dfrac{227.9 \text{ N}}{6.65 \times 10^{-4} \text{ kg/m}}} = 585.4 \text{ m/s}$, which gives $f_1 = \dfrac{585.4 \text{ m/s}}{2(0.600 \text{ m})} = 488 \text{ Hz}$.

EVALUATE: The wires have different fundamental frequencies because they have different tensions and different linear mass densities.

15.61. **IDENTIFY and SET UP:** The average power is given by $P_{av} = \frac{1}{2}\sqrt{\mu F}\,\omega^2 A^2$. Rewrite this expression in terms of v and λ in place of F and ω.

EXECUTE: (a) $P_{av} = \frac{1}{2}\sqrt{\mu F}\,\omega^2 A^2$

$v = \sqrt{F/\mu}$ so $\sqrt{F} = v\sqrt{\mu}$

$\omega = 2\pi f = 2\pi(v/\lambda)$

Using these two expressions to replace $\sqrt{F}$ and ω gives $P_{av} = 2\mu\pi^2 v^3 A^2/\lambda^2$;

$\mu = (6.00 \times 10^{-3} \text{ kg})/(8.00 \text{ m})$

$A = \left(\dfrac{2\lambda^2 P_{av}}{4\pi^2 v^3 \mu}\right)^{1/2} = 7.07 \text{ cm}$

(b) EVALUATE: $P_{av} \sim v^3$ so doubling v increases P_{av} by a factor of 8.
$P_{av} = 8(50.0 \text{ W}) = 400.0 \text{ W}$

15.63. **IDENTIFY:** The chain of beads vibrates in a standing wave pattern.

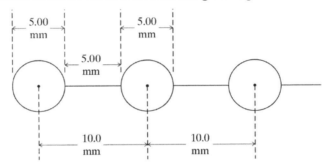

Figure 15.63

SET UP: First find the number of beads on a string 1.005 m long. Fig. 15.63 shows the first few beads. Starting at the left end, the string is made up of units each consisting of one bead of diameter 5.00 mm plus a space of 5.00 mm, so each unit is 10.0 mm long. To make a chain 1.005 m long, we need 100 of these units plus 5.00 mm for the end bead on the right. So the total mass of the 101 beads is 101 g = 0.101 kg. We use $y(x,t) = A_{SW} \sin kx \sin \omega t$, $v = \sqrt{\dfrac{F}{\mu}}$, $v = f\lambda$, $v_y = \dfrac{\partial y}{\partial t}$,
$\lambda_n = 2L/n$, $k = 2\pi/\lambda$, $\omega = 2\pi f$, and $F = kx$ (this k is the force constant).

EXECUTE: (a) $\mu = m/L = (0.101 \text{ kg})/(1.50 \text{ m}) = 0.0673$ kg/m.

(b) $T = kx = (28.8 \text{ N/m})(1.50 \text{ m} - 1.005 \text{ m}) = 14.3$ N.

(c) $v = \sqrt{\dfrac{F}{\mu}} = \sqrt{\dfrac{14.3 \text{ N}}{0.0673 \text{ kg/m}}} = 14.6$ m/s.

(d) We want the wave speed v. The wave pattern has 4 antinodes, so the chain is vibrating in its 4th harmonic, so $n = 4$. $\lambda_n = 2L/n = (3.00 \text{ m})/4 = 0.750$ m. Using $v = f\lambda$ gives $f = v/\lambda = (14.6 \text{ m/s})/(0.750 \text{ m}) = 19.5$ Hz.

(e) The motionless beads are at the nodes. In the 4th harmonic, nodes occur at the ends of the chain, in the middle, and ¼ and ¾ of the way along the chain. The beads at these locations are #1, 26, 51, 76, and 101.

(f) Bead #26 is at a node, so bead #13 is at an antinode, so its maximum speed is $v_{y,\text{max}} = A\omega$. Using $\omega = 2\pi f$, we get $A = \dfrac{v_{y,\text{max}}}{\omega} = \dfrac{v_{y,\text{max}}}{2\pi f} = \dfrac{7.54 \text{ m/s}}{2\pi(19.5 \text{ Hz})} = 0.0615 \text{ m} = 6.15 \text{ cm}$.

(g) When the chain is stretched, the beads are 15.0 mm apart center-to-center, with the origin at the center of the first bead. The first few coordinates are:
$x_1 = 0$ mm $= (1 - 1)(15.0$ mm$)$
$x_2 = 10.0$ mm $= (2 - 1)(15.0$ mm$)$
$x_3 = 20.0$ mm $= (3 - 1)(15.0$ mm$)$
$x_4 = 30.0$ mm $= (4 - 1)(15.0$ mm$)$
From this pattern we recognize that the nth bead is at $x_n = (n - 1)(15.0$ mm$)$, where $n = 1, 2, \ldots$.

(h) The 30th bead is at $x_{30} = (30 - 1)(15.0$ mm$) = 435$ mm $= 0.435$ m. The standing wave equation is $y(x,t) = A_{\text{SW}} \sin kx \sin \omega t$. Thus $v_y = \dfrac{\partial y}{\partial t} = \omega A_{\text{SW}} \sin kx \cos \omega t$, so at any position x, $v_{y\text{-max}} = \omega A_{\text{SW}} \sin kx = 2\pi f A_{\text{SW}} \sin kx$. From part (f) $A_{\text{SW}} = 0.0615$ m, and $k = 2\pi / \lambda = 2\pi/(0.750$ m$) = 8.378$ m^{-1}. Therefore $v_{y,\text{max}} = 2\pi(19.5$ Hz$)(0.0615$ m$) \sin[(8.3778$ m$^{-1})(0.435$ m$)] = -3.63$ m/s, so the maximum *speed* is 3.63 m/s.

EVALUATE: The 30th bead has a smaller maximum speed than the 13th bead. This is reasonable since the 13th bead is at an antinode but the 30th bead is not.

15.65. **IDENTIFY and SET UP:** $v = \sqrt{F/\mu}$ is the wave speed and $v_y = \dfrac{\partial y}{\partial t}$ is the transverse speed of a point on the string. $v = f\lambda$, $\lambda_n = 2L/n$, $\mu = m/L$, $v_{\text{max}} = \omega A$ (maximum v_y), $a_{\text{max}} = \omega^2 A$ (maximum a_y). The first overtone is the second harmonic ($n = 2$).

EXECUTE: (a) $v_{\text{max}} = \omega A = 2\pi f A$, which gives 28.0 m/s $= (0.0350$ m$)(2\pi f)$, so $f = 127.32$ Hz. $\lambda_n = 2L/n = 2L/2 = L = 2.50$ m. Using these results to get v gives $v = f\lambda = (127.32$ Hz$)(2.50$ m$) = 318.3$ m/s. Now combine $v = \sqrt{F/\mu}$ and $\mu = m/L$ to find m, giving
$m = LF/v^2 = (2.50$ m$)(90.0$ N$)/(318.3$ m/s$)^2 = 0.00222$ kg $= 2.22$ g.
(b) $a_{\text{max}} = \omega^2 A = A(2\pi f)^2 = (0.0350$ m$)[2\pi(127.32$ Hz$)]^2 = 22{,}400$ m/s^2.

EVALUATE: It is important to distinguish between the transverse velocity of a point on the string, $v_y = \dfrac{\partial y}{\partial t}$, and speed of the wave, $v = \sqrt{F/\mu}$. The wave speed is constant in time, but the transverse speed is not. The maximum acceleration of a point on the string is about 2300g!

15.67. **IDENTIFY:** The standing wave frequencies are given by $f_n = n\left(\dfrac{v}{2L}\right)$. $v = \sqrt{F/\mu}$. Use the density of steel to calculate μ for the wire.

SET UP: For steel, $\rho = 7.8 \times 10^3$ kg/m^3. For the first overtone standing wave, $n = 2$.

EXECUTE: $v = \dfrac{2Lf_2}{2} = (0.550$ m$)(311$ Hz$) = 171$ m/s. The volume of the wire is $V = (\pi r^2)L$. $m = \rho V$ so $\mu = \dfrac{m}{L} = \dfrac{\rho V}{L} = \rho \pi r^2 = (7.8 \times 10^3$ kg/m$^3)\pi(0.57 \times 10^{-3}$ m$)^2 = 7.96 \times 10^{-3}$ kg/m. The tension is $F = \mu v^2 = (7.96 \times 10^{-3}$ kg/m$)(171$ m/s$)^2 = 233$ N.

EVALUATE: The tension is not large enough to cause much change in length of the wire.

15.69. IDENTIFY: When the rock is submerged in the liquid, the buoyant force on it reduces the tension in the wire supporting it. This in turn changes the frequency of the fundamental frequency of the vibrations of the wire. The buoyant force depends on the density of the liquid (the target variable). The vertical forces on the rock balance in both cases, and the buoyant force is equal to the weight of the liquid displaced by the rock (Archimedes's principle).

SET UP: The wave speed is $v = \sqrt{\dfrac{F}{\mu}}$ and $v = f\lambda$. $B = \rho_{liq} V_{rock} g$. $\Sigma F_y = 0$.

EXECUTE: $\lambda = 2L = 6.00$ m. In air, $v = f\lambda = (42.0$ Hz$)(6.00$ m$) = 252$ m/s. $v = \sqrt{\dfrac{F}{\mu}}$ so

$\mu = \dfrac{F}{v^2} = \dfrac{164.0 \text{ N}}{(252 \text{ m/s})^2} = 0.002583$ kg/m. In the liquid, $v = f\lambda = (28.0$ Hz$)(6.00$ m$) = 168$ m/s.

$F = \mu v^2 = (0.002583$ kg/m$)(168$ m/s$)^2 = 72.90$ N. $F + B - mg = 0$.
$B = mg - F = 164.0$ N $- 72.9$ N $= 91.10$ N. For the rock,

$V = \dfrac{m}{\rho} = \dfrac{(164.0 \text{ N}/9.8 \text{ m/s}^2)}{3200 \text{ kg/m}^3} = 5.230 \times 10^{-3}$ m^3. $B = \rho_{liq} V_{rock} g$ and

$\rho_{liq} = \dfrac{B}{V_{rock} g} = \dfrac{91.10 \text{ N}}{(5.230 \times 10^{-3} \text{ m}^3)(9.8 \text{ m/s}^2)} = 1.78 \times 10^3$ kg/m^3.

EVALUATE: This liquid has a density 1.78 times that of water, which is rather dense but not impossible.

15.73. IDENTIFY AND SET UP: Assume that the mass M is large enough so that there no appreciable motion of the string at the pulley or at the oscillator. For a string fixed at both ends, $\lambda_n = 2L/n$. The node-to-node distance d is $\lambda/2$, so $d = \lambda/2$. $v = f\lambda = \sqrt{F/\mu}$.

EXECUTE: (a) Because it is essentially fixed at its ends, the string can vibrate in only wavelengths for which $\lambda_n = 2L/n$, so $d = \lambda/2 = L/n$, where $n = 1, 2, 3, \ldots$.

(b) $f\lambda = \sqrt{F/\mu}$ and $\lambda = 2d$. Combining these two conditions and squaring gives $f^2(4d^2) = T/\mu = Mg/\mu$. Solving for μd^2 gives $\mu d^2 = \left(\dfrac{g}{4f^2}\right) M$. Therefore the graph of μd^2 versus M should be a straight line having slope equal to $g/4f^2$. Figure 15.73 shows this graph.

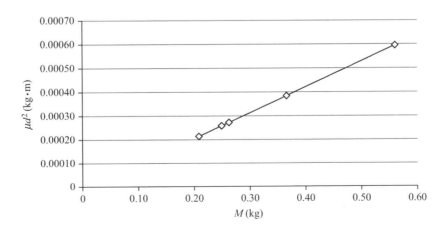

Figure 15.73

(c) The best fit straight line for the data has the equation $\mu d^2 = (0.001088 \text{ m})M - 0.00009074$ kg·m. The slope is $g/4f^2$, so $g/4f^2 = 0.001088$ m. Solving for f gives $f = 47.5$ Hz.

(d) For string A, $\mu = 0.0260$ g/cm $= 0.00260$ kg/m. We want the mass M for $\lambda = 48.0$ cm. Using $f\lambda = \sqrt{F/\mu}$ where $F = Mg$, squaring and solving for M, we get $M = \dfrac{\mu(f\lambda)^2}{g}$. Putting in the numbers gives $M = (0.00260 \text{ kg/m})[(47.5 \text{ Hz})(0.480 \text{ m})]^2/(9.80 \text{ m/s}^2) = 0.138$ kg $= 138$ g.

EVALUATE: In part (d), if the string is vibrating in its fundamental mode, $n = 1$, so $d = L = 48.0$ cm. The mass of the string in that case would be $m = \mu L = (0.00260 \text{ kg/m})(0.48 \text{ m}) = 0.00125$ kg $= 1.25$ g, so the string would be much lighter than the 138-g weight attached to it.

15.75. **IDENTIFY AND SET UP:** $P_{av} = \dfrac{1}{2}\sqrt{\mu F}\,\omega^2 A^2$, $v = \sqrt{F/\mu}$, and $\omega = 2\pi f$.

EXECUTE: Combining $P_{av} = \dfrac{1}{2}\sqrt{\mu F}\,\omega^2 A^2$ and $\omega = 2\pi f$ gives $P_{av} = 2\pi^2 \sqrt{\mu F}\,A^2 f^2$. Therefore a graph of P_{av} versus f^2 should be a straight line having slope $2\pi^2 \sqrt{\mu F}\,A^2$.

(b) Calculate the slope from the graph shown in the problem. Estimating the points (40,000 Hz^2, 12.2 W) and (10,000 Hz^2, 3 W), we get a slope of (9.2 W)/(30,000 Hz^2) = 3.07×10^{-4} W/Hz^2. (Answers here will vary, depending on the accuracy in reading the graph.) We can get F from the slope of the graph and then use $v = \sqrt{F/\mu}$ to calculate v. Using our measured slope, we have $2\pi^2\sqrt{\mu F}\,A^2$ = slope = 3.07×10^{-4} W/Hz^2. Solving for F gives $F = \dfrac{(\text{slope})^2}{4\pi^4 A^4 \mu}$. Putting this result into $v = \sqrt{F/\mu}$ gives us $v = \sqrt{\dfrac{F}{\mu}} = \sqrt{\dfrac{(\text{slope})^2}{4\pi^4 A^4 \mu^2}} = \dfrac{\text{slope}}{2\pi^2 A^2 \mu}$. This gives $v = (3.07\times 10^{-4} \text{ W/Hz}^2)/[2\pi^2(0.0040 \text{ m})^2(0.0035 \text{ kg/m})] = 280$ m/s.

(c) From the graph, $P = 10.0$ W corresponds to $f^2 = 33{,}000$ Hz^2, so $\omega = 2\pi f = 2\pi\sqrt{33{,}000 \text{ Hz}^2} = 1100$ rad/s.

EVALUATE: At 280 m/s and with an angular frequency of 1100 rad/s, the string is moving too fast to follow individual waves.

STUDY GUIDE FOR SOUND AND HEARING

Summary

In this chapter, we expand the concept of mechanical waves in order to understand sound and hearing. We'll begin with a description of longitudinal sound waves and their amplitudes, periods, frequencies, and wavelengths. We'll see how waves propagate through gases, liquids, and solids; how to determine the speed and intensity of sound waves; and how sound is produced by musical instruments. We'll also examine how the frequency of sound waves changes relative to the motion of the source and listener, summarized in the Doppler effect.

Objectives

After studying this chapter, you'll understand

- How sound waves are formed and propagate through media.
- How to apply the concepts of superposition, standing waves, nodes, and antinodes to sound waves.
- How to calculate the intensity of a sound wave.
- The allowed frequencies for longitudinal standing sound waves.
- The definition and how to calculate sound beats.
- How to apply the Doppler effect to moving sources and listeners.
- How to apply acoustics to a variety of systems.

Concepts and Equations

Term	Description
Sound Waves	Sound consists of longitudinal waves propagating through a medium. The pressure amplitude is given by $$p_{max} = BkA,$$ where B is the bulk modulus of the medium, k is the wave number, and A is the displacement amplitude. The speed of the sound wave depends on the medium: $$v = \sqrt{\frac{B}{\rho}} \quad \text{(longitudinal wave in a fluid);}$$ $$v = \sqrt{\frac{\gamma RT}{M}} \quad \text{(longitudinal wave in an ideal gas);}$$ $$v = \sqrt{\frac{Y}{\rho}} \quad \text{(longitudinal wave in a solid rod).}$$
Intensity	The intensity of a sound wave is the rate at which energy is transported per unit area per unit time. The intensity of a sinusoidal wave is given by $$I = \frac{1}{2}\sqrt{\rho B}\,\omega^2 A^2 = \frac{p_{max}^2}{2\rho v} = \frac{p_{max}^2}{2\sqrt{\rho B}}.$$ The intensity β of a sound wave is a logarithmic measure given by $$\beta = (10 \text{ dB})\log\frac{I}{I_0},$$ where I_0 is the reference intensity (10^{-12} W/m^2). The units of β are decibels (dB).
Standing Sound Waves	Standing sound waves that propagate in a fluid in a pipe can reflect and form longitudinal standing waves. The closed end of a pipe is a displacement node and a pressure antinode; the open end of a pipe is a displacement antinode and a pressure node. For a pipe of length L with an open end, the fundamental frequency and harmonics are $$f_n = n\frac{v}{2L} = nf_1 \quad (n = 1, 2, 3,\ldots).$$ For a pipe of length L with a closed end, the fundamental frequency and harmonics are $$f_n = n\frac{v}{4L} = nf_1 \quad (n = 1, 3, 5,\ldots).$$
Interference	When waves overlap in the same region of space, the waves are said to interfere. When the waves combine to form a wave with a larger amplitude, the waves interfere constructively, or reinforce one another. When the waves differ by a half cycle, their sum results in a wave with a smaller amplitude, and the waves interfere destructively, or cancel.
Beats	Beats are heard when two tones of slightly different frequencies are sounded together, creating a beat frequency that is the difference of the original two frequencies. The beat frequency given by $$f_{beat} = f_a - f_b.$$

Doppler Effect	The Doppler effect is the frequency shift that occurs when the listener is in motion relative to the source of sound. The listener's frequency f_L is related to the source frequency f_S by $$f_L = \frac{v + v_L}{v + v_S} f_S,$$ where v is the speed of sound and v_L and v_S are the *x*- components of the speed of the listener and source, respectively.

Key Concept 1: In a sound wave, the pressure amplitude (maximum pressure fluctuation) and displacement amplitude (maximum displacement of a particle in the medium) are proportional to each other. The proportionality constant depends on the wavelength of the sound and the bulk modulus of the medium.

Key Concept 2: When a sound wave travels from one medium into a different medium, the wave frequency and angular frequency remain the same. The wave number and wavelength can change, however, as can the pressure amplitude and displacement amplitude.

Key Concept 3: The speed of sound waves in a fluid depends on the fluid's bulk modulus and density. A sound wave of a given frequency has a longer wavelength in a medium that has a faster sound speed.

Key Concept 4: The speed of sound in a gas is determined by the temperature of the gas, its molar mass, and its ratio of heat capacities.

Key Concept 5: The intensity (power per unit area) of a sound wave is proportional to the square of the pressure amplitude of the wave. The proportionality constant depends on the density of the medium and the speed of sound in the medium.

Key Concept 6: If two sound waves in a given medium have the same intensity but different frequencies, the wave with the higher frequency has the greater *displacement* amplitude. The two waves have the same *pressure* amplitude, however.

Key Concept 7: To find the acoustic power output of a source of sound, multiply the area over which the emitted sound wave is distributed by the average intensity of the sound over that area.

Key Concept 8: The sound intensity level (in decibels, or dB) is a logarithmic measure of the intensity of a sound wave. Adding 10 dB to the sound intensity level corresponds to multiplying the intensity by a factor of 10.

Key Concept 9: The *difference* between the sound intensity levels of two sounds is proportional to the logarithm of the *ratio* of the intensities of those sounds.

Key Concept 10: In a standing sound wave, a pressure node is a displacement antinode, and vice versa. The sound is loudest at a pressure antinode; there is no sound at a pressure node.

Key Concept 11: For a pipe open at both ends (an "open pipe"), the normal-mode frequencies of a standing sound wave include both even and odd multiples of the pipe's fundamental frequency. For a pipe open at one end and closed at the other (a "stopped pipe"), the only normal-mode frequencies are the odd multiples of the pipe's fundamental frequency. The fundamental frequency of a stopped pipe is half that of an open pipe of the same length.

Key Concept 12: If you force or drive a mechanical system (such as a guitar string or the air in a pipe) to vibrate at a frequency *f*, the system will oscillate with maximum amplitude (or *resonate*) if *f* equals one of the normal-mode frequencies of the system.

Key Concept 13: Two sound waves of the same frequency interfere *constructively* at a certain point if the waves arrive there in phase. If the waves arrive at that point out of phase, they interfere *destructively*.

Key Concept 14: If a source of sound is moving through still air, a listener behind the source hears a sound of increased wavelength. A listener in front of the source hears a sound of decreased wavelength.

Key Concept 15: If a source of sound is moving through still air, a listener behind the source hears a sound of decreased frequency. A listener in front of the source hears a sound of increased frequency.

Key Concept 16: If a listener is moving away from a source of sound, the listener hears a sound of decreased frequency. If the listener is moving toward the source, the listener hears a sound of increased frequency.

Key Concept 17: When a listener and source of sound are both moving relative to the air, the listener hears a sound of decreased frequency if the listener and source are moving apart. The listener hears a sound of increased frequency if the listener and source are moving closer together.

Key Concept 18: If a source of sound moves relative to a wall (or other reflecting surface), there are two shifts in the frequency of the sound: The frequency received by and reflected from the wall is shifted compared to the sound emitted by the source, and the frequency received back at the source is shifted compared to the sound reflected from the wall. Both shifts increase the frequency if the source is approaching the wall, and both shifts decrease the frequency if the source is moving away from the wall.

Key Concept 19: An object moving through the air faster than the speed of sound continuously produces a cone-shaped shock wave. The angle of the cone depends on the object's Mach number (the ratio of its speed to the speed of sound).

Conceptual Questions

1: Threshold of pain

By what factor must you amplify the intensity of a normal conversation to make it reach the threshold of pain?

IDENTIFY, SET UP, AND EXECUTE The intensity of a normal conversation is 65 dB, and the intensity at the threshold of pain is 120 dB, according to Table 2 in the text. These two intensities differ by 55 dB, or $10^{5.5} = 3.1 \times 10^5$. Therefore, you must amplify the normal conversation by 3.1×10^5 to reach the threshold of pain.

EVALUATE A normal conversation is five orders of magnitude less than the pain threshold. Remember this fact the next time your physics professor lectures: His lectures are *not* painful, because his voice hasn't reached the sound pain threshold!

2: Explaining Doppler shift

Your younger brother asks you to explain why the sound from train whistles changes from a high pitch to a low pitch when a train passes. How do you explain the change?

IDENTIFY, SET UP, AND EXECUTE You first explain that sound comes from vibrations, or changing pressure. A high pitch comes from faster vibrations and a low pitch comes from slower vibrations. The train whistle produces a steady number of vibrations. When the train approaches, the vibrations become compressed, effectively increasing the number of vibrations that reach your ear per unit time. When the train leaves, the vibrations become expanded, effectively slowing the vibrations.

EVALUATE This description provides an alternative explanation of the Doppler shift. As the listener and source move toward each other, the wave fronts become closer together and the frequency increases.

3: An orchestra warming up

While you wait for an orchestral performance, you hear the musicians tuning their instruments. You observe that several musicians play the same note for a few seconds to check the tune. What are they doing?

IDENTIFY, SET UP, AND EXECUTE The musicians are trying to play the same frequency when they tune their instruments. If one or more instruments vibrate at a slightly different frequency, the different frequencies interfere and produce beats. When tuning, the musicians listen for beats and readjust their instruments until the beats are removed.

EVALUATE You can listen for beats when you hear music. Beats may be intentional; for example, some pipe organs have a slow beat to create an undulating effect.

4: Frequencies in a water bottle

Estimate the two lowest frequencies that you can achieve by blowing across the top of a plastic water bottle.

IDENTIFY, SET UP, AND EXECUTE We can treat the bottle as a pipe that is open at one end and closed at the other. The normal-mode frequencies of a closed-end pipe are given by

$$f_n = \frac{nv}{4L},$$

where we want frequencies corresponding to $n=1$ and 3. The water bottle is approximately 8", or 20.3 cm, long. Taking the speed of sound to be 344 m/s, we find that the two frequencies are 424 Hz and 1271 Hz.

EVALUATE Note that only the odd n's are valid for the closed-end pipe.
You can now build a pipe organ from recycled soda cans and soda bottles.

Problems

1: Finding a plug in a tube

In an attempt to find where a plug is in a tube containing air, a plumber blows air across the opening of the tube and hears a resonance at a frequency of 80 Hz. If this is the fundamental mode, how far away from the end of the pipe is the plug? Take the velocity of sound in air to be 345 m/s.

IDENTIFY We'll use the relationship between pipe length and normal-mode frequencies to find the distance the plug is from the end of the tube—the target variable.

SET UP We'll use the normal-mode relationship to find the fundamental frequency of a closed pipe, since the plug effectively closes the pipe.

EXECUTE The fundamental frequency of the closed pipe is

$$f_1 = \frac{v}{4L}.$$

We know the frequency and speed of sound, so we solve for the length:

$$L = \frac{v}{4f_1} = \frac{(345 \text{ m/s})}{4(80 \text{ Hz})} = 1.08 \text{ m}.$$

The plug is 1.08 from the end of the pipe.

KEY CONCEPT 11 EVALUATE This problem illustrates how to count overtones carefully and how to interpret integer results.

Practice Problem: What frequency would be heard if the plug was 1.26 m away? *Answer:* 68.5 Hz

Extra Practice: How far away does the plug need to be so that the frequency is too low to be heard? *Answer:* 4.31 m

2: Overtones in a pipe

An open pipe of length 1.5 m is played on a day when the speed of sound in air is 345 m/s. How many overtones can be heard by a person with good hearing?

IDENTIFY The number of overtones is the number of frequencies above the fundamental frequency. We'll use the relationship between pipe length and normal-mode frequencies to find the number of overtones—the target variable.

SET UP A person with good hearing can hear in the range from 20 Hz to 20,000 Hz. We'll find the number of frequencies in that range for the pipe, and then we'll subtract the fundamental frequency to solve the problem.

EXECUTE The frequencies of standing waves in an open pipe are

$$f_n = nf_1,$$

where n is an integer. The fundamental frequency of the pipe is

$$f_1 = \frac{v}{2L}.$$

For this pipe,

$$f_1 = \frac{v}{2L} = \frac{(345 \text{ m/s})}{2(1.5 \text{ m})} = 115 \text{ Hz}.$$

The fundamental frequency is above 20 Hz, so it can be heard. The highest frequency that can be heard by the human ear is 20,000 Hz. This frequency corresponds to

$$n = \frac{20{,}000 \text{ Hz}}{f_1} = \frac{20{,}000 \text{ Hz}}{115 \text{ Hz}} = 173.9.$$

Since we cannot hear nine-tenths of a frequency, we truncate n to 173. Thus, 173 frequencies can be heard: the fundamental frequency and 172 overtones.

KEY CONCEPT 11 **EVALUATE** This problem illustrates how to count overtones carefully and how to interpret integer results.

Practice Problem: What is the frequency of the 172nd overtone? *Answer:* 19,895 Hz

Extra Practice: How many overtones can be heard in a 2.0 m pipe? *Answer:* 230

3: Making notes

A 1-meter-long tube open at one end and closed at the other contains water to a depth d. Assuming that the sound waves have a displacement node at the water surface and an antinode at the open end, find the depth of liquid that makes the tube resonate at middle C (264 Hz) and one octave below middle C (132 Hz). Take the speed of sound in air to be 345 m/s.

IDENTIFY We'll use the relationship between pipe length and normal-mode frequencies to find the distance the water is from the top of the tube. We'll subtract that distance from the length of the pipe to find the height of the water—the target variable.

SET UP We'll use the normal-mode relationship to find the fundamental frequency of a closed pipe, since there is a displacement node at the water. We'll find the lowest mode, that corresponding to $n=1$.

EXECUTE The fundamental frequency of a closed pipe is

$$f_1 = \frac{v}{4L},$$

where L is the distance from the top of the tube to the top of the water. We solve for L for the two frequencies:

$$L_{\text{middle C}} = \frac{v}{4f_1} = \frac{(345 \text{ m/s})}{4(264 \text{ Hz})} = 0.327 \text{ m};$$

$$L_{\text{below C}} = \frac{v}{4f_1} = \frac{(345 \text{ m/s})}{4(132 \text{ Hz})} = 0.653 \text{ m}.$$

The water must be at a height of $1.0 \text{ m} - 0.327 \text{ m} = 0.673 \text{ m}$ for middle C and at a height of $1.0 \text{ m} - 0.653 \text{ m} = 0.347 \text{ m}$ for one octave below middle C.

KEY CONCEPT 11 EVALUATE This problem illustrates how to design a pipe organ that is made by filling several pipes of the same length with water. The pipe, however, would be a tough organ to tune, because the water will evaporate.

Practice Problem: What is the lowest frequency that can be heard in this pipe? *Answer:* 86.3 Hz

Extra Practice: What is the height of the water when you hear a frequency of 20,000 Hz? *Answer:* 0.996 m

4: Power at a concert

You are given the task of determining how much power is needed in the sound system at the new stadium. There is a single set of speakers on the stage. If the design calls for an intensity of 100 dB at the farthest seats (120 m from the speakers), how much power is required?

IDENTIFY AND SET UP In order to determine the intensity from the power, we assume that the sound is distributed over a sphere of radius 120 m. We'll use the definition of intensity to relate the design intensity to the power.

EXECUTE The intensity is given by

$$\beta = (10 \text{ dB})\log\frac{I}{I_0}.$$

Taking the logarithm of both sides gives

$$I = I_0 10^{(\beta/10 \text{ dB})}.$$

The intensity is the power per unit area, where the area in this case is that of a sphere $(4\pi r^2)$. Combining the various equations gives

$$P = AI = (4\pi r^2)(I_0 10^{(\beta/10\ \text{dB})}) = (4\pi(120\ \text{m})^2)((10^{212}\ \text{W/m}^2)10^{((100\ \text{dB})/10\ \text{dB})}) = 1.8\ \text{kW}.$$

The required power is 1.8 kW.

KEY CONCEPT 7 **EVALUATE** Our stadium requires a substantial sound system in order for all visitors to hear the concert. The amount of power required would harm the hearing of those near the speaker. Stadiums are designed with multiple speakers placed around the stadium and closer to the visitors, to reduce the maximum volume.

Practice Problem: What is the intensity of the sound for persons seated 20 m from the speakers? *Answer:* 116 dB, close to the threshold for permanent hearing damage.

Extra Practice: How far away can you be and still hear the speakers at an "average whisper" level? *Answer:* 1.2×10^6 m

5: Speed of approaching train

You are driving along a country road at 20.0 m/s. A train approaches on a rail that parallels the road. The train whistle blasts at a frequency of 800 Hz, but you hear a 950-Hz whistle. What is the speed of the approaching train?

IDENTIFY Our target variable is the speed of the approaching train.

SET UP The Doppler effect describes the frequency shift for moving sources, so we'll use the Doppler formula to determine the speed of the approaching train. Our coordinate system is shown in Figure 1; positive velocities are taken to be from the listener toward the source. Both the source and listener are moving. The listener's velocity is positive and the source's velocity is negative in our coordinate system. The speed of sound is taken to be 340 m/s.

Figure 1 Problem 5 sketch.

EXECUTE The listener's frequency is related to the source frequency by the Doppler shift,

$$f_L = \frac{v + v_L}{v + v_S} f_S,$$

where v is the speed of sound and v_L and v_S are the x-components of the speed of the listener and source, respectively. We can rearrange to solve for v_S:

$$v_S = \frac{v + v_L}{f_L} f_S - v.$$

In this case, v_L is +20.0 m/s, f_L is 950 Hz, f_S is 800 Hz, and v is 340 m/s. Substituting yields

$$v_S = \frac{v + v_L}{f_L} f_S - v = \frac{(340\ \text{m/s}) + (20.0\ \text{m/s})}{(950\ \text{Hz})}(800\ \text{Hz}) - (340\ \text{m/s}) = -36.8\ \text{m/s}.$$

The train is approaching at 36.8 m/s, or 132 kilometers per hour.

KEY CONCEPT 17 **EVALUATE** This problem illustrates how to use the Doppler shift to find the speed of an object. We expected and found a negative speed, indicating that the source was moving toward the listener. We see that proper Doppler-shift solutions require a coordinate system and careful interpretation of the directions of the velocities.

Practice Problem: What frequency would you hear if you were moving away from the train at 20.0 m/s? *Answer:* 844 Hz.

Extra Practice: What frequency would you hear in this case after the train passes you? *Answer:* 764 Hz

Try It Yourself!

1: Designing an organ pipe

You are asked to design an organ pipe that will produce a middle C on the "even-tempered scale." Middle C is equivalent to a frequency of 261.6 Hz. (a) If the tube is open at both ends, how long should it be? (b) If the tube is open at one end and closed at the other, how long should it be? Take the speed of sound in air to be 345 m/s.

Solution Checkpoints

IDENTIFY AND SET UP Use the normal-mode frequency relationships for open and closed tubes to find the lengths.

EXECUTE The fundamental frequency of an open pipe is

$$f_1 = \frac{v}{2L}.$$

The fundamental frequency of a closed pipe is

$$f_1 = \frac{v}{4L}.$$

The two lengths are 0.659 m and 0.330 m.

KEY CONCEPT 11 **EVALUATE** Which pipe is more desirable for a compact pipe organ?

Practice Problem: What lengths are required for a note that is one octave higher (twice the frequency)? *Answer:* 0.330 m and 0.165 m

Extra Practice: What length do you need for an open pipe playing a very low A (27.5 Hz)? *Answer:* 6.27 m

2: Designing a organ pipe, part 2

Find the allowed normal-mode frequencies of the two organ pipes in the previous problem. Take the speed of sound in air to be 345 m/s.

Solution Checkpoints

IDENTIFY AND SET UP Use the normal-mode frequency relationships for open and closed tubes to find the frequencies.

EXECUTE The normal-mode frequencies of an open pipe are

$$f_n = \frac{nv}{2L}.$$

Substituting $n = 1, 2, 3...$, we find that $f_1 = 261.6$ Hz, $f_2 = 523.2$ Hz, $f_3 = 784.8$ Hz,...
The fundamental frequency of a closed pipe is

$$f_1 = \frac{v}{4L}.$$

The normal-mode frequencies are integer multiples of the fundamental frequency; that is, $f_n = nf_1$. So, substituting $n = 1, 3, 5...$ we find that $f_1 = 261.6$ Hz, $f_3 = 784.8$ Hz, $f_5 = 1308$ Hz,...

KEY CONCEPT 11 **EVALUATE** Which pipe is more desirable for the number of frequencies it can produce?

Practice Problem: What is the highest harmonic that can be heard in the open pipe? *Answer:* 76th

Extra Practice: How many harmonics can be heard in the closed pipe? *Answer:* 38

3: Doppler-shift practice

Consider a source that produces a sound with a frequency of 500 Hz. If the speed of sound in air is 345 m/s, and the source and listener both move along the line joining them at speeds of 25 m/s, what frequencies can be heard by the listener for all possible directions of velocities?

Solution Checkpoints

IDENTIFY AND SET UP Use the Doppler-shift equation to find the possible frequencies while varying the direction, or sign, of the velocities.

EXECUTE The listener's frequency is related to the source frequency by the Doppler shift,

$$f_L = \frac{v + v_L}{v + v_S} f_S,$$

where v is the speed of sound and v_L and v_S are the speeds of the listener and source, respectively. When the two velocities are in the same direction, either positive or negative, there is no Doppler shift and the listener hears a sound with a frequency of 500 Hz.

If the source and listener are moving away from each other, (say, v_L is negative and v_S is positive,) then the frequency is reduced to 432 Hz. If the source and listener are moving toward each other (say, v_L is positive and v_S is negative), then the frequency is increased to 579 Hz.

KEY CONCEPT 17 **EVALUATE** This problem illustrates the amount of Doppler shift for four possible combinations of velocity between two objects.

Practice Problem: If the source moves at 15 m/s, then what possible frequencies can be heard if both the listener and the source are moving to the right? *Answer:* 485 Hz and 514 Hz

Extra Practice: What if they move in opposite directions? *Answer:* 444 Hz and 561 Hz

Key Example Variation Problems

Solutions to these problems are in Chapter 16 of the Student's Solutions Manual.

Be sure to review EXAMPLES 16.5, 16.6, 16.7, 16.8, and 16.9 (Section 16.3) before attempting these problems.

VP16.9.1 A 256 Hz sound wave in air (density 1.20 kg/m^3, speed of sound 344 m/s) has intensity 5.50×10^{-8} W/m^2. (a) What is the wave's pressure amplitude? (b) If the intensity remains the same but the frequency is doubled to 512 Hz, how does this affect the pressure amplitude?

VP16.9.2 At a certain distance from a fire alarm, the sound intensity level is 85.0 dB. (a) What is the intensity of this sound? (b) How many times greater is the intensity of this sound than that of a 67.0 dB sound?

VP16.9.3 A lion can produce a roar with a sound intensity level of 114 dB at a distance of 1.00 m. What is the sound intensity level at a distance of (a) 4.00 m and (b) 15.8 m from the lion? Assume that intensity obeys the inverse-square law.

VP16.9.4 The sound intensity level inside a typical modern airliner in flight is 66.0 dB. The air in the cabin has density 0.920 kg/m^3 (less than in the atmosphere at sea level) and speed of sound 344 m/s. (a) What is the pressure amplitude of this sound? (b) If the pressure amplitude were increased by a factor of 10, what would the new sound intensity level be?

Be sure to review EXAMPLES 16.11 and 16.12 (Sections 16.4 and 16.5) before attempting these problems.

VP16.12.1 A particular open organ pipe has a fundamental frequency of 220 Hz (known to musicians as A3 or "A below middle C") when the speed of sound waves in air is 344 m/s. (a) What is the length of this pipe? (b) The third harmonic of this pipe has the same frequency as the fundamental frequency of a stopped pipe. What is the length of this stopped pipe?

VP16.12.2 You have two organ pipes, one open and one stopped. Which harmonic (if any) of the stopped pipe has the same frequency as the third harmonic of the open pipe if the stopped pipe length is (a) 16, (b) 12, or (c) 13 that of the open pipe?

VP16.12.3 One of the strings of a bass viol is 0.680 m long and has a fundamental frequency of 165 Hz. (a) What is the speed of waves on this string? (b) When this string vibrates at its fundamental frequency, it causes the air in a nearby stopped organ pipe to vibrate at that pipe's fundamental frequency. The speed of sound in the pipe is 344 m/s. What is the length of this pipe?

VP16.12.4 A stopped pipe 1.00 m in length is filled with helium at 20°C (speed of sound 999 m/s). When the helium in this pipe vibrates at its third harmonic frequency, it causes the air at 20°C (speed of sound 344 m/s) in a nearby open pipe to vibrate at its fifth harmonic frequency. What are the frequency and wavelength of the sound wave (a) in the helium in the stopped pipe and (b) in the air in the open pipe? (c) What is the length of the open pipe?

Be sure to review EXAMPLES 16.14, 16.15, 16.16, 16.17, and 16.18 (Section 16.8) before attempting these problems.

VP16.18.1 The siren on an ambulance emits a sound of frequency 2.80×10^3 Hz. If the ambulance is traveling at 26.0 m/s (93.6 km/h, or 58.2 mi/h), the speed of sound is 340 m/s, and the air is still, what are the frequency and wavelength that you hear if you are standing (a) in front of the ambulance or (b) behind the ambulance?

VP16.18.2 A stationary bagpiper is playing a Highland bagpipe, in which one reed produces a continuous sound of frequency 440 Hz. The air is still and the speed of sound is 340 m/s. (a) What is the wavelength of the sound wave produced by the bagpipe? What are the frequency and wavelength of the sound wave that a bicyclist hears if she is (b) approaching the bagpiper at 10.0 m/s or (c) moving away from the bagpiper at 10.0 m/s?

VP16.18.3 A police car moving east at 40.0 m/s is chasing a speeding sports car moving east at 35.0 m/s. The police car's siren has frequency 1.20×10^3 Hz, the speed of sound is 340 m/s, and the air is still. (a) What is the frequency of sound that the driver of the speeding sports car hears? (b) If the speeding sports car were to turn around and drive west at 35.0 m/s toward the approaching police car, what frequency would the driver of the sports car hear?

VP16.18.4 For a scene in an action movie, a car drives at 25.0 m/s directly toward a wall. The car's horn is on continuously and produces a sound of frequency 415 Hz. (a) If the speed of sound is 340 m/s and the air is still, what is the frequency of the sound that the driver of the car hears reflected from the wall? (b) How fast would the car have to move for the reflected sound that the driver hears to have frequency 495 Hz?

16

STUDENT'S SOLUTIONS MANUAL FOR SOUND AND HEARING

VP16.9.1. **IDENTIFY:** We want to know the pressure amplitude of a sound wave.

SET UP: $I = \dfrac{p_{max}^2}{2\rho v}$

EXECUTE: **(a)** Solve for p_{max} giving $p_{max} = \sqrt{2\rho v I}$. Putting in the numbers we get
$p_{max} = \sqrt{2(1.20 \text{ kg/m}^3)(344 \text{ m/s})(5.50\times 10^{-8} \text{ W/m}^2)} = 6.74\times 10^{-3}$ Pa.
(b) Neither ρ nor v is affected by the frequency change, so p_{max} remain unchanged.
EVALUATE: Changing the air density would affect the pressure amplitude.

VP16.9.2. **IDENTIFY:** This problem deals with sound intensity and intensity level.

SET UP: $\beta = (10 \text{ dB}) \log \dfrac{I}{I_0}$

EXECUTE: **(a)** We know the sound level is 85.0 dB and want the sound intensity. Using
$\beta = (10 \text{ dB}) \log \dfrac{I}{I_0}$ gives $85.0 \text{ dB} = (10 \text{ dB}) \log \dfrac{I}{I_0}$. Solving for I gives
$I = 10^{8.5} I_0 = 10^{8.5} 10^{-12} \text{ W/m}^2 = 10^{-3.50} \text{ W/m}^2 = 3.16\times 10^{-4}$ W/m^2.
(b) Solve $\beta = (10 \text{ dB}) \log \dfrac{I}{I_0}$ for I, giving $I = I_0 10^{\beta/(10 \text{ dB})}$. Now take the ratio of the two intensities. $\dfrac{I_{85}}{I_{67}} = \dfrac{I_0 10^{(85.0 \text{ dB})/(10 \text{ dB})}}{I_0 10^{(67.0 \text{ dB})/(10 \text{ dB})}} = \dfrac{10^{8.50}}{10^{6.70}} = 63.1$, so I_{85} is 63.1 times greater than I_{67}.
EVALUATE: The ratio of the sound intensity levels is $85/67 = 1.27$, but the intensity ratio is *much* greater.

VP16.9.3. **IDENTIFY:** This problem involves sound intensity and intensity level.
SET UP: We know the sound intensity level of the lion's roar is 114 at 1.00 m, and we want to know what it is at 4.00 m and 15.8 m from the lion. The sound intensity obeys an inverse-square law, but the sound intensity level does not. We use $\beta = (10 \text{ dB}) \log \dfrac{I}{I_0}$ and $I_2 = I_1 \left(\dfrac{r_1}{r_2}\right)^2$.

EXECUTE: **(a)** <u>At 1.00 m</u>: $\beta_1 = (10 \text{ dB}) \log \dfrac{I_1}{I_0} = 114$ dB. Solve for I_1: $I_1 = 10^{11.4} I_0$.

At 4.00 m: $I_2 = I_1 \left(\dfrac{r_1}{r_2}\right)^2$ gives $I_4 = I_1 \left(\dfrac{1.00 \text{ m}}{4.00 \text{ m}}\right)^2 = \dfrac{I_1}{16.0} = \dfrac{10^{11.4} I_0}{16.0}$. Now find the sound intensity level: $\beta_4 = (10 \text{ dB}) \log \dfrac{I_4}{I_0} = (10 \text{ dB}) \log \left(\dfrac{10^{11.4} I_0}{16.0 I_0}\right) = 102$ dB.

(b) At 15.8 m we follow the same procedure as above, giving

$$\beta_{15.8} = (10 \text{ dB}) \log \dfrac{I_{16}}{I_0} = (10 \text{ dB}) \log \left(\dfrac{10^{11.4} I_0}{(15.8)^2 I_0}\right) = 90.0 \text{ dB}.$$

EVALUATE: The sound intensity obeys an inverse-square law, but sound intensity level does not.

VP16.9.4. **IDENTIFY:** We are investigating the relationship between sound intensity level and the pressure amplitude of a sound wave.

SET UP: We know that $I = \dfrac{p_{max}^2}{2\rho v}$, $\beta = (10 \text{ dB}) \log \dfrac{I}{I_0}$, and $I_0 = 10^{-12}$ W/m². We want to find the pressure amplitude of the sound wave and then see how changing it would affect the sound intensity level.

EXECUTE: **(a)** First use $\beta = (10 \text{ dB}) \log \dfrac{I}{I_0}$ to find the intensity I. $\beta = (10 \text{ dB}) \log \dfrac{I}{I_0} = 66.0$ dB, so $I = 10^{6.60} I_0 = 3.9811 \times 10^{-6}$ W/m². Now use I to find p_{max}. Solving $I = \dfrac{p_{max}^2}{2\rho v}$ for p_{max} gives $p_{max} = \sqrt{2\rho v I}$, which gives

$$p_{max} = \sqrt{2(0.920 \text{ kg/m}^3)(344 \text{ m/s})(3.9811 \times 10^{-6} \text{ W/m}^2)} = 5.02 \times 10^{-2} \text{ Pa}.$$

(b) Since $I \propto p_{max}^2$, increasing p_{max} by a factor of 10.0 will increase I by a factor of $10.0^2 = 100$, so $I_2 = 100 I_1$. Taking the ratio of the sound intensity levels gives
$\dfrac{\beta_2}{\beta_1} = \dfrac{(10 \text{ dB}) \log(I_2/I_0)}{(10 \text{ dB}) \log(I_1/I_0)} = \dfrac{\log(100 I_1/I_0)}{\log(I_1/I_0)} = \dfrac{2 + \log(I_1/I_0)}{\log(I_1/I_0)}$. But $\beta_1 = (10 \text{ dB}) \log(I_1/I_0)$, so $\log(I_1/I_0) = \dfrac{\beta_1}{10} = \dfrac{66.0 \text{ dB}}{10} = 6.60$. Therefore $\dfrac{\beta_2}{\beta_1} = \dfrac{2 + \dfrac{\beta_1}{10}}{\dfrac{\beta_1}{10}} = \dfrac{2 + 6.60}{6.60} = 1.303$. So we have $\beta_2 = 1.303 \beta_1 = (1.303)(66.0 \text{ dB}) = 86.0$ dB.

EVALUATE: Careful in handling logarithms because $\log(100 I_1/I_0)$ is *not* equal to $100 \log I_1/I_0$.

VP16.12.1. **IDENTIFY:** We are dealing with standing sound waves in open and stopped pipes.

SET UP: For an open pipe, $f_n = nv/2L$ ($n = 1, 2, 3, \ldots$), and for a stopped pipe $f_n = nv/4L$ ($n = 1, 3, 5, \ldots$).

EXECUTE: **(a)** We want the length L of this open pipe. For the fundamental frequency, $f_1 = v/2L$, so the length is $L = (344 \text{ m/s})/[2(220 \text{ Hz})] = 0.782$ m.

(b) For the open pipe in its 3rd harmonic, $n = 3$, so $f_3 = \dfrac{3v}{2L_3}$.

For the stopped pipe in its fundamental frequency, $f_1 = \dfrac{v}{4L_1}$.

The two frequencies are the same, so $\dfrac{3v}{2L_3} = \dfrac{v}{4L_1}$, so $L_1 = L_3/6 = (0.782 \text{ m})/6 = 0.130$ m.

EVALUATE: Careful on stopped pipes: n must be an *odd* integer, but on open pipes n can be odd and even integers.

VP16.12.2. IDENTIFY: We are investigating the harmonics of a stopped and an open organ pipe. We want to find what harmonics of a stopped pipe will resonate at the same frequency as the third harmonic of an open pipe.

SET UP: For an open pipe $f_n = nv/2L$ ($n = 1, 2, 3, \ldots$), and for a stopped pipe $f_n = nv/4L$ ($n = 1, 3, 5, \ldots$). We know that $f_3 = 3v/2L_o$ for the open pipe. We want to find values of n for the stopped pipe so that it will have the same frequency as the third harmonic of the open pipe. That is $f_n(\text{stopped}) = f_3(\text{open})$, where we want to find n for the stopped pipe. Equating the frequencies gives $\dfrac{nv}{4L_s} = \dfrac{3v}{2L_o}$, so $n = 6\dfrac{L_s}{L_o}$.

EXECUTE: **(a)** In this case $L_s = L_o/6$ so $n = 6\dfrac{L_s}{L_o} = 6\left(\dfrac{L_o}{6L_o}\right) = 1$.

(b) In this case, $L_s = L_o/2$, so $n = 6\dfrac{L_s}{L_o} = 6\left(\dfrac{L_o}{2L_o}\right) = 3$.

(c) In this case, $L_s = L_o/3$, so $n = 6\dfrac{L_s}{L_o} = 6\left(\dfrac{L_o}{3L_o}\right) = 2$. But n is only *odd* for a stopped pipe, so there are no harmonics of the stopped pipe having a frequency that is equal to the 3rd harmonic frequency of the open pipe.

EVALUATE: We cannot always make two pipes resonate at a given frequency.

VP16.12.3. IDENTIFY: A string vibrating in its fundamental frequency causes a nearby stopped organ pipe to vibrate in its fundamental frequency. So we are dealing with standing waves on a string and in a stopped pipe.

SET UP: For a string fixed at its ends, $f_n = nv_{\text{str}}/2L$ and $\lambda_n = 2L/n$ ($n = 1, 2, 3, \ldots$), and for a stopped pipe $f_n = nv/4L$ ($n = 1, 3, 5, \ldots$). For any wave $v = f\lambda$.

EXECUTE: **(a)** For the fundamental mode of the string, $\lambda_1 = 2L$ so $v_{\text{str}} = f\lambda = f(2L) =$ (165 Hz)(2)(0.680 m) = 224 m/s.

(b) The organ pipe in its fundamental mode is vibrating at the same frequency as the string. Using $f_n = nv/4L$ and solving for L gives $L = \dfrac{v}{4f_1} = \dfrac{344 \text{ m/s}}{4(165 \text{ Hz})} = 0.521$ m.

EVALUATE: The frequency of the sound wave is the same as the frequency at which the string is vibrating, but the wavelengths of the two waves are *not* the same because they have different speeds.

VP16.12.4. IDENTIFY: A stopped pipe 1.00 m long that is filled with helium is vibrating in its third harmonic, which causes a nearby open pipe to vibrate in its fifth harmonic.

SET UP: We want to know the frequency and wavelength of the sound in both pipes and the length of the open pipe. The speed of sound in the helium is $v_{\text{He}} = 999$ m/s, $v = f\lambda$, $\lambda_n = 4L/n$ ($n = 1, 3, 4, \ldots$) for a stopped pipe, and $f_n = nv/2L$ ($n = 1, 2, 3 \ldots$) for an open pipe.

EXECUTE: **(a)** In the stopped pipe, $\lambda_3 = 4L/3 = 4(1.00 \text{ m})/3 = 1.33$ m.

Using $v = f\lambda$ gives $f_3 = \dfrac{v_{\text{He}}}{\lambda_3} = \dfrac{999 \text{ m/s}}{1.33 \text{ m}} = 749$ Hz.

(b) In the open pipe, we know that $f_5 = 749$ Hz.

Using $v = f\lambda$ gives $\lambda_5 = \dfrac{v_{\text{air}}}{f_5} = \dfrac{344 \text{ m/s}}{749 \text{ Hz}} = 0.459$ m.

(c) Using $f_n = nv/2L$ in the open pipe gives $f_5 = \dfrac{5v_{\text{air}}}{2L}$. Solving for L gives

$$L = \dfrac{5v_{\text{air}}}{2f_5} = \dfrac{5(344 \text{ m/s})}{2(749 \text{ Hz})} = 1.15 \text{ m}.$$

EVALUATE: The speed of sound is different in the helium than it is in air because the density of the helium is not the same as that of air.

VP16.18.1. IDENTIFY: You do not hear the same frequency that the siren is emitting due to the motion of the ambulance. This is due to the Doppler effect.

SET UP: We have a moving source of sound and a stationary listener. The target variables are the frequency and wavelength of the sound heard by the listener, so use $f_L = \left(\dfrac{v+v_L}{v+v_S}\right)f_S$ and $v = f\lambda$.

EXECUTE: (a) Listener in front of source: $v_S = -26.0$ m/s. $f_L = \left(\dfrac{v+v_L}{v+v_S}\right)f_S$ gives

$$f_L = \left(\dfrac{340 \text{ m/s} + 0}{340 \text{ m/s} - 26.0 \text{ m/s}}\right)(2.80\times 10^3 \text{ Hz}) = 3.03\times 10^3 \text{ Hz}.$$

The wavelength is $\lambda = \dfrac{v}{f} = \dfrac{340 \text{ m/s}}{3.03\times 10^3 \text{ Hz}} = 0.112$ m.

(b) Listener behind source: $v_S = +26.0$ m/s.

$$f_L = \left(\dfrac{340 \text{ m/s} + 0}{340 \text{ m/s} + 26.0 \text{ m/s}}\right)(2.80\times 10^3 \text{ Hz}) = 2.60\times 10^3 \text{ Hz}.$$

The wavelength is $\lambda = \dfrac{v}{f} = \dfrac{340 \text{ m/s}}{2.60\times 10^3 \text{ Hz}} = 0.131$ m.

EVALUATE: When the source moves toward the listener, the wave crests are closer together so the frequency is increased. When the source moves away from the listener, the distance between wave crests is greater so the frequency is decreased.

VP16.18.2. IDENTIFY: The bike rider hears a sound different from what the bagpiper is emitting due to the bike's motion. This is due to the Doppler effect.

SET UP: The bagpiper is the stationary source and the bike rider is the moving observer. We want the frequency and wavelength of the sound the bike rider hears, so we use $v = f\lambda$ and $f_L = \left(\dfrac{v+v_L}{v+v_S}\right)f_S$ with $v_S = 0$.

EXECUTE: (a) $\lambda = \dfrac{v}{f} = \dfrac{340 \text{ m/s}}{440 \text{ Hz}} = 0.773$ m.

(b) <u>Biker approaching source</u>: $v_L = +10.0$ m/s. Using $f_L = \left(\dfrac{v+v_L}{v+v_S}\right)f_S$ gives

$$f_L = \left(\dfrac{340 \text{ m/s} + 10.0 \text{ m/s}}{340 \text{ m/s} + 0}\right)(440 \text{ Hz}) = 453 \text{ Hz}.$$

Now use $v = f\lambda$, but v is the speed of sound relative to the listener, which is 340 m/s + 10.0 m/s = 350 m/s. Therefore $\lambda = \dfrac{v}{f} = \dfrac{350 \text{ m/s}}{453 \text{ Hz}} = 0.773$ m.

Biker moving away from source: $v_L = -10.0$ m/s. Now we get

$$f_L = \left(\frac{340 \text{ m/s} - 10.0 \text{ m/s}}{340 \text{ m/s}}\right)(440 \text{ Hz}) = 427 \text{ Hz}.$$

Now use $v = f\lambda$, where $v = 340$ m/s $- 10.0$ m/s $= 330$ m/s. $\lambda = \dfrac{v}{f} = \dfrac{330 \text{ m/s}}{453 \text{ Hz}} = 0.773$ m.

EVALUATE: The frequency varies due to the speed of the listener because she runs into the waves at a higher (or lower) rate due to her motion. But the wavelength is just the distance between wave crests, so it is not affected by her motion and we get the same answer in both cases.

VP16.18.3. **IDENTIFY:** The police car and the sports car are both moving, so both of their motions affect the frequency that the listener receives. We need to use the Doppler effect.

SET UP: Our target variable is the frequency f_L of sound that the listener receives. We use $f_L = \left(\dfrac{v + v_L}{v + v_S}\right) f_S$. Figure VP16.18.3 shows both situations.

Figure VP16.18.3

EXECUTE: **(a)** The police car is moving in the direction of the listener and the listener is moving away from the police car, so $v_S = -40.0$ m/s and $v_L = -35.0$ m/s. $f_L = \left(\dfrac{v + v_L}{v + v_S}\right) f_S$ gives

$$f_L = \left(\frac{340 \text{ m/s} - 35.0 \text{ m/s}}{340 \text{ m/s} - 40.0 \text{ m/s}}\right)(1200 \text{ Hz}) = 1220 \text{ Hz}.$$

(b) The police car is moving in the direction of the listener and the listener is in the direction of the police car, so $v_S = -40.0$ m/s and $v_L = +35.0$ m/s. $f_L = \left(\dfrac{v + v_L}{v + v_S}\right) f_S$ gives

$$f_L = \left(\frac{340 \text{ m/s} + 35.0 \text{ m/s}}{340 \text{ m/s} - 40.0 \text{ m/s}}\right)(1200 \text{ Hz}) = 1500 \text{ Hz}.$$

EVALUATE: In part (a), the motion of the police car causes a higher frequency but the motion of the sports car causes a lower frequency, so there is a small change in the frequency heard by the listener compared to the emitted frequency. In part (b) both of their motions increase the frequency, so there is a large frequency change.

VP16.18.4. **IDENTIFY:** The motion of the car increases the sound frequency in front of it, and this motion also increases the frequency of the sound the driver hears reflected from the wall. There are two Doppler effects to consider.

SET UP: We need to break this problem into two steps: (1) the car is a moving source and the wall is a stationary listener and (2) the wall is a stationary source and the car is a moving listener. The wall reflects the same frequency it receives from the car. $f_L = \left(\dfrac{v + v_L}{v + v_S}\right) f_S$. In part (a) we want to find the frequency that the driver receives reflected from the wall. In part (b) we want to find how fast he most drive to hear sound of frequency 495 Hz coming from the wall.

EXECUTE: (a) Car as moving source and wall as stationary listener: $v_L = 0$ (the wall), $v_S = -25.0$ m/s. Use $f_L = \left(\dfrac{v + v_L}{v + v_S}\right) f_S$ to find the frequency f_w that the wall receives.

$$f_w = \left(\dfrac{340 \text{ m/s}}{340 \text{ m/s} - 25.0 \text{ m/s}}\right)(415 \text{ Hz}) = 447.94 \text{ Hz}.$$

Car as moving listener and wall as stationary source: $v_S = 0$ (the wall), $v_L = +25.0$ m/s, and $f_S = 447.94$ Hz. Use $f_L = \left(\dfrac{v + v_L}{v + v_S}\right) f_S$ to find the frequency f_d that the driver receives coming back from the wall. $f_d = \left(\dfrac{340 \text{ m/s} + 25.0 \text{ m/s}}{340 \text{ m/s}}\right)(447.94 \text{ Hz}) = 481$ Hz.

(b) We want to find v_S so that $f_c = 495$ Hz. We follow the same steps as in part (a).

Car as listener and wall as source: $v_L = v_c = ?$ (where v_c is the *magnitude* of the car's speed), $f_S = f_w = ?$, $f_L = f_c = 495$ Hz, $v_S = 0$ (wall). Using $f_L = \left(\dfrac{v + v_L}{v + v_S}\right) f_S$ gives $495 \text{ Hz} = \left(\dfrac{340 \text{ m/s} + v_c}{340 \text{ m/s}}\right) f_w$, so

$$f_w = \dfrac{(495 \text{ Hz})(340 \text{ m/s})}{340 \text{ m/s} + v_c}.$$

Car as source and wall as listener: $v_S = -v_c$, $v_L = 0$, $f_S = 415$ Hz, $f_L = f_w = ?$ so $f_L = \left(\dfrac{v + v_L}{v + v_S}\right) f_S$ gives

$$f_w = \left(\dfrac{340 \text{ m/s}}{340 \text{ m/s} - v_c}\right)(415 \text{ Hz}).$$

Now use the result for f_w that we just found in the previous step.

Equating these two equations for f_w and solving for v_c gives us $\dfrac{(495 \text{ Hz})(340 \text{ m/s})}{340 \text{ m/s} + v_c} = \left(\dfrac{340 \text{ m/s}}{340 \text{ m/s} - v_c}\right)(415 \text{ Hz})$, so $v_c = 29.9$ m/s.

EVALUATE: Our result in part (b) gives $v_c = 29.9$ m/s which is greater than the speed of 25.0 m/s in part (a). This result is reasonable because the frequency change in (b) is greater than the one in (a).

16.1. IDENTIFY and SET UP: $v = f\lambda$ gives the wavelength in terms of the frequency. Use $p_{max} = BkA$ to relate the pressure and displacement amplitudes.

EXECUTE: (a) $\lambda = v/f = (344 \text{ m/s})/1000 \text{ Hz} = 0.344$ m.

(b) $p_{max} = BkA$ and Bk is constant gives $p_{max1}/A_1 = p_{max2}/A_2$

$A_2 = A_1 \left(\dfrac{p_{max2}}{p_{max1}}\right) = 1.2 \times 10^{-8} \text{ m} \left(\dfrac{30 \text{ Pa}}{3.0 \times 10^{-2} \text{ Pa}}\right) = 1.2 \times 10^{-5}$ m.

(c) $p_{max} = BkA = 2\pi BA/\lambda$

$p_{max}\lambda = 2\pi BA = $ constant so $p_{max1}\lambda_1 = p_{max2}\lambda_2$ and $\lambda_2 = \lambda_1 \left(\dfrac{p_{max1}}{p_{max2}}\right) = (0.344 \text{ m}) \left(\dfrac{3.0 \times 10^{-2} \text{ Pa}}{1.5 \times 10^{-3} \text{ Pa}}\right)$

$= 6.9$ m

$f = v/\lambda = (344 \text{ m/s})/6.9 \text{ m} = 50$ Hz.

EVALUATE: The pressure amplitude and displacement amplitude are directly proportional. For the same displacement amplitude, the pressure amplitude decreases when the frequency decreases and the wavelength increases.

16.5. **IDENTIFY** and **SET UP:** Use $t = $ distance/speed. Calculate the time it takes each sound wave to travel the $L = 60.0$ m length of the pipe. Use $v = \sqrt{\dfrac{Y}{\rho}}$ to calculate the speed of sound in the brass rod.

EXECUTE: Wave in air: $t = (60.0 \text{ m})/(344 \text{ m/s}) = 0.1744$ s.

Wave in the metal: $v = \sqrt{\dfrac{Y}{\rho}} = \sqrt{\dfrac{9.0 \times 10^{10} \text{ Pa}}{8600 \text{ kg/m}^3}} = 3235$ m/s, so $t = \dfrac{60.0 \text{ m}}{3235 \text{ m/s}} = 0.01855$ s.

The time interval between the two sounds is $\Delta t = 0.1744 \text{ s} - 0.01855 \text{ s} = 0.156$ s.

EVALUATE: The restoring forces that propagate the sound waves are much greater in solid brass than in air, so v is much larger in brass.

16.7. **IDENTIFY:** $d = vt$ for the sound waves in air and in water.

SET UP: Use $v_{\text{water}} = 1482$ m/s at 20°C, as given in Table 16.1. In air, $v = 344$ m/s.

EXECUTE: Since along the path to the diver the sound travels 1.2 m in air, the sound wave travels in water for the same time as the wave travels a distance $22.0 \text{ m} - 1.20 \text{ m} = 20.8 \text{ m}$ in air. The depth of the diver is $(20.8 \text{ m})\dfrac{v_{\text{water}}}{v_{\text{air}}} = (20.8 \text{ m})\dfrac{1482 \text{ m/s}}{344 \text{ m/s}} = 89.6$ m. This is the depth of the diver; the distance from the horn is 90.8 m.

EVALUATE: The time it takes the sound to travel from the horn to the person on shore is $t_1 = \dfrac{22.0 \text{ m}}{344 \text{ m/s}} = 0.0640$ s. The time it takes the sound to travel from the horn to the diver is $t_2 = \dfrac{1.2 \text{ m}}{344 \text{ m/s}} + \dfrac{89.6 \text{ m}}{1482 \text{ m/s}} = 0.0035 \text{ s} + 0.0605 \text{ s} = 0.0640$ s. These times are indeed the same. For three figure accuracy the distance of the horn above the water can't be neglected.

16.9. **IDENTIFY:** $v = f\lambda$. The relation of v to gas temperature is given by $v = \sqrt{\dfrac{\gamma RT}{M}}$.

SET UP: Let $T = 22.0°\text{C} = 295.15$ K.

EXECUTE: At 22.0°C, $\lambda = \dfrac{v}{f} = \dfrac{325 \text{ m/s}}{1250 \text{ Hz}} = 0.260 \text{ m} = 26.0$ cm. $\lambda = \dfrac{v}{f} = \dfrac{1}{f}\sqrt{\dfrac{\gamma RT}{M}}$.

$\dfrac{\lambda}{\sqrt{T}} = \dfrac{1}{f}\sqrt{\dfrac{\gamma R}{M}}$, which is constant, so $\dfrac{\lambda_1}{\sqrt{T_1}} = \dfrac{\lambda_2}{\sqrt{T_2}}$.

$T_2 = T_1 \left(\dfrac{\lambda_2}{\lambda_1}\right)^2 = (295.15 \text{ K})\left(\dfrac{28.5 \text{ cm}}{26.0 \text{ cm}}\right)^2 = 354.6 \text{ K} = 81.4°\text{C}$.

EVALUATE: When T increases v increases and for fixed f, λ increases. Note that we did not need to know either γ or M for the gas.

16.13. IDENTIFY AND SET UP: We want the sound intensity level to increase from 20.0 dB to 60.0 dB. The previous problem showed that $\beta_2 - \beta_1 = (10\ \text{dB})\log\left(\dfrac{I_2}{I_1}\right)$. We also know that $\dfrac{I_2}{I_1} = \dfrac{r_1^2}{r_2^2}$.

EXECUTE: Using $\beta_2 - \beta_1 = (10\ \text{dB})\log\left(\dfrac{I_2}{I_1}\right)$, we have $\Delta\beta = +40.0$ dB. Therefore $\log\left(\dfrac{I_2}{I_1}\right) = 4.00$, so $\dfrac{I_2}{I_1} = 1.00\times 10^4$. Using $\dfrac{I_2}{I_1} = \dfrac{r_1^2}{r_2^2}$ and solving for r_2, we get

$r_2 = r_1\sqrt{\dfrac{I_1}{I_2}} = (15.0\ \text{m})\sqrt{\dfrac{1}{1.00\times 10^4}} = 15.0$ cm.

EVALUATE: A change of 10^2 in distance gives a change of 10^4 in intensity. Our analysis assumes that the sound spreads from the source uniformly in all directions.

16.17. IDENTIFY: Use $I = \dfrac{vp_{max}^2}{2B}$ to relate I and p_{max}. $\beta = (10\ \text{dB})\log(I/I_0)$. The equation $p_{max} = BkA$ says the pressure amplitude and displacement amplitude are related by $p_{max} = BkA = B\left(\dfrac{2\pi f}{v}\right)A$.

SET UP: At 20°C the bulk modulus for air is 1.42×10^5 Pa and $v = 344$ m/s. $I_0 = 1\times 10^{-12}$ W/m^2.

EXECUTE: (a) $I = \dfrac{vp_{max}^2}{2B} = \dfrac{(344\ \text{m/s})(6.0\times 10^{-5}\ \text{Pa})^2}{2(1.42\times 10^5\ \text{Pa})} = 4.4\times 10^{-12}$ W/m^2

(b) $\beta = (10\ \text{dB})\log\left(\dfrac{4.4\times 10^{-12}\ \text{W/m}^2}{1\times 10^{-12}\ \text{W/m}^2}\right) = 6.4$ dB

(c) $A = \dfrac{vp_{max}}{2\pi fB} = \dfrac{(344\ \text{m/s})(6.0\times 10^{-5}\ \text{Pa})}{2\pi(400\ \text{Hz})(1.42\times 10^5\ \text{Pa})} = 5.8\times 10^{-11}$ m

EVALUATE: This is a very faint sound and the displacement and pressure amplitudes are very small. Note that the displacement amplitude depends on the frequency but the pressure amplitude does not.

16.21. IDENTIFY: The intensity of sound obeys an inverse square law.

SET UP: $\dfrac{I_2}{I_1} = \dfrac{r_1^2}{r_2^2}$. $\beta = (10\ \text{dB})\log\left(\dfrac{I}{I_0}\right)$, with $I_0 = 1\times 10^{-12}$ W/m^2.

EXECUTE: (a) $\beta = 53$ dB gives $5.3 = \log\left(\dfrac{I}{I_0}\right)$ and $I = (10^{5.3})I_0 = 2.0\times 10^{-7}$ W/m^2.

(b) $r_2 = r_1\sqrt{\dfrac{I_1}{I_2}} = (3.0\ \text{m})\sqrt{\dfrac{4}{1}} = 6.0$ m.

(c) $\beta = \dfrac{53\ \text{dB}}{4} = 13.25$ dB gives $1.325 = \log\left(\dfrac{I}{I_0}\right)$ and $I = 2.1\times 10^{-11}$ W/m^2.

$r_2 = r_1\sqrt{\dfrac{I_1}{I_2}} = (3.0\ \text{m})\sqrt{\dfrac{2.0\times 10^{-7}\ \text{W/m}^2}{2.1\times 10^{-11}\ \text{W/m}^2}} = 290$ m.

EVALUATE: **(d)** Intensity obeys the inverse square law but noise level does not.

16.29. **IDENTIFY:** We are looking at the standing wave pattern in a pipe. The pattern has a displacement node at one end and a displacement antinode at the other end of the pipe.

SET UP: For a stopped pipe $\lambda_n = 4L/n$ (n an odd integer). $v = f\lambda$

EXECUTE: **(a)** With a displacement node at one end and an antinode at the other, this must be a stopped pipe.

(b) The nth harmonic has $\frac{n+1}{2}$ nodes, so for 5 nodes, $5 = \frac{n+1}{2}$, which gives $n = 5$. This pipe is resonating in its 5th harmonic.

(c) Using $v = f\lambda$ gives $\lambda_9 = \frac{v}{f_9} = \frac{344 \text{ m/s}}{1710 \text{ Hz}} = 0.20117$ m. Using $\lambda_n = 4L/n$ gives

$L = \frac{n\lambda_n}{4} = \frac{9(0.20117 \text{ m})}{4} = 0.453$ m.

(d) $f_9 = 9f_1$, so $f_1 = (1710 \text{ Hz})/9 = 190$ Hz.

EVALUATE: A stopped pipe does not have even harmonics because it has a displacement node at the closed end and an antinode at the open end.

16.33. **(a) IDENTIFY** and **SET UP:** Path difference from points A and B to point Q is $3.00 \text{ m} - 1.00 \text{ m} = 2.00$ m, as shown in Figure 16.33. Constructive interference implies path difference $= n\lambda$, $n = 1, 2, 3, \ldots$

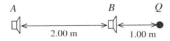

Figure 16.33

EXECUTE: $2.00 \text{ m} = n\lambda$ so $\lambda = 2.00$ m/n

$f = \frac{v}{\lambda} = \frac{nv}{2.00 \text{ m}} = \frac{n(344 \text{ m/s})}{2.00 \text{ m}} = n(172 \text{ Hz})$, $n = 1, 2, 3, \ldots$

The lowest frequency for which constructive interference occurs is 172 Hz.

(b) IDENTIFY and **SET UP:** Destructive interference implies path difference $= (n/2)\lambda$, $n = 1, 3, 5, \ldots$

EXECUTE: $2.00 \text{ m} = (n/2)\lambda$ so $\lambda = 4.00$ m/n

$f = \frac{v}{\lambda} = \frac{nv}{4.00 \text{ m}} = \frac{n(344 \text{ m/s})}{(4.00 \text{ m})} = n(86 \text{ Hz})$, $n = 1, 3, 5, \ldots$

The lowest frequency for which destructive interference occurs is 86 Hz.

EVALUATE: As the frequency is slowly increased, the intensity at Q will fluctuate, as the interference changes between destructive and constructive.

16.35. **IDENTIFY:** For constructive interference the path difference is an integer number of wavelengths and for destructive interference the path difference is a half-integer number of wavelengths.

SET UP: $\lambda = v/f = (344 \text{ m/s})/(688 \text{ Hz}) = 0.500$ m

EXECUTE: To move from constructive interference to destructive interference, the path difference must change by $\lambda/2$. If you move a distance x toward speaker B, the distance to B gets shorter by x and the distance to A gets longer by x so the path difference changes by $2x$. $2x = \lambda/2$ and $x = \lambda/4 = 0.125$ m.

EVALUATE: If you walk an additional distance of 0.125 m farther, the interference again becomes constructive.

16.37. **IDENTIFY:** For constructive interference, the path difference is an integer number of wavelengths. For destructive interference, the path difference is a half-integer number of wavelengths.

SET UP: One speaker is 4.50 m from the microphone and the other is 4.92 m from the microphone, so the path difference is 0.42 m. $f = v/\lambda$.

EXECUTE: **(a)** $\lambda = 0.42$ m gives $f = \dfrac{v}{\lambda} = 820$ Hz; $2\lambda = 0.42$ m gives $\lambda = 0.21$ m and $f = \dfrac{v}{\lambda} = 1640$ Hz; $3\lambda = 0.42$ m gives $\lambda = 0.14$ m and $f = \dfrac{v}{\lambda} = 2460$ Hz, and so on. The frequencies for constructive interference are $n(820$ Hz$)$, $n = 1, 2, 3, \ldots$.

(b) $\lambda/2 = 0.42$ m gives $\lambda = 0.84$ m and $f = \dfrac{v}{\lambda} = 410$ Hz; $3\lambda/2 = 0.42$ m gives $\lambda = 0.28$ m and $f = \dfrac{v}{\lambda} = 1230$ Hz; $5\lambda/2 = 0.42$ m gives $\lambda = 0.168$ m and $f = \dfrac{v}{\lambda} = 2050$ Hz, and so on. The frequencies for destructive interference are $(2n+1)(410$ Hz$)$, $n = 0, 1, 2, \ldots$.

EVALUATE: The frequencies for constructive interference lie midway between the frequencies for destructive interference.

16.39. **IDENTIFY:** The beat is due to a difference in the frequencies of the two sounds.

SET UP: $f_{\text{beat}} = f_1 - f_2$. Tightening the string increases the wave speed for transverse waves on the string and this in turn increases the frequency.

EXECUTE: **(a)** If the beat frequency increases when she raises her frequency by tightening the string, it must be that her frequency is 433 Hz, 3 Hz above concert A.

(b) She needs to lower her frequency by loosening her string.

EVALUATE: The beat would only be audible if the two sounds are quite close in frequency. A musician with a good sense of pitch can come very close to the correct frequency just from hearing the tone.

16.41. **IDENTIFY:** Apply the Doppler shift equation $f_L = \left(\dfrac{v + v_L}{v + v_S}\right) f_S$.

SET UP: The positive direction is from listener to source. $f_S = 1200$ Hz. $f_L = 1240$ Hz.

EXECUTE: $v_L = 0$. $v_S = -25.0$ m/s. $f_L = \left(\dfrac{v}{v + v_S}\right) f_S$ gives

$v = \dfrac{v_S f_L}{f_S - f_L} = \dfrac{(-25 \text{ m/s})(1240 \text{ Hz})}{1200 \text{ Hz} - 1240 \text{ Hz}} = 780$ m/s.

EVALUATE: $f_L > f_S$ since the source is approaching the listener.

16.43. **IDENTIFY:** Apply the Doppler shift equation $f_L = \left(\dfrac{v + v_L}{v + v_S}\right) f_S$.

SET UP: The positive direction is from listener to source. $f_S = 392$ Hz.

EXECUTE: **(a)** $v_S = 0$. $v_L = -15.0$ m/s.

$f_L = \left(\dfrac{v + v_L}{v + v_S}\right) f_S = \left(\dfrac{344 \text{ m/s} - 15.0 \text{ m/s}}{344 \text{ m/s}}\right)(392 \text{ Hz}) = 375$ Hz

(b) $v_S = +35.0$ m/s. $v_L = +15.0$ m/s. $f_L = \left(\dfrac{v + v_L}{v + v_S}\right) f_S = \left(\dfrac{344 \text{ m/s} + 15.0 \text{ m/s}}{344 \text{ m/s} + 35.0 \text{ m/s}}\right)(392 \text{ Hz}) = 371$ Hz

(c) $f_{beat} = f_1 - f_2 = 4$ Hz

EVALUATE: The distance between whistle A and the listener is increasing, and for whistle A $f_L < f_S$. The distance between whistle B and the listener is also increasing, and for whistle B $f_L < f_S$.

16.45. **IDENTIFY:** The distance between crests is λ. In front of the source $\lambda = \dfrac{v - v_S}{f_S}$ and behind the source $\lambda = \dfrac{v + v_S}{f_S}$. $f_S = 1/T$.

SET UP: $T = 1.6$ s. $v = 0.32$ m/s. The crest to crest distance is the wavelength, so $\lambda = 0.12$ m.

EXECUTE: **(a)** $f_S = 1/T = 0.625$ Hz. $\lambda = \dfrac{v - v_S}{f_S}$ gives

$v_S = v - \lambda f_S = 0.32$ m/s $- (0.12$ m$)(0.625$ Hz$) = 0.25$ m/s.

(b) $\lambda = \dfrac{v + v_S}{f_S} = \dfrac{0.32 \text{ m/s} + 0.25 \text{ m/s}}{0.625 \text{ Hz}} = 0.91$ m

EVALUATE: If the duck was held at rest but still paddled its feet, it would produce waves of wavelength $\lambda = \dfrac{0.32 \text{ m/s}}{0.625 \text{ Hz}} = 0.51$ m. In front of the duck the wavelength is decreased and behind the duck the wavelength is increased. The speed of the duck is 78% of the wave speed, so the Doppler effects are large.

16.51. **IDENTIFY:** Apply the Doppler shift formulas. We first treat the stationary police car as the source and then as the observer as he receives his own sound reflected from the on-coming car.

SET UP: $f_L = \left(\dfrac{v + v_L}{v + v_S}\right) f_S$.

EXECUTE: (a) Since the frequency is increased the moving car must be approaching the police car. Let v_c be the speed of the moving car. The speed v_p of the police car is zero. First consider the moving car as the listener, as shown in Figure 16.51(a).

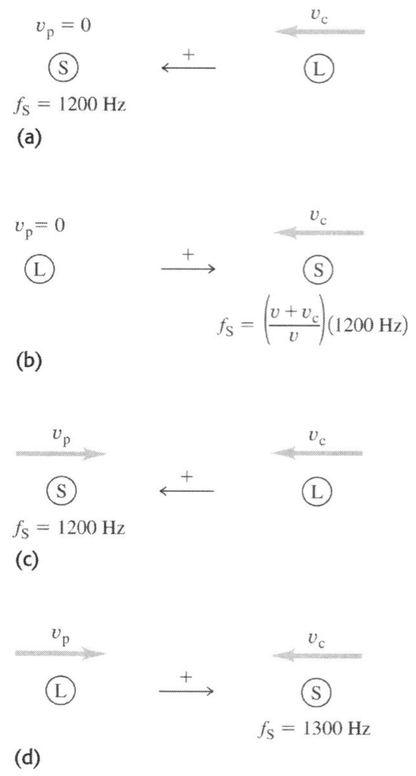

Figure 16.51

$$f_L = \left(\frac{v + v_L}{v + v_S}\right) f_S = \left(\frac{v + v_c}{v}\right)(1200 \text{ Hz})$$

Then consider the moving car as the source and the police car as the listener, as shown in Figure 16.51(b):

$$f_L = \left(\frac{v + v_L}{v + v_S}\right) f_S \text{ gives } 1250 \text{ Hz} = \left(\frac{v}{v - v_c}\right)\left(\frac{v + v_c}{v}\right)(1200 \text{ Hz}).$$

Solving for v_c gives

$$v_c = \left(\frac{50}{2450}\right) v = \left(\frac{50}{2450}\right)(344 \text{ m/s}) = 7.02 \text{ m/s}$$

(b) Repeat the calculation of part (a), but now $v_p = 20.0$ m/s, toward the other car.

Waves received by the car (Figure 16.51(c)):

$$f_L = \left(\frac{v + v_c}{v - v_p}\right) f_S = \left(\frac{344 \text{ m/s} + 7 \text{ m/s}}{344 \text{ m/s} - 20 \text{ m/s}}\right)(1200 \text{ Hz}) = 1300 \text{ Hz}$$

Waves reflected by the car and received by the police car (Figure 16.51(d)):

$$f_L = \left(\frac{v + v_p}{v - v_c}\right) f_S = \left(\frac{344 \text{ m/s} + 20 \text{ m/s}}{344 \text{ m/s} - 7 \text{ m/s}}\right)(1300 \text{ Hz}) = 1404 \text{ Hz}$$

EVALUATE: The cars move toward each other with a greater relative speed in (b) and the increase in frequency is much larger there.

16.53. **IDENTIFY:** Apply $\sin\alpha = v/v_S$ to calculate α. Use the method of Example 16.19 to calculate t.

SET UP: Mach 1.70 means $v_S/v = 1.70$.

EXECUTE: **(a)** In $\sin\alpha = v/v_S$, $v/v_S = 1/1.70 = 0.588$ and $\alpha = \arcsin(0.588) = 36.0°$.

(b) As in Example 16.19, $t = \dfrac{1250 \text{ m}}{(1.70)(344 \text{ m/s})(\tan\ 36.0°)} = 2.94$ s.

EVALUATE: The angle α decreases when the speed v_S of the plane increases.

16.57. **IDENTIFY:** We are investigating how the speed of sound depends on the air temperature as we look at higher and higher altitudes.

SET UP: Eq. 16.10: $v = \sqrt{\dfrac{\gamma RT}{M}}$, where $M = 0.0288$ kg/mol and $\gamma = 1.40$. First use the given information to sketch the graph of T versus y, as in Fig. 16.57. We use the slope-intercept equation of a straight line $T = my + b$. For this graph, $m = -\dfrac{20.0 \text{ K}}{300.0 \text{ m}} = -0.06667$ K/m and $b = 278$ K. Using these results, the speed v of sound as a function of altitude y is $v = \sqrt{\dfrac{\gamma R}{M}(my+b)}$. We want to find the time for the sound to travel 300.0 m straight upward, and we know $v_y = dy/dt$.

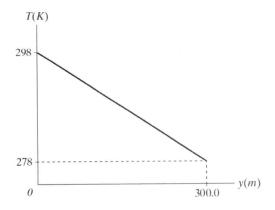

Figure 16.57

EXECUTE: **(a)** $v_y = dy/dt$ gives $dt = \dfrac{dy}{v}$, so we integrate, giving $\displaystyle\int_0^{t_{300}} dt = \int_0^{300 \text{ m}} \dfrac{dy}{v}$. The time integral gives simply t_{300}, but the y integral requires a bit more effort. Using our equation for v from above, the y integral becomes $\displaystyle\int_0^{300 \text{ m}} \dfrac{dy}{v} = \int_0^{300 \text{ m}} \dfrac{dy}{\sqrt{\dfrac{\gamma R}{M}}\sqrt{my+b}} = \sqrt{\dfrac{M}{\gamma R}}\int_0^{300 \text{ m}} \dfrac{dy}{(my+b)^{1/2}}$. To do the integral, let $u = my + b$, so $du = m\, dy$. The integral now becomes $\displaystyle\int \dfrac{(1/m)du}{u^{1/2}} = \dfrac{2}{m}u^{1/2}$. Returning to the original variables, the integral is $\sqrt{\dfrac{M}{\gamma R}}\displaystyle\int_0^{300 \text{ m}} \dfrac{dy}{(my+b)^{1/2}} = \sqrt{\dfrac{M}{\gamma R}}\dfrac{2}{m}(my+b)^{1/2}\Big|_0^{300 \text{ m}} = t_{300}$. Putting in the numbers for m, b, M, R, and γ gives $t_{300} = 0.879$ s.

(b) At 300.0 m, $T = 5.00°C = 298$ K, so using these numbers gives $v = \sqrt{\dfrac{\gamma RT}{M}} = 335.2$ m/s. The horizontal distance the sound travels is $x = vt_{300} = (335.2 \text{ m/s})(0.879 \text{ s}) = 295$ m.

EVALUATE: Since $v \propto \sqrt{T}$ (with T in K units), there is not much difference between the surface temperature and the temperature at 300 m, so v at 300 m is not very much different from v at the surface, as we see.

16.59. **IDENTIFY and SET UP:** The frequency of any harmonic is an integer multiple of the fundamental. For a stopped pipe only odd harmonics are present. For an open pipe, all harmonics are present. See which pattern of harmonics fits to the observed values in order to determine which type of pipe it is. Then solve for the fundamental frequency and relate that to the length of the pipe.

EXECUTE: **(a)** For an open pipe the successive harmonics are $f_n = nf_1$, $n = 1, 2, 3, \ldots$ For a stopped pipe the successive harmonics are $f_n = nf_1$, $n = 1, 3, 5, \ldots$. If the pipe is open and these harmonics are successive, then $f_n = nf_1 = 1372$ Hz and $f_{n+1} = (n+1)f_1 = 1764$ Hz. Subtract the first equation from the second: $(n+1)f_1 - nf_1 = 1764$ Hz $- 1372$ Hz. This gives $f_1 = 392$ Hz. Then $n = \dfrac{1372 \text{ Hz}}{392 \text{ Hz}} = 3.5$. But n must be an integer, so the pipe can't be open. If the pipe is stopped and these harmonics are successive, then $f_n = nf_1 = 1372$ Hz and $f_{n+2} = (n+2)f_1 = 1764$ Hz (in this case successive harmonics differ in n by 2). Subtracting one equation from the other gives $2f_1 = 392$ Hz and $f_1 = 196$ Hz. Then $n = 1372$ Hz/$f_1 = 7$ so 1372 Hz $= 7f_1$ and 1764 Hz $= 9f_1$. The solution gives integer n as it should; the pipe is stopped.

(b) From part (a) these are the seventh and ninth harmonics.

(c) From part (a) $f_1 = 196$ Hz.

For a stopped pipe $f_1 = \dfrac{v}{4L}$ and $L = \dfrac{v}{4f_1} = \dfrac{344 \text{ m/s}}{4(196 \text{ Hz})} = 0.439$ m.

EVALUATE: It is essential to know that these are successive harmonics and to realize that 1372 Hz is not the fundamental. There are other lower frequency standing waves; these are just two successive ones.

16.61. **IDENTIFY:** Destructive interference occurs when the path difference is a half-integer number of wavelengths. Constructive interference occurs when the path difference is an integer number of wavelengths.

SET UP: $\lambda = \dfrac{v}{f} = \dfrac{344 \text{ m/s}}{784 \text{ Hz}} = 0.439$ m

EXECUTE: **(a)** If the separation of the speakers is denoted h, the condition for destructive interference is $\sqrt{x^2 + h^2} - x = \beta\lambda$, where β is an odd multiple of one-half. Adding x to both sides, squaring, canceling the x^2 term from both sides, and solving for x gives $x = \dfrac{h^2}{2\beta\lambda} - \dfrac{\beta}{2}\lambda$. Using $\lambda = 0.439$ m and $h = 2.00$ m yields 9.01 m for $\beta = \tfrac{1}{2}$, 2.71 m for $\beta = \tfrac{3}{2}$, 1.27 m for $\beta = \tfrac{5}{2}$, 0.53 m for $\beta = \tfrac{7}{2}$, and 0.026 m for $\beta = \tfrac{9}{2}$. These are the only allowable values of β that give positive solutions for x.

(b) Repeating the above for integral values of β, constructive interference occurs at 4.34 m, 1.84 m, 0.86 m, 0.26 m. Note that these are between, but not midway between, the answers to part (a).

(c) If $h = \lambda/2$, there will be destructive interference at speaker B. If $\lambda/2 > h$, the path difference can never be as large as $\lambda/2$. (This is also obtained from the above expression for x, with $x = 0$ and $\beta = \frac{1}{2}$.) The minimum frequency is then $v/2h = (344 \text{ m/s})/(4.0 \text{ m}) = 86$ Hz.

EVALUATE: When f increases, λ is smaller and there are more occurrences of points of constructive and destructive interference.

16.63. **(a) IDENTIFY** and **SET UP:** Use $v = f\lambda$ to calculate λ.

EXECUTE: $\lambda = \dfrac{v}{f} = \dfrac{1482 \text{ m/s}}{18.0 \times 10^3 \text{ Hz}} = 0.0823$ m.

(b) IDENTIFY: Apply the Doppler effect equation, $f_L = \left(\dfrac{v + v_L}{v}\right) f_S = \left(1 + \dfrac{v_L}{v}\right) f_S$. The frequency of the directly radiated waves is $f_S = 18{,}000$ Hz. The moving whale first plays the role of a moving listener, receiving waves with frequency f_L'. The whale then acts as a moving source, emitting waves with the same frequency, $f_S' = f_L'$ with which they are received. Let the speed of the whale be v_W.

SET UP: Whale receives waves: (Figure 16.63a)

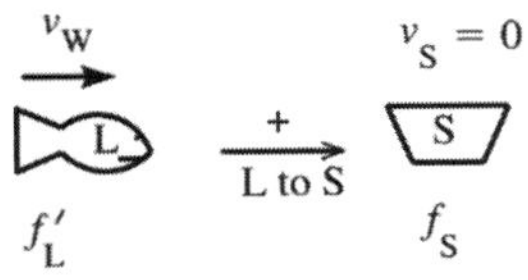

EXECUTE: $v_L = +v_W$

$$f_L' = f_S\left(\dfrac{v + v_L}{v + v_S}\right) = f_S\left(\dfrac{v + v_W}{v}\right)$$

Figure 16.63a

SET UP: Whale re-emits the waves: (Figure 16.63b)

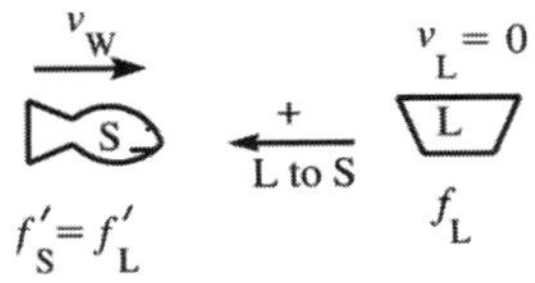

EXECUTE: $v_S = -v_W$

$$f_L = f_S\left(\dfrac{v + v_L}{v + v_S}\right) = f_S'\left(\dfrac{v}{v - v_W}\right)$$

Figure 16.63b

But $f_S' = f_L'$ so $f_L = f_S\left(\dfrac{v + v_W}{v}\right)\left(\dfrac{v}{v - v_W}\right) = f_S\left(\dfrac{v + v_W}{v - v_W}\right)$.

Then $\Delta f = f_S - f_L = f_S\left(1 - \dfrac{v + v_W}{v - v_W}\right) = f_S\left(\dfrac{v - v_W - v - v_W}{v - v_W}\right) = \dfrac{-2f_S v_W}{v - v_W}$.

$\Delta f = \dfrac{-2(1.80 \times 10^4 \text{ Hz})(4.95 \text{ m/s})}{1482 \text{ m/s} - 4.95 \text{ m/s}} = -120$ Hz.

EVALUATE: Δf is negative, which means that $f_L > f_S$. This is reasonable because the listener and source are moving toward each other so the frequency is raised.

16.67. **IDENTIFY:** Follow the method of Example 16.18 and apply the Doppler shift formula twice, once with the insect as the listener and again with the insect as the source.

SET UP: Let v_{bat} be the speed of the bat, v_{insect} be the speed of the insect, and f_i be the frequency with which the sound waves both strike and are reflected from the insect. The positive direction in each application of the Doppler shift formula is from the listener to the source.

EXECUTE: The frequencies at which the bat sends and receives the signals are related by

$$f_L = f_i \left(\frac{v + v_{bat}}{v - v_{insect}}\right) = f_S \left(\frac{v + v_{insect}}{v - v_{bat}}\right)\left(\frac{v + v_{bat}}{v - v_{insect}}\right).$$ Solving for v_{insect},

$$v_{insect} = v\left[\frac{1 - \frac{f_S}{f_L}\left(\frac{v + v_{bat}}{v - v_{bat}}\right)}{1 + \frac{f_S}{f_L}\left(\frac{v + v_{bat}}{v - v_{bat}}\right)}\right] = v\left[\frac{f_L(v - v_{bat}) - f_S(v + v_{bat})}{f_L(v - v_{bat}) + f_S(v + v_{bat})}\right].$$

Letting $f_L = f_{refl}$ and $f_S = f_{bat}$ gives the result.

(b) If $f_{bat} = 80.7$ kHz, $f_{refl} = 83.5$ kHz, and $v_{bat} = 3.9$ m/s, then $v_{insect} = 2.0$ m/s.

EVALUATE: $f_{refl} > f_{bat}$ because the bat and insect are approaching each other.

16.69. **IDENTIFY:** The sound from the speaker moving toward the listener will have an increased frequency, while the sound from the speaker moving away from the listener will have a decreased frequency. The difference in these frequencies will produce a beat.

SET UP: The greatest frequency shift from the Doppler effect occurs when one speaker is moving away and one is moving toward the person. The speakers have speed $v_0 = r\omega$, where $r = 0.75$ m.

$f_L = \left(\frac{v + v_L}{v + v_S}\right)f_S$, with the positive direction from the listener to the source. $v = 344$ m/s.

EXECUTE: **(a)** $f = \frac{v}{\lambda} = \frac{344 \text{ m/s}}{0.313 \text{ m}} = 1100$ Hz. $\omega = (75 \text{ rpm})\left(\frac{2\pi \text{ rad}}{1 \text{ rev}}\right)\left(\frac{1 \text{ min}}{60 \text{ s}}\right) = 7.85$ rad/s and

$v_0 = (0.75 \text{ m})(7.85 \text{ rad/s}) = 5.89$ m/s.

For speaker A, moving toward the listener: $f_{LA} = \left(\frac{v}{v - 5.89 \text{ m/s}}\right)(1100 \text{ Hz}) = 1119$ Hz.

For speaker B, moving toward the listener: $f_{LB} = \left(\frac{v}{v + 5.89 \text{ m/s}}\right)(1100 \text{ Hz}) = 1081$ Hz.

$f_{beat} = f_1 - f_2 = 1119 \text{ Hz} - 1081 \text{ Hz} = 38$ Hz.

(b) A person can hear individual beats only up to about 7 Hz and this beat frequency is much larger than that.

EVALUATE: As the turntable rotates faster the beat frequency at this position of the speakers increases.

16.71. **IDENTIFY and SET UP:** There is a node at the piston, so the distance the piston moves is the node to node distance, $\lambda/2$. Use $v = f\lambda$ to calculate v and $v = \sqrt{\frac{\gamma RT}{M}}$ to calculate γ from v.

EXECUTE: **(a)** $\lambda/2 = 37.5$ cm, so $\lambda = 2(37.5 \text{ cm}) = 75.0$ cm $= 0.750$ m.
$v = f\lambda = (500 \text{ Hz})(0.750 \text{ m}) = 375$ m/s

(b) Solve $v = \sqrt{\gamma RT/M}$ for γ: $\gamma = \frac{Mv^2}{RT} = \frac{(28.8 \times 10^{-3} \text{ kg/mol})(375 \text{ m/s})^2}{(8.3145 \text{ J/mol}\cdot\text{K})(350 \text{ K})} = 1.39$.

(c) EVALUATE: There is a node at the piston so when the piston is 18.0 cm from the open end the node is inside the pipe, 18.0 cm from the open end. The node to antinode distance is $\lambda/4 = 18.8$ cm, so the antinode is 0.8 cm beyond the open end of the pipe.
The value of γ we calculated agrees with the value given for air in Example 16.4.

16.73. **IDENTIFY:** The wire vibrating in its fundamental causes the tube to resonate with the same frequency in its fundamental. We are dealing with standing waves on a string and in an open pipe.

SET UP: Fig. 16.73a illustrates the information in the problem. The information we have is as follows:
wire: $\mu = 1.40$ g/m $= 0.00140$ kg/m, vibrating in its fundamental f_1.
pole: $M = 8.00$ kg, $L = 1.56$ m
tube: 39.0 cm long, $m = 4.00$ kg, hollow (open pipe), vibrating in its fundamental f_1
We want to find the frequency f_1 at which the wire and tube are vibrating and the height h in Fig. 16.73a. The equations we use for the tube are $f_n = nv_s/2L_t$, $\lambda_n = 2L_t/n$, $\lambda_1 = 2L_t$ and for the wire $f_n = nv/2L_w$, $\lambda_n = 2L_w/n$, $\lambda_1 = 2L_w$. We also use $v = f\lambda$.

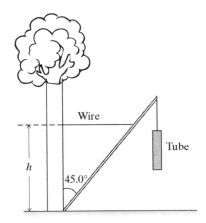

Figure 16.73a

EXECUTE: **(a)** For the tube in its fundamental mode, we have $\lambda_1 = 2L_t = 2(0.390 \text{ m}) = 0.780$ m. The frequency is $f_1 = \dfrac{v_s}{\lambda_t} = \dfrac{344 \text{ m/s}}{0.780 \text{ m}} = 441$ Hz. This is the frequency at which the wire and the tube are resonating.

(b) We want to find h in Fig. 16.73a. Now look at the wire. $f_w \lambda_w = v_w = \sqrt{\dfrac{F}{\mu}}$. In its fundamental mode $\lambda_1 = 2L_w$. From Fig. 16.73a we see that $L_w = h \tan 45.0° = h$, so $\lambda_1 = 2h$. From part (a) we know that $f_1 = 441$ Hz, so $f_w \lambda_w = v_w = \sqrt{\dfrac{F}{\mu}}$ gives $f_1(2h) = (441 \text{ Hz})(2h) = \sqrt{\dfrac{F}{\mu}}$, which becomes

$(882 \text{ Hz})h = \sqrt{\dfrac{F}{\mu}}$.   (Eq. 1)

We need to find F in order to find h. Make a free-body diagram of the pole as in Fig. 16.73b and apply $\Sigma \tau_z = 0$ about the hinge. Using Fig. 16.73 as a guide, we get

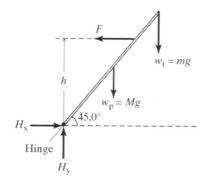

Figure 16.73b

$Fh - Mg\dfrac{L}{2}\cos 45.0° - mgL\cos 45.0° = 0$, which simplifies to $Fh = \left(\dfrac{M}{2} + m\right)gL\cos 45.0°$. Putting in $M = 8.00$ kg, $m = 4.00$ kg, and $L = 1.56$ m, and solving for F, we get $F = \dfrac{86.48 \text{ N} \cdot \text{m}}{h}$. Now return to Eq. 1. Square the equation and substitute for F, giving $(882 \text{ Hz})^2 h^2 = \dfrac{F}{\mu} = \dfrac{86.48 \text{ N} \cdot \text{m}}{h(0.00140 \text{ kg/m})}$. Solving for h gives $h = 0.430$ m.

EVALUATE: Changing h would change the length of the wire as well as the tension in it. These changes would affect the fundamental frequency of the wire.

Study Guide for Temperature and Heat

Summary

Heat will be introduced as a method of energy transfer due to temperature differences, and the rate of heat transfer will be calculated. We'll also learn about the amount of heat required to change the phase of matter and about the three types of heat transfer: conduction, convection, and radiation.

Objectives

After studying this chapter, you'll understand

- The definitions of *temperature* and *thermal equilibrium*.
- The three temperature scales and how to measure temperature.
- How thermal expansion describes the change in length and volume of materials due to temperature changes.
- About heat, phase changes, and calorimetry and how to apply these concepts to problems.
- How heat is transferred by conduction, convection, and radiation.

Concepts and Equations

Term	Description
Thermal Equilibrium	Two objects in thermal equilibrium have the same temperature.
Temperature Scales	The Celsius temperature scale defines 0°C as the freezing point of water and 100°C as the boiling point of water. The Fahrenheit temperature scale defines 32°F as the freezing point of water and 212°F as the boiling point of water. The Kelvin scale defines *absolute zero* as 0 K and uses the Celsius unit as its standard unit. Note that 0 K is −273.15°C.
Thermal Expansion and Thermal Stress	Materials change size, or thermally expand, due to changes in temperature. An object of length L_0 at temperature T_0 will have length L at temperature $T = T_0 + \Delta T$, or $$L = L_0 + \Delta L = L_0(1 + \alpha \Delta T),$$ where α is the coefficient of linear expansion with units K^{-1}. An object of volume V_0 at temperature T_0 will have volume V at temperature $T = T_0 + \Delta T$, or $$V = V_0 + \Delta V = V_0(1 + \beta \Delta T),$$ where β is the coefficient of volume expansion with units K^{-1}. When a material is heated or cooled while being held such that it cannot contract or expand, it is under tensile stress given by $$\frac{F}{A} = -Y\alpha \Delta T.$$
Heat	Heat is energy transferred from one object to another due to changes in temperature. The quantity of heat Q needed to raise the temperature of a mass m of material by an amount ΔT is $$Q = mc\Delta T,$$ where c is the specific heat capacity of the material. The SI unit of heat capacity is the joule per kilogram per kelvin (J/(kg K)).
Phase Change	A phase transition is the change from one phase of matter to another. Phases include solid, liquid, and gas. The heat of fusion, L_f, is the heat per unit mass required to change a solid material to liquid. The heat of vaporization, L_v, is the heat per unit mass required to change a liquid material to gas. The heat of sublimation, L_s, is the heat per unit mass required to change a solid material to gas.
Calorimetry	Calorimetry is the measurement of heat in a system. For an isolated system, the algebraic sum of the quantities of heat must add to zero: $$\Sigma Q = 0.$$
Heat Transfer	Heat may be transferred through conduction, convection, and radiation. *Conduction* is the transfer of energy within a material without bulk motion of the material. *Convection* is the transfer of energy due to the motion of mass from one region to another. *Radiation* is the transfer of energy through electromagnetic waves. The heat current H for an area A and length L through which the heat flows is given by $$H = \frac{dQ}{dt} = kA\frac{T_H - T_C}{L},$$

> where T_H and T_C are, respectively, the temperatures of the hot and cold sides of the material and k is the thermal conductivity. The heat current H due to radiation is
>
> $$H = Ae\sigma T^4,$$
>
> where A is the surface area, e is the emissivity of the surface (a pure number between 0 and 1), T is the absolute temperature, and σ is the Stefan–Boltzmann constant (5.6705×10^{-8} W/m^2/K^4).

Key Concept 1: The Celsius and Kelvin scales have different zero points, but differences in temperature are the same in both scales: Increasing the temperature by 37.00 °C is the same as increasing it by 37.00 K.

Key Concept 2: A change in temperature causes the *length* of an object to change by an amount that is approximately proportional to the object's initial length and to the temperature change ΔT.

Key Concept 3: A change in temperature causes the *volume* of an object to change by an amount that is approximately proportional to the object's initial volume and to the temperature change ΔT.

Key Concept 4: To keep the length of an object constant when the temperature changes, forces must be applied to both of its ends. The required stress (force per unit area) is proportional to the temperature change.

Key Concept 5: To find the amount of heat required to change the temperature of a mass m of material by an amount ΔT, multiply m by ΔT and by the specific heat c of the material. Heat is positive when it flows into an object and negative when it flows out.

Key Concept 6: Any energy flow (not just heat) into or out of a quantity of material can cause the temperature of the material to change. The rate of energy flow is equal to the mass times the specific heat of the material times the rate of temperature change.

Key Concept 7: In a calorimetry problem in which two objects at different temperatures interact by exchanging heat, energy is conserved: The sum of the heat flows (one positive, one negative) into the two objects is zero. The heat flow stops when the two objects reach the same temperature.

Key Concept 8: When heat flows between two objects and one or both of them change phase, your calculations must include the heat required to cause the phase change. This depends on the object's mass and material and on which phase change occurs.

Key Concept 9: In many calorimetry problems you won't know whether or not a phase change occurs. To find out, try working the problem three ways (assuming no phase change, assuming part of the object changes phase, and assuming all of it changes phase). The way that leads to a sensible result is the correct one.

Key Concept 10: To find the amount of heat released when a quantity of substance undergoes combustion, multiply the mass that combusts by the heat of combustion of the substance.

Key Concept 11: If a temperature difference is maintained between the two sides of an object of thickness L and cross-sectional area A, there will be a steady heat current due to conduction from the high-temperature side to the low-temperature side. This conduction heat current is proportional to the temperature difference and to the ratio A/L.

Key Concept 12: When there is a steady heat flow by conduction through two materials in succession, the heat current is the same in both materials: No energy is lost in going from one material to the next.

Key Concept 13: Even if two different objects have the same constant temperature difference between their ends, the heat current H through the two objects can be different. The value of H depends on the dimensions of the object and the thermal conductivity of the material of the object.

Key Concept 14: All objects emit energy in the form of electromagnetic radiation due to their temperature. The heat current of this radiation is proportional to the object's surface area, to the emissivity of its surface, and to the fourth power of the object's Kelvin temperature.

Key Concept 15: An object at Kelvin temperature T emits electromagnetic radiation but also absorbs radiation from its surroundings at Kelvin temperature T_S. The *net* heat current is proportional to the object's surface area, to the emissivity of its surface, and to the difference between T^4 and T_S^4.

Conceptual Questions

1: Do holes expand or contract?

Your younger brother knows that solids expand as they heat up. He thinks that the metal surrounding the hole in a cookie sheet will expand into the hole as the cookie sheet heats up. Is he right or wrong?

IDENTIFY, SET UP, AND EXECUTE Figure 1 shows a sketch of a cookie sheet with a hole in it. After considering the problem, you realize that the hole will enlarge as the cookie sheet heats up, since all dimensions of an object enlarge with temperature. The challenge is how to best explain this phenomenon to him.

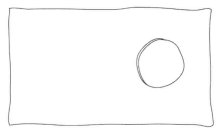

Figure 1 Question 1 sketch.

If you give the problem a bit more thought, you come up with a convincing argument. If the cookie sheet had no hole, the whole sheet would increase with temperature. If you punch out a hole in the same cookie sheet and consider the piece of metal that was removed, this piece expands as its temperature rises. Therefore, the hole in the cookie sheet must also expand, just as it did when the cookie sheet was holeless.

EVALUATE Thermal expansion must be considered carefully. Here, we see that a confusing point can be clarified by imagining what happens to the piece that was once the hole.

CAUTION Holes expand when heated! Keep the results of this problem in mind when you encounter similar problems. Holes don't shrink when heated.

2: Cooler after a shower

When you step out of the shower, you often feel cold. After drying off, you feel warmer, even though the room's temperature is the same as when you stepped out of the shower. Why?

IDENTIFY, SET UP, AND EXECUTE When you step out of the shower, water on your body evaporates. Evaporation requires heat energy (the heat of vaporization), much of which energy comes from heat leaving your body. You feel cold because your body is transferring its heat to evaporate the water. When you are dry, there is little heat lost due to evaporation.

EVALUATE Evaporation also explains why one feels cooler in a dry climate than in a humid climate: Your sweat evaporates more rapidly in a dry climate, taking away more heat, than in a humid climate.

3: Cold water versus cold air

Would you prefer to spend 10 minutes in a 40°F (4°F) room or in a 40°F pool? Why?

IDENTIFY, SET UP, AND EXECUTE Both the room and the pool are at the same temperature, but the 40°F room would be much more comfortable. The reason is that the specific heat of air is much less than the specific heat of water (i.e., air will carry away less heat from your body than the water would in any time interval). Since the air carries away less heat, you are more comfortable in the room.

EVALUATE Specific heat is the amount of heat needed to change the temperature of a material per unit mass and per unit temperature. Larger specific heats mean that more heat is carried away from an object.

Problems

1: Volume of a copper cup

A copper cup is filled to the brim with ethanol at 0°C. When the cup and ethanol are heated to 350°C, 4.7 cm³ of ethanol spills from the cup. What is the initial volume of the cup?

IDENTIFY We'll use temperature expansion to find the change in volume of the cup and the ethanol. Their difference will lead to the initial volume of the cup—the target variable.

SET UP Both the cup and the ethanol expand as the temperature rises; the difference in their expansion is equal to the volume of the spilled ethanol. We'll apply the volume expansion equation to both the cup and the ethanol, setting their difference equal to the volume of the spill. The coefficient of volume expansion is $5.1 \times 10^{-5}/\text{K}$ for copper and $7.5 \times 10^{-5}/\text{K}$ for ethanol.

EXECUTE For any material, the change in volume due to temperature is
$$\Delta V = \beta V_0 \Delta T.$$

We are given the volume of the spill, which is the change in volume of the ethanol minus the change in volume of the cup:
$$V_{spill} = \Delta V_{ethanol} - \Delta V_{cup}.$$

The initial volumes of the cup and the ethanol are the same. We'll call their common volume V_0. The temperature of both materials is 35°C. Replacing the changes in volumes yields
$$V_{spill} = \beta_{ethanol} V_0 \Delta T - \beta_{copper} V_0 \Delta T.$$

Solving for V_0, we obtain

$$V_0 = \frac{V_{spill}}{(\beta_{ethanol} - \beta_{copper})\Delta T} = \frac{(4.7 \text{ cm}^3)}{((75 \times 10^{-5}/\text{K}) - (5.1 \times 10^{-5}/\text{K}))(35°\text{C})} = 190 \text{ cm}^3.$$

The original volume of the cup is 190 cm³.

KEY CONCEPT 3 **EVALUATE** This is a straightforward application of volume thermal expansion. We did need to note carefully that both the copper and the ethanol expanded, so the spillage was the difference in the changes in volumes. We could almost ignore the change in volume of the copper, since the coefficient of thermal expansion is much smaller for copper than for ethanol.

Practice Problem: What is the final volume of the cup? *Answer:* 190.3 cm³.

Extra Practice: How much would overflow if it was a steel cup filled to the top with glycerin? *Answer:* 48.8 cm³

2: Stress in a wire

An aluminum wire is stretched across a large steel frame. Initially, the wire is at 20°C and is unstressed. The system (wire plus frame) is cooled by 50°C. If the area of contact for the wire is 9.0×10^{-6} m², what force is exerted on the wire?

IDENTIFY The differences in the expansion of the wire and frame will lead to tensile stress acting on the wire. The target variable is the force on the wire.

SET UP We'll first use temperature expansion to find the change in lengths of the wire and frame. We will then find the tensile stress on the wire. The coefficient of linear expansion is 2.4×10^{-5}/K for aluminum and 1.2×10^{-5}/K for steel. Young's modulus for the aluminum is 0.7×10^{11} N/m².

EXECUTE The change in length for a material due to temperature is
$$\Delta L = \alpha L_0 \Delta T.$$
When the aluminum is cooled by 50°C, the change in its length is
$$\left(\frac{\Delta L}{L_0}\right)_{al} = \alpha_{al} \Delta T = (2.4 \times 10^{-5})(50) = 1.2 \times 10^{-3}.$$
For the steel, the change in length is
$$\left(\frac{\Delta L}{L_0}\right)_{st} = \alpha_{st} \Delta T = (1.2 \times 10^{-5})(50) = 0.60 \times 10^{-3}.$$
Both of these changes are decreases, since the temperature has decreased. We see the aluminum changes more than the steel. Because the steel decreases less, stress is induced in the aluminum. The stress is given by
$$\frac{F}{A} = Y_{al}\left(\frac{\Delta L}{L_0}\right).$$

The stress is proportional to the net change in length of the wire. The frame shrinks, so the net change in length is the difference in the changes of the wire and the frame. The force is then

$$F = AY_{al}\left(\frac{\Delta L}{L_0}\right) = (9.0\times10^{2\,6}\text{ m}^2)(0.7\times10^{11}\text{ N/m}^2)(1.2\times10^{-3} - 0.60\times10^{-3}) = 380\text{ N}.$$

The stress on the wire is 380 N.

KEY CONCEPT 4 **EVALUATE** This is an application of linear thermal expansion. To find the stress, we did need to note carefully the difference in how the two materials contracted.

3: Ice to steam

A copper calorimeter of mass 2.0 kg initially contains 1.5 kg of ice at −10°C. How much heat energy must be added to convert all of the ice to water and then half of the water into steam?

IDENTIFY We'll use heat capacity and heat of fusion to determine how the ice melts and turns to steam. The target variable is the amount of heat needed to convert the ice to water and the water to steam.

SET UP We'll solve the problem in several steps. We'll first find the heat required to raise the temperature of the ice to 0°C, then find the heat required to melt the ice, then find the heat required to warm the water to 100°C, and finally find the heat required to vaporize half of the water. We know that the final temperature of the remaining water must be 100°C, since the water remains in equilibrium throughout the process. We must also add the heat required to heat the copper pot to 100°C.

EXECUTE The heat required to heat the ice to 0°C is

$$Q_1 = m_{ice}\,c_{ice}\,\Delta T = (1.5\text{ kg})(2100\text{ J/kg/K})(10.0°\text{C}) = 31{,}500\text{ J}.$$

We used the specific heat of ice (2010 J/kg/K) to find Q_1. The heat required to melt the ice is the heat of fusion for ice:

$$Q_2 = m_{ice}\,L_f = (1.5\text{ kg})(3.34\times10^5\text{ J/kg}) = 501{,}000\text{ J}.$$

The melted ice must warm to the final temperature (100.0°C):

$$Q_3 = m_{ice}\,c_{water}\,\Delta T = m_{ice}(4190\text{ J/kg/K})(100.0°\text{C}) = 628{,}500\text{ J}.$$

Here, we used the heat capacity of water (4190 J/kg/K). Half of the mass turns to steam. Using the heat of vaporization (2.256×10^6 J/kg), we find that the heat required is

$$Q_4 = \tfrac{1}{2}m_{ice}\,L_f = \tfrac{1}{2}(1.5\text{ kg})(2.256\times10^6\text{ J/kg}) = 1{,}692{,}000\text{ J}.$$

The copper pot also increases in temperature, from −10°C to 100°C. The heat required to bring about this increase is

$$Q_{copper} = m_{copper}\,c_{copper}\,\Delta T = (2.0\text{ kg})(390\text{ J/kg/K})(110.0°\text{C}) = 85{,}800\text{ J}.$$

The total heat is the sum of the five quantities of heat:

$$Q_t = Q_1 + Q_2 + Q_3 + Q_4 + Q_{copper} = 2.94\times10^6\text{ J}.$$

So 2.94×10^6 J are needed to heat the pot and ice from −10°C to 100°C and then to vaporize half of the water that is produced.

KEY CONCEPT 8 **EVALUATE** We see how we must solve calorimetry problems in multiple steps. We need to include the latent heat when materials change phase and the heat capacity when materials heat up.

> **CAUTION** **Temperature is not heat!** Temperature characterizes the state of an object. Heat is the flow of energy. The two terms may be synonymous in everyday language, but they are not synonymous in physics.

Practice Problem: How much more energy is required to end up with steam at 120°C? *Answer:* 1.75 MJ

Extra Practice: How much energy is required to boil 3.0 kg of ethanol at 50°C? *Answer:* 2.77 MJ

4: Cooling hot tea

You wish to chill your freshly brewed tea with the minimum amount of ice that will avoid watering it down too much. What is the minimum amount of ice you should add to 2.0 kg of freshly brewed tea at 95°C to cool it to 5.0°C? The ice is initially at a temperature of −5.0°C.

IDENTIFY We'll set the heat lost from the tea equal to the heat gained by the ice. The target variable is the amount of ice needed to cool the tea.

SET UP The amount of heat lost by the tea is given by the specific heat capacity equation, since the tea doesn't go through a phase change. The ice melts, so, in calculating the heat gain of the ice, we need to include the latent heat of fusion, plus the changes due to the ice warming to 0°C, and the changes due to the melted ice warming to 5.0°C.

EXECUTE The heat transfer from the hot tea as it cools to 5.0°C is negative:

$$Q_{tea} = m_{tea} c_{water} \Delta T_{tea} = (2.0 \text{ kg})(4190 \text{ J/kg/K})(5.0°C - 95°C) = -754,000 \text{ J}.$$

Here, we used the heat capacity of water (4190 J/kg/K) for the tea. The ice must warm to 0°C, then melt, and then heat to 5.0°C. We find the heat required for each segment of the ice warming. For the ice to heat to 0°C, we use the specific heat of ice (2010 J/kg/K):

$$Q_{ice} = m_{ice} c_{ice} \Delta T_{ice} = m_{ice} (2010 \text{ J/kg/K})[0.0°C - (-5.0°C)] = m_{ice}(10,000 \text{ J/kg}).$$

The heat needed to melt the ice is the heat of fusion for ice:

$$Q_{melt} = m_{ice} L_f = m_{ice}(3.34 \times 10^5 \text{ J/kg}).$$

The melted ice must warm to the final temperature (5.0°C):

$$Q_{melted\ ice} = m_{ice} c_{water} \Delta T_{melted\ ice} = m_{ice}(4190 \text{ J/kg/K})(5.0°C - 0.0°C) = m_{ice}(21,000 \text{ J/kg}).$$

The sum of these four quantities must be zero:

$$Q_{tea} + Q_{ice} + Q_{melt} + Q_{melted\ ice} = -754,000 \text{ J} + m_{ice}(10,000 \text{ J/kg})$$
$$+ m_{ice}(334,000 \text{ J/kg}) + m_{ice}(21,000 \text{ J/kg}) = 0.$$

Therefore,

$$m_{ice} = \frac{754,000 \text{ J}}{(10,000 \text{ J/kg}) + (334,000 \text{ J/kg}) + (21,000 \text{ J/kg})} = 2.1 \text{ kg}.$$

It takes a minimum of 2.1 kg of ice to cool the tea down.

KEY CONCEPT 7 **EVALUATE** Despite your best effort, the tea will be watery. Putting the ice in a bag will prevent the melted ice water from mixing with the tea. More importantly, we see how we must proceed stepwise through calorimetry problems.

Practice Problem: Repeat the problem using a frozen chunk of silver? *Answer:* 322 kg

Extra Practice: How much heat is required to melt that amount of silver, assuming it is already at the melting point? *Answer:* 28.5 MJ

5: Heat flow through three bars

A composite rod is made up of three equal lengths and cross sections of aluminum, brass, and copper. The free aluminum end is maintained at 100°C, and the free end of the copper rod is maintained at 0°C. If the surface of the rod is insulated to prevent radial heat flow, find the temperature at each junction.

IDENTIFY We'll use the heat current through the rods to find the temperatures at the rod junctions—the target variables.

SET UP A sketch of the rod is shown in Figure 2. The heat current is the same through each segment of the rod, so we'll write the heat equations for each segment and set them equal to each other to solve the problem.

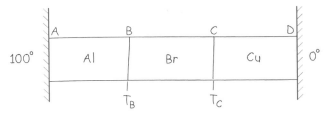

Figure 2 Problem 5 sketch.

EXECUTE The heat current through any rod is

$$H = kA \frac{T_H - T_C}{L}.$$

We can write the heat current per unit area times the length through the individual rods as

$$\frac{H_{AB} L}{A} = k_{Al}(T_H - T_C) = (205 \text{ W/m} \cdot \text{K})(100°\text{C} - T_B),$$

$$\frac{H_{BC} L}{A} = k_{Br}(T_H - T_C) = (109 \text{ W/m} \cdot \text{K})(T_B - T_C),$$

$$\frac{H_{CD} L}{A} = k_{Co}(T_H - T_C) = (385 \text{ W/m} \cdot \text{K})(T_C - 0°\text{C}).$$

These expressions must all be equal, since they each reference the same heat current, area, and length of the rod segment. Setting the last two equations equal to each other results in

$$(109 \text{ W/m} \cdot \text{K})T_B - (109 \text{ W/m} \cdot \text{K})T_C = (385 \text{ W/m} \cdot \text{K})T_C,$$

or

$$T_B = \frac{385+109}{109}T_C = 4.53T_C.$$

Setting the first and last equations equal to each other gives

$$(205 \text{ W/m} \cdot \text{K})(100°\text{C}) - (205 \text{ W/m} \cdot \text{K})T_B = (385 \text{ W/m} \cdot \text{K})T_C.$$

Replacing T_B yields

$$205(100°\text{C}) - 205\,(4.53T_C) = 385\,T_C,$$
$$T_C = 15.6°\text{C}.$$

Then

$$T_B = 70.7°\text{C}.$$

The aluminum/brass junction is at 70.7°C and the brass/copper junction is at 15.6°C.

KEY CONCEPT 11 EVALUATE Although one might expect the three equal segments to have equal temperature differences, we see that the varying thermal conductivities of the segments caused a nonuniform temperature distribution. The segment with the highest thermal conductivity (copper) had the smallest temperature difference between its ends, and the segment with the lowest thermal conductivity (brass) had the greatest temperature difference between its ends.

Practice Problem: What are the temperatures if the aluminum section is replaced with silver? *Answer:* 18.3°C and 82.7°C

Extra Practice: What if the brass section is now replaced with fiberglass? *Answer:* 0.01°C and 99.99°C

6: Time required to melt a block of ice

A long steel rod that is insulated to prevent heat loss along its sides is in perfect thermal contact with a large container of boiling water at one end and a 3.0-kg block of ice at the other. The steel rod is 1.2 m long with cross-sectional area 3.50 cm². How long does it take for the block of ice to melt? The ice block is initially at 0°C.

IDENTIFY We'll combine our knowledge of heat conduction with our knowledge of heat of fusion to solve this problem. The target variable is the time required for the ice to melt.

SET UP A sketch of the problem is shown in Figure 3. We begin by determining the heat required to melt the ice. We then find the rate of heat flow into the ice. With that information, we can find the time it takes for the ice to melt.

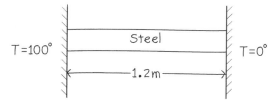

Figure 3 Problem 6 sketch.

EXECUTE The heat required to melt the ice is the heat of fusion for ice:

$$Q_{melt} = m_{ice} L_f = (3.0 \text{ kg})(3.34 \times 10^5 \text{ J/kg}) = 1.0 \times 10^6 \text{ J}.$$

The rate of heat flow is given by

$$H = \frac{\Delta Q}{\Delta t} = kA \frac{T_H - T_C}{L},$$

where k is the thermal conductivity, A and L are, respectively, the area and length of the bar, and T_H and T_C are, respectively, the temperatures of the hot and cold sides of the bar. We find that

$$H = \frac{\Delta Q}{\Delta t} = kA \frac{T_H - T_C}{L} = (50.2 \text{ W/(m} \cdot \text{K)})(6.5 \times 10^{-4} \text{ m}^2) \frac{(100°C) - (0°C)}{(1.2 \text{ m})} = 2.7 \text{ W},$$

where we used 50.2 W/m/K as the thermal conductivity of steel. The time required to melt the ice is

$$\Delta t = \frac{Q_{melt}}{H} = \frac{(10^6 \text{ J})}{(2.7 \text{ W})} = 370,000 \text{ s}.$$

The time required to melt the ice is 370,000 s, or 103 hours.

EVALUATE We see that the thin steel bar is a relatively poor conductor of heat. Replacing the steel with a copper bar would increase the rate by almost a factor of 8, due to the differences in thermal conductivity. Increasing the rod's diameter and shortening the rod would also increase the rate of melting.

Practice Problem: How much time does it take if we replace the steel with silver? *Answer:* 12.6 hours

Extra Practice: How much time does it take if we double the area of the silver bar and make it half as long? *Answer:* 3.16 hours

Try It Yourself!

1: Niagara Falls

Water flowing at a speed of 5.0 m/s falls over a 50-m-high waterfall into a still pool below. Calculate the approximate rise in water temperature due to the conversion of mechanical energy into thermal energy.

Solution Checkpoints

IDENTIFY AND SET UP Use energy conservation, equating the loss in mechanical energy to heat.

EXECUTE The water has kinetic energy and gravitational energy that together convert to heat. Equating the two forms of energy

$$\tfrac{1}{2}mv^2 + mgh = mc\Delta T.$$

The rise in temperature is $0.12°C$.

KEY CONCEPT 5 **EVALUATE** Did you need to know the mass of the water?

Practice Problem: How high does a waterfall need to be so that the 5.0 m/s water increases its temperature by 1°C? *Answer:* 426 m

Extra Practice: How fast (in mph) does water need to be moving so that a 50 m waterfall increases its temperature by 1°C? *Answer:* 192 mph

2: Melting ice, again

A copper calorimeter of mass 2.0 kg initially contains 1.5 kg of ice at −10°C. (a) What will the final temperature be if the heat added is 5×10^5 J? (b) What will the final temperature be if the heat added is 10^6 J?

Solution Checkpoints

IDENTIFY AND SET UP Use heat capacity and heat of fusion to determine how the temperature of the ice increases with the heat provided. Follow a series of steps and determine whether the heat at each step exceeds the heat provided.

EXECUTE (a) Examining Problem 4, we see that 5×10^5 J would be exhausted during the melting phase of the problem. The final temperature is 0°C.

(b) With 10^6 J of heat, all of the ice melts but the temperature doesn't reach 100°C. The temperature can be found by adding up the heat required in each step:

$$\Delta Q = m_{ice} c_{ice}(10°C) + m_{copper} c_{copper} \Delta T + m_{ice} c_{water}(\Delta T - 10°C) + m_{ice} L_f.$$

The final temperature is 64.8°C.

KEY CONCEPT 8 **EVALUATE** Why is −10°C required in the term expressing the heat capacity of water?

Practice Problem: What is the final temperature if there is twice as much ice initially? *Answer:* 0°C

Extra Practice: How much energy is needed to warm up and completely melt that much ice? *Answer:* 1.06 MJ

Key Example Variation Problems

Solutions to these problems are in Chapter 17 of the Student's Solutions Manual.

Be sure to review EXAMPLES 17.2, 17.3, and 17.4 (Section 17.4) before attempting these problems.

VP17.4.1 A metal rod is 0.500 m in length at a temperature of 15.0°C. When you raise its temperature to 37.0°C, its length increases by 0.220 mm. (a) What is the coefficient of linear expansion of the metal? (b) If a second rod of the same metal has length 0.300 m at 25.0°C, how will its length change if the temperature drops to −20.0°C?

VP17.4.2 A copper mug that can hold 250 cm³ of liquid is filled to the brim with ethanol at 20.0°C. If you lower the temperature of the mug and ethanol to −50.0°C, what is the maximum additional volume of ethanol you can add to the mug without spilling any? (See Table 17.2. Ethanol remains a liquid at temperatures down to −114°C.)

VP17.4.3 A cylindrical brass rod is 10.0 cm in length and 0.500 cm in radius at 25.0°C. How much force do you have to apply to each end of the rod to maintain its length when the temperature is decreased to 13.0°C? Are the required forces tensile or compressive? (See Table 17.1. Brass has Young's modulus 9×10^{10} Pa.)

VP17.4.4 A rod made of metal A is attached end to end to another rod made of metal B, making a combined rod of overall length L. The coefficients of linear expansion of metals A and B are α_A and α_B, respectively, and $\alpha_B > \alpha_A$. When the temperature of the combined rod is increased by ΔT, the overall length increases by ΔL. What was the initial length of the rod of metal A?

Be sure to review EXAMPLES 17.7, 17.8, and 17.9 (Section 17.6) before attempting these problems.

VP17.9.1 You place a piece of aluminum at 250.0°C in 5.00 kg of liquid water at 20.0°C. None of the water boils, and the final temperature of the water and aluminum is 22.0°C. What is the mass of the piece of aluminum? Assume no heat is exchanged with the container that holds the water. (See Table 17.3.)

VP17.9.2 You place an ice cube of mass 7.50×10^{-3} kg and temperature 0.00°C on top of a copper cube of mass 0.460 kg. All of the ice melts, and the final equilibrium temperature of the two substances is 0.00°C. What was the initial temperature of the copper cube? Assume no heat is exchanged with the surroundings. (See Tables 17.3 and 17.4.)

VP17.9.3 You have 1.60 kg of liquid ethanol at 28.0°C that you wish to cool. What mass of ice at initial temperature −5.00°C should you add to the ethanol so that all of the ice melts and the resulting ethanol−water mixture has temperature 10.0°C? Assume no heat is exchanged with the container that holds the ethanol. (See Tables 17.3 and 17.4.)

VP17.9.4 You put a silver ingot of mass 1.25 kg and initial temperature 315°C in contact with 0.250 kg of ice at initial temperature −8.00°C. Assume no heat is exchanged with the surroundings. (a) What is the final equilibrium temperature? (b) What fraction of the ice melts? (See Tables 17.3 and 17.4.)

Be sure to review EXAMPLES 17.11, 17.12, 17.13, 17.14, and 17.15 (Section 17.7) before attempting these problems.

VP17.15.1 A square pane of glass 0.500 m on a side is 6.00 mm thick. When the temperatures on the two sides of the glass are 25.0°C and −10.0°C, the heat current due to conduction through the glass is 1.10×10^3 W. (a) What is the thermal conductivity of the glass? (b) If the thickness of the glass is increased to 9.00 mm, what will be the heat current?

VP17.15.2 A brass rod and a lead rod, each 0.250 m long and each with cross-sectional area 2.00×10^{-4} m^2, are joined end to end to make a composite rod of overall length 0.500 m. The free end of the brass rod is maintained at a high temperature, and the free end of the lead rod is maintained at a low temperature. The temperature at the junction of the two rods is 185°C, and the heat current due to conduction through the composite rod is 6.00 W. What are the temperatures of (a) the free end of the brass rod and (b) the free end of the lead rod? (See Table 17.5.)

VP17.15.3 The emissivity of the surface of a star is approximately 1. The star Sirius A emits electromagnetic radiation at a rate of 9.7×10^{27} W and has a surface temperature of 9940 K. What is the radius of Sirius in meters and as a multiple of the sun's radius (6.96×10^8 m)?

VP17.15.4 A building in the desert is made of concrete blocks (emissivity 0.91) and has an exposed surface area of 525 m². If the building is maintained at 20.0°C but the temperature on a hot desert night is 35.0°C, what is the net rate at which the building absorbs energy by radiation?

STUDENT'S SOLUTIONS MANUAL FOR TEMPERATURE AND HEAT

VP17.4.1. **IDENTIFY:** The rods expand (or contract) when their temperature is changed, so we are dealing with thermal expansion.
SET UP: $\Delta L = \alpha L_0 \Delta T$
EXECUTE: **(a)** We want to find α, the coefficient of linear expansion. Using $\Delta L = \alpha L_0 \Delta T$ gives 2.20×10^{-4} m = $\alpha (0.500$ m$)(37.0°C - 15.0°C)$, so $\alpha = 2.00 \times 10^{-5}$ (C°)$^{-1}$ = 2.00×10^{-5} K^{-1}.
(b) We want the change in length. Using $\Delta L = \alpha L_0 \Delta T$ gives
$\Delta L = \left[2.00 \times 10^{-5} \text{ (C°)}^{-1}\right](0.300 \text{ m})(-20.0°C - 25.0°C) = -2.7 \times 10^{-4}$ m = -0.27 mm. Its length would *decrease* by 0.27 mm.
EVALUATE: Increasing temperature causes thermal expansion, while decreasing temperature causes thermal contraction. In both cases, the relative change in length is normally very small.

VP17.4.2. **IDENTIFY:** The mug and ethanol both decrease in volume as their temperature is decreased, so we are dealing with thermal contraction.
SET UP: We are dealing with volume contraction, so we use $\Delta V = \beta V_0 \Delta T$. If the volume of the ethanol and the mug decreased by the same amount, the ethanol would continue to fill the mug. In that case we could not add any additional ethanol to the mug. But $\beta_{\text{ethanol}} > \beta_{\text{mug}}$, so the ethanol contracts more than the mug, leaving room to add additional ethanol. The amount of empty space after the contraction will be $V_{\text{empy}} = \Delta V_{\text{ethanol}} - \Delta V_{\text{mug}}$, which is what we want to find. Table 17.2 tells us that $\beta_{\text{Cu}} = 5.1 \times 10^{-5}$ (C°)$^{-1}$ and $\beta_{\text{ethanol}} = 75 \times 10^{-5}$ (C°)$^{-1}$.
EXECUTE: $V_{\text{empy}} = \Delta V_{\text{ethanol}} - \Delta V_{\text{mug}} = \beta_e V_0 \Delta T - \beta_c V_0 \Delta T = (\beta_e - \beta_c) V_0 \Delta T$. Using the coefficients of volume expansion from Table 17.2, $V_0 = 250$ cm^3, and $\Delta T = -70.0°C$, we get $V_{\text{empty}} = -12$ cm^3. This is the net change in volume, and it is negative since the volumes have decreased. The available volume is 12 cm^3.
EVALUATE: If β_{ethanol} were less than β_{mug}, ethanol would spill out of the mug as the temperature fell because the mug would contract more than the ethanol.

VP17.4.3. **IDENTIFY:** The thermal stress is due to the contraction of the brass rod as its temperature is decreased.
SET UP: Thermal stress is $\dfrac{F}{A} = -Y\alpha\Delta T$. Table 17.1 gives $\alpha_{\text{brass}} = 2.0 \times 10^{-5}$ (C°)$^{-1}$ and Y_{brass} is given in the problem. The target variable is the thermal stress.

EXECUTE: The force is $F = -AY\alpha\Delta T = -\pi r^2 Y\alpha\Delta T$. Using the given values for Y and α, $r = 5.00\times 10^{-3}$ m, and $\Delta T = 13.0°C - 25.0°C = -12°C$, we get $F = 1700$ N.

EVALUATE: The bar tends to shrink when cooled, so the force prevent this. Therefore the forces are *tensile*.

VP17.4.4. **IDENTIFY:** The rods increase in length when their temperature is raised, so we are dealing with thermal expansion.

SET UP: The change in length of the combined rod is the sum of the changes of the two rods. So we know that $L = L_A + L_B$ and $\Delta L = \Delta L_A + \Delta L_B$, and we use $\Delta L = \alpha L_0 \Delta T$ for each rod. We want to solve for L_A in terms of the other quantities.

EXECUTE: $\Delta L = \Delta L_A + \Delta L_B = \alpha_A L_A \Delta T + \alpha_B (L - L_A)\Delta T$. Solve for L_A: $L_A = \dfrac{\alpha_B L - \dfrac{\Delta L}{\Delta T}}{\alpha_B - \alpha_A}$.

EVALUATE: The units may not look right. But realize that $\Delta L = \alpha L_0 \Delta T$ tells us that $\dfrac{\Delta L}{\Delta T} = \alpha L_0$ so the units in the numerator are those of αL, so the overall units are those of $\dfrac{\alpha L}{\alpha}$, which are units of length.

VP17.9.1. **IDENTIFY:** The heat lost by the aluminum is equal to the heat gained by the water, so the net heat transfer for the system is zero.

SET UP: The aluminum is hotter, so it loses heat and the water gains that heat. Use $Q = mc\Delta T$ for each substance. The target variable is the mass of the aluminum. From Table 17.3 we have $c_{Al} = 910$ J/kg·K and $c_w = 4190$ J/kg·K.

EXECUTE: Use $Q_{Al} + Q_w = 0$: $m_{Al} c_{Al} \Delta T_{Al} + m_w c_w \Delta T_w = 0$. Using the numbers gives $m_{Al}(910$ J/kg·K$)(-228.0$ K$) + (5.00$ kg$)(4190$ J/kg·K$)(2.0°C) = 0$, so $m_{Al} = 0.20$ kg.

EVALUATE: It is *not true* that $T_C = T_K$. But it *is true* that $\Delta T_C = \Delta T_K$ because the size of the Kelvin degree is the same as the size of the Celsius degree.

VP17.9.2. **IDENTIFY:** Heat is transferred to the copper cube to the ice cube, and this heat melts the ice. We are dealing with temperature changes and a change of phase for the ice from solid to liquid.

SET UP: The net heat transfer between the ice and copper is zero because the heat lost by the copper is gained by the ice cube. For temperature changes we use $Q = mc\Delta T$ and for the phase change we use $Q = mL_f$. The target variable is the initial temperature of the copper.

EXECUTE: Use $Q_{ice} + Q_{Cu} = 0$ and call T the initial temperature of the copper. This gives $m_{Cu} c_{Cu} \Delta T_{Cu} + m_i L_i = 0$. Using c_{Cu} from Table 17.3 and L_f from Table 17.4, we have $(0.460$ kg$)(390$ J/kg·K$)(0.00°C - T) + (7.50\times 10^{-3}$ kg$)(334\times 10^3$ J/kg$) = 0$, so $T = 14.0°C$.

EVALUATE: Just the melting a small ice cube can produce a large temperature change for the copper because c_{Cu} is small and L_f is large for water.

VP17.9.3. **IDENTIFY:** We mix ice and ethanol and wait until the mixture reaches an equilibrium temperature. The heat lost by the ethanol is equal to the heat gained by the ice, so the net heat transferred is zero.

SET UP: The target variable is the mass of the ice, m_i. For temperature changes we use $Q = mc\Delta T$ and for the phase change we use $Q = mL_f$. The heat transferred to the ice causes three changes: (1) increase in the ice temperature from $-5.00°C$ to $0.00°C$, (2) melt the ice at $0.00°C$, and (3) increase the temperature of the melted ice from $0.00°C$ to $10.0°C$.

EXECUTE: Use $Q_{ice} + Q_{ethanol} = 0$ and include the three changes for the ice listed above. This gives $m_i c_i \Delta T_i + m_i L_i + m_i c_w \Delta T_w + m_e c_e \Delta T_e = 0$. We know that $m_e = 1.60$ kg, and from Tables 17.3 and

17.4, we have $c_i = 2100$ J/kg·K, $c_w = 4190$ J/kg·K, $c_e = 2428$ J/kg·K, and $L_f = 334 \times 10^3$ J/kg. The temperature changes are $\Delta T_i = 5.00°C$, $\Delta T_w = 10.0°C$, and $\Delta T_e = -18.0°C$. Putting in the numbers and solving for m_i gives $m_i = 0.181$ kg.
EVALUATE: We must treat the ice in three stages. In addition to melting at 0°C, the ice and liquid water have different specific heats so we cannot treat the ice change from –5.00°C to 10.0°C in a single step.

VP17.9.4. **IDENTIFY:** The ice cools a silver ingot. Several possible outcomes are possible depending on the relative masses of the ice and silver and their initial temperatures: the ice could remain ice but at a higher temperature, the ice could partially melt and remain at 0.00°C, the ice could all melt and remain at 0.00°C, or the ice could all melt and increase its temperature above 0.00°C.
SET UP: For temperature changes use $Q = mc\Delta T$ and for the phase change use $Q = mL_f$. First see if there is enough heat in the silver to bring the ice up to its melting point temperature of 0.00°C and to melt it at 0.00°C. Use quantities from Tables 17.3 and 17.4 for water, ice, and silver. The target variable is the final equilibrium temperature T of the system.
EXECUTE: The heat to cool the silver down to 0.00°C is $Q = mc\Delta T$, which gives
$Q = (1.25 \text{ kg})(234 \text{ J/kg·K})(315°C) = 9.214 \times 10^4$ J. The heat needed to melt all the ice is
$Q = mc\Delta T + mL_f = (0.250 \text{ kg})(2100 \text{ J/kg·K})(8.00°C) + (0.250 \text{ kg})(334 \times 10^3 \text{ J/kg}) = 8.77 \times 10^4$ J.
As we see, there is enough heat in the silver ingot to melt all the ice. Now we use the same procedure as in the previous problem except we have silver instead of ethanol. This gives
$m_i c_i \Delta T_i + m_i L_f + m_i c_w \Delta T_w + m_s c_s \Delta T_s = 0$. The temperature changes are $\Delta T_i = 8.00°C$,
$\Delta T_w = T - 00.0°C$, and $\Delta T_s = T - 315°C$, $m_i = 0.250$ kg, and $m_s = 1.25$ kg. The result is $T = 3.31°C$, and all the ice melts.
EVALUATE: We must treat the ice in three stages. In addition to melting at 0°C, the ice and liquid water have different specific heats so we cannot treat the ice change from –8.00°C to 3.31°C in a single step.

VP17.15.1. **IDENTIFY:** The heat flows through the pane of glass, so we are dealing with heat conduction.
SET UP: The rate of heat flow is $H = kA\dfrac{T_H - T_C}{L}$.

EXECUTE: (a) We want the thermal conductivity k of the glass. Solve $H = kA\dfrac{T_H - T_C}{L}$ for k,
which gives $k = \dfrac{LH}{A(T_H - T_C)} = \dfrac{(0.00600 \text{ m})(1100 \text{ W})}{(0.500 \text{ m})^2 (35.0 \text{ C°})} = 0.754$ W/m·K.

(b) $H \propto 1/L$, so $\dfrac{H_9}{H_6} = \dfrac{1/9}{1/6} = \dfrac{2}{3}$, which gives $H_9 = \dfrac{2}{3}(1100 \text{ W}) = 733$ W.

EVALUATE: Other ways to decrease the heat current would be to decrease the area of the pane or to use a double-pane window which traps a layer of air between two panes of glass. The air has a *much* lower thermal conductivity than glass.

VP17.15.2. **IDENTIFY:** Heat current flows through the two end-to-end rods (brass and lead), so we are dealing with heat conduction.
SET UP: Use $H = kA\dfrac{T_H - T_C}{L}$. Table 17.5 gives $k_{brass} = 109.0$ W/m·K and $k_{lead} = 34.7$ W/m·K.
The heat current is the same (6.00 W) in both rods. The target variables are the temperatures of the free ends of the two-rod system.

EXECUTE: (a) <u>Brass</u>: Using $H = kA\dfrac{T_H - T_C}{L}$ and putting in the numbers gives

$6.00 \text{ W} = (109.0 \text{ W/m} \cdot \text{K})(2.00 \times 10^{-4} \text{ m})\left(\dfrac{T_{\text{brass}} - 185°\text{C}}{0.250 \text{ m}}\right)$, so $T_{\text{brass}} = 254°\text{C}$.

(b) <u>Lead</u>: We follow the same procedure as for brass, giving

$6.00 \text{ W} = (34.7 \text{ W/m} \cdot \text{K})(2.00 \times 10^{-4} \text{ m})\left(\dfrac{185°\text{C} - T_{\text{lead}}}{0.250 \text{ m}}\right)$, so $T_{\text{lead}} = -31°\text{C}$.

EVALUATE: From $H = kA\dfrac{T_H - T_C}{L}$ we see that $\Delta T \propto \dfrac{1}{k}$. So, if all else is the same, an object with a small k should have a large temperature difference compared to one with a large k. Our results agree with this since $\Delta T_{\text{lead}} > \Delta T_{\text{brass}}$ and $k_{\text{lead}} < k_{\text{brass}}$.

VP17.15.3. IDENTIFY: We are dealing the radiation from a star.

SET UP: $H = Ae\sigma T^4$, where T must be in K, $e = 1$ for the star, and $A = 4\pi R^2$.

EXECUTE: $H = Ae\sigma T^4 = 4\pi R^2 \sigma T^4$, so solve for R, giving $R = \sqrt{\dfrac{H}{4\pi\sigma T^4}} = \dfrac{1}{T^2}\sqrt{\dfrac{H}{4\pi\sigma}}$. Putting in the numbers gives $R = \dfrac{1}{(9940 \text{ K})^2}\sqrt{\dfrac{9.7 \times 10^{27} \text{ W}}{4\pi(5.67 \times 10^{-8} \text{ W/m}^2 \cdot \text{K}^4)}} = 1.2 \times 10^9$ m.

$\dfrac{R_{\text{Sirius}}}{R_{\text{sun}}} = \dfrac{1.2 \times 10^9 \text{ m}}{6.96 \times 10^8 \text{ m}} = 1.7$.

EVALUATE: The surface area of Sirius is $(1.7)^2 = 2.9$ times as great as our sun, so Sirius is a luminous star.

VP17.15.4. IDENTIFY: The building radiates heat into the air, but the hot air also radiates heat back into the building. So we are looking for the *net* radiation, which will be *into* the building because its surface is less hot than the outside air.

SET UP: $H_{\text{net}} = Ae\sigma\left(T^4 - T_s^4\right)$, where temperatures must be in K and $e = 0.91$.

EXECUTE: $H_{\text{net}} = (525 \text{ m}^2)(0.91)(5.67 \times 10^{-8} \text{ W/m}^2 \cdot \text{K}^4)\left[(293 \text{ K})^4 - (308 \text{ K})^4\right] = -4.4 \times 10^4$ W.

The minus sign tells us that net heat flows *into* the building from the hot desert air at a rate of 4.4×10^4 W $= 44$ kW.

EVALUATE: The heat flow into the building would be even greater during the day when the outside temperature could be 40°C or even higher.

17.5. IDENTIFY: Convert ΔT in kelvins to C° and to F°.

SET UP: $1 \text{ K} = 1 \text{ C°} = \frac{9}{5}\text{F°}$

EXECUTE: (a) $\Delta T_F = \frac{9}{5}\Delta T_C = \frac{9}{5}(-10.0 \text{ C°}) = -18.0$ F°

(b) $\Delta T_C = \Delta T_K = -10.0$ C°

EVALUATE: Kelvin and Celsius degrees are the same size. Fahrenheit degrees are smaller, so it takes more of them to express a given ΔT value.

17.9. IDENTIFY and SET UP: Fit the data to a straight line for $p(T)$ and use this equation to find T when $p = 0$.

EXECUTE: (a) If the pressure varies linearly with temperature, then $p_2 = p_1 + \gamma(T_2 - T_1)$.

$$\gamma = \frac{p_2 - p_1}{T_2 - T_1} = \frac{6.50 \times 10^4 \text{ Pa} - 4.80 \times 10^4 \text{ Pa}}{100°\text{C} - 0.01°\text{C}} = 170.0 \text{ Pa/C°}$$

Apply $p = p_1 + \gamma(T - T_1)$ with $T_1 = 0.01°\text{C}$ and $p = 0$ to solve for T.

$0 = p_1 + \gamma(T - T_1)$

$$T = T_1 - \frac{p_1}{\gamma} = 0.01°\text{C} - \frac{4.80 \times 10^4 \text{ Pa}}{170 \text{ Pa/C°}} = -282°\text{C}.$$

(b) Let $T_1 = 100°\text{C}$ and $T_2 = 0.01°\text{C}$; use $T_2/T_1 = p_2/p_1$ to calculate p_2, where T is in kelvins.

$$p_2 = p_1\left(\frac{T_2}{T_1}\right) = 6.50 \times 10^4 \text{ Pa}\left(\frac{0.01 + 273.15}{100 + 273.15}\right) = 4.76 \times 10^4 \text{ Pa; this differs from the } 4.80 \times 10^4 \text{ Pa}$$

that was measured so $T_2/T_1 = p_2/p_1$ is not precisely obeyed.

EVALUATE: The answer to part (a) is in reasonable agreement with the accepted value of $-273°\text{C}$.

17.15. IDENTIFY: Apply $\Delta V = V_0 \beta \Delta T$.

SET UP: For copper, $\beta = 5.1 \times 10^{-5}$ $(\text{C}°)^{-1}$. $\Delta V/V_0 = 0.150 \times 10^{-2}$.

EXECUTE: $\Delta T = \dfrac{\Delta V/V_0}{\beta} = \dfrac{0.150 \times 10^{-2}}{5.1 \times 10^{-5} \ (\text{C}°)^{-1}} = 29.4 \text{ C}°$. $T_\text{f} = T_\text{i} + \Delta T = 49.4°\text{C}$.

EVALUATE: The volume increases when the temperature increases.

17.17. IDENTIFY: Apply $\Delta V = V_0 \beta \Delta T$ to the volume of the flask and to the mercury. When heated, both the volume of the flask and the volume of the mercury increase.

SET UP: For mercury, $\beta_\text{Hg} = 18 \times 10^{-5}$ $(\text{C}°)^{-1}$.

8.95 cm³ of mercury overflows, so $\Delta V_\text{Hg} - \Delta V_\text{glass} = 8.95 \text{ cm}^3$.

EXECUTE: $\Delta V_\text{Hg} = V_0 \beta_\text{Hg} \Delta T = (1000.00 \text{ cm}^3)(18 \times 10^{-5} \ (\text{C}°)^{-1})(55.0 \text{ C}°) = 9.9 \text{ cm}^3$.

$\Delta V_\text{glass} = \Delta V_\text{Hg} - 8.95 \text{ cm}^3 = 0.95 \text{ cm}^3$.

$$\beta_\text{glass} = \frac{\Delta V_\text{glass}}{V_0 \Delta T} = \frac{0.95 \text{ cm}^3}{(1000.00 \text{ cm}^3)(55.0 \text{ C}°)} = 1.7 \times 10^{-5} \ (\text{C}°)^{-1}.$$

EVALUATE: The coefficient of volume expansion for the mercury is larger than for glass. When they are heated, both the volume of the mercury and the inside volume of the flask increase. But the increase for the mercury is greater and it no longer all fits inside the flask.

17.21. IDENTIFY: Apply $\Delta L = L_0 \alpha \Delta T$ and stress $= F/A = -Y\alpha \Delta T$.

SET UP: For steel, $\alpha = 1.2 \times 10^{-5}$ $(\text{C}°)^{-1}$ and $Y = 2.0 \times 10^{11}$ Pa.

EXECUTE: **(a)** $\Delta L = L_0 \alpha \Delta T = (12.0 \text{ m})(1.2 \times 10^{-5} \ (\text{C}°)^{-1})(42.0 \text{ C}°) = 0.0060 \text{ m} = 6.0 \text{ mm}$.

(b) stress $= -Y\alpha \Delta T = -(2.0 \times 10^{11} \text{ Pa})(1.2 \times 10^{-5} \ (\text{C}°)^{-1})(42.0 \text{ C}°) = -1.0 \times 10^8$ Pa. The minus sign means the stress is compressive.

EVALUATE: Commonly occurring temperature changes result in very small fractional changes in length but very large stresses if the length change is prevented from occurring.

17.23. IDENTIFY: We are dealing with the thermal expansion of two rods.

SET UP: Use $\Delta L = \alpha L_0 \Delta T$. From Table 17.1 we get the coefficients of linear expansion for aluminum and Invar. We know that $\Delta L_{Al} = 2\Delta L_{Invar}$ and $\Delta T_{Al} = \frac{1}{3}\Delta T_{Invar}$. We want L_{Al}/L_{Invar}.

EXECUTE: Apply $\Delta L = \alpha L_0 \Delta T$ to each bar and take the ratio of the length changes, giving $\frac{\Delta L_{Al}}{\Delta L_{Invar}} = \frac{\alpha_{Al} L_{Al} \Delta T_{Al}}{\alpha_{Invar} L_{Invar} \Delta T_{Invar}}$. Solving for the length ratio gives $\frac{L_{Al}}{L_{Invar}} = \frac{\alpha_{Invar} \Delta L_{Al} \Delta T_{Invar}}{\alpha_{Al} \Delta L_{Invar} \Delta T_{Al}} = \left(\frac{0.09}{2.4}\right)\left(\frac{\Delta L_{Al}}{\Delta L_{Invar}}\right)\left(\frac{3\Delta T_{Al}}{\Delta T_{Al}}\right) = 0.225$, which we round to 0.2. This also gives $\frac{L_{Invar}}{L_{Al}} = \frac{1}{0.225} = 4$.

EVALUATE: The coefficient of linear expansion for Invar is much less than that of aluminum, so the Invar rod needs to be much longer than the aluminum rod to expand half as much with only 1/3 the temperature change.

17.25. IDENTIFY and SET UP: Apply $Q = mc\,\Delta T$ to the kettle and water.

EXECUTE: kettle
$Q = mc\Delta T$, $c = 910$ J/kg·K (from Table 17.3)
$Q = (1.10 \text{ kg})(910 \text{ J/kg·K})(85.0°C - 20.0°C) = 6.5065 \times 10^4$ J

water
$Q = mc\Delta T$, $c = 4190$ J/kg·K (from Table 17.3)
$Q = (1.80 \text{ kg})(4190 \text{ J/kg·K})(85.0°C - 20.0°C) = 4.902 \times 10^5$ J

Total $Q = 6.5065 \times 10^4$ J $+ 4.902 \times 10^5$ J $= 5.55 \times 10^5$ J.

EVALUATE: Water has a much larger specific heat capacity than aluminum, so most of the heat goes into raising the temperature of the water.

17.31. IDENTIFY and SET UP: Set the change in gravitational potential energy equal to the quantity of heat added to the water.

EXECUTE: The change in mechanical energy equals the decrease in gravitational potential energy, $\Delta U = -mgh$; $|\Delta U| = mgh$. $Q = |\Delta U| = mgh$ implies $mc\Delta T = mgh$

$\Delta T = gh/c = (9.80 \text{ m/s}^2)(225 \text{ m})/(4190 \text{ J/kg·K}) = 0.526$ K $= 0.526$ C°

EVALUATE: Note that the answer is independent of the mass of the object. Note also the small change in temperature that corresponds to this large change in height!

17.33. IDENTIFY: Some of the kinetic energy of the bullet is transformed through friction into heat, which raises the temperature of the water in the tank.

SET UP: Set the loss of kinetic energy of the bullet equal to the heat energy Q transferred to the water. $Q = mc\Delta T$. From Table 17.3, the specific heat of water is 4.19×10^3 J/kg·C°.

EXECUTE: The kinetic energy lost by the bullet is
$K_i - K_f = \frac{1}{2}m(v_i^2 - v_f^2) = \frac{1}{2}(15.0 \times 10^{-3} \text{ kg})[(865 \text{ m/s})^2 - (534 \text{ m/s})^2] = 3.47 \times 10^3$ J, so for the water
$Q = 3.47 \times 10^3$ J. $Q = mc\Delta T$ gives $\Delta T = \frac{Q}{mc} = \frac{3.47 \times 10^3 \text{ J}}{(13.5 \text{ kg})(4.19 \times 10^3 \text{ J/kg·C°})} = 0.0613$ C°.

EVALUATE: The heat energy required to change the temperature of ordinary-size objects is very large compared to the typical kinetic energies of moving objects.

17.35. **IDENTIFY** and **SET UP:** Heat comes out of the metal and into the water. The final temperature is in the range $0 < T < 100°C$, so there are no phase changes. $Q_{\text{system}} = 0$.

(a) EXECUTE: $Q_{\text{water}} + Q_{\text{metal}} = 0$

$m_{\text{water}} c_{\text{water}} \Delta T_{\text{water}} + m_{\text{metal}} c_{\text{metal}} \Delta T_{\text{metal}} = 0$

$(1.00 \text{ kg})(4190 \text{ J/kg} \cdot \text{K})(2.0 \text{ C°}) + (0.500 \text{ kg})(c_{\text{metal}})(-78.0 \text{ C°}) = 0$

$c_{\text{metal}} = 215 \text{ J/kg} \cdot \text{K}$

(b) EVALUATE: Water has a larger specific heat capacity so stores more heat per degree of temperature change.

(c) If some heat went into the styrofoam then Q_{metal} should actually be larger than in part (a), so the true c_{metal} is larger than we calculated; the value we calculated would be smaller than the true value.

17.37. **IDENTIFY:** The amount of heat lost by the iron is equal to the amount of heat gained by the water. The water must first be heated to 100°C and then vaporized.

SET UP: The relevant equations are $Q = mc\Delta T$ and $Q = L_v m$. The specific heat of iron is $c_{\text{iron}} = 0.47 \times 10^3 \text{ J/(kg} \cdot \text{K)}$, the specific heat of water is $c_{\text{water}} = 4.19 \times 10^3 \text{ J/(kg} \cdot \text{K)}$, and the heat of vaporization of water is $L_v = 2256 \times 10^3 \text{ J/kg}$.

EXECUTE: The iron cools: $Q_{\text{iron}} = m_i c_i \Delta T_i$.

The water warms and vaporizes: $Q_{\text{water}} = c_w m_w \Delta T_w + m_w L_{v_w} = m_w (c_w \Delta T_w + L_{v_w})$.

Assume that all of the heat lost by the iron is gained by the water so that $Q_{\text{water}} = -Q_{\text{iron}}$. Equating the respective expressions for each Q and solving for m_w we obtain

$$m_w = \frac{-m_i c_i \Delta T_i}{c_w \Delta T_w + L_{v_w}} = \frac{-(1.20 \text{ kg})(0.47 \times 10^3 \text{ J/kg} \cdot \text{K})(120.0°\text{C} - 650.0°\text{C})}{(4.19 \times 10^3 \text{ J/kg} \cdot \text{K})(100.0°\text{C} - 15.0°\text{C}) + 2256 \times 10^3 \text{ J/kg}} = 0.114 \text{ kg}.$$

EVALUATE: Note that only a relatively small amount of water is required to cause a very large temperature change in the iron. This is due to the high heat of vaporization and specific heat of water, and the relatively low specific heat capacity of iron.

17.41. **IDENTIFY:** By energy conservation, the heat lost by the copper is gained by the ice. This heat must first increase the temperature of the ice from −20.0°C to the melting point of 0.00°C, then melt some of the ice. At the final thermal equilibrium state, there is ice and water, so the temperature must be 0.00°C. The target variable is the initial temperature of the copper.

SET UP: For temperature changes, $Q = mc\Delta T$ and for a phase change from solid to liquid $Q = mL_F$.

EXECUTE: For the ice,

$Q_{\text{ice}} = (2.00 \text{ kg})[2100 \text{ J/(kg} \cdot \text{C°)}](20.0\text{C°}) + (0.80 \text{ kg})(3.34 \times 10^5 \text{ J/kg}) = 3.512 \times 10^5 \text{ J}$. For the copper, using the specific heat from the table in the text gives

$Q_{\text{copper}} = (6.00 \text{ kg})[390 \text{ J/(kg} \cdot \text{C°)}](0°\text{C} - T) = -(2.34 \times 10^3 \text{ J/C°})T$. Setting the sum of the two heats equal to zero gives $3.512 \times 10^5 \text{ J} = (2.34 \times 10^3 \text{ J/C°})T$, which gives $T = 150°\text{C}$.

EVALUATE: Since the copper has a smaller specific heat than that of ice, it must have been quite hot initially to provide the amount of heat needed.

17.45. **IDENTIFY and SET UP:** The heat that must be added to a lead bullet of mass m to melt it is $Q = mc\Delta T + mL_f$ ($mc\Delta T$ is the heat required to raise the temperature from 25°C to the melting point of 327.3°C; mL_f is the heat required to make the solid → liquid phase change.) The kinetic energy of the bullet if its speed is v is $K = \tfrac{1}{2}mv^2$.

EXECUTE: $K = Q$ says $\tfrac{1}{2}mv^2 = mc\Delta T + mL_f$

$v = \sqrt{2(c\Delta T + L_f)}$

$v = \sqrt{2[(130 \text{ J/kg} \cdot \text{K})(327.3°\text{C} - 25°\text{C}) + 24.5 \times 10^3 \text{ J/kg}]} = 357$ m/s

EVALUATE: This is a typical speed for a rifle bullet. A bullet fired into a block of wood does partially melt, but in practice not all of the initial kinetic energy is converted to heat that remains in the bullet.

17.47. **IDENTIFY:** Use $Q = Mc\Delta T$ to find Q for a temperature rise from 34.0°C to 40.0°C. Set this equal to $Q = mL_v$ and solve for m, where m is the mass of water the camel would have to drink.

SET UP: $c = 3480$ J/kg·K and $L_v = 2.42 \times 10^6$ J/kg. For water, 1.00 kg has a volume 1.00 L. $M = 400$ kg is the mass of the camel.

EXECUTE: The mass of water that the camel saves is

$m = \dfrac{Mc\Delta T}{L_v} = \dfrac{(400 \text{ kg})(3480 \text{ J/kg} \cdot \text{K})(6.0 \text{ K})}{(2.42 \times 10^6 \text{ J/kg})} = 3.45$ kg which is a volume of 3.45 L.

EVALUATE: This is nearly a gallon of water, so it is an appreciable savings.

17.49. **IDENTIFY:** The asteroid's kinetic energy is $K = \tfrac{1}{2}mv^2$. To boil the water, its temperature must be raised to 100.0°C and the heat needed for the phase change must be added to the water.

SET UP: For water, $c = 4190$ J/kg·K and $L_v = 2256 \times 10^3$ J/kg.

EXECUTE: $K = \tfrac{1}{2}(2.60 \times 10^{15} \text{ kg})(32.0 \times 10^3 \text{ m/s})^2 = 1.33 \times 10^{24}$ J. $Q = mc\Delta T + mL_v$.

$m = \dfrac{Q}{c\Delta T + L_v} = \dfrac{1.33 \times 10^{22} \text{ J}}{(4190 \text{ J/kg} \cdot \text{K})(90.0 \text{ K}) + 2256 \times 10^3 \text{ J/kg}} = 5.05 \times 10^{15}$ kg.

EVALUATE: The mass of water boiled is 2.5 times the mass of water in Lake Superior.

17.51. **IDENTIFY and SET UP:** Heat flows out of the water and into the ice. The net heat flow for the system is zero. The ice warms to 0°C, melts, and then the water from the melted ice warms from 0°C to the final temperature.

EXECUTE: $Q_{\text{system}} = 0$; calculate Q for each component of the system: (Beaker has small mass says that $Q = mc\Delta T$ for beaker can be neglected.)

0.250 kg of water: cools from 75.0°C to 40.0°C

$Q_{\text{water}} = mc\Delta T = (0.250 \text{ kg})(4190 \text{ J/kg} \cdot \text{K})(40.0°\text{C} - 75.0°\text{C}) = -3.666 \times 10^4$ J.

ice: warms to 0°C; melts; water from melted ice warms to 40.0°C

$Q_{\text{ice}} = mc_{\text{ice}}\Delta T + mL_f + mc_{\text{water}}\Delta T$.

$Q_{\text{ice}} = m[(2100 \text{ J/kg} \cdot \text{K})(0°\text{C} - (-20.0°\text{C})) + 334 \times 10^3 \text{ J/kg} + (4190 \text{ J/kg} \cdot \text{K})(40.0°\text{C} - 0°\text{C})]$.

$Q_{ice} = (5.436 \times 10^5 \text{ J/kg})m$. $Q_{system} = 0$ says $Q_{water} + Q_{ice} = 0$.

$-3.666 \times 10^4 \text{ J} + (5.436 \times 10^5 \text{ J/kg})m = 0$. $m = \dfrac{3.666 \times 10^4 \text{ J}}{5.436 \times 10^5 \text{ J/kg}} = 0.0674$ kg.

EVALUATE: Since the final temperature is 40.0°C we know that all the ice melts and the final system is all liquid water. The mass of ice added is much less than the mass of the 75°C water; the ice requires a large heat input for the phase change.

17.53. IDENTIFY: We mix liquids at different temperatures, so this is a problem in calorimetry.

SET UP: The unknown liquid cools down from 30.0°C to 14.0°C. The ice all melts and the resulting water increases from 0.0°C to 14.0°C. The heat gained by the ice all comes from the unknown liquid, so the net heat change for the mixture is zero. The ice goes through two changes: melting at 0.0°C followed by a temperature increase to 14.0°C. We use $Q = mc\Delta T$ and want to find the specific heat c of the unknown liquid.

EXECUTE: $Q_{ice} = Q_{melt} + Q_{increase\,temp} = m_{ice}L_f + m_{ice}c_{water}\Delta T_{water}$ and $Q_{unknown} = mc\Delta T_{unknown}$. Using $Q_{ice} + Q_{unknown} = 0$ gives $m_{ice}L_f + m_{ice}c_{water}\Delta T_{water} + mc\Delta T_{unknown} = 0$. Using the given masses and temperatures as well as $L_f = 334 \times 10^3$ J/kg and $c_{water} = 4190$ J/kg·K, we get $c = 2370$ J/kg·K.

EVALUATE: From Table 17.3 we see that this value is close to the specific heats of ethylene and glycol, so it is a reasonable result.

17.55. IDENTIFY: A heat current flows through the two bars, so we are dealing with thermal conduction.

SET UP: Both bars have the same length and cross-sectional area, and we want to find the thermal conductivity of the unknown metal. The graph plots T versus T_H, so we need to relate these quantities to interpret the graph. For this we use $H = kA\dfrac{T_H - T_C}{L}$. At steady state, H is the same in both bars.

EXECUTE: For the copper bar $H = k_{Cu}A\dfrac{T_H - T}{L}$, and for the unknown bar $H = kA\dfrac{T - 0.0°C}{L}$. Equating the heat currents and simplifying gives $k_{Cu}(T_H - T) = kT$. Solving for T gives $T = \dfrac{k_{Cu}}{k + k_{Cu}}T_H$, so a graph of T versus T_H should be a straight line having slope $\dfrac{k_{Cu}}{k + k_{Cu}}$. Solving for k gives $k = k_{Cu}\left(\dfrac{1}{\text{slope}} - 1\right)$. From Table 17.5, we have $k_{Cu} = 385$ W/mol·K, so

$k = (385 \text{ W/m·K})\left(\dfrac{1}{0.710} - 1\right) = 157$ W/m·K.

EVALUATE: From Table 17.5, $k = 157$ W/m·K is between that of brass and aluminum, so our result is reasonable.

17.57. IDENTIFY and SET UP: Call the temperature at the interface between the wood and the styrofoam T. The heat current in each material is given by $H = kA(T_H - T_C)/L$.

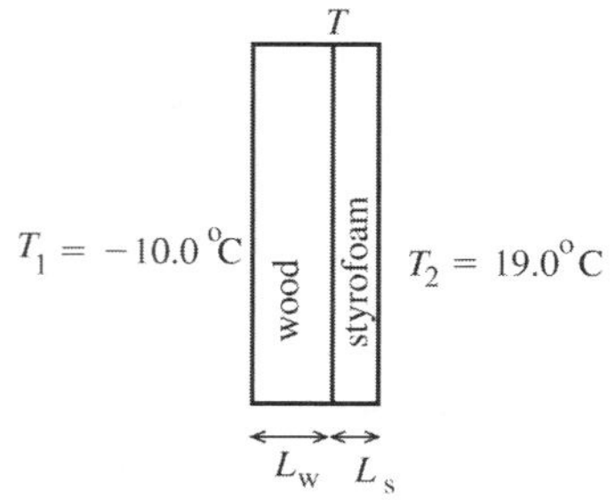

See Figure 17.57.
Heat current through the wood:
$$H_w = k_w A(T - T_1)L_w$$
Heat current through the styrofoam:
$$H_s = k_s A(T_2 - T)/L_s$$

Figure 17.57

In steady-state heat does not accumulate in either material. The same heat has to pass through both materials in succession, so $H_w = H_s$.

EXECUTE: (a) This implies $k_w A(T - T_1)/L_w = k_s A(T_2 - T)/L_s$

$k_w L_s (T - T_1) = k_s L_w (T_2 - T)$

$$T = \frac{k_w L_s T_1 + k_s L_w T_2}{k_w L_s + k_s L_w} = \frac{-0.0176 \text{ W} \cdot \text{°C/K} + 0.01539 \text{ W} \cdot \text{°C/K}}{0.00257 \text{ W/K}} = -0.86°C.$$

EVALUATE: The temperature at the junction is much closer in value to T_1 than to T_2. The styrofoam has a very small k, so a larger temperature gradient is required for than for wood to establish the same heat current.

(b) IDENTIFY and SET UP: Heat flow per square meter is $\dfrac{H}{A} = k\left(\dfrac{T_H - T_C}{L}\right)$. We can calculate this either for the wood or for the styrofoam; the results must be the same.

EXECUTE: Wood: $\dfrac{H_w}{A} = k_w \dfrac{T - T_1}{L_w} = (0.080 \text{ W/m} \cdot \text{K})\dfrac{-0.86°C - (-10.0°C)}{0.030 \text{ m}} = 24 \text{ W/m}^2$.

Styrofoam: $\dfrac{H_s}{A} = k_s \dfrac{T_2 - T}{L_s} = (0.027 \text{ W/m} \cdot \text{K})\dfrac{19.0°C - (-0.86°C)}{0.022 \text{ m}} = 24 \text{ W/m}^2$.

EVALUATE: H must be the same for both materials and our numerical results show this. Both materials are good insulators and the heat flow is very small.

17.61. IDENTIFY and SET UP: The heat conducted through the bottom of the pot goes into the water at 100°C to convert it to steam at 100°C. We can calculate the amount of heat flow from the mass of material that changes phase. Then use $H = kA(T_H - T_C)/L$ to calculate T_H, the temperature of the lower surface of the pan.

EXECUTE: $Q = mL_v = (0.390 \text{ kg})(2256 \times 10^3 \text{ J/kg}) = 8.798 \times 10^5 \text{ J}$

$H = Q/t = 8.798 \times 10^5 \text{ J}/180 \text{ s} = 4.888 \times 10^3 \text{ J/s}$

Then $H = kA(T_H - T_C)/L$ says that $T_H - T_C = \dfrac{HL}{kA} = \dfrac{(4.888 \times 10^3 \text{ J/s})(8.50 \times 10^{-3} \text{ m})}{(50.2 \text{ W/m} \cdot \text{K})(0.150 \text{ m}^2)} = 5.52 \text{ C°}$

$T_H = T_C + 5.52 \text{ C°} = 100°C + 5.52 \text{ C°} = 105.5°C$.

EVALUATE: The larger $T_H - T_C$ is the larger H is and the faster the water boils.

17.65. IDENTIFY: Use $H = Ae\sigma T^4$ to calculate A.

SET UP: $H = Ae\sigma T^4$ so $A = H/e\sigma T^4$
150-W and all electrical energy consumed is radiated says $H = 150$ W.
EXECUTE:
$$A = \frac{150 \text{ W}}{(0.35)(5.67 \times 10^{-8} \text{ W/m}^2 \cdot \text{K}^4)(2450 \text{ K})^4} = 2.1 \times 10^{-4} \text{ m}^2 (1 \times 10^4 \text{ cm}^2/1 \text{ m}^2) = 2.1 \text{ cm}^2$$
EVALUATE: Light-bulb filaments are often in the shape of a tightly wound coil to increase the surface area; larger A means a larger radiated power H.

17.73. IDENTIFY and SET UP: Use the temperature difference in M° and in C° between the melting and boiling points of mercury to relate M° to C°. Also adjust for the different zero points on the two scales to get an equation for T_M in terms of T_C.

(a) EXECUTE: normal melting point of mercury: $-39°\text{C} = 0.0°\text{M}$
normal boiling point of mercury: $357°\text{C} = 100.0°\text{M}$
$100.0 \text{ M}° = 396 \text{ C}°$ so $1 \text{ M}° = 3.96 \text{ C}°$

Zero on the M scale is -39 on the C scale, so to obtain T_C multiply T_M by 3.96 and then subtract $39°$: $T_C = 3.96 T_M - 39°$

Solving for T_M gives $T_M = \frac{1}{3.96}(T_C + 39°)$

The normal boiling point of water is $100°\text{C}$; $T_M = \frac{1}{3.96}(100° + 39°) = 35.1°\text{M}$.

(b) $10.0 \text{ M}° = 39.6 \text{ C}°$
EVALUATE: A M° is larger than a C° since it takes fewer of them to express the difference between the boiling and melting points for mercury.

17.75. IDENTIFY and SET UP: Use $\Delta V = V_0 \beta \Delta T$ for the volume expansion of the oil and of the cup. Both the volume of the cup and the volume of the olive oil increase when the temperature increases, but β is larger for the oil so it expands more. When the oil starts to overflow,
$\Delta V_{\text{oil}} = \Delta V_{\text{glass}} + (3.00 \times 10^{-3} \text{ m})A$, where A is the cross-sectional area of the cup.

EXECUTE: $\Delta V_{\text{oil}} = V_{0,\text{oil}} \beta_{\text{oil}} \Delta T = (9.7 \text{ cm}) A \beta_{\text{oil}} \Delta T$.
$\Delta V_{\text{glass}} = V_{0,\text{glass}} \beta_{\text{glass}} \Delta T = (10.0 \text{ cm}) A \beta_{\text{glass}} \Delta T$.
$(9.7 \text{ cm}) A \beta_{\text{oil}} \Delta T = (10.0 \text{ cm}) A \beta_{\text{glass}} \Delta T + (0.300 \text{ cm}) A$. The A divides out. Solving for ΔT gives $\Delta T = 47.4 \text{ C}°$. $T_2 = T_1 + \Delta T = 69.4°\text{C}$.
EVALUATE: If the expansion of the cup is neglected, the olive oil will have expanded to fill the cup when $(0.300 \text{ cm})A = (9.7 \text{ cm}) A \beta_{\text{oil}} \Delta T$, so $\Delta T = 45.5 \text{ C}°$ and $T_2 = 77.5°\text{C}$. Our result is somewhat higher than this. The cup also expands but not as much since $\beta_{\text{glass}} \ll \beta_{\text{oil}}$.

17.77. IDENTIFY and SET UP: Call the metals A and B. Use the data given to calculate α for each metal.
EXECUTE: $\Delta L = L_0 \alpha \Delta T$ so $\alpha = \Delta L/(L_0 \Delta T)$

metal A: $\alpha_A = \dfrac{\Delta L}{L_0 \Delta T} = \dfrac{0.0650 \text{ cm}}{(30.0 \text{ cm})(100 \text{ C}°)} = 2.167 \times 10^{-5} \text{ (C}°)^{-1}$

metal B: $\alpha_B = \dfrac{\Delta L}{L_0 \Delta T} = \dfrac{0.0350 \text{ cm}}{(30.0 \text{ cm})(100 \text{ C}°)} = 1.167 \times 10^{-5} \text{ (C}°)^{-1}$

EVALUATE: L_0 and ΔT are the same, so the rod that expands the most has the larger α.

IDENTIFY and **SET UP:** Now consider the composite rod (Figure 17.77). Apply $\Delta L = L_0 \alpha \Delta T$. The target variables are L_A and L_B, the lengths of the metals A and B in the composite rod.

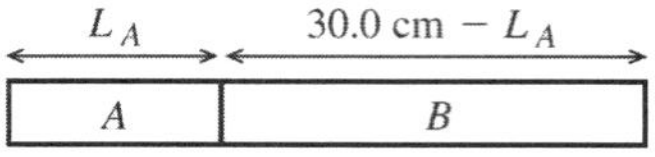

$\Delta T = 100\ C°$
$\Delta L = 0.058\ \text{cm}$

Figure 17.77

EXECUTE: $\Delta L = \Delta L_A + \Delta L_B = (\alpha_A L_A + \alpha_B L_B)\Delta T$

$\Delta L/\Delta T = \alpha_A L_A + \alpha_B (0.300\ \text{m} - L_A)$

$$L_A = \frac{\Delta L/\Delta T - (0.300\ \text{m})\alpha_B}{\alpha_A - \alpha_B} = \frac{(0.058 \times 10^{-2}\ \text{m})/(100\ C°) - (0.300\ \text{m})(1.167 \times 10^{-5}(C°)^{-1})}{1.00 \times 10^{-5}\ (C°)^{-1}} = 23.0\ \text{cm}$$

$L_B = 30.0\ \text{cm} - L_A = 30.0\ \text{cm} - 23.0\ \text{cm} = 7.0\ \text{cm}$

EVALUATE: The expansion of the composite rod is similar to that of rod A, so the composite rod is mostly metal A.

17.81. **(a) IDENTIFY** and **SET UP:** The diameter of the ring undergoes linear expansion (increases with T) just like a solid steel disk of the same diameter as the hole in the ring. Heat the ring to make its diameter equal to 2.5020 in.

EXECUTE: $\Delta L = \alpha L_0 \Delta T$ so $\Delta T = \dfrac{\Delta L}{L_0 \alpha} = \dfrac{0.0020\ \text{in.}}{(2.5000\ \text{in.})(1.2 \times 10^{-5}(C°)^{-1})} = 66.7\ C°$

$T = T_0 + \Delta T = 20.0°C + 66.7\ C° = 87°C$

(b) IDENTIFY and **SET UP:** Apply the linear expansion equation to the diameter of the brass shaft and to the diameter of the hole in the steel ring.

EXECUTE: $L = L_0(1 + \alpha \Delta T)$

Want L_s (steel) $= L_b$ (brass) for the same ΔT for both materials: $L_{0s}(1 + \alpha_s \Delta T) = L_{0b}(1 + \alpha_b \Delta T)$ so $L_{0s} + L_{0s}\alpha_s \Delta T = L_{0b} + L_{0b}\alpha_b \Delta T$

$\Delta T = \dfrac{L_{0b} - L_{0s}}{L_{0s}\alpha_s - L_{0b}\alpha_b} = \dfrac{2.5020\ \text{in.} - 2.5000\ \text{in.}}{(2.5000\ \text{in.})(1.2 \times 10^{-5}(C°)^{-1}) - (2.5020\ \text{in.})(2.0 \times 10^{-5}(C°)^{-1})}$

$\Delta T = \dfrac{0.0020}{3.00 \times 10^{-5} - 5.00 \times 10^{-5}}\ C° = -100\ C°$

$T = T_0 + \Delta T = 20.0°C - 100\ C° = -80°C$

EVALUATE: Both diameters decrease when the temperature is lowered but the diameter of the brass shaft decreases more since $\alpha_b > \alpha_s$; $|\Delta L_b| - |\Delta L_s| = 0.0020$ in.

17.83. **IDENTIFY:** We mix an unknown liquid at 90.0°C with water at 0.0°C and measure the final equilibrium temperature T. The heat lost by the unknown liquid is equal to the heat gained by the water, so $Q_{net} = 0$.

SET UP: $Q_{net} = Q_{water} + Q_{unknown}$. We need to relate the mass of the water m_w to T in order to interpret the graph of m_w versus T^{-1}. Let x refer to the unknown liquid. We want to find c_x. For temperature changes, we use $Q = mc\Delta T$.

EXECUTE: Using $Q_{net} = Q_{water} + Q_{unknown} = 0$ gives $m_w c_w \Delta T_w + m_x c_x \Delta T_x = 0$. Solve for m_w as a function of T^{-1}. $m_w c_w (T - 0.0°C) = m_x c_x (90.0°C - T)$, so $m_w = \left(\dfrac{m_x c_x}{c_w}\right)\left(\dfrac{90.0°C}{T} - 1\right)$. The graph of m_w versus T^{-1} should be a straight line having slope equal to $\dfrac{m_x c_x 90.0°C}{c_w}$. Solving for c_x gives

$$c_x = \dfrac{c_w (\text{slope})}{m_x(90.0°C)} = \dfrac{(4190 \text{ J/kg} \cdot \text{K})(2.15 \text{ kg} \cdot \text{C°})}{(0.050 \text{ kg})(90.0°C)} = 2000 \text{ J/kg} \cdot \text{K}.$$

EVALUATE: From Table 17.3 we see that c_x is about half that of water, but close to that of ice, ethanol, and ethylene glycol, so we get a reasonable result.

17.85. **IDENTIFY and SET UP:** To calculate Q, use $Q = mc\,\Delta T$ in the form $dQ = nC\,dT$ and integrate, using $C(T)$ given in the problem. C_{av} is obtained from $C = \dfrac{1}{n}\dfrac{dQ}{dT}$ using the finite temperature range instead of an infinitesimal dT.

EXECUTE: (a) $dQ = nC\,dT$

$$Q = n\int_{T_1}^{T_2} C\,dT = n\int_{T_1}^{T_2} k(T^3/\theta^3)dT = (nk/\theta^3)\int_{T_1}^{T_2} T^3\,dt = (nk/\theta^3)\left(\tfrac{1}{4}T^4\Big|_{T_1}^{T_2}\right)$$

$$Q = \dfrac{nk}{4\theta^3}(T_2^4 - T_1^4) = \dfrac{(1.50 \text{ mol})(1940 \text{ J/mol} \cdot \text{K})}{4(281 \text{ K})^3}\left[(40.0 \text{ K})^4 - (10.0 \text{ K})^4\right] = 83.6 \text{ J}.$$

(b) $C_{av} = \dfrac{1}{n}\dfrac{\Delta Q}{\Delta T} = \dfrac{1}{1.50 \text{ mol}}\left(\dfrac{83.6 \text{ J}}{40.0 \text{ K} - 10.0 \text{ K}}\right) = 1.86$ J/mol·K

(c) $C = k(T/\theta)^3 = (1940 \text{ J/mol} \cdot \text{K})(40.0 \text{ K}/281 \text{ K})^3 = 5.60$ J/mol·K

EVALUATE: C is increasing with T, so C at the upper end of the temperature integral is larger than its average value over the interval.

17.89. **IDENTIFY:** The energy generated in the body is used to evaporate water, which prevents the body from overheating.

SET UP: Energy is (power)(time); calculate the heat energy Q produced in one hour. The mass m of water that vaporizes is related to Q by $Q = mL_v$. 1.0 kg of water has a volume of 1.0 L.

EXECUTE: (a) $Q = (0.80)(500 \text{ W})(3600 \text{ s}) = 1.44 \times 10^6$ J. The mass of water that evaporates each hour is $m = \dfrac{Q}{L_v} = \dfrac{1.44 \times 10^6 \text{ J}}{2.42 \times 10^6 \text{ J/kg}} = 0.60$ kg.

(b) (0.60 kg/h)(1.0 L/kg) = 0.60 L/h. The number of bottles of water is

$$\dfrac{0.60 \text{ L/h}}{0.750 \text{ L/bottle}} = 0.80 \text{ bottles/h}.$$

EVALUATE: It is not unreasonable to drink 8/10 of a bottle of water per hour during vigorous exercise.

17.93. **IDENTIFY:** The heat lost by the water is equal to the amount of heat gained by the ice. First calculate the amount of heat the water could give up if it is cooled to 0.0°C. Then see how much heat it would take to melt all of the ice. If the heat to melt the ice is less than the heat the water would give up, the ice all melts and then the resulting water is heated to some final temperature.

SET UP: $Q = mc\,\Delta T$ and $Q = mL_f$.

EXECUTE: (a) Heat from water if cooled to 0.0°C: $Q = mc\,\Delta T$

$Q = mc\,\Delta T = (1.50 \text{ kg})(4190 \text{ J/kg} \cdot \text{K})(28.0 \text{ K}) = 1.760 \times 10^5$ J

Heat to melt all of the ice: $Q = mc\,\Delta T + mL_f = m(c\Delta T + L_f)$

$Q = (0.600 \text{ kg})[(2100 \text{ J/kg} \cdot \text{K})(22.0 \text{ K}) + 3.34 \times 10^5 \text{ J/kg}] = 2.276 \times 10^5$ J

Since the heat required to melt all the ice is greater than the heat available by cooling the water to 0.0°C, not all the ice will melt.

(b) Since not all the ice melts, the final temperature of the water (and ice) will be 0.0°C. So the heat from the water will melt only part of the ice. Call m the mass of the melted ice. Therefore $Q_{\text{from water}} = 1.760 \times 10^5$ J $= (0.600$ kg$)(2100$ J/kg·K$)(22.0$ K$) + m(3.34 \times 10^5$ J/kg$)$, which gives $m = 0.444$ kg, which is the amount of ice that melts. The mass of ice remaining is 0.600 kg – 0.444 kg = 0.156 kg. The final temperature will be 0.0°C since some ice remains in the water.

EVALUATE: An alternative approach would be to assume that all the ice melts and find the final temperature of the water in the container. This actually comes out to be negative, which is not possible if all the ice melts. Therefore not all the ice could have melted. Once you know this, proceed as in part (b).

17.95. IDENTIFY and SET UP: Assume that all the ice melts and that all the steam condenses. If we calculate a final temperature T that is outside the range 0°C to 100°C then we know that this assumption is incorrect. Calculate Q for each piece of the system and then set the total $Q_{\text{system}} = 0$.

EXECUTE: (a) <u>Copper can</u> (changes temperature from 0.0° to T; no phase change):
$Q_{\text{can}} = mc\Delta T = (0.446$ kg$)(390$ J/kg·K$)(T - 0.0°C) = (173.9$ J/K$)T$
<u>Ice</u> (melting phase change and then the water produced warms to T):
$Q_{\text{ice}} = +mL_f + mc\Delta T = (0.0950$ kg$)(334 \times 10^3$ J/kg$) + (0.0950$ kg$)(4190$ J/kg·K$)(T - 0.0°C)$
$Q_{\text{ice}} = 3.173 \times 10^4$ J $+ (398.0$ J/K$)T$.
<u>Steam</u> (condenses to liquid and then water produced cools to T):
$Q_{\text{steam}} = -mL_v + mc\Delta T = -(0.0350$ kg$)(2256 \times 10^3$ J/kg$) + (0.0350$ kg$)(4190$ J/kg·K$)(T - 100.0°C)$
$Q_{\text{steam}} = -7.896 \times 10^4$ J $+ (146.6$ J/K$)T - 1.466 \times 10^4$ J $= -9.362 \times 10^4$ J $+ (146.6$ J/K$)T$
$Q_{\text{system}} = 0$ implies $Q_{\text{can}} + Q_{\text{ice}} + Q_{\text{steam}} = 0$.
$(173.9$ J/K$)T + 3.173 \times 10^4$ J $+ (398.0$ J/K$)T - 9.362 \times 10^4$ J $+ (146.6$ J/K$)T = 0$
$(718.5$ J/K$)T = 6.189 \times 10^4$ J
$T = \dfrac{6.189 \times 10^4 \text{ J}}{718.5 \text{ J/K}} = 86.1°C$.

(b) No ice, no steam, and 0.0950 kg + 0.0350 kg = 0.130 kg of liquid water.

EVALUATE: This temperature is between 0°C and 100°C so our assumptions about the phase changes being complete were correct.

17.99. IDENTIFY: Apply $H = kA\dfrac{T_H - T_C}{L}$.

SET UP: For the glass use $L = 12.45$ cm, to account for the thermal resistance of the air films on either side of the glass.

EXECUTE: (a) $H = (0.120$ W/m·K$)(2.00 \times 0.95$ m$^2)\left(\dfrac{28.0 \text{ C°}}{5.0 \times 10^{-2} \text{ m} + 1.8 \times 10^{-2} \text{ m}}\right) = 93.9$ W.

(b) The heat flow through the wood part of the door is reduced by a factor of
$1 - \dfrac{(0.50)^2}{(2.00 \times 0.95)} = 0.868$, so it becomes 81.5 W. The heat flow through the glass is

$H_{glass} = (0.80 \text{ W/m} \cdot \text{K})(0.50 \text{ m})^2 \left(\dfrac{28.0 \text{ C°}}{12.45 \times 10^{-2} \text{ m}} \right) = 45.0$ W, and so the ratio is

$\dfrac{81.5 + 45.0}{93.9} = 1.35$.

EVALUATE: The single-pane window produces a significant increase in heat loss through the door. (See Problem 17.101).

17.101. IDENTIFY and SET UP: Use H written in terms of the thermal resistance R: $H = A\Delta T/R$, where $R = L/k$ and $R = R_1 + R_2 + \ldots$ (additive).

EXECUTE: <u>single pane:</u> $R_s = R_{glass} + R_{film}$, where $R_{film} = 0.15 \text{ m}^2 \cdot \text{K/W}$ is the combined thermal resistance of the air films on the room and outdoor surfaces of the window.

$R_{glass} = L/k = (4.2 \times 10^{-3} \text{ m})/(0.80 \text{ W/m} \cdot \text{K}) = 0.00525 \text{ m}^2 \cdot \text{K/W}$

Thus $R_s = 0.00525 \text{ m}^2 \cdot \text{K/W} + 0.15 \text{ m}^2 \cdot \text{K/W} = 0.1553 \text{ m}^2 \cdot \text{K/W}$.

<u>double pane:</u> $R_d = 2R_{glass} + R_{air} + R_{film}$, where R_{air} is the thermal resistance of the air space between the panes. $R_{air} = L/k = (7.0 \times 10^{-3} \text{ m})/(0.024 \text{ W/m} \cdot \text{K}) = 0.2917 \text{ m}^2 \cdot \text{K/W}$

Thus $R_d = 2(0.00525 \text{ m}^2 \cdot \text{K/W}) + 0.2917 \text{ m}^2 \cdot \text{K/W} + 0.15 \text{ m}^2 \cdot \text{K/W} = 0.4522 \text{ m}^2 \cdot \text{K/W}$

$H_s = A\Delta T/R_s$, $H_d = A\Delta T/R_d$, so $H_s/H_d = R_d/R_s$ (since A and ΔT are same for both)

$H_s/H_d = (0.4522 \text{ m}^2 \cdot \text{K/W})/(0.1553 \text{ m}^2 \cdot \text{K/W}) = 2.9$

EVALUATE: The heat loss is about a factor of 3 less for the double-pane window. The increase in R for a double pane is due mostly to the thermal resistance of the air space between the panes.

17.103. IDENTIFY: The jogger radiates heat but the air radiates heat back into the jogger.

SET UP: The emissivity of a human body is taken to be 1.0. In the equation for the radiation heat current, $H_{net} = Ae\sigma(T^4 - T_s^4)$, the temperatures must be in kelvins.

EXECUTE: **(a)** $P_{jog} = (0.80)(1300 \text{ W}) = 1.04 \times 10^3$ J/s.

(b) $H_{net} = Ae\sigma(T^4 - T_s^4)$, which gives

$H_{net} = (1.85 \text{ m}^2)(1.00)(5.67 \times 10^{-8} \text{ W/m}^2 \cdot \text{K}^4)([306 \text{ K}]^4 - [313 \text{ K}]^4) = -87.1$ W. The person gains 87.1 J of heat each second by radiation.

(c) The total excess heat per second is 1040 J/s + 87 J/s = 1130 J/s.

(d) In 1 min = 60 s, the runner must dispose of $(60 \text{ s})(1130 \text{ J/s}) = 6.78 \times 10^4$ J. If this much heat goes to evaporate water, the mass m of water that evaporates in one minute is given by $Q = mL_v$, so

$m = \dfrac{Q}{L_v} = \dfrac{6.78 \times 10^4 \text{ J}}{2.42 \times 10^6 \text{ J/kg}} = 0.028$ kg = 28 g.

(e) In a half-hour, or 30 minutes, the runner loses $(30 \text{ min})(0.028 \text{ kg/min}) = 0.84$ kg. The runner must drink 0.84 L, which is $\dfrac{0.84 \text{ L}}{0.750 \text{ L/bottle}} = 1.1$ bottles.

EVALUATE: The person *gains* heat by radiation since the air temperature is greater than his skin temperature.

17.105. **(a) IDENTIFY** and **EXECUTE:** Heat must be conducted from the water to cool it to 0°C and to cause the phase transition. The entire volume of water is not at the phase transition temperature, just the upper surface that is in contact with the ice sheet.

(b) IDENTIFY: The heat that must leave the water in order for it to freeze must be conducted through the layer of ice that has already been formed.
SET UP: Consider a section of ice that has area A. At time t let the thickness be h. Consider a short time interval t to $t+dt$. Let the thickness that freezes in this time be dh. The mass of the section that freezes in the time interval dt is $dm = \rho\, dV = \rho A\, dh$. The heat that must be conducted away from this mass of water to freeze it is $dQ = dm L_f = (\rho A L_f)dh$. $H = dQ/dt = kA(\Delta T/h)$, so the heat dQ conducted in time dt throughout the thickness h that is already there is $dQ = kA\left(\dfrac{T_H - T_C}{h}\right)dt$. Solve for dh in terms of dt and integrate to get an expression relating h and t.
EXECUTE: Equate these expressions for dQ.

$$\rho A L_f dh = kA\left(\dfrac{T_H - T_C}{h}\right)dt$$

$$h\, dh = \left(\dfrac{k(T_H - T_C)}{\rho L_f}\right)dt$$

Integrate from $t=0$ to time t. At $t=0$ the thickness h is zero.

$$\int_0^h h\, dh = [k(T_H - T_C)/\rho L_f]\int_0^t dt$$

$$\tfrac{1}{2}h^2 = \dfrac{k(T_H - T_C)}{\rho L_f}t \text{ and } h = \sqrt{\dfrac{2k(T_H - T_C)}{\rho L_f}}\sqrt{t}$$

The thickness after time t is proportional to $\sqrt{t}$.
(c) The expression in part (b) gives

$$t = \dfrac{h^2 \rho L_f}{2k(T_H - T_C)} = \dfrac{(0.25\text{ m})^2(920\text{ kg/m}^3)(334\times 10^3\text{ J/kg})}{2(1.6\text{ W/m}\cdot\text{K})(0°\text{C}-(-10°\text{C}))} = 6.0\times 10^5\text{ s}$$

$t = 170$ h.

(d) Find t for $h = 40$ m. t is proportional to h^2, so $t = (40\text{ m}/0.25\text{ m})^2(6.00\times 10^5\text{ s}) = 1.5\times 10^{10}$ s. This is about 500 years. With our current climate this will not happen.
EVALUATE: As the ice sheet gets thicker, the rate of heat conduction through it decreases. Part (d) shows that it takes a very long time for a moderately deep lake to totally freeze.

17.109. **IDENTIFY:** The latent heat of fusion L_f is defined by $Q = mL_f$ for the solid → liquid phase transition. For a temperature change, $Q = mc\Delta T$.

SET UP: At $t = 1$ min the sample is at its melting point and at $t = 2.5$ min all the sample has melted.
EXECUTE: **(a)** It takes 1.5 min for all the sample to melt once its melting point is reached and the heat input during this time interval is $(1.5\text{ min})(10.0\times 10^3\text{ J/min}) = 1.50\times 10^4$ J. $Q = mL_f$.

$$L_f = \dfrac{Q}{m} = \dfrac{1.50\times 10^4\text{ J}}{0.500\text{ kg}} = 3.00\times 10^4\text{ J/kg}.$$

(b) The liquid's temperature rises 30 C° in 1.5 min. $Q = mc\Delta T$.

$$c_{\text{liquid}} = \dfrac{Q}{m\Delta T} = \dfrac{1.50\times 10^4\text{ J}}{(0.500\text{ kg})(30\text{ C°})} = 1.00\times 10^3\text{ J/kg}\cdot\text{K}.$$

The solid's temperature rises 15 C° in 1.0 min.

$$c_{\text{solid}} = \frac{Q}{m\Delta T} = \frac{1.00 \times 10^4 \text{ J}}{(0.500 \text{ kg})(15 \text{ C°})} = 1.33 \times 10^3 \text{ J/kg} \cdot \text{K}.$$

EVALUATE: The specific heat capacities for the liquid and solid states are different. The values of c and L_f that we calculated are within the range of values in Tables 17.3 and 17.4.

17.111. IDENTIFY: At steady state, the heat current in both bars is the same when they are connected end-to-end. The heat to melt the ice is the heat conducted through the bars.

SET UP: $Q = mL$ and $H = kA\dfrac{T_H - T_C}{L}$.

EXECUTE: With bar A alone: 0.109 kg of ice melts in 45.0 min = (45.0)(60) s. Therefore the heat current is $H = mL_f/t = (0.109 \text{ kg})(334 \times 10^5 \text{ J/kg})/[(45.0)(60) \text{ s}] = 13.48 \text{ J/s} = 13.48 \text{ W}$. Applying this result to the heat flow in bar A gives $H = kA\dfrac{T_H - T_C}{L}$. Solving for k_A gives $k_A = HL/A(T_H - T_C)$. Numerically we get
$k_A = (13.48 \text{ W})(0.400 \text{ m})/[(2.50 \times 10^{-4} \text{ m}^2)(100 \text{ C°})] = 215.7 \text{ W/m} \cdot \text{K}$, which rounds to 216 W/m·K.

With the two bars end-to-end: The heat current is the same in both bars, so $H_A = H_B$. Using $H = kA\dfrac{T_H - T_C}{L}$ for each bar, we get $\dfrac{k_A A(100°C - 62.4°C)}{L} = \dfrac{k_B A(62.4°C - 0°C)}{L}$. Using our result for k_A and canceling A and L, we get $k_B = 130$ W/m·K.

EVALUATE: $k_A = 216$ W/m·K, which is slightly larger than that of aluminum, and $k_B = 130$ W/m·K, which is between that of aluminum and brass. Therefore these results are physically reasonable.

17.115. IDENTIFY: Apply the equation $H = kA(T_H - T_C)/L$. For a cylindrical surface, the area A in this equation is a function of the distance r from the central axis, and the material must be considered as a series of shells with thickness dr and a temperature difference dT between the inside and outside of the shell. The heat current will be a constant, and must be found by integrating a differential equation.

SET UP: The surface area of the curved side of a cylinder is $2\pi rL$. When $x \ll 1$, $\ln(1+x) \approx x$.

EXECUTE: **(a)** For a cylindrical shell, $H = kA(T_H - T_C)/L$ becomes
$H = k(2\pi rL)\dfrac{dT}{dr}$ or $\dfrac{Hdr}{2\pi r} = kLdT$. Between the limits $r = a$ and $r = b$, this integrates to
$\dfrac{H}{2\pi}\ln(b/a) = kL(T_2 - T_1)$, or $H = \dfrac{2\pi kL(T_2 - T_1)}{\ln(b/a)}$.

(b) Using $H = k(2\pi rL)\dfrac{dT}{dr}$ from part (a), we integrate: $\int_a^r \dfrac{Hdr}{k2\pi rL} = -\int_{T_2}^T dT$, which gives
$\dfrac{H}{2\pi kL}\ln(r/a) = -(T - T_2) = T_2 - T$. Solving for T and using H from part (a) gives
$T = T_2 - \dfrac{H}{2\pi kL}\ln(r/a) = \dfrac{(T_2 - T_1)2\pi kL\ln(r/a)}{2\pi kL \ln(b/a)} = T_2 - \dfrac{(T_2 - T_1)\ln(r/a)}{\ln(b/a)}$, which can also be written as
$T(r) = T_2 + (T_1 - T_2)\dfrac{\ln(r/a)}{\ln(b/a)}$.

(c) For a thin-walled cylinder, $a \approx b$, so $\dfrac{b-a}{a} \ll 1$. We can write $\dfrac{b}{a} = 1 + \dfrac{b-a}{a}$, and the log term becomes $\ln\left(\dfrac{b}{a}\right) = \ln\left(1 + \dfrac{b-a}{a}\right) \approx \dfrac{b-a}{a}$ using $\ln(1+x) \approx x$ for $x \ll 1$. Therefore the rate of heat flow becomes $H = \dfrac{(T_2 - T_1) 2\pi k L}{\dfrac{b-a}{a}} = k(2\pi a L)\dfrac{T_2 - T_1}{b-a}$. This is equivalent to Eq. (17.21) in which $A = 2\pi a L$, which is the surface area of the thin cylindrical shell of radius a and length L.

(d) For steady-state heat flow, the rate of flow through the cork is the same as through the Styrofoam. Using our result from part (a), we have $H_C = H_S$, which gives
$(140°C - T)2\pi k_C L/(\ln 4/2) = (T - 15°C)2\pi k_S L/(\ln 6/4)$
Cancelling $2\pi L$ and using the given values for k_C and k_S, we get $T = 73°C$.

(e) Use the result of part (a) for H.
$$H = \dfrac{(140°C - 73°C)2\pi(0.0400 \text{ W/m}\cdot\text{K})(2.00 \text{ m})}{\ln(4/2)} = 49 \text{ W}.$$

EVALUATE: As a check, calculate H in the Styrofoam. $\Delta T = 73°C - 15°C = 58 \text{ C}°$.
$$H_S = \dfrac{2\pi(2.00 \text{ m})(0.027 \text{ W/m}\cdot\text{K})}{\ln(6.00/4.00)}(58 \text{ C}°) = 49 \text{ W}.$$ This is the same as we just found, which it should be for steady-state flow.

18

STUDY GUIDE FOR THERMAL PROPERTIES OF MATTER

Summary

In this chapter, we extend our investigation into thermodynamics, viewing systems from both the macroscopic and microscopic perspectives and building links between the two perspectives. We'll learn about equations of state for materials and examine the ideal-gas equation as one such equation. This investigation will allow us to build a model for the kinetic energy of individual molecules and predict the behavior of gases.

Objectives

After studying this chapter, you'll understand

- How to define the *mole* and *Avogadro's number*.
- How to define *equations of state* and how to apply the ideal gas equation.
- How to determine the kinetic energy of gases and how to apply that energy to individual particles.
- The origins of molar heat capacities for materials and gases.

Concepts and Equations

Term	Description
Mole	One mole (mol) is the amount of substance that contains the same number of elementary units as there are atoms in 0.012 kg of carbon 12. The number of molecules in a mole is Avogadro's number, $N_A = 6.022 \times 10^{23}$ molecules per mole. The molar mass is the mass of 1 mole of a substance.
Equation of State	An equation of state expresses the relation among pressure, temperature, and volume of a certain amount of a substance in equilibrium. The pressure p, volume V, and absolute temperature T are the state variables.
Ideal-Gas Equation	The ideal-gas equation is the equation of state for an ideal gas that approximates the behavior of a real gas at a low pressure and a high temperature. The pressure p, temperature T, volume V, and number of moles, n, of the gas are related by $$pV = nRT,$$ where R is the ideal-gas constant. In SI units, when pressure is given in Pa and volume is given in m³, $R = 8.3145$ J/(mol·K).
Kinetic Theory of Gases	The total translational kinetic energy K_{tr} of all the molecules in an ideal gas is proportional to the temperature T and quantity of gas, n, in moles. Expressed as an equation, the total translational kinetic energy is $$K_{tr} = \tfrac{3}{2}nRT.$$ For a single molecule, the average translational kinetic energy is $$K_{av} = \tfrac{3}{2}kT,$$ where $k = R/N_A = 1.381 \times 10^{-23}$ J/(molecule·K) is the Boltzmann constant. The mean free path of molecules in an ideal gas is given by $$\lambda = v t_{mean} = \frac{V}{4\pi\sqrt{2}\, r^2 N}.$$
Molar Heat Capacity	The amount of heat Q needed for a temperature change ΔT is $$Q = nC\Delta T,$$ where n is the number of moles of the substance and C is the molar heat capacity. The molar heat capacity at constant volume is given in certain cases by $$C_V = \tfrac{3}{2}R \quad \text{(monatomic gas)},$$ $$C_V = \tfrac{5}{2}R \quad \text{(diatomic gas)},$$ $$C_V = 3R \quad \text{(monatomic solid)}.$$
Molecular Speeds	The speeds of molecules in an ideal gas are given by the Maxwell–Boltzmann distribution: $$f(v) = 4\pi \left(\frac{m}{2\pi kT}\right)^{3/2} v^2 e^{-mv^2/2kT}.$$ The quantity $f(v)\,dv$ describes the fraction of molecules with speeds between v and $v + dv$.
Phases of Matter	Ordinary matter exists in solid, liquid, and gas phases. A phase diagram shows the conditions under which two phases can coexist in phase equilibrium. All three phases can coexist at the triple point.

Key Concept 1: The ideal-gas equation, Eq. (18.3), relates the pressure, volume, absolute temperature, and number of moles for a quantity of an ideal gas.

Key Concept 2: For a fixed quantity of an ideal gas, the pressure p, volume V, and absolute temperature T may all change, but the quantity pV/T remains constant.

Key Concept 3: You can determine the mass of a quantity of an ideal gas from its pressure, volume, and absolute temperature, and the molar mass of the gas.

Key Concept 4: The ideal-gas equation can also be expressed as a relationship among the pressure, density, molar mass, and absolute temperature of an ideal gas [Eq. (12.4)].

Key Concept 5: To find the mass of a single molecule of a substance, divide the molar mass of that substance by Avogadro's number (the number of molecules in a mole).

Key Concept 6: The average translational kinetic energy of a molecule in an ideal gas at absolute temperature T is $\frac{3}{2}kT$, where k is the Boltzmann constant, no matter what the mass of the molecules. However, the root-mean-square speed v_{rms} of molecules (that is, the square root of the average value of v^2) in an ideal gas does depend on the mass per molecule m: $v_{rms} = \sqrt{3kT/m}$.

Key Concept 7: In general, for any collection of numbers, the root-mean-square value is *not* equal to the average value. For molecules in an ideal gas, the root-mean-square speed is always greater than the average speed.

Key Concept 8: The mean free path and mean free time are, respectively, the average distance and average time that a molecule in a gas travels between collisions with other molecules. Both depend on the temperature and pressure of the gas and the radius of a molecule. The mean free time also depends on the speed of the molecule; the mean free path does not.

Conceptual Questions

1: Don't hold your breath

Explain why scuba divers are taught not to hold their breath as they ascend to the surface from depths under the water.

IDENTIFY, SET UP, AND EXECUTE We know from fluid statics that pressure increases with depth in water. The ideal-gas equation states that pressure and volume are inversely proportional for a given temperature and quantity of gas. Ascending to the surface reduces the ambient pressure, causing an increase in volume. (We assume that the temperature is constant in the water.) By holding her breath, a scuba diver traps a quantity of air inside her lungs. As the pressure decreases upon her ascent, her lungs expand, possibly damaging some lung tissue. If the diver exhales during the ascent, the pressure cannot build to dangerous levels.

EVALUATE This problem combines our knowledge of fluid statics and our knowledge of ideal gases and helps illustrate the relation between pressure and volume. High-altitude weather balloons also expand as they rise, so they are partially filled at the ground to prevent the balloons from bursting as they ascend.

2: Atmosphere on the earth and moon

Why does the earth, but not the moon, have an atmosphere?

IDENTIFY, SET UP, AND EXECUTE The escape velocity of molecules on the earth is about 11 km/s, much higher than the average rms speed of molecules in the atmosphere. Without sufficient speed, the molecules remain near earth, thus creating an atmosphere.

The gravitational potential on the moon's surface is about 20 times weaker than that on the earth's surface, so the escape speed is about 20 times less, or 2400 m/s. This lesser escape speed greatly enhances the probability that molecules of whatever atmosphere the moon might have had have all escaped into space.

EVALUATE The problem illustrates how molecular motion can help explain common physics phenomena.

Problems

1: Changing volume in a diving bell

A diving bell (a circular cylinder 3.0 m high, open at the bottom) is lowered into a lake. By how much does the water rise as the bell is lowered 75 m? The surface temperature of the lake is $25°C$ and the temperature at the 75 m depth is $15°C$.

IDENTIFY We assume that the gas is ideal, so we use the ideal-gas equation to relate the surface values of pressure, temperature, and volume to the values at depth. The target value is the height of the water in the diving bell at depth.

SET UP Figure 1 shows a diagram of the situation. We'll use fluid statics to relate the pressure at the surface to the pressure at depth. These two relations will be combined to find the final height of water in the bell.

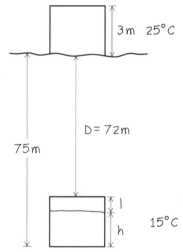

Figure 1 Problem 1 sketch.

EXECUTE The same amount of gas is trapped inside the bell both at the surface and at depth; therefore,

$$\frac{p_s V_s}{T_s} = \frac{p_D V_D}{T_D} = \text{constant},$$

where the subscript S indicates the value at the surface and subscript D indicates the value at depth. Substituting the known values gives

$$\frac{(1.01 \times 10^5 \text{ Pa})(A)(3.0 \text{ m})}{(298 \text{ K})} = \frac{p_D A l}{(288 \text{ K})},$$

where we replaced the volume of the cylinder with its area times its height. There two unknowns in this equation:

$$p_D l = (288 \text{ K}) \frac{(1.01 \times 10^5 \text{ Pa})(3.0 \text{ m})}{(298 \text{ K})} = 2.93 \times 10^5 \text{ Pa m}.$$

We can find the pressure at depth from fluid statics, using

$$p_D = p_S + \rho g (D + l).$$

Substituting the known values yields

$$p_D = (1.01 \times 10^5 \text{ Pa}) + (1 \times 10^3 \text{ kg/m}^3)(9.8 \text{ m/s}^2)(72 \text{ m} + l),$$

$$p_D = (8.07 \times 10^5 \text{ Pa}) + (7.06 \times 10^5 \text{ Pa/m})l.$$

This equation also has two unknowns. Combining the two equations to eliminate p_D results in

$$p_D = \frac{(2.93 \times 10^5 \text{ Pa m})}{l} = (8.07 \times 10^5 \text{ Pa}) + (7.06 \times 10^5 \text{ Pa/m})l.$$

Rearranging terms gives

$$(7.06/\text{m}^2)l^2 + (8.07/\text{m})l - 2.93 = 0.$$

This is a quadratic equation with solutions $l = 0.290, -1.43$ m. The negative root is nonphysical, so the correct l is 0.29 m. The water rises 3.0 m − 0.29 m, or 2.7 m, as the bell descends.

KEY CONCEPT 2 **EVALUATE** This problem illustrates how the increased pressure at depth reduces the volume of gas in the diving bell. If you consider the reverse process, you can see how the volume would increase as the bell rises to the surface, as we discussed in Conceptual Question 1. You can also try both situations by using a bucket of water. Submerge an inverted glass in a bucket of water, and see how the water level in the glass rises as the glass is lowered. Then use a hose to add air to the bottom of an inverted glass at the bottom of the bucket. As you raise the glass, you should see air leaving it.

2: Spacing between hydrogen molecules

Find the average spacing between H_2 molecules, assuming that the molecules are at the vertices of a fictitious cubic structure, in (a) gaseous H_2 at STP, (b) liquid H_2 at 20 K, where the density is 41,060 mol/m^3, and (c) solid H_2 at 4.2 K, where the molar volume is 22.91×10^{-6} m^3/mol.

IDENTIFY We'll use the definitions of *density, Avogadro's number,* and *molar mass* to find the average spacing.

SET UP We'll assign a spacing of a to the distance between H_2 molecules, giving a volume of a^3 per molecule. One mole of gas occupies 22.4 L at STP, and 1 L is 0.001 m^3.

EXECUTE (a) For the gaseous H_2, we multiply the volume of one molecule by N_A to get the volume at STP:

$$N_A a^3 = 22.4 \text{ L/mol} = 22.4 \times 10^{-3} \text{ m}^3/\text{mol}.$$

Solving for a gives

$$a = \sqrt[3]{\frac{22.4 \times 10^{-3} \text{ m}^3/\text{mol}}{N_A}} = \sqrt[3]{\frac{22.4 \times 10^{-3} \text{ m}^3/\text{mol}}{6.203 \times 10^{23}}} = 3.34 \times 10^{-9} \text{ m}.$$

(b) We are given that each cubic meter contains 41,060 moles of liquid H_2. In equation form, this is

$$1 \text{ m}^3 = 41,060 \, N_A a^3.$$

Solving for a results in

$$a = \sqrt[3]{\frac{1 \text{ m}^3}{41,060 \, N_A}} = \sqrt[3]{\frac{1 \text{ m}^3}{(41,060)(6.203 \times 10^{23})}} = 3.44 \times 10^{-10} \text{ m}.$$

(c) We are given the molar volume:

$$N_A a^3 = 22.91 \times 10^{-6} \text{ m}^3/\text{mol}.$$

Solving for a produces

$$a = \sqrt[3]{\frac{22.91 \times 10^{-6} \text{ m}^3/\text{mol}}{N_A}} = \sqrt[3]{\frac{22.91 \times 10^{-6} \text{ m}^3/\text{mol}}{6.203 \times 10^{23}}} = 3.37 \times 10^{-10} \text{ m}.$$

The average spacing is 3.34×10^{-9} m for the gaseous H_2, 3.44×10^{-10} m for the liquid H_2, and 3.37×10^{-10} m for the solid H_2.

EVALUATE We see that the average spacing for the liquid and solid H_2 is similar and about 10 times closer than the spacing for the gaseous H_2.

3: Spacing in a vacuum

A common type of laboratory vacuum pump produces an ultimate pressure of 10 microns. How many molecules (in moles) of gas are present in a volume of 0.15 m^3 reduced to that pressure at 300 K?

IDENTIFY We'll use the ideal-gas law to solve for the number of molecules in the volume—the target variable.

SET UP To use the ideal-gas law, we need to know the pressure in standard units. One micron is a measure of pressure in terms of the height of mercury. One atmosphere is 760 mm of mercury; one micron is one-thousandth of a millimeter of mercury.

EXECUTE First convert the pressure to atmospheres, using the information provided:

$$10 \text{ microns} = (10 \times 10^{-6})\left(\frac{1 \text{ atm}}{0.760 \text{ m}}\right) = 1.32 \times 10^{-5} \text{ atm}.$$

The ideal-gas law is

$$pV = nRT.$$

Solving for n and using the molar gas constant yields

$$n = \frac{pV}{RT} = \frac{(1.32 \times 10^{-5} \text{ atm})(1.5 \times 10^2 \text{ m}^3)}{(0.08206 \text{ L} \cdot \text{atm/mol} \cdot \text{K})(300 \text{ K})} = 8.04 \times 10^{-5} \text{ mol.}$$

There is 8.04×10^{-5} of a mole of molecules in the container after evacuating.

EVALUATE This may seem like a small number, but there are almost 5×10^{19} molecules remaining in the volume. One of the toughest challenges in physics research is creating ultrahigh vacuums to study the behavior of small numbers of particles. Physicists don't want collisions with remaining air molecules to interfere with the molecules whose behavior they are studying.

KEY CONCEPT 1   Did we assume that the gas was an ideal gas? Yes, we took the gas to be an ideal gas. This assumption is valid because the gas is at low pressure.

Practice Problem: How many moles are present if the temperature is 200 K? *Answer:* 1.2×10^{-4} moles

Extra Practice: How many moles are present if the temperature is 400 K? *Answer:* 6.0×10^{-5}

4: Mixing gases

A 1-liter flask at 293 K contains a mixture of 3 g of N_2 and 3 g of H_2 gas. Assuming that the gases behave as ideal gases, (a) calculate the partial pressure exerted by both gases and (b) calculate the rms speeds of the two gases.

IDENTIFY We'll use the ideal-gas law and kinetic theory to solve the problem. The target variables are the partial pressures and rms speeds of the two gases.

SET UP Partial pressure is the pressure exerted by each gas separately. We'll use the ideal-gas law to find the partial pressure of each gas. The rms speed may be calculated by an expression found in the text.

EXECUTE (a) The volume and temperature are given, so we need to convert the amount of each gas to moles. We have

$$n_{N_2} = \frac{3 \text{ g}}{28 \text{ g/mol}} = \frac{3}{28} \text{ mol,}$$

$$n_{H_2} = \frac{3 \text{ g}}{2 \text{ g/mol}} = \frac{3}{2} \text{ mol.}$$

The ideal-gas law is

$$pV = nRT.$$

Solving for p for each gas yields

$$p_{N_2} = \frac{nRT}{V} = \frac{(\frac{3}{28} \text{ mol})(8.314 \text{ J/mol} \cdot \text{K})(293 \text{ K})}{10^{-3} \text{ m}^3} = 2.61 \times 10^5 \text{ Pa} = 2.6 \text{ atm,}$$

$$p_{H_2} = \frac{nRT}{V} = \frac{(\frac{3}{2} \text{ mol})(8.314 \text{ J/mol} \cdot \text{K})(293 \text{ K})}{10^{-3} \text{ m}^3} = 3.65 \times 10^6 \text{ Pa} = 36 \text{ atm.}$$

(b) The rms speed is found from

$$v_{\text{rms}} = \sqrt{\frac{3RT}{M}}.$$

Solving for our gases gives

$$v_{N_2} = \sqrt{\frac{3RT}{M}} = \sqrt{\frac{3(8.314 \text{ J/mol} \cdot \text{K})(293 \text{ K})}{(28 \times 10^{-3} \text{ kg/mol})}} = 511 \text{ m/s},$$

$$v_{H_2} = \sqrt{\frac{3RT}{M}} = \sqrt{\frac{3(8.314 \text{ J/mol} \cdot \text{K})(293 \text{ K})}{(2 \times 10^{-3} \text{ kg/mol})}} = 1910 \text{ m/s}.$$

KEY CONCEPTS 1, 6

EVALUATE We see that both the pressure and the rms speed are higher for the H_2 gas, consistent with its smaller mass.

Practice Problem: What is the partial pressure if there are 6 g of hydrogen? *Answer:* 72 atm

Extra Practice: What is the rms speed if there are 6 g of hydrogen? *Answer:* 1910 m/s

5: Escape velocities of the sun and earth

Calculate the escape velocities at the surface of the sun and the earth, and determine whether any molecules have rms thermal speeds comparable with these escape velocities at 300 K.

IDENTIFY We'll use energy conservation to determine the escape velocities and kinetic theory to find the rms speeds. The target variables are the escape velocities at the surface of the sun and earth and the rms speeds for gases at 300 K.

SET UP For a molecule to escape from the sun or the earth, all of the molecule's gravitational potential energy must convert to kinetic energy. We use that relation to solve for the minimum escape velocity. The rms speed is given by an expression found in the text.

EXECUTE The escape velocity is found from energy conservation:

$$\frac{GMm}{R} = \frac{1}{2}mv^2.$$

Solving for the velocity gives

$$v = \sqrt{\frac{2GM}{R}}.$$

Solving for the escape velocities of the sun and the earth yields

$$v_{\text{sun}} = \sqrt{\frac{2GM_{\text{sun}}}{R_{\text{sun}}}} = \sqrt{\frac{2(6.67 \times 10^{-11} \text{ N} \cdot \text{m}^2/\text{kg}^2)(2 \times 10^{30} \text{ kg})}{7.0 \times 10^8 \text{ m}}} = 6.2 \times 10^5 \text{ m/s},$$

$$v_{\text{earth}} = \sqrt{\frac{2GM_{\text{earth}}}{R_{\text{earth}}}} = \sqrt{\frac{2(6.67 \times 10^{-11} \text{ N} \cdot \text{m}^2/\text{kg}^2)(6.0 \times 10^{24} \text{ kg})}{6.38 \times 10^6 \text{ m}}} = 1.12 \times 10^4 \text{ m/s}.$$

The rms speed is found from

$$v_{\text{rms}} = \sqrt{\frac{3RT}{M}}.$$

The hydrogen atom has the highest rms speed of any gaseous atom, so we calculate the speed for hydrogen. For a hydrogen atom at 300 K,

$$v_H = \sqrt{\frac{3RT}{M}} = \sqrt{\frac{3(8.314 \text{ J/mol} \cdot \text{K})(300 \text{ K})}{(1 \times 10^{-3} \text{ kg/mol})}} = 2700 \text{ m/s}.$$

The rms speed is 2700 m/s, less than the escape velocity of the sun or the earth.

KEY CONCEPT 6 **EVALUATE** We see that it is unlikely for any gas to have sufficient rms speed to escape the sun or the earth. The sun's average surface temperature is roughly 6000 K, corresponding to 12,000 m/s, a speed still too low for escape. Why does the moon not have an atmosphere?

Practice Problem: Find the escape velocity of a particle on the moon. *Answer:* 2370 m/s, enough for most gases to escape.

Extra Practice: What is the escape velocity of Jupiter? *Answer:* 6.02×10^4 m/s

Try It Yourself!

1: Helium gas

Helium gas is admitted to a volume of 200 cm³ at a temperature of 77 K until the pressure is equal to 1 atm. (a) If the temperature of the container is raised to 20°C, what will the pressure inside the container be? (b) If the system has a relief valve that will not permit the pressure to exceed 1 atm, what fraction of gas remains at 20°C?

Solution Checkpoints

IDENTIFY AND SET UP Use the ideal-gas law to solve the problem. Is that law a valid approximation?

EXECUTE (a) For the closed system, of the number of moles, pressure, temperature, and volume, which changes as the temperature rises? Only pressure and temperature change, so their ratio remains constant:

$$\frac{p_1}{T_1} = \frac{p_2}{T_2}.$$

The final pressure is 3.8 atm.

(b) For the valved system, what remains constant? Both the moles and temperature change, yielding

$$n_1 RT_1 = n_2 RT_2.$$

26.3% of the gas remains.

KEY CONCEPT 1 **EVALUATE** How could the system be changed so that both the pressure and the number of moles remain constant?

Practice Problem: What pressure exists if the final temperature is 37°C? *Answer:* 4.03 atm

Extra Practice: What temperature is required to obtain a final pressure of 6.0 atm? *Answer:* 189°C

2: Nitrogen gas

(a) Calculate the volume occupied by 1 mole of nitrogen gas at the critical temperature (162.2 K) and critical pressure (33.9×10^5 N/m^2). Express your answer as a ratio of the volume to the known critical volume (90.1×10^{-6} m^3). (b) Calculate the pressure at the critical volume and temperature.

Solution Checkpoints

IDENTIFY AND SET UP Treat the nitrogen as an ideal gas and use the ideal-gas law to solve the problem.

EXECUTE (a) The ideal-gas law can be used to find the volume:

$$V = \frac{nRT}{p}.$$

The volume is $3.44 V_C$.

(b) The critical pressure is $3.44 p_c$.

KEY CONCEPT 2 **EVALUATE** What do the critical pressure, temperature, and volume refer to? Do you think the results are valid?

3: RMS speed values

In a gas at 300 K that is a mixture of the diatomic molecules H$_2$ (2 g/mol) and D$_2$ (4 g/mol), find the rms speeds of the molecules

Solution Checkpoints

IDENTIFY AND SET UP Kinetic theory gives the rms speed of molecules.

EXECUTE The rms speed is found from

$$v = \sqrt{\frac{3RT}{M}}.$$

The rms speed is 1930 m/s for hydrogen and 1370 m/s for deuterium (D$_2$).

KEY CONCEPT 6 **EVALUATE** Could you devise a method for separating these two molecules by using the differences in their rms speeds? Explain.

Practice Problem: What are the partial pressures if there are 4 g of each gas in a 1-liter flask? *Answer:* 49 atm and 25 atm

Extra Practice: What are the partial pressures if we double the amount of D$_2$? *Answer:* 49 atm and 49 atm

Key Example Variation Problems

Solutions to these problems are in Chapter 18 of the Student's Solutions Manual.

Be sure to review EXAMPLES 18.1, 18.2, 18.3, and 18.4 (Section 18.1) before attempting these problems.

VP18.4.1 When the temperature is 30.0°C, the pressure of the air inside a bicycle tire of fixed volume 1.40×10^{-3} m^3 is 5.00×10^5 Pa. (a) What will be the pressure inside the tire when the temperature drops to 10.0°C? (b) How many moles of air are inside the tire?

VP18.4.2 When a research balloon is released at sea level, where the temperature is 15.0°C and the atmospheric pressure is 1.01×10^5 Pa, the helium in it has volume 13.0 m^3. (a) When the balloon reaches an altitude of 32.0 km, where the temperature is –44.5°C and the pressure is 868 Pa, what is the volume of the helium? (b) If this balloon is spherical, how many times larger is its radius at 32.0 km than at sea level?

VP18.4.3 The dwarf planet Pluto has a very thin atmosphere made up almost entirely of nitrogen (N_2, molar mass 2.8×10^{-2} kg/mol). At Pluto's surface the temperature is 42 K and the atmospheric pressure is 1.0 Pa. At the surface, (a) how many moles of gas are there per cubic meter of atmosphere, and (b) what is the density of the atmosphere in kg/m^3? (For comparison, the values are 42 mol/m^3 and 1.2 kg/m^3 at the earth's surface.)

VP18.4.4 When the pressure on n moles of helium gas is suddenly changed from an initial value of p_1 to a final value of p_2, the density of the gas changes from its initial value of ρ_1 to a final value of $\rho_2 = \rho_1(p_2/p_1)^{3/5}$. (a) If the initial absolute temperature of the gas is T_1, what is its final absolute temperature T_2 in terms of T_1, p_1, and p_2? (b) If the final pressure is 0.500 times the initial pressure, what are the ratio of the final density to the initial density and the ratio of the final temperature to the initial temperature? (c) Repeat part (b) if the final pressure is 2.00 times the initial pressure.

Be sure to review EXAMPLES 18.6 and 18.7 (Section 18.3) before attempting these problems.

VP18.7.1 At what temperature (in °C) is the rms speed of helium atoms (molar mass 4.00 g/mol) the same as the rms speed of nitrogen molecules (molar mass 28.0 g/mol) at 20.0°C? (Note that helium remains a gas at temperatures above –269°C.)

VP18.7.2 For the first 380,000 years after the Big Bang, the temperature of the matter in the universe was too high for nuclei and electrons to form atoms. The first hydrogen atoms (mass per atom 1.67×10^{-27} kg) did not form until the temperature had dropped to about 3000 K. (a) What was the average translational kinetic energy of hydrogen atoms when the temperature was 3.0×10^3 K ? (b) What was the rms speed of these atoms?

VP18.7.3 The air in a room with dimensions 5.0 m $\times$ 5.0 m $\times$ 2.4 m is at temperature 20.0°C and pressure 1.00 atm $= 1.01\times10^5$ Pa. (a) How many air molecules are in this room? (b) What is the total translational kinetic energy of these molecules? (c) How fast would a car of mass 1.5×10^3 kg have to move to have the same translational kinetic energy?

VP18.7.4 (a) What is the average value of the first 10 integers (1 through 10)? (b) What is the average value of the squares of the first 10 integers? (c) What is the rms value of the first 10 integers?

Be sure to review EXAMPLE 18.8 (Section 18.3) before attempting these problems.

VP18.8.1 At the surface of Mars the atmosphere has average pressure 6.0×10^2 Pa and average temperature $-63°C$. (a) What is the mean free path of atmospheric molecules (assumed to be spheres of radius 2.0×10^{-10} m) at the surface of Mars? (b) How many times greater is your answer in part (a) than the mean free path of atmospheric molecules at the earth's surface, where the average pressure is 1.01×10^5 Pa and the average temperature is $15°C$?

VP18.8.2 A cubical box 1.00 m on a side contains air at 20.0°C. (a) What would the pressure inside the box have to be in order for the mean free path of air molecules (assumed to be spheres of radius 2.0×10^{-10} m) to be 1.00 m, so that a typical molecule suffers no collisions with other molecules as it travels the width of the box? (b) How many moles of air would be inside the box at the pressure calculated in part (a)? (At 1 atm pressure, this box would contain 41.4 mol of air.)

VP18.8.3 A cylinder for storing helium (molar mass 4.00 g/mol) has an interior volume of $50.0 \text{ L} = 5.00\times10^{-25}$ m^3. The cylinder holds 4.00×10^2 mol of helium under pressure at 27.0°C. If you assume the helium atoms are spheres of radius 3.1×10^{-11} m, what are (a) the mean free path of a helium atom in the tank and (b) the mean free time for a helium atom moving at the rms speed?

VP18.8.4 A gas with molecules of radius r and mass per molecule m is at temperature T and pressure p. (a) Write an expression for the mean free time for a molecule moving at the rms speed for this gas. (b) Which single change would have the greatest effect on the mean free time: doubling the radius r, doubling the pressure p, or doubling the temperature T?

STUDENT'S SOLUTIONS MANUAL FOR THERMAL PROPERTIES OF MATTER

VP18.4.1. **IDENTIFY:** An ideal gas inside a tire undergoes a temperature change at constant volume, so the ideal gas law applies.
SET UP: Use $pV = nRT$, with $T_1 = 30.0°C = 303$ K and $T_2 = 10.0°C = 283$ K.
EXECUTE: **(a)** We want the new pressure p_2. At fixed volume, p/T is constant, so $\dfrac{p_1}{T_1} = \dfrac{p_2}{T_2}$ which

gives $p_2 = p_1 \dfrac{T_2}{T_1} = (5.00\times 10^5 \text{ Pa})\left(\dfrac{283 \text{ K}}{303 \text{ K}}\right) = 4.67\times 10^5$ Pa.

(b) We want the number n of moles in the tire. Solve $pV = nRT$ for n, giving $n = pV/RT$. Using $p = 5.00\times 10^5$ Pa, $T = 303$ K, and $V = 1.40\times 10^{-3}$ m^3 gives $n = 0.278$ mol.
EVALUATE: The temperature T must *always* be in Kelvins when using the ideal gas law.

VP18.4.2. **IDENTIFY:** As the balloon rises, the temperature and pressure decrease and the volume increases. The ideal gas law applies.
SET UP: Use $pV = nRT$. Since n remains constant, pV/T is constant.
At sea level: $T_1 = 15.0°C = 288$ K, $p_1 = 1.01\times 10^5$ Pa, $V_1 = 13.0$ m^3.
At 32.0 km: $T_2 = -44.5°C = 228.5$ K, $p_2 = 868$ Pa, $V_2 = $?
EXECUTE: **(a)** We want the volume at 32.0 km altitude. Since pV/T is constant, we have
$\dfrac{p_1 V_1}{T_1} = \dfrac{p_2 V_2}{T_2}$, so $V_2 = V_1 \dfrac{T_2}{T_1}\dfrac{p_1}{p_2}$. Putting in the numbers shown above, we get $V_2 = 1.20\times 10^3$ m^3.

(b) We want the ratio of the radii R_2/R_1 of the balloons. Using the volume of a sphere gives us

$\dfrac{V_2}{V_1} = \dfrac{\frac{4}{3}\pi R_2^3}{\frac{4}{3}\pi R_1^3} = \left(\dfrac{R_2}{R_1}\right)^3$, so $\dfrac{R_2}{R_1} = \left(\dfrac{V_2}{V_1}\right)^{1/3} = \left(\dfrac{1.20\times 10^3 \text{ m}^3}{13.0 \text{ m}^3}\right)^{1/3} = 4.52$, so the radius is 4.52 times

greater at 32.0 km than it is at sea level.
EVALUATE: The volume at 32.0 km is 1200/13 = 92 times as great as at sea level.

VP18.4.3. **IDENTIFY:** The pressure and temperature of Pluto's atmosphere are much lower than they are in Earth's atmosphere. The ideal gas law applies.
SET UP: Use $pV = nRT$, where $T = 42$ K and $p = 1.0$ Pa on Pluto. The molar mass of N_2 is 0.0028 kg/mol.
EXECUTE: **(a)** We want the molar density of Pluto's atmosphere in mol/m^3. Using $pV = nRT$ gives $n/V = p/RT$. Using $p = 1.0$ Pa and $T = 42$ K, we get $n/V = 2.9 \times 10^{-3}$ mol/m^3.
(b) We want the mass density of Pluto's atmosphere in kg/m^3. Convert the molar density to mass density using the fact that the molar density of N_2 is 0.028 kg/mol. This gives
$(2.9 \times 10^{-3} \text{ mol/m})(0.028 \text{ kg/mol}) = 8.0 \times 10^{-5}$ kg/m^3.
EVALUATE: Compare our results on Pluto to those on Earth.
$\dfrac{(n/V)_P}{(n/V)_E} = \dfrac{2.9 \times 10^{-3} \text{ mol/m}^3}{42 \text{ mol/m}^3} = 6.8 \times 10^{-5}$. $\dfrac{(m/V)_P}{(m/V)_E} = \dfrac{8.0 \times 10^{-5} \text{ kg/m}^3}{1.2 \text{ kg/m}^3} = 6.7 \times 10^{-5}$. The ratios are nearly the same, and both are very small due to Pluto's extremely low pressure.

VP18.4.4. **IDENTIFY:** We are investigating the effects of changes in an ideal gas using the ideal gas law.
SET UP: $pV = nRT$, and for constant n we have pV/T is constant. $\rho = m/V$ so $V = m/\rho$. We are given that $\rho_2 = \rho_1(p_2/p_1)^{3/5}$.
EXECUTE: **(a)** We want T_2. $\dfrac{p_1 V_1}{T_1} = \dfrac{p_2 V_2}{T_2}$, so $T_2 = T_1 \dfrac{V_2}{V_1} \dfrac{p_2}{p_1} = T_1 \dfrac{p_2}{p_1} \dfrac{m/\rho_1}{m/\rho_2} = T_1 \dfrac{p_2}{p_1} \dfrac{\rho_1}{\rho_2}$. Using $\rho_2 = \rho_1(p_2/p_1)^{3/5}$ gives $T_2 = T_1 \dfrac{p_2}{p_1} \dfrac{\rho_1}{\rho_1 \left(\dfrac{p_2}{p_1}\right)^{3/5}} = T_1 \left(\dfrac{p_2}{p_1}\right)\left(\dfrac{p_1}{p_2}\right)^{3/5} = T_1 \left(\dfrac{p_2}{p_1}\right)^{2/5}$.

(b) We want ρ_2/ρ_1 and T_2/T_1 when $p_2 = 0.500 p_1$. Using $\rho_2 = \rho_1(p_2/p_1)^{3/5}$ we get
$\dfrac{\rho_2}{\rho_1} = \left(\dfrac{p_2}{p_1}\right)^{3/5} = \left(\dfrac{0.500 p_1}{p_1}\right)^{3/5} = 0.660$.
Using our result from part (a) gives $\dfrac{T_2}{T_1} = \left(\dfrac{p_2}{p_1}\right)^{2/5} = \left(\dfrac{0.500 p_1}{p_1}\right)^{2/5} = 0.758$.

(c) We want ρ_2/ρ_1 and T_2/T_1 when $p_2 = 2.00 p_1$. Using $\rho_2 = \rho_1(p_2/p_1)^{3/5}$ we get
$\dfrac{\rho_2}{\rho_1} = \left(\dfrac{p_2}{p_1}\right)^{3/5} = \left(\dfrac{2.00 p_1}{p_1}\right)^{3/5} = 1.52$.
Using our result from part (a) gives $\dfrac{T_2}{T_1} = \left(\dfrac{p_2}{p_1}\right)^{2/5} = \left(\dfrac{2.00 p_1}{p_1}\right)^{2/5} = 1.32$.

EVALUATE: In part (b) our result says that decreasing p decreases ρ, and in part (c) it says that increasing p increases ρ. These are reasonable results.

VP18.7.1. **IDENTIFY:** We are comparing molecular speeds of different molecules.
SET UP and EXECUTE: Use $v_{rms} = \sqrt{\dfrac{3RT}{M}}$.

N_2 molecule at 20.0°C: $v_{rms} = \sqrt{\dfrac{3RT}{M}} = \sqrt{\dfrac{3R(293 \text{ K})}{28.0 \text{ g/mol}}}$.

He atom with the same v_{rms}: $v_{rms} = \sqrt{\dfrac{3RT}{4.00 \text{ g/mol}}}$

Equate the speeds and solve for T: $\sqrt{\dfrac{3RT}{4.00 \text{ g/mol}}} = \sqrt{\dfrac{3R(293 \text{ K})}{28.0 \text{ g/mol}}}$, $T = 41.86 \text{K} = -231°\text{C}$.

EVALUATE: It was not necessary to use R or convert g to kg since they divided out. It is best to save numerical calculations until the end to minimize tedious work.

VP18.7.2. IDENTIFY: We are looking at atomic kinetic energy and speed.

SET UP: $v_{rms} = \sqrt{\dfrac{3kT}{m}}$, $K_{tr} = \dfrac{3}{2}kT$. We want the average kinetic energy of an atom and the rms speed.

EXECUTE: (a) $K_{av} = \dfrac{3}{2}kT = \dfrac{3}{2}(1.381 \times 10^{-23} \text{ J/K})(3000 \text{ K}) = 6.2 \times 10^{-20}$ J.

(b) Use $v_{rms} = \sqrt{\dfrac{3kT}{m}}$ with $T = 3000$ K and $m = 1.67 \times 10^{-27}$ kg, giving $v_{rms} = 8600$ m/s.

EVALUATE: The speed in part (b) is about 19,000 mph!

VP18.7.3. IDENTIFY: We want to find out how many molecules are in a typical room and how much kinetic energy they have.

SET UP: Use $pV = nRT$, $K_{tr} = \dfrac{3}{2}kT$, $K_{tot} = NK_{tr}$, and $N = nN_A$. The target variables are N and K_{tr}.

EXECUTE: (a) Solve $pV = nRT$ to get $n = pV/RT$. Using $V = (5.0 \text{ m})(5.0 \text{ m})(2.4 \text{ m}) = 612 \text{ m}^3$, $p = 1.01 \times 10^5$ Pa, and $T = 293$ K gives $n = 2488$ mol. The number of molecules in 2488 mol is $N = nN_A = (2488 \text{ mol})(6.022 \times 10^{23} \text{ molecules/mol}) = 1.5 \times 10^{27}$ molecules.

(b) $K_{tot} = NK_{tr} = N\dfrac{3}{2}kT$. Using $T = 293$ K and $N = 1.5 \times 10^{27}$ molecules gives $K_{tot} = 9.1 \times 10^6$ J.

(c) We want the speed of the car so its kinetic energy would be 9.1×10^6 J. $K_{car} = K_{tot}$ gives $\dfrac{1}{2}mv^2 = K_{tot}$, so $v = \sqrt{\dfrac{2K_{tot}}{m}} = \sqrt{\dfrac{2(9.1 \times 10^6 \text{ J})}{1500 \text{ kg}}} = 110$ m/s.

EVALUATE: The speed in part (c) is about 250 mph or 400 km/h.

VP18.7.4. IDENTIFY: We want to compare the average, average of squares, and rms for the first 10 integers.

SET UP and EXECUTE: (a) The average is $n_{av} = \dfrac{1 + 2 + 3 + \cdots + 10}{10} = \dfrac{55}{10} = 5.5$.

(b) The average of the squares is $(n^2)_{av} = \dfrac{1^2 + 2^2 + 3^2 + \cdots + 10^2}{10} = \dfrac{385}{10} = 38.5$.

(c) The rms value is $n_{rms} = \sqrt{(n^2)_{av}} = \sqrt{38.5} = 6.20$.

EVALUATE: Note that the rms value is *not* the same as the average value.

VP18.8.1. IDENTIFY: We are comparing the mean free path for molecules on Mars with those on Earth.

SET UP: We use $\lambda = \dfrac{kT}{4\pi\sqrt{2}r^2 p}$, with $T_M = -63°\text{C} = 210$ K and $T_E = 15°\text{C} = 288$ K, $p_M = 6.0 \times 10^2$ Pa. The molecular radius is 2.0×10^{10} m.

EXECUTE: (a) We want the mean free path on Mars. Using $\lambda = \dfrac{kT}{4\pi\sqrt{2}\,r^2 p}$ with values given above we get $\lambda_M = 6.8\times10^{-6}$ m.

(b) We want to compare the mean free path on Mars to that on Earth by finding λ_M/λ_E. Taking the ratio gives $\dfrac{\lambda_M}{\lambda_E} = \dfrac{T_M/p_M}{T_E/p_E} = \left(\dfrac{T_M}{T_E}\right)\left(\dfrac{p_E}{p_M}\right) = \left(\dfrac{210\text{ K}}{288\text{ K}}\right)\left(\dfrac{1.01\times10^5\text{ Pa}}{600\text{ Pa}}\right) = 120$. The mean free path on Mars is 120 times greater than it is on Earth.

EVALUATE: The low pressure on Mars means that molecules are far apart compared to those on Earth, so they can travel much farther on Mars before running into each other.

VP18.8.2. **IDENTIFY:** We are investigating the mean free path of molecules.

SET UP: Use $\lambda = \dfrac{kT}{4\pi\sqrt{2}\,r^2 p}$ and $pV = nRT$.

EXECUTE: (a) We want to find the pressure so that the mean free path is 1.00 m. Solve $\lambda = \dfrac{kT}{4\pi\sqrt{2}\,r^2 p}$ for p, giving $p = \dfrac{kT}{4\pi\sqrt{2}\,r^2 \lambda}$. Using $T = 293$ K, $r = 2.0\times10^{-10}$ m, and $\lambda = 1.00$ m, we get $p = 5.7\times10^{-3}$ Pa.

(b) Our target variable is the number of moles n inside the box. Solving $pV = nRT$ for n gives $n = pV/RT$. Using the pressure from part (a) and the same values for the other variables gives $n = 2.3\times10^{-6}$ mol.

EVALUATE: Even at this low pressure, the number of air molecules in the box is $N = nN_A = (2.3\times10^{-6}\text{ mol})(6.022\times10^{23}\text{ molecules/mol}) = 1.4\times10^{18}$ molecules.

VP18.8.3. **IDENTIFY:** We want to find the mean free path and mean free time for helium in a cylindrical tank.

SET UP: We use $\lambda = \dfrac{kT}{4\pi\sqrt{2}\,r^2 p}$, $v_{\rm rms} = \sqrt{\dfrac{3RT}{M}}$, and $pV = nRT$. $v_{\rm rms} t_{\rm mean} = \lambda$. We know that $T = 27.0°\text{C} = 300.0$ K, $V = 5.00\times10^{-2}$ m³, $n = 400$ mol, and $r = 3.1\times10^{-11}$ m, so $\lambda = 1.2\times10^{-8}$ m.

EXECUTE: (a) We want the mean free path. First use $pV = nRT$ to find p, so $p = nRT/V$. Using this result gives $\lambda = \dfrac{kT}{4\pi\sqrt{2}\,r^2 p} = \dfrac{kT}{4\pi\sqrt{2}\,r^2\left(\dfrac{nRT}{V}\right)} = \dfrac{kV}{4\pi\sqrt{2}\,r^2 nR}$. Using the given values for V, n, and r gives $\lambda = 1.2\times10^{-8}$ m.

(b) We want the mean free time. $v_{\rm rms} = \sqrt{\dfrac{3RT}{M}}$ and $t_{\rm mean} = \lambda/v_{\rm rms}$. Combining these equations gives $t_{\rm mean} = \dfrac{\lambda}{v_{\rm rms}} = \dfrac{\lambda}{\sqrt{\dfrac{3RT}{M}}} = \lambda\sqrt{\dfrac{M}{3RT}}$. Using $M = 4.00\times10^{-3}$ kg/mol, $\lambda = 1.2\times10^{-8}$ m, and $T = 300$ K, we get $t_{\rm mean} = 8.9\times10^{-12}$ s.

EVALUATE: The number of collisions per second would be $1/t_{\rm mean} = 1.1\times10^{11}$ collision/s.

VP18.8.4. **IDENTIFY:** We want to find the mean free time for gas molecules.

SET UP: We use $\lambda = \dfrac{kT}{4\pi\sqrt{2}\,r^2 p}$, $v_{\rm rms} = \sqrt{\dfrac{3kT}{m}}$, $v_{\rm rms} t_{\rm mean} = \lambda$.

EXECUTE: (a) Combining the above equations gives $t_{mean} = \dfrac{\lambda}{v_{rms}} = \dfrac{\dfrac{kT}{4\pi\sqrt{2}r^2 p}}{\sqrt{\dfrac{3kT}{m}}}$, which simplifies to $t_{mean} = \dfrac{1}{4\pi r^2 p}\sqrt{\dfrac{mkT}{6}}$.

(b) From part (a), we see that $t_{mean} \propto \dfrac{\sqrt{T}}{r^2 p}$. From this result we can see the following.

Doubling p would cut t_{mean} in half.
Doubling T would increase t_{mean} by a factor of $\sqrt{2}$.
Doubling r would decrease t_{mean} by a factor of $\dfrac{1}{2^2} = \dfrac{1}{4}$.

Therefore doubling r would have the greatest effect on the mean free time.

EVALUATE: If we doubled all of these quantities at once, the effect would reduce t_{mean} by a factor of $\dfrac{\sqrt{2}}{8} \approx 0.177$.

18.3. IDENTIFY: $pV = nRT$.

SET UP: T is constant.

EXECUTE: nRT is constant so $p_1 V_1 = p_2 V_2$.

$p_2 = p_1 \left(\dfrac{V_1}{V_2}\right) = (0.355 \text{ atm})\left(\dfrac{0.110 \text{ m}^3}{0.390 \text{ m}^3}\right) = 0.100 \text{ atm}.$

EVALUATE: For T constant, p decreases as V increases.

18.7. IDENTIFY: We are asked to compare two states. Use the ideal gas law to obtain T_2 in terms of T_1 and ratios of pressures and volumes of the gas in the two states.

SET UP: $pV = nRT$ and n, R constant implies $pV/T = nR = $ constant and $p_1 V_1 / T_1 = p_2 V_2 / T_2$.

EXECUTE: $T_1 = (27 + 273)\text{K} = 300 \text{ K}$

$p_1 = 1.01 \times 10^5 \text{ Pa}$

$p_2 = 2.72 \times 10^6 \text{ Pa} + 1.01 \times 10^5 \text{ Pa} = 2.82 \times 10^6 \text{ Pa}$ (in the ideal gas equation the pressures must be absolute, not gauge, pressures)

$T_2 = T_1 \left(\dfrac{p_2}{p_1}\right)\left(\dfrac{V_2}{V_1}\right) = 300 \text{ K} \left(\dfrac{2.82 \times 10^6 \text{ Pa}}{1.01 \times 10^5 \text{ Pa}}\right)\left(\dfrac{46.2 \text{ cm}^3}{499 \text{ cm}^3}\right) = 776 \text{ K}$

$T_2 = (776 - 273)°\text{C} = 503°\text{C}.$

EVALUATE: The units cancel in the V_2/V_1 volume ratio, so it was not necessary to convert the volumes in cm^3 to m^3. It was essential, however, to use T in kelvins.

18.9. IDENTIFY: $pV = nRT$.

SET UP: $T_1 = 300 \text{ K}$, $T_2 = 430 \text{ K}$.

EXECUTE: (a) n, R are constant so $\dfrac{pV}{T} = nR = $ constant. $\dfrac{p_1 V_1}{T_1} = \dfrac{p_2 V_2}{T_2}$.

$$p_2 = p_1 \left(\frac{V_1}{V_2}\right)\left(\frac{T_2}{T_1}\right) = (7.50\times 10^3 \text{ Pa})\left(\frac{0.750 \text{ m}^3}{0.410 \text{ m}^3}\right)\left(\frac{430 \text{ K}}{300 \text{ K}}\right) = 1.97\times 10^4 \text{ Pa}.$$

EVALUATE: Since the temperature increased while the volume decreased, the pressure must have increased. In $pV = nRT$, T must be in kelvins, even if we use a ratio of temperatures.

18.15. IDENTIFY: We are looking at the gauge pressure in a gas, so we apply the ideal gas law.

SET UP: $pV = nRT$, $p_{tot} = p_{atm} + p_{gauge}$. The pressure we need to use is the *total* pressure since that is the pressure the gas exerts in the container, so $p_{tot} = 0.876$ atm + 1.000 atm = 1.876 atm.

EXECUTE: Using the numbers given in the problem, $n = pV/RT$. Putting in the numbers gives

$$n = \frac{(1.876 \text{ atm})(5.43 \text{ L})}{(0.08206 \text{ L}\cdot\text{atm/mol}\cdot\text{K})(295.2 \text{ K})} = 0.421 \text{ mol}.$$

EVALUATE: We use the gauge pressure because that is the actual pressure in the tank. A pressure gauge reads the gauge pressure, but that is not the tank pressure since it is only the pressure above one atmosphere.

18.17. IDENTIFY: We know the volume, pressure, and temperature of the gas and want to find its mass and density.

SET UP: $V = 3.00\times 10^{-3}$ m^3. $T = 295$ K. $p = 2.03\times 10^{-8}$ Pa. The ideal gas law, $pV = nRT$, applies.

EXECUTE: **(a)** $pV = nRT$ gives

$$n = \frac{pV}{RT} = \frac{(2.03\times 10^{-8} \text{ Pa})(3.00\times 10^{-3} \text{ m}^3)}{(8.315 \text{ J/mol}\cdot\text{K})(295 \text{ K})} = 2.48\times 10^{-14} \text{ mol}.$$ The mass of this amount of gas is

$m = nM = (2.48\times 10^{-14} \text{ mol})(28.0\times 10^{-3} \text{ kg/mol}) = 6.95\times 10^{-16}$ kg.

(b) $\rho = \dfrac{m}{V} = \dfrac{6.95\times 10^{-16} \text{ kg}}{3.00\times 10^{-3} \text{ m}^3} = 2.32\times 10^{-13}$ kg/m^3.

EVALUATE: The density at this level of vacuum is 13 orders of magnitude less than the density of air at STP, which is 1.20 kg/m^3.

18.21. IDENTIFY: Use $pV = nRT$ to calculate the number of moles and then the number of molecules would be $N = nN_A$.

SET UP: 1 atm $= 1.013\times 10^5$ Pa. 1.00 cm$^3 = 1.00\times 10^{-6}$ m^3. $N_A = 6.022\times 10^{23}$ molecules/mol.

EXECUTE: **(a)**

$$n = \frac{pV}{RT} = \frac{(9.00\times 10^{-14} \text{ atm})(1.013\times 10^5 \text{ Pa/atm})(1.00\times 10^{-6} \text{ m}^3)}{(8.314 \text{ J/mol}\cdot\text{K})(300.0 \text{ K})} = 3.655\times 10^{-18} \text{ mol}.$$

$N = nN_A = (3.655\times 10^{-18} \text{ mol})(6.022\times 10^{23}$ molecules/mol$) = 2.20\times 10^6$ molecules.

(b) $N = \dfrac{pVN_A}{RT}$ so $\dfrac{N}{p} = \dfrac{VN_A}{RT} =$ constant and $\dfrac{N_1}{p_1} = \dfrac{N_2}{p_2}$.

$N_2 = N_1\left(\dfrac{p_2}{p_1}\right) = (2.20\times 10^6 \text{ molecules})\left(\dfrac{1.00 \text{ atm}}{9.00\times 10^{-14} \text{ atm}}\right) = 2.44\times 10^{19}$ molecules.

EVALUATE: The number of molecules in a given volume is directly proportional to the pressure. Even at the very low pressure in part (a) the number of molecules in 1.00 cm^3 is very large.

18.23. IDENTIFY: Use $pV = nRT$ and $n = \dfrac{N}{N_A}$ with $N = 1$ to calculate the volume V occupied by 1 molecule. The length l of the side of the cube with volume V is given by $V = l^3$.

SET UP: $T = 27°C = 300$ K. $p = 1.00$ atm $= 1.013 \times 10^5$ Pa. $R = 8.314$ J/mol·K. $N_A = 6.022 \times 10^{23}$ molecules/mol.

The diameter of a typical molecule is about 10^{-10} m. 0.3 nm $= 0.3 \times 10^{-9}$ m.

EXECUTE: (a) $pV = nRT$ and $n = \dfrac{N}{N_A}$ gives

$$V = \dfrac{NRT}{N_A p} = \dfrac{(1.00)(8.314 \text{ J/mol·K})(300 \text{ K})}{(6.022 \times 10^{23} \text{ molecules/mol})(1.013 \times 10^5 \text{ Pa})} = 4.09 \times 10^{-26} \text{ m}^3.$$

$l = V^{1/3} = 3.45 \times 10^{-9}$ m.

(b) The distance in part (a) is about 10 times the diameter of a typical molecule.

(c) The spacing is about 10 times the spacing of atoms in solids.

EVALUATE: There is space between molecules in a gas whereas in a solid the atoms are closely packed together.

18.25. IDENTIFY: The ideal gas law applies. The translational kinetic energy of a gas depends on its absolute temperature.

SET UP: $pV = nRT$, $K_{\text{tr}} = 3/2 \, nRT$, $K = \tfrac{1}{2} mv^2$.

EXECUTE: (a) From $pV = nRT$, we have $n = pV/RT$. Putting this into $K_{\text{tr}} = 3/2 \, nRT$, we have $K_{\text{tr}} = 3/2 \, (pV/RT)(RT) = 3/2 \, pV = (3/2)(1.013 \times 10^5 \text{ Pa})(8.00 \text{ m})(12.00 \text{ m})(4.00 \text{ m}) = 5.83 \times 10^7$ J.

(b) $K = \tfrac{1}{2} mv^2$: $\tfrac{1}{2}(2000 \text{ kg})v^2 = 5.83 \times 10^7$ J, gives $v = 242$ m/s.

EVALUATE: No automobile can travel this fast! Obviously the molecules in the room have a great deal of kinetic energy because there are so many of them.

18.27. IDENTIFY: We make several measurements of the pressure and temperature of a gas. Using a graph of the pressure versus the temperature, we want to determine the number N of gas molecules in the container.

SET UP: We use the ideal gas law to find a relationship between p and T_C so we can interpret the graph. The gas volume is $V = 80.0$ cm^3 = 0.0800 L. We use $pV = nRT$ and $N = nN_A$. The target variable is N.

EXECUTE: $pV = nRT = nR(T_C + 273)$, so $p = \dfrac{nR(T_C + 273)}{V} = \dfrac{nR}{V} T_C + \dfrac{273nR}{V}$. Therefore a graph of p versus T_C should be a straight line having slope equal to nR/V. Putting this in terms of N gives slope $= \dfrac{nR}{V} = \dfrac{(N/N_A)R}{V}$, so $N = \dfrac{N_A V(\text{slope})}{R}$. Putting in the appropriate numbers gives

$$N = \dfrac{(6.022 \times 10^{23} \text{ molec/mol})(0.0800 \text{ L})(1.10 \text{ atm/C°})}{0.08206 \text{ L·atm/mol·K}} = 6.46 \times 10^{23} \text{ molecules}.$$

EVALUATE: The number of moles is $\dfrac{6.46 \times 10^{23}}{6.022 \times 10^{23}} = 1.07$ mol.

18.31. IDENTIFY and SET UP: Apply the analysis of Section 18.3.

EXECUTE: **(a)** $\frac{1}{2}m(v^2)_{av} = \frac{3}{2}kT = \frac{3}{2}(1.38\times10^{-23}$ J/molecule·K$)(300$ K$) = 6.21\times10^{-21}$ J.

(b) We need the mass m of one molecule:

$$m = \frac{M}{N_A} = \frac{32.0\times10^{-3} \text{ kg/mol}}{6.022\times10^{23} \text{ molecules/mol}} = 5.314\times10^{-26} \text{ kg/molecule.}$$

Then $\frac{1}{2}m(v^2)_{av} = 6.21\times10^{-21}$ J (from part (a)) gives

$$(v^2)_{av} = \frac{2(6.21\times10^{-21} \text{ J})}{m} = \frac{2(6.21\times10^{-21} \text{ J})}{5.314\times10^{-26} \text{ kg}} = 2.34\times10^5 \text{ m}^2/\text{s}^2.$$

(c) $v_{rms} = \sqrt{(v^2)_{rms}} = \sqrt{2.34\times10^4 \text{ m}^2/\text{s}^2} = 484$ m/s.

(d) $p = mv_{rms} = (5.314\times10^{-26}$ kg$)(484$ m/s$) = 2.57\times10^{-23}$ kg·m/s.

(e) Time between collisions with one wall is $t = \dfrac{0.20 \text{ m}}{v_{rms}} = \dfrac{0.20 \text{ m}}{484 \text{ m/s}} = 4.13\times10^{-4}$ s.

In a collision $\vec{v}$ changes direction, so $\Delta p = 2mv_{rms} = 2(2.57\times10^{-23}$ kg·m/s$) = 5.14\times10^{-23}$ kg·m/s

$F = \dfrac{dp}{dt}$ so $F_{av} = \dfrac{\Delta p}{\Delta t} = \dfrac{5.14\times10^{-23} \text{ kg·m/s}}{4.13\times10^{-4} \text{ s}} = 1.24\times10^{-19}$ N.

(f) pressure $= F/A = 1.24\times10^{-19}$ N/(0.10 m)$^2 = 1.24\times10^{-17}$ Pa (due to one molecule).

(g) pressure $= 1$ atm $= 1.013\times10^5$ Pa.

Number of molecules needed is 1.013×10^5 Pa/$(1.24\times10^{-17}$ Pa/molecule$) = 8.17\times10^{21}$ molecules.

(h) $pV = NkT$ (Eq. 18.18), so

$$N = \frac{pV}{kT} = \frac{(1.013\times10^5 \text{ Pa})(0.10 \text{ m})^3}{(1.381\times10^{-23} \text{ J/molecule·K})(300 \text{ K})} = 2.45\times10^{22} \text{ molecules.}$$

(i) From the factor of $\frac{1}{3}$ in $(v_x^2)_{av} = \frac{1}{3}(v^2)_{av}$.

EVALUATE: This exercise shows that the pressure exerted by a gas arises from collisions of the molecules of the gas with the walls.

18.33. IDENTIFY and SET UP: Use equal v_{rms} to relate T and M for the two gases. $v_{rms} = \sqrt{3RT/M}$, so $v_{rms}^2/3R = T/M$, where T must be in kelvins. Same v_{rms} so same T/M for the two gases and $T_{N_2}/M_{N_2} = T_{H_2}/M_{H_2}$.

EXECUTE: $T_{N_2} = T_{H_2}\left(\dfrac{M_{N_2}}{M_{H_2}}\right) = [(20+273)\text{K}]\left(\dfrac{28.014 \text{ g/mol}}{2.016 \text{ g/mol}}\right) = 4.071\times10^3$ K

$T_{N_2} = (4071-273)°C = 3800°C$.

EVALUATE: A N_2 molecule has more mass so N_2 gas must be at a higher temperature to have the same v_{rms}.

18.35. **IDENTIFY:** We add heat energy to a gas in a sealed rigid container and want to find the new rms speed of the molecules due to this added energy.

SET UP: We know that at constant volume $Q = nC_V \Delta T$ where $C_V = \frac{3}{2}R$ for a monatomic gas, and that $v_{rms} = \sqrt{\frac{3RT}{M}}$. First use $v_{rms} = \sqrt{\frac{3RT}{M}}$ to get the original temperature. Then use $Q = nC_V \Delta T$ to get ΔT, and finally use $v_{rms} = \sqrt{\frac{3RT}{M}}$ again to get the new v_{rms}.

EXECUTE: Get the original temperature T_1. Square $v_{rms} = \sqrt{\frac{3RT}{M}}$ to get $T_1 = \frac{Mv_{rms}^2}{3R}$. Using the numbers gives $T_1 = \frac{(0.00400 \text{ kg/mol})(900 \text{ m/s})^2}{3(8.314 \text{ J/mol} \cdot \text{K})} = 129.9$ K. Now get the new temperature T_2. Using $Q = nC_V \Delta T = n\left(\frac{3}{2}R\right)\Delta T$ gives 2400 J = (3.00 mol)(3/2)(8.314 J/mol·K) ΔT, so we have ΔT = 64.15 K. Thus $T_2 = T_1 + \Delta T$ = 129.9 K + 64.15 K = 194.05 K. Finally use $v_{rms} = \sqrt{\frac{3RT}{M}}$ to calculate the new rms speed using $T_2 = 194.05$ K and $M = 0.00400$ kg/mol. The result is v_{rms} = 1100 m/s.

EVALUATE: Since $v_{rms} \propto \sqrt{T}$, when our fractional increase in T is $\frac{194.05}{129.9} = 1.494$, the fractional increase in v_{rms} is $\sqrt{1.494} = 1222$. Thus we would expect that $v_2 = 1.222 v_1 = (1.222)(900$ m/s$) = $ 1100 m/s, which is exactly what we found.

18.37. **IDENTIFY:** Use $dQ = nC_V dT$ applied to a finite temperature change.

SET UP: $C_V = 5R/2$ for a diatomic ideal gas and $C_V = 3R/2$ for a monatomic ideal gas.

EXECUTE: **(a)** $Q = nC_V \Delta T = n\left(\frac{5}{2}R\right) \Delta T$. $Q = (1.80$ mol$)\left(\frac{5}{2}\right)(8.314$ J/mol·K$)(50.0$ K$) = 1870$ J.

(b) $Q = nC_V \Delta T = n\left(\frac{3}{2}R\right) \Delta T$. $Q = (1.80$ mol$)\left(\frac{3}{2}\right)(8.314$ J/mol·K$)(50.0$ K$) = 1120$ J.

EVALUATE: More heat is required for the diatomic gas; not all the heat that goes into the gas appears as translational kinetic energy, some goes into energy of the internal motion of the molecules (rotations).

18.41. **IDENTIFY:** Apply $v_{mp} = \sqrt{2kT/m}$, $v_{av} = \sqrt{8kT/\pi m}$, and $v_{rms} = \sqrt{3kT/m}$.

SET UP: Note that $\frac{k}{m} = \frac{R/N_A}{M/N_A} = \frac{R}{M}$. $M = 44.0 \times 10^{-3}$ kg/mol.

EXECUTE: **(a)** $v_{mp} = \sqrt{2(8.3145 \text{ J/mol} \cdot \text{K})(300 \text{ K})/(44.0 \times 10^{-3} \text{ kg/mol})} = 3.37 \times 10^2$ m/s.

(b) $v_{av} = \sqrt{8(8.3145 \text{ J/mol} \cdot \text{K})(300 \text{ K})/(\pi(44.0 \times 10^{-3} \text{ kg/mol}))} = 3.80 \times 10^2$ m/s.

(c) $v_{rms} = \sqrt{3(8.3145 \text{ J/mol} \cdot \text{K})(300 \text{ K})/(44.0 \times 10^{-3} \text{ kg/mol})} = 4.12 \times 10^2$ m/s.

EVALUATE: The average speed is greater than the most probable speed and the rms speed is greater than the average speed.

18.45. IDENTIFY: Refer to the phase diagram in Figure 18.24 in the textbook.

SET UP: For water the triple-point pressure is 610 Pa and the critical-point pressure is 2.212×10^7 Pa.

EXECUTE: (a) To observe a solid to liquid (melting) phase transition the pressure must be greater than the triple-point pressure, so $p_1 = 610$ Pa. For $p < p_1$ the solid to vapor (sublimation) phase transition is observed.

(b) No liquid to vapor (boiling) phase transition is observed if the pressure is greater than the critical-point pressure. $p_2 = 2.212 \times 10^7$ Pa. For $p_1 < p < p_2$ the sequence of phase transitions is solid to liquid and then liquid to vapor.

EVALUATE: Normal atmospheric pressure is approximately 1.0×10^5 Pa, so the solid to liquid to vapor sequence of phase transitions is normally observed when the material is water.

18.49. IDENTIFY: We can model the atmosphere as a fluid of constant density, so the pressure depends on the depth in the fluid, as we saw in Section 12.2.

SET UP: The pressure difference between two points in a fluid is $\Delta p = \rho g h$, where h is the difference in height of two points.

EXECUTE: (a) $\Delta p = \rho g h = (1.2 \text{ kg/m}^3)(9.80 \text{ m/s}^2)(1000 \text{ m}) = 1.18 \times 10^4$ Pa.

(b) At the bottom of the mountain, $p = 1.013 \times 10^5$ Pa. At the top, $p = 8.95 \times 10^4$ Pa.

$pV = nRT =$ constant so $p_b V_b = p_t V_t$ and $V_t = V_b \left(\dfrac{p_b}{p_t} \right) = (0.50 \text{ L}) \left(\dfrac{1.013 \times 10^5 \text{ Pa}}{8.95 \times 10^4 \text{ Pa}} \right) = 0.566$ L.

EVALUATE: The pressure variation with altitude is affected by changes in air density and temperature and we have neglected those effects. The pressure decreases with altitude and the volume increases. You may have noticed this effect: bags of potato chips "puff up" when taken to the top of a mountain.

18.51. IDENTIFY: The buoyant force on the balloon must be equal to the weight of the load plus the weight of the gas.

SET UP: The buoyant force is $F_B = \rho_{air} V g$. A lift of 290 kg means $\dfrac{F_B}{g} - m_{hot} = 290$ kg, where m_{hot} is the mass of hot air in the balloon. $m = \rho V$.

EXECUTE: $m_{hot} = \rho_{hot} V$. $\dfrac{F_B}{g} - m_{hot} = 290$ kg gives $(\rho_{air} - \rho_{hot})V = 290$ kg.

Solving for ρ_{hot} gives $\rho_{hot} = \rho_{air} - \dfrac{290 \text{ kg}}{V} = 1.23 \text{ kg/m}^3 - \dfrac{290 \text{ kg}}{500.0 \text{ m}^3} = 0.65 \text{ kg/m}^3$. $\rho_{hot} = \dfrac{pM}{RT_{hot}}$.

$\rho_{air} = \dfrac{pM}{RT_{air}}$. $\rho_{hot} T_{hot} = \rho_{air} T_{air}$ so

$T_{hot} = T_{air} \left(\dfrac{\rho_{air}}{\rho_{hot}} \right) = (288 \text{ K}) \left(\dfrac{1.23 \text{ kg/m}^3}{0.65 \text{ kg/m}^3} \right) = 545$ K $= 272°C$.

EVALUATE: This temperature is well above normal air temperatures, so the air in the balloon would need considerable heating.

18.53. IDENTIFY: We are asked to compare two states. Use the ideal-gas law to obtain m_2 in terms of m_1 and the ratio of pressures in the two states. Apply $pV = \dfrac{m_{\text{total}}}{M} RT$ to the initial state to calculate m_1.

SET UP: $pV = nRT$ can be written $pV = (m/M)RT$
T, V, M, R are all constant, so $p/m = RT/MV = $ constant.
So $p_1/m_1 = p_2/m_2$, where m is the mass of the gas in the tank.

EXECUTE: $p_1 = 1.30 \times 10^6$ Pa $+ 1.01 \times 10^5$ Pa $= 1.40 \times 10^6$ Pa
$p_2 = 3.40 \times 10^5$ Pa $+ 1.01 \times 10^5$ Pa $= 4.41 \times 10^5$ Pa
$m_1 = p_1 VM/RT$; $V = hA = h\pi r^2 = (1.00 \text{ m})\pi(0.060 \text{ m})^2 = 0.01131 \text{ m}^3$

$$m_1 = \frac{(1.40 \times 10^6 \text{ Pa})(0.01131 \text{ m}^3)(44.1 \times 10^{-3} \text{ kg/mol})}{(8.3145 \text{ J/mol} \cdot \text{K})((22.0 + 273.15)\text{K})} = 0.2845 \text{ kg}$$

Then $m_2 = m_1 \left(\dfrac{p_2}{p_1}\right) = (0.2845 \text{ kg}) \left(\dfrac{4.41 \times 10^5 \text{ Pa}}{1.40 \times 10^6 \text{ Pa}}\right) = 0.0896$ kg.

m_2 is the mass that remains in the tank. The mass that has been used is
$m_1 - m_2 = 0.2845$ kg $- 0.0896$ kg $= 0.195$ kg.

EVALUATE: Note that we have to use absolute pressures. The absolute pressure decreases by a factor of approximately 3 and the mass of gas in the tank decreases by a factor of approximately 3.

18.57. IDENTIFY: Air and water can be pumped in and out of a diving bell.

SET UP and EXECUTE: The average density can be changed by pumping water in and out of the bell. We use $\rho = m/V$, $p = p_0 + \rho gh$, and $pV = nRT$.

(a) We want to know what volume of seawater to add to the bell so it is neutrally buoyant. The average density of the bell and all its contents must be equal to the density of seawater. Call m_b the mass of the bell, m_d the mass of the diver, and m_w the mass of the added water. Using $\rho_{\text{bell}} = \rho_{\text{seawater}}$ gives $\dfrac{m_b + m_d + m_w}{V_{\text{bell}}} = \rho_{\text{seawater}}$. Using $V_{\text{bell}} = \pi r^2 L$ and solving for m_w gives
$m_w = \pi r^2 L \rho_{\text{seawater}} - m_b - m_d$. Using $\rho_{\text{seawater}} = 1025$ kg/m^3, $r = 0.750$ m, $L = 2.50$ m, $m_b = 4350$ kg, and $m_d = 80.0$ kg, we get $m_w = 98.31$ kg. The volume of this water is $V = m/\rho$ so
$$V = \frac{98.31 \text{ kg}}{1025 \text{ kg/m}^3} = 0.0959 \text{ m}^3 = 95.9 \text{ L}.$$

(b) We want to know the rate at which we should release air from the tank to maintain a constant pressure as the bell descends at a steady 1.0 m/s, so we want dn/dt. Combining $p = p_0 + \rho gh$ and $pV = nRT$ gives $p_0 + \rho gh = (RT/V)n$. Taking the time derivative of both sides of this equation gives
$\rho g \dfrac{dy}{dt} = \left(\dfrac{RT}{V}\right) \dfrac{dn}{dt}$. Solving for dn/dt gives $\dfrac{dn}{dt} = \dfrac{\rho g V}{RT} \dfrac{dy}{dt} = \dfrac{\rho g V v}{RT}$. V is the volume of air in the bell which is $V = V_b - V_w = \pi r^2 L - V_w = \pi(0.750 \text{ m})^2(2.50 \text{ m}) - 0.0959 \text{ m}^3 = 4.322 \text{ m}^3$. Using this value for V plus $\rho = 1025$ kg/m^3, $T = 293$ K, $v = 1.00$ m/s, $R = 8.314$ J/mol$\cdot$K, and $L = 2.50$ m, we get $dn/dt = 17.8$ mol/s.

(c) The tank contains 600 ft^3 of gas, which is 1.6992×10^4 L. It was loaded at standard conditions of 1.0 atm at 0°C = 273 K, so $n = pV/RT$. Putting in the numbers gives $n = 758.5$ mol. The tank releases gas at a rate of 17.8 mol/s, so (17.8 mol/s)$t = 758.5$ mol, which gives $t = 42.6$ s. During this time the bell is descending at a steady 1.00 m/s, so the distance it travels is (1.00 m/s)(42.6 s) = 42.6 m.

EVALUATE: At 42.6 m the water pressure would be $p = p_0 + \rho g h = p_0 + (1025 \text{ kg/m}^3)(g)(42.6 \text{ m})$ = 5.3 atm, so the diving bell should not have any leaks!

18.59. **(a) IDENTIFY:** Consider the gas in one cylinder. Calculate the volume to which this volume of gas expands when the pressure is decreased from $(1.20 \times 10^6 \text{ Pa} + 1.01 \times 10^5 \text{ Pa}) = 1.30 \times 10^6 \text{ Pa}$ to $1.01 \times 10^5 \text{ Pa}$. Apply the ideal-gas law to the two states of the system to obtain an expression for V_2 in terms of V_1 and the ratio of the pressures in the two states.

SET UP: $pV = nRT$

n, R, T constant implies $pV = nRT = $ constant, so $p_1 V_1 = p_2 V_2$.

EXECUTE: $V_2 = V_1 (p_1 / p_2) = (1.90 \text{ m}^3) \left(\dfrac{1.30 \times 10^6 \text{ Pa}}{1.01 \times 10^5 \text{ Pa}} \right) = 24.46 \text{ m}^3$

The number of cylinders required to fill a 750 m³ balloon is $750 \text{ m}^3 / 24.46 \text{ m}^3 = 30.7$ cylinders.

EVALUATE: The ratio of the volume of the balloon to the volume of a cylinder is about 400. Fewer cylinders than this are required because of the large factor by which the gas is compressed in the cylinders.

(b) IDENTIFY: The upward force on the balloon is given by Archimedes's principle: $B = $ weight of air displaced by balloon $= \rho_{\text{air}} V g$. Apply Newton's second law to the balloon and solve for the weight of the load that can be supported. Use the ideal-gas equation to find the mass of the gas in the balloon.

SET UP: The free-body diagram for the balloon is given in Figure 18.59.

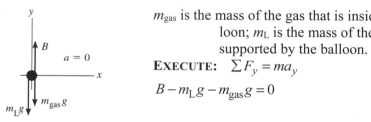

m_{gas} is the mass of the gas that is inside the balloon; m_{L} is the mass of the load that is supported by the balloon.

EXECUTE: $\Sigma F_y = m a_y$

$B - m_{\text{L}} g - m_{\text{gas}} g = 0$

Figure 18.59

$\rho_{\text{air}} V g - m_{\text{L}} g - m_{\text{gas}} g = 0$

$m_{\text{L}} = \rho_{\text{air}} V - m_{\text{gas}}$

Calculate m_{gas}, the mass of hydrogen that occupies 750 m³ at 15°C and $p = 1.01 \times 10^5$ Pa.

$pV = nRT = (m_{\text{gas}} / M) RT$

gives $m_{\text{gas}} = pVM / RT = \dfrac{(1.01 \times 10^5 \text{ Pa})(750 \text{ m}^3)(2.02 \times 10^{-3} \text{ kg/mol})}{(8.3145 \text{ J/mol} \cdot \text{K})(288 \text{ K})} = 63.9$ kg.

Then $m_{\text{L}} = (1.23 \text{ kg/m}^3)(750 \text{ m}^3) - 63.9 \text{ kg} = 859$ kg, and the weight that can be supported is

$w_{\text{L}} = m_{\text{L}} g = (859 \text{ kg})(9.80 \text{ m/s}^2) = 8420$ N.

(c) $m_{\text{L}} = \rho_{\text{air}} V - m_{\text{gas}}$

$m_{\text{gas}} = pVM / RT = (63.9 \text{ kg})((4.00 \text{ g/mol}) / (2.02 \text{ g/mol})) = 126.5$ kg (using the results of part (b)).

Then $m_{\text{L}} = (1.23 \text{ kg/m}^3)(750 \text{ m}^3) - 126.5 \text{ kg} = 796$ kg.

$w_{\text{L}} = m_{\text{L}} g = (796 \text{ kg})(9.80 \text{ m/s}^2) = 7800$ N.

EVALUATE: A greater weight can be supported when hydrogen is used because its density is less.

18.61. IDENTIFY: Apply Bernoulli's equation to relate the efflux speed of water out the hose to the height of water in the tank and the pressure of the air above the water in the tank. Use the ideal-gas equation to relate the volume of the air in the tank to the pressure of the air.

SET UP: Points 1 and 2 are shown in Figure 18.61.

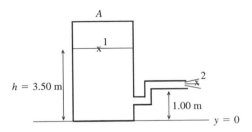

$p_1 = 4.20 \times 10^5$ Pa
$p_2 = p_{air} = 1.00 \times 10^5$ Pa
large tank implies $v_1 \approx 0$

Figure 18.61

EXECUTE: **(a)** $p_1 + \rho g y_1 + \frac{1}{2}\rho v_1^2 = p_2 + \rho g y_2 + \frac{1}{2}\rho v_2^2$

$\frac{1}{2}\rho v_2^2 = p_1 - p_2 + \rho g(y_1 - y_2)$

$v_2 = \sqrt{(2/\rho)(p_1 - p_2) + 2g(y_1 - y_2)}$

$v_2 = 26.2$ m/s.

(b) $h = 3.00$ m
The volume of the air in the tank increases so its pressure decreases. $pV = nRT = $ constant, so $pV = p_0 V_0$ (p_0 is the pressure for $h_0 = 3.50$ m and p is the pressure for $h = 3.00$ m)

$p(4.00 \text{ m} - h)A = p_0(4.00 \text{ m} - h_0)A$

$p = p_0\left(\dfrac{4.00 \text{ m} - h_0}{4.00 \text{ m} - h}\right) = (4.20 \times 10^5 \text{ Pa})\left(\dfrac{4.00 \text{ m} - 3.50 \text{ m}}{4.00 \text{ m} - 3.00 \text{ m}}\right) = 2.10 \times 10^5$ Pa.

Repeat the calculation of part (a), but now $p_1 = 2.10 \times 10^5$ Pa and $y_1 = 3.00$ m.

$v_2 = \sqrt{(2/\rho)(p_1 - p_2) + 2g(y_1 - y_2)}$

$v_2 = 16.1$ m/s

$h = 2.00$ m

$p = p_0\left(\dfrac{4.00 \text{ m} - h_0}{4.00 \text{ m} - h}\right) = (4.20 \times 10^5 \text{ Pa})\left(\dfrac{4.00 \text{ m} - 3.50 \text{ m}}{4.00 \text{ m} - 2.00 \text{ m}}\right) = 1.05 \times 10^5$ Pa

$v_2 = \sqrt{(2/\rho)(p_1 - p_2) + 2g(y_1 - y_2)}$

$v_2 = 5.44$ m/s.

(c) $v_2 = 0$ means $(2/\rho)(p_1 - p_2) + 2g(y_1 - y_2) = 0$

$p_1 - p_2 = -\rho g(y_1 - y_2)$

$y_1 - y_2 = h - 1.00$ m

$p = p_0\left(\dfrac{0.50 \text{ m}}{4.00 \text{ m} - h}\right) = (4.20 \times 10^5 \text{ Pa})\left(\dfrac{0.50 \text{ m}}{4.00 \text{ m} - h}\right)$. This is p_1, so

$(4.20 \times 10^5 \text{ Pa})\left(\dfrac{0.50 \text{ m}}{4.00 \text{ m} - h}\right) - 1.00 \times 10^5 \text{ Pa} = (9.80 \text{ m/s}^2)(1000 \text{ kg/m}^3)(1.00 \text{ m} - h)$

$(210/(4.00 - h)) - 100 = 9.80 - 9.80h$, with h in meters.

$210 = (4.00 - h)(109.8 - 9.80h)$

$9.80h^2 - 149h + 229.2 = 0$ and $h^2 - 15.20h + 23.39 = 0$

quadratic formula: $h = \frac{1}{2}\left(15.20 \pm \sqrt{(15.20)^2 - 4(23.39)}\right) = (7.60 \pm 5.86)$ m

h must be less than 4.00 m, so the only acceptable value is $h = 7.60$ m $- 5.86$ m $= 1.74$ m.

EVALUATE: The flow stops when $p + \rho g(y_1 - y_2)$ equals air pressure. For $h = 1.74$ m, $p = 9.3 \times 10^4$ Pa and $\rho g(y_1 - y_2) = 0.7 \times 10^4$ Pa, so $p + \rho g(y_1 - y_2) = 1.0 \times 10^5$ Pa, which is air pressure.

18.65. IDENTIFY: The mass of one molecule is the molar mass, M, divided by the number of molecules in a mole, N_A. The average translational kinetic energy of a single molecule is $\frac{1}{2}m(v^2)_{av} = \frac{3}{2}kT$. Use $pV = NkT$ to calculate N, the number of molecules.

SET UP: $k = 1.381 \times 10^{-23}$ J/molecule·K. $M = 28.0 \times 10^{-3}$ kg/mol. $T = 295.15$ K. The volume of the balloon is $V = \frac{4}{3}\pi(0.250 \text{ m})^3 = 0.0654 \text{ m}^3$. $p = 1.25$ atm $= 1.27 \times 10^5$ Pa.

EXECUTE: (a) $m = \dfrac{M}{N_A} = \dfrac{28.0 \times 10^{-3} \text{ kg/mol}}{6.022 \times 10^{23} \text{ molecules/mol}} = 4.65 \times 10^{-26}$ kg.

(b) $\frac{1}{2}m(v^2)_{av} = \frac{3}{2}kT = \frac{3}{2}(1.381 \times 10^{-23}$ J/molecule·K$)(295.15$ K$) = 6.11 \times 10^{-21}$ J.

(c) $N = \dfrac{pV}{kT} = \dfrac{(1.27 \times 10^5 \text{ Pa})(0.0654 \text{ m}^3)}{(1.381 \times 10^{-23} \text{ J/molecule·K})(295.15 \text{ K})} = 2.04 \times 10^{24}$ molecules.

(d) The total average translational kinetic energy is

$N\left(\frac{1}{2}m(v^2)_{av}\right) = (2.04 \times 10^{24}$ molecules$)(6.11 \times 10^{-21}$ J/molecule$) = 1.25 \times 10^4$ J.

EVALUATE: The number of moles is $n = \dfrac{N}{N_A} = \dfrac{2.04 \times 10^{24} \text{ molecules}}{6.022 \times 10^{23} \text{ molecules/mol}} = 3.39$ mol.

$K_{tr} = \frac{3}{2}nRT = \frac{3}{2}(3.39 \text{ mol})(8.314 \text{ J/mol·K})(295.15 \text{ K}) = 1.25 \times 10^4$ J, which agrees with our results in part (d).

18.69. IDENTIFY: The equipartition principle says that each atom has an average kinetic energy of $\frac{1}{2}kT$ for each degree of freedom. There is an equal average potential energy.

SET UP: The atoms in a three-dimensional solid have three degrees of freedom and the atoms in a two-dimensional solid have two degrees of freedom.

EXECUTE: (a) In the same manner that $C_V = 3R$ was obtained, the heat capacity of the two-dimensional solid would be $2R = 16.6$ J/mol·K.

(b) The heat capacity would behave qualitatively like those in Figure 18.21 in the textbook, and the heat capacity would decrease with decreasing temperature.

EVALUATE: At very low temperatures the equipartition theorem doesn't apply. Most of the atoms remain in their lowest energy states because the next higher energy level is not accessible.

18.71. (a) IDENTIFY and SET UP: Apply conservation of energy $K_1 + U_1 + W_{other} = K_2 + U_2$, where $U = -Gmm_p/r$. Let point 1 be at the surface of the planet, where the projectile is launched, and let point 2 be far from the earth. Just barely escapes says $v_2 = 0$.

EXECUTE: Only gravity does work says $W_{other} = 0$.

$U_1 = -Gmm_p/R_p$; $r_2 \to \infty$ so $U_2 = 0$; $v_2 = 0$ so $K_2 = 0$.

The conservation of energy equation becomes $K_1 - Gmm_p / R_p = 0$ and $K_1 = Gmm_p / R_p$.

But $g = Gm_p / R_p^2$ so $Gm_p / R_p = R_p g$ and $K_1 = mgR_p$, as was to be shown.

EVALUATE: The greater gR_p is, the more initial kinetic energy is required for escape.

(b) IDENTIFY and SET UP: Set K_1 from part (a) equal to the average kinetic energy of a molecule as given by the equation $\frac{1}{2}m(v^2)_{av} = \frac{3}{2}kT$. $\frac{1}{2}m(v^2)_{av} = mgR_p$ (from part (a)). But also, $\frac{1}{2}m(v^2)_{av} = \frac{3}{2}kT$, so $mgR_p = \frac{3}{2}kT$.

EXECUTE: $T = \dfrac{2mgR_p}{3k}$

Nitrogen:

$m_{N_2} = (28.0 \times 10^{-3} \text{ kg/mol})/(6.022 \times 10^{23} \text{ molecules/mol}) = 4.65 \times 10^{-26}$ kg/molecule

$T = \dfrac{2mgR_p}{3k} = \dfrac{2(4.65 \times 10^{-26} \text{ kg/molecule})(9.80 \text{ m/s}^2)(6.37 \times 10^6 \text{ m})}{3(1.381 \times 10^{-23} \text{ J/molecule} \cdot \text{K})} = 1.40 \times 10^5$ K

Hydrogen:

$m_{H_2} = (2.02 \times 10^{-3} \text{ kg/mol})/(6.022 \times 10^{23} \text{ molecules/mol}) = 3.354 \times 10^{-27}$ kg/molecule

$T = \dfrac{2mgR_p}{3k} = \dfrac{2(3.354 \times 10^{-27} \text{ kg/molecule})(9.80 \text{ m/s}^2)(6.37 \times 10^6 \text{ m})}{3(1.381 \times 10^{-23} \text{ J/molecule} \cdot \text{K})} = 1.01 \times 10^4$ K

(c) $T = \dfrac{2mgR_p}{3k}$

Nitrogen:

$T = \dfrac{2(4.65 \times 10^{-26} \text{ kg/molecule})(1.63 \text{ m/s}^2)(1.74 \times 10^6 \text{ m})}{3(1.381 \times 10^{-23} \text{ J/molecule} \cdot \text{K})} = 6370$ K

Hydrogen:

$T = \dfrac{2(3.354 \times 10^{-27} \text{ kg/molecule})(1.63 \text{ m/s}^2)(1.74 \times 10^6 \text{ m})}{3(1.381 \times 10^{-23} \text{ J/molecule} \cdot \text{K})} = 459$ K

(d) EVALUATE: The "escape temperatures" are much less for the moon than for the earth. For the moon a larger fraction of the molecules at a given temperature will have speeds in the Maxwell-Boltzmann distribution larger than the escape speed. After a long time most of the molecules will have escaped from the moon.

18.75. IDENTIFY: $f(v)dv$ is the probability that a particle has a speed between v and $v + dv$. The equation for the Maxwell-Boltzmann distribution gives $f(v)$. v_{mp} is given by $v_{mp} = \sqrt{2kT/m}$.

SET UP: For O_2, the mass of one molecule is $m = M / N_A = 5.32 \times 10^{-26}$ kg.

EXECUTE: (a) $f(v)dv$ is the fraction of the particles that have speed in the range from v to $v + dv$. The number of particles with speeds between v and $v + dv$ is therefore $dN = Nf(v)dv$ and

$\Delta N = N \displaystyle\int_{v}^{v + \Delta v} f(v)dv.$

(b) Setting $v = v_{mp} = \sqrt{\dfrac{2kT}{m}}$ in $f(v)$ gives $f(v_{mp}) = 4\pi\left(\dfrac{m}{2\pi kT}\right)^{3/2}\left(\dfrac{2kT}{m}\right)e^{-1} = \dfrac{4}{e\sqrt{\pi}\,v_{mp}}$. For oxygen gas at 300 K, $v_{mp} = 3.95\times10^{2}$ m/s and $f(v)\Delta v = 0.0421$.

(c) Increasing v by a factor of 7 changes f by a factor of $7^{2}e^{-48}$, and $f(v)\Delta v = 2.94\times10^{-21}$.

(d) Multiplying the temperature by a factor of 2 increases the most probable speed by a factor of $\sqrt{2}$, and the answers are decreased by $\sqrt{2}$: 0.0297 and 2.08×10^{-21}.

(e) Similarly, when the temperature is one-half what it was in parts (b) and (c), the fractions increase by $\sqrt{2}$ to 0.0595 and 4.15×10^{-21}.

EVALUATE: **(f)** At lower temperatures, the distribution is more sharply peaked about the maximum (the most probable speed), as is shown in Figure 18.23a in the textbook.

18.77. IDENTIFY: At equilibrium the net upward force of the gas on the piston equals the weight of the piston. When the piston moves upward the gas expands, the pressure of the gas drops and there is a net downward force on the piston. For simple harmonic motion the net force has the form $F_y = -ky$, for a displacement y from equilibrium, and $f = \dfrac{1}{2\pi}\sqrt{\dfrac{k}{m}}$.

SET UP: $pV = nRT$. T is constant.

EXECUTE: **(a)** The difference between the pressure, inside and outside the cylinder, multiplied by the area of the piston, must be the weight of the piston. The pressure in the trapped gas is $p_0 + \dfrac{mg}{A} = p_0 + \dfrac{mg}{\pi r^2}$.

(b) When the piston is a distance $h + y$ above the cylinder, the pressure in the trapped gas is $\left(p_0 + \dfrac{mg}{\pi r^2}\right)\left(\dfrac{h}{h+y}\right)$ and for values of y small compared to h, $\dfrac{h}{h+y} = \left(1+\dfrac{y}{h}\right)^{-1} \sim 1 - \dfrac{y}{h}$. The net force, taking the positive direction to be upward, is then

$$F_y = \left[\left(p_0 + \dfrac{mg}{\pi r^2}\right)\left(1 - \dfrac{y}{h}\right) - p_0\right](\pi r^2) - mg = -\left(\dfrac{y}{h}\right)(p_0\pi r^2 + mg).$$

This form shows that for positive h, the net force is down; the trapped gas is at a lower pressure than the equilibrium pressure, and so the net force tends to restore the piston to equilibrium.

(c) The angular frequency of small oscillations would be given by

$$\omega^2 = \dfrac{(p_0\pi r^2 + mg)/h}{m} = \dfrac{g}{h}\left(1 + \dfrac{p_0\pi r^2}{mg}\right). \quad f = \dfrac{\omega}{2\pi} = \dfrac{1}{2\pi}\sqrt{\dfrac{g}{h}\left(1 + \dfrac{p_0\pi r^2}{mg}\right)^{1/2}}.$$

If the displacements are not small, the motion is not simple harmonic. This can be seen be considering what happens if $y \sim -h$; the gas is compressed to a very small volume, and the force due to the pressure of the gas would become unboundedly large for a finite displacement, which is not characteristic of simple harmonic motion. If $y \gg h$ (but not so large that the piston leaves the cylinder), the force due to the pressure of the gas becomes small, and the restoring force due to the atmosphere and the weight would tend toward a constant, and this is not characteristic of simple harmonic motion.

EVALUATE: The assumption of small oscillations was made when $\dfrac{h}{h+y}$ was replaced by $1 - y/h$; this is accurate only when y/h is small.

18.79. IDENTIFY: The measurement gives the dew point. Relative humidity is defined in Problem 18.46, and the vapor pressure table is given with the problem in the text.

SET UP: relative humidity = $\dfrac{\text{partial pressure of water vapor at temperature } T}{\text{vapor pressure of water at temperature } T}$. At 28.0°C the vapor pressure of water is 3.78×10^3 Pa.

EXECUTE: **(a)** The experiment shows that the dew point is 16.0°C, so the partial pressure of water vapor at 30.0°C is equal to the vapor pressure at 16.0°C, which is 1.81×10^3 Pa.

Thus the relative humidity = $\dfrac{1.81 \times 10^3 \text{ Pa}}{4.25 \times 10^3 \text{ Pa}} = 0.426 = 42.6\%$.

(b) For a relative humidity of 35%, the partial pressure of water vapor is $(0.35)(3.78 \times 10^3 \text{ Pa}) = 1.323 \times 10^3$ Pa. This is close to the vapor pressure at 12°C, which would be at an altitude $(30°C - 12°C)/(0.6 \text{ C}°/100 \text{ m}) = 3$ km above the ground.

(c) For a relative humidity of 80%, the vapor pressure will be the same as the water pressure at around 24°C, corresponding to an altitude of about 1 km.

EVALUATE: The lower the dew point is compared to the air temperature, the smaller the relative humidity. Clouds form at a lower height when the relative humidity at the surface is larger.

18.83. IDENTIFY: The equation $\lambda = \dfrac{V}{4\pi\sqrt{2}r^2 N}$ gives the mean free path λ. In the equation $t_{\text{mean}} = \dfrac{V}{4\pi\sqrt{2}r^2 vN}$, use $v_{\text{rms}} = \sqrt{\dfrac{3RT}{M}}$ in place of v. $pV = nRT = NkT$. The escape speed is $v_{\text{escape}} = \sqrt{\dfrac{2GM}{R}}$.

SET UP: For atomic hydrogen, $M = 1.008 \times 10^{-3}$ kg/mol.

EXECUTE: **(a)** From $\lambda = \dfrac{V}{4\pi\sqrt{2}r^2 N}$, we have

$\lambda = [4\pi\sqrt{2}r^2(N/V)]^{-1} = [4\pi\sqrt{2}(5.0 \times 10^{-11} \text{ m})^2 (50 \times 10^6 \text{ m}^{-3})]^{-1} = 4.5 \times 10^{11}$ m.

(b) $v_{\text{rms}} = \sqrt{3RT/M} = \sqrt{3(8.3145 \text{ J/mol} \cdot \text{K})(20 \text{ K})/(1.008 \times 10^{-3} \text{ kg/mol})} = 703$ m/s, and the time between collisions is then $(4.5 \times 10^{11} \text{ m})/(703 \text{ m/s}) = 6.4 \times 10^8$ s, about 20 yr. Collisions are not very important.

(c) $p = (N/V)kT = (50/1.0 \times 10^{-6} \text{ m}^3)(1.381 \times 10^{-23} \text{ J/K})(20 \text{ K}) = 1.4 \times 10^{-14}$ Pa.

(d) $v_{\text{escape}} = \sqrt{\dfrac{2GM}{R}} = \sqrt{\dfrac{2G(Nm/V)(4\pi R^3/3)}{R}} = \sqrt{(8\pi/3)G(N/V)mR^2}$

$v_{\text{escape}} = \sqrt{(8\pi/3)(6.673 \times 10^{-11} \text{ N} \cdot \text{m}^2/\text{kg}^2)(50 \times 10^6 \text{ m}^{-3})(1.67 \times 10^{-27} \text{ kg})(10 \times 9.46 \times 10^{15} \text{ m})^2}$

$v_{\text{escape}} = 650$ m/s. This is lower than v_{rms} and the cloud would tend to evaporate.

(e) In equilibrium (clearly not *thermal* equilibrium), the pressures will be the same; from $pV = NkT$, $kT_{\text{ISM}}(N/V)_{\text{ISM}} = kT_{\text{nebula}}(N/V)_{\text{nebula}}$ and the result follows.

(f) With the result of part (e),

$T_{\text{ISM}} = T_{\text{nebula}}\left(\dfrac{(N/V)_{\text{nebula}}}{(N/V)_{\text{ISM}}}\right) = (20 \text{ K})\left(\dfrac{50 \times 10^6 \text{ m}^3}{(200 \times 10^{-6} \text{ m}^3)^{-1}}\right) = 2 \times 10^5$ K, more than three times the temperature of the sun. This indicates a high average kinetic energy, but the thinness of the ISM means that a ship would not burn up.

EVALUATE: The temperature of a gas is determined by the average kinetic energy per atom of the gas. The energy density for the gas also depends on the number of atoms per unit volume, and this is very small for the ISM.

Study Guide for The First Law of Thermodynamics

Summary

In this chapter, we investigate and quantify thermodynamic processes—processes that exchange heat and do work. We'll examine thermodynamic systems and energy in these systems. Our examination will lead us to the first law of thermodynamics and thermodynamic processes. Four common thermodynamics processes will be highlighted, and the implications of those processes for ideal gases will be examined.

Objectives

After studying this chapter, you'll understand

- How to define *thermodynamic processes*.
- How heat is transferred and work is done in a thermodynamic process.
- The definition of, and how to apply, the *first law of thermodynamics*.
- How a path between initial and final states affects a thermodynamic process.
- The four common thermodynamic processes (adiabatic, isochoric, isobaric, and isothermal).
- How to apply the common thermodynamic processes to find the changes in heat, work, and internal energy of thermodynamic systems.
- How ideal gases are described in the common thermodynamic processes.

Concepts and Equations

Term	Description
Heat and Work in Thermodynamic Processes	A thermodynamic system may exchange energy with its surroundings by heat transfer or by mechanical work. The work done by the system is given by $$W = \int_{V_1}^{V_2} p\,dV$$ $$= p(V_2 - V_1) \quad \text{(constant pressure only)}.$$ In any thermodynamic process, the heat added to the system and the work done by the system depend on the steps the system takes from its initial to its final states, as well as on the initial and final states themselves.
First Law of Thermodynamics	The first law of thermodynamics states that when heat Q is added to a system while work W is performed by the system, the internal energy U changes by $$\Delta U = Q - W.$$ For infinitesimal changes, $$dU = dQ - dW.$$ The internal energy of any thermodynamic system depends only on its state. The change in internal energy in any process depends only on the initial and final states.
Thermodynamic Processes	Common thermodynamic processes include the adiabatic process, in which no heat flows into or out of the system ($Q = 0$); the isochoric process, in which the volume remains constant ($W = 0$); the isobaric process, in which the pressure remains constant [$W = p(V_2 - V_1)$]; and the isothermal process, in which the temperature remains constant.
Properties of an Ideal Gas	The internal energy of an ideal gas depends only on its temperature, not its pressure or volume. The molar heat capacity at constant volume (C_V) and the molar heat capacity at constant pressure (C_p) for an ideal gas are related by $$C_p = C_V + R.$$ For an adiabatic process in an ideal gas, both $TV^{\gamma-1}$ and pV^γ are constant, where $\gamma = C_p/C_V$. The work done by an ideal gas during an adiabatic expansion is given by $$W = nC_V(T_1 - T_2)$$ $$= \frac{C_V}{R}(p_1V_1 - p_2V_2)$$ $$= \frac{1}{\gamma - 1}(p_1V_1 - p_2V_2).$$

Key Concept 1: To find the amount of work a system does on its surroundings during a volume change, you can either calculate the integral of pressure p with respect to volume V, or find the area under the curve of p versus V. The work is positive when the volume increases, and negative when the volume decreases.

From Chapter 19 of Student's Study Guide to accompany *University Physics with Modern Physics, Volume 1*, Fifteenth Edition. Hugh D. Young and Roger A. Freedman. Copyright © 2020 by Pearson Education, Inc. All rights reserved.

Key Concept 2: The first law of thermodynamics states that the change in internal energy ΔU of a system equals the heat flow Q into the system minus the work W that the system does on its surroundings. Q is positive if heat flows into the system, negative if heat flows out; W is positive if the system does work on its surroundings, negative if the surroundings do work on the system.

Key Concept 3: In a cyclic thermodynamic process a system returns to the same state it was in initially, so the net change ΔU in internal energy is zero. The net work W done by the system in a cyclic process equals the area enclosed by the path that the system follows on a pV diagram. Since $\Delta U = 0$, the net heat flow Q into the system in a cyclic process is equal to W.

Key Concept 4: When a system starts in one state and ends in a different state, the internal energy change ΔU of the system is the same no matter what path the system takes between the two states. However, the heat flow Q into the system and the work W done by the system *do* depend on the path taken.

Key Concept 5: For a system that undergoes a thermodynamic process, if you can calculate any two of the quantities ΔU (internal energy change), Q (heat flow into the system), and W (work done by the system), you can calculate the third using the first law of thermodynamics, $\Delta U = Q - W$.

Key Concept 6: When n moles of an ideal gas undergo a thermodynamic process in which the temperature changes by ΔT, the change in internal energy is equal to the product of n, ΔT, and the molar heat capacity at constant volume C_V: $\Delta U = nC_V\Delta T$. This is true whether or not the volume of the system remains constant.

Key Concept 7: In an adiabatic process there is no heat flow into or out of the system, so $Q = 0$ and the internal energy equals the negative of the work done by the system: $\Delta U = -W$. If the system is an ideal gas, the temperature T, volume V, and pressure p all change in the process, but the quantities $TV^{\gamma-1}$ and P^γ remain constant (γ is the ratio of specific heats for the gas).

Conceptual Questions

1: *pV* diagrams

One mole of helium gas is placed in a sealed container and undergoes an isochoric process that results in a doubling of the helium's pressure. Next, the gas undergoes an adiabatic process until the volume of the container is tripled. It then undergoes an isobaric expansion which results in a volume that is four times its original volume. Finally, the helium undergoes an isothermal compression that leaves the container with the same volume that it had after the first process. Sketch a pV diagram for this combined process.

IDENTIFY, SET UP, AND EXECUTE Figure 1 shows the resulting pV diagram. The diagram starts at point *a* with initial pressure p_0 and volume V_0. Then comes the isochoric process, at constant volume, represented by a vertical line to point *b*, where the pressure has doubled. Next is the adiabatic process, in which no heat is exchanged. This process follows the path to point *c*, where the volume has tripled. Next is the isobaric process, carried out at constant pressure and represented by a horizontal line to point *d*, where the volume has increased by V_0 from point *c*. Finally, in the isothermic process, the segment follows the path to point *e*, where the pressure has increased to $2p_0$.

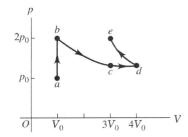

Figure 1 Question 1.

EVALUATE This problem helps clarify the differences in the four common thermodynamic processes. pV diagrams are a valuable aid in solving problems involving these processes. The diagrams will help lead us through a particular problem, as well as provide a check on our results.

2: Internal energy in a thermodynamic process

A container of argon gas undergoes a multistep process. First, it undergoes an isobaric expansion that triples its volume. Next, it goes through an isochoric process that results in a doubling of the argon's pressure. Then it cools adiabatically by 50 K. After that, it undergoes a second isochoric process that doubles its volume. Finally, it undergoes isobaric compression that leaves it at its initial temperature. Find the total change in internal energy.

IDENTIFY, SET UP, AND EXECUTE The change in the internal energy of an ideal gas depends only upon temperature. Argon is an ideal gas. Since the final temperature is the same as the initial temperature, the total change in internal energy is zero.

EVALUATE This complicated scenario shows that we need to focus on the important parts of the process to interpret the results properly. Although it is trivial to find the change in internal energy, it would be cumbersome to find the heat added to the system.

Problems

1: Adiabatic compression of helium

Helium gas is expanded adiabatically from a 12-liter volume at STP to a 33-liter volume. Find the final temperature and pressure of the gas and the work done on the gas.

IDENTIFY Since the helium is expanded adiabatically, both pV^γ and $TV^{\gamma-1}$ are constant during the process. The target variables are the final temperature and pressure of the gas and the work done on the gas.

SET UP For helium, $\gamma = 1.67$ (from Table 1 in the text). *Standard temperature and pressure (STP)* refers to a temperature of 273 K and a pressure of 1 atmosphere.

EXECUTE To find the final pressure, we use the relation

$$pV^\gamma = \text{constant} = p_1 V_1^\gamma = p_2 V_2^\gamma,$$

where the subscripts 1 and 2 indicate "before" and "after," respectively. Rearranging terms to find the final pressure, we obtain

$$p_2 = \frac{p_1 V_1^\gamma}{V_2^\gamma} = \frac{(1.01\times 10^5 \text{ Pa})(12 \text{ L})^{1.67}}{(33 \text{ L})^{1.67}} = 1.86\times 10^4 \text{ Pa}.$$

To find the final temperature, we use the relation

$$TV^{\gamma-1} = \text{constant} = T_1 V_1^{\gamma-1} = T_2 V_2^{\gamma-1}.$$

Rearranging terms in the preceding equation gives a final temperature of

$$T_2 = \frac{T_1 V_1^{\gamma-1}}{V_2^{\gamma-1}} = \frac{(273 \text{ K})(12 \text{ L})^{0.67}}{(33 \text{ L})^{0.67}} = 139 \text{ K}.$$

The work done by an ideal gas in an adiabatic process is

$$W = nC_V (T_1 - T_2).$$

For helium, C_V is 12.47, from Table 1. The number of moles is

$$n = \frac{12 \text{ L}}{22.4 \text{ L}} = 0.536 \text{ mol}.$$

The work done by the gas is

$$W = nC_V(T_1 - T_2) = (0.536 \text{ mol})(12.47 \text{ J/mol} \cdot \text{K})(273 \text{ K} - 139 \text{ K}) = 895 \text{ J}.$$

The work done *on* the gas is the opposite, or -895 J. The final pressure is 1.86×10^4 Pa and the final temperature is 139 K.

KEY CONCEPT 7 **EVALUATE** We see that both the temperature and the pressure decreased in this adiabatic expansion. That makes sense, since no heat was transferred into or out of the system, so having a larger volume required a lower pressure and temperature.
   Would you find the same final temperature if you used the ideal-gas equation? If you check, you'll find that you indeed do find the same final temperature.

Practice Problem: What is the final temperature if oxygen is used instead of helium?
Answer: 182 K

Extra Practice: What is the work done on the oxygen? *Answer:* 1020 J

2: Isochoric and isobaric process with helium

Two moles of helium gas are taken from point *a* to point *c* in the diagram shown in Figure 2. Find the change in internal energy along path *abc*.

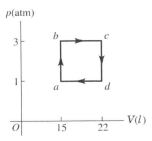

Figure 2 Problem 2.

IDENTIFY The target variable is the change in internal energy.

SET UP We break the process up into two segments, one from a to b and one from b to c. The first segment is an isochoric process (carried out at constant volume) and the second is an isobaric process (carried out at constant pressure). We'll use the relations for those segments to determine the work, heat, and temperature changes. We'll combine the work and heat changes to find the change in internal energy.

EXECUTE We find the change in internal energy along ab by first finding the change in temperature along ab:

$$\Delta T_{ab} = \frac{(p_b - p_a)V_a}{nR} = \frac{(3.03 \times 10^5 \text{ Pa} - 1.01 \times 10^5 \text{ Pa})(15 \text{ L})(10^{-3} \text{ m}^3/\text{L})}{(2 \text{ mol})(8.31 \text{ J/mol/K})} = 182 \text{ K}.$$

The heat transferred during ab is then

$$Q_{ab} = nC_V \Delta T_{ab} = (2 \text{ mol})(12.47 \text{ J/mol/K})(182 \text{ K}) = 4500 \text{ J},$$

where we used the molar heat capacity at constant volume for helium ($C_V = 12.47$ J/mol/K, from Table 1). The work done during segment ab is zero, since that segment is isochoric. The change in internal energy for segment ab is then

$$\Delta U_{ab} = Q_{ab} - W_{ab} = 4500 \text{ J} - 0 = 4500 \text{ J}.$$

For segment bc, we follow the same procedure. The change in temperature along bc is

$$\Delta T_{bc} = \frac{p_b(V_c - V_b)}{nR} = \frac{(3.03 \times 10^5 \text{ Pa})(2.2 \times 10^{-2} \text{ m}^3 - 1.5 \times 10^{-2} \text{ m}^3)}{(2 \text{ mol})(8.31 \text{ J/mol/K})} = 127 \text{ K}.$$

The heat transferred during bc is

$$Q_{bc} = nC_P \Delta T_{bc} = (2 \text{ mol})(20.78 \text{ J/mol/K})(127 \text{ K}) = 5300 \text{ J},$$

where we used the molar heat capacity at constant pressure for helium ($C_P = 20.78$ J/mol/K, from Table 1). The work done during segment bc is

$$W_{bc} = p_b \Delta V_{bc} = (3.03 \times 10^5 \text{ Pa})((2.2 \times 10^{-3} \text{ m}^3) - (1.5 \times 10^{-3} \text{ m}^3)) = 2100 \text{ J}.$$

The change in internal energy during segment bc is

$$\Delta U_{bc} = Q_{bc} - W_{bc} = 5300 \text{ J} - 2100 \text{ J} = 3200 \text{ J}.$$

The total change in internal energy is

$$\Delta U = \Delta U_{ab} + \Delta U_{bc} = 4500 \text{ J} + 3200 \text{ J} = 7700 \text{ J}.$$

The total change in internal energy in the system is 7700 J.

KEY CONCEPT 5 **EVALUATE** We see that by breaking up a process into segments, determining the type of process that takes place during each segment, and knowing which variables change and which remain constant during each segment, one can easily find the change in internal energy.

How should the change in internal energy along path adc compare with the change along path abc? They should be the same for helium, an ideal gas.

Practice Problem: Find the change in internal energy along segments ad and dc, and compare their sums with the change in internal energy along path abc. Answer: $\Delta U_{ad} = 1100$ J, $\Delta U_{dc} = 6600$ J, $\Delta U_{adc} = 7700$ J.

Extra Practice: What is the change in internal energy if there is only one mole of helium? *Answer:* 7700 J

3: Monatomic gas process

One mole of an ideal monatomic gas starts at point A in Figure 3 ($T = 273$ K, $p = 1$ atm) and undergoes an adiabatic expansion to point B, where the volume of the gas has doubled. These two processes are followed by an isothermal compression to the original volume at point C and an isobaric increase to the original point A. Find (a) the temperature at point B, (b) the pressure at point C, and (c) the total work done for the entire cycle. Take γ to be 5/3.

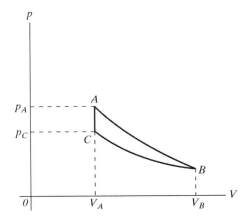

Figure 3 Problem 3.

IDENTIFY The target variables are the temperature at point B, the pressure at point C, and the total work done during the process.

SET UP We'll break the process up into the three segments shown in the figure. We'll use the relations for those segments to determine the temperature, pressure, and work.

EXECUTE (a) For the adiabatic process, we use the relation

$$TV^{\gamma-1} = \text{constant} = T_1 V_1^{\gamma-1} = T_2 V_2^{\gamma-1}.$$

Rearranging terms to find the temperature at point B gives

$$T_2 = \frac{T_1 V_1^{\gamma-1}}{V_2^{\gamma-1}} = \frac{(273 \text{ K})(V)^{0.67}}{(2V)^{0.67}} = 172 \text{ K}.$$

(b) The temperature at point C is the same as at point B, since the second step is isothermal. Also, the volume is the same at point C as it was at point A. We use the standard ideal-gas law:

$$\frac{p_1 V_1}{T_1} = \frac{p_3 V_3}{T_3}.$$

Solving for p_3 gives

$$p_3 = p_1 \frac{T_3}{T_1} = (1 \text{ atm}) \frac{(172 \text{ K})}{(273 \text{ K})} = 0.630 \text{ atm}.$$

(c) The total work done by the gas during the entire cycle is the sum of the separate amounts of work done during each cycle. The path AB is adiabatic, so no heat is exchanged and the work is

$$W_{AB} = 2\,C_V(T_2 - T_1) = 2\,(\tfrac{3}{2}(8.314 \text{ J/mol} \cdot \text{K}))(172 \text{ K} - 273 \text{ K}) = 1260 \text{ J}.$$

For segment BC, the temperature is constant. The work is given by the integral

$$\begin{aligned}
W &= \int_{V_1}^{V_2} p\,dV \\
&= \int_{V_2}^{V_3} \frac{RT}{V} dV = RT \ln V \Big|_{V_2}^{V_3} \\
&= RT \ln(V_3/V_2) = (8.314 \text{ J/mol} \cdot \text{K})(172 \text{ K})\ln(V/2V) \\
&= -991 \text{ J}.
\end{aligned}$$

No work is done during segment CA, since the system changes isobarically. The total work done is the sum, 268 J.

KEY CONCEPT 1 **EVALUATE** We solved this problem by breaking the process into segments and working through each segment to find the final state variables. Can you use the area inside the cycle to find the amount of work? The area is greater than zero, indicating positive work.

Practice Problem: What is the temperature at point B if the volume has tripled there?
Answer: 131 K

Extra Practice: What pressure do you now get at point C? *Answer:* 0.479 atm

4: Expansion process for argon

One mole of argon is initially at 25°C and occupies a volume of 35 liters. The argon is first expanded at constant pressure until the volume is doubled and then expanded adiabatically until the temperature returns to 25°C. Find the total change in internal energy, the total work done by the argon, and the final volume and pressure of the argon.

IDENTIFY Sketch the process and then use the definitions to solve. The target variables are the total change in internal energy, the total work done, and the final volume and pressure.

SET UP Figure 4 shows the pV diagram for the process. We'll break the process up into two segments, one from a to b and one from b to c. The first segment is an isobaric process (carried out under constant pressure) and the second is adiabatic (no heat exchanged). We'll use the relations for those segments to determine the work, heat, and temperature changes. We'll combine these results to find the quantities of interest.

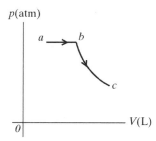

Figure 4 Problem 4 sketch.

EXECUTE The total change in internal energy is zero, since the internal energy of an ideal gas depends only on temperature and the final temperature is equal to the initial temperature.

Next, we find the temperature at point b. Segment ab is at constant pressure, so
$$\frac{V_a}{T_a} = \frac{V_b}{T_b}.$$

The temperature at b is
$$T_b = \frac{V_b}{V_a}T_a = \frac{(70 \text{ L})}{(35 \text{ L})}(273+25)\text{ K} = 596 \text{ K}.$$

The heat supplied during ab is
$$Q_{ab} = nC_P\Delta T_{ab} = (1 \text{ mol})(20.78 \text{ J/mol/K})(596 \text{ K} - 298 \text{ K}) = 6190 \text{ J},$$

where we used the molar heat capacity at constant pressure for argon ($C_P = 20.78$ J/mol/K, from Table 1). The heat transferred in bc is zero, since segment bc is adiabatic. The total heat supplied in the complete process is 6190 J. Because the total internal energy change is zero, the heat supplied must be equal to the work done by the argon. The work done by the argon is 6190 J.

Next, we find the final volume and pressure. To find the final volume in the adiabatic process (bc), we use the relation
$$TV^{\gamma-1} = \text{constant} = T_b V_b^{\gamma-1} = T_c V_c^{\gamma-1},$$

where $\gamma = 1.67$ for argon. Rearranging terms to find the final volume gives
$$V_c = \sqrt[\gamma-1]{\frac{T_b V_b^{\gamma-1}}{T_c}} = \sqrt[0.67]{\frac{(596 \text{ K})(70 \text{ L})^{0.67}}{(298 \text{ K})}} = 197 \text{ L}.$$

We can use the equation of state for an ideal gas to find the final pressure:
$$p_c = \frac{nRT_c}{V_c} = \frac{(1 \text{ mol})(8.31 \text{ J/mol/K})(298 \text{ K})}{(197 \times 10^{-3} \text{ m}^3)} = 1.26 \times 10^4 \text{ Pa}.$$

The final pressure is 12,600 Pa and the final volume is 197 liters.

KEY CONCEPT 7 **EVALUATE** We again see that we need to break the process up into segments and work through each segment to find the final state variables. We also see that the pV graph is useful in solving problems involving thermodynamic processes.

Practice Problem: What is the final volume if the volume has tripled at point b? *Answer:* 541 L

Extra Practice: What is the pressure now at point c? *Answer:* 4,580 Pa

Try It Yourself!

1: Ideal-gas process

Consider n moles of an ideal gas that undergo the constant-volume and constant-pressure processes along the paths shown in Figure 5 from the initial state a to b, then from b to c, then from c to d, and then back from a to d. For each of these processes, calculate (a) the work done by the system and (b) the heat taken in by the system.

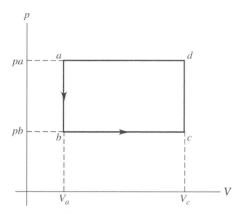

Figure 5 Try It Yourself 1.

Solution Checkpoints

IDENTIFY AND SET UP We break the process up into segments and use the relations for isochoric and isobaric processes.

EXECUTE (a) No work is done by the gas during the isochoric processes. For the isobaric processes, the work done by the gas is the pressure times the change in volume:

$$W_{b \to c} = p_b (V_c - V_a),$$
$$W_{d \to a} = p_a (V_a - V_c).$$

The work done from b to c is a positive quantity, and the work done from d to a is negative, resulting in negative total work. Work is done *on* the gas.

(b) The heat taken in by the ideal gas will be the number of moles, times the change in temperature, times the heat capacity at constant volume or constant pressure. This relationship gives

$$Q_{a \to b} = nC_V (T_b - T_a),$$
$$Q_{b \to c} = nC_p (T_c - T_b),$$
$$Q_{c \to d} = nC_V (T_d - T_c),$$
$$Q_{d \to a} = nC_p (T_a - T_d).$$

EVALUATE How does the heat taken in by the gas compare with the work done by the gas?

2: Ideal-gas process, version 2

A total of n moles of an ideal monatomic gas is taken from point 1 on the T_1 isotherm to point 3 on the T_2 isotherm along the path $1 \to 2 \to 3$ as shown in Figure 6. (a) Calculate the change in internal energy of the gas and the heat that must be added to it in this process.
(b) Suppose the path $1 \to 4 \to 3$ is followed instead. Calculate the heat added along this path.

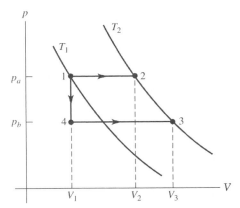

Figure 6 Try It Yourself 2.

Solution Checkpoints

IDENTIFY AND SET UP We break the process up into segments and use the relations for isothermic and isobaric processes.

EXECUTE (a) In going from 1 to 3, the change in internal energy depends on the temperature differences and not the path taken. This gives

$$\Delta U = nC_V(T_3 - T_1).$$

The heat taken in from 1 to 3 is

$$\Delta Q = nC_p(T_2 - T_1) + nRT \ln\left(\frac{V_3}{V_2}\right).$$

(b) The change in internal energy is the same as in (a). The heat is

$$\Delta Q = nC_V(T_2 - T_1) + p_3(V_3 - V_1).$$

KEY CONCEPT 4 **EVALUATE** How does the heat exchanged on the two paths differ. Why?

Key Example Variation Problems

Solutions to these problems are in Chapter 19 of the Student's Solutions Manual.

Be sure to review EXAMPLES 19.2, 19.3, 19.4, and 19.5 (Section 19.4) before attempting these problems.

VP19.5.1 Consider the following thermodynamic processes. (a) The internal energy of a quantity of gas increases by 2.5×10^3 J as the gas does 3.25×10^3 J of work on its surroundings. How much heat flows into the gas? (b) Ice enclosed in a container does 2.50×10^4 J of work on its container as 7.00×10^3 J of heat flows out of the ice. What is the internal energy change of the ice? (c) The internal energy of a metal block increases by 4.25×10^3 J as 2.40×10^3 J of heat flows into the block. How much work did the block do on its surroundings?

VP19.5.2 How much work does a quantity of gas (volume V, pressure p) do in each of these processes? (a) V increases from 2.00×10^{-3} m^3 to 4.50×10^{-3} m^3 while p is a constant 6.20×10^4 Pa. (b) V is a constant 2.00×10^{-3} m^3 while p increases from

6.00×10^4 Pa to 9.00×10^4 Pa. (c) V decreases from 6.00×10^{-3} m^3 to 3.00×10^{-3} m^3 while p is a constant 1.25×10^5 Pa. (d) V increases from 2.00×10^{-3} m^3 to 4.50×10^{-3} m^3 while p increases from 1.50×10^5 Pa to 5.50×10^5 Pa. The graph of this process on a pV-diagram is a straight line.

VP19.5.3 A quantity of gas is enclosed in a cylinder that has a movable piston. The gas has volume $V = 3.60\times10^{-3}$ m^3 and pressure $p = 2.00\times10^5$ Pa. (a) First V decreases to $V = 2.40\times10^{-3}$ m^3 while p remains constant; then V remains constant while p increases to 6.00×10^5 Pa. During this process 1.56×10^3 J of heat flows into the gas. For this process, how much work does the gas do, and what is its internal energy change? (b) The gas returns to the same initial volume and pressure. Now p increases to 6.00×10^5 Pa while V remains constant; then V decreases to 2.40×10^{-3} m^3 while p remains constant. For this process, how much work does the gas do, and how much heat flows into the gas?

VP19.5.4 At an atmospheric pressure of 1.01×10^5 Pa, carbon dioxide (CO_2) sublimes at a temperature of -78.5°C. The heat of sublimation is 5.71×10^5 J/kg. You let heat flow into 6.0 kg of solid CO_2 (volume 4.0×10^{-3} m^3) at atmospheric pressure and a constant -78.5°C until it has completely sublimed to gas at -78.5°C. The CO_2 gas occupies a volume of 3.4 m^3. For this process, calculate (a) how much heat flows into the CO_2, (b) how much work the CO_2 does, and (c) the internal energy change of the CO_2.

Be sure to review EXAMPLE 19.6 (Section 19.7) before attempting these problems.

VP19.6.1 For neon gas, $C_V = \frac{3}{2}R$ and $C_p = \frac{5}{2}R$. Calculate the change in internal energy of 4.00 mol of neon if its temperature (a) increases from 20.0°C to 40.0°C at constant volume; (b) increases from 15.0°C to 30.0°C at constant pressure; (c) increases from 12.0°C to 24.0°C while both the pressure and volume increase.

VP19.6.2 For nitrogen gas, $C_V = \frac{5}{2}R$ and $C_p = \frac{7}{2}R$. You have 2.10 mol of nitrogen gas at temperature 27.0°C and pressure $p = 1.00\times10^5$ Pa. (a) Calculate the initial volume V of the gas. (b) Calculate the final temperature (in °C) of the gas and its internal energy change if (i) p doubles while V remains constant, (ii) V doubles while p remains constant, and (iii) both p and V double.

VP19.6.3 A quantity of argon gas ($C_V = \frac{3}{2}R$ and $C_p = \frac{5}{2}R$) is at pressure $p = 1.20\times10^5$ Pa and occupies volume $V = 0.250$ m^3. Calculate the change in internal energy of the gas if the final pressure and volume of the gas are
(a) $p = 2.40\times10^5$ Pa, $V = 0.250$ m^3; (b) $p = 1.20\times10^5$ Pa, $V = 0.125$ m^3;
(c) $p = 1.80\times10^5$ Pa, $V = 0.600$ m^3.

VP19.6.4 A quantity of n moles of oxygen gas ($C_V = \frac{5}{2}R$ and $C_p = \frac{7}{2}R$) is at absolute temperature T. You increase the absolute temperature to $2T$. Find the change in internal energy of the gas, the heat flow into the gas, and the work done by the gas if the process you used to increase the temperature is (a) isochoric, (b) isobaric, or (c) adiabatic.

Be sure to review EXAMPLE 19.7 (Section 19.8) before attempting these problems.

VP19.7.1 A quantity of argon gas ($C_V = \frac{3}{2}R$ and $C_p = \frac{5}{2}R$) is at pressure 4.00×10^5 Pa and occupies volume 2.00×10^{-3} m³. Then the gas expands adiabatically to a new volume 6.00×10^{-3} m³. Calculate (a) the value of γ for argon, (b) the final pressure of the gas, and (c) the work done by the gas as it expands.

VP19.7.2 A quantity of oxygen gas ($C_V = \frac{5}{2}R$ and $C_p = \frac{7}{2}R$) is at absolute temperature 325 K, is under pressure 1.00×10^5 Pa, and occupies volume 6.50×10^{-3} m³. After you compress this gas adiabatically, its absolute temperature is 855 K. Calculate (a) the value of γ for oxygen, (b) the final volume of the gas, (c) the final pressure of the gas, and (d) the work done by the gas as it is compressed.

VP19.7.3 Initially 5.00 mol of neon gas ($C_V = \frac{3}{2}R$ and $\gamma = \frac{5}{2}R$) are at absolute temperature 305 K and occupy volume 4.00×10^{-2} m³. Then the gas expands adiabatically to a new volume of 9.00×10^{-2} m³. Calculate (a) the initial pressure of the gas, (b) the final pressure of the gas, (c) the final temperature of the gas, and (d) the work done by the gas as it expands.

VP19.7.4 You have 1.25 mol of hydrogen gas ($C_V = \frac{5}{2}R$ and $\gamma = \frac{7}{2}R$) at absolute temperature 325 K. You allow the gas to expand adiabatically to a final temperature of 195 K. (a) How much work does the gas do while being compressed? (b) What is the ratio of its final volume to its initial volume? (c) What is the ratio of the final gas pressure to the initial gas pressure?

STUDENT'S SOLUTIONS MANUAL FOR THE FIRST LAW OF THERMODYNAMICS

VP19.5.1. **IDENTIFY:** We investigate several thermodynamic processes and apply the first law of thermodynamics.
SET UP: The first law of thermodynamics is $\Delta U = Q - W$.
EXECUTE: **(a)** $\Delta U = Q - W$: 2.50×10^3 J $= Q - 3.25 \times 10^3$ J. $Q = 5.75 \times 10^3$ J.
(b) $\Delta U = Q - W$: $\Delta U = -7.00 \times 10^3$ J $- 2.50 \times 10^4$ J $= -3.20 \times 10^4$ J.
(c) $\Delta U = Q - W$: 4.25×10^3 J $= 2.40 \times 10^3$ J $- W$. $W = -1.85 \times 10^3$ J.
EVALUATE: Careful! Q is the heat put *into* the gas and W is the work done *by* the gas. If heat comes *out* of the gas, Q will be negative. If work is done *on* the gas, W will be negative.

VP19.5.2. **IDENTIFY:** We want to calculate the work that a gas does in various thermodynamic processes.
SET UP: $W = p\Delta V$ if pressure is constant. In general $W = \int p dV$. The work done by a gas is equal to the area under the curve on a pV-diagram.
EXECUTE: **(a)** The pressure is constant, so we use $W = p\Delta V = p(V_2 - V_1) =$
$(6.20 \times 10^4$ Pa$)(4.50 \times 10^{-3}$ m$^3 - 2.00 \times 10^{-3}$ m$^3) = 155$ J.
(b) $\Delta V = 0$ so $W = 0$.
(c) Pressure is constant so $W = p\Delta V = (1.25 \times 10^5$ Pa$)(-3.00 \times 10^{-3}$ m$^3) = -375$ J.
(d) First sketch a pV-diagram of the process, as shown in Fig. VP19.5.2. The work done by the gas is the area under the curve. This area is made up of a triangle above a rectangle. The work is $W = A_{\text{triangle}} + A_{\text{rectangle}}$ which gives
$W = \frac{1}{2}(2.50 \times 10^{-3}$ m$^3)(4.00 \times 10^5$ Pa$) + (2.50 \times 10^{-3}$ m$^3)(1.50 \times 10^5$ Pa$) = 875$ J.

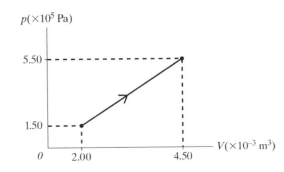

Figure VP19.5.2

EVALUATE: For more complicated curves, the area under the curve can be replaced by the definite integral $W = \int p\,dV$.

VP19.5.3. **IDENTIFY:** A gas undergoes various thermodynamic changes. We want to determine the work it does and the heat energy put into it. The ideal gas law and the first law of thermodynamics apply.

SET UP: We apply $\Delta U = Q - W$, $W = p\Delta V$ for constant pressure, and $pV = nRT$. We know that $p_1 = 2.00 \times 10^5$ Pa and $V_1 = 3.60 \times 10^{-3}$ m^3.

EXECUTE: **(a)** We want the work the gas does and its internal energy change. First find the work W done by the gas during the two processes 1 and 2. $W = W_1 + W_2$. This gives

$W = p_1 \Delta V_1 + p \Delta V_2 = p_1 \Delta V_1 + 0 = (2.00 \times 10^5 \text{ Pa})(-1.20 \times 10^{-3} \text{ m}^3) = -2.40 \times 10^2$ J.

Now use $\Delta U = Q - W = 1560 \text{ J} - (-240 \text{ J}) = 1.80 \times 10^3$ J.

(b) We want to find the work the gas does and how much heat flows into the gas. As before,

$W = W_1 + W_2 = p_1 \Delta V_1 + p_2 \Delta V_2 = 0 + p_2 \Delta V_2 = (6.00 \times 10^5 \text{ Pa})(-1.20 \times 10^{-3} \text{ m}^3) = -7.20 \times 10^2$ J.

Now apply $\Delta U = Q - W$ but we first need to find ΔU.

For the process in (a): $p_1 V_1 = (2.00 \times 10^5 \text{ Pa})(3.60 \times 10^{-3} \text{ m}^3) = 720$ J

$p_2 V_2 = (6.00 \times 10^5 \text{ Pa})(2.40 \times 10^{-3} \text{ m}^3) = 1440$ J

For the process in (b): $p_1 V_1 = (2.00 \times 10^5 \text{ Pa})(3.60 \times 10^{-3} \text{ m}^3) = 720$ J

$p_2 V_2 = (6.00 \times 10^5 \text{ Pa})(2.40 \times 10^{-3} \text{ m}^3) = 1440$ J

In both cases, the *change* in pV is $\Delta(pV) = 720$ J. From $pV = nRT$ we see that $nR\Delta T = \Delta(pV)$. Therefore for these two processes, ΔT is the same because $\Delta(pV)$ is the same, and if ΔT is the same, ΔU must also be the same. So $\Delta U = 1.80 \times 10^3$ J as we found in part (a). Now we use

$\Delta U = Q - W$: 1.80×10^3 J $= Q - (-7.20 \times 10^2$ J$)$, which gives $Q = 1.08 \times 10^3$ J.

EVALUATE: Note that ΔU depends only on ΔT, so if ΔT is the same ΔU must be the same, no matter what the process. ΔU is independent of path.

VP19.5.4. **IDENTIFY:** Heat flows into the solid CO_2 causing the gas to do work and change its internal energy. The first law of thermodynamics applies.

SET UP: $\Delta U = Q - W$, for sublimation $Q = mL_{sub}$, where $L_{sub} = 5.71 \times 10^5$ J/kg. At constant pressure $W = p\Delta V$.

EXECUTE: **(a)** $Q = mL_{sub} = (6.0 \text{ kg})(5.71 \times 10^5 \text{ J/kg}) = 3.4 \times 10^6$ J.

(b) We want the work. The pressure is constant so the work done is

$W = p\Delta V = (1.01 \times 10^5 \text{ Pa})(3.4 \text{ m}^3 - 0.0040 \text{ m}^3) = 3.4 \times 10^5$ J.

(c) We want the change in internal energy. $\Delta U = Q - W = 3.4 \times 10^6$ J $- 3.4 \times 10^5$ J $= 3.1 \times 10^6$ J.

EVALUATE: The internal energy change is due to an increase in the electrical potential energy of CO_2 molecules.

VP19.6.1. **IDENTIFY:** This problem involves molar heat capacity and internal energy for the monatomic gas neon.

SET UP: $Q = nC_V \Delta T$, $C_V = \frac{3}{2}R$ for a monatomic gas, $\Delta U = Q - W$. The target variable is the change in internal energy ΔU.

EXECUTE: **(a)** At constant volume, $W = 0$, so $\Delta U = Q = nC_V \Delta T = n\frac{3}{2}R\Delta T$. Using $n = 4.00$ mol and $\Delta T = 20.0$ K gives $\Delta U = 998$ J.

(b) Since ΔU is independent of path, it is the same for constant volume and constant pressure processes (or any other process) if ΔT is the same. Therefore $\Delta U = Q_V = nC_V\Delta T = n\frac{3}{2}R\Delta T$. Using $n = 4.00$ mol and $\Delta T = 15.0$ K gives $\Delta U = 748$ J.
(c) Use the same reasoning as in part (b) since ΔU is independent of path. We have $\Delta U = n\frac{3}{2}R\Delta T$, and using $n = 4.00$ mol and $\Delta T = 12.0$ K gives $\Delta U = 599$ J.
EVALUATE: For *any* ideal gas processes, if ΔT is the same, ΔU is the same.

VP19.6.2. IDENTIFY: We are investigating the behavior of ideal N_2 gas. The ideal gas law and the first law of thermodynamics both apply.
SET UP: Use $pV = nRT$, $\Delta U = Q - W$, $Q = nC_V\Delta T$, and $C_V = \frac{3}{2}R$.
EXECUTE: (a) We want the initial volume of the gas. Solving $pV = nRT$ for V gives $V = nRT/p$. Using $n = 2.10$ mol, $T = 300$ K, and $p = 1.00\times10^5$ Pa gives $V = 5.24\times10^{-2}$ m^3.
(b) (i) We want T_2 and ΔU when the pressure doubles at constant volume. For constant volume, $pV = nRT$ tells us that if p doubles, T also doubles, so $T_2 = 600$ K $= 327°$C.
For constant volume, $W = 0$, so $\Delta U = Q = nC_V\Delta T = n\frac{5}{2}R\Delta T$. Using $n = 2.10$ mol and $\Delta T = 300$ K gives $\Delta U = 1.31\times10^4$ J.
(ii) We want T_2 and ΔU when the volume doubles at constant pressure. For constant pressure, $pV = nRT$ tells us that if V doubles, T also doubles, so $T_2 = 600$ K $= 327°$C. By the4 same reasoning as in part (ii), $\Delta U = 1.31\times10^4$ J.
(iii) We want T_2 and ΔU when both pressure and volume double. From $pV = nRT$ we see that if both p and V double, the product pV increases by a factor of 4. Therefore T also increases by a factor of 4, so $T_2 = 4T_1 = 4(300$ K$) = 1200$ K $= 927°$C.
Now $\Delta T = 1200$ K $- 300$ K $= 900$ K, which is 3 times the change in parts (i) and (ii), so ΔU is 3 times as large as before. Therefore $\Delta U = 3(1.31\times10^4$ J$) = 3.93\times10^4$ J.
EVALUATE: Note that if we quadruple the Kelvin temperature we do *not* quadruple the Celsius temperature.

VP19.6.3. IDENTIFY: Ideal monatomic argon gas goes through changes in its temperature and pressure. We want to find the resulting change in its internal energy. The ideal gas law and first law of thermodynamics both apply.
SET UP: We use $pV = nRT$, $\Delta U = Q - W$, $\Delta U = nC_V\Delta T$ for *any* process, and $C_V = \frac{3}{2}R$ for a monatomic gas. $pV = nRT$ tells us that $\Delta(pV) = nR\Delta T$, so $\Delta T = \frac{\Delta(pV)}{nR}$. This gives $\Delta U = nC_V\Delta T = n\left(\frac{3}{2}R\right)\frac{\Delta(pV)}{nR} = \frac{3}{2}\Delta(pV) = \frac{3}{2}(p_2V_2 - p_1V_1)$. We want to find ΔU for each process. We know that $p_1 = 1.20\times10^5$ Pa and $V_1 = 0.250$ m^3 for each process.
EXECUTE: (a) $V_1 = V_2 = 0.250$ m^3 and $p_2 = 2.4\times10^5$ Pa. Use $\Delta U = \frac{3}{2}(p_2V_2 - p_1V_1)$ which becomes $\Delta U = \frac{3}{2}V_1(p_2 - p_1) = \frac{3}{2}(0.250$ m$^3)(1.20\times10^5$ Pa$) = 4.50\times10^4$ J.

(b) $p_1 = p_2 = 1.20 \times 10^5$ Pa and $V_2 = 0.125$ m³. Use $\Delta U = \frac{3}{2}(p_2V_2 - p_1V_1)$ which becomes $\Delta U = \frac{3}{2}p_1(V_2 - V_1) = \frac{3}{2}(1.20 \times 10^5 \text{ Pa})(0.125 \text{ m}^3 - 0.250 \text{ m}^3) = -2.25 \times 10^4$ J.

(c) $p_2 = 1.80 \times 10^5$ Pa and $V_2 = 0.600$ m³. Use $\Delta U = \frac{3}{2}(p_2V_2 - p_1V_1)$ which gives $\Delta U = 1.17 \times 10^5$ J.

EVALUATE: Even though only one of the processes was at constant volume, we could use $\Delta U = nC_V \Delta T$ to calculate ΔU because ΔU is the same *as if* the process were at constant volume. Remember that ΔU is independent of path.

VP19.6.4. **IDENTIFY:** We increase the temperature of a gas from T to $2T$ by isochoric, isobaric, and adiabatic processes. In each case we want to find the change in the internal energy of the gas, the heat flow in-to it, and work done by the gas. The first law of thermodynamics applies.

SET UP: $\Delta U = Q - W$, $\Delta U = nC_V\Delta T$, at constant pressure $Q = nC_p\Delta T$, and for a diatomic gas $C_V = \frac{5}{2}R$ and $C_p = \frac{7}{2}R$. The target variable is ΔU for each process.

EXECUTE: **(a)** The *volume* is constant for an isochoric process, so $W = 0$. Therefore $\Delta U = Q - W$ gives $\Delta U = Q = nC_V\Delta T$. Since the temperature increases from T to $2T$, $\Delta T = 2T - T = T$. Thus $\Delta U = Q = nC_V\Delta T = n\left(\frac{5}{2}R\right)T = \frac{5}{2}nRT$.

(b) The *pressure* is constant for an isobaric process. $Q = nC_p\Delta T = n\left(\frac{7}{2}R\right)T = \frac{7}{2}nRT$.

$\Delta U = nC_V\Delta T = \frac{5}{2}nRT$, the same as in part (a). From $\Delta U = Q - W$ we have $W = Q - \Delta U = \frac{7}{2}nRT - \frac{5}{2}nRT = nRT$.

(c) For an adiabatic process, no heat enters or leaves the gas, so $Q = 0$. From part (a) we know that $\Delta U = nC_V\Delta T = \frac{5}{2}nRT$. $\Delta U = Q - W$ gives $W = Q - \Delta U = 0 - \Delta U = -\frac{5}{2}nRT$.

EVALUATE: As we have seen here, doubling the gas temperature can have different results depending on how the process is carried out.

VP19.7.1. **IDENTIFY:** We are investigating monatomic argon gas during an adiabatic expansion.

SET UP: pV^γ is constant during an adiabatic process, $\gamma = C_p / C_V$, $W = \frac{C_V}{R}(p_1V_1 - p_2V_2)$.

EXECUTE: **(a)** $\gamma = \frac{C_p}{C_V} = \frac{5/2R}{3/2R} = 5/3$.

(b) We want the final pressure p_2. Since pV^γ is constant, we have $p_1V_1^\gamma = p_2V_2^\gamma$ which gives $p_2 = p_1\left(\frac{V_1}{V_2}\right)^\gamma$. Using $p_1 = 4.00 \times 10^5$ Pa, $V_1 = 2.00 \times 10^{-3}$ m³, $V_2 = 6.00 \times 10^{-3}$ m³, and $\gamma = 5/3$, we have $p_2 = 6.41 \times 10^4$ Pa.

(c) We want the work W. $W = \frac{C_V}{R}(p_1V_1 - p_2V_2) = \frac{3/2R}{R}(p_1V_1 - p_2V_2) = \frac{3}{2}(p_1V_1 - p_2V_2)$. Using the pressures and volumes above, we get $W = 623$ J.

EVALUATE: Careful! An adiabatic process is *not* the same as an isothermal process. The temperature can change during an adiabatic process but not during an isothermal process.

VP19.7.2. IDENTIFY: We are investigating diatomic oxygen O_2 gas during an adiabatic compression.

SET UP: $TV^{\gamma-1}$ and pV^γ are constant during an adiabatic process, $\gamma = C_p/C_V$,

$$W = \frac{1}{\gamma-1}(p_1V_1 - p_2V_2).$$

EXECUTE: (a) $\gamma = \dfrac{C_p}{C_V} = \dfrac{7/2R}{5/2R} = 7/5.$

(b) We want the final volume V_2. Since $TV^{\gamma-1}$ is constant, we have $T_1V_1^{\gamma-1} = T_2V_2^{\gamma-1}$ which gives $V_2^{\gamma-1} = V_1^{\gamma-1}\dfrac{T_1}{T_2}$. Taking the $\gamma-1$ root gives $V_2 = V_1\left(\dfrac{T_1}{T_2}\right)^{1/(\gamma-1)} = V_1\left(\dfrac{T_1}{T_2}\right)^{5/2}$. Using $p_1 = 1.00\times10^5$ Pa, $V_1 = 6.50\times10^{-3}$ m^3, $T_1 = 325$ K, and $T_2 = 855$ K, we have $V_2 = 5.79\times10^{-4}$ m^3.

(c) We want the final pressure p_2. pV^γ is constant so $p_1V_1^\gamma = p_2V_2^\gamma$. Solving for p_2 gives

$$p_2 = p_1\left(\frac{V_1}{V_2}\right)^\gamma.$$ Using $p_1 = 1.00\times10^5$ Pa, $V_1 = 6.50\times10^{-3}$ m^3, and $V_2 = 5.79\times10^{-4}$ m^3, we get $p_2 = 2.95\times10^6$ Pa.

(d) We want the work done by the gas during this compression. Use $W = \dfrac{1}{\gamma-1}(p_1V_1 - p_2V_2)$ and put in the numbers from above, we get $W = -2650$ J. The minus sign tells us that work is done *on* the gas to compress it.

EVALUATE: Regardless of the process, the gas must still obey the ideal gas law, so pV/T should be the same for states 1 and 2. Using the values for the pressure, volume, and temperature of each state, we get $p_1V_1/T_1 = 2.00$ J/K and $p_2V_2/T_2 = 2.00$ J/K, so our results are consistent with the ideal gas law.

VP19.7.3. IDENTIFY: Monatomic neon gas expands adiabatically.

SET UP: $TV^{\gamma-1}$ and pV^γ are constant during an adiabatic process and $\gamma = C_p/C_V = 5/3$. We also have $\Delta U = Q - W$, $\Delta U = nC_V\Delta T$, $pV = nRT$. We know $T_1 = 305$ K, $V_1 = 0.0400$ m^3, and $V_2 = 0.0900$ m^3, and $n = 5.00$ mol. $Q = 0$ for an adiabatic process.

EXECUTE: (a) We want the initial pressure p_1. Solving $pV = nRT$ gives $p_1 = nRT_1/V_1$. Using the numbers above gives $p_1 = 3.17\times10^5$ Pa.

(b) We want the final pressure p_2. Since pV^γ is constant, $p_1V_1^\gamma = p_2V_2^\gamma$. Solving for p_2 gives

$$p_2 = p_1\left(\frac{V_1}{V_2}\right)^\gamma.$$ Using $\gamma = 5/3$ and the values for the volumes and pressure, we get $p_2 = 8.20\times10^4$ Pa.

(c) We want the final temperature T_2. Since $TV^{\gamma-1}$ is constant, $T_1V_1^{\gamma-1} = T_2V_2^{\gamma-1}$. Solving for T_2 gives $T_2 = T_1\left(\dfrac{V_1}{V_2}\right)^{\gamma-1}$. Using $T_1 = 305$ K and the values for the volumes, we have $T_2 = 178$ K.

(d) We want the work done by the gas. For an adiabatic process, $Q = 0$, so $\Delta U = Q - W$ gives $W = -\Delta U = -nC_V \Delta T = -n\left(\frac{3}{2}R\right)\Delta T$. Using $\Delta T = 178$ K $- 305$ K and $n = 5.00$ mol, we get $W = 7940$ J.

EVALUATE: We can check the answer in part (d) by using $W = \frac{1}{\gamma - 1}(p_1V_1 - p_2V_2)$. Putting in the numbers from above gives $W = 7950$ J. The slight difference in the last digit is from rounding.

VP19.7.4. **IDENTIFY:** Diatomic hydrogen gas H_2 expands adiabatically.

SET UP: $TV^{\gamma-1}$ and pV^γ are constant during an adiabatic process and $\gamma = C_p/C_V = 7/5$. We also have $\Delta U = Q - W$, $\Delta U = nC_V\Delta T$, $pV = nRT$. We know that $T_1 = 325$ K, $T_2 = 195$ K, and $n = 1.25$ mol. $Q = 0$ for an adiabatic process.

EXECUTE: **(a)** We want the work done by the gas. Using $\Delta U = Q - W$ with $Q = 0$ gives

$$W = -\Delta U = -nC_V\Delta T = -n\left(\frac{5}{2}R\right)\Delta T = -(1.25 \text{ mol})(5/2)(8.314 \text{ J/mol}\cdot\text{K})(130 \text{ K}), \text{ so}$$

$W = -3.38 \times 10^3$ J.

(b) We want V_2/V_1. Since $TV^{\gamma-1}$ is constant, $T_1V_1^{\gamma-1} = T_2V_2^{\gamma-1}$, which gives Using the numbers gives $\dfrac{V_2}{V_1} = \left(\dfrac{325 \text{ K}}{195 \text{ K}}\right)^{5/2} = 3.59$.

(c) We want p_2/p_1. Solving $p_1V_1^\gamma = p_2V_2^\gamma$ for p_2/p_1 gives $\dfrac{p_2}{p_1} = \left(\dfrac{V_1}{V_2}\right)^\gamma = \left(\dfrac{1}{3.59}\right)^{7/5} = 0.167$.

EVALUATE: The gas must also obey the ideal gas law, so we check the ratio p_2/p_1 ratio. Using $pV = nRT$ this ratio is $\dfrac{p_2}{p_1} = \dfrac{nRT_2/V_2}{nRT_1/V_1} = \left(\dfrac{T_2}{T_1}\right)\left(\dfrac{V_1}{V_2}\right) = \left(\dfrac{195 \text{ K}}{325 \text{ K}}\right)\left(\dfrac{1}{3.59}\right) = 0.167$, which agrees with our result in part (c).

19.5. **IDENTIFY:** For an isothermal process $W = nRT\ln(p_1/p_2)$. Solve for p_1.

SET UP: For a compression (V decreases) W is negative, so $W = -392$ J. $T = 295.15$ K.

EXECUTE: **(a)** $\dfrac{W}{nRT} = \ln\left(\dfrac{p_1}{p_2}\right)$. $\dfrac{p_1}{p_2} = e^{W/nRT}$.

$\dfrac{W}{nRT} = \dfrac{-392 \text{ J}}{(0.305 \text{ mol})(8.314 \text{ J/mol}\cdot\text{K})(295.15 \text{ K})} = -0.5238$.

$p_1 = p_2 e^{W/nRT} = (1.76 \text{ atm})e^{-0.5238} = 1.04$ atm.

(b) In the process the pressure increases and the volume decreases. The pV-diagram is sketched in Figure 19.5.

EVALUATE: W is the work done by the gas, so when the surroundings do work on the gas, W is negative. The gas was compressed at constant temperature, so its pressure must have increased, which means that $p_1 < p_2$, which is what we found.

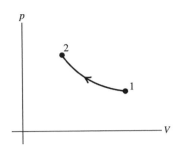

Figure 19.5

19.9. IDENTIFY: $\Delta U = Q - W$. For a constant pressure process, $W = p\Delta V$.

SET UP: $Q = +1.15 \times 10^5$ J, since heat enters the gas.

EXECUTE: (a) $W = p\Delta V = (1.65 \times 10^5 \text{ Pa})(0.320 \text{ m}^3 - 0.110 \text{ m}^3) = 3.47 \times 10^4$ J.

(b) $\Delta U = Q - W = 1.15 \times 10^5 \text{ J} - 3.47 \times 10^4 \text{ J} = 8.04 \times 10^4$ J.

EVALUATE: (c) $W = p\Delta V$ for a constant pressure process and $\Delta U = Q - W$ both apply to any material. The ideal gas law wasn't used and it doesn't matter if the gas is ideal or not.

19.11. IDENTIFY: Part ab is isochoric, but bc is not any of the familiar processes.

SET UP: $pV = nRT$ determines the Kelvin temperature of the gas. The work done in the process is the area under the curve in the pV diagram. Q is positive since heat goes into the gas.
1 atm $= 1.013 \times 10^5$ Pa. 1 L $= 1 \times 10^{-3}$ m^3. $\Delta U = Q - W$.

EXECUTE: (a) The lowest T occurs when pV has its smallest value. This is at point a, and

$$T_a = \frac{p_a V_a}{nR} = \frac{(0.20 \text{ atm})(1.013 \times 10^5 \text{ Pa/atm})(2.0 \text{ L})(1.0 \times 10^{-3} \text{ m}^3/\text{L})}{(0.0175 \text{ mol})(8.315 \text{ J/mol} \cdot \text{K})} = 278 \text{ K}.$$

(b) a to b: $\Delta V = 0$ so $W = 0$.

b to c: The work done by the gas is positive since the volume increases. The magnitude of the work is the area under the curve so $W = \frac{1}{2}(0.50 \text{ atm} + 0.30 \text{ atm})(6.0 \text{ L} - 2.0 \text{ L})$ and

$W = (1.6 \text{ L} \cdot \text{atm})(1 \times 10^{-3} \text{ m}^3/\text{L})(1.013 \times 10^5 \text{ Pa/atm}) = 162$ J.

(c) For abc, $W = 162$ J. $\Delta U = Q - W = 215$ J $- 162$ J $= 53$ J.

EVALUATE: 215 J of heat energy went into the gas. 53 J of energy stayed in the gas as increased internal energy and 162 J left the gas as work done by the gas on its surroundings.

19.17. IDENTIFY: For a constant pressure process, $W = p\Delta V$, $Q = nC_p \Delta T$, and $\Delta U = nC_V \Delta T$. $\Delta U = Q - W$ and $C_p = C_V + R$. For an ideal gas, $p\Delta V = nR\Delta T$.

SET UP: From Table 19.1, $C_V = 28.46$ J/mol $\cdot$ K.

EXECUTE: (a) The pV diagram is shown in Figure 19.17 (next page).

(b) $W = pV_2 - pV_1 = nR(T_2 - T_1) = (0.250 \text{ mol})(8.3145 \text{ J/mol} \cdot \text{K})(100.0 \text{ K}) = 208$ J.

(c) The work is done on the piston.

(d) Since $\Delta U = nC_V \Delta T$ holds for any process, we have
$\Delta U = nC_V \Delta T = (0.250 \text{ mol})(28.46 \text{ J/mol} \cdot \text{K})(100.0 \text{ K}) = 712$ J.

(e) Either $Q = nC_p \Delta T$ or $Q = \Delta U + W$ gives $Q = 920$ J to three significant figures.

(f) The lower pressure would mean a correspondingly larger volume, and the net result would be that the work done would be the same as that found in part (b).

EVALUATE: $W = nR\Delta T$, so W, Q and ΔU all depend only on ΔT. When T increases at constant pressure, V increases and $W > 0$. ΔU and Q are also positive when T increases.

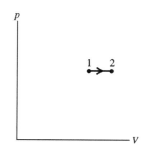

Figure 19.17

19.19. IDENTIFY: For constant volume, $Q = nC_V \Delta T$. For constant pressure, $Q = nC_p \Delta T$.

SET UP: From Table **19.1 in the text**, $C_V = 20.76$ J/mol·K and $C_p = 29.07$ J/mol·K.

EXECUTE: (a) Using $Q = nC_V \Delta$, $\Delta T = \dfrac{Q}{nC_V} = \dfrac{645 \text{ J}}{(0.185 \text{ mol})(20.76 \text{ J/mol·K})} = 167.9$ K and $T = 948$ K.

The pV-diagram is sketched in Figure 19.19a.

(b) Using $Q = nC_p \Delta T$, $\Delta T = \dfrac{Q}{nC_p} = \dfrac{645 \text{ J}}{(0.185 \text{ mol})(29.07 \text{ J/mol·K})} = 119.9$ K and $T = 900$ K.

The pV-diagram is sketched in Figure 19.19b.

EVALUATE: At constant pressure some of the heat energy added to the gas leaves the gas as expansion work and the internal energy change is less than if the same amount of heat energy is added at constant volume. ΔT is proportional to ΔU.

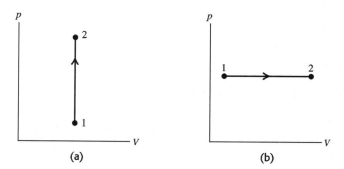

Figure 19.19

19.21. IDENTIFY: $\Delta U = Q - W$. For an ideal gas, $\Delta U = C_V \Delta T$, and at constant pressure, $W = p \Delta V = nR \Delta T$.

SET UP: $C_V = \tfrac{3}{2} R$ for a monatomic gas.

EXECUTE: $\Delta U = n\left(\tfrac{3}{2} R\right)\Delta T = \tfrac{3}{2} p\Delta V = \tfrac{3}{2} W$. Then $Q = \Delta U + W = \tfrac{5}{2} W$, so $W/Q = \tfrac{2}{5}$.

EVALUATE: For diatomic or polyatomic gases, C_V is a different multiple of R and the fraction of Q that is used for expansion work is different.

19.23. IDENTIFY: $\Delta U = Q - W$. Apply $Q = nC_p \Delta T$ to calculate C_p. Apply $\Delta U = nC_V \Delta T$ to calculate C_V. $\gamma = C_p/C_V$.

SET UP: $\Delta T = 15.0\ \text{C}° = 15.0\ \text{K}$. Since heat is added, $Q = +970\ \text{J}$.

EXECUTE: (a) $\Delta U = Q - W = 1970\ \text{J} - 223\ \text{J} = 747\ \text{J}$.

(b) $C_p = \dfrac{Q}{n\Delta T} = \dfrac{970\ \text{J}}{(1.75\ \text{mol})(15.0\ \text{K})} = 37.0\ \text{J/mol·K}$.

$C_V = \dfrac{\Delta U}{n\Delta T} = \dfrac{747\ \text{J}}{(1.75\ \text{mol})(15.0\ \text{K})} = 28.5\ \text{J/mol·K}$. $\gamma = \dfrac{C_p}{C_V} = \dfrac{37.0\ \text{J/mol·K}}{28.5\ \text{J/mol·K}} = 1.30$.

EVALUATE: The value of γ we calculated is similar to the values given in Tables 19.1 for polyatomic gases.

19.27. IDENTIFY: For an adiabatic process of an ideal gas, $p_1 V_1^\gamma = p_2 V_2^\gamma$, $W = \dfrac{1}{\gamma - 1}(p_1 V_1 - p_2 V_2)$, and $T_1 V_1^{\gamma-1} = T_2 V_2^{\gamma-1}$.

SET UP: For a monatomic ideal gas $\gamma = 5/3$.

EXECUTE: (a) $p_2 = p_1 \left(\dfrac{V_1}{V_2}\right)^\gamma = (1.50 \times 10^5\ \text{Pa}) \left(\dfrac{0.0800\ \text{m}^3}{0.0400\ \text{m}^3}\right)^{5/3} = 4.76 \times 10^5\ \text{Pa}$.

(b) This result may be substituted into $W = \dfrac{1}{\gamma - 1} p_1 V_1 (1 - (V_1/V_2)^{\gamma-1})$, or, substituting the above form for p_2,

$W = \dfrac{1}{\gamma - 1} p_1 V_1 (1 - (V_1/V_2)^{\gamma-1}) = \dfrac{3}{2}(1.50 \times 10^5\ \text{Pa})(0.0800\ \text{m}^3)\left[1 - \left(\dfrac{0.0800}{0.0400}\right)^{2/3}\right] = -1.06 \times 10^4\ \text{J}$.

(c) From $T_1 V_1^{\gamma-1} = T_2 V_2^{\gamma-1}$, $(T_2/T_1) = (V_2/V_1)^{\gamma-1} = (0.0800/0.0400)^{2/3} = 1.59$, and since the final temperature is higher than the initial temperature, the gas is heated.

EVALUATE: In an adiabatic compression $W < 0$ since $\Delta V < 0$. $Q = 0$ so $\Delta U = -W$. $\Delta U > 0$ and the temperature increases.

19.31. IDENTIFY: Combine $T_1 V_1^{\gamma-1} = T_2 V_2^{\gamma-1}$ with $pV = nRT$ to obtain an expression relating T and p for an adiabatic process of an ideal gas.

SET UP: $T_1 = 299.15\ \text{K}$.

EXECUTE: $V = \dfrac{nRT}{p}$ so $T_1 \left(\dfrac{nRT_1}{p_1}\right)^{\gamma-1} = T_2 \left(\dfrac{nRT_2}{p_2}\right)^{\gamma-1}$ and $\dfrac{T_1^\gamma}{p_1^{\gamma-1}} = \dfrac{T_2^\gamma}{p_2^{\gamma-1}}$.

$T_2 = T_1 \left(\dfrac{p_2}{p_1}\right)^{(\gamma-1)/\gamma} = (299.15\ \text{K}) \left(\dfrac{0.850 \times 10^5\ \text{Pa}}{1.01 \times 10^5\ \text{Pa}}\right)^{0.4/1.4} = 284.8\ \text{K} = 11.6°\text{C}$.

EVALUATE: For an adiabatic process of an ideal gas, when the pressure decreases the temperature decreases.

19.33. **IDENTIFY:** As helium undergoes an adiabatic process it's Kelvin temperature doubles.

SET UP: Treat helium as an ideal monatomic gas, so $\gamma = 5/3$. We also have $pV = nRT$ and $p_1V_1^\gamma = p_2V_2^\gamma$, and we know that $T_2 = 2T_1$. We want the factor by which p changes during this process, so we want p_2/p_1.

EXECUTE: From $pV = nRT$ we have $V = nRT/p$. Taking the ratio of the volumes gives
$$\frac{V_1}{V_2} = \frac{nRT_1/p_1}{nRT_2/p_2} = \left(\frac{T_1}{T_2}\right)\left(\frac{p_2}{p_1}\right).$$ Now use $p_1V_1^\gamma = p_2V_2^\gamma$, which gives $\dfrac{p_2}{p_1} = \left(\dfrac{V_1}{V_2}\right)^\gamma$. Use the previous result for the volume ratio to get $\dfrac{p_2}{p_1} = \left(\dfrac{T_1}{T_2} \cdot \dfrac{p_2}{p_1}\right)^\gamma$, which can be arranged to get

$\dfrac{p_2}{p_1}\left(\dfrac{p_2}{p_1}\right)^{-\gamma} = \left(\dfrac{T_1}{T_2}\right)^\gamma$. Using $\gamma = 5/3$ and $T_1/T_2 = \frac{1}{2}$, this becomes $\dfrac{p_2}{p_1} = \left(\dfrac{1}{2}\right)^{-5/2} = 4\sqrt{2}$.

EVALUATE: Look at the volume ratio V_2/V_1, which is $\dfrac{V_2}{V_1} = \dfrac{nRT_2/p_2}{nRT_1/p_1} = \left(\dfrac{T_2}{T_1}\right)\left(\dfrac{p_1}{p_2}\right)$, so

$\dfrac{V_2}{V_1} = (2)\left(\dfrac{1}{4\sqrt{2}}\right) = \dfrac{\sqrt{2}}{4} \approx 0.354$, which means that the volume decreases. This is reasonable because the only way to increase the gas temperature adiabatically is to compress it.

19.35. **IDENTIFY and SET UP:** For an ideal gas, $pV = nRT$. The work done is the area under the path in the pV-diagram.

EXECUTE: (a) The product pV increases and this indicates a temperature increase.
(b) The work is the area in the pV plane bounded by the blue line representing the process and the verticals at V_a and V_b. The area of this trapezoid is
$\frac{1}{2}(p_b + p_a)(V_b - V_a) = \frac{1}{2}(2.40 \times 10^5 \text{ Pa})(0.0400 \text{ m}^3) = 4800 \text{ J}.$

EVALUATE: The work done is the average pressure, $\frac{1}{2}(p_1 + p_2)$, times the volume increase.

19.37. **IDENTIFY:** We can read the values from the pV-diagram and apply the ideal gas law and the first law of thermodynamics.

SET UP: At each point $pV = nRT$, with $T = 85 \text{ K} + 273 \text{ K} = 358 \text{ K}$. For an isothermal process of an ideal gas, $W = nRT \ln(V_2/V_1)$. $\Delta U = nC_V \Delta T$ for any ideal gas process.

EXECUTE: (a) At point b, $p = 0.200$ atm $= 2.026 \times 10^4$ Pa and $V = 0.100$ m^3.
$$n = \frac{pV}{RT} = \frac{(2.026 \times 10^4 \text{ Pa})(0.100 \text{ m}^3)}{(8.315 \text{ J/mol} \cdot \text{K})(358 \text{ K})} = 0.681 \text{ moles}.$$

(b) n, R, and T are constant so $p_aV_a = p_bV_b$.
$$V_a = V_b\left(\frac{p_b}{p_a}\right) = (0.100 \text{ m}^3)\left(\frac{0.200 \text{ atm}}{0.600 \text{ atm}}\right) = 0.0333 \text{ m}^3.$$

(c) $W = nRT \ln(V_b/V_a) = (0.681 \text{ mol})(8.315 \text{ J/mol} \cdot \text{K})(358 \text{ K})\ln\left(\dfrac{0.100 \text{ m}^3}{0.0333 \text{ m}^3}\right) = 2230 \text{ J} = 2.23 \text{ kJ}.$

W is positive and corresponds to work done by the gas.
(d) $\Delta U = nC_V \Delta T$ so for an isothermal process ($\Delta T = 0$), $\Delta U = 0$.

EVALUATE: W is positive when the volume increases, so the area under the curve is positive. For *any* isothermal process, $\Delta U = 0$.

19.39. IDENTIFY: Use $\Delta U = Q - W$ and the fact that ΔU is path independent.

$W > 0$ when the volume increases, $W < 0$ when the volume decreases, and $W = 0$ when the volume is constant. $Q > 0$ if heat flows into the system.

SET UP: The paths are sketched in Figure **19.39**.

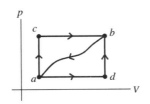

$Q_{acb} = +90.0$ J (positive since heat flows in)
$W_{acb} = +60.0$ J (positive since $\Delta V > 0$)

Figure 19.39

EXECUTE: (a) $\Delta U = Q - W$

ΔU is path independent; Q and W depend on the path.
$\Delta U = U_b - U_a$
This can be calculated for any path from a to b, in particular for path acb:
$\Delta U_{a \to b} = Q_{acb} - W_{acb} = 90.0 \text{ J} - 60.0 \text{ J} = 30.0 \text{ J}$.
Now apply $\Delta U = Q - W$ to path adb; $\Delta U = 30.0$ J for this path also.
$W_{adb} = +15.0$ J (positive since $\Delta V > 0$)
$\Delta U_{a \to b} = Q_{adb} - W_{adb}$ so $Q_{adb} = \Delta U_{a \to b} + W_{adb} = 30.0 \text{ J} + 15.0 \text{ J} = +45.0 \text{ J}$.

(b) Apply $\Delta U = Q - W$ to path ba: $\Delta U_{b \to a} = Q_{ba} - W_{ba}$
$W_{ba} = -35.0$ J (negative since $\Delta V < 0$)
$\Delta U_{b \to a} = U_a - U_b = -(U_b - U_a) = -\Delta U_{a \to b} = -30.0$ J
Then $Q_{ba} = \Delta U_{b \to a} + W_{ba} = -30.0 \text{ J} - 35.0 \text{ J} = -65.0 \text{ J}$.
($Q_{ba} < 0$; the system liberates heat.)

(c) $U_a = 0$, $U_d = 8.0$ J
$\Delta U_{a \to b} = U_b - U_a = +30.0$ J, so $U_b = +30.0$ J.
Process $a \to d$:
$\Delta U_{a \to d} = Q_{ad} - W_{ad}$
$\Delta U_{a \to d} = U_d - U_a = +8.0$ J
$W_{adb} = +15.0$ J and $W_{adb} = W_{ad} + W_{db}$. But the work W_{db} for the process $d \to b$ is zero since $\Delta V = 0$ for that process. Therefore $W_{ad} = W_{adb} = +15.0$ J.
Then $Q_{ad} = \Delta U_{a \to d} + W_{ad} = +8.0 \text{ J} + 15.0 \text{ J} = +23.0$ J (positive implies heat absorbed).
Process $d \to b$:
$\Delta U_{d \to b} = Q_{db} - W_{db}$
$W_{db} = 0$, as already noted.
$\Delta U_{d \to b} = U_b - U_d = 30.0 \text{ J} - 8.0 \text{ J} = +22.0$ J.
Then $Q_{db} = \Delta U_{d \to b} + W_{db} = +22.0$ J (positive; heat absorbed).

EVALUATE: The signs of our calculated Q_{ad} and Q_{db} agree with the problem statement that heat is absorbed in these processes.

19.43. IDENTIFY: Segment *ab* is isochoric, *bc* is isothermal, and *ca* is isobaric.

SET UP: For *bc*, $\Delta T = 0$, $\Delta U = 0$, and $Q = W = nRT \ln(V_c/V_b)$. For ideal H_2 (diatomic), $C_V = \frac{5}{2}R$ and $C_p = \frac{7}{2}R$. $\Delta U = nC_V \Delta T$ for any process of an ideal gas.

EXECUTE: **(a)** $T_b = T_c$. For states *b* and *c*, $pV = nRT =$ constant so $p_b V_b = p_c V_c$ and

$$V_c = V_b \left(\frac{p_b}{p_c}\right) = (0.20 \text{ L})\left(\frac{2.0 \text{ atm}}{0.50 \text{ atm}}\right) = 0.80 \text{ L}.$$

(b) $T_a = \dfrac{p_a V_a}{nR} = \dfrac{(0.50 \text{ atm})(1.013 \times 10^5 \text{ Pa/atm})(0.20 \times 10^{-3} \text{ m}^3)}{(0.0040 \text{ mol})(8.315 \text{ J/mol} \cdot \text{K})} = 305 \text{ K}.$ $V_a = V_b$ so for states *a*

and *b*, $\dfrac{T}{p} = \dfrac{V}{nR} =$ constant so $\dfrac{T_a}{p_a} = \dfrac{T_b}{p_b}$. $T_b = T_c = T_a\left(\dfrac{p_b}{p_a}\right) = (305 \text{ K})\left(\dfrac{2.0 \text{ atm}}{0.50 \text{ atm}}\right) = 1220 \text{ K};$

$T_c = 1220$ K.

(c) *ab*: $Q = nC_V \Delta T = n(\frac{5}{2}R) \Delta T$, which gives

$Q = (0.0040 \text{ mol})(\frac{5}{2})(8.315 \text{ J/mol} \cdot \text{K})(1220 \text{ K} - 305 \text{ K}) = +76$ J. Q is positive and heat goes into the gas.

ca: $Q = nC_p \Delta T = n(\frac{7}{2}R) \Delta T$, which gives

$Q = (0.0040 \text{ mol})(\frac{7}{2})(8.315 \text{ J/mol} \cdot \text{K})(305 \text{ K} - 1220 \text{ K}) = -107$ J. Q is negative and heat comes out of the gas.

bc: $Q = W = nRT \ln(V_c/V_b)$, which gives

$Q = (0.0040 \text{ mol})(8.315 \text{ J/mol} \cdot \text{K})(1220 \text{ K}) \ln(0.80 \text{ L}/0.20 \text{ L}) = 56$ J. Q is positive and heat goes into the gas.

(d) *ab*: $\Delta U = nC_V \Delta T = n(\frac{5}{2}R)\Delta T$, which gives

$\Delta U = (0.0040 \text{ mol})(\frac{5}{2})(8.315 \text{ J/mol} \cdot \text{K})(1220 \text{ K} - 305 \text{ K}) = +76$ J. The internal energy increased.

bc: $\Delta T = 0$ so $\Delta U = 0$. The internal energy does not change.

ca: $\Delta U = nC_V \Delta T = n(\frac{5}{2}R)\Delta T$, which gives

$\Delta U = (0.0040 \text{ mol})(\frac{5}{2})(8.315 \text{ J/mol} \cdot \text{K})(305 \text{ K} - 1220 \text{ K}) = -76$ J. The internal energy decreased.

EVALUATE: The net internal energy change for the complete cycle $a \to b \to c \to a$ is $\Delta U_{\text{tot}} = +76 \text{ J} + 0 + (-76 \text{ J}) = 0$. For any complete cycle the final state is the same as the initial state and the net internal energy change is zero. For the cycle the net heat flow is $Q_{\text{tot}} = +76 \text{ J} + (-107 \text{ J}) + 56 \text{ J} = +25$ J. $\Delta U_{\text{tot}} = 0$ so $Q_{\text{tot}} = W_{\text{tot}}$. The net work done in the cycle is positive and this agrees with our result that the net heat flow is positive.

19.45. IDENTIFY AND SET UP: We have information on the pressure and volume of the gas during the process, but we know almost nothing else about the gas. We do know that the first law of thermodynamics must apply to the gas during this process, so $Q = \Delta U + W$, and the work done by the gas is

$W = \int_{V_1}^{V_2} p \, dV$. If W is positive, the gas does work, but if W is negative, work is done on the gas.

EXECUTE: **(a)** Figure 19.45 shows the pV-diagram for this process. On the pV-diagram, we see that the graph is a closed figure; the gas begins and ends in the same state.

Student's Solutions Manual for The First Law of Thermodynamics

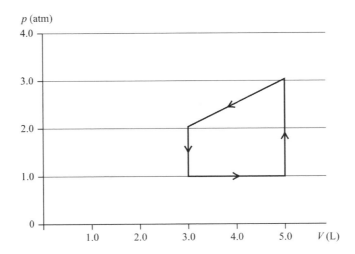

Figure 19.45

(b) Applying $Q = \Delta U + W$, we see that $\Delta U = 0$ because the gas ends up at the same state from which it began. Therefore $Q = W$. $W = \int_{V_1}^{V_2} p\, dV$, so the work is the area under the curve on a pV-diagram. For a closed cycle such as this one, the work is the area enclosed within the diagram. We calculate this work geometrically: $|W|$ = area (rectangle) + area (triangle) = (2.0 L)(1.0 atm) + $\frac{1}{2}$ (2.0 L)(1.0 atm) = 3.0 L·atm = 300 J. But the net work is negative, so $Q = -3.0$ L·atm $= -300$ J. Since Q is negative, heat flows out of the gas.
EVALUATE: We know that the work is negative because in the upper part of the diagram, the volume is decreasing, which means that the gas is being compressed.

19.47. IDENTIFY: $pV = nRT$. For an isothermal process $W = nRT \ln(V_2/V_1)$. For a constant pressure process, $W = p\Delta V$.

SET UP: $1 \text{ L} = 10^{-3} \text{ m}^3$.
EXECUTE: **(a)** The pV-diagram is sketched in Figure 19.47.
(b) At constant temperature, the product pV is constant, so
$V_2 = V_1(p_1/p_2) = (1.5 \text{ L})\left(\dfrac{1.00 \times 10^5 \text{ Pa}}{2.50 \times 10^4 \text{ Pa}}\right) = 6.00$ L. The final pressure is given as being the same as $p_3 = p_2 = 2.5 \times 10^4$ Pa. The final volume is the same as the initial volume, so $T_3 = T_1(p_3/p_1) = 75.0$ K.
(c) Treating the gas as ideal, the work done in the first process is
$W = nRT \ln(V_2/V_1) = p_1 V_1 \ln(p_1/p_2)$. $W = (1.00 \times 10^5 \text{ Pa})(1.5 \times 10^{-3} \text{ m}^3) \ln\left(\dfrac{1.00 \times 10^5 \text{ Pa}}{2.50 \times 10^4 \text{ Pa}}\right) = 208$ J.

For the second process, $W = p_2(V_3 - V_2) = p_2(V_1 - V_2) = p_2 V_1 [1 - (p_1/p_2)]$.

$W = (2.50 \times 10^4 \text{ Pa})(1.5 \times 10^{-3} \text{ m}^3)\left(1 - \dfrac{1.00 \times 10^5 \text{ Pa}}{2.50 \times 10^4 \text{ Pa}}\right) = -113$ J.

The total work done is $208 \text{ J} - 113 \text{ J} = 95$ J.
(d) Heat at constant volume. No work would be done by the gas or on the gas during this process.
EVALUATE: When the volume increases, $W > 0$. When the volume decreases, $W < 0$.

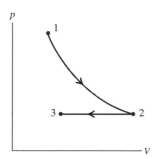

Figure 19.47

19.49. **IDENTIFY:** For an adiabatic process of an ideal gas, $T_1 V_1^{\gamma-1} = T_2 V_2^{\gamma-1}$. $pV = nRT$.

SET UP: For air, $\gamma = 1.40 = \frac{7}{5}$.

EXECUTE: **(a)** As the air moves to lower altitude its density increases; under an adiabatic compression, the temperature rises. If the wind is fast-moving, Q is not as likely to be significant, and modeling the process as adiabatic (no heat loss to the surroundings) is more accurate.

(b) $V = \dfrac{nRT}{p}$, so $T_1 V_1^{\gamma-1} = T_2 V_2^{\gamma-1}$ gives $T_1^\gamma p_1^{1-\gamma} = T_2^\gamma p_2^{1-\gamma}$. The temperature at the higher pressure is $T_2 = T_1 (p_1/p_2)^{(\gamma-1)/\gamma} = (258.15 \text{ K}) \left[(8.12 \times 10^4 \text{ Pa})/(5.60 \times 10^4 \text{ Pa})\right]^{2/7} = 287.1 \text{ K} = 13.9°\text{C}$ so the temperature would rise by 11.9 C°.

EVALUATE: In an adiabatic compression, $Q = 0$ but the temperature rises because of the work done on the gas.

19.51. **IDENTIFY:** Assume that the gas is ideal and that the process is adiabatic. Apply $T_1 V_1^{\gamma-1} = T_2 V_2^{\gamma-1}$ and $p_1 V_1^\gamma = p_2 V_2^\gamma$ to relate pressure and volume and temperature and volume. The distance the piston moves is related to the volume of the gas. Use $W = nC_V (T_1 - T_2)$ to calculate W.

(a) SET UP: $\gamma = C_p / C_V = (C_V + R)/C_V = 1 + R/C_V = 1.40$. The two positions of the piston are shown in Figure 19.51.

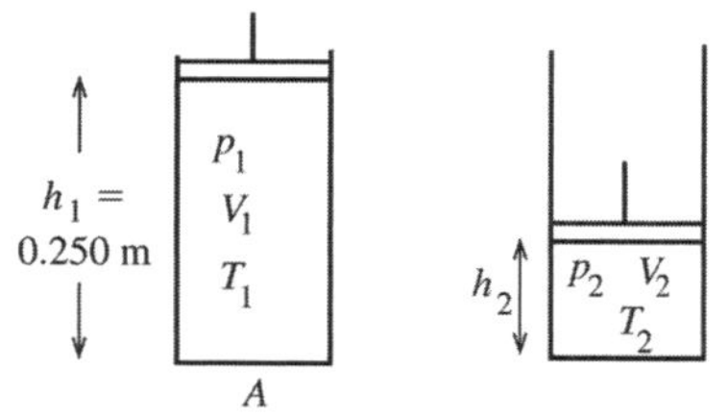

$p_1 = 1.01 \times 10^5$ Pa
$p_2 = 3.80 \times 10^5$ Pa $+ p_{\text{air}} = 4.81 \times 10^5$ Pa
$V_1 = h_1 A$
$V_2 = h_2 A$

Figure 19.51

EXECUTE: For an adiabatic process of an ideal gas, $p_1 V_1^\gamma = p_2 V_2^\gamma$.

$p_1 h_1^\gamma A^\gamma = p_2 h_2^\gamma A^\gamma$

$h_2 = h_1 \left(\dfrac{p_1}{p_2}\right)^{1/\gamma} = (0.250 \text{ m}) \left(\dfrac{1.01 \times 10^5 \text{ Pa}}{4.81 \times 10^5 \text{ Pa}}\right)^{1/1.40} = 0.08199$ m

The piston has moved a distance $h_1 - h_2 = 0.250$ m $- 0.08199$ m $= 0.168$ m.

(b) SET UP: $T_1 V_1^{\gamma-1} = T_2 V_2^{\gamma-1}$

$T_1 h_1^{\gamma-1} A^{\gamma-1} = T_2 h_2^{\gamma-1} A^{\gamma-1}$

EXECUTE: $T_2 = T_1 \left(\dfrac{h_1}{h_2}\right)^{\gamma-1} = 300.1 \text{ K}\left(\dfrac{0.250 \text{ m}}{0.08199 \text{ m}}\right)^{0.40} = 468.7 \text{ K} = 196°C.$

(c) SET UP AND EXECUTE: $W = nC_V(T_1 - T_2)$ gives

$W = (20.0 \text{ mol})(20.8 \text{ J/mol} \cdot \text{K})(300.1 \text{ K} - 468.7 \text{ K}) = -7.01 \times 10^4 \text{ J} = -70.1 \text{ kJ}.$ This is the work done *by* the gas. The work done *on* the gas by the pump is +70.1 kJ.

EVALUATE: In an adiabatic compression of an ideal gas the temperature increases. In any compression the work W done by the gas is negative.

19.55. IDENTIFY and SET UP: Use the ideal gas law, the first law of thermodynamics, and expressions for Q and W for specific types of processes.

EXECUTE: (a) initial expansion (state 1 → state 2)
$p_1 = 2.40 \times 10^5$ Pa, $T_1 = 355$ K, $p_2 = 2.40 \times 10^5$ Pa, $V_2 = 2V_1$
$pV = nRT$; $T/V = p/nR =$ constant, so $T_1/V_1 = T_2/V_2$ and $T_2 = T_1(V_2/V_1) = 355 \text{ K}(2V_1/V_1) = 710 \text{ K}$
$\Delta p = 0$ so $W = p\Delta V = nR\Delta T = (0.250 \text{ mol})(8.3145 \text{ J/mol} \cdot \text{K})(710 \text{ K} - 355 \text{ K}) = +738 \text{ J}$
$Q = nC_p\Delta T = (0.250 \text{ mol})(29.17 \text{ J/mol} \cdot \text{K})(710 \text{ K} - 355 \text{ K}) = +2590 \text{ J}$
$\Delta U = Q - W = 2590 \text{ J} - 738 \text{ J} = 1850 \text{ J}$

(b) At the beginning of the final cooling process (cooling at constant volume), $T = 710$ K. The gas returns to its original volume and pressure, so also to its original temperature of 355 K.
$\Delta V = 0$ so $W = 0$
$Q = nC_V\Delta T = (0.250 \text{ mol})(20.85 \text{ J/mol} \cdot \text{K})(355 \text{ K} - 710 \text{ K}) = -1850 \text{ J}$
$\Delta U = Q - W = -1850 \text{ J}.$

(c) For any ideal gas process $\Delta U = nC_V \Delta T$. For an isothermal process $\Delta T = 0$, so $\Delta U = 0$.

EVALUATE: The three processes return the gas to its initial state, so $\Delta U_{\text{total}} = 0$; our results agree with this.

19.59. IDENTIFY: For an adiabatic process, no heat enters or leaves the gas. An isochoric process takes place at constant volume, and an isobaric process takes place at constant pressure. The first law of thermodynamics applies.

SET UP: For any process, including an isochoric process, $Q = nC_V\Delta T$, and for an isobaric process, $Q = nC_p \Delta T$. $Q = \Delta U + W$.

EXECUTE: (a) Process a is adiabatic since no heat goes into or out of the system. In processes b and c, the temperature change is the same, but more heat goes into the gas for process c. Since the change in internal energy is the same for both b and c, some of the heat in c must be doing work, but not in b. Therefore b is isochoric and c is isobaric. To summarize: a is adiabatic, b is isochoric, c is isobaric.

(b) $Q_b = nC_V\Delta T$ and $Q_c = nC_p \Delta T$. Subtracting gives
$Q_c - Q_b = nC_p \Delta T - nC_V\Delta T = n(C_p - C_V)\Delta T = nR\Delta T = 20$ J. Solving for ΔT gives
$\Delta T = (20 \text{ J})/nR = (20 \text{ J})/[(0.300 \text{ mol})(8.314 \text{ J/mol} \cdot \text{K})] = 8.0 \text{ C}°$, so $T_2 = 20.0°C + 8.0°C = 28.0°C.$

(c) $\dfrac{Q_c}{Q_b} = \dfrac{nC_p\Delta T}{nC_V\Delta T} = \dfrac{C_p}{C_V} = \gamma = \dfrac{50 \text{ J}}{30 \text{ J}} = \dfrac{5}{3}.$ Since $\gamma = 5/3$, the gas must be monatomic, in which case we have $C_V = 3/2 R$ and $C_p = 5/2 R$. Therefore
Process a: $Q = \Delta U + W$ gives $0 = nC_V\Delta T + W.$
$W = -n(3/2 R) \Delta T = -(0.300 \text{ mol})(3/2)(8.314 \text{ J/mol} \cdot \text{K})(8.0 \text{ K}) = -30 \text{ J}.$

Process b: The volume is constant, so $W = 0$.

Process c: $Q = \Delta U + W$. ΔU is the same as for process a because ΔT is the same, so we have 50 J = 30 J + W, which gives W = 20 J.

(d) The greatest work has the greatest volume change. Using the results of part (c), process a has the greatest amount of work and hence the greatest volume change.

(e) The volume is increasing if W is positive. Therefore

Process a: W is negative, so the volume decreases.
Process b: W = 0 so the volume stays the same.
Process c: W is positive, so the volume increases.

EVALUATE: In Process a, no heat enters the gas, yet its temperature increases. This means that work must have been done on the gas, as we found.

19.61. IDENTIFY: The air in a cylinder can be compressed by a moveable piston at one end. It goes through a cycle that ends in the same state at which it began. The compression ratio v is defined as $V_{max}/V_{min} = v$.

SET UP: Summarize the steps of the cycle and make a pV-diagram of the process, shown in Fig. 19.61.

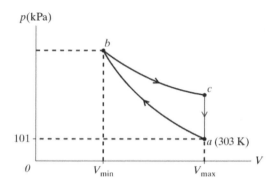

Figure 19.61

<u>Segment ab</u>: Beginning at ambient temperature, an adiabatic compression quickly increases the air temperature.
<u>Segment bc</u>: Isothermal expansion to maximum volume
<u>Segment ca</u>: Isochoric cooling back to ambient temperature
We know the following things about this system: For diatomic air $C_V = 20.8$ J/mol·K and $\gamma = 1.40$. The volume of the air in the cylinder is $V = AL$, where A is the area of the faces and L is the length of the gas-containing part of the cylinder. $L_{max} = 30.0$ cm $= 0.300$ m and $L_{min} = L_{max}/v = (0.300 \text{ m})/v$. The cycle starts at point a with $T_a = 30.0°C = 303$ K, $p_a = 101$ kPa, and $V_a = L_{max}A = (0.300 \text{ m})A$, where $A = \pi r^2$.

EXECUTE: (a) We want the work done by the air for a complete cycle abc. We need to break this process up into its three segments, so $W_{tot} = W_{ab} + W_{bc} + W_{ca}$. Segment ab is an adiabatic compression, so $W_{ab} = \frac{1}{\gamma-1}(p_aV_a - p_bV_b)$. Using $p_aV_a^\gamma = p_bV_b^\gamma$ with $V_b = V_a/v$, this becomes

$$W_{ab} = \frac{1}{\gamma-1}\left[p_aV_a - p_aV_b\left(\frac{V_a}{V_b}\right)^\gamma\right] = \frac{p_a}{\gamma-1}\left[V_a - v^\gamma(V_a/v)\right] = \frac{p_aV_a}{\gamma-1}\left(1 - v^\gamma/v\right) = \frac{p_aV_a}{\gamma-1}\left(1 - v^{\gamma-1}\right).$$

Segment bc is isothermal, so T = constant = T_b. Using $pV = nRT$ and $W_{bc} = \int_b^c pdV$, we have

$$W_{bc} = \int_b^c \frac{nRT_b}{V}dV = nRT_b \ln(V_c/V_b).$$ But $V_c = V_a$ and $V_b = V_a/v$, so $V_c/V_b = v$, so W_{bc} is

$W_{bc} = nRT_b \ln v = p_b V_b \ln v$. From $p_a V_a^\gamma = p_b V_b^\gamma$ we have $p_b = p_a \left(\dfrac{V_a}{V_b}\right)^\gamma$, and we also know that $V_b = V_a/v$, so $W_{bc} = p_a \left(\dfrac{V_a}{V_b}\right)^\gamma \dfrac{V_a}{v} \ln v = p_a V_a v^\gamma \dfrac{1}{v} \ln v = p_a V_a v^{\gamma-1} \ln v$. The work during segment ca is 0 because the volume is constant. $V_a = (0.300 \text{ m})A = (0.300 \text{ m})\pi(0.0150 \text{ m})^2 = 2.1205\times 10^{-4}$ m^3. Using the values for p_a, V_a, and γ we get $W_{ab} = (53.5 \text{ J})(1 - v^{0.40})$ and $W_{bc} = (21.4 \text{ J})v^{0.40} \ln v$.

$W_{\text{tot}} = 53.5 \text{ J} + v^{0.40}[(21.42 \text{ J})\ln v - 53.5 \text{ J}]$.

(b) We want the maximum integer value of v so that maximum temperature is no greater than 400°C. The maximum temperature occurs along segment bc, so the maximum that T_b can be is 400°C = 673 K. Using $T_b V_b^{\gamma-1} = T_a V_a^{\gamma-1}$ gives $T_b = T_a \left(\dfrac{V_a}{V_b}\right)^{\gamma-1} = T_a \left(\dfrac{vV_b}{V_b}\right)^{\gamma-1} = T_a v^{\gamma-1}$. This gives $v^{0.40} = 673/303$, so $v = (673/303)^{1/0.400} = 7.35$. Since v must be an integer, $v_{\max} = 7$. The temperature at b for this value of v is $T_b = T_a v^{\gamma-1} = (303 \text{ K})7^{0.40} = 660 \text{ K} = 387°C$. (Note that this is *not* 400°C, but the requirement was that the maximum temperature be *no greater than* 400°C, not that it be *equal to* 400°C.)

(c) We want the minimum value of v so that the gas does at least 25.0 J of work per cycle. Using $W_{\text{tot}} = 53.5 \text{ J} + v^{0.40}[(21.42 \text{ J})\ln v - 53.5 \text{ J}] = 25.0$ J gives $-28.55 \text{ J} = v^{0.40}[(21.42 \text{ J})\ln v - 53.5 \text{ J}]$. The LHS of this equation is negative, so the RHS must also be negative. The only way this is possible is for $[(21.42 \text{ J})\ln v - 53.5 \text{ J}]$ to be negative. This tells us that $\ln v < 53.5/21.42$, so $v < 12.18$. Since v must be an integer, $v < 12$. But in part (b) we saw that $v \leq 7$. We try $v = 7$ to see how much work is done. Using $v = 7$ in $W_{\text{tot}} = 53.5 \text{ J} + v^{0.40}[(21.42 \text{ J})\ln v - 53.5 \text{ J}]$ gives $= 27.7$ J, which is acceptable since it is more than 25.0 J. Now try $v = 6$, which gives $W_{\text{tot}} = 22.5$ J. This is not acceptable since it is less than 25.0 J, and no smaller integers would be acceptable either. Therefore $v_{\min} = 7$.

(d) $v \geq 7$ and $v \leq 7$, so $v = 7$.

(e) We want the heat that leaves the gas during the isochoric segment, which is ca. The work is zero, so $\Delta U = Q - W = Q$. Therefore $Q = \Delta U = nC_V \Delta T$. First we need n. Using $pV = nRT$ gives $n = p_a V_a/RT_a$, which gives $n = 8.50\times 10^{-3}$ mol. Now we can use $Q = \Delta U = nC_V \Delta T$. This gives $Q = (8.50\times 10^{-3} \text{ mol})(20.8 \text{ J/mol}\cdot\text{K})(303 \text{ K} - 660 \text{ K}) = -63.1$ J. The minus sign tells us that the heat comes *out of* the air.

EVALUATE: This device does 27.7 J of work per cycle operating between a high temperature of 660 K (387°C) and low temperature of 303 K (30°C).

Study Guide for The Second Law of Thermodynamics

Summary

In this chapter, we'll complete our investigation of thermodynamics, examining thermodynamic processes and the second law of thermodynamics. Heat engines and refrigerators transform heat into work or energy in cyclic processes. *Thermal efficiency* and *performance coefficients* for engines and refrigerators will be defined. The second law of thermodynamics limits the efficiency of engines and has profound implications in many physical processes. The second law can be quantified in terms of entropy, a measure of disorder. We'll examine several common cyclic processes to aid our understanding of thermodynamics.

Objectives

After studying this chapter, you'll understand
- How to define and identify *reversible processes*.
- How to analyze heat engines and refrigeration cycles.
- How to apply the second law of thermodynamics.
- How to calculate entropy for a variety of systems.
- How to apply thermodynamic principles to a variety of engine and refrigeration cycles.

Concepts and Equations

Term	Description
Directions of Thermodynamic Processes	Heat flows spontaneously from hotter objects to cooler objects in thermodynamic processes. A reversible or equilibrium process is a process that can be reversed by infinitesimal changes in the conditions of the process and in which the system is always in, or very close to, thermal equilibrium. All other thermodynamic processes are irreversible.
Heat Engine	A heat engine takes heat Q_H from a source, converts part of the heat to work W, and discards the remaining heat $\|Q_C\|$ at a lower temperature. The heat engine's thermal efficiency e is $$e = \frac{W}{Q_H} = 1 + \frac{Q_C}{Q_H} = 1 - \frac{|Q_C|}{|Q_H|}.$$
Otto Cycle	A gasoline engine operating in the Otto cycle has a theoretical maximum thermal efficiency given by $$e = 1 - \frac{1}{r^{\gamma-1}},$$ where r is the compression ratio.
Refrigerator	A refrigerator takes heat Q_C from a cold source, performs work W, and discards the heat $\|Q_H\|$ to a warmer source. The performance coefficient K is $$K = \frac{Q_C}{W} = \frac{|Q_C|}{|Q_H| - |Q_C|}.$$
The Second Law of Thermodynamics	The second law of thermodynamics states that it is impossible for any cyclic system to convert heat completely into work. It also states that no cyclic process can transfer heat from a cold place to a hot place without any input of work.
Carnot Cycle	The Carnot cycle operates between two heat reservoirs and represents the most efficient heat engine. The Carnot cycle combines the reversible adiabatic and isothermal expansion and contraction between two heat reservoirs at temperatures T_H and T_C, respectively. The efficiency of the Carnot cycle is $$e_{\text{Carnot}} = 1 - \frac{T_C}{T_H} = \frac{T_H - T_C}{T_H}.$$
Entropy	Entropy is a quantitative measure of the disorder of a system. The entropy change in a reversible thermodynamic system is $$\Delta S = \int_1^2 \frac{dQ}{T}.$$ The second law of thermodynamics can be stated as "The entropy of an isolated system may increase, but not decrease. The total entropy of a system interacting with its surroundings may never decrease."

Key Concept 1: A heat engine takes in energy in the form of heat from a hot reservoir, uses some of that energy to do work, and rejects the remaining energy as heat that goes into a cold reservoir. The fraction of the heat from the hot reservoir that is converted into work is called the thermal efficiency of the engine.

Key Concept 2: An engine that utilizes the Carnot cycle, which consists of only reversible isothermal and reversible adiabatic processes, has the greatest efficiency possible consistent with the second law of thermodynamics. The smaller the ratio of the temperature of the cold reservoir to the temperature of the hot reservoir, the greater the Carnot efficiency.

Key Concept 3: When analyzing any of the steps taken by a heat engine as it goes through a thermodynamic cycle, if you can calculate any two of the quantities ΔU, Q, and W, you can calculate the third using the first law of thermodynamics, $\Delta U = Q - W$. The efficiency of the engine equals the sum of the work done in all steps divided by the heat taken in from the hot reservoir.

Key Concept 4: A refrigerator takes heat in from a cold reservoir and rejects heat to a hot reservoir. Work must be done on the refrigerator to make this happen. The coefficient of performance K of the refrigerator equals the amount of heat that is rejected to the hot reservoir divided by the amount of work that must be done on the refrigerator. The greater the value of K, the more heat can be removed from the cold reservoir for a given expenditure of work.

Key Concept 5: To find the entropy change of a system that goes through a reversible thermodynamic process at a constant temperature T, divide the heat Q that flows into the system during the process by T. If heat flows *in* (Q is positive), the entropy and disorder of the system *increase;* if heat flows *out* (Q is negative), the entropy and disorder of the system *decrease.*

Key Concept 6: The entropy change ΔS of a thermodynamic system does not depend on the path taken between the initial and final states of the system. You can calculate ΔS for any initial and final states by analyzing a *reversible* process that takes the system between these states; in general this involves evaluating an integral.

Key Concept 7: In the special case where a system undergoes a reversible adiabatic process, the system undergoes *no* change in entropy.

Key Concept 8: To calculate the entropy change of a system in an *irreversible* process such as a free expansion, you must consider a *reversible* process that takes the system from the same initial state to the same final state.

Key Concept 9: For the special case of a heat engine that uses the Carnot cycle, the net entropy change of the engine and its environment (the hot and cold reservoirs) in a complete cycle is zero. This is not the case for other, less ideal cycles.

Key Concept 10: When heat flows irreversibly from an object at higher temperature to one at lower temperature, the entropy of the high-temperature object decreases and the entropy of the low-temperature object increases. The *net* entropy of the two objects together always increases.

Key Concept 11: An alternative way to calculate entropy changes is in terms of the number of microscopic states available to the system. If there are w possible microscopic states of the system for a given macroscopic state, the entropy for the macroscopic state equals the Boltzmann constant k multiplied by the natural logarithm of w.

Conceptual Questions

1: Cleaning your room

Your parents are always nagging you about cleaning your room. After learning about the second law of thermodynamics, you explain to your parents that it is impossible to clean your room, since cleaning would reduce the entropy inside your room and violate the second law of thermodynamics. Your mother recalls her college physics course and convinces you that you can clean your room without violating the second law. How does she convince you?

IDENTIFY, SET UP, AND EXECUTE Your mother agrees with you that the entropy of a closed system can never decrease. But she notes that when you clean your room, the system consists of you plus your belongings. You can decrease the entropy of your belongings in your room by increasing the entropy of your body, as long as the total entropy increases. You can certainly clean your room!

EVALUATE This problem shows how the entropy of isolated components in a system may decrease as long as the system's total entropy increases. It also shows that you shouldn't argue with your mother, although you may want to try the argument on your father, who doesn't remember his physics course.

2: Leaving a refrigerator door open to cool a room

When the air-conditioning system at your house fails, your younger brother suggests leaving the refrigerator door open to cool the house. Is this method effective?

IDENTIFY, SET UP, AND EXECUTE A refrigerator cools its contents by taking heat away from the contents, performing work, and expelling heat to a warmer region. The heat expelled is always greater than the heat removed from the contents. The refrigerator must add net heat to its surroundings. Opening the refrigerator will result in a warmer room, so it is not an effective method of cooling the room.

EVALUATE Can opening the refrigerator warm the house on a cold day? Yes, since it must expel heat to operate. It wouldn't be a very efficient heat source, but it would provide some heat to the room.

3: Water as a fuel

Some people have suggested using water as a clean fuel. The idea is to break apart water molecules into hydrogen and oxygen. Then, when the hydrogen (combined with water) is burned, it produces energy without pollution. How does the second law of thermodynamics relate to this idea?

IDENTIFY, SET UP, AND EXECUTE Because the breaking apart of the water and the burning of hydrogen constitutes a reversible cycle, the net entropy must increase. The process may actually create *more* pollution, since it takes more energy to dissociate the water than is recovered by burning the hydrogen. For example, if gasoline is used to generate the hydrogen, it would take more gasoline to generate an equivalent amount of hydrogen-based power than if the gasoline were used to operate the vehicle directly.

EVALUATE There is the possibility that pollution would be reduced. If the hydrogen-generating plants installed high-quality pollution filters, than there could be less pollution generated overall by the plant compared with the pollution generated by many cars. However, a hydrogen-burning car will always require more total energy to operate. Hydrogen should probably be considered an alternative energy storage method.

Problems

1: Work in a heat engine

A heat engine carries 0.2 mol of argon through the cyclic process shown in Figure 1. Process *ab* is isochoric, process *bc* is adiabatic, and process *ca* is isobaric at a pressure of 2.0 atm. Find the net work done by the gas in the complete cycle. The temperatures at the of endpoints of the process are $T_a = 290$ K, $T_b = 650$ K, and $T_c = 440$ K.

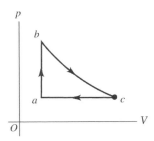

Figure 1 Problem 1.

IDENTIFY Use the principles of heat engines to find the work done.

SET UP We'll break the cycle up into processes and find the work done during each process. We'll need to find the pressure and volume at the three points before finding the work. Argon is an ideal gas, so we'll use the ideal-gas relations.

EXECUTE Starting at point a, we find the volume V_a from the ideal-gas equation:

$$V_a = \frac{nRT_a}{p_a} = \frac{(0.2 \text{ mol})(8.31 \text{ J/mol/K})(290 \text{ K})}{(2.02 \times 10^5 \text{ Pa})} = 2.39 \times 10^{-3} \text{ m}^3.$$

At point b, the volume is the same as at point a. We find the pressure at b:

$$p_b = \frac{nRT_b}{V_b} = \frac{(0.2 \text{ mol})(8.31 \text{ J/mol/K})(650 \text{ K})}{(2.39 \times 10^{-3} \text{ m}^3)} = 4.52 \times 10^5 \text{ Pa}.$$

We find the volume V_c at c:

$$V_c = \frac{nRT_c}{p_c} = \frac{(0.2 \text{ mol})(8.31 \text{ J/mol/K})(440 \text{ K})}{(2.02 \times 10^5 \text{ Pa})} = 3.62 \times 10^{-3} \text{ m}^3.$$

With these values, we find the work done during each process. There is no work done during the process ab, since it is an isochoric process. Process bc is adiabatic and there is no heat exchanged. The work is opposite the change in internal energy, or

$$W_{bc} = -\Delta U_{bc} = -nC_V(T_c - T_b) = -(0.2 \text{ mol})(12.47 \text{ J/mol/K})(440 \text{ K} - 650 \text{ K}) = 524 \text{ J},$$

where we used the molar heat capacity at constant volume for argon $(C_V = 12.47 = \text{J/mol/K},)$. Process ca is isobaric and the work is

$$W_{ca} = p_c \Delta V_{ca} = (2.02 \times 10^5 \text{ Pa})((2.39 \times 10^{-3} \text{ m}^3) - (3.62 \times 10^{-3} \text{ m}^3)) = -248 \text{ J}.$$

Note that the work done by the gas is negative, since it is compressed in ca. The work done during the complete cycle is the sum of the separate amounts of work done during each of the three processes:

$$W = W_{ab} + W_{bc} + W_{ca} = 0 + 524 \text{ J} - 248 \text{ J} = 276 \text{ J}.$$

The gas does 276 J of work in one cycle.

KEY CONCEPT 1 **EVALUATE** Since the area inside the pV cycle diagram is equal to the work, our positive result is in agreement with the area shown in the diagram.

Practice Problem: What is the net work done if there is 1 mole of argon? *Answer:* 1410 J

Extra Practice: What is the net work done if there is 1 mole of oxygen? *Answer:* 3170 J

2: Efficiency of a heat engine

Find the thermal efficiency of an engine that operates in accordance with the cycle shown in Figure 2, in which 2 moles of helium stored at 2.0 atm in a 10-liter vessel starts at point *a*, undergoes an isochoric process to quadruple its pressure at point *b*, triples in volume in an isobaric expansion to point *c*, reduces its pressure to one-fourth its pressure at point *c* through an isochoric process at point *d*, and goes through an isobaric compression reducing its volume by one-third to return to point *a*.

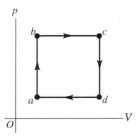

Figure 2 Problem 2.

IDENTIFY The target variable is the thermal efficiency of the engine.

SET UP We need to know the work and heat of the cycle to find the efficiency of the engine. We break the cycle up into processes and use our knowledge of isochoric (constant-volume) and isobaric (constant-pressure) processes. We are given the changes in pressure and volume, so we can proceed immediately to calculating the work done during each of the four processes. Helium is an ideal gas, so we can use the ideal-gas relations.

EXECUTE We start by determining the pressure and volume at the points *b*, *c*, and *d*. The pressure at points *b* and *c* is 8 atm $(4P_a)$ and the pressure at *d* is 2 atm (P_a). The volume at *b* is 10 liters (V_a) and the volume at *c* and *d* is 30 liters $(3V_a)$. Next, we find the work for each process. No work is done during processes *ab* and *cd*, since they are isochoric. Process *bc* is isobaric and the work done is

$$W_{bc} = p_b \Delta V_{bc} = (8.08 \times 10^5 \text{ Pa})((30 \times 10^{-3} \text{ m}^3) - (10 \times 10^{-3} \text{ m}^3)) = 16,200 \text{ J}.$$

Process *da* is also isobaric and the work done is

$$W_{da} = p_d \Delta V_{da} = (2.02 \times 10^5 \text{ Pa})((10 \times 10^{-3} \text{ m}^3) - (30 \times 10^{-3} \text{ m}^3)) = -4040 \text{ J}.$$

The total work is the sum of the separate amounts of work done during each of the four processes:

$$W = W_{ab} + W_{bc} + W_{cd} + W_{da} = 0 + 16,200 \text{ J} + 0 - 4040 \text{ J} = 12,100 \text{ J}.$$

We need to find the heat flowing into the engine. Heat flows into the engine during processes *ab* and *bc*. To find the heat flow into the engine, we need to know the temperature at points *a*, *b*, and *c*. The ideal-gas equation gives

$$T_a = \frac{p_a V_a}{nR} = \frac{(2.02 \times 10^5 \text{ Pa})(10 \times 10^{-3} \text{ m}^3)}{(2 \text{ mol})(8.31 \text{ J/mol/K})} = 122 \text{ K},$$

$$T_b = \frac{p_b V_b}{nR} = \frac{(8.08 \times 10^5 \text{ Pa})(10 \times 10^{-3} \text{ m}^3)}{(2 \text{ mol})(8.31 \text{ J/mol/K})} = 486 \text{ K},$$

$$T_c = \frac{p_c V_c}{nR} = \frac{(8.08 \times 10^5 \text{ Pa})(30 \times 10^{-3} \text{ m}^3)}{(2 \text{ mol})(8.31 \text{ J/mol/K})} = 1460 \text{ K}.$$

The heat flow in process ab (carried out at constant volume) is

$$Q_{ab} = nC_V \Delta T_{ab} = (2 \text{ mol})(12.47 \text{ J/mol/K})(486 \text{ K} - 122 \text{ K}) = 9080 \text{ J},$$

where we used the molar heat capacity at constant volume for helium ($C_V = 12.47$ J/mol/K,).
The heat flow in process bc (carried out at constant pressure) is

$$Q_{bc} = nC_P \Delta T_{bc} = (2 \text{ mol})(20.78 \text{ J/mol/K})(1460 \text{ K} - 486 \text{ K}) = 40{,}500 \text{ J},$$

where we used the molar heat capacity at constant pressure for helium ($C_P = 20.78 =$ J/mol/K).
The total heat flowing into the engine in one cycle is therefore

$$Q_H = Q_{ab} + Q_{bc} = 40{,}500 \text{ J} + 9080 \text{ J} = 49{,}600 \text{ J}.$$

The efficiency of the engine is

$$e = \frac{W}{Q_H} = \frac{12{,}100 \text{ J}}{49{,}600 \text{ J}} = 24.4\%.$$

KEY CONCEPT 3 **EVALUATE** We see that the engine is 24.4% efficient. Note that, to find the efficiency, we started by determining the state variables for the points on the pV diagram. Then we found the work and heat flow in the cycle and combined these two pieces of information to solve the problem.

Practice Problem: What is the net work done if the initial pressure is 1 atm? *Answer:* 6060 J

Extra Practice: What is the heat flow into the engine in this case? *Answer:* 24,700 J

3: Efficiency of a diesel engine

Find the thermal efficiency of the diesel cycle shown in Figure 3 for an engine having a compression ratio $r = 18$, an expansion ratio $E = 6 = rV/V_C$, and $C_P/C_V = 1.4$.

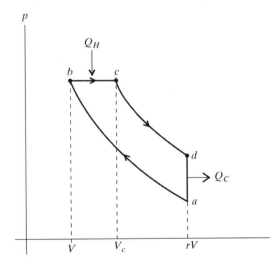

Figure 3 Problem 3.

IDENTIFY The target variable is the thermal efficiency of the engine. The efficiency involves the heat into and out of the system.

SET UP We start by finding the heat into and out of the system. Along the adiabatic lines, no heat is transferred. We'll be able to find the heat in terms of the changes in temperature and then solve for the temperatures. We'll substitute the compression and expansion factors where appropriate.

EXECUTE The efficiency is given by

$$e = 1 + \frac{Q_C}{Q_H}.$$

The heat entering the cycle between points b and c is given by

$$Q_{bc} = nC_p(T_c - T_b).$$

The heat leaving the cycle between points d and a is given by

$$Q_{da} = nC_V(T_a - T_d).$$

No heat is exchanged between any other pairs of points, as they lie upon adiabatic lines. The efficiency is then

$$e = 1 + \frac{T_a - T_d}{\gamma(T_c - T_b)},$$

where we replaced the ratio of the specific heats with γ. To find the efficiency, we need to find the temperatures. We seek relations between the temperatures that should cancel out. Points b and c are at the same pressure, so we have

$$\frac{V}{T_b} = \frac{V_c}{T_c}.$$

The ratio of the volumes can be substituted for the volumes:

$$T_c = T_b \frac{r}{E}.$$

To relate these temperatures to the temperature at a, we use the adiabatic relation

$$T_b V^{\gamma-1} = T_a V_a^{\gamma-1} = T_a(rV)^{\gamma-1}.$$

So we have

$$T_b = T_a r^{\gamma-1}, \qquad T_c = T_a r^{\gamma-1} \frac{r}{E}.$$

We find the temperature at point d by using the other adiabatic line:

$$T_c V_c^{\gamma-1} = T_d V_d^{\gamma-1} = T_d(rV)^{\gamma-1}.$$

Solving for T_d yields

$$T_d = \left(\frac{r}{E}\right)^{\gamma} T_a.$$

We now replace the temperatures with their expressions in terms of T_a to find the efficiency:

Study Guide for The Second Law of Thermodynamics

$$e = 1 + \frac{T_a - T_d}{\gamma(T_c - T_b)}$$

$$= 1 + \frac{(1 - (r/E)^\gamma)}{\gamma(r^\gamma/E - r^{\gamma-1})}$$

$$= 1 + \frac{(1 - (18/6)^{1.4})}{1.4(18^{1.4}/6 - 18^{0.4})} = 0.59.$$

The efficiency is 59%.

EVALUATE We see that the diesel engine is 59% efficient. Diesel engines are generally more efficient than gasoline engines.

4: Entropy change in melting ice

A heat reservoir at 50°C is used to melt 25 kg of ice at 0°C. What is the entropy change in the melted ice? What is the entropy change in the reservoir? What is the total entropy change in the system?

IDENTIFY The target variables are the entropy changes for the ice, reservoir, and total system.

SET UP Entropy change in a reversible process is equal to the heat transferred divided by the temperature of the material. We can find the heat required to melt the ice, which must be equal to the heat provided by the reservoir.

EXECUTE The heat required to melt the ice is given by the heat of fusion relation and is

$$Q = mL_f = (25 \text{ kg})(335 \times 10^3 \text{ J/kg}) = 8.375 \times 10^6 \text{ J},$$

where we used the latent heat of fusion for ice (335×10^3 J/kg). The change in entropy for the water is

$$\Delta S_{\text{ice}} = \frac{Q}{T} = \frac{8.375 \times 10^6 \text{ J}}{273 \text{ K}} = 30,700 \text{ J/K}.$$

The change in entropy for the reservoir is equal to the heat leaving the reservoir divided by the temperature of the reservoir. The reservoir loses as much heat as the ice gains. The entropy change is

$$\Delta S_{\text{reservoir}} = \frac{-Q}{T} = \frac{-8.375 \times 10^6 \text{ J}}{323 \text{ K}} = 25,900 \text{ J/K}.$$

The total change in entropy is the sum of the entropies for the ice and reservoir:

$$\Delta S_{\text{total}} = \Delta S_{\text{ice}} = \Delta S_{\text{reservoir}} = 30,700 \text{ J/K} - 25,900 \text{ J/K} = 4800 \text{ J/K}.$$

The entropy of the ice increases by 30,700 J/K, the entropy of the reservoir decreases by 25,900 J/K, and the total entropy of the system increases by 4800 J/K.

KEY CONCEPT 5 **EVALUATE** The entropy of the reservoir decreased, but this does not violate the second law, since the system is the combination of the ice, the water, and the reservoir. The total change in entropy is positive, as expected.

Practice Problem: What is the total change in entropy of the system if the heat reservoir is at 20°C?
Answer: 2120 J/K

Extra Practice: What temperature does the reservoir need to be at so that the total change in entropy is 500 J/K? *Answer:* 4. °C

5: Entropy change in isothermal expansion

Find the change in entropy for an ideal gas that undergoes an isothermal expansion from an initial volume to a final volume that is twice the initial volume.

IDENTIFY The target variable is the entropy change for the gas.

SET UP Entropy change is related to the work and the temperature of the material. For an ideal gas, the internal energy doesn't change. We'll use the standard definitions to find the entropy change in the process.

EXECUTE Writing the change in entropy in terms of work and temperature gives
$$TdS = dQ = 0 + dW,$$
since the change in internal energy is zero. The work done by an ideal gas undergoing an isothermal process is given by
$$dW = pdV = \frac{nRT}{V}dV.$$
Combining the two equations yields an expression for the change in entropy:
$$dS = \frac{nR}{V}dV.$$
We integrate to find the change:
$$S = \int_{v_1}^{v_2} \frac{nR}{V}dV = nR \ln V \Big|_{v}^{2v} = nR \ln 2.$$
The change in entropy is $R \ln 2$ for each mole of gas.

KEY CONCEPT 6 **EVALUATE** The entropy of the gas increased, as expected.

Try It Yourself!

1: Efficiency of a heat engine

Following the cycle shown in Figure 4, find the efficiency of a heat engine using an ideal monatomic gas as its working substance. Take $\gamma = 1.4$ and $r = V_b/V_a = 2.5$.

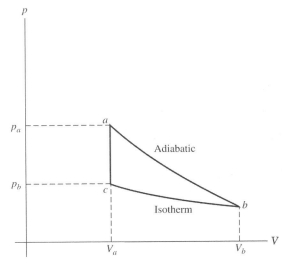

Figure 4 Try It Yourself 1.

Solution Checkpoints

IDENTIFY AND SET UP Break the process up into segments and the heat exchanged during each segment.

EXECUTE No heat is exchanged in segment ab. (Why?) Along segment bc, heat is removed from the system and is equivalent to

$$Q = nRT_c \ln(V_a/V_b).$$

Along segment ca, heat is put into the system and is equivalent to

$$Q = nC_V(T_a - T_c).$$

The temperatures are written in terms of T_c to find the efficiency, which is

$$e = 1 - \frac{((\gamma - 1) \ln r)}{r^{\gamma - 1} - 1}.$$

EVALUATE What effect does increasing r have on the efficiency?

Practice Problem: What is the efficiency when $r = 4.0$? *Answer:* 25.2%

Extra Practice: What is the efficiency when $r = 18$? *Answer:* 46.9%

2: Entropy in mixed water

Equal volumes of water at 80°C and 20°C are mixed together. Find the increase in entropy for a total volume of 1.0 m^3.

Solution Checkpoints

IDENTIFY AND SET UP The final temperature of the mixture should be halfway between the initial temperatures of the two volumes of water if the heat capacities are constant and independent of temperature. No process is noted, so we can choose any reversible process. Assume that the process keeps the volume constant.

EXECUTE The change in entropy for the system is

$$\Delta S = \int_1^2 \frac{dQ}{T} = C_V \int_{T_1}^{T_2} \frac{dT}{T} = C_V \ln\left(\frac{T_2}{T_1}\right),$$

where the temperatures are in kelvins. The entropy of the hot water decreases, while the entropy of the cold water increases. The total change in entropy is 3.62×10^4 J/K.

KEY CONCEPT 10 EVALUATE How did you determine C_V?

Practice Problem: What is the increase in entropy if the temperatures are 60°C and 40°C? *Answer:* 4020 J/K

Extra Practice: What is the increase in entropy if the temperatures are 99°C and 1°C? *Answer:* 97,600 J/K

Key Example Variation Problems

Solutions to these problems are in Chapter 20 of the Student's Solutions Manual.

Be sure to review EXAMPLE 20.1 (Section 20.2) before attempting these problems.

VP20.1.1 A diesel engine has efficiency 0.180. (a) In order for this engine to do 1.24×10^4 J of work, how many joules of heat must it take in? (b) How many joules of this heat is discarded?

VP20.1.2 In one cycle a heat engine absorbs 3.82×10^4 J of heat from the hot reservoir and rejects 3.16×10^4 J of heat to the cold reservoir. What is the efficiency of this engine?

VP20.1.3 Measurements of a gasoline engine show that it has an efficiency of 0.196 and that it exhausts 4.96×10^8 J of heat during 20 minutes of operation. During that time, (a) how much heat does the engine take in and (b) how much work does the engine do?

VP20.1.4 An aircraft piston engine that burns gasoline (heat of combustion 5.0×10^7 J/kg) has a power output of 1.10×10^5 W. (a) How much work does this engine do in 1.00 h? (b) This engine burns 34 kg of gasoline per hour. How much heat does the engine take in per hour? (c) What is the efficiency of the engine?

Be sure to review EXAMPLES 20.2, 20.3, and 20.4 (Section 20.6) before attempting these problems.

VP20.4.1 In one cycle a Carnot engine takes in 8.00×10^4 J of heat and does 1.68×10^4 J of work. The temperature of the engine's cold reservoir is 25.0°C. (a) What is the efficiency of this engine? (b) How much heat does this engine exhaust per cycle? (c) What is the temperature (in °C) of the hot reservoir?

VP20.4.2 For the Carnot cycle described in Example 20.3, you change the temperature of the cold reservoir from 27°C to −73°C. The initial pressure and volume at point a are unchanged, the volume still doubles during the isothermal expansion $a \rightarrow b$, and the volume still decreases by one-half during the isothermal compression $c \to d$. For this modified cycle, calculate (a) the new efficiency of the cycle and (b) the amount of work done in each of the four steps of the cycle.

VP20.4.3 A Carnot refrigerator has a cold reservoir at $-10.0°C$ and a hot reservoir at $25.0°C$. (a) What is its coefficient of performance? (b) How much work input does this refrigerator require to remove 4.00×10^6 J of heat from the cold reservoir?

VP20.4.4 A Carnot engine uses the expansion and compression of n moles of argon gas, for which $C_V = \frac{3}{2}R$. This engine operates between temperatures T_C and T_H. During the isothermal expansion $a \to b$, the volume of the gas increases from V_a to $V_b = 2V_a$. (a) Calculate the work W_{ab} done during the isothermal expansion $a \to b$. Give your answer in terms of n, R, and T_H. (b) Calculate the work W_{bc} done during the adiabatic expansion $b \to c$. Give your answer in terms of n, R, T_C and T_H. (c) For this engine, $W_{ab} = W_{bc}$. Find the ratio T_C/T_H and the efficiency of the engine.

Be sure to review EXAMPLES 20.6, 20.7, 20.8, 20.9, and 20.10 (Section 20.7) before attempting these problems.

VP20.10.1 Ethanol melts at 159 K (heat of fusion 1.042×10^5 J/kg) and boils at 351 K (heat of vaporization 8.54×10^5 J/kg). Liquid ethanol has a specific heat of 2428 J/kg·K (which we assume does not depend on temperature). If you have 1.00 kg of ethanol originally in the solid state at 159 K, calculate the change in entropy of the ethanol when it (a) melts at 159 K, (b) increases in temperature as a liquid from 159 K to 351 K, and (c) boils at 351 K.

VP20.10.2 Initially 5.00 mol of helium (which we can treat as an ideal gas) occupies volume 0.120 m3 and is at temperature $20.0°C$. You allow the helium to expand so that its final volume is 0.360 m^3 and its final temperature is also $20.0°C$. Calculate the net change in entropy of the helium if (a) you make the helium expand isothermally from 0.120 m^3 to 0.360 m^3 and (b) you first increase the temperature of the helium so that it expands at constant pressure from 0.120 m^3 to 0.360 m^3, then cool the helium at constant volume to $20.0°C$.

VP20.10.3 In one cycle of operation, an engine that does *not* use the Carnot cycle takes in 8.00×10^4 J of heat from a reservoir at $260°C$, does 1.60×10^4 J of work, and rejects 6.40×10^4 J of heat to a reservoir at $20°C$. At the end of the cycle, the engine is in the same state as at the beginning of the cycle. Calculate the change in entropy in one cycle for (a) the engine, (b) the reservoir at $260°C$, (c) the reservoir at $20°C$, and (d) the system of engine and two reservoirs. Does the net entropy of the system increase, decrease, or stay the same?

VP20.10.4 You combine 1.00 kg of ice (heat of fusion 3.34×10^5 J/K) at 0.0°C and 0.839 kg of liquid water (specific heat 4.19×10^3 J/K)) at 95.0°C. When the system comes to equilibrium, all of the ice has melted and the temperature of the mixture is 0.0°C. Calculate the change in entropy for (a) the ice as it melts, (b) the water that was initially at 95.0°C, and (c) the combination of the two. Does the net entropy of the system increase, decrease, or stay the same?

STUDENT'S SOLUTIONS MANUAL FOR THE SECOND LAW OF THERMODYNAMICS

VP20.1.1. IDENTIFY: We are dealing the efficiency of a diesel engine.
SET UP: Efficiency = $e = W/Q_H$ and $Q_H = W + Q_C$.
EXECUTE: **(a)** We want the heat taken in, which is Q_H. Solve $e = W/Q_H$ for Q_H giving

$$\frac{1.16K_B}{K_B} = \frac{T_C(A)}{T_C(B)} \frac{\Delta T_B}{\Delta T_A} = \frac{T_C(A)}{T_C(B)} \frac{1.30\Delta T_A}{\Delta T_A}$$

(b) $Q_C = W + Q_H = Q = mc\Delta T. - 1.24 \times 10^4$ J $= 5.65 \times 10^4$ J.
EVALUATE: Only 18% of the heat taken in is converted to work. This is around the efficiency of automobile engines.

VP20.1.2. IDENTIFY: We are dealing with the efficiency of a heat engine.
SET UP and EXECUTE: $e = 1 - \left|\frac{Q_C}{Q_H}\right| = 1 - \frac{3.16 \times 10^4 \text{ J}}{3.82 \times 10^4 \text{ J}} = 0.173$.
EVALUATE: Only 17.3% of the heat input is converted to work; the rest is rejected from the engine as waste heat.

VP20.1.3. IDENTIFY: We are dealing with the efficiency of a gasoline heat engine.
SET UP: The efficiency is $e = 0.196$ and $Q_C = 4.96 \times 10^8$ J in 20 min.
EXECUTE: **(a)** We want to find Q_H. $e = 1 - \left|\frac{Q_C}{Q_H}\right| = 1 - \frac{4.96 \times 10^8 \text{ J}}{Q_H} = 0.196$. $Q_H = 6.17 \times 10^8$ J.
(b) We want the work W. $W = Q_H - Q_C = 6.17 \times 10^8$ J $- 4.96 \times 10^8$ J $= 1.21 \times 10^8$ J.
EVALUATE: The power output is $P = W/t = (1.21 \times 10^8$ J$)/[(20)(60)$ s$] = Q = W. = 101$ kW.

VP20.1.4. IDENTIFY: A heat engine has a power output of 1.10×10^5 W and gets its heat input from the combustion of gasoline.
SET UP and EXECUTE: The heat of combustion of gasoline is 5.0×10^7 J/kg.
(a) We want to see how much work this engine does in one hour. The power output is the rate at which the engine does work, so the work it does in one hour is $W = Pt = (1.10 \times 10^5$ W$)(3600$ s$) = 3.96 \times 10^8$ J.
(b) We want the heat input. $Q_H = (34$ kg$)(5.0 \times 10^7$ J/kg$) = \frac{Q_C}{Q_H} = -\frac{T_C}{T_H}$.
(c) We want the efficiency. $C_V = R/(\gamma - 1) = 20.79$ J/mol·K;

EVALUATE: The heat wasted each hour is $Q_H - W =$
$Q = nC_V\Delta T = (0.350 \text{ mol})(20.79 \text{ J/mol}\cdot\text{K})(600\text{ K} - 300\text{ K}) = 2180\text{ J} - 3.96\times10^8\text{ J} =$
$W = p\Delta V = (1.013\times10^5 \text{ Pa})(8.62\times10^{-3}\text{ m}^3 - 14.1\times10^{-3}\text{ m}^3) = -560\text{ J}$

VP20.4.1. **IDENTIFY:** Th3is problem deals with a Carnot heat engine.

SET UP: For any heat engine $e = W/Q_H$ and $Q_H = W + Q_C$, and for a Carnot engine $e = 1 - \dfrac{T_C}{T_H}$.

EXECUTE: **(a)** We want the efficiency.
$\Delta U = nC_V\Delta T = (0.350\text{ mol})(20.79\text{ J/mol}\cdot\text{K})(300\text{ K} - 492\text{ K}) = -1400\text{ J}$,

(b) We want Q_C. $Q_H = W + Q_C$, so $Q_C = 8.00\times10^4\text{ J} - 1.68\times10^4\text{ J} = 6.32\times10^4\text{ J}$.
(c) We want T_H. $C_p = C_V + R = 7R/2$ gives $\Delta V = 0$ $T_H = 377\text{ K} = 104°\text{C}$.

EVALUATE: The temperature must be in Kelvins to use these formulas.

VP20.4.2. **IDENTIFY:** We are analyzing a Carnot heat engine.

SET UP and EXECUTE: **(a)** We want the efficiency. $e = 1 - \dfrac{T_C}{T_H} = 1 - (200\text{ K})/(500\text{ K}) = 0.60$.

(b) Segment ab is the same as in Example 20.3, so $W_{ab} = 576$ J.
$W_{bc} = nC_V(T_H - T_C) = (0.200\text{ mol})(20.8\ \Delta p = 0)\ (500\text{ K} - 200\text{ K}) = 1250\text{ J}$.
$W_{cd} = nRT_C\ln(V_d/V_c) = (0.200\text{ mol})(8.314\text{ J/mol}\cdot\text{K})(200\text{ K})\ln(1/2) = -231\text{ J}$.
$W_{da} = nC_V(T_C - T_H) = (0.200\text{ mol})(20.8\text{ J/mol}\cdot\text{K})(200\text{ K} - 500\text{ K}) = -1250\text{ J}$.
EVALUATE: The total work is 576 J + 1250 J − 231 J − 1250 J = 345 J. This is much more work than in Example 20.3 due to the greater temperature difference between the hot and cold reservoirs. We get much more work from the same heat input of 576 J.

VP20.4.3. **IDENTIFY:** This problem is about a refrigerator operating on a Carnot cycle. $T_C = -10.0°\text{C} = 263.15\text{ K}$ and $T_H = 25.0°\text{C} = 298.15\text{ K}$.
SET UP and EXECUTE: **(a)** We want the coefficient of performance K.
$K = \dfrac{T_C}{T_H - T_C} = \dfrac{263.15\text{ K}}{298\text{ K} - 263\text{ K}} = 7.52$.

(b) We want the work input. $K = \dfrac{|Q_C|}{|Q_H| - |Q_C|} = \dfrac{|Q_C|}{W}$. $W = \dfrac{|Q_C|}{K} = \dfrac{4.00\times10^6\text{ J}}{7.52} = 5.32\times10^5\text{ J}$.

EVALUATE: $Q_H = W + |Q_C| = W = 0 + 4.00\times10^6\text{ J} = 4.53\times10^6\text{ J}$. Each cycle this refrigerator discharges 4.53×10^6 J of heat into the room.

VP20.4.4. **IDENTIFY:** We are dealing with a Carnot heat engine. We know that $V_b = 2V_a$.
SET UP and EXECUTE: **(a)** We want the work done during the isothermal expansion ab.
$e = \dfrac{W}{Q_H} = \dfrac{610\text{ J}}{7.01\times10^3\text{ J}} = 0.087 = 8.7\%$.

(b) We want the work done during the adiabatic expansion bc. $Q = 0$, so $W_{bc} = Q_{bc}$. So
$W_{bc} = nC_V\Delta T_{bc} = n\left(\dfrac{3}{2}R\right)(T_H - T_C) = \dfrac{3}{2}nR(T_H - T_C)$.

(c) Given that $W_{ab} = W_{bc}$, we want e and the ratio of the cold to hot temperatures. Using the results from parts (a) and (b) gives $nRT_H\ln 2 = \dfrac{3}{2}nR(T_H - T_C)$, which gives $\dfrac{T_C}{T_H} = 1 - \dfrac{2}{3}\ln 2$. The efficiency is $e = 1 - \dfrac{279.15\text{K}}{300.15\text{K}} = 7.0\%. = 1 - \left(1 - \dfrac{2}{3}\ln 2\right) = \dfrac{2}{3}\ln 2 = 0.462$.

EVALUATE: This is quite an efficient engine. It is not necessarily true that $W_{ab} = W_{bc}$ for any Carnot engine.

VP20.10.1. IDENTIFY: We want to calculate the entropy changes due to melting, boiling, and temperature change.

SET UP and EXECUTE: For constant temperature $\Delta S = \dfrac{Q}{T}$, and for changing temperature ΔU We want the entropy change in each case.

(a) The temperature stays constant, so $\Delta S = \dfrac{Q}{T} = \dfrac{mL_f}{T} = \dfrac{(1.00 \text{ kg})(1.042 \times 10^5 \text{ J/kg})}{159 \text{ K}} = 655 \text{ J/K}$.

(b) The temperature increases, so $\Delta S = \int \dfrac{dQ}{T} = \int_{T_1}^{T_2} \dfrac{mc\,dT}{T} = mc\ln(T_2/T_1) = 1920 \text{ J/K}$ using $T_1 = 159$ K, $T_2 = 159$ K, $m = 1.00$ kg, and $c = 2428$ J/kg·K.

(c) The temperature stays constant, so $\Delta S = \dfrac{Q}{T} = \dfrac{mL_v}{T} = \dfrac{(1.00 \text{ kg})(8.54 \times 10^5 \text{ J/kg})}{351 \text{ K}} = 2430 \text{ J/K}$.

EVALUATE: The greatest increase is for boiling because $L_v \gg L_f$.

VP20.10.2. IDENTIFY: This problem involves the entropy change due to helium expansion by two different processes. We want to compare the entropy change in the two cases.

SET UP and EXECUTE: For constant temperature $\Delta S = \dfrac{Q}{T}$, and for changing temperature $\Delta S = \int \dfrac{dQ}{T}$.

(a) For an isothermal expansion we use $\Delta S = \dfrac{Q}{T}$. Since T is constant, $\Delta U = 0$, so $Q = W$. This gives

$$Q = W = \int_{V_1}^{V_2} p\,dV = \int_{V_1}^{V_2} \dfrac{nRT}{V}dV = nRT\ln(V_2/V_1). \quad \Delta S = \dfrac{Q}{T} = \dfrac{nRT\ln(V_2/V_1)}{T} = nR\ln(V_2/V_1).$$

Putting in the numbers gives $\Delta S = (5.00 \text{ mol})(8.314 \text{ J/mol·K})\ln\left(\dfrac{0.360 \text{ m}^3}{0.120 \text{ m}^3}\right) = 45.7 \text{ J/K}$.

(b) Start with a pV-diagram showing the process (Fig. VP20.10.2). Break the process into its two segments: ab is isobaric and bc is isochoric. We know that $T_a = T_c = 20.0°C$ and $V_c = V_b = 3V_a$.

<u>Segment ab (constant pressure):</u> The temperature is not constant, so we use $\Delta S = \int \dfrac{dQ}{T}$. The first law of thermodynamics (in its differential form) gives $dQ = dU + dW$. The pressure is constant, so $pV = nRT$ gives $p\,dV = nR\,dT$. Using this in the entropy calculation gives

$$dS_{ab} = \dfrac{dQ}{T} = \dfrac{dU + dW}{T} = \dfrac{nC_V dT}{T} + \dfrac{p\,dV}{T} = \dfrac{nC_V dT}{T} + \dfrac{nR\,dT}{T}.$$ Integrating gives

$$\Delta S_{ab} = \int_{T_a}^{T_b} \left(\dfrac{nC_V}{T} + \dfrac{nR}{T}\right)dT = nC_V \ln(T_b/T_a) + nR\ln(T_b/T_a).$$

<u>Segment bc (constant volume):</u> Since the volume is constant, $dW = 0$, so $dQ = dU = nC_V dT$. Therefore $\Delta S_{bc} = \int \dfrac{dQ}{T} = \int \dfrac{dU}{T} = \int_{T_b}^{T_c} \dfrac{nC_V dT}{T} = nC_V \ln(T_c/T_b)$.

The total entropy change is $\Delta S_{ac} = \Delta S_{ab} + \Delta S_{bc} = nC_V \ln(T_b/T_a) + nR\ln(T_b/T_a) + nC_V\ln(T_c/T_b)$. But $T_a = T_c = 20.0°C$, so $T_c/T_b = T_a/T_b$ which means that the first and third terms in the last sum are equal but have opposite signs, so they cancel. This leaves $\Delta S_{ac} = nR\ln(T_b/T_a)$. Using $pV = nRT$ for constant pressure tells us that $\dfrac{T_b}{T_a} = \dfrac{V_b}{V_a} = 3$. Thus the total entropy change is $\Delta S_{ac} = nR\ln 3 = (5.00 \text{ mol})(8.314 \text{ J/mol·K})\ln 3 = 45.7 \text{ J/K}$.

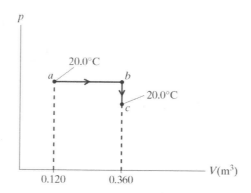

Figure VP20 10.2

EVALUATE: Notice that the entropy change is the same for both paths between the same initial and final states.

VP20.10.3. IDENTIFY: We are dealing the entropy change of a heat engine that is *not* using a Carnot cycle.

SET UP: For a constant temperature process $\Delta S = \dfrac{Q}{T}$. For one cycle, we want the entropy changes of the gas in the engine, the hot reservoir, and the cold reservoir. The two reservoirs are assumed to be so large that they do not change their temperatures as the engine runs. $T_H = 260°C = 533.15$ K and $T_C = 20°C = 293.15$ K.

EXECUTE: (a) $\Delta S = 0$ for a reversible cycle.

(b) The engine takes a quantity of heat Q_H *out* of the hot reservoir, so

$\Delta S_{\text{hot res}} = \dfrac{Q}{T} = \dfrac{-Q_H}{T_H} = \dfrac{-8.00\times 10^4 \text{ J}}{533.15 \text{ K}} = -150 \text{ J/K}$.

(b) The engine transfers a quantity of heat Q_C *into* the hot reservoir, so

$\Delta S_{\text{cold res}} = \dfrac{Q}{T} = \dfrac{+Q_C}{T_H} = \dfrac{+6.40\times 10^4 \text{ J}}{293.15 \text{ K}} = +218 \text{ J/K}$.

(c) $\Delta S_{\text{tot}} = \Delta S_{\text{engine}} + \Delta S_{\text{hot res}} + \Delta S_{\text{cold res}} = 0 - 150 \text{ J/K} + 218 \text{ J/K} = +68 \text{ J/K}$. The entropy of the system *increases*.

EVALUATE: The entropy of the universe (that is, the total entropy) increases.

VP20.10.4. IDENTIFY: We are looking at the entropy change for an *irreversible* process.

SET UP: For ice melting at constant temperature $\Delta S = \dfrac{Q}{T}$, and for water changing temperature $\Delta S = \int \dfrac{dQ}{T}$. We want the entropy change of the ice and of the water and the total entropy change for the entire system. $T_1 = 0.0°C = 273$ K and $T_2 = 95.0°C = 368$ K.

EXECUTE: (a) $\Delta S_{\text{ice}} = \dfrac{Q}{T} = \dfrac{mL_f}{T} = \dfrac{(1.00 \text{ kg})(3.34\times 10^5 \text{ J/kg})}{273.15 \text{ K}} = +1220 \text{ J/K}$.

(b) $\Delta S_{\text{water}} = \int \frac{dQ}{T} = \int_{368\text{ K}}^{273\text{ K}} \frac{mcdT}{T} = (0.839\text{ kg})(4190\text{ J/kg}\cdot\text{K})\ln\left(\frac{273\text{ K}}{368\text{ K}}\right) = -1050\text{ J/K}.$

(c) $\Delta S_{\text{system}} = +1220\text{ J/K} - 1050\text{ J/K} = +170\text{ J/K}.$ The entropy increases.

EVALUATE: This is an irreversible process so the entropy should increase, as we've found.

20.3. IDENTIFY and **SET UP:** The problem deals with a heat engine. $W = +3700$ W and $Q_H = +16,100$ J. Use $e = \frac{W}{Q_H} = 1 - \left|\frac{Q_C}{Q_H}\right|$ to calculate the efficiency e and $W = |Q_H| - |Q_C|$ to calculate $|Q_C|$. Power $= W/t$.

EXECUTE: (a) $e = \frac{\text{work output}}{\text{heat energy input}} = \frac{W}{Q_H} = \frac{3700\text{ J}}{16,100\text{ J}} = 0.23 = 23\%.$

(b) $W = Q = |Q_H| - |Q_C|$
Heat discarded is $|Q_C| = |Q_H| - W = 16,100\text{ J} - 3700\text{ J} = 12,400\text{ J}.$

(c) Q_H is supplied by burning fuel; $Q_H = mL_c$ where L_c is the heat of combustion.
$m = \frac{Q_H}{L_c} = \frac{16,100\text{ J}}{4.60\times 10^4\text{ J/g}} = 0.350\text{ g}.$

(d) $W = 3700$ J per cycle
In $t = 1.00$ s the engine goes through 60.0 cycles.
$P = W/t = 60.0(3700\text{ J})/1.00\text{ s} = 222$ kW
$P = (2.22\times 10^5\text{ W})(1\text{ hp}/746\text{ W}) = 298$ hp

EVALUATE: $Q_C = -12,400$ J. In one cycle $Q_{\text{tot}} = Q_C + Q_H = 3700$ J. This equals W_{tot} for one cycle.

20.5. IDENTIFY: This cycle involves adiabatic (ab), isobaric (bc), and isochoric (ca) processes.

SET UP: ca is at constant volume, ab has $Q = 0$, and bc is at constant pressure. For a constant pressure process $W = p\Delta V$ and $Q = nC_p\,\Delta T$. $pV = nRT$ gives $n\Delta T = \frac{p\Delta V}{R}$, so $Q = \left(\frac{C_p}{R}\right)p\Delta V$.

If $\gamma = 1.40$ the gas is diatomic and $C_p = \frac{7}{2}R$. For a constant volume process $W = 0$ and $Q = nC_V\,\Delta T$. $pV = nRT$ gives $n\Delta T = \frac{V\Delta p}{R}$, so $Q = \left(\frac{C_V}{R}\right)V\Delta p$. For a diatomic ideal gas $C_V = \frac{5}{2}R$. 1 atm $= 1.013\times 10^5$ Pa.

EXECUTE: (a) $V_b = 9.0\times 10^{-3}\text{ m}^3$, $p_b = 1.5$ atm and $V_a = 2.0\times 10^{-3}\text{ m}^3$. For an adiabatic process $p_a V_a^\gamma = p_b V_b^\gamma$. $p_a = p_b\left(\frac{V_b}{V_a}\right)^\gamma = (1.5\text{ atm})\left(\frac{9.0\times 10^{-3}\text{ m}^3}{2.0\times 10^{-3}\text{ m}^3}\right)^{1.4} = 12.3$ atm.

(b) Heat enters the gas in process ca, since T increases.
$Q = \left(\frac{C_V}{R}\right)V\Delta p = \left(\frac{5}{2}\right)(2.0\times 10^{-3}\text{ m}^3)(12.3\text{ atm} - 1.5\text{ atm})(1.013\times 10^5\text{ Pa/atm}) = 5470$ J.
$Q_H = 5470$ J.

(c) Heat leaves the gas in process bc, since T decreases.

$Q = \left(\dfrac{C_p}{R}\right) p\Delta V = \left(\dfrac{7}{2}\right)(1.5\text{ atm})(1.013\times 10^5\text{ Pa/atm})(-7.0\times 10^{-3}\text{ m}^3) = -3723$ J. $Q_C = -3723$ J.

(d) $W = Q_H + Q_C = +5470$ J $+ (-3723$ J$) = 1747$ J.

(e) $e = \dfrac{W}{Q_H} = \dfrac{1747\text{ J}}{5470\text{ J}} = 0.319 = 31.9\%$.

EVALUATE: We did not use the number of moles of the gas.

20.9. IDENTIFY: For the Otto-cycle engine, $e = 1 - r^{1-\gamma}$.

SET UP: r is the compression ratio.

EXECUTE: **(a)** $e = 1 - (8.8)^{-0.40} = 0.581$, which rounds to 58%.

(b) $e = 1 - (9.6)^{20.40} = 0.595$ an increase of 1.4%.

EVALUATE: An increase in r gives an increase in e.

20.11. IDENTIFY: The heat $Q = mc\Delta T$ that comes out of the water to cool it to 5.0°C is Q_C for the refrigerator.

SET UP: For water 1.0 L has a mass of 1.0 kg and $c = 4.19\times 10^3$ J/kg·C°. $P = \dfrac{|W|}{t}$. The coefficient of performance is $K = \dfrac{|Q_C|}{|W|}$.

EXECUTE: $Q = mc\Delta T = (12.0\text{ kg})(4.19\times 10^3\text{ J/kg·C°})(5.0°\text{C} - 31°\text{C}) = -1.31\times 10^6$ J.

$|Q_C| = 1.31\times 10^6$ J. $K = \dfrac{|Q_C|}{|W|} = \dfrac{|Q_C|}{Pt}$ so $t = \dfrac{|Q_C|}{PK} = \dfrac{1.31\times 10^6\text{ J}}{(135\text{ W})(2.25)} = 4313$ s $= 71.88$ min $= 1.20$ h.

EVALUATE: 1.2 h seems like a reasonable time to cool down the dozen bottles.

20.13. IDENTIFY: The hot reservoir of a Carnot engine is 72.0 C° higher than the cold reservoir and this engine is 12.5% efficient. We want the temperature of the two reservoirs.

SET UP: For a Carnot engine $e = 1 - \dfrac{T_C}{T_H}$. $T_C = T_H - 72.0$ K.

EXECUTE: $e = 1 - \dfrac{T_C}{T_H} = 1 - \dfrac{T_H - 72.0\text{ K}}{T_H} = 0.125$ gives $T_H = 576$ K, so $T_C = 576$ K $- 72$ K $= 504$ K.

EVALUATE: Careful! The 72.0 C° is a temperature *difference*. Since Kelvin and Celsius degrees are the same size, so $\Delta T_K = \Delta T_C$. We should *not* convert the 72.0 C° to get 345 K.

20.19. IDENTIFY: $|Q_H| = |W| + |Q_C|$. $Q_H < 0$, $Q_C > 0$. $K = \dfrac{|Q_C|}{|W|}$. For a Carnot cycle, $\dfrac{Q_C}{Q_H} = -\dfrac{T_C}{T_H}$.

SET UP: $T_C = 270$ K, $T_H = 320$ K. $|Q_C| = 415$ J.

EXECUTE: **(a)** $Q_H = -\left(\dfrac{T_H}{T_C}\right) Q_C = -\left(\dfrac{320\text{ K}}{270\text{ K}}\right)(415\text{ J}) = -492$ J.

(b) For one cycle, $|W| = |Q_H| - |Q_C| = 492$ J $- 415$ J $= 77$ J. $P = \dfrac{(165)(77\text{ J})}{60\text{ s}} = 212$ W.

(c) $K = \dfrac{|Q_C|}{|W|} = \dfrac{415\text{ J}}{77\text{ J}} = 5.4$.

EVALUATE: The amount of heat energy $|Q_H|$ delivered to the high-temperature reservoir is greater than the amount of heat energy $|Q_C|$ removed from the low-temperature reservoir.

20.21. IDENTIFY and SET UP: We are looking at two refrigerators, A and B, operating on a Carnot cycle. The coefficient of performance of a Carnot refrigerator is $K_{Carnot} = \dfrac{T_C}{T_H - T_C} = \dfrac{T_C}{\Delta T}$. We know that $K_A = K_B + 0.16 K_B = 1.16 K_B$, $\Delta T_B = \Delta T_A + 0.30 \Delta T_A = 1.30 \Delta T_A$, and $T_C(B) = 180$ K. We want to find the cold reservoir temperature for A, $T_C(A)$.

EXECUTE: $K_A = \dfrac{T_C(A)}{\Delta T_A}$ and $K_B = \dfrac{T_C(B)}{\Delta T_B}$, so $\dfrac{K_A}{K_B} = \dfrac{\frac{T_C(A)}{\Delta T_A}}{\frac{T_C(B)}{\Delta T_B}}$. Using the given information, we have $\dfrac{1.16 K_B}{K_B} = \dfrac{T_C(A)}{T_C(B)} \dfrac{\Delta T_B}{\Delta T_A} = \dfrac{T_C(A)}{T_C(B)} \dfrac{1.30 \Delta T_A}{\Delta T_A}$. This gives $1.16 = \dfrac{1.30 T_C(A)}{180 \text{ K}}$, so $T_C(A) = 161$ K.

EVALUATE: All temperatures must be in Kelvins.

20.25. IDENTIFY: The process is at constant temperature, so $\Delta S = \dfrac{Q}{T}$. $\Delta U = Q - W$.

SET UP: For an isothermal process of an ideal gas, $\Delta U = 0$ and $Q = W$. For a compression, $\Delta V < 0$ and $W < 0$.

EXECUTE: $Q = W = -1850$ J. $\Delta S = \dfrac{-1850 \text{ J}}{293 \text{ K}} = -6.31$ J/K.

EVALUATE: The entropy change of the gas is negative. Heat must be removed from the gas during the compression to keep its temperature constant and therefore the gas is not an isolated system.

20.27. IDENTIFY: Each phase transition occurs at constant temperature and $\Delta S = \dfrac{Q}{T}$. $Q = mL_v$.

SET UP: For vaporization of water, $L_v = 2256 \times 10^3$ J/kg.

EXECUTE: (a) $\Delta S = \dfrac{Q}{T} = \dfrac{mL_v}{T} = \dfrac{(1.00 \text{ kg})(2256 \times 10^3 \text{ J/kg})}{(373.15 \text{ K})} = 6.05 \times 10^3$ J/K. Note that this is the change of entropy of the water as it changes to steam.
(b) The magnitude of the entropy change is roughly five times the value found in Example 20.5.

EVALUATE: Water is less ordered (more random) than ice, but water is far less random than steam; a consideration of the density changes indicates why this should be so.

20.29. IDENTIFY: For a free expansion, $\Delta S = nR \ln(V_2/V_1)$.

SET UP: $V_1 = 2.40$ L $= 2.40 \times 10^{-3}$ m^3.

EXECUTE: $\Delta S = (0.100 \text{ mol})(8.314 \text{ J/mol} \cdot \text{K}) \ln\left(\dfrac{425 \text{ m}^3}{2.40 \times 10^{-3} \text{ m}^3}\right) = 10.0$ J/K.

EVALUATE: $\Delta S_{system} > 0$ and the free expansion is irreversible.

20.31. IDENTIFY: The total work that must be done is $W_{tot} = mg\Delta y$. $|W| = |Q_H| - |Q_C|$. $Q_H > 0$, $W > 0$ and $Q_C < 0$. For a Carnot cycle, $\dfrac{Q_C}{Q_H} = -\dfrac{T_C}{T_H}$.

SET UP: $T_C = 373$ K, $T_H = 773$ K. $|Q_H| = 250$ J.

EXECUTE: (a) $Q_C = -Q_H \left(\dfrac{T_C}{T_H}\right) = -(250 \text{ J})\left(\dfrac{373 \text{ K}}{773 \text{ K}}\right) = -121$ J.

(b) $|W| = 250$ J $- 121$ J $= 129$ J. This is the work done in one cycle.

$W_{tot} = (500 \text{ kg})(9.80 \text{ m/s}^2)(100 \text{ m}) = 4.90 \times 10^5$ J. The number of cycles required is

$\dfrac{W_{tot}}{|W|} = \dfrac{4.90 \times 10^5 \text{ J}}{129 \text{ J/cycle}} = 3.80 \times 10^3$ cycles.

EVALUATE: In $\dfrac{Q_C}{Q_H} = -\dfrac{T_C}{T_H}$, the temperatures must be in kelvins.

20.33. IDENTIFY: We know the efficiency of this Carnot engine, the heat it absorbs at the hot reservoir and the temperature of the hot reservoir.

SET UP: For a heat engine $e = \dfrac{W}{|Q_H|}$ and $Q_H + Q_C = W$. For a Carnot cycle, $\dfrac{Q_C}{Q_H} = -\dfrac{T_C}{T_H}$. $Q_C < 0$, $W > 0$, and $Q_H > 0$. $T_H = 135°C = 408$ K. In each cycle, $|Q_H|$ leaves the hot reservoir and $|Q_C|$ enters the cold reservoir. The work done on the water equals its increase in gravitational potential energy, mgh.

EXECUTE: (a) $e = \dfrac{W}{Q_H}$ so $W = eQ_H = (0.220)(410 \text{ J}) = 90.2$ J.

(b) $Q_C = W - Q_H = 90.2$ J $- 410$ J $= -319.85$ J, which rounds to -320 J.

(c) $\dfrac{Q_C}{Q_H} = -\dfrac{T_C}{T_H}$ so $T_C = -T_H\left(\dfrac{Q_C}{Q_H}\right) = -(408 \text{ K})\left(\dfrac{-319.8 \text{ J}}{410 \text{ J}}\right) = 318$ K $= 45°$C.

(d) $\Delta S = \dfrac{-|Q_H|}{T_H} + \dfrac{|Q_C|}{T_C} = \dfrac{-410 \text{ J}}{408 \text{ K}} + \dfrac{319.8 \text{ J}}{318 \text{ K}} = 0$. The Carnot cycle is reversible and $\Delta S = 0$.

(e) $W = mgh$ so $m = \dfrac{W}{gh} = \dfrac{90.2 \text{ J}}{(9.80 \text{ m/s}^2)(35.0 \text{ m})} = 0.263$ kg $= 263$ g.

EVALUATE: The Carnot cycle is reversible so $\Delta S = 0$ for the world. However some parts of the world gain entropy while other parts lose it, making the sum equal to zero.

20.35. IDENTIFY: The same amount of heat that enters the person's body also leaves the body, but these transfers of heat occur at different temperatures, so the person's entropy changes.

SET UP: 1 food calorie = 1000 cal = 4186 J. The heat enters the person's body at $37°C = 310$ K and leaves at a temperature of $30°C = 303$ K. $\Delta S = \dfrac{Q}{T}$.

EXECUTE: $|Q| = (0.80)(2.50 \text{ g})(9.3 \text{ food calorie/g})\left(\dfrac{4186 \text{ J}}{1 \text{ food calorie}}\right) = 7.79 \times 10^4$ J.

$\Delta S = \dfrac{+7.79 \times 10^4 \text{ J}}{310 \text{ K}} + \dfrac{-7.79 \times 10^4 \text{ J}}{303 \text{ K}} = -5.8$ J/K. Your body's entropy decreases.

EVALUATE: The entropy of your body can decrease without violating the second law of thermodynamics because you are not an isolated system.

20.37. IDENTIFY: $pV = nRT$, so pV is constant when T is constant. Use the appropriate expression to calculate Q and W for each process in the cycle. $e = \dfrac{W}{Q_H}$.

SET UP: For an ideal diatomic gas, $C_V = \tfrac{5}{2}R$ and $C_p = \tfrac{7}{2}R$.

EXECUTE: (a) $p_a V_a = 2.0 \times 10^3$ J. $p_b V_b = 2.0 \times 10^3$ J. $pV = nRT$ so $p_a V_a = p_b V_b$ says $T_a = T_b$.

(b) For an isothermal process, $Q = W = nRT \ln(V_2/V_1)$. ab is a compression, with $V_b < V_a$, so $Q < 0$ and heat is rejected. bc is at constant pressure, so $Q = nC_p \Delta T = \dfrac{C_p}{R} p\Delta V$. ΔV is positive, so $Q > 0$ and heat is absorbed. ca is at constant volume, so $Q = nC_V \Delta T = \dfrac{C_V}{R} V\Delta p$. Δp is negative, so $Q < 0$ and heat is rejected.

(c) $T_a = \dfrac{p_a V_a}{nR} = \dfrac{2.0 \times 10^3 \text{ J}}{(1.00)(8.314 \text{ J/mol} \cdot \text{K})} = 241$ K. $T_b = \dfrac{p_b V_b}{nR} = T_a = 241$ K.

$T_c = \dfrac{p_c V_c}{nR} = \dfrac{4.0 \times 10^3 \text{ J}}{(1.00)(8.314 \text{ J/mol} \cdot \text{K})} = 481$ K.

(d) $Q_{ab} = nRT \ln\left(\dfrac{V_b}{V_a}\right) = (1.00 \text{ mol})(8.314 \text{ J/mol} \cdot \text{K})(241 \text{ K}) \ln\left(\dfrac{0.0050 \text{ m}^3}{0.010 \text{ m}^3}\right) = -1.39 \times 10^3$ J.

$Q_{bc} = nC_p \Delta T = (1.00)\left(\dfrac{7}{2}\right)(8.314 \text{ J/mol} \cdot \text{K})(241 \text{ K}) = 7.01 \times 10^3$ J.

$Q_{ca} = nC_V \Delta T = (1.00)\left(\dfrac{5}{2}\right)(8.314 \text{ J/mol} \cdot \text{K})(-241 \text{ K}) = -5.01 \times 10^3$ J.

$Q_{net} = Q_{ab} + Q_{bc} + Q_{ca} = 610$ J. $W_{net} = Q_{net} = 610$ J.

(e) $e = \dfrac{W}{Q_H} = \dfrac{610 \text{ J}}{7.01 \times 10^3 \text{ J}} = 0.087 = 8.7\%$.

EVALUATE: We can calculate W for each process in the cycle. $W_{ab} = Q_{ab} = -1.39 \times 10^3$ J. $W_{bc} = p\Delta V = (4.0 \times 10^5 \text{ Pa})(0.0050 \text{ m}^3) = 2.00 \times 10^3$ J. $W_{ca} = 0$. $W_{net} = W_{ab} + W_{bc} + W_{ca} = 610$ J, which does equal Q_{net}.

20.39. IDENTIFY: $T_b = T_c$ and is equal to the maximum temperature. Use the ideal gas law to calculate T_a. Apply the appropriate expression to calculate Q for each process. $e = \dfrac{W}{Q_H}$. $\Delta U = 0$ for a complete cycle and for an isothermal process of an ideal gas.

SET UP: For helium, $C_V = 3R/2$ and $C_p = 5R/2$. The maximum efficiency is for a Carnot cycle, and $e_{\text{Carnot}} = 1 - T_C/T_H$.

EXECUTE: (a) $Q_{in} = Q_{ab} + Q_{bc}$. $Q_{out} = Q_{ca}$. $T_{max} = T_b = T_c = 327°C = 600$ K.

$\dfrac{p_a V_a}{T_a} = \dfrac{p_b V_b}{T_b} \rightarrow T_a = \dfrac{p_a}{p_b} T_b = \dfrac{1}{3}(600 \text{ K}) = 200$ K.

$p_b V_b = nRT_b \rightarrow V_b = \dfrac{nRT_b}{p_b} = \dfrac{(2 \text{ moles})(8.31 \text{ J/mol} \cdot \text{K})(600 \text{ K})}{3.0 \times 10^5 \text{ Pa}} = 0.0332 \text{ m}^3$.

$$\frac{p_b V_b}{T_b} = \frac{p_c V_c}{T_c} \rightarrow V_c = V_b \frac{p_b}{p_c} = (0.0332 \text{ m}^3)\left(\frac{3}{1}\right) = 0.0997 \text{ m}^3 = V_a.$$

$$Q_{ab} = nC_V \Delta T_{ab} = (2 \text{ mol})\left(\frac{3}{2}\right)(8.31 \text{ J/mol} \cdot \text{K})(400 \text{ K}) = 9.97 \times 10^3 \text{ J}.$$

$$Q_{bc} = W_{bc} = \int_b^c p\, dV = \int_b^c \frac{nRT_b}{V} dV = nRT_b \ln \frac{V_c}{V_b} = nRT_b \ln 3.$$

$Q_{bc} = (2.00 \text{ mol})(8.31 \text{ J/mol} \cdot \text{K})(600 \text{ K})\ln 3 = 1.10 \times 10^4 \text{ J}.$ $Q_{\text{in}} = Q_{ab} + Q_{bc} = 2.10 \times 10^4 \text{ J}.$

$$Q_{\text{out}} = Q_{ca} = nC_p \Delta T_{ca} = (2.00 \text{ mol})\left(\frac{5}{2}\right)(8.31 \text{ J/mol} \cdot \text{K})(400 \text{ K}) = 1.66 \times 10^4 \text{ J}.$$

(b) $Q = \Delta U + W = 0 + W \rightarrow W = Q_{\text{in}} - Q_{\text{out}} = 2.10 \times 10^4 \text{ J} - 1.66 \times 10^4 \text{ J} = 4.4 \times 10^3 \text{ J}.$

$$e = W/Q_{\text{in}} = \frac{4.4 \times 10^3 \text{ J}}{2.10 \times 10^4 \text{ J}} = 0.21 = 21\%.$$

(c) $e_{\max} = e_{\text{Carnot}} = 1 - \frac{T_C}{T_H} = 1 - \frac{200 \text{ k}}{600 \text{ k}} = 0.67 = 67\%.$

EVALUATE: The thermal efficiency of this cycle is about one-third of the efficiency of a Carnot cycle that operates between the same two temperatures.

20.43. IDENTIFY: Use $pV = nRT$. Apply the expressions for Q and W that apply to each type of process. $e = \dfrac{W}{Q_H}.$

SET UP: For O_2, $C_V = 20.85$ J/mol·K and $C_p = 29.17$ J/mol·K.

EXECUTE: (a) $p_1 = 2.00$ atm, $V_1 = 4.00$ L, $T_1 = 300$ K.

$p_2 = 2.00$ atm. $\dfrac{V_1}{T_1} = \dfrac{V_2}{T_2}.$ $V_2 = \left(\dfrac{T_2}{T_1}\right) V_1 = \left(\dfrac{450 \text{ K}}{300 \text{ K}}\right)(4.00 \text{ L}) = 6.00$ L.

$V_3 = 6.00$ L. $\dfrac{p_2}{T_2} = \dfrac{p_3}{T_3}.$ $p_3 = \left(\dfrac{T_3}{T_2}\right) p_2 = \left(\dfrac{250 \text{ K}}{450 \text{ K}}\right)(2.00 \text{ atm}) = 1.11$ atm.

$V_4 = 4.00$ L. $p_3 V_3 = p_4 V_4.$ $p_4 = p_3 \left(\dfrac{V_3}{V_4}\right) = (1.11 \text{ atm})\left(\dfrac{6.00 \text{ L}}{4.00 \text{ L}}\right) = 1.67$ atm.

These processes are shown in Figure 20.43.

(b) $n = \dfrac{p_1 V_1}{RT_1} = \dfrac{(2.00 \text{ atm})(4.00 \text{ L})}{(0.08206 \text{ L} \cdot \text{atm/mol} \cdot \text{K})(300 \text{ K})} = 0.325$ mol

Process $1 \rightarrow 2$: $W = p\Delta V = nR\Delta T = (0.325 \text{ mol})(8.315 \text{ J/mol} \cdot \text{K})(150 \text{ K}) = 405$ J.
$Q = nC_p \Delta T = (0.325 \text{ mol})(29.17 \text{ J/mol} \cdot \text{K})(150 \text{ K}) = 1422$ J.

Process $2 \rightarrow 3$: $W = 0.$ $Q = nC_V \Delta T = (0.325 \text{ mol})(20.85 \text{ J/mol} \cdot \text{K})(-200 \text{ K}) = -1355$ J.

Process $3 \rightarrow 4$: $\Delta U = 0$ and

$$Q = W = nRT_3 \ln\left(\frac{V_4}{V_3}\right) = (0.325 \text{ mol})(8.315 \text{ J/mol} \cdot \text{K})(250 \text{ K})\ln\left(\frac{4.00 \text{ L}}{6.00 \text{ L}}\right) = -274 \text{ J}.$$

Process $4 \rightarrow 1$: $W = 0.$ $Q = nC_V \Delta T = (0.325 \text{ mol})(20.85 \text{ J/mol} \cdot \text{K})(50 \text{ K}) = 339$ J.

(c) $W = 405 \text{ J} - 274 \text{ J} = 131$ J.

(d) $e = \dfrac{W}{Q_H} = \dfrac{131 \text{ J}}{1422 \text{ J} + 339 \text{ J}} = 0.0744 = 7.44\%.$

$e_{\text{Carnot}} = 1 - \dfrac{T_C}{T_H} = 1 - \dfrac{250 \text{ K}}{450 \text{ K}} = 0.444 = 44.4\%$; e_{Carnot} is much larger.

EVALUATE: $Q_{\text{tot}} = +1422 \text{ J} + (-1355 \text{ J}) + (-274 \text{ J}) + 339 \text{ J} = 132 \text{ J}$. This is equal to W_{tot}, apart from a slight difference due to rounding. For a cycle, $W_{\text{tot}} = Q_{\text{tot}}$, since $\Delta U = 0$.

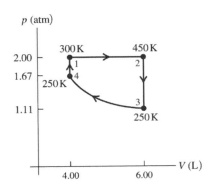

Figure 20.43

20.45. IDENTIFY: The efficiency of the composite engine is $e_{12} = \dfrac{W_1 + W_2}{Q_{H1}}$, where Q_{H1} is the heat input to the first engine and W_1 and W_2 are the work outputs of the two engines. For any heat engine, $W = Q_C + Q_H$, and for a Carnot engine, $\dfrac{Q_{\text{low}}}{Q_{\text{high}}} = -\dfrac{T_{\text{low}}}{T_{\text{high}}}$, where Q_{low} and Q_{high} are the heat flows at the two reservoirs that have temperatures T_{low} and T_{high}.

SET UP: $Q_{\text{high},2} = -Q_{\text{low},1}$. $T_{\text{low},1} = T'$, $T_{\text{high},1} = T_H$, $T_{\text{low},2} = T_C$ and $T_{\text{high},2} = T'$.

EXECUTE: $e_{12} = \dfrac{W_1 + W_2}{Q_{H1}} = \dfrac{Q_{\text{high},1} + Q_{\text{low},1} + Q_{\text{high},2} + Q_{\text{low},2}}{Q_{\text{high},1}}$. Since $Q_{\text{high},2} = -Q_{\text{low},1}$, this reduces

to $e_{12} = 1 + \dfrac{Q_{\text{low},2}}{Q_{\text{high},1}}$. $Q_{\text{low},2} = -Q_{\text{high},2} \dfrac{T_{\text{low},2}}{T_{\text{high},2}} = Q_{\text{low},1} \dfrac{T_C}{T'} = -Q_{\text{high},1} \left(\dfrac{T_{\text{low},1}}{T_{\text{high},1}}\right)\dfrac{T_C}{T'} = -Q_{\text{high},1}\left(\dfrac{T'}{T_H}\right)\dfrac{T_C}{T'}$.

This gives $e_{12} = 1 - \dfrac{T_C}{T_H}$. The efficiency of the composite system is the same as that of the original engine.

EVALUATE: The overall efficiency is independent of the value of the intermediate temperature T'.

20.47. (a) IDENTIFY and SET UP: Calculate e from $e = 1 - 1/(r^{\gamma-1})$, Q_C from $e = (Q_H + Q_C)/Q_H$, and then W from $W = Q_C + Q_H$.

EXECUTE: $e = 1 - 1/(r^{\gamma-1}) = 1 - 1/(10.6^{0.4}) = 0.6111$

$e = (Q_H + Q_C)/Q_H$ and we are given $Q_H = 200$ J; calculate Q_C.

$Q_C = (e-1)Q_H = (0.6111 - 1)(200 \text{ J}) = -78$ J. (negative, since corresponds to heat leaving)

Then $W = Q_C + Q_H = -78 \text{ J} + 200 \text{ J} = 122$ J. (positive, in agreement with Figure 20.6 in the text)

EVALUATE: Q_H, $W > 0$, and $Q_C < 0$ for an engine cycle.

(b) IDENTIFY and SET UP: The stoke times the bore equals the change in volume. The initial volume is the final volume V times the compression ratio r. Combining these two expressions gives an equation for V. For each cylinder of area $A = \pi(d/2)^2$ the piston moves 0.0864 m and the volume changes from rV to V, as shown in Figure 20.47a.

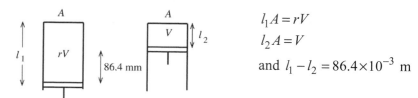

$l_1 A = rV$
$l_2 A = V$
and $l_1 - l_2 = 86.4 \times 10^{-3}$ m

Figure 20.47a

EXECUTE: $l_1 A - l_2 A = rV - V$ and $(l_1 - l_2)A = (r-1)V$

$$V = \frac{(l_1 - l_2)A}{r-1} = \frac{(86.4 \times 10^{-3} \text{ m})\pi(41.25 \times 10^{-3} \text{ m})^2}{10.6 - 1} = 4.811 \times 10^{-5} \text{ m}^3$$

At point a the volume is $rV = 10.6(4.811 \times 10^{-5} \text{ m}^3) = 5.10 \times 10^{-4} \text{ m}^3$.

(c) IDENTIFY and **SET UP:** The processes in the Otto cycle are either constant volume or adiabatic. Use the Q_H that is given to calculate ΔT for process bc. Use $T_1 V_1^{\gamma-1} = T_2 V_2^{\gamma-1}$ and $pV = nRT$ to relate p, V and T for the adiabatic processes ab and cd.

EXECUTE: Point a: $T_a = 300$ K, $p_a = 8.50 \times 10^4$ Pa and $V_a = 5.10 \times 10^{-4}$ m^3.

Point b: $V_b = V_a/r = 4.81 \times 10^{-5}$ m^3. Process $a \to b$ is adiabatic, so $T_a V_a^{\gamma-1} = T_b V_b^{\gamma-1}$.

$T_a (rV)^{\gamma-1} = T_b V^{\gamma-1}$

$T_b = T_a r^{\gamma-1} = 300$ K$(10.6)^{0.4} = 771$ K

$pV = nRT$ so $pV/T = nR = $ constant, so $p_a V_a / T_a = p_b V_b / T_b$

$p_b = p_a (V_a / V_b)(T_b / T_a) = (8.50 \times 10^4$ Pa$)(rV/V)(771$ K$/300$ K$) = 2.32 \times 10^6$ Pa

Point c: Process $b \to c$ is at constant volume, so $V_c = V_b = 4.81 \times 10^{25}$ m^3

$Q_H = nC_V \Delta T = nC_V (T_c - T_b)$. The problem specifies $Q_H = 200$ J; use to calculate T_c. First use the p, V, T values at point a to calculate the number of moles n.

$$n = \frac{pV}{RT} = \frac{(8.50 \times 10^4 \text{ Pa})(5.10 \times 10^{-4} \text{ m}^3)}{(8.3145 \text{ J/mol} \cdot \text{K})(300 \text{ K})} = 0.01738 \text{ mol}$$

Then $T_c - T_b = \dfrac{Q_H}{nC_V} = \dfrac{200 \text{ J}}{(0.01738 \text{ mol})(20.5 \text{ J/mol} \cdot \text{K})} = 561.3$ K, and

$T_c = T_b + 561.3$ K $= 771$ K $+ 561$ K $= 1332$ K

$p/T = nR/V = $ constant so $p_b / T_b = p_c / T_c$

$p_c = p_b (T_c / T_b) = (2.32 \times 10^6$ Pa$)(1332$ K$/771$ K$) = 4.01 \times 10^6$ Pa

Point d: $V_d = V_a = 5.10 \times 10^{-4}$ m^3

Process $c \to d$ is adiabatic, so $T_d V_d^{\gamma-1} = T_c V_c^{\gamma-1}$

$T_d (rV)^{\gamma-1} = T_c V^{\gamma-1}$

$T_d = T_c / r^{\gamma-1} = 1332$ K$/10.6^{0.4} = 518$ K

$p_c V_c / T_c = p_d V_d / T_d$

$p_d = p_c (V_c / V_d)(T_d / T_c) = (4.01 \times 10^6$ Pa$)(V/rV)(518$ K$/1332$ K$) = 1.47 \times 10^5$ Pa

EVALUATE: Can look at process $d \to a$ as a check.

$Q_C = nC_V(T_a - T_d) = (0.01738 \text{ mol})(20.5 \text{ J/mol} \cdot \text{K})(300 \text{ K} - 518 \text{ K}) = -78$ J, which agrees with part (a). The cycle is sketched in Figure 20.47b.

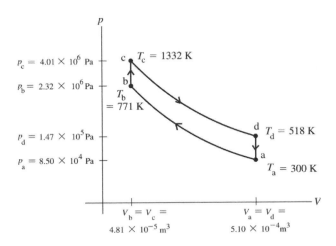

Figure 20.47b

(d) IDENTIFY and SET UP: The Carnot efficiency is given by $e_{\text{Carnot}} = 1 - \dfrac{T_C}{T_H}$. T_H is the highest temperature reached in the cycle and T_C is the lowest.

EXECUTE: From part (a) the efficiency of this Otto cycle is $e = 0.611 = 61.1\%$.
The efficiency of a Carnot cycle operating between 1332 K and 300 K is
$e_{\text{Carnot}} = 1 - T_C/T_H = 1 - 300 \text{ K}/1332 \text{ K} = 0.775 = 77.5\%$, which is larger.

EVALUATE: The second law of thermodynamics requires that $e \leq e_{\text{Carnot}}$, and our result obeys this law.

20.49. **IDENTIFY and SET UP:** A refrigerator is like a heat engine run in reverse. In the pV-diagram shown with the figure, heat enters the gas during parts ab and bc of the cycle, and leaves during ca. Treating H$_2$ as a diatomic gas, we know that $C_V = \tfrac{5}{2}R$ and $C_p = \tfrac{7}{2}R$. Segment bc is isochoric, so $Q_{bc} = nC_V \Delta T$. Segment ca is isobaric, so $Q_{ca} = nC_p \Delta T$. Segment ab is isothermal, so $Q_{ab} = nRT \ln(V_b/V_a)$. The coefficient of performance of a refrigerator is $K = \dfrac{|Q_C|}{|W|} = \dfrac{|Q_C|}{|Q_H| - |Q_C|}$, and $pV = nRT$ applies. Calculate the values for Q_C and Q_H and use the definition of K. Use 1000 L = 1 m^3 and work in units of L·atm.

EXECUTE: Use $pV = nRT$ to find p_b. Since ab is isothermal, $p_aV_a = p_bV_b$, which gives $p_b = (0.700 \text{ atm})(0.0300 \text{ m}^3)/(0.100 \text{ m}^3) = 0.210$ atm.
$Q_C = Q_{ab} + Q_{bc}$, so we need to calculate these quantities.
$Q_{ab} = nRT \ln(V_b/V_a) = p_aV_a \ln(V_b/V_a) = (0.700 \text{ atm})(30.0 \text{ L}) \ln[(100 \text{ L})/(30 \text{ L})] = 25.2834$ L·atm.
$Q_{bc} = nC_V \Delta T_{bc} = n(\tfrac{5}{2}R) \Delta T_{bc} = \tfrac{5}{2} V_b \Delta p_{bc} = (5/2)(100 \text{ L})(0.700 \text{ atm} - 0.210 \text{ atm}) = 122.5$ L·atm.
Therefore $Q_C = Q_{ab} + Q_{bc} = 25.2834$ L·atm $+ 122.500$ L·atm $= 147.7834$ L·atm.
$Q_H = Q_{ca} = nC_p\Delta T_{ca} = n\left(\tfrac{7}{2}R\right)\Delta T_{ca} = \tfrac{7}{2}p_c\Delta T_{ca} = (7/2)(0.700 \text{ atm})(30.0 \text{ L} - 100.0 \text{ L}) = -171.500$ L·atm.

Now get K: $K = \dfrac{|Q_C|}{|Q_H| - |Q_C|} = \dfrac{147.7834 \text{ L} \cdot \text{atm}}{(171.500 \text{ L} \cdot \text{atm} - 147.7834 \text{ L} \cdot \text{atm})} = 6.23$.

EVALUATE: K is greater than 1, which it must be. Efficiencies are less than 1.

20.53. IDENTIFY and SET UP: The most efficient heat engine operating between any two given temperatures is the Carnot engine, and its efficiency is $e_{\text{Carnot}} = 1 - \dfrac{T_C}{T_H}$.

EXECUTE: **(a)** For prototype A, $e_{\max} = e_{\text{Carnot}} = 1 - \dfrac{T_C}{T_H} = 1 - (320\text{ K})/(450\text{ K}) = 0.289 = 28.9\%$. By similar calculations, we get the following:
A: $e_{\max} = 0.289 = 28.9\%$
B: $e_{\max} = 0.383 = 38.3\%$
C: $e_{\max} = 0.538 = 53.8\%$
D: $e_{\max} = 0.244 = 24.4\%$
(b) Engine C claims a maximum efficiency of 56%, which is greater than the maximum possible for its temperature range, so it is impossible.
(c) We get the following ratios:
A: $e_{\text{claimed}}/e_{\max} = 0.21/0.289 = 0.73$
B: $e_{\text{claimed}}/e_{\max} = 0.35/0.383 = 0.90$
D: $e_{\text{claimed}}/e_{\max} = 0.20/0.244 = 0.82$
In decreasing order, we have B, D, A.
EVALUATE: Engine B is not only the most efficient, it is also closest to its maximum possible efficiency for its temperature range. Engines A and D have nearly the same efficiency, but D comes somewhat closer to its theoretical maximum than does A.

20.57. IDENTIFY and SET UP: Once you cut through all the extraterrestrial description, this problem boils down to an ordinary Carnot heat engine. First list all the known information and then summarize it on a pV-diagram as shown in Fig. 20.57. We know the following information:

$T_d = 123$ K = ambient temperature = T_a
$p_b = 20.3$ kPa
p_d = ambient pressure because the gauge pressure is zero at d
Trigger volume is V_a.
Combustion starts at point b.
$r_a = 1/3\ r_d \rightarrow r_d = 3 r_a \rightarrow V_d = 27 V_a$
$V_b = \tfrac{1}{2} V_a \rightarrow V_a = 2 V_b$
$\gamma = 1.30$

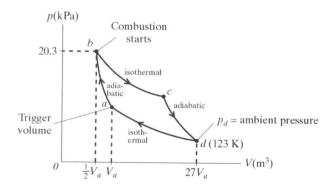

Figure 20.57

EXECUTE: **(a)** We want the combustion temperature, which is T_b. Since da is isothermal, $T_a = T_d = 123$ K. Since ab is adiabatic, we know that $T_a V_a^{\gamma-1} = T_b V_b^{\gamma-1}$. Using $V_b = \tfrac{1}{2} V_a$, we have

$$T_b = T_a (V_a / V_b)^{\gamma-1} = (123 \text{ K})\left(\frac{2V_b}{V_b}\right)^{1.30-1} = 151 \text{ K}.$$

(b) We want the ambient pressure, which is p_d. Since ab is adiabatic, we use $p_a V_a^\gamma = p_b V_b^\gamma$ to get p_a. Then use $p_a V_a = p_d V_d$ to find p_d since da is isothermal.

Find p_a: $p_a = p_b (V_b / V_a)^\gamma = (20.3 \text{ kPa})\left(\dfrac{V_b}{2V_b}\right)^{1.30} = 8.244 \text{ kPa}.$

Find p_d: $p_d = p_a \left(\dfrac{V_a}{V_d}\right) = (8.244 \text{ kPa})\left(\dfrac{V_a}{27V_a}\right) = 0.305$ kPa $= 305$ Pa. At point d the gauge pressure is zero, so 305 Pa is the ambient pressure.

(c) We want the heat input Q_{in} each minute. The work that is done is $W = 60$ kJ/h $= 1.00$ kJ/min. For a Carnot engine $e = \dfrac{W}{Q_{\text{in}}} = 1 - \dfrac{T_C}{T_H} = 1 - \dfrac{T_d}{T_b}$. Using the result from part (a) and the given information,

we have $\dfrac{1.00 \text{ kJ}}{Q_{\text{in}}} = 1 - \dfrac{123 \text{ K}}{151 \text{ K}}$, so $Q_{\text{in}} = 5.39$ kJ.

(d) We want the heat rejected Q_{out} each minute. $Q_{\text{in}} = W + Q_{\text{out}}$, so 5.39 kJ = 1.00 kJ + Q_{out}, which gives $Q_{\text{out}} = 4.39$ kJ.

EVALUATE: The efficiency of this engine is $e = W/Q_{\text{in}} = (1.00 \text{ kJ})/(5.39 \text{ kJ}) = 0.185$. Sporons are about as efficient at doing work as automobile engines on Earth.